T0211445

QUANTUM CHROMODYNAMICS

Perturbative and Nonperturbative Aspects

Aimed at graduate students and researchers in theoretical physics, this book presents the modern theory of strong interaction: quantum chromodynamics (QCD).

The book exposes various perturbative and nonperturbative approaches to the theory, including chiral effective theory, the problems of anomalies, vacuum tunnel transitions, and the problem of divergence of the perturbative series. The QCD sum rules approach is exposed in detail. A great variety of hadronic properties (masses of mesons and baryons, magnetic moments, formfactors, quark distributions in hadrons, etc.) have been found using this method. The evolution of hadronic structure functions is presented in detail, together with polarization phenomena. The problem of jets in QCD is treated through theoretical description and experimental observation. The connection with Regge theory is emphasized. The book covers many aspects of theory which are not discussed in other books, such as CET, QCD sum rules, and BFKL.

B. L. IOFFE is Head of the Laboratory of Theoretical Physics at ITEP, Moscow. He discovered the origin of baryon masses in QCD, the light-cone dominance in deep inelastic scattering. He predicted that parity violation in weak interactions should be accompanied by violation of charge symmetry or time reversal invariance, and the observation of parity violating spin-momentum correlations would mean the charge symmetry violation. He has won several prizes, including the Alexander von Humboldt Award.

V. S. FADIN is Head of the Theoretical Department of the Budker Institute of Nuclear Physics. He is one of the authors of the BFKL equation for parton distributions, which gives the most common basis for the theoretical description of semi-hard processes. He is also one of the discoverers of the coherence effect in gluon emission, which is crucial for inclusive spectra of hadrons in their soft region.

L. N. LIPATOV is Director of the Theoretical Physics Department at the Institute for Nuclear Physics, St. Petersburg. He is one of the authors of the DGLAP and BFKL equations for the parton distributions used in the theoretical description of the deep inelastic ep scattering and other processes in QCD. He has been awarded prizes from the ITEP and FIAN institutes for outstanding scientific achievements.

CAMBRIDGE MONOGRAPHS ON
PARTICLE PHYSICS,
NUCLEAR PHYSICS AND COSMOLOGY

General Editors: T. Ericson, P. V. Landshoff

1. K. Winter (ed.): *Neutrino Physics*
2. J. F. Donoghue, E. Golowich and B. R. Holstein: *Dynamics of the Standard Model*
3. E. Leader and E. Predazzi: *An Introduction to Gauge Theories and Modern Particle Physics, Volume 1: Electroweak Interactions, the 'New Particles' and the Parton Model*
4. E. Leader and E. Predazzi: *An Introduction to Gauge Theories and Modern Particle Physics, Volume 2: CP-Violation, QCD and Hard Processes*
5. C. Grupen: *Particle Detectors*
6. H. Grosse and A. Martin: *Particle Physics and the Schrödinger Equation*
7. B. Anderson: *The Lund Model*
8. R. K. Ellis, W. J. Stirling and B. R. Webber: *QCD and Collider Physics*
9. I. I. Bigi and A. I. Sanda: *CP Violation*
10. A. V. Manohar and M. B. Wise: *Heavy Quark Physics*
11. R. K. Bock, H. Grote, R. Frühwirth and M. Regler: *Data Analysis Techniques for High-Energy Physics, Second edition*
12. D. Green: *The Physics of Particle Detectors*
13. V. N. Gribov and J. Nyiri: *Quantum Electrodynamics*
14. K. Winter (ed.): *Neutrino Physics, Second edition*
15. E. Leader: *Spin in Particle Physics*
16. J. D. Walecka: *Electron Scattering for Nuclear and Nucleon Scattering*
17. S. Narison: *QCD as a Theory of Hadrons*
18. J. F. Letessier and J. Rafelski: *Hadrons and Quark-Gluon Plasma*
19. A. Donnachie, H. G. Dosch, P. V. Landshoff and O. Nachtmann: *Pomeron Physics and QCD*
20. A. Hoffmann: *The Physics of Synchroton Radiation*
21. J. B. Kogut and M. A. Stephanov: *The Phases of Quantum Chromodynamics*
22. D. Green: *High P_T Physics at Hadron Colliders*
23. K. Yagi, T. Hatsuda and Y. Miake: *Quark-Gluon Plasma*
24. D. M. Brink and R. A. Broglia: *Nuclear Superfluidity*
25. F. E. Close, A. Donnachie and G. Shaw: *Electromagnetic Interactions and Hadronic Structure*
26. C. Grupen and B. A. Shwartz: *Particle Detectors, Second edition*
27. V. Gribov: *Strong Interactions of Hadrons at High Energies*
28. I. I. Bigi and A. I. Sanda: *CP Violation, Second edition*
29. P. Jaranowski and A. Królak: *Analysis of Gravitational Wave Data*
30. B. L. Ioffe, V. S. Fadin and L. N. Lipatov: *Quantum Chromodynamics: Perturbative and Nonperturbative Aspects*

QUANTUM CHROMODYNAMICS
Perturbative and
Nonperturbative Aspects

B. L. IOFFE
A. I. Alikhanov Institute of Theoretical and Experimental Physics, Moscow

V. S. FADIN
G. I. Budker Institute of Nuclear Physics, Novosibirsk

L. N. LIPATOV
St. Petersburg Nuclear Physics Institute, Gatchina

CAMBRIDGE
UNIVERSITY PRESS

CAMBRIDGE
UNIVERSITY PRESS

University Printing House, Cambridge CB2 8BS, United Kingdom

Published in the United States of America by Cambridge University Press, New York

Cambridge University Press is part of the University of Cambridge.

It furthers the University's mission by disseminating knowledge in the pursuit of education, learning and research at the highest international levels of excellence.

www.cambridge.org
Information on this title: www.cambridge.org/9781107424753

First published 2010
First paperback edition 2014

A catalogue record for this publication is available from the British Library

Library of Congress Cataloguing in Publication data
Ioffe, B. L. (Boris Lazarevich), 1926–
Quantum chromodynamics : perturbative and nonperturbative aspects / B. L. Ioffe, V. S. Fadin, L. N. Lipatov.
p. cm.
Includes bibliographical references and index.
ISBN 978-0-521-63148-8
1. Quantum chromodynamics. I. Fadin, V. S. (Viktor Sergeevich), 1942–
II. Lipatov, L. N. (Lev Nikolaevich), 1940– III. Title.
QC793.3.Q35195   2010
539.7'548—dc22
2009021695

ISBN 978-0-521-63148-8 Hardback
ISBN 978-1-107-42475-3 Paperback

Contents

Preface   *page* ix

1   General properties of QCD   1
   1.1   QCD Lagrangian   1
   1.2   Quantization of the QCD Lagrangian   2
   1.3   The Gribov ambiguity   7
   1.4   Feynman rules   10
   1.5   Regularization   14
   1.6 γ_5 problem   15
   1.7   Renormalization   18
   1.8   One-loop calculations   21
   1.9   Renormalization group   27
   1.10   Asymptotic freedom in QCD   35
   1.11   The renormalization scheme and scale ambiguity   38
   1.12   Anomalous dimensions of twist-2 operators   42
   1.13   Colour algebra   45
       References   49

2   Chiral symmetry and its spontaneous violation   53
   2.1   The general properties of QCD at low energies   53
   2.2   The masses of the light quarks   54
   2.3   Spontaneous violation of chiral symmetry. Quark condensate   59
   2.4   Goldstone theorem   61
   2.5   Chiral effective theory (CET) at low energies   65
   2.6   Low-energy sum rules in CET   74
   2.7   The nucleon and pion–nucleon interaction in CET   76
       Problems   77
       References   79

3   Anomalies   82
   3.1   Generalities   82
   3.2   The axial anomaly   82

3.3	The axial anomaly and the scattering of polarized electron (muon) on polarized gluon	93
3.4	The scale anomaly	95
3.5	The infrared power-like singularities in photon-photon, photon-gluon, and gluon-gluon scattering in massless QED and QCD. Longitudinal gluons in QCD	100
	Problems	104
	References	105

4 Instantons and topological quantum numbers — 107

4.1	Tunneling in quantum mechanics	108
4.2	Instantons and the topological current	115
4.3	Instantons in Minkowski space-time	127
4.4	Fermions in the instanton field. Atiyah–Singer theorem	130
4.5	The vacuum structure in QCD	133
4.6	The pre-exponential factor of the instanton action. The dilute gas instanton model	134
4.7	Quark propagator in the instanton field	139
4.8	Appendix	141
	Problems	142
	References	142

5 Divergence of perturbation series — 145

5.1	Renormalization group approach to renormalons	147
5.2	High-order estimates in zero-dimensional models	152
5.3	Zero charge and asymptotic freedom in scalar models	155
5.4	Renormalized strongly nonlinear scalar model	159
5.5	Functional approach to the high-order estimates	165
5.6	Series divergence in models with gauge interactions	174
5.7	Asymptotic estimates in the pure Yang–Mills theory	182
5.8	Asymptotic estimates in quantum electrodymamics	186
5.9	Applications of high-order estimates	190
	References	193

6 QCD sum rules — 195

6.1	Operator product expansion	195
6.2	Condensates	199
6.3	Condensates, induced by external fields	202
6.4	QCD sum rules method	204
6.5	Determination of $\alpha_s(Q^2)$ and the condensates from low-energy data	207
6.6	Calculations of light meson masses and coupling constants	229
6.7	Sum rules for baryon masses	239

6.8	Calculation technique	252
6.9	Static properties of hadrons	259
6.10	Three-point functions and formfactors at intermediate momentum transfers	279
6.11	Valence quark distributions in hadrons	285
	Problems	299
	References	302

7   Evolution equations   309

7.1	Introduction	309
7.2	Parton model in QCD	311
7.3	Evolution equations for parton distributions	316
7.4	Splitting kernels in the Born approximation	324
7.5	OPE on light cone and parton model	332
7.6	Evolution equations for fragmentation functions	334
7.7	Parton distributions in QCD in LLA	338
7.8	Hard processes beyond the LLA	345
7.9	Parton-number correlators	353
7.10	Deep inelastic electron scattering off the polarized proton	359
7.11	Parton distributions in polarized nucleon	367
7.12	Evolution equations for quasipartonic operators	372
7.13	Q^2-dependence of the twist-3 structure functions for the polarized target	379
7.14	Infrared evolution equations at small x	383
	References	397

8   QCD jets   402

8.1	Total cross section of e^+e^--annihilation into hadrons	402
8.2	Jet production	405
8.3	Two-jet events	406
8.4	Three-jet events	412
8.5	Event shape	414
8.6	Inclusive spectra	415
8.7	Colour coherence	417
8.8	Soft-gluon approximations	423
8.9	Soft-gluon distributions	429
8.10	Hump-backed shape of parton spectra	433
8.11	Multiplicity distributions and KNO scaling	437
8.12	Moments of fragmentation functions at small $j-1$	438
8.13	Modified Leading Logarithmic Approximation (MLLA)	441
	References	444

9   **BFKL approach**   448

    9.1   Introduction   448
    9.2   Gluon reggeization   450
    9.3   Reggeon vertices and trajectory   454
    9.4   BFKL equation   469
    9.5   BFKL pomeron   476
    9.6   Bootstrap of the gluon reggeization   491
    9.7   Next-to-leading order BFKL   504
        References   527

10   **Further developments in high-energy QCD**   533

    10.1   Effective-action approach   533
    10.2   BFKL dynamics and integrability   542
    10.3   The odderon in QCD   550
    10.4   Baxter–Sklyanin representation   561
    10.5   Maximal transcendentality and anomalous dimensions   570
    10.6   Discussion of obtained results   575
        References   575

Notations   579
Index   580

Preface

Quantum Chromodynamics becomes now – moving into the second decade of the twenty-first century – a very extended theory with many branches. Therefore, it is impossible to expose in one book the whole QCD with all its branches and applications. On the other hand, it is a rapidly developing theory and for the authors of a book on QCD it is a serious danger to go too close to the border between that which is known with certainty and the domain of unknown. For these reasons, we decided to present in this book some aspects of QCD that are not very closely related to one another. In the selection of these aspects, surely, the individual preferences of authors also played a role.

What is not discussed in this book includes the theory of heavy quark systems – charm, bottom, top mesons and baryons; QCD at finite temperature and density; the problem of phase transitions in QCD; and lattice calculations. The QCD corrections to weak interactions, particularly to the production of W, Z, and Higgs bosons, are also beyond the scope of the book. We intentionally avoided various model approaches related to QCD, restricting ourself to exposition of the results following from the first principles of the theory. The phenomenology of hard processes and of the parton model was only lightly touched on in this book. We refer the reader to our previous book: B. L. Ioffe, V. A. Khoze and L. N. Lipatov, *Hard Process*, North Holland, 1984, where these topics were considered in detail.

We are greatly indebted to W. von Schlippe, who read the manuscript, performed the enormous work of improving the English, and gave very significant advice. We are very thankful to N. S. Libova, who also participated in improving the English and partly printed the text, and to M. N. Markina, who did the hard work of printing and preparing the manuscript. We acknowledge A. N. Sidorov, who kindly presented the data of parton distributions, A. Samsonov and A. Oganesian for drawing the figures, and S. Bass, who kindly sent his book to one of us (B.I.).

B. Ioffe, V. Fadin, L. Lipatov

1

General properties of QCD

1.1 QCD Lagrangian

As in any gauge theory, the quantum chromodynamics (QCD) Lagrangian can be derived with the help of the gauge invariance principle from the free matter Lagrangian. Since quark fields enter the QCD Lagrangian additively, let us consider only one quark flavour. We will denote the quark field $\psi(x)$, omitting spinor and colour indices [$\psi(x)$ is a three-component column in colour space; each colour component is a four-component spinor]. The free quark Lagrangian is:

$$\mathcal{L}_q = \overline{\psi}(x)(i\,\partial\!\!\!/ - m)\psi(x), \tag{1.1}$$

where m is the quark mass,

$$\partial\!\!\!/ = \partial_\mu \gamma_\mu = \frac{\partial}{\partial x_\mu}\gamma_\mu = \gamma_0\frac{\partial}{\partial t} + \gamma\frac{\partial}{\partial \boldsymbol{r}}. \tag{1.2}$$

The Lagrangian $\mathcal{L}_q$ is invariant under global (x–independent) gauge transformations

$$\psi(x) \to U\psi(x), \quad \overline{\psi}(x) \to \overline{\psi}(x)U^+, \tag{1.3}$$

with unitary and unimodular matrices U

$$U^+ = U^{-1}, \quad |U| = 1, \tag{1.4}$$

belonging to the fundamental representation of the colour group $SU(3)_c$. The matrices U can be represented as

$$U \equiv U(\theta) = \exp(i\theta^a t^a), \tag{1.5}$$

where θ^a are the gauge transformation parameters; the index a runs from 1 to 8; t^a are the colour group generators in the fundamental representation; and $t^a = \lambda^a/2$, λ^a are the Gell-Mann matrices.

Invariance under the global gauge transformations (1.3) can be extended to local (x-dependent) ones, i.e. to those where θ^a in the transformation matrix (1.5) is x-dependent. This can be achieved by introducing gluon fields $A_\mu^a(x)$ which transform according to

$$A_\mu(x) \to A_\mu^{(\theta)}(x) = U A_\mu(x)U^{-1} + \frac{i}{g}(\partial_\mu U)U^{-1}, \tag{1.6}$$

where g is a coupling constant,

$$A_\mu(x) = A_\mu^a(x)t^a, \tag{1.7}$$

and the partial derivative ∂_μ in (1.1) is replaced by the covariant derivative ∇_μ

$$\nabla_\mu \equiv \partial_\mu + igA_\mu. \tag{1.8}$$

The transformation law of the covariant derivative,

$$\nabla_\mu^{(\theta)} = \partial_\mu + igA_\mu^{(\theta)} = U\nabla_\mu U^+, \tag{1.9}$$

ensures the gauge invariance of the Lagrangian

$$\mathcal{L}_q + \mathcal{L}_{qg} = \overline{\psi}(x)(i\,\slashed{\nabla} - m)\psi(x) = \mathcal{L}_q - g\overline{\psi}(x)\slashed{A}(x)\psi(x). \tag{1.10}$$

The gauge field Lagrangian

$$\mathcal{L}_g = -\frac{1}{2}\,\mathrm{Tr}\left[G_{\mu\nu}(x)G_{\mu\nu}(x)\right] \tag{1.11}$$

is expressed in terms of the field intensities $G_{\mu\nu}$, which are built from the covariant derivatives ∇_μ

$$G_{\mu\nu}(x) = -\frac{i}{g}\left[\nabla_\mu, \nabla_\nu\right] = \partial_\mu A_\nu(x) - \partial_\nu A_\mu(x) + ig[A_\mu(x), A_\nu(x)]. \tag{1.12}$$

For the components

$$G_{\mu\nu}^a(x) = 2\,\mathrm{Tr}\left(t^a G_{\mu\nu}(x)\right) \tag{1.13}$$

we have

$$G_{\mu\nu}^a(x) = \partial_\mu A_\nu^a(x) - \partial_\nu A_\mu^a(x) - igT_{bc}^a A_\mu^b(x)A_\nu^c(x), \tag{1.14}$$

where $T_{bc}^a = -if^{abc}$ are group generators of the adjoint representation; f^{abc} are the group structure constants. Invariance of the Lagrangian (1.11) follows from the transformation law for the field strengths

$$G_{\mu\nu}^{(\theta)} = -\frac{i}{g}\left[\nabla_\mu^{(\theta)}, \nabla_\nu^{(\theta)}\right] = UG_{\mu\nu}U^+. \tag{1.15}$$

Thus, the QCD Lagrangian is

$$\mathcal{L}_{QCD} = \mathcal{L}_g + \mathcal{L}_q + \mathcal{L}_{qg}. \tag{1.16}$$

1.2 Quantization of the QCD Lagrangian

The Lagrangian (1.16) is invariant under the transformations (1.3)–(1.6). However, because of this invariance, a canonical quantization (i.e. exploiting operators with commutation laws $[p_i q_j] = -i\delta_{ij}$) of this Lagrangian is impossible, since the momentum, canonically conjugate to coordinate $A_0^a(x)$, is zero. The reason is that the gauge invariance means the presence of superfluous, nonphysical fields (degrees of freedom) in the Lagrangian. In other words, massless gluons have only two possible helicities: ± 1. Only a two-component field is required to describe them, but we have introduced the four-component field $A_\mu^a(x)$. This problem is similar to those that arise in quantum electrodynamics (QED), but the solution in QCD is more complicated because the symmetry group $SU(3)_c$ is non-abelian, unlike the abelian $U(1)$ in QED.

The quantization problem can be solved in several ways. One is elimination of superfluous degrees of freedom with the help of constraint equations (the Lagrange–Euler equation for the A_0^a-component does not contain the second-order time derivative of A_0^a and is therefore a constraint equation) and some gauge condition (for example, the Coulomb gauge condition $\partial_i A_i^a = 0$), which can be imposed due to gauge invariance of the Lagrangian. Clearly, this leads to the loss of an explicitly relativistic invariance of calculations. Of course, physical values obtained as a result of these calculations have correct transformation properties (because of their gauge invariance); however, the loss of explicit invariance of intermediate calculations can produce a sense of aesthetic protest; moreover, it can lead to more cumbersome calculations. This method is applied as a rule in cases where a noncovariant approach has evident advantages (as the Coulomb gauge in nonrelativistic problems).

In another approach, a term $\mathcal{L}_{fix}$, which allows one to perform the canonical quantization, is added to the Lagrangian (for example, $\mathcal{L}_{fix} = -\frac{1}{2}(\partial_\mu A_\mu^a)^2$). In classical theory, this term is neutralized by the corresponding gauge condition (in our example, the Lorentz condition $\partial_\mu A_\mu^a = 0$). In quantum theory, such a condition cannot be imposed since components A_μ^a are quantized independently. Therefore, the total space contains states with arbitrary numbers of pseudogluons, i.e. massless particles with nonphysical polarization vectors – timelike and longitudinal. In such an approach, the canonical quantization leads to an indefinite metric, i.e. a not positive definite norm of states, which clearly is incompatible with the probabilistic interpretation. Of course, the physical subspace contains only gluons with transverse polarizations, and the indefinite metric does not show up there. But the S-matrix built using $\mathcal{L}_{QCD} + \mathcal{L}_{fix}$ as a total Lagrangian has matrix elements between physical and nonphysical states. Recall that the same applies to QED. However, in QED, the corresponding S-matrix is unitary in the physical subspace, so that it can be successfully used for calculation of physical observables. This is not quite so in the QCD case because of self-interaction of gluons. Here, the unitarity in the physical subspace can be achieved only by embedding ghost fields with an unusual property: They must be quantized as fermions with spin zero. Originally, this was pointed out by Feynman [1], who suggested the method of resolving the problem in the one-loop approximation. Later the method was developed in more detail by DeWitt [2]–[4]. Now the most common method of quantization of gauge fields is based on the functional integral formalism. In this formalism, the impossibility of canonical quantization becomes apparent as the divergence of the functional integral $\int \mathcal{D}A \exp(iS_{QCD})$, where $S_{QCD} = \int d^4x \mathcal{L}_{QCD}$ is the classical action for the gluon fields. The reason of the divergence is that, since both the integration measure

$$\mathcal{D}A = \prod_{a,\mu,x} dA_\mu^a(x), \tag{1.17}$$

and S_{QCD} are invariant under the gauge transformations (1.3), (1.6), the integral contains as a factor the volume of the gauge group

$$\int \mathcal{D}\theta = \int \prod_{a,x} d\theta^a(x).$$ (1.18)

However, it is sufficient to integrate only over a subspace of gauge nonequivalent fields $A^a_\mu(x)$, i.e. fields not interconnected by the transformations (1.6). Such a subspace can be defined by the gauge condition

$$(\hat{G}A)^a(x) = B^a(x),$$ (1.19)

with some operator $\hat{G}$ acting on the fields A^c_μ. The factor (1.18) can be extracted explicitly using the trick suggested in [5], namely by exploiting the equality

$$\det \hat{M}(A) \int \mathcal{D}\theta \prod_{a,x} \delta((\hat{G}A^{(\theta)})^a(x) - B^a(x)) = 1,$$ (1.20)

where the operator $\hat{M}(A)$ is defined by its matrix elements

$$(M(A))^{ab}(x, y) = \left(\frac{\delta(\hat{G}A^{(\theta)})^a(x)}{\delta\theta^b(y)} \right),$$ (1.21)

and $\det \hat{M}(A)$ is called the Faddeev–Popov determinant. Here $\delta/\delta\theta$ is the functional derivative, which is defined for any functional $\mathcal{F}(\theta)$ as

$$\frac{\delta\mathcal{F}(\theta)}{\delta\theta^b(y)} = \lim_{\epsilon \to 0} \frac{\mathcal{F}(\theta^a(x) + \epsilon\delta^{ab}\delta(x-y)) - \mathcal{F}(\theta^a(x))}{\epsilon},$$ (1.22)

and the values of θ in (1.21) are determined by the solution of the equation $(\hat{G}A^{(\theta)})^a(x) = B^a(x)$ with given A and B. It is supposed that this solution is unique.

From the definition (1.20), it follows that $\det \hat{M}(A)$ is gauge invariant, since $(A^{(\tilde{\theta})})^\theta = A^{(\tilde{\theta}+\theta)}$. Therefore, we have

$$\int \mathcal{D}A \, e^{iS_{QCD}} = \int \mathcal{D}A \, e^{iS_{QCD}} \det \hat{M}(A) \int \mathcal{D}\theta \prod_{a,x} \delta\left((\hat{G}A^{(\theta)})^a(x) - B^a(x)\right)$$

$$= \int \mathcal{D}\theta \int \mathcal{D}A^{(\theta)} \det \hat{M}(A^{(\theta)}) \prod_{a,x} \delta\left((\hat{G}A^{(\theta)})^a(x) - B^a(x)\right) e^{iS_{QCD}}$$

$$= \int \mathcal{D}\theta \int \mathcal{D}A \det \hat{M}(A) \prod_{a,x} \delta\left((\hat{G}A)^a(x) - B^a(x)\right) e^{iS_{QCD}},$$ (1.23)

where $\det \hat{M}(A)$ is now given by (1.21) at $\theta = 0$, so that the integrand is θ-independent and we can omit the factor $\int \mathcal{D}\theta$. Therefore, the functional integral over the gauge field can be defined as the last expression in (1.23) without $\int \mathcal{D}\theta$. Moreover, since (1.23) does not depend on B, one can integrate it over B with an appropriate weight factor and define the functional integral as the result of integration. The commonly used definition is

$$\int \mathcal{D}B e^{-\frac{i}{2\xi} \int d^4x (B^a(x))^2} \int \mathcal{D}A \det \hat{M}(A) \prod_{a,x} \delta\left((\hat{G}A)^a(x) - B^a(x)\right) e^{iS_{QCD}}$$

$$= \int \mathcal{D}A \det \hat{M}(A) \, e^{iS_{QCD} - \frac{i}{2\xi} \int d^4x ((\hat{G}A)^a(x))^2}. \tag{1.24}$$

The latter exponential can be written as $\exp(i \int d^4x (\mathcal{L}_{QCD} + \mathcal{L}_{fix}))$, where the gauge fixing term is given by

$$\mathcal{L}_{fix} = -\frac{1}{2\xi}\left((\hat{G}A)^a(x)\right)^2. \tag{1.25}$$

The factor $\det \hat{M}(A)$ can also be presented in the Lagrangian form using nonphysical ghost fields $\phi^a(x)$ obeying Fermi–Dirac statistic. The functional integral formalism can be extended to the case of fermion fields, considering them as anticommuting Grassmann variables. For a finite number of such variables $\psi_i, i = 1 \div N$, $[\psi_i \psi_j]_+ \equiv \psi_i \psi_j + \psi_j \psi_i = 0$ for all i, j, the set of monomials

$$1, \ \psi_i, \ \psi_i \psi_j, \ldots, \psi_1 \psi_2 \ldots \psi_N \tag{1.26}$$

gives the basis of the linear space of the Grassmann algebra. There are left and right derivatives in this space that are defined by the rules

$$\frac{\partial}{\partial \psi_j}\left(\psi_{i_1} \psi_{i_2} \cdots \psi_{i_n}\right) = \begin{cases} (-1)^{k-1} \psi_{i_1} \cdots \psi_{i_{k-1}} \psi_{i_{k+1}} \cdots \psi_{i_n}, & \text{if } i_k = j, \\ 0, & \text{if } i_l \neq j, l = 1, 2, \ldots, n, \end{cases} \tag{1.27}$$

$$\left(\psi_{i_1} \psi_{i_2} \cdots \psi_{i_n}\right) \frac{\overleftarrow{\partial}}{\partial \psi_j} = \begin{cases} (-1)^{n-k} \psi_{i_1} \cdots \psi_{i_{k-1}} \psi_{i_{k+1}} \cdots \psi_{i_n}, & \text{if } i_k = j, \\ 0, & \text{if } i_l \neq j, l = 1, 2, \ldots, n. \end{cases} \tag{1.28}$$

Note that the derivatives are anticommuting:

$$\frac{\partial^2}{\partial \psi_i \partial \psi_j} = -\frac{\partial^2}{\partial \psi_j \partial \psi_i}. \tag{1.29}$$

The integrals over the Grassman variables are defined by the rules

$$\int d\psi_i = 0, \quad \int d\psi_i \, \psi_j = \delta_{ij}, \tag{1.30}$$

and

$$[d\psi_i, \psi_j]_+ = [d\psi_i, d\psi_j]_+ = 0. \tag{1.31}$$

These rules give the important relation

$$\int d\psi_1 d\psi_2 \cdots d\psi_N d\bar{\psi}_1 d\bar{\psi}_2 \cdots d\bar{\psi}_N \exp\left(-\sum_{i,j} \bar{\psi}_i M_{ij} \psi_j\right) = (-1)^{\frac{N(N+1)}{2}} \det M, \tag{1.32}$$

where M_{ij} are matrix elements (which can be complex numbers) of the matrix M, and $\bar{\psi}_i$ are the Grassmann variables that can be considered either as complex conjugate to ψ_i or as independent Grassman variables. In any case, they are supposed to anticommute with ψ_i. Recall that for ordinary complex numbers ϕ_i instead of (1.32) one has

$$\int d\phi_1 d\phi_2 \cdots d\phi_N d\phi_1^* d\phi_2^* \cdots d\phi_N^* \exp\left(-\sum_{i,j} \phi_i^* M_{ij}\phi_j\right) = (2\pi)^N \det^{-1} M. \quad (1.33)$$

The relation (1.32) allows $\det \hat{M}(A)$ in (1.24) as the functional integral over the Grassman variables $\phi^a(x)$ and $\phi^{a\dagger}(x)$ [where $\phi^{a\dagger}(x)$ can be the complex conjugate of $\phi^a(x)$ or can have nothing to do with $\phi^a(x)$]:

$$\det \hat{M}(A) = C \int \mathcal{D}\phi \mathcal{D}\phi^\dagger \exp\left[i \int d^4x d^4 y \phi^{\dagger a}(x) (M(A))^{ab}(x,y)\phi^b(y)\right], \quad (1.34)$$

where C is an inessential numerical constant. Usually, the operator $\hat{G}$ in (1.19) is local, so that one can write

$$(M(A))^{ab}(x,y)) = \hat{R}^{ab}(x)\delta(x-y), \quad (1.35)$$

and

$$\det \hat{M}(A) = C \int \mathcal{D}\phi \mathcal{D}\phi^\dagger \exp\left[i \int d^4x \phi^{\dagger a}(x)\hat{R}^{ab}(x)\phi^b(x)\right]$$

$$= C \int \mathcal{D}\phi \mathcal{D}\phi^\dagger \exp\left[i \int d^4x \mathcal{L}_{ghost}\right], \quad (1.36)$$

where $\mathcal{L}_{ghost}$ is the Lagrangian of the fields $\phi^a(x)$, which are called Faddeev–Popov ghosts. Therefore, one can introduce the QCD Lagrangian unitary in physical space

$$\mathcal{L}_{eff} = \mathcal{L}_{QCD} + \mathcal{L}_{fix} + \mathcal{L}_{ghost} = \bar{\psi}(x)(i\,\slashed{\nabla} - m)\psi(x) - \frac{1}{4}G^a_{\mu\nu}(x)G^a_{\mu\nu}(x)$$

$$- \frac{1}{2\xi}\left((\hat{G}A)^a(x)\right)^2 + \phi^{\dagger a}(x)\hat{R}^{ab}(x)\phi^b(x), \quad (1.37)$$

and consider it as a quantum Lagrangian, where the ghosts are fermions. We also can use this Lagrangian to describe several quark flavours; in this case ψ and $\bar{\psi}$, besides being Dirac bispinors and colour triplets, have also the flavour indices, $\slashed{\nabla}$ is multiplied by the unit matrix in flavour space and m is the mass matrix.

The Lagrangian (1.37) also can be used to construct the generating functional

$$Z[J, \eta, \bar{\eta}] = \int \mathcal{D}A\mathcal{D}\psi\mathcal{D}\bar{\psi}\mathcal{D}\phi\mathcal{D}\phi^\dagger \exp\left\{i \int d^4x \left(\mathcal{L}_{eff} + A^a_\mu(x)J^a_\mu(x)\right.\right.$$

$$\left.\left. + \bar{\psi}(x)\eta(x) + \bar{\eta}(x)\psi(x)\right)\right\}, \quad (1.38)$$

where $J^a_\mu(x)$, $\eta(x)$, and $\bar{\eta}(x)$ are sources for the gluon fields $A^a_\mu(x)$ and quark fields $\bar{\psi}(x)$ and $\psi(x)$, respectively. All sources and fields here are considered as classical, but whereas $J^a_\mu(x)$ and $A^a_\mu(x)$ are usual c–numbers, $\eta(x)$, $\bar{\eta}(x)$ and $\psi(x)$, $\bar{\psi}(x)$ are anticommuting

Grassman variables. This property has to be taken into account in defining Green functions with participation of quarks and antiquarks. For example, the quark propagator is defined as

$$\langle 0|T\left(\psi_\alpha(x)\overline{\psi}_\beta(y)\right)|0\rangle = -\frac{(-i)^2\delta^2 Z[J,\eta,\bar{\eta}]}{Z[0,0,0]\delta\bar{\eta}_\alpha(x)\delta\eta_\beta(y)}\bigg|_{J=\eta=\bar{\eta}=0}, \tag{1.39}$$

where $\psi_\alpha(x)$ and $\overline{\psi}_\beta(y)$ are Heisenberg operators, and α and β symbolize both spinor and colour indices. Green functions for any number of particles can be defined in a similar way as corresponding functional derivatives of the generating functional (1.38).

1.3 The Gribov ambiguity

When deriving (1.37), it was supposed that at any fixed A and B the equation $(\hat{G}A^{(\theta)})^a(x) = B^a(x)$ has a unique solution with respect to θ, i.e. the absence of any solution or the existence of several solutions were excluded. There are no examples for the first possibility (i.e. the absence of any solution), however, the existence of many solutions (i.e. many gauge-equivalent fields obeying the same gauge condition) is an ordinary case, as was pointed out by Gribov [6],[7].

The simplest example presented by Gribov [6] corresponds to the Coulomb gauge defined by the equation

$$\partial_i A_i(x) = 0, \tag{1.40}$$

for the case of the $SU(2)$ colour group, where $t^a = \tau^a/2$ and τ^a are the Pauli matrices. The existence of gauge-equivalent fields means that the equality

$$\partial_i A_i^{(\theta)}(x) = 0, \tag{1.41}$$

where

$$A_i^{(\theta)}(x) = UA_i(x)U^{-1} + \frac{i}{g}(\partial_i U)U^{-1}, \quad U = \exp(i\theta^a(x)t^a), \quad \partial_i A_i(x) = 0, \tag{1.42}$$

is valid for nontrivial $\theta^a(x)$ and the matrix U tending to the identity matrix at $|\mathbf{r}| \to \infty$. Taking account of the equalities

$$U^{-1}\partial_i U = -(\partial_i U^{-1})U, \quad \partial_i(\partial_i U)U^{-1} = -U\left(\partial_i(\partial_i U^{-1})U\right)U^{-1}, \tag{1.43}$$

it follows from (1.41), (1.42), that

$$\left[\nabla_i(A),(\partial_i U^{-1})U\right] = 0, \tag{1.44}$$

where $\nabla_i(A) = \partial_i + igA_i$.

Let us consider first the case $A = 0$. It can be shown, using (1.42), that in this case the solutions of (1.44) are transverse fields, which are gauge equivalent to $A = 0$. It is easy to demonstrate the existence of such solutions for the gauge group $SU(2)$. One can look for spherically symmetric solutions of the form

$$U = e^{i\theta(r)\mathbf{n}\tau} = \cos\theta + i\mathbf{n}\tau\sin\theta, \tag{1.45}$$

where $n = r/r$. Equation (1.44) follows from the extremum condition for the functional

$$W = \int d^3x \, \mathrm{Tr}\left((\partial_i U)(\partial_i U^{-1}) - 2ig A_i(\partial_i U^{-1})U\right). \tag{1.46}$$

For $A = 0$ and U given by (1.45), W is equal to

$$W = 2\int d^3x \left(\left(\frac{d\theta}{dr}\right)^2 + \frac{2\sin^2\theta}{r^2}\right), \tag{1.47}$$

so that Equation (1.44) gives

$$r\frac{d^2}{dr^2}(\theta r) - \sin(2\theta) = 0. \tag{1.48}$$

With $t = \ln r$, the equation turns into the equation of motion for a particle in the potential $-\sin^2\theta$ with friction:

$$\frac{d^2\theta}{dt^2} + \frac{d\theta}{dt} - \sin(2\theta) = 0. \tag{1.49}$$

The fields $A_i^{(\theta)}$ (1.42) are expressed through the solutions of (1.49) as:

$$A_{ai}^{(\theta)} = -\frac{2}{g}\left(n_a n_i \frac{d\theta}{dr} + (\delta_{ai} - n_a n_i)\frac{\sin(2\theta)}{2r} + \epsilon_{aib} n_b \frac{\sin^2\theta}{r}\right). \tag{1.50}$$

Fields nonsingular at $r = 0$ $(t = -\infty)$ correspond to motion of the particle which is at unstable equilibrium $\theta = 0$ at $t \to -\infty$ [points $\theta = n\pi$ are equivalent because the fields (1.50) do not change at $\theta \to \theta \pm \pi$]

$$\theta_{r\to 0} = ce^t = cr \tag{1.51}$$

where the constant c defines the motion. Because of friction $-\pi < \theta < \pi$ at any t and

$$\theta|_{r\to\infty} = \pm\frac{\pi}{2} + \frac{c_1}{\sqrt{r}}\cos\left(\frac{\sqrt{7}}{2}\ln r + c_2\right), \tag{1.52}$$

where the constants $c_{1,2}$ are determined by c in (1.51). The corresponding fields given by (1.50)

$$A_{ai}^{(\theta)}|_{r\to\infty} = -\frac{2}{g}\epsilon_{aib} n_b \frac{1}{r} \tag{1.53}$$

decrease as $1/r$.

Thus, there is a family of pure gauge transverse fields, i.e. nonzero fields $A^{(\theta)}$, which up to gauge transformations are equivalent to zero fields and satisfy the gauge condition (1.41). It is a particular case of a general statement that the gauge condition (1.41) does not fix uniquely the field from a family of gauge-equivalent fields.

The existence of a solution of (1.44) for an arbitrary field A_i^a means the existence of local extremes of the action (1.46), and can be easily understood for large fields. First, it is clear that $W = 0$ at $U = 1$ ($\theta^a = 0$), and it is the absolute minimum of the contribution of the first term in (1.46). At small θ^a, the functional W (1.46) becomes

$$W_0 = \int d^3x \frac{1}{2}\left(\theta^a(x)\hat{R}^{ab}(x)\theta^b(x)\right),\tag{1.54}$$

where $\hat{R}^{ab}(x) = -\partial_i D_i^{ab}(A)$ is the Faddeev–Popov operator (1.35) in the case of Coulomb gauge, $D_i^{ab}(A) = \partial_i\delta^{ab} + gf^{cab}A_i^c$. The operator $\hat{R}$ is analogous to the nonrelativistic Hamiltonian for a particle with spin in a velocity-dependent field. Evidently, if the field A_i^a is small (more precisely, the product of a typical magnitude of the field on a typical length of the space occupied by the field is small), then $\hat{R}$ has only positive eigenvalues. But if the field is sufficiently large and has an appropriate configuration (with a deep potential well), the appearance of negative eigenvalues is quite natural. Negative eigenvalues of $\hat{R}$ mean negative values of the functional W_0 (1.54). Since $\min W \leq W_0$ and $W = 0$ at $U = 1$, for such fields there is a nontrivial extremum of W (1.46), i.e. a solution of (1.41). Thus, for such fields the Coulomb gauge is not uniquely defined. Note that the extremum is reached for $U \to 1$ at $r \to \infty$, since the contribution of the first term in (1.46) is non-negative.

The existence of solutions of (1.42) can be demonstrated explicitly in the case when

$$A_i^a(x) = \frac{1}{g}\epsilon_{iaj}\frac{x_j}{r^2}f(r).\tag{1.55}$$

It is easy to see that $\partial_i A_i^a(x) = 0$. For such fields with the parameterization (1.45), the functional (1.46) takes the form of

$$W = 2\int d^3x\left(\left(\frac{d\theta}{dr}\right)^2 + \frac{2\sin^2\theta}{r^2}(1 - f(r))\right),\tag{1.56}$$

and instead of (1.49) one has

$$\frac{d^2\theta}{dt^2} + \frac{d\theta}{dt} - \sin(2\theta)(1 - f(r)) = 0.\tag{1.57}$$

Besides solutions of the same type as for $f = 0$, for sufficiently large f there can be other solutions corresponding to the particle in positions of unstable equilibrium $\theta = n\pi$ at $t \to \infty$ as well at $t \to -\infty$ (with or without change of n). The corresponding fields $A_i^{(\theta)}(x)$ (1.42) rapidly decrease with r if $f(r)$ decreases.

The problem of the existence of many gauge equivalent fields (Gribov copies) satisfying the same gauge condition is inherent not only in the Coulomb gauge, but for covariant gauge conditions as well. Thus, for the gauge $\partial A_\mu/\partial x_\mu = 0$, the existence of Gribov copies follows from the same line of argument as for (1.41), if the theory is formulated in the four-dimensional Euclidean space. On the contrary, axial and planar gauges were found to be free of Gribov ambiguity [8].

The existence of Gribov copies means that Eqs. (1.24) and (1.38) have to be improved. Gribov suggested [7] that the problem of copies can be solved if the integration in the functional space is restricted by the potentials for which the Faddeev–Popov determinant is positive (in Euclidean space). This restriction does not concern small fields, and therefore is not significant for perturbation theory. In particular, it is not significant for hard processes, where perturbation theory is applied. But the region of sufficiently large fields, where the operator $\hat{R}$ could have negative eigenvalues, is excluded from the integral (1.38).

The existence of Gribov copies is evidently important for lattice QCD. For investigation of their influence see, for instance, [9]–[11] and references therein.

1.4 Feynman rules

Feynman diagrams and Feynman rules are defined by the effective QCD Lagrangian (1.37). The rules for external lines are the same as in QED apart from evident colour indices. The rules for internal lines can be divided in two parts: independent and dependent of a choice of gauge. The quark propagator belongs to the first part. Using α, β for the Lorentz indices and i, j for colour indices, we have for the propagator

$$i\delta_{ij}\frac{(\not{p}+m)_{\alpha\beta}}{p^2-m^2+i0}. \qquad (1.58)$$

We will omit quark indices in the following. Then we have

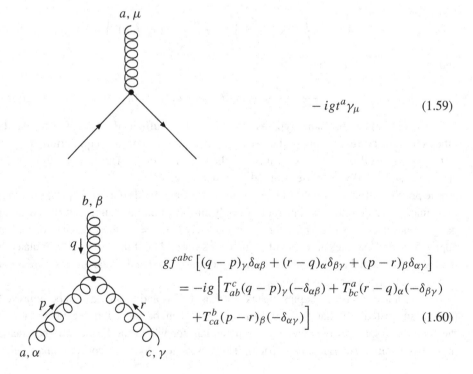

$$-igt^a\gamma_\mu \qquad (1.59)$$

$$gf^{abc}\left[(q-p)_\gamma\delta_{\alpha\beta}+(r-q)_\alpha\delta_{\beta\gamma}+(p-r)_\beta\delta_{\alpha\gamma}\right]$$
$$=-ig\left[T^c_{ab}(q-p)_\gamma(-\delta_{\alpha\beta})+T^a_{bc}(r-q)_\alpha(-\delta_{\beta\gamma})\right.$$
$$\left.+T^b_{ca}(p-r)_\beta(-\delta_{\alpha\gamma})\right] \qquad (1.60)$$

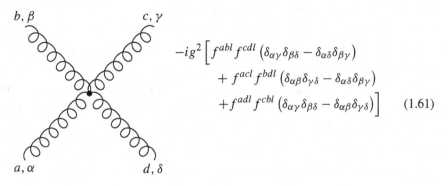

$$-ig^2 \left[f^{abl} f^{cdl} \left(\delta_{\alpha\gamma}\delta_{\beta\delta} - \delta_{\alpha\delta}\delta_{\beta\gamma} \right) \right.$$
$$+ f^{acl} f^{bdl} \left(\delta_{\alpha\beta}\delta_{\gamma\delta} - \delta_{\alpha\delta}\delta_{\beta\gamma} \right)$$
$$\left. + f^{adl} f^{cbl} \left(\delta_{\alpha\gamma}\delta_{\beta\delta} - \delta_{\alpha\beta}\delta_{\gamma\delta} \right) \right] \qquad (1.61)$$

The gluon propagator in any gauge can be presented as

$$i\delta^{ab} \frac{d_{\mu\nu}}{k^2 + i0} \qquad (1.62)$$

with gauge dependent $d_{\mu\nu}$.

A gauge is called covariant if $\mathcal{L}_{fix}$ does not depend on external vectors. A commonly used choice is

$$\mathcal{L}_{fix} = -\frac{1}{2\xi} \left(\partial_\mu A_\mu^a \right)^2. \qquad (1.63)$$

In this case

$$d_{\mu\nu} = -\delta_{\mu\nu} + (1 - \xi) \frac{k_\mu k_\nu}{k^2}, \qquad (1.64)$$

and one needs to introduce ghosts with the Lagrangian (note that it does not depend on the gauge parameter ξ)

$$\mathcal{L}_{ghost} = \left(\partial_\mu \phi^\dagger \right)^a \left(D_\mu \phi \right)^a = \left(\partial_\mu \phi^\dagger \right)^a \left(\partial_\mu \phi \right)^a - g f^{abc} \partial_\mu \phi^{\dagger a} A_\mu^b \phi^c, \qquad (1.65)$$

where

$$\left(D_\mu \phi \right)^a = D_\mu^{ab} \phi^b, \quad D_\mu^{ab} = \partial_\mu \delta^{ab} + ig A_\mu^c T_{ab}^c, \quad T_{ab}^c = -i f^{abc}. \qquad (1.66)$$

Therefore we need to add to the diagram elements presented above the ghost propagator

$$i\delta^{ab} \frac{1}{r^2 + i0}, \qquad (1.67)$$

and the ghost-gluon interaction vertex

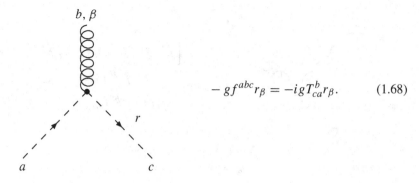

$$- g f^{abc} r_\beta = -i g T^b_{ca} r_\beta. \qquad (1.68)$$

Among the non-covariant gauges the most popular are the Coulomb, axial [12]–[16], and planar [17] gauges. The Coulomb gauge is a particular case of gauges with

$$\mathcal{L}_{fix} = -\frac{1}{2\xi} \left(\bar{\partial}_\mu A^a_\mu \right)^2, \qquad (1.69)$$

where $\bar{\partial}_\mu = \partial_\mu - n_\mu (n\partial)$, n is a timelike vector, $n^2 = 1$. With (1.69) we have

$$d_{\mu\nu} = -\delta_{\mu\nu} + \left(1 - \xi \frac{k^2}{k^2 - (kn)^2} \right) \frac{k_\mu k_\nu}{k^2 - (kn)^2} - \frac{kn}{k^2 - (kn)^2} (k_\mu n_\nu + n_\mu k_\nu), \qquad (1.70)$$

and

$$\mathcal{L}_{ghost} = \left(\bar{\partial}_\mu \phi^\dagger \right)^a \left(\bar{D}_\mu \phi \right)^a = \left(\bar{\partial}_\mu \phi^\dagger \right)^a \left(\bar{\partial}_\mu \phi \right)^a - g f^{abc} \left(\bar{\partial}_\mu \phi^{\dagger a} \right) A^b_\mu \phi^c. \qquad (1.71)$$

It follows from (1.71) that the Coulomb gauge requires the substitution $r \to \bar{r} = r - n(nr)$ in eqs. (1.67) and (1.68) for the ghost propagator and the ghost-gluon vertex.

The Coulomb gauge can be obtained as the limit of $\xi \to 0$, so that

$$d_{\mu\nu} = -\delta_{\mu\nu} + \frac{k_\mu k_\nu}{k^2 - (kn)^2} - \frac{kn}{k^2 - (kn)^2} (k_\mu n_\nu + n_\mu k_\nu). \qquad (1.72)$$

In the Lorentz frame, where $n = 0$, we have

$$\frac{d_{00}}{k^2} = \frac{1}{\mathbf{k}^2}, \quad \frac{d_{ij}}{k^2} = \left(\delta_{ij} - \frac{k_i k_j}{\mathbf{k}^2} \right) \frac{1}{k^2}. \qquad (1.73)$$

It is clear from (1.73), that the 00-component of the propagator corresponds to the Coulomb interaction, while the ij-components describe propagation of transverse gluons. Therefore, in contrast to covariant gauges, the Coulomb gauge allows a clear physical interpretation. On the other hand, an evident disadvantage of this gauge is its noncovariance. Moreover, it contains long-range action, that, taken literally, violates causality. Indeed, causality is violated in contributions of separate diagrams. But, of course, for physical values, which are represented by gauge invariant sets of diagrams, causality is restored.

The Coulomb gauge is convenient for calculations of nonrelativistic observables. Note that the ghost fields in this gauge are not actually dynamical variables, which is easily seen in the Lorentz frame where $n = (1, 0, 0, 0)$. Instead of using the ghost Lagrangian (1.71), one can rewrite the Faddeev–Popov determinant as

$$\det \left(\bar{\partial}_\mu \bar{D}_\mu \right) = \det \left(\bar{\partial}_\mu \bar{\partial}_\mu \right) \exp \left\{ \mathrm{Tr} \ln \left(1 - \frac{i}{\bar{\partial}_\mu \bar{\partial}_\mu} g \bar{\partial}_\nu A_\nu \right) \right\}, \qquad (1.74)$$

where $A_\mu = A_\mu^a t^a$, and expand the exponential function.

Axial and planar gauges are ghost-free. Generally, for the axial gauges

$$\mathcal{L}_{fix} = -\frac{1}{2\xi} \left(n_\mu A_\mu^a \right)^2, \qquad (1.75)$$

$$d_{\mu\nu} = -\delta_{\mu\nu} - \frac{k_\mu k_\nu}{(kn)^2} (n^2 + \xi k^2) + \frac{k_\mu n_\nu + n_\mu k_\nu}{(kn)}. \qquad (1.76)$$

These gauges are called *physical* gauges because of the absence of ghosts. The gauges with $\xi \to 0$ and $n^2 = 0$ are called light-cone gauges. In this case, we have

$$d_{\mu\nu} = \sum_{\lambda=1}^{2} e^*_{\lambda\,\mu} e_{\lambda\,\nu} + k^2 \frac{n_\mu n_\nu}{(kn)^2}, \qquad (1.77)$$

where the polarization vectors e_λ are transverse both to k and n. The light-cone gauges are widely employed, in particular, in perturbative analysis of deep inelastic scattering, where their use allows reduction of the analysis to effective ladder diagrams. Physical gauges are convenient because vertices for emission of a gluon with momentum k and physical polarization by a particle with momentum p in the massless limit vanish at zero-emission angle ϑ_{kp}. Consequently, for real gluons, the corresponding matrix elements behave as ϑ_{kp}^{-1}. Taking account of the angular integration measure $\vartheta d\vartheta$, this means that interference terms in cross sections have no collinear singularities (singularities at $\vartheta_{kp} = 0$). For virtual gluons, collinear singularities appear only in contributions of diagrams where a gluon line connects two lines with the same momentum.

However, it must be noted that physical gauges also have an evident inconvenience related to the use of external vectors. In particular, they require special renormalization procedures (see, for instance, Refs. [18]–[20]). Therefore, these gauges are used mainly for qualitative analyses and calculations of leading contributions. Moreover, even in one-loop calculations in the axial gauges, there is the problem of definition of the unphysical singularity at $(kn) = 0$. There are different prescriptions for treatment of this singularity, which seems in the one-loop case to be equivalent between themselves and to covariant gauges, but a strict proof of such statement for any number of loops is absent. Consequently, many-loop calculations are performed in covariant gauges.

Another example of a ghost-free gauge is the planar gauge [17], with

$$\mathcal{L}_{fix} = -\frac{1}{2n^2} (n_\mu A_\mu^a) \partial^2 (n_\nu A_\nu^a) \qquad (1.78)$$

and

$$d_{\mu\nu} = -\delta_{\mu\nu} + \frac{k_\mu n_\nu + n_\mu k_\nu}{(kn)}. \qquad (1.79)$$

1.5 Regularization

Calculating radiative corrections in field theories results in ultraviolet divergences in integrals over momenta of virtual particles. In renormalizable theories, these divergences can be hidden by transition from the bare charges and masses, appearing as parameters of an initial Lagrangian, to renormalized charges and masses. All physical quantities must not contain the divergences being expressed in terms of the renormalized charges and masses. But, in intermediate calculations – which are necessary in order to express the physical quantities through the renormalized parameters – one has to work with divergent integrals. In order to make the calculations work, one needs first to regularize the theory, i.e. to make the integrals convergent by introducing some regularization parameters.

Physically most transparent is the regularization by an ultraviolet cut-off Λ. This was widely used in quantum electrodynamics (see, for instance, [21] and references therein). Unfortunately, such regularization violates both Lorentz and gauge invariance, thus its use requires great care. Other methods of regularization worth mentioning include the Pauli–Villars method [22], where convergence of integrals is achieved by means of introduction of auxiliary fields with large (tending to infinity) mass and wrong statistics; the analytical regularization [23], consisting of change of the power of propagator denominators; the higher covariant derivative method [24],[25]; and the zeta-function method [26].

The most suitable technically and currently most popular method is the dimensional regularization [27]–[32], where the space-time dimension $D = 4 - 2\epsilon$ is kept different from the physical value 4. Positive values of ϵ remove ultraviolet divergences, while negative values make integrals convergent in the infrared region. Results of integration are presented as analytical functions of D. Therefore, the dimensional regularization can be used for both ultraviolet and infrared divergences. Both of them become apparent as poles at $\epsilon = 0$.

The only change of the generating functional (1.38) in dimensional regularization is the space-time dimension: $d^4x \rightarrow d^D x$. All required properties such as gauge invariance, unitarity, and so on, are preserved. Note, however, that values related to the space-time dimension are changed. Thus, canonical dimensions of fields (in mass units) become

$$[\psi] = [m]^{\frac{D-1}{2}}, \quad [A^a_\mu] = [m]^{\frac{D-2}{2}}, \quad [\phi] = [m]^{\frac{D-2}{2}}, \tag{1.80}$$

and the coupling g in the QCD Lagrangian (1.16) acquires the dimension

$$[g] = [m]^{\frac{4-D}{2}}. \tag{1.81}$$

The anticommutation relations of the Dirac matrices have the same form

$$\{\gamma_\mu, \gamma_\nu\} \equiv \gamma_\mu\gamma_\nu + \gamma_\nu\gamma_\mu = 2\delta_{\mu\nu}. \tag{1.82}$$

The normalization of Tr $(\gamma_\mu\gamma_\nu)$ is ambiguous. We will use also the same form as at $D = 4$:

$$\mathrm{Tr}\,(\gamma_\mu\gamma_\nu) = 4\delta_{\mu\nu}. \tag{1.83}$$

Since $\delta_{\mu\mu} = \delta_{\mu\nu}\delta_{\mu\nu} = D$, we have

$$\gamma_\mu\gamma_\nu\gamma_\mu = (2 - D)\gamma_\nu, \quad \gamma_\mu\gamma_\nu\gamma_\rho\gamma_\mu = 4\delta_{\nu\rho} + (D - 4)\gamma_\nu\gamma_\rho,$$
$$\gamma_\mu\gamma_\nu\gamma_\rho\gamma_\sigma\gamma_\mu = -2\gamma_\sigma\gamma_\rho\gamma_\nu - (D - 4)\gamma_\nu\gamma_\rho\gamma_\sigma. \tag{1.84}$$

The basic integral is

$$\int \frac{d^D k}{(2\pi)^D i} \frac{1}{(k^2 - 2(pk) - r^2 + i0)^\alpha} = \frac{(-1)^\alpha}{(4\pi)^{\frac{D}{2}}} \frac{\Gamma(\alpha - \frac{D}{2})}{\Gamma(\alpha)} \frac{1}{(r^2 + p^2 - i0)^{\alpha - \frac{D}{2}}}, \quad (1.85)$$

where $\Gamma(x)$ is the Euler gamma-function. Integrals with k in the numerator can be obtained from (1.85) by taking derivatives with respect to p. The most frequently encountered integrals are

$$\int \frac{d^D k}{(2\pi)^D i} \frac{k_\mu}{(k^2 - 2(pk) - r^2 + i0)^\alpha} = \frac{(-1)^\alpha}{(4\pi)^{\frac{D}{2}}} \frac{\Gamma(\alpha - \frac{D}{2})}{\Gamma(\alpha)} \frac{p_\mu}{(r^2 + p^2 - i0)^{\alpha - \frac{D}{2}}}, \quad (1.86)$$

$$\int \frac{d^D k}{(2\pi)^D i} \frac{k_\mu k_\nu}{(k^2 - 2(pk) - r^2 + i0)^\alpha} = \frac{(-1)^\alpha}{(4\pi)^{\frac{D}{2}}} \frac{\Gamma(\alpha - \frac{D}{2})}{\Gamma(\alpha)} \frac{1}{(r^2 + p^2 - i0)^{\alpha - \frac{D}{2}}}$$

$$\times \left(p_\mu p_\nu - \delta_{\mu\nu} \frac{(r^2 + p^2)}{2(\alpha - 1 - \frac{D}{2})} \right). \quad (1.87)$$

The case of several propagator denominators can be reduced to (1.85) with the help of the generalized Feynman parametrization

$$\frac{1}{a_1^{\alpha_1} a_2^{\alpha_2} \dots a_n^{\alpha_n}} = \frac{\Gamma(\alpha)}{\Gamma(\alpha_1)\Gamma(\alpha_2)\dots\Gamma(\alpha_n)} \int_0^1 dx_1 \int_0^1 dx_2 \cdots \int_0^1 dx_n \delta(1 - x_1 - x_2 - \cdots - x_n)$$

$$\times \frac{x_1^{\alpha_1 - 1} x_2^{\alpha_2 - 1} \dots x_n^{\alpha_n - 1}}{(a_1 x_1 + a_2 x_2 + \cdots + a_n x_n)^\alpha}, \quad (1.88)$$

where $\alpha = \alpha_1 + \alpha_2 + \cdots + \alpha_n$. This equation solves the problem when all the denominators are quadratic in k. In axial gauges there are denominators linear in k. In this case, the problem can be solved applying (1.88) to quadratic and linear denominators separately and then using the equality

$$\frac{1}{a^\alpha b^\beta} = \frac{\Gamma(\alpha + \beta)}{\Gamma(\alpha)\Gamma(\beta)} \int_0^\infty dz \frac{z^{\beta - 1}}{(a + bz)^{\alpha + \beta}}, \quad (1.89)$$

where a and b mean denominators quadratic and linear in k respectively.

1.6 γ5 problem

At the space-time dimension $D = 4$, the Dirac matrix γ_5 is defined as

$$\gamma_5 = \frac{-i}{4!} \epsilon_{\mu\nu\rho\sigma} \gamma_\mu \gamma_\nu \gamma_\rho \gamma_\sigma, \quad (1.90)$$

where $\epsilon_{0123} = 1$, and has the properties

$$\text{Tr}\left(\gamma_5 \gamma_\mu \gamma_\nu \gamma_\rho \gamma_\sigma \right) = 4i\epsilon_{\mu\nu\rho\sigma}, \quad (1.91)$$

$$\{\gamma_\mu \gamma_5\} = 0. \quad (1.92)$$

A well-known shortcoming of dimensional regularization is the absence of a natural generalization of the γ_5 to noninteger values $D \neq 4$ (see [33] for a review). This shortcoming seems to be unimportant in QCD, which is a parity conserving theory, so that, at first sight, the problem of definition of γ_5 at $D \neq 4$ is irrelevant for QCD. But it arises in calculations of QCD corrections to electro-weak processes. Moreover, different generalizations of γ_5 lead to discrepancies in calculations of radiative corrections to pure QCD processes. This problem also arises at the consideration of polarization phenomena, interactions with external axial currents, and so forth. In particular, the problem appears [34] because of the absence of a unique representation of 4-quark operators in connection with violation of the identity

$$\frac{1}{2}(\gamma_\mu\gamma_\nu\gamma_\rho - \gamma_\rho\gamma_\nu\gamma_\mu) \otimes \frac{1}{2}(\gamma_\mu\gamma_\nu\gamma_\rho - \gamma_\rho\gamma_\nu\gamma_\mu) = 6\gamma_5\gamma_\mu \otimes \gamma_5\gamma_\mu \qquad (1.93)$$

in dimensional regularization.

The definition (1.90) cannot be unambiguously continued to D dimensions. The problem is that the totally antisymmetric ϵ-tensor is a purely 4-dimensional object and cannot be self-consistently continued to D dimensions. In dimensional regularization, only computational rules are meaningful; the invariant tensors and gamma-matrices are considered formal objects obeying certain algebraic identities and no explicit values should be given to their indices [35],[36]. But a set of rules must be complete, consistent, and must lead to a satisfactory renormalization scheme. Although usability for practical calculations is not necessary, it is very desirable, and it is particularly desirable to conserve the anticommutativity of γ_5 with γ_μ. Unfortunately, this cannot be done, at least in a nonsophisticated way. Thus, generalization of (1.92) to D dimensions, together with $\delta_{\mu\mu} = D$ and the cyclic property of the trace operation, leads to

$$2D \operatorname{Tr} \gamma_5 = \operatorname{Tr}(\{\gamma_5\gamma_\alpha\}\gamma_\alpha) = 0,$$
$$2(D - 2) \operatorname{Tr}(\gamma_5\gamma_\mu\gamma_\nu) = -\operatorname{Tr}(\{\gamma_5\gamma_\alpha\}\gamma_\mu\gamma_\alpha\gamma_\nu) = 0,$$
$$2(D - 2)(D - 4) \operatorname{Tr}(\gamma_5\gamma_\mu\gamma_\nu\gamma_\rho\gamma_\sigma) = (D - 2) \operatorname{Tr}(\{\gamma_5\gamma_\alpha\}\gamma_\mu\gamma_\nu\gamma_\alpha\gamma_\rho\gamma_\sigma)$$
$$- 4(D - 2)\delta_{\mu\nu} \operatorname{Tr}(\gamma_5\gamma_\rho\gamma_\sigma) - 4(D - 2)\delta_{\rho\sigma} \operatorname{Tr}(\gamma_5\gamma_\mu\gamma_\nu) = 0, \qquad (1.94)$$

so that instead of (1.91) the trace $\gamma_5\gamma_\mu\gamma_\nu\gamma_\rho\gamma_\sigma$ should vanish at $D \neq 4$. But this contradicts the general requirement that every self-consistent regularization must be smooth at $D = 4$. Therefore, γ_5 anticommuting with γ_μ is forbidden in the usual dimensional regularization.

This discrepancy can be avoided within the framework of the modified form of dimensional regularization (called also dimensional reduction), as proposed by Siegel [37] for supersymmetric theories, where the Dirac algebra and vector fields are considered in four dimensions. Then, $\gamma_\mu\gamma_\mu = 4$ and it seems at first sight that (1.91) is allowed. Unfortunately, for the $4-D$-dimensional parts $\tilde{\gamma}_\mu$ of gamma matrices (which participate in loop traces), one comes to the same inconsistency as in dimensional regularization. A similar inconsistency [38] arises when one tries to generalize the 4–dimensional equality

$$\epsilon_{\mu_1\mu_2\mu_3\mu_4}\epsilon_{\nu_1\nu_2\nu_3\nu_4} = -\sum_{\mathcal{P}\in S_4} \operatorname{sign}\mathcal{P} \prod_{i=1}^{4} \delta_{\mu_i\nu_{\mathcal{P}(i)}}, \qquad (1.95)$$

where $\mathcal{P}$ is an element of the permutation group S_4. The reason for this inconsistency is related to the equality

$$\delta_{\mu\nu}\tilde{\delta}_{\nu\rho} = \tilde{\delta}_{\mu\rho}, \tag{1.96}$$

where $\delta_{\mu\nu}$ and $\tilde{\delta}_{\mu\nu}$ are metric tensors in D- and 4-dimensional spaces (so that $\{\gamma_\mu, \gamma_\nu\} = 2\tilde{\delta}_{\mu\nu}$). In turn, (1.96) follows from the requirement $\not{p}^2 = p^2$, and, consequently,

$$p_\mu p_\nu(\delta_{\mu\nu} - \tilde{\delta}_{\mu\nu}) = 0 \tag{1.97}$$

for any p.

Since (1.92) greatly simplifies spinorial calculations, there are many suggestions of ways to conserve it in sophisticated variants of dimensional reduction (with 4-dimensional Dirac algebra), including:

- Modifying (1.91) to

$$\text{Tr}\left(\gamma_5\gamma_\mu\gamma_\nu\gamma_\rho\gamma_\sigma\right) = 4i\epsilon_{\mu\nu\rho\sigma} + \mathcal{O}(D-4), \tag{1.98}$$

  where $\mathcal{O}(D-4)$ is some D-dependent constant which must be fixed order by order in perturbation theory by Ward identities and Bose symmetry [39].
- Rejecting the cyclicity of the the trace operation [40].
- Redefining the trace operation for the traces with odd numbers of γ-matrices [41],[42].

For other suggestions see, e.g. [43]–[45]. Such proposals were criticized in [33], with the conclusion that regularization through dimensional reduction is inconsistent. Nevertheless, the method can be used – with caution – in specific calculations. Thus, one can use (1.92) if traces with an odd number of γ_5 are not involved.

In some cases, the γ_5 problem can be solved by calculation of the imaginary part of the amplitude, which can be done in four dimensions, and restoration of the whole amplitude by using dispersion relations.

A consistent definition of γ_5 in D dimensions is possible if (1.92) is rejected. An explicit version of such definition was proposed by 't Hooft and Veltman [29] and elaborated in [46]. The general discussion is done in [36]. Introducing the "4-dimensional" metric tensor $\tilde{\delta}_{\mu\nu}$ with the properties

$$\tilde{\delta}_{\mu\nu} = \tilde{\delta}_{\nu\mu}, \quad \tilde{\delta}_{\mu\nu}\tilde{\delta}_{\nu\rho} = \tilde{\delta}_{\mu\rho}, \quad \delta_{\mu\nu}\tilde{\delta}_{\nu\rho} = \tilde{\delta}_{\mu\nu}\delta_{\nu\rho} = \tilde{\delta}_{\mu\rho}, \quad \tilde{\delta}_{\mu\mu} = 4, \tag{1.99}$$

and the tensor $\epsilon_{\mu\nu\rho\sigma}$ satisfying

$$\epsilon_{\mu_1\mu_2\mu_3\mu_4}\epsilon_{\nu_1\nu_2\nu_3\nu_4} = -\sum_{\mathcal{P}\in S_4}\text{sign}\,\mathcal{P}\prod_{i=1}^{4}\tilde{\delta}_{\mu_i\nu_{\mathcal{P}(i)}}, \tag{1.100}$$

γ_5 is defined by (1.90) and satisfies the equalities

$$\{\tilde{\gamma}_\mu\gamma_5\} = 0, \quad \{\gamma_\mu\gamma_5\} = 2\gamma_5(\gamma_\mu - \tilde{\gamma}_\mu), \tag{1.101}$$

$$\text{Tr}\,\gamma_5 = Tr(\gamma_5\gamma_\mu\gamma_\nu) = \text{Tr}(\gamma_5\gamma_{\mu_1}\gamma_{\mu_2}\cdots\gamma_{\mu_{2n+1}}) = 0,$$

$$\text{Tr}\left(\gamma_5\gamma_\mu\gamma_\nu\gamma_\rho\gamma_\sigma\right) = 4i\epsilon_{\mu\nu\rho\sigma}, \quad (\gamma_5)^2 = 1, \tag{1.102}$$

where $\tilde{\gamma}_\mu = \tilde{\delta}_{\mu\nu}\gamma_\nu$. The relations (1.101) correspond to the definition used by 't Hooft and Veltman [29] (γ_5 is anticommuting with γ_μ if $\mu = 0, 1, 2, 3$ and commuting in other cases), but do not explicitly violate relativistic invariance picking out the first four dimensions.

1.7 Renormalization

The bare charge g and mass m entering the QCD Lagrangian (1.16) are not physical values. The transition from the bare parameters to physical ones, i.e. to the quantities that can be (at least in principle) experimentally measured, is called renormalization. It is worthwhile to recall the following circumstances related to this procedure.[1]

First, the physical coupling constants and masses depend on the scale. This important fact was first realized in the famous papers by Landau, Abrikosov and Khalatnikov [48] and by Gell-Mann and Low [49].

Second, renormalization would make sense even if ultraviolet divergences were absent. Although in this case the renormalization would be finite, so that physical quantities could be expressed through bare parameters, it would be more convenient to express the former in terms of experimentally measured ones. In the actual case of the existence of ultraviolet divergences, renormalization is not only convenient but necessary.

Third, the renormalization procedure is not unique, that is clear from its physical meaning. Indeed, it is a question of our definition, i.e. of how we choose the parameters called the renormalized charge and mass. Moreover, instead of these we could use other physical quantities, for example, cross sections of two physical processes. The choice of the renormalization procedure is called the renormalization scheme. Note that in QCD, in contrast to QED, renormalization cannot result in transition from bare to physical (observable) charge and mass. The reason is that observable hadrons are colourless objects that are supposed to consist of coloured quarks and gluons; quarks and gluons are confined in hadrons and are not observed as isolated states.

Fourth, in theories with ultraviolet divergences, the renormalization is closely related to regularization. Evidently, connections between bare and renormalized quantities in such theories make sense only in the presence of some regularization. The choice of the regularization method is called regularization scheme. In what follows we will assume dimensional regularization. The renormalization scheme is closely related to the regularization scheme, and usually they are mentioned together. A complete definition of any theory requires not only knowledge of the Lagrangian, but also of the regularization and renormalization schemes. Of course, all regularization and renormalization schemes must lead to the same values for physical observables. Note, however, that although this statement seems obvious, it is inapplicable in any fixed order of perturbation theory (see Section 1.11 for details).

For physical quantities, such as probabilities and cross sections, which are related to S-matrix elements, the renormalization procedure is reduced to the transition from

---

[1] For more detail see, for instance, [47]

bare charges and masses to the renormalized ones (below the subscript 0 means bare quantities):

$$g_0 = Z_g \mu^\epsilon g, \quad m_0 = Z_m m. \tag{1.103}$$

Here, the dimensional regularization with $D = 4 - 2\epsilon$ is assumed and the factor μ^ϵ, where μ is the renormalization scale, is introduced in order to make the renormalized coupling dimensionless. Regularizations with $D = 4$ do not require this factor. Note that all renormalization constants Z_i are dimensionless.

However, the most general objects of field theories are Green functions. Their renormalization requires additional renormalization constants. Thus, for the quark and gluon propagators

$$G_0(p) = Z_2 G(p), \quad G_{0\,\mu\nu}^{(tr)ab}(k) = Z_3 G_{\mu\nu}^{(tr)ab}(k), \tag{1.104}$$

where the superscript (tr) means the transverse part, Z_2 and Z_3 are the renormalization constants. Renormalization is required also for the interaction vertices; omitting the Lorentz and colour structures, one can write

$$\Gamma_0^{(\bar{q}gq)}(p_i) = Z_{1q}^{-1}\Gamma^{(\bar{q}gq)}(p_i), \quad \Gamma_0^{(3g)}(k_i) = Z_1^{-1}\Gamma^{(3g)}(k_i), \quad \Gamma_0^{(4g)}(k_i) = Z_4^{-1}\Gamma^{(4g)}(k_i), \tag{1.105}$$

where $\Gamma^{(\bar{q}gq)}$, $\Gamma^{(3g)}$, and $\Gamma^{(4g)}$ are the renormalized one-particle irreducible vertices and Z_i are the renormalization constants. The renormalizability of the theory means that the renormalized quantities, being expressed in terms of the renormalized charges and masses, remain finite in the limit when the regularization is removed.

In the covariant gauges (1.63), one also needs to renormalize the ghost propagator, the vertex of the ghost-gluon interaction, and the gauge parameter ξ. The renormalization constants for the first two items are denoted $\tilde{Z}_3$ and $\tilde{Z}_1$, respectively; for the gauge parameter, the constant is usually taken equal to that of the gluon field:

$$\xi = Z_3^{-1}\xi_0. \tag{1.106}$$

The renormalized coupling constant can be defined using any of the interaction vertices. Its universality required by gauge invariance means that

$$Z_g^{-1} = Z_3^{1/2}Z_2 Z_{1q}^{-1} = Z_3^{3/2}Z_1^{-1} = Z_3 Z_4^{-1/2} = Z_3^{1/2}\tilde{Z}_3\tilde{Z}_1^{-1}. \tag{1.107}$$

Note that the definition of the charge renormalization constant Z_g in terms of the ghost-gluon interaction appears to be the most suitable for its calculation. Eq. (1.107) means that the renormalization constants are not independent. The relations

$$\frac{Z_3}{Z_1} = \frac{Z_2}{Z_{1q}} = \frac{Z_3^{1/2}}{Z_4^{1/2}} = \frac{\tilde{Z}_3}{\tilde{Z}_1} \tag{1.108}$$

are fulfilled as a consequence of gauge invariance. These relations appear as a particular case of the Slavnov–Taylor identities [50],[51], which correspond to the Ward identities in QED and guarantee the universality of the renormalized coupling constant g.

It is convenient to construct perturbation theory using renormalized quantities. The effective Lagrangian (1.37) is expressed in terms of the bare quantities, which must now carry the subscript 0. In terms of the renormalized fields

$$\psi = Z_2^{-1/2}\psi_0, \quad A_\mu^a = Z_3^{-1/2}A_{0\,\mu}^a, \quad \phi = \tilde{Z}_2^{-1/2}\phi_0, \qquad (1.109)$$

the Lagrangian (1.37) with the gauge-fixing term (1.63) takes the form of

$$\mathcal{L}_{eff} = Z_2\bar{\psi}\left(i\,\slashed{\partial} - Z_3 Z_g \mu^\epsilon g\,\slashed{A}^a t^a - Z_m m\right)\psi$$
$$- \frac{1}{4}Z_3\left(\partial_\mu A_\nu^a - \partial_\nu A_\mu^a - Z_3 Z_g \mu^\epsilon g f^{abc} A_\mu^b(x)A_\nu^c(x)\right)^2$$
$$- \frac{1}{2\xi}(\partial_\mu A_\mu^a)^2 + \tilde{Z}_2\left(\partial_\mu \phi^\dagger\right)^a \left(\delta^{ac}\partial_\mu - Z_3 Z_g \mu^\epsilon g f^{abc} A_\mu^b\right)\phi^c. \qquad (1.110)$$

The renormalized perturbation theory is obtained from (1.110) by writing the renormalized constants in the form of $Z_i = 1 + \Delta Z_i$ and considering the terms with ΔZ_i as perturbations. These terms are called the counterterms. Here, each of ΔZ_i is written as a series in g^2, and the coefficients of these series are determined by the renormalization conditions, order by order, of perturbation theory.

The renormalization in noncovariant gauges has specific peculiarities. In particular, in the planar gauge [see (1.78) and (1.79)] the field renormalization is nonmultiplicative [52]:

$$A_{0\,\mu}^a = z_3^{1/2}\left[A_\mu^a - n_\mu \frac{(nA^a)}{n^2}(1 - \tilde{z}_3^{-1})\right], \qquad (1.111)$$

where z_3 is a renormalization constant for physical field components but $\tilde{z}_3$ is related to the change of the gauge group under renormalization.

The renormalization constants are different in different renormalization schemes. The schemes that are commonly used can be divided into three classes: on-shell, off-shell, and minimal subtraction (MS)-like schemes.

In on-shell scheme, the renormalization constants are defined in such a way that the inverse renormalized propagators, together with their first derivatives and the renormalized vertices, are equal to the corresponding Born values on the mass shell. Such schemes usually are used in QED, but even in QED on-shell schemes have a certain inconvenience related to the infrared divergences. On-shell schemes are not used in QCD. The technical reason for this is that infrared divergences are much more severe in QCD than in QED. The physical reason is the absence of the mass shell for quarks and gluons due to confinement.

In the off-shell schemes, analogous conditions on the propagators and vertices are imposed at some off-shell values of momenta in the space-like region [53]–[56]. These schemes also are called momentum-space subtraction (MOM) schemes; values of momenta where the renormalization conditions are imposed are called renormalization scale or subtraction point.

The on-shell and the off-shell renormalization schemes can be applied to any regularization scheme but, MS-like schemes cannot because MS-like schemes assume dimensional regularization. In the MS scheme [57], μ in Eq. (1.103) is called renormalization scale, and renormalized quantities are obtained from the corresponding nonrenormalized quanties by subtraction of pole terms in $1/\epsilon$. Schemes that differ from the MS scheme by a shift of the renormalization scale μ are called MS-like schemes. The most frequently used is the $\overline{\text{MS}}$ scheme [58] which is obtained from the MS scheme by the replacement

$$\mu^2 \to \bar{\mu}^2 = \frac{\mu^2}{4\pi} e^{\gamma_E}, \tag{1.112}$$

where $\gamma_E = 0.57721\ldots$ is the Euler constant. The reason is that in one-loop calculations $1/\epsilon$ always appears in the combination

$$\frac{1}{\epsilon} - \gamma_E + \ln 4\pi \equiv \frac{1}{\bar{\epsilon}}. \tag{1.113}$$

Note that, unlike the MOM schemes, renormalization in the MS-like schemes does not mean imposing the renormalization conditions at some values of momenta, so that μ does not have the sense of a renormalization or subtraction point.

1.8 One-loop calculations

One-loop calculations are straightforward in the covariant gauges with dimensional regularization. The simplest object is the ghost self-energy part $M^2(p)$. One has

$$-i M^2(p)\delta^{ab} =$$

$$= \delta^{ab} \frac{g^2 C_A}{(2\pi)^D} \int \frac{d^D k \mu^{2\epsilon} p_\mu (p-k)_\nu d_{\mu\nu}(k)}{(k^2+i0)((k-p)^2+i0)}, \tag{1.114}$$

where $d_{\mu\nu}$ is the numerator of the gluon propagator (1.64), $C_A = N_c$. Using (1.88) and (1.85) we get

$$
\begin{aligned}
M^2(p) &= \frac{g^2 C_A p^2}{4(2\pi)^D i} \int \frac{d^D k \mu^{2\epsilon}}{(k^2+i0)((k-p)^2+i0)} \left(2 - (1-\xi)\frac{p^2}{k^2}\right) \\
&= \frac{g^2 C_A p^2 \Gamma(\epsilon)}{4(4\pi)^{\frac{D}{2}}} \left(\frac{\mu^2}{-p^2}\right)^\epsilon \int_0^1 dx (x(1-x))^{-\epsilon} \left(2 - \epsilon\frac{(1-\xi)}{x}\right) \\
&= \frac{g^2 C_A p^2 \Gamma(\epsilon)}{4(4\pi)^{\frac{D}{2}}} \left(\frac{\mu^2}{-p^2}\right)^\epsilon \frac{\Gamma^2(1-\epsilon)}{\Gamma(1-2\epsilon)} \left(\frac{2}{1-2\epsilon} + (1-\xi)\right).
\end{aligned} \tag{1.115}
$$

In the limit of $\epsilon \to 0$ and neglecting terms that vanish in this limit we obtain

$$M^2(p) = \frac{g^2 C_A p^2}{4(4\pi)^2} \left[\left(\frac{1}{\bar{\epsilon}} - \ln\left(\frac{-p^2}{\mu^2}\right)\right)(3-\xi) + 4\right]. \tag{1.116}$$

It follows from (1.116) that the ghost renormalization constant $\tilde{Z}_3$ in the MS scheme is

$$\tilde{Z}_3 = 1 + \frac{g^2 C_A}{4(4\pi)^2} (3-\xi)\frac{1}{\epsilon}. \tag{1.117}$$

For the quark self-energy part $\Sigma(p)$ one has:

$$-i\Sigma(p) = \frac{g^2 C_F}{(2\pi)^D} \int \frac{d^D k \mu^{2\epsilon} \gamma_\mu (\not{p} - \not{k} + m)\gamma_\nu d_{\mu\nu}}{(k^2 + i0)((k-p)^2 - m^2 + i0)}, \tag{1.118}$$

where $C_F = (N_c^2 - 1)/(2N_c)$. Using (1.84), (1.88) and (1.85), it is easy to get

$$\Sigma(p) = g^2 C_F \int_0^1 dx \int \frac{d^D k \mu^{2\epsilon}}{(2\pi)^D i} \left[\frac{\gamma_\mu(\not{p} - \not{k} + m)\gamma_\mu - (1-\xi)(m - \not{p})}{(k^2 - 2x(kp) + x(p^2 - m^2) + i0)^2} \right.$$

$$\left. - 2(1-\xi)\frac{(1-x)\,\not{k}(p^2 - m^2)}{(k^2 - 2x(kp) + x(p^2 - m^2) + i0)^3} \right] = \frac{g^2 C_F \Gamma(\epsilon)\mu^{2\epsilon}}{(4\pi)^{\frac{D}{2}}} \int_0^1 dx\, x^{-\epsilon}$$

$$\times \left[\frac{\gamma_\mu(\not{p}(1-x)+m)\gamma_\mu - (1-\xi)(m - \not{p})}{(m^2 - p^2(1-x))^\epsilon} + \epsilon(1-\xi)\frac{(1-x)\,\not{p}(p^2 - m^2)}{(m^2 - p^2(1-x))^{1+\epsilon}} \right], \tag{1.119}$$

hence in the limit of $\epsilon \to 0$

$$\Sigma(p) = \frac{g^2 C_F}{(4\pi)^2} \left[\frac{1}{\epsilon}(3m + \xi(m - \not{p})) + \not{p} - 2m \right.$$

$$+ \int_0^1 dx \left((2\,\not{p}(1-x) - 4m + (1-\xi)(m - \not{p})) \ln\left(\frac{x(m^2 - p^2(1-x))}{\mu^2} \right) \right.$$

$$\left. \left. + (1-\xi)\frac{(1-x)\,\not{p}(p^2 - m^2)}{(m^2 - p^2(1-x))} \right) \right]. \tag{1.120}$$

From (1.120) it follows that the renormalization constants Z_2 and Z_m in the MS scheme are:

$$Z_2 = 1 - \frac{g^2 C_F \xi}{(4\pi)^2} \frac{1}{\epsilon},$$

$$Z_m = 1 - \frac{3g^2 C_F}{(4\pi)^2} \frac{1}{\epsilon}, \tag{1.121}$$

so that Z_m is gauge independent and Z_2 turns out to be unity in the Landau gauge ($\xi = 0$).

The gluon polarization operator $\delta^{ab}\Pi_{\mu\nu}(k)$ contains contributions of quark, gluon, and ghost loops. In the first of these,

$$i\delta^{ab}\Pi^f_{\mu\nu}(k) \quad = \quad \text{(1.122)}$$

the tensor $\Pi^f_{\mu\nu}$ differs from the quark contribution to the photon polarization operator only by the coupling constant and the coefficient $1/2$ coming from the trace of colour matrices:

$$i\Pi^f_{\mu\nu}(k) = -\frac{g^2\mu^{2\epsilon}}{2}\sum_f \int \frac{d^D p}{(2\pi)^D} \frac{\text{Tr}\,[\gamma_\mu(\not{p}+m_f)\gamma_\nu(\not{p}-\not{k}+m_f)]}{(p^2-m_f^2+i0)((p-k)^2-m_f^2+i0)}, \quad \text{(1.123)}$$

where the sum goes over quark flavours. This part is gauge invariant. On the contrary, the part related to gluons is gauge dependent. In the covariant gauges, together with the proper gluon contribution,

$$i\delta^{ab}\Pi^g_{\mu\nu}(k) \quad = \quad \text{(1.124)}$$

it contains the ghost contribution

$$i\delta^{ab}\Pi^{gh}_{\mu\nu}(k) \quad = \quad . \quad \text{(1.125)}$$

For the tensors $\Pi^i_{\mu\nu}$ ($i = g, gh$), one has

$$i\Pi^i_{\mu\nu}(k) = \frac{g^2\mu^{2\epsilon}N_c}{2}\int \frac{d^D q\, d^D r\, \delta^D(k+q+r)}{(2\pi)^D(q^2+i0)(r^2+i0)}T^i_{\mu\nu}(k;q,r), \quad \text{(1.126)}$$

where

$$T^g_{\mu\nu}(k;q,r) = \gamma_{\mu\alpha\beta}(k,q,r)d_{\alpha\gamma}(q)d_{\beta\sigma}(r)\gamma_{\nu\gamma\sigma}(k,q,r), \quad \text{(1.127)}$$

$$T^{gh}_{\mu\nu}(k;q,r) = -2q_\mu(q+k)_\nu. \quad \text{(1.128)}$$

Here, $\gamma_{\mu\alpha\beta}$ represents the three-gluon vertex,

$$\gamma_{\mu\alpha\beta}(k, q, r) = (q - k)_\beta \, \delta_{\mu\alpha} + (r - q)_\mu \delta_{\alpha\beta} + (k - r)_\alpha \delta_{\mu\beta}, \tag{1.129}$$

and hence

$$
T^g_{\mu\nu}(k, q, r) = \frac{\delta_{\mu\nu}}{2} \left(9k^2 + (q - r)^2 \right) - \frac{9}{2} k_\mu k_\nu + \left(D - \frac{3}{2} \right) (q - r)_\mu (q - r)_\nu
$$
$$
- \frac{(1 - \xi)}{q^2} \left[\delta_{\mu\nu}(k^2 - r^2)^2 + k_\mu k_\nu q^2 - (k_\mu q_\nu + q_\mu k_\nu)(q^2 + 3kq) - q_\mu q_\nu (r^2 - 2k^2) \right]
$$
$$
+ \frac{(1 - \xi)}{r^2} \left[\delta_{\mu\nu}(k^2 - q^2)^2 + k_\mu k_\nu r^2 - (k_\mu r_\nu + r_\mu k_\nu)(r^2 + 3kr) - r_\mu r_\nu (q^2 - 2k^2) \right]
$$
$$
+ \frac{(1 - \xi)^2}{r^2 q^2} \left[r_\mu r_\nu (kq)^2 + q_\mu q_\nu (kr)^2 - (q_\mu r_\nu + r_\mu q_\nu)(rk)(qk) \right]. \tag{1.130}
$$

Integration over loop momenta is easily performed using the Feynman parametrization. The result is

$$
\Pi^f_{\mu\nu}(k) = \left(k_\mu k_\nu - k^2 \delta_{\mu\nu} \right) \Pi_f(k^2),
$$
$$
\Pi_f(k^2) = \frac{4g^2 \mu^{2\epsilon} \Gamma(\epsilon)}{(4\pi)^{D/2}} \sum_f \int_0^1 dx \, \frac{x(1 - x)}{(m_f^2 - k^2 x(1 - x) - i0)^\epsilon}; \tag{1.131}
$$
$$
\Pi^g_{\mu\nu}(k) = \frac{g^2 N_c}{2(4\pi)^{D/2}} \left(\frac{\mu^2}{-k^2 - i0} \right)^\epsilon \frac{\Gamma(\epsilon)\Gamma^2(1 - \epsilon)(1 - \epsilon)}{\Gamma(D)} \Big[k^2 \delta_{\mu\nu}(6D - 5)
$$
$$
- k_\mu k_\nu (7D - 6) + (1 - \xi)(k^2 \delta_{\mu\nu} - k_\mu k_\nu)
$$
$$
\times \left(2(D - 1)(2D - 7) - \frac{1 - \xi}{2}(D - 1)(D - 4) \right) \Big]; \tag{1.132}
$$

and

$$
\Pi^{gh}_{\mu\nu}(k) = \frac{g^2 N_c}{2(4\pi)^{D/2}} \left(\frac{\mu^2}{-k^2 - i0} \right)^\epsilon \frac{\Gamma(\epsilon)\Gamma^2(1 - \epsilon)(1 - \epsilon)}{\Gamma(D)} \Big[k^2 \delta_{\mu\nu} + k_\mu k_\nu (D - 2) \Big]. \tag{1.133}
$$

In contrast to the quark part, which is transverse, $\Pi^g_{\mu\nu}(k)$ and $\Pi^{gh}_{\mu\nu}(k)$ separately are not transverse. Nevertheless, their sum is transverse, as is required by gauge invariance.

For $\epsilon \to 0$ we have

$$
\Pi_f(k^2) = \frac{4g^2}{(4\pi)^2} \left(\frac{n_f}{6} \frac{1}{\bar\epsilon} - \sum_f \int_0^1 dx \, x(1 - x) \ln \left(\frac{m_f^2 - k^2 x(1 - x) - i0}{\mu^2} \right) \right), \tag{1.134}
$$

where n_f is the number of quark flavours,

$$
\Pi^g_{\mu\nu}(k) = \frac{g^2 N_c}{12(4\pi)^2} \left(\frac{1}{\bar\epsilon} - \ln \left(\frac{-k^2 - i0}{\mu^2} \right) + \frac{8}{3} \right) \Big[k^2 \delta_{\mu\nu}(19 - 12\epsilon) - k_\mu k_\nu (22 - 14\epsilon)
$$
$$
+ (1 - \xi)(k^2 \delta_{\mu\nu} - k_\mu k_\nu) \Big(2(3 - 14\epsilon) + (1 - \xi)3\epsilon \Big) \Big]; \tag{1.135}
$$

and

$$\Pi_{\mu\nu}^{gh}(k) = \frac{g^2 N_c}{12(4\pi)^2}\left(\frac{1}{\bar\epsilon} - \ln\left(\frac{-k^2 - i0}{\mu^2}\right) + \frac{8}{3}\right)\left[k^2\delta_{\mu\nu} + 2k_\mu k_\nu(1 - \epsilon)\right]. \quad (1.136)$$

Denoting the part related to gluons $\Pi_{\mu\nu}^G$, we have

$$\Pi_{\mu\nu}^G(k) = \Pi_{\mu\nu}^g(k) + \Pi_{\mu\nu}^{gh}(k) = (k_\mu k_\nu - k^2\delta_{\mu\nu})\Pi_G(k^2), \quad (1.137)$$

where

$$\Pi_G(k^2) = -\frac{g^2 N_c}{12(4\pi)^2}\left(\frac{1}{\bar\epsilon} - \ln\left(\frac{-k^2 - i0}{\mu^2}\right) + \frac{8}{3}\right)$$
$$\times \left[20 - 12\epsilon + (1 - \xi)\left(2(3 - 14\epsilon) + (1 - \xi)3\epsilon\right)\right]. \quad (1.138)$$

The transverse part of the gluon propagator is

$$G_{\mu\nu}^{tr}(k) = \frac{i}{k^2(1 + \Pi(k^2))}\left(-\delta_{\mu\nu} + \frac{k_\mu k_\nu}{k^2}\right), \quad (1.139)$$

where $\Pi(k^2) = \Pi_f(k^2) + \Pi_G(k^2)$. From (1.104), (1.134), (1.138) it follows that in the MS scheme

$$Z_3 = 1 - \frac{g^2}{(4\pi)^2}\frac{1}{\epsilon}\left[N_c\left(-\frac{13}{6} + \frac{\xi}{2}\right) + \frac{2}{3}n_f\right]. \quad (1.140)$$

Note that in the photon case,

$$\text{Im }\Pi_\gamma(s) = \frac{s}{4\pi\alpha}\sigma_{e^+e^-\to anything}^\gamma(s), \quad (1.141)$$

where $\sigma_{e^+e^-\to anything}^\gamma$ is the total cross section of one-photon e^+e^- annihilation. This means that $\text{Im }\Pi_\gamma$ is positive definite. This is evidently true for the quark contribution to the gluon polarization operator. From (1.134), we have

$$\text{Im }\Pi_f(k^2) = \frac{g^2}{(4\pi)}\frac{1}{6}\sum_f \frac{k^2 + 2m_f^2}{k^2}\sqrt{\frac{k^2 - 4m_f^2}{k^2}}\,\theta(k^2 - 4m_f^2). \quad (1.142)$$

Instead, we see from (1.138) that, first of all, in the gluon case $\text{Im }\Pi_G$ is gauge dependent:

$$\text{Im }\Pi_G(k^2) = -\frac{\pi g^2 N_c}{6(4\pi)^2}\left[10 + 3(1 - \xi)\right], \quad (1.143)$$

and is negative at $\xi < 13/3$. Separate contributions of ghost fields and gluon fields to $\Pi_{\mu\nu}$ are not transverse. Nevertheless, it is possible to give them some sense considering their convolutions with polarization vectors $e(k)$ satisfy the transversality and normalization conditions

$$e_\mu k_\mu = 0, \quad e_\mu e_\mu^* = -1. \quad (1.144)$$

Then we have

$$e_\mu e_\nu^* \, \text{Im} \, \Pi_{\mu\nu}^g (k^2) = -\frac{\pi g^2 N_c}{12(4\pi)^2} k^2 \left[19 + 6(1 - \xi) \right], \tag{1.145}$$

$$e_\mu e_\nu^* \, \text{Im} \, \Pi_{\mu\nu}^{gh} (k^2) = -\frac{\pi g^2 N_c}{12(4\pi)^2} k^2, \tag{1.146}$$

and

$$e_\mu e_\nu^* \, \text{Im} \, \Pi_{\mu\nu}^G (k^2) = -\frac{\pi g^2 N_c}{6(4\pi)^2} k^2 \left[10 + 3(1 - \xi) \right]. \tag{1.147}$$

Evidently, the negative sign of (1.146) is related to "wrong statistics" of the ghosts. As for (1.145), it receives contributions from intermediate gluons both with physical and unphysical polarizations; the first contribution is positive, whereas the last one is gauge dependent. To find the first contribution we can use the Cutkosky rule. The discontinuity of (1.135) is obtained by the substitution

$$\frac{1}{r^2 + i0} \to -2\pi i \delta(r^2), \qquad \frac{1}{q^2 + i0} \to -2\pi i \delta(q^2). \tag{1.148}$$

After that we can perform summation over physical polarizations by substitution in (1.127):

$$- d_{\mu\nu} \to \delta_{\mu\nu} - \frac{q_\mu r_\nu + r_\mu q_\nu}{(qr)}. \tag{1.149}$$

As a result, we obtain

$$\text{Im} \, \Pi_{\mu\nu}^{g \, phys} = \frac{\pi g^2 N_c}{3(4\pi)^2} \left(k_\mu k_\nu - k^2 \delta_{\mu\nu} \right), \tag{1.150}$$

so that the contribution of physical gluons in the intermediate state to the polarization tensor is transverse, and

$$e_\mu e_\nu^* \, \text{Im} \, \Pi_{\mu\nu}^{g \, phys} = \frac{\pi g^2 N_c}{3(4\pi)^2} k^2, \tag{1.151}$$

the contribution to the cross section, is positive.

Calculations of the one-loop vertex parts is considerably more complicated. We do not present them here. Instead, let us summarize the one-loop renormalization constants in the covariant gauges. Denoting

$$Z_i = 1 - \frac{g^2}{(4\pi)^2} \frac{1}{\epsilon} a_i, \quad \tilde{Z}_i = 1 - \frac{g^2}{(4\pi)^2} \frac{1}{\epsilon} \tilde{a}_i, \tag{1.152}$$

in the theory with n_f quark flavours we have

$$a_m = 3C_F, \quad a_2 = C_F \xi, \quad a_3 = N_c \left(-\frac{13}{6} + \frac{\xi}{2} \right) + \frac{2}{3} n_f, \quad a_1 = N_c \left(-\frac{17}{12} + \frac{3\xi}{4} \right) + \frac{2}{3} n_f,$$

$$a_{1q} = N_c \frac{3 + \xi}{4} + C_F \xi, \quad a_4 = N_c \left(-\frac{2}{3} + \xi \right) + \frac{2}{3} n_f, \quad \tilde{a}_3 = N_c \frac{-3 + \xi}{4}, \quad \tilde{a}_1 = N_c \frac{\xi}{2}. \tag{1.153}$$

From (1.152) and (1.153) it is easy to see that the relations (1.108) are fulfilled and Eq. (1.107) gives

$$Z_g = 1 - \frac{g^2}{(4\pi)^2} \frac{1}{\epsilon} \left(\frac{11}{6} N_c - \frac{1}{3} n_f \right), \tag{1.154}$$

independent of ξ, as it should be.

1.9 Renormalization group

As we have discussed, there is a great arbitrariness in the renormalization. First, one can use different renormalization schemes. Then, after choosing a renormalization scheme, there is an arbitrariness in the renormalization scale μ. The renormalization can be carried out at any scale. The parameters of the theory renormalized at two different scales are connected with each other by finite renormalizations. The set of finite renormalizations constitutes the renormalization group and obeys functional equations characteristic of the group. For an infinitesimal change of the scale μ, the functional equations are reduced to differential equations that are called renormalization group equations [49], [59]–[65].

It is worth mentioning again that the difference between MOM and MS-like schemes is that in the MS-like schemes the renormalization scale μ is not really a renormalization point. In the MS scheme, only pole terms are subtracted, so the renormalization constants in this scheme are mass-independent (when expressed in terms of g they depend neither on the quark masses m_q nor on the renormalization scale μ). In the MOM schemes, renormalization constants contain dependence on m_q and μ. This circumstance leads, in particular, to a difference in the renormalization group equations.

In the MS scheme, indicating explicitly the dependence of the renormalized charge g on the renormalization scale μ, we have

$$g(\mu) = \mu^{-\epsilon} Z_g^{-1} g_0. \tag{1.155}$$

The renormalization constant Z_g in this scheme takes the form

$$Z_g = 1 + \sum_{n=1}^{\infty} \frac{Z_g^{(n)}}{\epsilon^n}, \quad Z_g^{(n)} = \sum_{k=n}^{\infty} c_{kn} g^{2k}. \tag{1.156}$$

Since g_0 is independent of μ, we get

$$\frac{d \ln g}{d \ln \mu} = -\epsilon - g \frac{d \ln Z_g}{dg} \frac{d \ln g}{d \ln \mu}, \tag{1.157}$$

so that

$$\frac{d \ln g}{d \ln \mu} = \frac{-\epsilon}{1 + g \, d \ln Z_g / dg} = -\epsilon + g \frac{dZ_g^{(1)}}{dg}. \tag{1.158}$$

The last equality follows from finiteness of the left-hand side in the limit $\epsilon \to 0$, which means cancelation of all pole terms in its Taylor series in g. Defining in this limit

$$\frac{d \ln g}{d \ln \mu} = \beta(g), \tag{1.159}$$

we have

$$\beta(g) = g \frac{dZ_g^{(1)}}{dg}. \tag{1.160}$$

Note that whereas (1.160) relates to the MS-scheme, (1.159) serves by the definition of the β-function in any scheme with renormalization scale μ. It follows from this definition that all schemes that differ by a shift of the renormalization scale, as MS-like schemes, have the same β-function; in particular, β-functions of the MS and $\overline{\text{MS}}$ schemes coincide. Let us stress that everything stated above applies equally to both massless and massive quarks, since in the MS-like schemes the renormalization constants are mass-independent.

The mass must be renormalized too. In the MS scheme the renormalization is written as

$$m_0 = Z_m \, m(\mu), \tag{1.161}$$

where m_0 and $m(\mu)$ are the bare and the renormalized masses, respectively, and Z_m is the mass renormalization constant, which is expanded similarly to (1.156):

$$Z_m = 1 + \sum_{n=1}^{\infty} \frac{Z_m^{(n)}}{\epsilon^n}, \quad Z_m^{(n)} = \sum_{k=n}^{\infty} d_{kn} g^{2k}. \tag{1.162}$$

By definition,

$$\frac{d \ln m(\mu)}{d \ln \mu} = -\gamma_m(g(\mu)), \tag{1.163}$$

where γ_m is the mass anomalous dimension. Using μ–independence of m_0 and Eqs. (1.158) and (1.160), one has

$$\frac{d \ln m(\mu)}{d \ln \mu} = -\frac{d \ln Z_m}{dg} \frac{dg}{d \ln \mu} = -\frac{d \ln Z_m}{dg} g \big(-\epsilon + \beta(g) \big). \tag{1.164}$$

From finiteness of the left-hand side it follows in the limit of $\epsilon \to 0$

$$\gamma_m(g) = -g \frac{dZ_m^{(1)}}{dg}. \tag{1.165}$$

It is worth noting that the mass-independent functions $\beta(g)$ and $\gamma_m(g)$ in the MS-like schemes are also gauge-independent [66],[67].

Let us write $\beta(g)$ as

$$\beta(g) = -\beta_0 \frac{\alpha_s}{4\pi} - \beta_1 \frac{\alpha_s^2}{(4\pi)^2} - \beta_2 \frac{\alpha_s^3}{(4\pi)^3} - \beta_3 \frac{\alpha_s^4}{(4\pi)^4} \cdots, \tag{1.166}$$

where $\alpha_s = g^2/(4\pi)$. As follows from (1.160), in the MS-scheme the coefficients β_i are determined by the residues of the first-order poles in ϵ in the renormalization constant Z_g. From (1.154), we come to the result obtained in [68],[69]:

$$\beta_0 = \frac{11}{3} N_c - \frac{2}{3} n_f = 11 - \frac{2}{3} n_f. \tag{1.167}$$

The positive sign of β_0 (at a not very large n_f) is the striking phenomenon providing the remarkable property of QCD – asymptotic freedom. Due just to this property, QCD has become the theory of strong interactions.

The two-loop beta-function obtained in [70],[71],[72] is presented as

$$\beta_1 = \frac{34}{3}C_A^2 - 4C_F T_F n_f - \frac{20}{3}C_A T_F n_f = \frac{34}{3}N_c^2 - \left(\frac{13}{3}N_c - \frac{1}{N_c}\right)n_f$$

$$= 102 - \frac{38}{3}n_f, \tag{1.168}$$

where

$$C_A = N_c, \quad C_F = \frac{N_c^2 - 1}{2N_c}, \quad T_F = \frac{1}{2}. \tag{1.169}$$

The three-loop result [73],[74], valid for an arbitrary semi-simple compact Lie group, is:

$$\beta_2 = \frac{2857}{54}C_A^3 + 2C_F^2 T_F n_f - \frac{205}{9}C_F C_A T_F n_f - \frac{1415}{27}C_A^2 T_F n_f$$

$$+ \frac{44}{9}C_F T_F^2 n_f^2 + \frac{158}{27}C_A T_F^2 n_f^2 = \frac{2857}{2} - \frac{5033}{18}n_f + \frac{325}{54}n_f^2. \tag{1.170}$$

The coefficient β_3 in the MS-like schemes was found [75] also for an arbitrary semi-simple compact Lie group. In the case of QCD, it takes the form of

$$\beta_3 = \left(\frac{149753}{6} + 3564\zeta(3)\right) - \left(\frac{1078361}{162} + \frac{6508}{27}\zeta(3)\right)n_f$$

$$+ \left(\frac{50065}{162} + \frac{6472}{81}\zeta(3)\right)n_f^2 + \frac{1093}{729}n_f^3, \tag{1.171}$$

where $\zeta(n)$ is the Riemann *zeta*-function. Numerically for QCD,

$$\beta_0 = 11 - 0.66667n_f,$$
$$\beta_1 = 102 - 12.6667n_f,$$
$$\beta_2 = 1428.50 - 279.611n_f + 6.01852n_f^2,$$
$$\beta_3 = 29243.0 - 6946.30n_f + 405.089n_f^2 + 1.49931n_f^3. \tag{1.172}$$

As already noted, multi-loop calculations are performed in covariant gauges. In particular, the result (1.171) was obtained in [75] using a general covariant gauge (1.63). The explicit cancellation of the gauge dependence in the β-function was used as an important check of the correctness of the calculations. The charge renormalization constant and β-function were obtained by calculating the renormalization constants for the ghost-ghost-gluon vertex, for the ghost propagator, and for the gluon propagator.

According to (1.159), the charges renormalized at different scales are connected by the equation

$$\ln\left(\frac{\mu}{\mu_0}\right) = \int\limits_{g(\mu_0)}^{g(\mu)} \frac{dg}{g\beta(g)}. \tag{1.173}$$

In the one-loop approximation for the coupling constant we have

$$\alpha_s(\mu) = \frac{\alpha_s(\mu_0)}{1 + \frac{\alpha_s(\mu_0)}{4\pi} \beta_0 \ln\left(\frac{\mu^2}{\mu_0^2}\right)}. \tag{1.174}$$

Usually, instead of an arbitrary point μ_0, the position Λ_{QCD} of the infrared pole of $g(\mu)$ is used:

$$\ln\left(\frac{\mu}{\Lambda_{QCD}}\right) = -\int_{g(\mu)}^{\infty} \frac{dg}{g\beta(g)}. \tag{1.175}$$

In the one-loop approximation this gives

$$\alpha_s(\mu) = \frac{4\pi}{\beta_0 \ln\left(\frac{\mu^2}{\Lambda_{QCD}^2}\right)}. \tag{1.176}$$

In higher orders, the definition (1.175) becomes unhandy; it is more convenient to write the solution of (1.159) as

$$\beta_0 \ln\left(\frac{\mu^2}{\Lambda^2}\right) = L\left(\frac{\alpha_s(\mu)}{4\pi}\right), \tag{1.177}$$

with

$$L(a) = \frac{1}{a} + \frac{\beta_1}{\beta_0} \ln(\beta_0 a) + \int_0^a dx \left[\frac{1}{x^2} - \frac{\beta_1}{\beta_0}\frac{1}{x} + \frac{\beta_0}{x\beta(4\pi\sqrt{x})}\right], \tag{1.178}$$

where $\beta(g)$ is given by (1.166). One can check by direct differentiation of (1.178) that (1.177) provides the correct evolution of α_s with μ. The parameter Λ in (1.177) is the integration constant (different from Λ_{QCD}). From the expansion

$$\frac{\beta_0}{x\beta(4\pi\sqrt{x})} = -\frac{1}{x^2} + \frac{\beta_1}{\beta_0 x} + \mathcal{O}(1) \tag{1.179}$$

one sees that the integrand in (1.178) is not singular at $x = 0$ and the integral can be expanded in powers of a, that makes the representation (1.177) more suitable than (1.175). In the three-loop approximation for the beta-function (1.166) this representation gives

$$\alpha_s(\mu) = \frac{4\pi}{\beta_0 \ln\left(\frac{\mu^2}{\Lambda^2}\right)} \left[1 - \frac{\beta_1}{\beta_0^2} \frac{\ln\left(\ln\left(\frac{\mu^2}{\Lambda^2}\right)\right)}{\ln\left(\frac{\mu^2}{\Lambda^2}\right)} + \frac{\beta_1^2}{\beta_0^4 \ln^2\left(\frac{\mu^2}{\Lambda^2}\right)} \right.$$
$$\left. \times \left(\left(\ln\left(\ln\left(\frac{\mu^2}{\Lambda^2}\right)\right) - \frac{1}{2}\right)^2 + \frac{\beta_2\beta_0}{\beta_1^2} - \frac{5}{4}\right)\right]. \tag{1.180}$$

The mass anomalous dimension (1.163) is known now also in four loops [76]. Writing it as

$$\gamma_m = \gamma_0 \frac{\alpha_s}{4\pi} + \gamma_1 \frac{\alpha_s^2}{(4\pi)^2} + \gamma_2 \frac{\alpha_s^3}{(4\pi)^3} + \gamma_3 \frac{\alpha_s^4}{(4\pi)^4} \cdots, \tag{1.181}$$

we have

$$\gamma_0 = 6C_F,$$

$$\gamma_1 = 3C_F^2 + \frac{97}{3}C_F C_A - \frac{20}{3}C_F T_F n_f,$$

$$\gamma_2 = 129C_F^3 - \frac{129}{2}C_F^2 C_A + \frac{11413}{54}C_A^2 C_F + 4C_F^2 T_F n_f (24\zeta(3) - 23)$$

$$- 8C_F C_A T_F n_f \left(12\zeta(3) + \frac{139}{27}\right) - \frac{280}{27}C_F T_F^2 n_f^2,$$

$$\gamma_3 = \frac{4603055}{81} + \frac{271360}{27}\zeta(3) - 17600\zeta(5) + 2\left(\frac{18400}{9}\zeta(5) + 880\zeta(4) - \frac{34192}{9}\zeta(3) - \frac{91723}{27}\right)n_f$$

$$+ 2\left(\frac{5242}{243} + \frac{800}{9}\zeta(3) - \frac{164}{3}\zeta(4)\right)n_f^2 + 2\left(\frac{64}{27}\zeta(3) - \frac{332}{243}\right)n_f^3. \tag{1.182}$$

Here, the first three coefficients are given for an arbitrary semi-simple compact Lie group and the fourth for the case of QCD. The one-loop contribution can be obtained from (1.121) and the two- and three-loop contributions were found in [77] and [78],[79], respectively. Numerically for QCD

$$\gamma_0 = 8,$$

$$\gamma_1 = 134.667 - 4.44445 \, n_f,$$

$$\gamma_2 = 2498 - 292.367 \, n_f - 3.45679 \, n_f^2,$$

$$\gamma_3 = 50659 - 9783.04 \, n_f - 141.395 \, n_f^2 + 2.96613 \, n_f^3. \tag{1.183}$$

From Eqs. (1.163) and (1.159) one obtains

$$m(\mu) = m(\mu_0) \exp\left(-\int_{g(\mu_0)}^{g(\mu)} \frac{dg}{g} \frac{\gamma_m(g)}{\beta(g)}\right). \tag{1.184}$$

Expansion of the ratio $\gamma_m(g)/\beta(g)$ in powers of g^2 gives

$$m(\mu) = m(\mu_0) \left(\frac{g(\mu)}{g(\mu_0)}\right)^{\gamma_0/\beta_0} \frac{\left(1 + \sum_{n=1}^{\infty} A_n \left(\alpha_s(\mu)/\pi\right)^n\right)}{\left(1 + \sum_{n=1}^{\infty} A_n \left(\alpha_s(\mu_0)/\pi\right)^n\right)}. \tag{1.185}$$

The first three coefficients A_n are determined by expansions (1.166) and (1.181). In particular,

$$A_1 = \frac{1}{8}\left(\frac{\gamma_1}{\beta_0} - \frac{\beta_1\gamma_0}{\beta_0^2}\right). \tag{1.186}$$

Numerically $A_n \sim 1$ for $n = 1, 2, 3$.

In the renormgroup equations for Green functions, both their mass and gauge dependence need to be taken into account. The connection of the bare Γ_0 and renormalized Γ functions is written as

$$\Gamma_0(p_i; g_0, m_0, \xi_0) = Z_\Gamma^{-1} \Gamma(p_i; g, m, \xi, \mu), \tag{1.187}$$

where

$$m = Z_m^{-1} m_0, \quad \xi = Z_3^{-1} \xi_0. \tag{1.188}$$

Independence of the unrenormalized function of μ leads to the equation

$$\left(\mu \frac{\partial}{\partial \mu} + g\beta(g) \frac{\partial}{\partial g} + \gamma_\Gamma(g, \xi) - \gamma_m(g)m \frac{\partial}{\partial m} + \delta(g, \xi)\xi \frac{\partial}{\partial \xi} \right) \Gamma(p_i; g, m, \xi, \mu) = 0, \tag{1.189}$$

where

$$\delta(g, \xi) = \frac{d\ln \xi}{d \ln \mu} = -\frac{d\ln Z_3}{d \ln \mu}, \tag{1.190}$$

and

$$\gamma_\Gamma(g, \xi) = -\frac{d\ln Z_\Gamma}{d \ln \mu} \tag{1.191}$$

is the anomalous dimension of the Green function. The derivatives in (1.190) and (1.191) are taken at fixed values of g_0, m_0, and ξ_0.

In MS-like schemes, $\delta(g, \xi)$ and $\gamma_\Gamma(g, \xi)$ are expressed in terms of the derivatives with respect to g of the coefficients $Z_3^{(1)}(g, \xi)$ and $Z_\Gamma^{(1)}(g, \xi)$ of the expansions

$$Z_3 = 1 + \sum_{n=1}^{\infty} \frac{Z_3^{(n)}(g, \xi)}{\epsilon^n}, \quad Z_\Gamma = 1 + \sum_{n=1}^{\infty} \frac{Z_\Gamma^{(n)}(g, \xi)}{\epsilon^n} \tag{1.192}$$

in the same way as it was done for the beta-function [see (1.157)–(1.160)] and for the mass anomalous dimension [see (1.163)–(1.165)]. Thus, we have

$$\frac{d\ln \xi}{d \ln \mu} = -\frac{d \ln Z_3}{d \ln \mu} = -\frac{\partial \ln Z_3}{\partial \ln g}(-\epsilon + \beta(g)) - \frac{\partial \ln Z_3}{\partial \ln \xi}\frac{d\ln \xi}{d \ln \mu}, \tag{1.193}$$

so that

$$\delta(g, \xi) = \frac{d\ln \xi}{d \ln \mu} = -\frac{\partial \ln Z_3/(\partial \ln g)(-\epsilon + \beta(g))}{1 + \partial \ln Z_3/(\partial \ln \xi)}, \tag{1.194}$$

and in the limit of $\epsilon \to 0$ using finiteness of the left-hand side we obtain

$$\delta(g, \xi) = g\frac{\partial Z_3^{(1)}}{\partial g}. \tag{1.195}$$

Accordingly,

$$\gamma_\Gamma(g, \xi) = g\frac{\partial Z_\Gamma^{(1)}}{\partial g}. \tag{1.196}$$

If the notation d_Γ is used for the canonical dimension of Γ,

$$\left(\mu \frac{\partial}{\partial \mu} + m \frac{\partial}{\partial m} + \lambda \frac{\partial}{\partial \lambda} \right) \Gamma(\lambda p_i; g, m, \xi, \mu) = d_\Gamma \, \Gamma(p_i; g, m, \xi, \mu), \tag{1.197}$$

then (1.189) can be rewritten as:

$$\left(\lambda\frac{\partial}{\partial\lambda} - g\beta(g)\frac{\partial}{\partial g} + (1 + \gamma_m(g))\, m\frac{\partial}{\partial m} - \delta(g,\xi)\xi\frac{\partial}{\partial\xi} - \gamma_\Gamma(g,\xi) - d_\Gamma\right)$$
$$\times \Gamma(\lambda p_i; g, m, \xi, \mu) = 0. \qquad (1.198)$$

Therefore, the renormalization group defines the transformation law of Γ at the simultaneous change of scale of all momenta p_i. For the case $\xi = 0$ in the MS-like schemes γ_Γ, as well as β and γ_m, depends only on g, so that Eq. (1.198) can be easily solved:

$$\Gamma(\lambda p_i; g, m, 0, \mu) = \lambda^{d_\Gamma}\Gamma(p_i; g(\lambda\mu), \tilde{m}(\lambda), 0, \mu)\exp\left[\int\limits_g^{g(\lambda\mu)}\frac{dx}{x\beta(x)}\gamma_\Gamma(x)\right], \quad (1.199)$$

where $\tilde{m}(\lambda)$ is defined by the equation

$$\frac{\lambda d\ln\tilde{m}(\lambda)}{d\lambda} = -(1 + \gamma_m(g(\lambda\mu))) \qquad (1.200)$$

with the initial condition $\tilde{m}(1) = m \equiv m(\mu)$. Note that since $g(\lambda\mu)$ and $\tilde{m}(\lambda)$ depend on $g \equiv g(\mu)$ and m as on initial conditions, we have

$$\frac{\partial\ln g(\lambda\mu)}{\partial\ln g} = \frac{\beta(g(\lambda\mu))}{\beta(g)}, \quad \frac{\partial m(\lambda)}{\partial m} = \frac{\tilde{m}(\lambda)}{m},$$
$$\frac{\partial\tilde{m}(\lambda)}{\partial g} = \frac{\tilde{m}(\lambda)}{\beta(g)}[\gamma_m(g(\lambda\mu)) - \gamma_m(g)]. \qquad (1.201)$$

The important point is that the Green function on the right-hand side (1.199) depends on the coupling constant at the point $\lambda\mu$. Due to the asymptotic freedom at large λ it can be calculated in perturbation theory. In the leading approximation one can use the Born value.

It is worth noting that one has to refer with caution to the statements about scheme- and gauge-independence of the first two coefficients of the β-function. In fact, even the first coefficient is scheme-dependent; in particular, in the MOM schemes this coefficient depends on quark masses. Thus, one can talk about the independence only in mass-independent schemes, such as MS schemes (or in the limit of negligibly small masses). Here, the independence is proved by presenting the ratio $g_2(\mu)/g_1(\mu)$ of the couplings in two schemes in the form of $1 + cg_1^2+$ terms of higher orders in g_1^2 and calculating $\beta^{(2)} = d\ln g_2/(d\ln\mu)$. It is easy to see that the first two coefficients of $\beta^{(2)}$ and $\beta^{(1)}$ coincide if c does not depend on μ. The scheme-independence of the first two coefficients leads, in particular, to their gauge-independence, because of the existence of physical schemes, where g is defined in terms of physical, i.e. gauge invariant, values.

However, as is shown in [80], there is a class of schemes, where c depends on μ because of gauge parameter dependence. In this case, only β_0 is scheme- and gauge-independent.

The MS-like schemes result in simpler calculations. The renormgroup equations in these schemes have the simplest form. However, sometimes virtues turn into shortcomings, and the mass-independence of the β- and γ-functions is just such a case. It is quite unnatural that these functions contain contributions from all quarks independently of their masses. It is also in evident contradiction with the decoupling theorem [81], according to which a heavy quark Q of mass M can be ignored in processes with typical momentum scale $k^2 \ll M^2$.

The point is that the decoupling theorem was proved using the MOM schemes. Actually, the proof is very simple. As already mentioned, in these schemes the renormalization can be performed by subtractions at the renormalization point μ^2. Let us take $\mu^2 \ll M^2$. After these subtractions all integrals become convergent. In the absence of quark loops they converge at values of virtual momenta $p^2 \sim \mu^2$ or $p^2 \sim k^2$. But integrals over momenta of Q-quark loops converge at $p^2 \sim M^2$, giving some negative power of M, which means heavy quark contributions vanish in the limit $M \to \infty$.

This proof is not valid in the MS-like schemes because, as already mentioned, the renormalization in these schemes cannot be performed by subtractions at some point of momentum space. Whereas in the MOM schemes the renormalization for Q quark loops looks like

$$\int_0^\infty dp^2 \left(\frac{1}{p^2 + k^2 + M^2} - \frac{1}{p^2 + \mu^2 + M^2} \right) \sim \mathcal{O}\left(\frac{k^2}{M^2}, \frac{\mu^2}{M^2} \right), \qquad (1.202)$$

in the MS-like schemes it is similar to

$$\int_0^\infty \frac{dp^2}{p^2 + k^2 + M^2} \left(\frac{\mu^2}{p^2} \right)^\epsilon - \int_{\mu^2}^\infty \frac{dp^2}{p^2} \left(\frac{\mu^2}{p^2} \right)^\epsilon \sim \mathcal{O} \ln \left(\frac{M^2}{\mu^2} \right). \qquad (1.203)$$

Therefore, the decoupling theorem is not applicable in its direct sense, and β_0 (1.167) contains contributions of arbitrary heavy quarks, which is quite unnatural. This means, in particular, that in the MS-like schemes physical amplitudes with typical momentum scale $k^2 \ll M^2$, being expressed in terms of $\alpha_s(\mu^2)$ with $\mu^2 \sim k^2$, contain large logarithms $\ln(M^2/\mu^2)$.

The standard solution of this problem is to provide decoupling explicitly [82],[83], using the notion of *active* flavours. In the region $k^2 < M^2$, the quark Q with mass M becomes inactive. In this region, one considers the dynamical degrees of freedom involving Q quarks as integrated out and works with the effective Lagrangian including only the remaining active quarks. That means, in particular, that only active quarks contribute to the β- and γ_m-functions that determine the dependence of $g(\mu)$ and $m(\mu)$. The requirement that such theory be consistent with the theory where the Q quark is also active leads to matching conditions between $g(\mu)$ and $m(\mu)$ in two theories at $\mu \sim M$, which are nontrivial in the two-loop approximation [84]–[87]. These conditions can be formulated in the language of finite renormalizations of the coupling constant and mass. The corresponding renormalization constants are found in [88] with α_s^3 accuracy.

1.10 Asymptotic freedom in QCD

Asymptotic freedom means decreasing α_s at small distances. This behaviour is opposite of that known in QED, where the effective charge increases when distances decrease. The discovery of asymptotic freedom in nonabelian gauge theories [68],[69] was a great break-through in the development of quantum field theories. A common belief before was that in any reasonable field theory the effective charge had the same behaviour as in QED, namely, it increased at small distances and decreased at large distances. Such phenomenon is called zero-charge, because the effective charge at large distances (*physical charge*) tends to zero when the scale where the effective charge is supposed to take a fixed value (*bare charge*) tends to infinity.

In QED, this phenomenon has a classical physical interpretation as the shielding of a trial charge source due to polarization by virtual pairs of charged particles created in the field of this source. Of course, a similar shielding of the colour charge takes place in QCD as well. This is clear from the quark contribution to the β-function. There is an analogous gluon contribution [the contribution of physical gluons to the polarization tensor, see (1.151)]. However, there is a new physical phenomenon related to gluons that leads to asymptotic freedom. The origin of this phenomenon is that, contrary to photons, the gluons themselves carry the colour charge and thereby spread out the colour charge of the source.

The QED behaviour of the effective charge also can be easily understood from the point of view of dispersion relations. Indeed, in QED, due to the Ward identity $Z_1 = Z_2$, the charge renormalization is determined by the photon polarization operator $\Pi_\gamma(s)$. The imaginary part of $\Pi_\gamma(s)$ is positive definite [see (1.141)]. Actually the positivity is not accidental but is dictated by unitarity. The positivity of Im Π and (1.139) leads to the zero-charge phenomenon.

In QCD in covariant gauges, the identity $Z_1 = Z_2$ is absent (as well as the positivity of Im Π). Nevertheless, at first sight it seems that the zero-charge phenomenon should remain. Indeed, the effective charge can be defined in terms of the interaction of two classical trial charges, and one can believe that in physical gauges the same line of arguments as in QED should work. But this impression is wrong. Moreover, the first calculation of the invariant charge [89] – even before the discovery of asymptotic freedom – actually was carried out just in a physical, namely the Coulomb, gauge and exhibited a decrease of the invariant charge at small distances. Unfortunately, it was not recognized at the time that the calculated value was just the invariant charge.

Using physical gauges makes clear the difference between QCD and QED and explains asymptotic freedom as a result of spreading of the colour charge of the trial source by gluons [90]. It is well known that the interaction of two distributed overlapping charges is weaker than that between two point charges of the same strength.

Most convenient for the elucidation is the Hamiltonian formulation of QCD in the Coulomb gauge. One can come to this formulation starting from the functional integral (1.24), with

$$(\hat{G}A)^a(x) = \partial_i A_i^a(x); \quad \hat{M}(A) = \partial_i D_i^{ab}(A),$$

$$e^{-\frac{i}{2\xi}\int d^4x((\hat{G}A)^a(x))^2}|_{\xi \to 0} \to \delta(\partial_i A_i^a(x)), \quad S_{QCD} \to 2\int d^4x \ \text{Tr}\left[-\frac{1}{4}G_{\mu\nu}G_{\mu\nu} + A_\mu J_\mu\right],$$

(1.204)

where $J_\mu = J_\mu^a t^a$, $J_\mu^a = (\rho_f^a(r), \vec{0})$ ($\rho_f^a(r)$ is the fixed colour charge source), and rewriting it in the following form:

$$\int \mathcal{D}A \det \hat{M}(A) \, \delta(\partial_i A_i^a(x)) \exp\left\{2i \int d^4x \ \text{Tr}\left[-\frac{1}{4}G_{\mu\nu}G_{\mu\nu} + A_\mu J_\mu\right]\right\}$$

$$= \int \mathcal{D}A\mathcal{D}A_0\mathcal{D}P \det \hat{M}(A) \, \delta(\partial_i A_i^a(x)) \exp\left\{2i \int d^4x\right.$$

$$\left.\times \text{Tr}\left[P_i \partial_0 A_i - \frac{1}{2}P_i P_i - \frac{1}{4}G_{ij}G_{ij} + A_0([\nabla_i P_i] + \rho_f)\right]\right\}.$$

(1.205)

Here $P_i = P_i^a t^a$. This equality is easily checked by integration over $\mathcal{D}P$. Now, defining ϕ by

$$P = P_\perp + \partial\phi, \quad \partial P_\perp = 0, \quad \partial^2\phi = \partial P, \quad \mathcal{D}P = \mathcal{D}P_\perp\mathcal{D}\phi,$$

(1.206)

we have, taking account of $\delta(\partial_i A_i^a(x))$,

$$[\nabla_i P_i] = \partial_i[\nabla_i \phi] + \rho,$$

(1.207)

where $\rho = ig[A_i P_{\perp i}]$, so that

$$\int \mathcal{D}A_0 \det \exp\left\{2i \ \text{Tr}\left[A_0\left([\nabla_i P_i] + \rho_f\right)\right]\right\} = \delta\left(\partial_i[\nabla_i\phi] + \rho + \rho_f\right)$$

$$= \delta\left(\hat{M}(A)\phi + \rho + \rho_f\right),$$

(1.208)

and therefore the determinant of $\hat{M}(A)$ in (1.205) is cancelled when integration over ϕ is performed. Now, using

$$\int d^4x \ \text{Tr}\left[P_i \partial_0 A_i - \frac{1}{2}P_i P_i\right] = \int d^4x \ \text{Tr}\left[P_{\perp i}\partial_0 A_i - \frac{1}{2}P_{\perp i}P_{\perp i} + \frac{1}{2}\phi\partial^2\phi\right], \quad (1.209)$$

we obtain after integration over ϕ and $A_\parallel$

$$\int \mathcal{D}A \det \hat{M}(A) \, \delta(\partial_i A_i^a(x)) \exp\left\{2i \int d^4x \ \text{Tr}\left[-\frac{1}{4}G_{\mu\nu}G_{\mu\nu} + A_\mu J_\mu\right]\right\}$$

$$= \int \mathcal{D}A_\perp\mathcal{D}P_\perp \exp\left\{2i \int d^4x \ \text{Tr}\left[P_{\perp i}\partial_0 A_{\perp i} - \frac{1}{2}P_{\perp i}P_{\perp i} - \frac{1}{4}G_{ij}G_{ij} + \frac{1}{2}\phi\partial^2\phi\right]\right\},$$

(1.210)

where

$$\phi = -\hat{M}(A)^{-1}\left(ig[A_{\perp i}P_{\perp i}] + \rho_f\right).$$

(1.211)

Thus, we come to the original Feynman formulation [91] of the path integral. The exponent in (1.210) is therefore $i \int d^4x (P^a_{\perp i} \partial_0 A^a_{\perp i}) - \mathcal{H}$, so that

$$\mathcal{H} = 2 \, \text{Tr} \left[\frac{1}{2} P_{\perp i} P_{\perp i} + \frac{1}{4} G_{ij} G_{ij} - \frac{1}{2} \phi \partial^2 \phi \right]. \tag{1.212}$$

Let $\rho_f = \rho_1 + \rho_2$, where ρ_1 and ρ_2 are two fixed charges forming a colour singlet. The energy of their interaction $E_{\rho_1 \rho_2}$ is given by the term containing the product $\rho_1 \rho_2$ in the correction to the energy of the vacuum state $|0\rangle$. Up to terms $\sim g^2$, it consists of two pieces:

$$E_{\rho_1 \rho_2} = E_1 + E_2 \tag{1.213}$$

having different physical meaning. The first of them is the first-order correction of perturbation theory where the perturbation is the term with ϕ in (1.212):

$$E_1 = -2 \int d^3x \, \text{Tr} \left[\rho_1(x) \langle 0 | \hat{M}(A)^{-1} \partial^2 \hat{M}(A)^{-1} | 0 \rangle \rho_2(x) \right]. \tag{1.214}$$

Such a term is absent in the QED case. Just this correction diminishes the interaction energy at small distances. Indeed, up to terms $\sim g^2$

$$\hat{M}(A)^{-1} \partial^2 \hat{M}(A)^{-1} f = \frac{1}{\partial^2} f - 2ig \frac{1}{\partial^2} \left[A_i, \frac{\partial_i}{\partial^2} f \right] - 3g^2 \frac{1}{\partial^2} \left[A_i, \frac{\partial_i}{\partial^2} \left[A_j, \frac{\partial_j}{\partial^2} f \right] \right], \tag{1.215}$$

and since

$$\langle 0 | A^a_i(x) | 0 \rangle = 0, \quad \langle 0 | A^a_i(x) A^b_j(y) | 0 \rangle = \delta^{ab} \int \frac{d^3k}{(2\pi)^3 2|k|} \left(\delta_{ij} - \frac{k_i k_j}{k^2} \right) e^{ik(x-y)}, \tag{1.216}$$

for $\rho_{1,2} = g \chi^a_{1,2} t^a \delta(x - x_{1,2})$ we obtain

$$E_1 = -2 \int d^3x \, \text{Tr} \left[\rho_1(x) \frac{1}{\partial^2} \rho_2(x) - 3g^2 \langle 0 | \left[A_i, \left(\frac{\partial_i}{\partial^2} \rho_1 \right) \right] \frac{1}{\partial^2} \left[A_j, \left(\frac{\partial_i}{\partial^2} \rho_2 \right) \right] | 0 \rangle \right]$$

$$= g^2 (\chi_1 \chi_2) \int \frac{d^3q}{(2\pi)^3} \frac{e^{i q (x_1 - x_2)}}{q^2} \left[1 + \frac{3g^2 C_A}{(2\pi)^3} \int \frac{d^3k}{2|k|(q-k)^2} \left(1 - \frac{(kq)^2}{k^2 q^2} \right) \right]. \tag{1.217}$$

The latter integral is logarithmically divergent. With the ultraviolet cut-off Λ^2 it gives

$$E_1 = g^2 (\chi_1 \chi_2) \int \frac{d^3q}{(2\pi)^3} \frac{e^{i q (x_1 - x_2)}}{q^2} \left[1 + \frac{g^2 C_A}{4\pi^2} \ln \left(\frac{\Lambda^2}{q^2} \right) \right]. \tag{1.218}$$

The logarithmic term in (1.218) increases when effective values of q^2 decrease, or the charge separation $|x_1 - x_2|$ increases. This means that the correction E_1 (1.218) gives antishielding.

On the contrary, the second term in (1.213) gives normal shielding. This is clear without calculations, since it comes from the second-order correction to the ground state, which is always negative. To find this, in (1.212) make the replacement:

$$\phi \partial^2 \phi \rightarrow 2ig [A_{\perp i} \partial_0 A_{\perp i}] \frac{1}{\partial^2} \rho_f. \tag{1.219}$$

The intermediate states in the second-order correction are two-gluon states $|g_1g_2\rangle$. Let k_1 and k_2 denote gluon momenta. Then

$$E_2 = -4g^2 \int \frac{d^3k_1}{(2\pi)^3 2\omega_1} \frac{d^3k_2}{(2\pi)^3 2\omega_2} \sum \int d^3x \langle 0| \, \mathrm{Tr} \left(\rho_1(x) \frac{1}{\partial^2}[A_{\perp i}(x), \partial_0 A_{\perp i}(x)] \right) |g_1g_2\rangle$$

$$\times \frac{1}{\omega_1 + \omega_2} \int d^3y \langle g_1g_2| \, \mathrm{Tr} \left([\partial_0 A_{\perp i}(y), A_{\perp i}(y)] \right) \frac{1}{\partial^2} \rho_2(y) |0\rangle, \tag{1.220}$$

where the sum goes over polarizations and colours of the produced gluons and their identity is taken into account. A straightforward calculation gives

$$E_2 = g^2(\chi_1\chi_2) \int \frac{d^3q}{(2\pi)^3} \frac{e^{i\,\boldsymbol{q}(\boldsymbol{x}_1-\boldsymbol{x}_2)}}{q^2} \frac{(-g^2C_A)}{4(2\pi)^3} \int \frac{d^3k \,(|\boldsymbol{k}| - |\boldsymbol{q} - \boldsymbol{k}|)^2}{|\boldsymbol{k}||\boldsymbol{q} - \boldsymbol{k}|(|\boldsymbol{k}| + |\boldsymbol{q} - \boldsymbol{k}|)q^2}$$

$$\times \left(1 + \frac{\boldsymbol{k}(\boldsymbol{q} - \boldsymbol{k})^2}{k^2(\boldsymbol{q} - \boldsymbol{k})^2} \right). \tag{1.221}$$

With logarithmic accuracy, we obtain

$$E_2 = g^2(\chi_1\chi_2) \int \frac{d^3q}{(2\pi)^3} \frac{e^{i\,\boldsymbol{q}(\boldsymbol{x}_1-\boldsymbol{x}_2)}}{q^2} \frac{(-g^2C_A)}{48\pi^2} \ln\left(\frac{\Lambda^2}{q^2}\right). \tag{1.222}$$

The minus sign provides for normal shielding. But the dominant correction is given by E_1, so that

$$E = g^2(\chi_1\chi_2) \int \frac{d^3q}{(2\pi)^3} \frac{e^{i\,\boldsymbol{q}(\boldsymbol{x}_1-\boldsymbol{x}_2)}}{q^2} \left[1 + \frac{(11g^2C_A)}{48\pi^2} \ln\left(\frac{\Lambda^2}{q^2}\right) \right], \tag{1.223}$$

which means antishielding and thus asymptotic freedom.

1.11 The renormalization scheme and scale ambiguity

All regularization and renormalization schemes must lead to the same values for physical observables. This statement seems to be obvious, but it is true only in principle; in practice, however, when a finite number of terms of perturbation theory is used, different schemes lead to different results. Let us consider a dimensionless physical quantity $V(Q)$ that depends on one variable $Q = \sqrt{|q^2|}$ (q is a four-momentum), with the expansion

$$V(Q) = a(\mu) \left[1 + \sum_{k=1}^{\infty} v_k \left(\frac{Q}{\mu} \right) a^k(\mu) \right], \tag{1.224}$$

where

$$a(\mu) = \frac{\alpha_s(\mu)}{4\pi}. \tag{1.225}$$

For simplicity, we consider the case when $V(Q)$ is proportional to the first power of α_s, but it is clear from the following that the consideration can be easily generalized to higher powers. The running coupling constant $\alpha_s(\mu)$ depends on its definition (renormalization scheme) and on the scale μ. For two different schemes we have

$$a_2(\mu) = a_1(\mu) \left[1 + \sum_{k=1}^{\infty} c_k a_1^k(\mu) \right], \tag{1.226}$$

where the subscripts 1 and 2 denote the schemes; the coefficients c_k depend on schemes, although this dependence is not indicated explicitly. In general, the coefficients c_k, as well as v_k in (1.224), can depend on μ through the renormalized gauge parameter [see discussion after (1.201)]. We confine ourselves to the case when they are gauge-independent.

Being a physical quantity, $V(Q)$ must not depend on the definition of the running coupling $a(\mu)$ (renormalization scheme) or on the scale μ. This means that the coefficients v_n are scheme- and scale-dependent; their dependence compensates the scheme- and scale-dependence of $\alpha_s(\mu)$. But this compensation is exact only in the infinite series (1.224). Truncation of the series leads to scheme- and scale-dependence. Denoting

$$V^{(n)}(Q; \mu) = a(\mu) \left[1 + \sum_{k=1}^{n-1} v_k \left(\frac{Q}{\mu} \right) a^k(\mu) \right], \tag{1.227}$$

we have $V_1^{(n)}(Q; \mu) \neq V_2^{(n)}(Q; \mu)$ and $V_i^{(n)}(Q; \mu_1) \neq V_i^{(n)}(Q; \mu_2)$ for $i = 1, 2$. Therefore, any fixed-order perturbative expansion contains a scheme and scale ambiguity. Of course, the difference of $V_i^{(n)}$ is of order α_s^{n+1}, and formally any choice of scheme and scale is admissible. But a wrong choice leads to a bad convergence of the series (1.224) and therefore to a large deviation of $V^{(n)}$ from $V(Q)$. On the contrary, a good choice puts $V^{(n)}$ closer to $V(Q)$. There are a few ways to resolve scheme-scale ambiguity.

The simplest way has no strictly defined rules. Some renormalization schemes are accepted without serious grounds and the appropriate scale is guessed. There are, however, at least three approaches that are based on definite principles. In the first of these [92],[93], called the method of effective charges (MEC) or fastest apparent convergence (FAC), $V(Q)$ is considered to be the physical coupling constant: $V(Q) \equiv a_v(Q) = \alpha_v(Q)/(4\pi)$. Its evolution with Q is described by the equation

$$\frac{da_v(Q)}{d \ln Q^2} = a_v(Q) \beta_v(g_v(Q)), \quad g_v = 4\pi \sqrt{a_v}. \tag{1.228}$$

Differentiating (1.224) with respect to $\ln Q^2$ having set $\mu = Q$, using the evolution equation for $g(Q)$ and comparing the result with the expansion

$$a_v \beta_v(g_v) = -\beta_0^v a_v^2 - \beta_1^v a_v^3 - \beta_2^v a_v^4 - \cdots . \tag{1.229}$$

gives equations connecting β_i^v with the coefficients β_i (1.166) and v_i (1.224). It is easy to see that, in accordance with the discussion after (1.201), the first two β_i^v coincide with β_i, $\beta_0^v = \beta_0$, $\beta_1^v = \beta_1$, and the next β_k^v are expressed in terms of β_l and $v_l(1)$ with $l \leq k$. In particular,

$$\beta_2^v = \beta_2 - \beta_1 v_1(1) - \beta_0 v_1^2(1) + \beta_0 v_2(1). \tag{1.230}$$

Thus, knowledge of β_l and $v_l(1)$ up to $l = k$ determines the function β_v up to the terms a_v^{k+1} [in the $(k + 1)$–loop approximation]. Then $V(Q)$ is given by the solution of the following equation [cf. (1.177)]:

$$\beta_0 \ln \left(\frac{Q^2}{\Lambda_v^2} \right) = L_v(V(Q)), \tag{1.231}$$

where $L_v(a)$ is defined as the right-hand side of (1.178) with the substitution $\beta \to \beta_v$ and Λ_v is the integration constant, which can be expressed [54],[55], in terms of the constant Λ for the running coupling $a(\mu)$ and the coefficient v_1 in (1.224):

$$\beta_0 \ln \left(\frac{\Lambda_v^2}{\Lambda^2} \right) = v_1(1). \tag{1.232}$$

The last relation, as well as the relations among β_i^v, β_i and v_i, can be obtained taking the difference of (1.177) and (1.231):

$$\beta_0 \ln \left(\frac{\Lambda_v^2}{\Lambda^2} \right) = L(a(Q)) - L_v(V(Q)) = \int_0^{a(Q)} dx \frac{\beta_0}{x\beta(4\pi\sqrt{x})} - \int_0^{a_v(Q)} dx \frac{\beta_0}{x\beta_v(4\pi\sqrt{x})}, \tag{1.233}$$

using (1.224) at $\mu = Q$ and expanding the right-hand member in powers of $a(Q)$.

In the two-loop approximation, because of the coincidence of the β–functions, $a_v(\mu)$ and $a(\mu)$ differ only by the scales, so that in this approximation

$$V_{FAC}^{(2)} = V^{(2)}(Q; \mu_c) = a(\mu_c), \tag{1.234}$$

where

$$\ln \left(\frac{Q}{\mu_c} \right) = \frac{v_1(1)}{2\beta_0}, \quad \mu_c = Q \exp \left(-\frac{v_1(1)}{2\beta_0} \right). \tag{1.235}$$

A second principle is the principle of minimal sensitivity (PMS) [94],[95], which means that the scheme and scale must be chosen so as to minimize the sensitivity of $V^{(n)}$ to their small variations. To realize this principle *in corpore* is not an easy task. One has to introduce some parametrization of schemes (for example, using coefficients of corresponding β-functions as the parameters) and to solve the problem of minimization in the many-dimensional space of these parameters and the scale μ. But in the two-loop approximation the only parameter is μ. From the PMS requirement $\mu \partial V^{(n)}(Q; \mu)/\partial \mu = 0$, using (1.159), (1.166), one obtains

$$a^2(\mu) \left[\frac{d}{d \ln \mu} v_1 \left(\frac{Q}{\mu} \right) - 2\beta_0 \right] - 2a^3(\mu) \left[\beta_1 + 2v_1 \left(\frac{Q}{\mu} \right) (\beta_0 + \beta_1 a(\mu)) \right] = 0. \tag{1.236}$$

The a^2 terms cancel each other because of the scale-invariance of $V(Q)$ (1.224), as $V(Q) - V^{(2)}(Q; \mu) \sim a^3(\mu)$. One can write

$$v_1 \left(\frac{Q}{\mu} \right) = -2\beta_0 \ln \left(\frac{Q}{\mu} \right) + v_1(1). \tag{1.237}$$

Then (1.236) gives the equation for the PMS scale μ_s:

$$\beta_1 + 2v_1 \left(\frac{Q}{\mu_s} \right) (\beta_0 + \beta_1 a(\mu_s)) = 0, \tag{1.238}$$

where $v_1 \left(\frac{Q}{\mu_s} \right)$ is defined by (1.237) and $a(\mu_s)$ by (1.177), (1.178). The two-loop PMS approximation for $V(Q)$ is obtained from $V^{(2)}(Q; \mu)$ putting $\mu = \mu_s$ and using (1.238) to express $v_1 \left(\frac{Q}{\mu_s} \right)$ through $a(\mu_s)$:

$$V^{(2)}_{PMS}(Q) = a(\mu_s) \left[1 - \frac{\beta_1 a(\mu_s)}{2 (\beta_0 + \beta_1 a(\mu_s))} \right]. \tag{1.239}$$

The approximation (1.239) without expansion in $a(\mu_s)$ is considered to be the most reliable one. With such expansion the approximation is simplified:

$$V^{(2)}_{PMS}(Q) = V^{(2)}(Q; \mu_s) = a(\mu_s) \left[1 - \frac{\beta_1}{2\beta_0} a(\mu_s) \right], \tag{1.240}$$

where for the scale μ_s we have from (1.238), taking account of (1.237)

$$\ln \left(\frac{Q}{\mu_s} \right) = \frac{\beta_1}{4\beta_0^2} + \frac{v_1(1)}{2\beta_0}, \quad \mu_s = Q \exp \left(-\frac{\beta_1}{4\beta_0^2} - \frac{v_1(1)}{2\beta_0} \right). \tag{1.241}$$

The FAC and PMS optimization procedures were used to estimate the coefficients v_n in the expansion (1.224) from the known coefficients v_k and β_k at $k < n$ (see, for example, [96] and references therein). The idea is that because corrections to the optimized $V^{(n)}$ are supposed to be small, most of the coefficients v_n come from their expansion in terms of $a(\mu)$. Thus, for $n = 2$, i.e. in the two-loop approximation, the estimates of v_2 are given by the coefficients at $a^3(\mu)$ in the expansions of (1.234) and (1.239).

Both in FAC and PMS approaches in the two-loop approximation, the problem is reduced to the scale choice. Here, both FAC and PMS scales μ_c (1.235) and μ_s (1.241) have a distant relation to typical virtualities for the processes under consideration. Rather, they are defined so as to absorb the correction v_1; as a result, the FAC correction becomes zero and the PMS one is small and process-independent, even in the case when v_1 is large. This property was criticized in [97], where it was pointed out that a large correction v_1 can be a consequence of bad convergence of the perturbation expansion rather than of a bad choice of an expansion parameter. The simplest example is the orthopositronium decay width in QED, where the expansion looks like

$$\Gamma = \Gamma_0 \left(1 - 10.3 \frac{\alpha}{\pi} + \cdots \right), \tag{1.242}$$

and the first-order correction should not be absorbed into a redefinition of α since it is not running at these energies. It was argued in [97] that the methods based on the FAC and PMS give wrong results for processes like Υ-decay, where the higher-order corrections are very large.

In contrast to these methods, the scale is defined in [97] such as to absorb only the part of v_1 related to charge renormalization. This can be done straightforwardly for processes without gluon-gluon interactions in the lowest order. For such processes, the dependence of v_1 on the number of flavours n_f comes only from the quark vacuum polarization. Since the polarization is inseparably linked with charge renormalization, all terms proportional to n_f are absorbed into the scale. If

$$v_1(1) = e_1 n_f + e_2, \tag{1.243}$$

then, in order to absorb the term $e_1 n_f$, one has to take the scale equal to

$$\mu_p = Q \exp\left(\frac{3e_1}{4}\right). \tag{1.244}$$

This procedure is called the Brodsky-Lepage-Mackenzie (BLM) scale setting. With this scale for $V(Q)$ one obtains in the second order:

$$V_{BLM}(Q) = V^{(2)}(Q; \mu_p) = a(\mu_p)\left[1 + e_p a(\mu_p)\right], \tag{1.245}$$

where

$$e_p = \left(n_f + \frac{3}{2}\beta_0\right)e_1 + e_2 = \frac{11}{2}N_c e_1 + e_2. \tag{1.246}$$

Thus, the principal difference between the approaches is that, in the first two, the scale is determined by the total correction, while in the third approach, the scale is determine by only the part of the correction that relates to charge renormalization. It should be noted that in the BLM approach a definite renormalization scheme is assumed and different schemes give different expansions. However, if two schemes differ only by an n_f-independent rescaling (as, for example, MS and $\overline{\text{MS}}$) they give the same result. Moreover, for two different schemes connected by (1.226) with $c_1(1) = \beta_0 c + d$, where d is n_f-independent, the coefficients e_p (1.246) will differ in d for all processes. This means in particular that the difference of e_p for different processes is scheme-independent. Therefore, for a bad scheme choice the coefficients e_p are large and of the same sign for most processes, which gives a possibility to eliminate unsuitable schemes.

1.12 Anomalous dimensions of twist-2 operators

As an example, let us calculate the anomalous dimensions of the nonsinglet twist-2 operator:

$$\hat{O}^i_{\mu_1\mu_2\ldots\mu_n}(x) = 2i^{n-1}\mathcal{S}\left[\bar{\psi} t^i_f \gamma_{\mu_1}\frac{\overset{\leftrightarrow}{\nabla}_{\mu_2}}{2}\frac{\overset{\leftrightarrow}{\nabla}_{\mu_3}}{2}\cdots\frac{\overset{\leftrightarrow}{\nabla}_{\mu_n}}{2}\psi\right]. \tag{1.247}$$

Here, ψ is supposed to describe three quark flavours; t^i_f are the generators of the flavour group in the fundamental representation; ∇_μ are the covariant derivatives (1.8); the sign $\leftrightarrow$ means difference of their actions to the right and to the left (neglecting the total derivatives it is possible to replace $\overset{\leftrightarrow}{\nabla}_\mu/2$ by ∇_μ); $\mathcal{S}$ denotes the symmetrization of indices $\mu_1\mu_2\cdots\mu_n$; and the subtractions which turn into zero all convolutions of $\hat{O}^i$ with $\delta_{\mu\nu}$.

The symmetrization and the subtractions result in definite Lorentz spin (equal to n) of the operator $\hat{\mathcal{O}}^i$. By definition, a twist τ of an operator $\hat{\mathcal{O}}$ is the difference between the canonical dimension d (in mass units) of the operator and its Lorentz spin n. It is not difficult to see that the minimal twist of operators built from quark and gluon fields is equal to 2. The operators of twist 2 play a major role in theoretical analysis of deep inelastic scattering.

To simplify the calculation let us perform it for the operator

$$\hat{\mathcal{O}}_n^i = c_{\mu_1} c_{\mu_2} \cdots c_{\mu_n} \hat{\mathcal{O}}_{\mu_1 \mu_2 \ldots \mu_n}^i = 2i^{n-1} \bar{\psi} \, \slashed{c} \left(\frac{c \overleftrightarrow{\nabla}}{2} \right)^{n-1} t_f^i \psi, \qquad (1.248)$$

with $c^2 = 0$, evidently having the same anomalous dimension as $\mathcal{O}_{\mu_1 \mu_2 \cdots \mu_n}^{(f)i}$. In order to find the dimension, it is sufficient to consider only matrix elements with zero momentum transfer, where $\overleftrightarrow{\nabla}_\mu / 2$ can be replaced by ∇_μ. In this case, in Born approximation, the operator $\hat{\mathcal{O}}_n^i$ has the following vertices:

$$= 2 \, \slashed{c}(cp)^{n-1} t_f^i = \Gamma_{n(qq)}^{i(B)}, \qquad (1.249)$$

$$= -2(n-1) \, \slashed{c} t^a t_f^i c_\mu \sum_{j=1}^{n-1} (cp_1)^{j-1} (cp_2)^{n-j-1}$$

$$= \Gamma_{n(qqg)}^{i(B)}, \qquad (1.250)$$

and so on. According to (1.196), the anomalous dimension is given in one-loop approximation by the doubled residue in the pole $\epsilon = 0$ of the renormalization constant $\mathcal{Z}_{\mathcal{O}}$ defined by

$$\hat{\mathcal{O}}_{n\,0}^i = \mathcal{Z}_{\mathcal{O}}^{-1} \hat{\mathcal{O}}_n^i. \qquad (1.251)$$

Here, the subscript 0 in the first member means, as usual, a bare quantity, and $\hat{\mathcal{O}}_n^i$ is the renormalized operator with finite matrix elements in the limit $\epsilon \to 0$. Let us consider the matrix elements between single-quark states; then we have $\mathcal{Z}_{\mathcal{O}} = Z_2^{-1} Z_n^f$, where Z_2 and Z_n^f are the renormalization constants of the quark propagator (1.104) and of the one-particle irreducible $\bar{q} \mathcal{O}_n^i q$ vertex (cf. (1.105)):

$$\Gamma_{n(qq)\,0}^i = (Z_n^f)^{-1} \Gamma_{n(qq)}^i. \qquad (1.252)$$

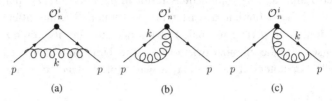

Fig. 1.1. The one-loop diagrams for the calculation of the renormalization constant for the operator $\mathcal{O}^{(f)i}_{\mu_1\mu_2\cdots\mu_n}$.

The renormalization constant Z_2 is given by (1.121). For the calculation of Z_n^f in g^2-order it is sufficient to find divergent contributions of the one-loop diagrams shown in Fig. 1.1.

As is seen from (1.121), Z_2 is gauge-dependent. The same is true for Z_n^f. On the contrary, $\mathcal{Z}_\mathcal{O}$ is gauge-independent, because of gauge invariance of the operator $\hat{\mathcal{O}}^i_n$. Therefore, we can use any gauge for its calculation. Eq. (1.121) gives the constant Z_2, in particular in the Feynman gauge, so that we will use this gauge. Since the divergent parts are independent of quark masses, we put $m = 0$. Then with the required accuracy for the contribution of the diagram Fig. 1.1a

$$\Gamma^{i(a)}_{n(qq)} = 2g^2 C_F t_f^i \int \frac{d^D k}{(2\pi)^D i} (ck)^{n-1} \frac{\gamma_\mu \not{k} \not{c} \not{k} \gamma_\mu}{(k^2 + i0)^2((p-k)^2 + i0)} \tag{1.253}$$

we get, using $\gamma_\mu \not{k} \not{c} \not{k} \gamma_\mu = 2(k^2 \not{c} - 2(kc) \not{k})$,

$$
\begin{aligned}
\Gamma^{i(a)}_{n(qq)} &= 4g^2 C_F t_f^i \int_0^1 dx \int \frac{d^D k}{(2\pi)^D i} \left[\frac{\not{c}(ck)^{n-1}}{[(k-xp)^2 + p^2 x(1-x) + i0]^2} - \frac{4(1-x)\not{k}(ck)^n}{[(k-xp)^2 + p^2 x(1-x) + i0]^3} \right] \\
&= 4g^2 C_F t_f^i \int_0^1 dx \int \frac{d^D k}{(2\pi)^D i} \left[\frac{\not{c}(cp)^{n-1} x^{n-1}}{[k^2 + p^2 x(1-x) + i0]^2} - \frac{4n x^{n-1}(1-x)(cp)^{n-1} \not{k}(ck)}{[k^2 + p^2 x(1-x) + i0]^3} \right] \\
&= \Gamma^{i(B)}_{n(qq)} \frac{g^2}{(4\pi)^2} \frac{C_F}{\epsilon} 2 \int_0^1 dx \left[x^{n-1} - n x^{n-1}(1-x) \right].
\end{aligned}
\tag{1.254}
$$

The diagrams Fig. 1.1b and Fig. 1.1c give equal contributions. In the same way, we obtain

$$
\begin{aligned}
\Gamma^{i(b+c)}_{n(qq)} &= -4g^2 C_F t_f^i \sum_{j=1}^{n-1} \int \frac{d^D k}{(2\pi)^D i} \frac{(ck)^{j-1}(cp)^{n-j-1} \not{c} \not{k} \not{c}}{(k^2 + i0)^2((p-k)^2 + i0)} \\
&= \Gamma^{i(B)}_{n(qq)} \frac{-g^2}{(4\pi)^2} \frac{C_F}{\epsilon} 4 \sum_{j=1}^{n-1} \int_0^1 dx\, x^j.
\end{aligned}
\tag{1.255}
$$

From (1.254) and (1.255), it follows that in the MS scheme

$$Z_n^{(f)} = 1 - \frac{g^2}{(4\pi)^2} \frac{C_F}{\epsilon} \left[\frac{2}{n(n+1)} - 4 \sum_{j=1}^{n-1} \frac{1}{j+1} \right]. \tag{1.256}$$

Using the result (1.104) for Z_2, we obtain for the anomalous dimension of the operator $\hat{O}^i_{\mu_1 \mu_2 \cdots \mu_n}$

$$
\begin{aligned}
\gamma_{O^i_n} &= -\frac{\alpha_s C_F}{\pi} \left[\frac{3}{2} + \frac{1}{n(n+1)} - 2 \sum_{j=1}^{n} \frac{1}{j} \right] \\
&= -\frac{\alpha_s C_F}{\pi} \left[\frac{3}{2} + \frac{1}{n(n+1)} - 2\psi(n+1) - 2\gamma_E \right].
\end{aligned}
\tag{1.257}
$$

Here,

$$
\psi(x) = (\ln \Gamma(x))' = \frac{\Gamma'(x)}{\Gamma(x)}, \quad \psi(x+1) = \frac{1}{x} + \psi(x), \quad \psi(1) = -\gamma_E,
\tag{1.258}
$$

$\gamma_E = .5772 \cdots$ is the Euler constant.

1.13 Colour algebra

The generators $\hat{T}^a$ of the colour group $SU(N_c)$ obey the commutation relations

$$
[\hat{T}^a \hat{T}^b] = i f^{abc} \hat{T}^c,
\tag{1.259}
$$

where the group structure constants f^{abc} are antisymmetric under the interchange of any two indices. In the fundamental representation the generators are denoted t^a. They have the properties

$$
\operatorname{Tr} t^a = 0, \quad t^a = t^{a\dagger}, \quad \operatorname{Tr} t^a t^b = \frac{1}{2} \delta^{ab}
\tag{1.260}
$$

and together with the identity matrix I create a complete set of $N_c \times N_c$ matrices. The completeness condition can be written as

$$
(t^a)^\alpha_{\ \beta} (t^a)^\gamma_{\ \delta} = \frac{1}{2} \delta^\alpha_\delta \delta^\gamma_\beta - \frac{1}{2N_c} \delta^\alpha_\beta \delta^\gamma_\delta.
\tag{1.261}
$$

The generators of the adjoint representation are denoted usually T^a, hence

$$
\left(T^a\right)_{bc} = -i f^{adc}.
\tag{1.262}
$$

From the completeness condition (1.261), one can easily obtain

$$
t^a t^a = C_F I, \quad t^a t^b t^a = \left(C_F - \frac{C_A}{2} \right) t^b, \quad t^a t^b t^c t^a = \frac{1}{4} \delta^{bc} I + \left(C_F - \frac{C_A}{2} \right) t^b t^c,
\tag{1.263}
$$

where C_F and C_A are the values of the Casimir operators in the fundamental and adjoint representations:

$$
C_F = \frac{N_c^2 - 1}{2N_c}, \quad C_A = N_c.
\tag{1.264}
$$

Taking account of (1.259) and (1.260), the completeness gives

$$
t^a t^b = \frac{\delta^{ab}}{2N_c} + \frac{1}{2} \left(d^{abc} + i f^{abc} \right) t^c,
\tag{1.265}
$$

where the d^{abc} are symmetric under the interchange of any two indices. Thus,

$$if^{abc} = 2\,\mathrm{Tr}([t^a t^b]t^c) = 2\,\mathrm{Tr}([t^b t^c]t^a), \quad d^{abc} = 2\,\mathrm{Tr}(\{t^a t^b\}t^c) = 2\,\mathrm{Tr}(\{t^b t^c\}t^a). \tag{1.266}$$

For $N_c \times N_c$ matrices T^a and D^a with matrix elements

$$T^a_{bc} = -if^{abc}, \quad D^a_{bc} = d^{abc} \tag{1.267}$$

we have the following useful identities, which result from (1.266) and (1.261):

$$\left[T^a, T^b\right] = if^{abc}T^c, \quad \left[T^a, D^b\right] = if^{abc}D^c, \tag{1.268}$$

$$T^a T^a = C_A I, \quad T^a T^b T^a = \frac{C_A}{2}T^b, \tag{1.269}$$

$$\mathrm{Tr}\left(T^a\right) = \mathrm{Tr}\left(D^a\right) = \mathrm{Tr}\left(T^a D^b\right) = 0, \tag{1.270}$$

$$\mathrm{Tr}\left(T^a T^b\right) = N_c \delta^{ab}, \quad \mathrm{Tr}\left(D^a D^b\right) = \frac{N_c^2 - 4}{N_c}\delta^{ab}, \tag{1.271}$$

$$\mathrm{Tr}\left(T^a T^b T^c\right) = i\frac{N_c}{2}f^{abc}, \quad \mathrm{Tr}\left(T^a T^b D^c\right) = \frac{N_c}{2}d^{abc}, \tag{1.272}$$

$$\mathrm{Tr}\left(D^a D^b T^c\right) = i\frac{N_c^2 - 4}{2N_c}f^{abc}, \quad \mathrm{Tr}\left(D^a D^b D^c\right) = \frac{N_c^2 - 12}{2N_c}d^{abc}, \tag{1.273}$$

$$\mathrm{Tr}\left(T^a T^b T^c T^d\right) = \delta^{ad}\delta^{bc} + \frac{1}{2}\left(\delta^{ab}\delta^{cd} + \delta^{ac}\delta^{bd}\right) + \frac{N_c}{4}\left(f^{adi}f^{bci} + d^{adi}d^{bci}\right), \tag{1.274}$$

$$\mathrm{Tr}\left(T^a T^b T^c D^d\right) = i\frac{N_c}{4}\left(d^{adi}f^{bci} - f^{adi}d^{bci}\right), \tag{1.275}$$

$$\mathrm{Tr}\left(T^a T^b D^c D^d\right) = \frac{1}{2}\left(\delta^{ab}\delta^{cd} - \delta^{ac}\delta^{bd}\right) + \frac{N_c^2 - 8}{4N_c}f^{adi}f^{bci} + \frac{N_c}{4}d^{adi}d^{bci}, \tag{1.276}$$

$$\mathrm{Tr}\left(T^a D^b T^c D^d\right) = -\frac{1}{2}\left(\delta^{ab}\delta^{cd} - \delta^{ac}\delta^{bd}\right) + \frac{N_c}{4}\left(f^{adi}f^{bci} + d^{adi}d^{bci}\right), \tag{1.277}$$

$$\mathrm{Tr}\left(T^a D^b D^c D^d\right) = i\frac{2}{N_c}f^{adi}d^{bci} + i\frac{N_c^2 - 8}{4N_c}f^{abi}d^{cdi} + i\frac{N_c}{2}d^{abi}f^{cdi}, \tag{1.278}$$

$$\mathrm{Tr}\left(D^a D^b D^c D^d\right) = \frac{N_c^2 - 4}{N_c^2}\delta^{ad}\delta^{bc} + \frac{1}{2}\delta^{ac}\delta^{bd} + \frac{N_c^2 - 8}{2N_c^2}\delta^{ab}\delta^{cd}$$

$$+ \frac{N_c}{4}f^{adi}f^{bci} + \frac{N_c^2 - 16}{4N_c}d^{adi}d^{bci} - \frac{4}{N_c}d^{abi}d^{cdi}, \tag{1.279}$$

$$f^{adi}f^{bci} + d^{adi}d^{bci} - f^{abi}f^{cdi} - d^{abi}d^{cdi} + \frac{2}{N_c}\left(\delta^{ad}\delta^{bc} - \delta^{ab}\delta^{cd}\right) = 0. \tag{1.280}$$

These relations are valid for arbitrary N_c. For $N_c = 3$, the following extra relation exists:

$$d^{bb'c}d^{aa'c} + d^{ba'c}d^{b'ac} + d^{bac}d^{a'b'c} = \frac{1}{3}\left(\delta^{bb'}\delta^{aa'} + \delta^{ba'}\delta^{b'a} + \delta^{ba}\delta^{a'b'}\right). \quad (1.281)$$

Sometimes it is convenient to use colour diagrams to perform colour algebra. Colour factors of Feynman diagrams in QCD are expressed in terms of the colour propagators of quarks and gluons

$$\beta \xrightarrow{\quad\alpha\quad} = \delta_{\alpha\beta}, \qquad b\,\text{\textonionnnn}\,a = \delta^{ab}, \qquad (1.282)$$

and the vertices

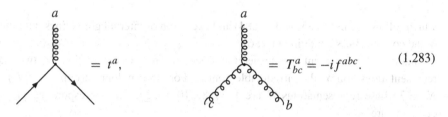

$$= t^a, \qquad = T^a_{bc} = -if^{abc}. \qquad (1.283)$$

Then the first and the third of Eqs. (1.260) are presented as

$$\text{\Large\bigcirc} = 0, \qquad \text{\Large\bigcirc} = T_F\,\text{\textonionnnn}, \qquad (1.284)$$

where $T_F = 1/2$ is the trace normalization of the fundamental representation. The commutation relations are depicted as

$$\qquad (1.285)$$

and the completeness condition (1.261) as

$$= T_F \qquad\qquad -\frac{T_F}{N_c} \qquad\qquad (1.286)$$

The sequential algorithm for calculation of colour factors corresponding to Feynman diagrams consists of the successive exclusion of the three-gluon vertices with the help of the equality

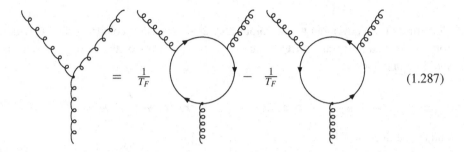

$$= \frac{1}{T_F} \qquad - \frac{1}{T_F} \qquad (1.287)$$

which follows from (1.285) and (1.284) and exclusion of internal gluon lines with the help of the completeness condition (1.286).

Finally, let us present the projection operators for decomposition of two adjoint representations onto the irreducible ones. For the colour group $SU(N_c)$ with $N_c = 3$, these representations $\mathcal{R}$ are $\underline{1}, \underline{8_a}, \underline{8_s}, \underline{10}, \overline{\underline{10}}, \underline{27}$. The corresponding projection operators are

$$\langle bb' | \hat{P}_1 | aa' \rangle = \frac{\delta^{bb'} \delta^{aa'}}{N_c^2 - 1}, \qquad (1.288)$$

$$\langle bb' | \hat{P}_{8_a} | aa' \rangle = \frac{f^{bb'c} f^{aa'c}}{N_c}, \qquad (1.289)$$

$$\langle bb' | \hat{P}_{8_s} | aa' \rangle = d^{bb'c} d^{aa'c} \frac{N_c}{N_c^2 - 4}, \qquad (1.290)$$

$$\langle bb' | \hat{P}_{10} | aa' \rangle = \frac{1}{4} \left[\delta^{ba} \delta^{b'a'} - \delta^{ba'} \delta^{b'a} - \frac{2}{N_c} f^{bb'c} f^{aa'c} + i f^{ba'c} d^{b'ac} + i d^{ba'c} f^{b'ac} \right], \quad (1.291)$$

$$\langle bb' | \hat{P}_{\overline{10}} | aa' \rangle = \frac{1}{4} \left[\delta^{ba} \delta^{b'a'} - \delta^{ba'} \delta^{b'a} - \frac{2}{N_c} f^{bb'c} f^{aa'c} - i f^{ba'c} d^{b'ac} - i d^{ba'c} f^{b'ac} \right], \quad (1.292)$$

$$\langle bb' | \hat{P}_{27} | aa' \rangle = \frac{1}{4} \left[\left(1 + \frac{2}{N_c} \right) \left(\delta^{ba} \delta^{b'a'} + \delta^{ba'} \delta^{b'a} \right) - \frac{2(N_c + 2)}{N_c(N_c + 1)} \delta^{bb'} \delta^{aa'} \right.$$

$$\left. - \left(1 + \frac{2}{N_c + 2} \right) d^{bb'c} d^{aa'c} + d^{bac} d^{b'a'c} + d^{b'ac} d^{ba'c} \right]. \qquad (1.293)$$

For generality, we do not put $N_c = 3$ here, so that the above expressions are valid for the $SU(N_c)$ group with arbitrary N_c. Corresponding representations in this case have dimensions

$$n_1 = 1, \quad n_{8_a} = n_{8_s} = N_c^2 - 1, \quad n_{10} = n_{\overline{10}} = \frac{(N_c^2 - 4)(N_c^2 - 1)}{4},$$

$$n_{27} = \frac{(N_c + 3)N_c^2(N_c - 1)}{4}. \tag{1.294}$$

However, for $N_c > 3$ there is an additional representation with dimension

$$n_{N_c > 3} = \frac{(N_c + 1)N_c^2(N_c - 3)}{4} \tag{1.295}$$

and projection operator

$$\langle bb'|\hat{P}_{N_c>3}|aa'\rangle = \frac{1}{4}\left[\left(1 - \frac{2}{N_c}\right)\left(\delta^{ba}\delta^{b'a'} + \delta^{ba'}\delta^{b'a}\right) + \frac{2(N_c - 2)}{N_c(N_c - 1)}\delta^{bb'}\delta^{aa'}\right.$$
$$\left. + \left(1 - \frac{2}{N_c - 2}\right)d^{bb'c}d^{aa'c} - d^{bac}d^{b'a'c} - d^{ba'c}d^{b'ac}\right]. \tag{1.296}$$

In $SU(3)$ this projection operator turns into zero due to the equality (1.281).

References

[1] Feynman, R. P., *Acta Phys. Polon.* **24** (1963) 697.

[2] DeWitt, B., *Phys. Rev.* **160** (1967) 1113.

[3] DeWitt, B., *Phys. Rev.* **162** (1967) 1195.

[4] DeWitt, B., *Phys. Rev.* **162** (1967) 1239.

[5] Faddeev, L. D. and Popov, V. N., *Phys. Lett.* **B25** (1967) 29.

[6] Gribov, V. N., In *Proceedings of the 12th Winter LNPI School on Nuclear and Elementary Particle Physics,* Leningrad (1977) 147–162. (V. N. Gribov, in Gauge Theories and Quark Confinement, Phasis, Moscow, 2002, pp. 257–270.)

[7] Gribov, V. N., *Nucl. Phys.* **B139** (1978) 1.

[8] Bassetto, A., Lazzizzera I. and Soldati, R., *Phys. Lett.* **B131** (1983) 177.

[9] Sobreiro, R. F. and Sorella, S. P., *JHEP* **0506** (2005) 054 [arXiv:hep-th/0506165].

[10] Ilgenfritz, E. M., Muller-Preussker, M., Sternbeck, A., Schiller, A. and Bogolubsky, I. L., arXiv:hep-lat/0609043.

[11] Maas, A., Cucchieri, A. and Mendes, T., arXiv:hep-lat/0610006.

[12] Arnowitt, R. L. and Ficler, S. I., *Phys. Rev.* **127** (1962) 1821.

[13] Schwinger, J., *Phys. Rev.* **130** (1963) 402.

[14] Fradkin, E. S. and Tyutin, I. V., *Phys. Rev.* **D2** (1970) 2841.

[15] Tomboulis, E., *Phys. Rev.* **D8** (1973) 2736.

[16] Delbourgo, R., Salam, A. and Strathdee, J., *Nuovo Cim.* **A32** (1974) 237.

[17] Dokshitzer, Yu. L., Dyakonov, D. I., Troyan, S. I., *Phys. Rep.* **58** (1980) 269.

[18] Frenkel, J., *Phys. Rev.* **D13** (1976) 2325.

[19] Kallosh, R. E. and Tyutin, I. V., *Yad. Fiz.* **17** (1973) 190 [*Sov. J. Nucl. Phys.* **17** (1973) 98].

50 *General properties of QCD*

[20] Konetschny, W. and Kummer, W., *Nucl. Phys.* **B100**, (1975) 106.

[21] Akhiezer, A. I. and Berestetsky, V. B., *Quantum electrodynamics*, Moscow, Nauka, 1981, (in Russian) Wiley-Interscience, N.Y., 1965.

[22] Pauli, W. and Villars, F., *Rev. Mod. Phys.* **21** (1949) 434.

[23] Speer, E. R., *J. Math. Phys.* **9** (1968) 1408.

[24] Lee, B. W. and Zinn Justin, J., *Phys. Rev.* **D5** (1972) 3137.

[25] Faddeev, L. D. and Slavnov, A. A., *Gauge Fields, introduction to quantum theory*, Benjamin Reading, 1980.

[26] Corrigan, E., Goddard, P., Osborn, H. and Templeton, S., *Nucl. Phys.* **B159** (1979) 469.

[27] Bollini, C. G. and Giambiagi, J. J., *Phys. Lett.* **B40** (1972) 566.

[28] Bollini, C. G. and Giambiagi, J. J., *Nuovo Cim.* **B12** (1972) 20.

[29] 't Hooft, G. and Veltman, M. J. G., *Nucl. Phys.* **B44** (1972) 189.

[30] 't Hooft, G. and Veltman, M. J. G., preprint CERN-73-09, Sep 1973.

[31] Ashmore, J. F., *Lett. Nuovo Cim.* **4** (1972) 289.

[32] Cicuta G. M. and Montaldi E., *Lett. Nuovo Cim.* **4** (1972) 329.

[33] Bonneau, G., *Int. J. Mod. Phys.* **A5** (1990) 3831.

[34] Adam, L. E. and Chetyrkin, K. G., *Phys. Lett.* **B329** (1994) 129 [arXiv:hep-ph/9404331].

[35] Wilson, K. G., *Phys. Rev.* **D7** (1973) 2911.

[36] Breitenlohner, P. and Maison, D., *Commun. Math. Phys.* **52** (1977) 11.

[37] Siegel, W., *Phys. Lett.* **B84** (1979) 193.

[38] Siegel, W., *Phys. Lett.* **B94** (1980) 37.

[39] Chanowitz, M. S., Furman, M. and Hinchliffe, I., *Phys. Lett.* **B78** (1978) 285; *Nucl. Phys.* **B159** (1979) 225.

[40] Nicolai, H. and Townsend, P. K., *Phys. Lett.* **B93** (1980) 111.

[41] Kreimer, D., *Phys. Lett.* **B237** (1990) 59.

[42] Korner, J. G., Kreimer, D. and Schilcher, K., *Z. Phys.* **C54** (1992) 503.

[43] Thompson, G. and Yu, H. L., *Phys. Lett.* **B151** (1985) 119.

[44] Abdelhafiz, M. I. and Zralek, M., *Acta Phys. Polon.* **B18** (1987) 21.

[45] Avdeev, L. V., Chochia, G. A. and Vladimirov, A. A., *Phys. Lett.* **B105** (1981) 272.

[46] Akyeampong, D. A. and Delbourgo, R., *Nuovo Cim.* **A17** (1973) 578; **18** (1973) 94; *Nuovo Cim.* **A19** (1974) 219.

[47] Collins, J. C., *Renormalization*, Cambridge University Press, 1984.

[48] Landau, L. D., Abrikosov, A. A. and Khalatnikov, I. M., *Dokl. Akad. Nauk USSR*, **95** (1954) 497, 773, 1117.

[49] Gell-Mann, M. and Low, F. E., *Phys. Rev.* **95** (1954) 1300.

[50] Slavnov, A. A., *Theor. Math. Phys.* **10** (1972) 99 [*Teor. Mat. Fiz.* **10** (1972) 153].

[51] Taylor, J. C., *Nucl. Phys.* **B33** (1971) 436.

[52] Milshtein, A. I. and Fadin, V. S., *Yad. Fiz.* **34** (1981) 1403.

[53] Georgi, H. and Politzer, H. D., *Phys. Rev.* **D14** (1976) 1829.

[54] Celmaster, W. and Gonsalves, R. J., *Phys. Rev. Lett.* **42** 1435 (1979).

[55] Celmaster, W. and Gonsalves, R. J., *Phys. Rev.* **D20** (1979) 1420.

[56] Barbieri, R., Caneschi, L., Curci, G. and d'Emilio, E., *Phys. Lett.* **B81** (1979) 207.

[57] 't Hooft, G., *Nucl. Phys.* **B61** (1973) 455.

[58] Bardeen, W. A., Buras, A. J., Duke, D. W. and Muta, T., *Phys. Rev.* **D18** (1978) 3998.

[59] Stueckelberger, E. C. G. and Petermann, A., *Helv. Phys. Acta* **26** (1953) 499.

[60] Bogolyubov, N. N. and Shirkov, D. V., *Doklady Akad. Nauk USSR*, **103** (1955) 203, 391.

[61] Shirkov, D. V., *Doklady Akad. Nauk USSR*, **105** (1955) 970.

[62] Logunov, A. A., *ZhETF* **30** (1956) 793.

[63] Ovsiannikov, L. V., *Doklady Akad. Nauk USSR*, **109** (1956) 1112.

[64] Symanzik, K., *Commun. Math. Phys.* **18** (1970) 227.

[65] Callan, C. G., *Phys. Rev.* **D2** (1970) 1541.

[66] Caswell, W. E. and Wilczek, F., *Phys. Lett.* **B49** (1974) 291.

[67] Gross, D., *Methods in Field theory*, Les Houches 1975, North Holland (1976), Chapter 4.

[68] Gross, D. J. and Wilczek, F., *Phys. Rev. Lett.* **30** (1973) 1343.

[69] Politzer, H. D., *Phys. Rev. Lett.* **30** (1973) 1346.

[70] Caswell, W. E., *Phys. Rev. Lett.* **33** (1974) 244.

[71] Jones, D. R. T., *Nucl. Phys.* **B75** (1974) 531.

[72] Egorian, E. S. and Tarasov, O. V., *Theor. Mat. Fiz.* **41** (1979) 26.

[73] Tarasov, O. V., Vladimirov, A. A. and Zharkov, A. Yu., *Phys. Lett.* **B93** (1980) 429.

[74] Larin, S. A. and Vermaseren, J. A. M., *Phys. Lett.* **B303** (1993) 334.

[75] van Ritbergen, T., Vermaseren, J. A. M. and Larin, S. A., *Phys. Lett.* **B400** (1997) 379.

[76] Vermaseren, J. A. M., Larin, S. A. and van Ritbergen, T., *Phys. Lett.* **B405** (1997) 327 [arXiv:hep-ph/9703284].

[77] Tarrach, R., *Nucl. Phys.* **B183** (1981) 384.

[78] Tarasov, O. V., preprint JINR P2-82-900 (Dubna, 1982), unpublished.

[79] Larin, S. A., in: *Proc. of the Int. School "Particles and Cosmology,"* April 1993, Baksan Neutrino Observatory of JINR, eds. V. A. Matveev, Kh.S. Nirov and V. A. Rubakov, World Scientific, Singapore, 1994.

[80] Espriu, D. and Tarrach, R., *Phys. Rev.* **D25** (1982) 1073.

[81] Appelquist, T. and Carazzone, J., *Phys. Rev.* **D11** (1975) 2856.

[82] Weinberg, S., *Phys. Lett.* **B91** (1980) 51.

[83] Ovrut, B. A. and Schnitzer, H. J., *Phys. Lett.* **B100** (1981) 403.

[84] Wetzel, W., *Nucl. Phys.* **B196** (1982) 259.

[85] Bernreuther, W. and Wetzel, W., *Nucl. Phys.* **B197** (1982) 228 [Erratum-ibid. **B513** (1998) 758].

[86] Bernreuther, W., *Z. Phys.* **B20** (1983) 331.

[87] Bernreuther, W., *Annals Phys.* **151** (1983) 127.

[88] Chetyrkin, K. G., Kniehl, B. A. and Steinhauser, M., *Nucl. Phys.* **B510** (1998) 61 [arXiv:hep-ph/9708255].

[89] Khriplovich, I. B., *Yad. Fiz.* **10** (1969) 409.

[90] Drell, S. D., *Transactions of the N.Y.Acad. of Sci.*, Series II, Vol. **40** (1980) 76. SLAC-PUB-2694.

[91] Feynman, R. P., *Rev. Mod. Phys.* **20** (1948) 367.

[92] Grunberg, G., *Phys. Lett.* **B95** (1980) 70 [Erratum-ibid. **B110** (1982) 501].

[93] Grunberg, G., *Phys. Rev.* **B29** (1984) 2315.

[94] Stevenson, P. M., *Phys. Rev.* **D23** (1981) 2916.

[95] Stevenson, P.M., *Phys. Lett.* **B100** (1981) 61.

[96] Kataev, A. L. and Starshenko, V. V., *Mod. Phys. Lett.* **A10** (1995) 235.

[97] Brodsky, S. J., Lepage, G. P. and Mackenzie, P. B., *Phys. Rev.* **D28** (1983) 228.

2
Chiral symmetry and its spontaneous violation

2.1 The general properties of QCD at low energies

The asymptotic freedom of QCD, i.e. the logarithmic decrease of the QCD coupling constant $\alpha_s(Q^2) \sim 1/\ln Q^2$ at large momentum transfers $Q^2 \to \infty$ (or, equivalently, the decrease of α_s at small distances, $\alpha_s(r) \sim 1/\ln r$) allows one to perform reliable theoretical calculations of hard processes, using perturbation theory.[1] However, the same property of the theory implies an increase of the running coupling constant in QCD at small momentum transfer, i.e. at large distances. Furthermore, this increase is unlimited within the framework of perturbation theory. Physically such growth is natural and is even needed, because otherwise the theory would not be a theory of *strong* interactions.

QCD possesses two remarkable properties. The first is the property of confinement: quarks and gluons cannot leave the region of their strong interaction and cannot be observed as real physical objects. Physical objects, observed experimentally at large distances, are hadrons – mesons and baryons. The second important property of QCD is the spontaneous violation of chiral symmetry. The masses of light u, d, s quarks that enter the QCD Lagrangian, especially the masses of u and d quarks from which the usual (non-strange) hadrons are built, are very small as compared to the characteristic QCD mass scale $M \sim 1$ GeV ($m_u, m_d < 10$ MeV, $m_s \sim 150$ MeV).[2] In QCD, the quark interaction is due to the exchange of vector gluonic field. Thus, if light quark masses are neglected, the QCD Lagrangian (its light quark part) becomes chirally symmetric, i.e. not only vector, but also axial currents are conserved. In this approximation the left-hand and right-hand chirality quark fields do not transform into each other. However, this chiral symmetry is not realized in the spectrum of hadrons and their low energy interactions. Indeed, in a chirally symmetric theory fermion states must be either massless or degenerate in parity. It is evident that baryons (particularly, the nucleon) do not possess such properties. This means that the chiral symmetry of the QCD Lagrangian is spontaneously broken. According to the Goldstone theorem, the spontaneous breaking of symmetry leads to the appearance of massless particles in the spectrum of physical states – the Goldstone bosons. In QCD, Goldstone bosons can be identified with the triplet of π mesons in the limit $m_u, m_d \to 0$,

---

[1] Here and in what follows the same notation is used for functions defined in coordinate and momentum spaces.
[2] By the characteristic mass scale we mean the scale at which the interaction in QCD becomes strong.

$m_s \neq 0$ ($SU(2)$ symmetry) and with the octet of pseudoscalar mesons (π, K, η) in the limit $m_u, m_d, m_s \to 0$ ($SU(3)$ symmetry). The local $SU(2)_L \times SU(2)_R$ symmetry (here L and R mean left-hand and right-hand quark currents) or $SU(3)_L \times SU(3)_R$ symmetry of strong interactions makes it possible to construct an effective low energy chiral theory of Goldstone bosons and their interactions with baryons.

The initial version of the approach was developed before QCD and was called the theory of partial conservation of axial current (PCAC). The effective Lagrangian of this theory represents the nonlinear interaction of pions with themselves and with nucleons and corresponds to the first term in the expansion in powers of momenta in the modern chiral effective theory. (The review of the PCAC theory at this stage can be found in [1]). When QCD was developed, it was proved that the appearance of Goldstone bosons is a consequence of spontaneous breaking of chiral symmetry of the QCD vacuum which leads to vacuum condensates violating the chiral symmetry. It was also established that baryon masses are expressed through the same vacuum condensates. Nowadays, one can formulate the chiral effective theory (CET) of hadrons as a succesive expansion of physical observables in powers of particle momenta and quark (or Goldstone boson) masses not only in tree approximation, as in PCAC, but also by taking into account loop corrections. (CET is often called chiral perturbation theory – ChPT.)

In this chapter we present foundations, basic ideas and concepts of CET as well as their connection with QCD, paying much attention to the general properties of pion interactions.

2.2 The masses of the light quarks

In what follows, u, d, s quarks will be called light quarks and all other quarks heavy quarks. The reason is that the masses of light quarks are small compared with the characteristic mass of strong interaction $M \sim 0.5 - 1.0\,\text{GeV}$ or m_ρ. The symmetry of strong interactions is $SU(3)_L \times SU(3)_R \times U(1)$. Its group generators are the charges corresponding to the left $(V - A)$ and right $(V + A)$ light quark chiral currents and $U(1)$ corresponds to the baryonic charge current. The experiment shows that the accuracy of $SU(3)_L \times SU(3)_R$ symmetry is of the same order as that of $SU(3)$ symmetry: the small parameter characterizing the chiral symmetry violation in strong interactions is generally of order $\sim 1/5 - 1/10$. If we restrict ourselves to the consideration of u and d quarks and hadrons which are built from them, then the symmetry of strong interaction is $SU(2)_L \times SU(2)_R \times U(1)$ and its accuracy is of order $(m_u + m_d)/M \sim 1\%$, i.e. of the same order as the isospin symmetry violation which arises from electromagnetic interactions.

The approximate validity of chiral symmetry means that not only divergences of the vector currents $\partial_\mu j_\mu^q$ are zero or small, but also those of the axial currents $\partial_\mu j_{\mu 5}^q$. (Here $q = u, d, s$. This statement refers to flavor nonsinglet axial currents. The divergence of the singlet axial current is determined by the anomaly and is nonzero even for massless quarks (see Chapter 3). The divergences of nonsinglet axial currents in QCD are proportional to the quark masses. Therefore the existence of chiral symmetry can be justified if the quark masses are small [2, 3, 4]. However, the baryon masses are by no means small – the chiral symmetry is not realized on the hadronic mass spectrum in a trivial way by vanishing of all the fermion masses. Chiral symmetry is broken spontaneously and massless particles – Goldstone bosons – appear in the physical spectrum. These Goldstone bosons belong to a pseudoscalar octet. They are massless if light quark masses are put to zero. The nonvanishing light quark masses lead to the explicit violation of chiral symmetry and provide the nonzero masses of the pseudoscalar meson octet. For this reason this octet of Goldstone bosons plays a special role in QCD.

Heavy quarks are decoupled in the low energy domain (this statement is called the Appelquist–Carazzone theorem) [5]. We ignore them in this chapter where QCD at low energies is considered.

The QCD Hamiltonian can be split into two pieces

$$H = H_0 + H_1, \tag{2.1}$$

where

$$H_1 = \int d^3x \left(m_u \bar{u}u + m_d \bar{d}d + m_s \bar{s}s \right). \tag{2.2}$$

Evidently, because of vector gluon–quark interaction the first term in the Hamiltonian, H_0, is $SU(3)_L \times SU(3)_R$ invariant and the only source of $SU(3)_L \times SU(3)_R$ violation is H_1. The quark masses m_q, $q = u, d, s$ in (2.2), are not renormalization invariant: they are scale dependent. It is possible to write

$$m_q(M) = Z_q(M/\mu)m_q(\mu), \tag{2.3}$$

where M characterizes the scale, μ is some fixed normalization point and $Z_q(M/\mu)$ are renormalization factors. If the light quark masses are small and can be neglected, the renormalization factors are flavour-independent. This means that the ratios

$$\frac{m_{q1}(M)}{m_{q2}(M)} = \frac{m_{q1}(\mu)}{m_{q2}(\mu)} \tag{2.4}$$

are scale-independent and have a definite physical meaning. (This relation holds if M is greater than the Goldstone mass m_K: its validity in the domain $M \sim m_K$ may be doubtful.) The u, d, s quark masses had been first estimated by Gasser and Leutwyler in 1975 [2] (see also their review [4]). In 1977, Weinberg [3] using PCAC and the Dashen theorem [6] to account for electromagnetic self-energies of mesons has proved that the ratios m_u/m_d and m_s/m_d could be expressed in terms of the K and π meson masses. In order to find the ratios m_u/m_d and m_s/m_d following Weinberg consider the axial currents

$$j_{\mu5}^{\bar{-}} = \bar{d}\gamma_\mu\gamma_5 u,$$
$$j_{\mu5}^3 = \left[\bar{u}\gamma_\mu\gamma_5 u - \bar{d}\gamma_\mu\gamma_5 d \right]/\sqrt{2}, \tag{2.5}$$
$$j_{\mu5}^{s\bar{-}} = \bar{s}\gamma_\mu\gamma_5 u, \quad j_{\mu5}^{s0} = \bar{s}\gamma_\mu\gamma_5 d, \tag{2.6}$$

and their matrix elements between the vacuum and π or K meson states:

$$\langle 0 \mid j_{\mu 5}^- \mid \pi^+ \rangle = i f_{\pi^+} p_\mu,$$

$$\langle 0 \mid j_{\mu 5}^3 \mid \pi^0 \rangle = i f_{\pi^0} p_\mu,$$

$$\langle 0 \mid j_{\mu 5}^{s-} \mid K^+ \rangle = i f_{K^+} p_\mu,$$

$$\langle \mid j_{\mu 5}^{s0} \mid K^0 \rangle = i f_{K^0} p_\mu, \tag{2.7}$$

where p_μ are π or K momenta. In the limit of exact $SU(3)$ symmetry all constants in the right-hand sides of (2.7) are equal: $f_{\pi^+} = f_{\pi^0} = f_{K^+} = f_{K^0}$, $SU(2)$ – isotopic symmetry results in equalities $f_{\pi^+} = f_{\pi^0}$, $f_{K^+} = f_{K^0}$. The constants $f_{\pi^+} \equiv f_\pi$ and $f_{K^+} \equiv f_K$ are the coupling constants of the decays $\pi^+ \to \mu^+ \nu$ and $K^+ \to \mu^+ \nu$. Their experimental values are: $f_\pi = 131$ MeV, $f_K = 160$ MeV. The ratio $f_K/f_\pi = 1.22$ characterizes the accuracy of $SU(3)$ symmetry. Multiply (2.7) by p_μ. Using the equality for the divergence of the axial current following from the QCD Lagrangian

$$\partial_\mu \left[\bar{q}_1(x) \gamma_\mu \gamma_5 q_2(x) \right] = i(m_{q_1} + m_{q_2}) \bar{q}_1(x) \gamma_5 q_2(x), \tag{2.8}$$

we get

$$i(m_u + m_d)\langle 0 \mid \bar{d}\gamma_5 u \mid \pi^+ \rangle \ = \ f_{\pi^+} m_{\pi^+}^2,$$

$$(i/\sqrt{2}) \left[(m_u + m_d)\langle 0 \mid \bar{u}\gamma_5 u - \bar{d}\gamma_5 d \mid \pi^0 \rangle + (m_u - m_d)\langle 0 \mid \bar{u}\gamma_5 u + \bar{d}\gamma_5 d \mid \pi^0 \rangle \right]$$
$$= f_{\pi^0} m_{\pi^0}^2,$$

$$i(m_s + m_u)\langle 0 \mid \bar{s}\gamma_5 u \mid K^+ \rangle \ = \ f_{K^+} m_{K^+}^2,$$

$$i(m_s + m_d)\langle 0 \mid \bar{s}\gamma_5 d \mid K^0 \rangle \ = \ f_{K^0} m_{K^0}^2. \tag{2.9}$$

Neglect electromagnetic (and weak) interactions and assume that isotopic invariance can be used for the matrix elements on the left-hand side of (2.9). Then

$$\langle 0 \mid \bar{u}\gamma_5 u + \bar{d}\gamma_5 d \mid \pi^0 \rangle = 0,$$

$$\langle 0 \mid \bar{d}\gamma_5 u \mid \pi^+ \rangle = \frac{1}{\sqrt{2}} \langle 0 \mid \bar{u}\gamma_5 u - \bar{d}\gamma_5 d \mid \pi^0 \rangle \tag{2.10}$$

and, as follows from (2.9), the $\pi^\pm$ and π^0 masses are equal in this approximation even when $m_u \neq m_d$. Hence the experimentally observed mass difference $\Delta m_\pi = m_{\pi^+} - m_{\pi^0} = 4.6$ MeV is caused by the electromagnetic interaction only. The sign of the K meson mass difference $\Delta m_K = m_{K^+} - m_{K^0} = -4.0$ MeV is opposite to that of the pion one. The electromagnetic kaon and pion mass differences in QCD or in the quark model are determined by the same Feynman diagrams and must be, at least, of the same sign. This means, in accord with (2.9), that $m_d > m_u$.

Assuming $SU(3)$ invariance of the matrix elements in (2.9) it is easy to get from (2.9) and (2.10)

$$\frac{m_u}{m_d} = \frac{\tilde{m}_\pi^2 - (\tilde{m}_{K^0}^2 - \tilde{m}_{K^+}^2)}{\tilde{m}_\pi^2 + (\tilde{m}_{K^0}^2 - \tilde{m}_{K^+}^2)}, \tag{2.11}$$

$$\frac{m_s}{m_d} = \frac{\tilde{m}_{K^0}^2 + \tilde{m}_{K^+}^2 - \tilde{m}_\pi^2}{\tilde{m}_{K^0}^2 - \tilde{m}_{K^+}^2 + \tilde{m}_\pi^2}. \tag{2.12}$$

The tildas in (2.11),(2.12) mean that the pion and kaon masses here are not physical, but are the masses in the limit when the electromagnetic interaction is switched off. In order to relate $\tilde{m}_\pi^2$, $\tilde{m}_K^2$ to physical masses, let us again use $SU(3)$ symmetry. In $SU(3)$ symmetry the photon is a U singlet and π^+ and K^+ belong to the U doublet.[3] Therefore, the electromagnetic corrections to $m_{\pi^+}^2$ and $m_{K^+}^2$ are equal

$$(\delta m_{\pi^+}^2)_{el} = (\delta m_{K^+}^2)_{el}. \tag{2.13}$$

It is possible to show, that in the limit m_π^2, $m_K^2 \to 0$, the electromagnetic corrections to the π^0 and K^0 masses tend to zero,

$$(\delta m_{\pi^0}^2)_{el} = (\delta m_{K^0}^2)_{el} = 0. \tag{2.14}$$

Eqs. (2.13), (2.14) can be rewritten in the form of the Dashen relation [6]

$$(m_{\pi^+}^2 - m_{\pi^0}^2)_{el} = (m_{K^+}^2 - m_{K^0}^2)_{el}. \tag{2.15}$$

From (2.14), (2.15) we have

$$\tilde{m}_\pi^2 = m_{\pi^0}^2,$$

$$\tilde{m}_{K^+}^2 - \tilde{m}_{K^0}^2 = m_{K^+}^2 - m_{K^0}^2 - (m_{\pi^+}^2 - m_{\pi^0}^2). \tag{2.16}$$

Substitution of (2.16) into (2.10), (2.11) leads to:

$$\frac{m_u}{m_d} = \frac{2m_{\pi^0}^2 - m_{\pi^+}^2 - (m_{K^0}^2 - m_{K^+}^2)}{m_{K^0}^2 - m_{K^+}^2 + m_{\pi^+}^2}, \tag{2.17}$$

$$\frac{m_s}{m_d} = \frac{m_{K^0}^2 + m_{K^+}^2 - m_{\pi^0}^2}{m_{K^0}^2 - m_{K^+}^2 + m_{\pi^+}^2}. \tag{2.18}$$

Numerically, this gives

$$\frac{m_u}{m_d} = 0.56, \qquad \frac{m_s}{m_d} = 20.1. \tag{2.19}$$

A strong violation of isotopic invariance, as well as the large difference between u, d and s quark masses, i.e. the violation of $SU(3)$ flavor symmetry, is evident from (2.19). (A more detailed analysis shows that the results (2.19) are practically independent of the assumption of $SU(3)$ symmetry of the corresponding matrix elements used in their derivation.) This seems to be in contradiction with the well-established isospin symmetry of strong interactions, as well as with the approximate $SU(3)$ symmetry. The resolution of

---

[3] The description of T, U, V subgroups of $SU(3)$ is given, e.g. in [7].

this puzzle is that the quark masses are small in comparison with the scale $M \sim m_\rho$ of strong interaction: the parameter characterizing isospin violation is $(m_d - m_u)/M$ and the parameter characterizing the $SU(3)$ symmetry violation is m_s/M.

The large m_s/m_d ratio explains the large mass splitting in the pseudoscalar meson octet. For $m_{K^+}^2/m_{\pi^+}^2$ we have from (2.9) $[\bar{m} = (m_u + m_d)/2]$

$$\frac{m_{K^+}^2}{m_{\pi^+}^2} = \frac{m_s + \bar{m}}{2\bar{m}} = 13 \tag{2.20}$$

in perfect agreement with experiment. The ratio m_η^2/m_π^2 expressed in terms of quark mass ratios is also in a good agreement with experiment.

The ratios (2.17), (2.18) were obtained in the first order in quark masses. Therefore, their accuracy is of order of the $SU(3)$ symmetry accuracy, i.e. about 10–20%.

Gasser and Leutwyler demonstrated that there is a relation valid in the second order in quark masses [8]

$$\left(\frac{m_u}{m_d}\right)^2 + \frac{1}{Q^2}\left(\frac{m_s}{m_d}\right)^2 = 1. \tag{2.21}$$

Using the Dashen theorem for electromagnetic self-energies of π and K mesons one can express Q as

$$Q_D^2 = \frac{(m_{K^0}^2 + m_{K^+}^2 - m_{\pi^+}^2 + m_{\pi^0}^2)(m_{K^0}^2 + m_{K^+}^2 - m_{\pi^+}^2 - m_{\pi^0}^2)}{4m_{\pi^0}^2(m_{K^0}^2 - m_{K^+}^2 + m_{\pi^+}^2 - m_{\pi^0}^2)}. \tag{2.22}$$

Numerically Q_D is equal to 24.2. However, the Dashen theorem is valid only in the first order in quark masses. The electromagnetic mass difference of K mesons calculated in [9] using the Cottingham formula [10] within the framework of the large N_c approach [11] increased $\Delta m_K = (M_{K^+} - M_{K^0})_{e.m.}$ from its Dashen value $\Delta m_K = 1.27\,\mathrm{MeV}$ to $\Delta m_K = 2.6\,\mathrm{MeV}$ and, correspondingly, decreased Q_D to $Q = 22.0 \pm 0.6$. The other way to find Q is from $\eta \to \pi^+\pi^-\pi^0$ decay, using the chiral effective theory. Unfortunately, the next two leading corrections are large in this approach [12], which makes the accuracy of the results uncertain. From the $\eta \to \pi^+\pi^-\pi^0$ decay data with account of interactions in the final state it was found that $Q = 22.4 \pm 0.9$ [13], $Q = 22.7 \pm 0.8$ [14] and $Q = 22.8 \pm 0.4$ [15] (the latter from the Dalitz plot). So, the final conclusion is that Q is in the interval $21.5 < Q < 23.5$. (See [16] for a review.) The ratio $\gamma = m_u/m_d$ can also be found from the ratio of $\psi' \to (J/\psi)\eta$ and $\psi' \to (J/\psi)\pi^0$ decays [17]. In [17] it was proved that

$$r = \frac{\Gamma(\psi' \to J/\psi + \pi^0)}{\Gamma(\psi' \to J/\psi + \eta)} = 3\left(\frac{1-\gamma}{1+\gamma}\right)^2 \left(\frac{m_\pi}{m_\eta}\right)^4 \left(\frac{p_\pi}{p_\eta}\right)^3, \tag{2.23}$$

where p_π and p_η are the pion and η momenta in the ψ' rest frame. Eq. (2.23) is valid in the first order in quark masses. The Particle Data Group [18] gives

$$r_{\mathrm{exp}} = (4.08 \pm 0.43) \cdot 10^{-2}. \tag{2.24}$$

Assuming the theoretical uncertainty in (2.23) as 30% and adding the theoretical and experimental errors in quadratures, we get from (2.23)

$$\gamma = \frac{m_u}{m_d} = 0.385 \pm 0.060. \tag{2.25}$$

A value close to (2.25) was found recently in [19]. The substitution of (2.25) into (2.21) with account of the above-mentioned uncertainty of Q results in

$$\frac{m_s}{m_d} = 20.8 \pm 1.3. \tag{2.26}$$

The value (2.25) is slightly less than the lowest-order result (2.19); (2.26) agrees with it. The values (2.25),(2.26) are in an agreement with lattice calculations [20].

The calculation of absolute values of quark masses is a more subtle problem. As was mentioned above, the masses are scale dependent. In perturbation theory, their scale dependence is given by the renormalization group equation (see Chapter 1):

$$\frac{dm(\mu)}{m(\mu)} = -\gamma[\,\alpha_s(\mu)\,]\frac{d\mu^2}{\mu^2} = -\sum_{r=1}^{\infty}\gamma_r a^r(\mu^2)\frac{d\mu^2}{\mu^2}. \tag{2.27}$$

In (2.27) $a = \alpha_s/\pi$, $\gamma_1 = 1$, $\gamma_2 = 91/24$, $\gamma_3 = 12.42$ for 3 flavours in the $\overline{\text{MS}}$ scheme [21]. In the first order in α_s it follows from (2.27) that:

$$\frac{m(Q^2)}{m(\mu^2)} = \left[\frac{\alpha_s(\mu^2)}{\alpha_s(Q^2)}\right]^{\gamma_m}, \tag{2.28}$$

where $\gamma_m = -4/9$ is the quark mass anomalous dimension. The recent calculations of m_s by QCD sum rules [22], from τ decay data [23, 24] and on lattice [20, 25], are not in good agreement with one another. The mean value estimated in [18] is: $m_s(2\text{ GeV}) \approx$ 105 MeV with an accuracy of about 20%. By taking $m_s(1\text{ GeV})/m_s(2\text{ GeV})$=1.40 we have then: $m_s(1\text{ GeV}) \approx 147$ MeV and, according to (2.25), (2.26), $m_d(1\text{ GeV}) = 7.1$ MeV, $m_u(1\text{ GeV}) = 2.9$ MeV. The light quark mass difference $m_d - m_u$ is equal to $m_d - m_u = 4.2 \pm 1.0$ MeV. This value agrees with that found by QCD sum rules from baryon octet mass splitting [26] and D and D^* isospin mass differences [27], $m_d - m_u = 3 \pm 1$ MeV. For the sum of the quark masses we have $m_u + m_d = 10.0 \pm 2.5$ MeV compared with $m_u + m_d = 12.8 \pm 2.5$ MeV found in [28]. For completeness the value of $m_c(m_c)$ is also presented here (see Chapter 6, Section 6.5.4):

$$m_c(m_c) = 1.275 \pm 0.015 \text{ GeV.} \tag{2.29}$$

2.3 Spontaneous violation of chiral symmetry. Quark condensate

As has been already mentioned, the large value of baryon masses indicate that chiral symmetry in QCD is violated. Generally, there are two possible mechanisms of chiral symmetry violation in quantum field theory. In the first (soft) mechanism, the symmetry is violated by the presence of fermion masses in the Lagrangian. (In the case of QCD by the presence of quark masses.) In the second mechanism, the chiral symmetry is broken spontaneously: the Lagrangian of the theory in chiral symmetric, but the spectrum of physical states is not. In QCD we deal with spontaneously broken chiral symmetry.[4]

In all known examples of field theories, spontaneous violation of global symmetry manifests itself in the modification of the properties of the ground state – the vacuum. Let us show that such phenomenon takes place also in QCD.

Consider the matrix element

$$iq_\mu(m_u + m_d) \int d^4x e^{iqx} \langle 0 \mid T\{j_{\mu5}^-(x), \ \bar{u}(0)\gamma_5 d(0)\} \mid 0 \rangle \qquad (2.30)$$

in the limit of massless u and d quarks (except for the overall factor $m_u + m_d$). Put q_μ inside the integral, integrate by parts, and use conservation of the axial current. Then only the term with the equal time commutator will remain

$$-(m_u + m_d) \int d^4x e^{iqx} \langle 0 \mid \delta(x_0)[\ j_{05}^-(x), \ \bar{u}(0)\gamma_5 d(0) \ ] \mid 0 \rangle$$

$$= (m_u + m_d)\langle 0 \mid \bar{u}u + \bar{d}d \mid 0 \rangle. \qquad (2.31)$$

Let us go now to the limit $q_\mu \to 0$ in (2.31) and perform summation over all intermediate states. The nonvanishing contribution comes only from the one-pion intermediate state since in this approximation the pion should be considered as massless. This contribution is equal to

$$q_\mu \langle 0 \mid j_{\mu5}^- \mid \pi^+ \rangle \frac{-1}{q^2} \langle \pi^+ \mid (m_u + m_d)\bar{u}\gamma_5 d \mid 0 \rangle = -f_\pi^2 m_\pi^2, \qquad (2.32)$$

where (2.7) and (2.9) were substituted when going to the right-hand side. Putting (2.32) in the left-hand side of (2.31) we get

$$\langle 0 \mid \bar{q}q \mid 0 \rangle = -\frac{1}{2} \frac{m_\pi^2 f_\pi^2}{m_u + m_d}, \qquad (2.33)$$

where $q = u$ or d and $SU(2)$ invariance of the QCD vacuum was used. Eq. (2.33) is the Gell-Mann–Oakes–Renner (GMOR) relation [29]. It can be derived also in another way. Assume the quark masses to be nonzero but small. Then the pion is massive and (2.30) tends to zero in the limit $q_\mu \to 0$. However, when we insert q_μ inside the integral, a term with the axial current divergence will appear in addition to the equal time commutator term (2.31). The account of this term, saturated by the one-pion intermediate state, results in the same Eq. (2.33). Numerically, with the quark mass values given in Section 2.2, $m_u + m_d = 10.0 \pm 2.5$ MeV we have

$$\langle 0 \mid \bar{q}q \mid 0 \rangle = -(257 \text{ MeV})^3 = -(1.70 \pm 0.42) \cdot 10^{-2} \text{ GeV}^3. \qquad (2.34)$$

As follows from (2.33), the product $(m_u + m_d)\langle 0 \mid \bar{q}q \mid 0 \rangle$ is scale independent, while $\langle 0 \mid \bar{q}q \mid 0 \rangle$ depends on the scale and the numerical value (2.34) refers to 1 GeV. Eq. (2.33)

---

[4] In principle, the chiral symmetry in baryonic states could be realized in a way that all baryonic states would be degenerate in parity with a splitting of the order of $m_u + m_d$. This is evidently not the case.

is valid in the first order in m_u, m_d, m_s. Therefore its accuracy is of order 20%. (This can be seen explicitly from the error shown in (2.34)). A more precise value of $\langle 0 \mid \bar{q}q \mid 0 \rangle$, following from the overall fit to the data in framework of QCD, will be given in Chapter 6. The quantity $\langle 0 \mid \bar{q}q \mid 0 \rangle$, called vacuum quark condensate can be also represented as

$$\langle 0 \mid \bar{q}q \mid 0 \rangle = \langle 0 \mid \bar{q}_L q_R + \bar{q}_R q_L \mid 0 \rangle, \tag{2.35}$$

where q_L and q_R are left and right quark fields $q_L = (1/2)(1+\gamma_5)q$, $q_R = (1/2)(1-\gamma_5)q$. It is evident from (2.34) that quark condensate violates chiral invariance and its numerical value (2.34) has a characteristic hadronic scale. The chiral invariance is violated globally, because $\langle 0 \mid \bar{q}q \mid 0 \rangle$ is noninvariant under global transformations $q \rightarrow e^{i\alpha\gamma_5}q$ with a constant α.

In perturbative QCD with massless quarks the quark condensate is zero in any order of perturbation theory. Therefore, the nonzero and nonsmall value of the quark condensate may arise only due to nonperturbative effects. The conclusion is that the nonperturbative field fluctuations which violate the chiral invariance of the Lagrangian are present and essential in QCD. Quark condensate plays a special role because its lowest dimension: d equals 3.

2.4 Goldstone theorem

Two arguments were presented in favor of chiral symmetry, approximately valid in QCD because of small u, d, s quark masses being spontaneously broken. These arguments were: the existence of large baryon masses and the appearance of a chiral symmetry violating quark condensate. Let us go to the limit of massless u, d, s quarks and show now that the direct consequence of each of these arguments is the appearance of massless pseudoscalar bosons in the hadronic spectrum.

Consider the matrix element of the axial current $j_{\mu5}^+ = \bar{u}\gamma_\mu\gamma_5 d$ between neutron and proton states. The general form of this matrix element is:

$$\langle p \mid j_{\mu5}^+ \mid n \rangle = \bar{v}_p(p')\left[\gamma_\mu\gamma_5 F_1(q^2) + q_\mu\gamma_5 F_2(q^2)\right]v_n(p), \tag{2.36}$$

where p and p' are the neutron and proton momentum, $q = p' - p$, $v_p(p')$, $v_n(p)$ are proton and neutron spinors and $F_1(q^2)$, $F_2(q^2)$ are formfactors. (Conservation of G parity was exploited in the derivation of (2.36)). Multiply (2.36) by q_μ and go to the limit $q^2 \rightarrow 0$, but $q_\mu \neq 0$. After multiplication, the left-hand side of (2.36) vanishes owing to axial current conservation. In the right-hand side using the Dirac equations for proton and neutron spinors, we have:

$$\bar{v}_p(p')\left[2mg_A + q^2 F_2(q^2)\right]\gamma_5 v_n(p), \tag{2.37}$$

where $g_A = F_1(0)$ is the neutron β-decay coupling constant, $g_A = 1.26$ and m is the nucleon mass (assumed to be equal for proton and neutron). The only way to avoid the

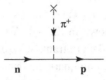

Fig. 2.1. The diagram describes the nucleon–axial current interaction by exchange of an intermediate pion: the solid line corresponds to the nucleon, the dashed line to the pion, and the cross means the interaction with external axial current.

discrepancy with the vanishing left-hand side of (2.36) is to assume that $F_2(q^2)$ has a pole at $q^2 = 0$:

$$F_2(q^2)_{q^2 \to 0} = -2mg_A \frac{1}{q^2}. \tag{2.38}$$

The pole in $F_2(q^2)$ corresponds to the appearance of a massless particle with pion quantum numbers. Then at small q^2 the matrix element in (2.36) has the form of:

$$\langle p \mid j_{\mu 5}^+ \mid n \rangle = g_A \bar{v}_p(p') \left(\delta_{\mu\nu} - \frac{q_\mu q_\nu}{q^2} \right) \gamma_\nu \gamma_5 v_n(p), \tag{2.39}$$

where conservation of the axial current is evident. The second term on the right-hand side of (2.39) describes the interaction of the nucleon with axial current by exchange of an intermediate pion, when the axial current creates a virtual π^+ and then the π^+ is absorbed by the neutron (Fig. 2.1). The low-energy pion–nucleon interaction can be parametrized phenomenologically by the Lagrangian

$$L_{\pi NN} = ig_{\pi NN} \, \bar{v}_N \gamma_5 \tau^a v_N \varphi^a, \tag{2.40}$$

where τ^a are the isospin Pauli matrices and $g_{\pi NN}$ is the πNN coupling constant, $g_{\pi NN}^2 / 4\pi \approx 14$. Using (2.7) and (2.40) the second term in (2.36) can be represented as

$$- \sqrt{2} \, g_{\pi NN} f_\pi \bar{v}_p \gamma_5 v_n \frac{q_\mu}{q^2}. \tag{2.41}$$

Comparison with (2.38) gives the Goldberger–Treiman relation [30]

$$g_{\pi NN} f_\pi = \sqrt{2} \, m g_A. \tag{2.42}$$

Experimentally, the Goldberger–Treiman relation is satisfied with 5% accuracy which strongly supports the hypothesis of spontaneous chiral symmetry violation in QCD. The main modification of (2.39) which arises from the nonvanishing pion mass is the replacement of the pion propagator: $q^2 \to q^2 - m_\pi^2$. Then the contribution of the second term vanishes at $q_\mu \to 0$ and becomes very small in the case of neutron β decay.

Since the only assumption in the above discussion was the conservation of the axial current, it can be generalized to any other component of the isospin 1 axial current if $SU(2)$ flavour symmetry is assumed, and to any octet axial current in the case of $SU(3)$ flavour symmetry. In the latter case we come to the conclusion that the octet of pseudoscalar mesons is massless in the limit of massless u, d, s quarks.

The massless bosons which arise by spontaneous symmetry breaking are Goldstone bosons and the theorem which states their appearance is the Goldstone theorem [31] (see also [32]). The proof of the Goldstone theorem presented above was based on the existence of massive baryons and on the nonvanishing nucleon β decay constant g_A. Before proceeding to another proof based on the existence of quark condensate in QCD, let us formulate some general features of spontaneously broken theories.

Let the Hamiltonian of the theory under consideration be invariant under some Lie group G, i.e. let the group generators Q_i commute with the Hamiltonian:

$$[Q_i, H] = 0, \quad i = 1, \ldots n. \tag{2.43}$$

The symmetry is spontaneously broken if the ground state is not invariant under G and a subset of $Q_l, l \leq m, 1 \leq m \leq n$ exists such that

$$Q_l \mid 0\rangle \neq 0. \tag{2.44}$$

Denote: $\mid B_l\rangle = Q_l \mid 0\rangle$. As follows from (2.43)

$$H \mid B_l\rangle = 0 \tag{2.45}$$

the states $\mid B_l\rangle$ have the same energy as the vacuum. These states may be considered as massless bosons at rest, i.e. as Goldstone bosons.[5] The operators $Q_j, j = m + 1, \ldots n$ generate a subgroup $K \subset G$, since from

$$Q_j \mid 0\rangle = 0 \tag{2.46}$$

it follows that

$$[Q_j, Q_{j'}] \mid 0\rangle = 0 \quad j, j' = m + 1, \ldots n. \tag{2.47}$$

In the case of QCD the group G is $SU(3)_L \times SU(3)_R$, which is spontaneously broken to $SU(3)_V$, i.e. to the group, where generators are the octet of vector charges. Q_l are the octet of axial charges and $\mid B_l\rangle$ are the octet of pseudoscalar mesons. (If only u, d quarks are considered as massless, all said above may be repeated, but relative to the $SU(2)_L \times SU(2)_R$ group.)

Strictly speaking, the states $\mid B_l\rangle$ are not well defined, they have an infinite norm. Indeed,

$$\langle B_l \mid B_l\rangle = \langle 0 \mid Q_l Q_l \mid 0\rangle = \int d^3x \langle 0 \mid j_l(\mathbf{x}, t) Q_l(t) \mid 0\rangle, \tag{2.48}$$

where $j_l(x)$ is the charge density operator corresponding to the generator Q_l. Extracting the $\mathbf{x}$ dependence of $j_l(\mathbf{x}, t)$ and using the fact that vacuum and intermediate states in (2.48) have zero momenta, we have

$$\langle B_l \mid B_l\rangle = \int d^3x \langle 0 \mid j_l(0, t) Q_l(t) \mid 0\rangle = V \langle 0 \mid j_l(0, t) Q_l(t) \mid 0\rangle, \tag{2.49}$$

where V is the total volume, $V \to \infty$. Physically, the infinite norm is well understood, since the massless Goldstone boson with zero momentum is distributed over the whole

---

[5] The statement that Q_l are operators of a continuous Lie group is essential: the theorem is not correct for discrete symmetry generators.

space. The prescription how to treat the problem is evident: give a small mass to the boson. In what follows, when the commutators will be considered, the problem can be circumvented by performing first the commutation resulting in δ functions, and after that the integration over the entire three-dimensional space.

Let us demonstrate now explicitly how this general theorem works in QCD. Go back to Eq. (2.31), which at $q = 0$ can be rewritten as

$$\langle 0 \mid [Q_5^-, \bar{u}\gamma_5 d] \mid 0 \rangle = -\langle 0 \mid \bar{u}u + \bar{d}d \mid 0 \rangle, \tag{2.50}$$

where

$$Q_5^- = \int d^3x j_{05}^-(x) \tag{2.51}$$

is the axial charge generator. It is evident from (2.50) that Q_5^- does not annihilate the vacuum, i.e. it belongs to the set of (2.44) generators. It is clear that the same property is inherent to all members of the octet of axial charges in $SU(3)$ symmetry (or to the isovector axial charges in $SU(2)$ symmetry). Applying the general considerations of Goldstone, Salam, and Weinberg [33] to our case, consider the vacuum commutator

$$\langle 0 \mid \left[j_{\mu 5}^-(x), \bar{u}(0)\gamma_5 d(0) \right] \mid 0 \rangle \tag{2.52}$$

in coordinate space. This expression can be written via the Lehmann–Källen representation

$$\langle 0 \mid [j_{\mu 5}^-(x), \bar{u}(0)\gamma_5 d(0)] \mid 0 \rangle = \frac{\partial}{\partial x_\mu} \int d\kappa^2 \Delta(x, \kappa^2) \rho^-(\kappa^2), \tag{2.53}$$

where $\Delta(x, \kappa^2)$ is the Pauli–Jordan (causal) function for a scalar particle with mass κ

$$\left(\partial_\mu^2 + \kappa^2 \right) \Delta(x, \kappa^2) = 0 \tag{2.54}$$

and $\rho^-(\kappa^2)$ is the spectral function, defined by

$$(2\pi)^{-3} p_\mu \theta(p_0) \rho^-(p^2) = -\sum_n \delta^4(p - p_n)\langle 0 \mid j_{\mu 5}^-(0) \mid n \rangle\langle n \mid \bar{u}(0)\gamma_5 d(0) \mid 0 \rangle. \tag{2.55}$$

Axial current conservation and (2.54) imply that

$$\kappa^2 \rho^-(\kappa^2) = 0, \tag{2.56}$$

hence

$$\rho^-(\kappa^2) = N\delta(\kappa^2), \tag{2.57}$$

where N is a constant.

Substitution of (2.57) into (2.53) gives

$$\langle 0 \mid \left[j_{\mu 5}^-(x), \bar{u}(0)\gamma_5 d(0) \right] \mid 0 \rangle = \frac{\partial}{\partial x_\mu} D(x)N, \tag{2.58}$$

where $D(x) = \Delta(x, 0)$. Put $\mu = 0$, $t = 0$, integrate (2.58) over three-dimensional space and use the equality $\partial D(x)/\partial t \mid_{t=0} = -\delta^3(x)$. Comparison of this result with (2.31) shows that N is proportional to the quark condensate and is nonzero. This means that the spectrum of physical states contains a massless Goldstone boson which gives a nonzero contribution to ρ^-. Its quantum numbers are those of π^+. It is easy to perform a similar calculation for other members of the pion multiplet in the case of $SU(2)$ symmetry or for the pseudoscalar meson octet in the case of $SU(3)$ symmetry. Obviously, the proof can be repeated for any other operator whose commutator with axial charges has a nonvanishing vacuum expectation value.

The two proofs presented above cannot be considered to be rigorous like a mathematical theorem, where the presence of Goldstone bosons in QCD is proved starting from the QCD Lagrangian and using the first principles of the theory. Indeed, in the first proof the existence of massive nucleons was taken as an experimental fact. In the second proof, the appearance of the nonvanishing quark condensate in QCD was exploited. The latter was proved (see Eqs. (2.30)–(2.32)) based on Ward identities which, as was demonstrated, became self-consistent only in the case of the existence of massless pions. Therefore, these proofs may be treated as a convincing physical argument, but not as a mathematical theorem (cf. [34]).

The two arguments mentioned above in favour of spontaneously broken chiral symmetry in QCD, namely the existence of large baryon masses and the appearance of a chiral symmetry violating quark condensate, are deeply interconnected. If it is believed that the origin of baryon masses in QCD is the spontaneous violation of chiral symmetry in the QCD vacuum, then one may expect that baryon masses can be expressed through chiral symmetry violating QCD vacuum condensates. Calculations performed within the framework of QCD sum rules have given support to this idea. Particularly, it was found that the proton mass is approximately equal to [35] (see Chapter 6)

$$m_p = [-2(2\pi)^2 \langle 0 \mid \bar{q}q \mid 0 \rangle]^{1/3}. \tag{2.59}$$

This formula demonstrates the fundamental fact that the appearance of the proton mass is caused by spontaneous violation of chiral invariance: the presence of the quark condensate. (Numerically, (2.59) gives the experimental value of the proton mass with an accuracy of about 15%). A similar formula applies to other baryons; see Chapter 6.

2.5 Chiral effective theory (CET) at low energies

An effective chiral theory based on QCD and exploiting the existence and properties of the Goldstone bosons can be formulated. This theory is an effective low-energy theory and is valid in terms of an expansion in powers of particle momenta (or in the derivatives of fields in coordinate space). The Lagrangian is represented as a series of terms with increasing powers of momenta. The theory breaks down at sufficiently high momenta, the characteristic parameters are $|p_i|/M$, where p_i are the spatial momenta of the Goldstone bosons entering the process under consideration and M is the characteristic scale of strong interaction. (Since p_i depend on the reference frame, some care must be taken when choosing the most suitable frame in each particular case.) The physical basis of the theory is the fact that in the limit of vanishing (or small enough) quark masses the spectrum of Goldstone bosons is separated by the gap from the spectrum of other hadrons. The chiral effective theory which works in the domain $|\boldsymbol{p}_i|/M \ll 1$ is a self-consistent theory and not a phenomenological model. Such theory can be formulated on the basis of $SU(2)_L \times SU(2)_R$ symmetry with pions as (quasi) Goldstone bosons. Then one may expect the accuracy of the theory to be of the order of isospin symmetry violation, i.e. of a few percent. Or the theory can be based on $SU(3)_L \times SU(3)_R$ symmetry with an octet of pseudoscalar bosons π, K, η as (quasi) Goldstone bosons. In this case the accuracy of the theory is of the order of $SU(3)$ symmetry violation, i.e. of order $m_s/M \sim 10 - 20\%$. Corrections of the order of m_q/M can be accounted for at the price of introducing additional parameters in the theory. For definiteness, the main part of this section deals with the case of $SU(2)_L \times SU(2)_R$.

The heuristic arguments for the formulation of the chiral effective theory are the following. In the limit of the quark and pion masses going to zero, Eq. (2.7) can be generalized to the operator form as a field equation

$$j^i_{\mu 5} = -\left(f_\pi/\sqrt{2}\right)\partial_\mu \varphi^i_\pi, \tag{2.60}$$

$$j^i_{\mu 5} = \bar{q}\gamma_\mu \gamma_5(\tau^i/2)q, \quad q = u, d, \tag{2.61}$$

where φ^i_π is the pion field, τ^i are the Pauli matrices, and $i = 1, 2, 3$ is the isospin index. (Normalization of the current $j^i_{\mu 5}$ is changed compared with (2.7) in order to have the standard commutation relations of current algebra). Taking the divergence of (2.60) we have

$$\partial_\mu j^i_{\mu 5} = \left(f_\pi/\sqrt{2}\right) m^2_\pi \varphi^i_\pi. \tag{2.62}$$

Eqs. (2.60) and (2.62) are correct near the pion mass shell.

Since the pion state is separated by a gap from the other massive states in the channel with pion quantum numbers these equations can be treated as field equations valid in the low-energy region (usually they are called the equations of partial conservation of axial current, PCAC).

A direct consequence of (2.62) is the Adler self-consistency condition [36]. Consider the amplitude of the process $A \to B + \pi$, where A and B are arbitrary hadronic states, in the limit of vanishing pion momentum p. The matrix element of this process can be written as

$$M_i(2\pi)^4 \delta^4(p_A - p - p_B) = \int d^4x\, e^{ipx}(\partial^2_\mu + m^2_\pi)\langle B \mid \varphi^i_\pi(x) \mid A\rangle. \tag{2.63}$$

The substitution of (2.62) gives

$$M_i = \frac{i(p^2 - m^2_\pi)}{(f_\pi/\sqrt{2})m^2_\pi} p_\mu \langle B \mid j^i_{\mu 5}(0) \mid A\rangle. \tag{2.64}$$

Going to the limit $p_\mu \to 0$ we get

$$M_i(A \to B\pi)_{p\to 0} \to 0 \tag{2.65}$$

which is the Adler condition. In deriving (2.65), it was implicitly assumed that the matrix element $\langle B \mid j^i_{\mu 5} \mid A\rangle$ does not contain pole terms where the axial current interacts with an external line. Generally, the Adler theorem does not work in such cases.

The chiral theory is based on the following principles:

1. The pion field transforms under some nonlinear representation of the group $G = SU(2)_L \times SU(2)_R$.
2. The action is invariant under these transformations.
3. After breaking the $SU(2)_L \times SU(2)_R$ symmetry reduces to $SU(2)_V$, i.e. to the symmetry generated by the isovector vector current.
4. In the lowest order the field equations (2.60), (2.62) are fulfilled.

The pion field may be represented by the 2×2 unitary matrix $U(x)$, $U^{-1} = U^+(x)$, that depends on $\varphi_\pi^i(x)$. The condition $|\det U| = 1$ is imposed on $U(x)$. Therefore the number of degrees of freedom of the matrix U is equal to that of three pion fields $\varphi_\pi^i(x)$. The transformation law under the group G transformations is given by

$$U'(x) = V_L U(x) V_R^+, \tag{2.66}$$

where V_L and V_R are unitary matrices of $SU(2)_L$ and $SU(2)_R$ transformations. (2.66) satisfies the necessary condition that after breaking, when G reduces to $SU(2)$ and $V_L = V_R = V$, the transformation law reduces to

$$U' = VU(x)V^{-1}, \tag{2.67}$$

i.e. to the transformation induced by the vector current.

It can be shown that the general form of the lowest-order effective Lagrangian, where only the terms up to p^2 are kept and the breaking arising from the pion mass is neglected, is: [8],[37]–[39]

$$L_{eff} = k \, \mathrm{Tr} \left(\partial_\mu U \cdot \partial_\mu U^+ \right), \tag{2.68}$$

where k is some constant.

The conserved vector and axial currents (Noether currents) which correspond to Lagrangian (2.68) can be found by applying to (2.68) the transformations (2.66) with

$$V_L = V_R = 1 + i\varepsilon\tau/2 \tag{2.69}$$

in the case of vector current and

$$V_L = V_R^+ = 1 + i\varepsilon\tau/2 \tag{2.70}$$

in the case of axial current. (Here ε is an infinitesimal isovector). The results are:

$$j_\mu^i = ik \, \mathrm{Tr} \left(\tau_i [\partial_\mu U, U^+] \right),$$
$$j_{\mu 5}^i = ik \, \mathrm{Tr} \left(\tau_i \{\partial_\mu U, U^+\} \right). \tag{2.71}$$

One may use various realizations of the matrix field $U(x)$ in terms of pionic fields $\varphi_\pi^i(x)$. All of them are equivalent and lead to the same physical consequences [40, 41]. Mathematically, this is provided by the statement that one realization differs from the other by a unitary (nonlinear) transformation (2.66). One of the useful realizations is

$$U(x) = \exp(i\alpha\tau\varphi_\pi(x)), \tag{2.72}$$

where α is a constant. Substitution of (2.72) into (2.68) and expansion in powers of the pion field up to the 4th order gives

$$L_{eff} = 2k\alpha^2(\partial_\mu\varphi_\pi)^2 + \frac{2}{3}k\alpha^4\left[(\varphi_\pi\partial_\mu\varphi_\pi)^2 - \varphi_\pi^2\cdot(\partial_\mu\varphi_\pi)^2\right] + \cdots \qquad (2.73)$$

From the requirement that the first term in (2.73), the kinetic energy, has the standard from, we have

$$k\alpha^2 = \frac{1}{4}. \qquad (2.74)$$

Substitution of (2.72) into (2.71) in the first nonvanishing order in the pionic field and account of (2.74) results in

$$j_\mu^i = \varepsilon_{ikl}\,\varphi_\pi^k\,\frac{\partial\varphi_\pi^l}{\partial x_\mu},$$

$$j_{\mu 5}^i = -2\sqrt{k}\,\frac{\partial\varphi_\pi^i}{\partial x_\mu}. \qquad (2.75)$$

The formula for the vector current – the first in Eqs. (2.75) – is the standard formula for the pion isovector current. Comparison of the second equation (2.75) with (2.60) finally fixes the constant k and, because of (2.74), α

$$k = \frac{1}{8}\,f_\pi^2, \qquad \alpha = \frac{\sqrt{2}}{f_\pi}. \qquad (2.76)$$

Therefore, the effective Lagrangian (2.68) as well as $U(x)$ are expressed in terms of one parameter – the pion decay constant f_π, which plays the role of the coupling constant in the theory. On dimensional grounds it is then clear that the expansions in powers of the momenta or in powers of the pion field are in fact the expansions in p^2/f_π^2 and φ^2/f_π^2. Particularly, the expansion of the effective Lagrangian (2.73) takes the form of

$$L_{eff} = \frac{1}{2}(\partial_\mu\varphi_\pi)^2 + \frac{1}{3}\frac{1}{f_\pi^2}\left[(\varphi_\pi\partial_\mu\varphi_\pi)^2 - \varphi_\pi^2(\partial_\mu\varphi_\pi)^2\right] + \cdots \qquad (2.77)$$

Turn now to the symmetry breaking term in the chiral effective theory Lagrangian. This term is proportional to the quark mass matrix

$$\mathcal{M} = \begin{pmatrix} m_u & 0 \\ 0 & m_d \end{pmatrix}. \qquad (2.78)$$

In the QCD Lagrangian, the corresponding term transforms under $SU(2)_L \times SU(2)_R$ transformations according to the representation $\left(\frac{1}{2}, \frac{1}{2}\right)$. This statement may be transferred to chiral theory by the requirement that in chiral theory the mass matrix (2.78) transforms according to

$$\mathcal{M}' = V_L\mathcal{M}V_R^+. \qquad (2.79)$$

The term in the Lagrangian linear in $\mathcal{M}$, of the lowest (zero) order in pion momenta and invariant under $SU(2)_L \times SU(2)_R$ transformation, has the form of

$$L' = \frac{f_\pi^2}{4}\{B \operatorname{Tr}(\mathcal{M}U^+) + B^*\operatorname{Tr}(\mathcal{M}U)\}, \tag{2.80}$$

where B is a constant and the factor f_π^2 is introduced for convenience. Impose the requirement of T invariance on the Lagrangian (2.80). The pion field is odd under T-inversion: $T(\varphi_\pi^i) = -\varphi_\pi^i$, so $TU = U^+$ and as a consequence $B = B^*$ and

$$L' = \frac{f_\pi^2}{4} B \operatorname{Tr}[\mathcal{M}(U + U^*)]. \tag{2.81}$$

In the lowest orders of the expansion in the pion field (2.81) reduces to

$$L' = \frac{1}{2} B(m_u + m_d)\left[f_\pi^2 - \varphi_\pi^2 + \frac{1}{6f_\pi^2}(\varphi_\pi^2)^2\right]. \tag{2.82}$$

The first term in square brackets gives a shift in vacuum energy resulting from symmetry breaking, the second corresponds to the pion mass term $(m_\pi^2/2)\varphi_\pi^2$ in the Lagrangian. With this identification we can determine the constant B:

$$B = \frac{m_\pi^2}{m_u + m_d} = -\frac{2}{f_\pi^2}\langle 0 \mid \bar{q}q \mid 0\rangle, \tag{2.83}$$

where the Gell-Mann–Oakes–Renner relation (2.33) was used. The relation (2.83) can be obtained also in another way. From the QCD Lagrangian we have

$$\frac{\partial}{\partial m_u}\langle 0 \mid L \mid 0\rangle = -\langle 0 \mid \bar{u}u \mid 0\rangle. \tag{2.84}$$

Differentiating (2.82) we get:

$$\frac{1}{2}Bf_\pi^2 = -\langle 0 \mid \bar{u}u \mid 0\rangle, \tag{2.85}$$

which coincides with (2.83). (The equality $\langle 0|\bar{u}u|0\rangle = \langle 0|\bar{d}d|0\rangle$ has been exploited).

As the simplest application of the effective Lagrangians (2.77), (2.82), calculate the pion–pion scattering amplitude to the first order in $1/f_\pi^2$. The result is [42]:

$$T = \delta^{ik}\delta^{lm}A(s, t, u) + \delta^{il}\delta^{km}A(t, s, u) + \delta^{im}\delta^{kl}A(u, t, s), \tag{2.86}$$

where

$$A(s, t, u) = \frac{2}{f_\pi^2}(s - m_\pi^2), \tag{2.87}$$

$$s = (p_1 + p_2)^2, \quad t = (p_1 - p_3)^2, \quad u = (p_1 - p_4)^2, \tag{2.88}$$

p_1, p_2 are the initial and p_3, p_4 are the final pion momenta. The isospin indices i, k refer to initial pions, l, m to the final ones. For example, for the $\pi^+\pi^0 \to \pi^+\pi^0$ scattering amplitude we get [42]

$$T = \frac{2}{f_\pi^2}(t - m_\pi^2), \tag{2.89}$$

where T is related to the centre of mass scattering amplitude $f_{c.m.}$ by

$$f_{c.m.} = \frac{1}{16\pi} \frac{1}{E} T, \tag{2.90}$$

and E is the energy of π^+ in centre of mass system.

Another realization often used instead of (2.72) is

$$U(x) = \frac{\sqrt{2}}{f_\pi} [\sigma(x) + i\tau\varphi_\pi(x)] \tag{2.91}$$

supplemented by the constraint

$$\sigma^2 + \varphi^2 = \frac{1}{2} f_\pi^2. \tag{2.92}$$

It can be shown by direct calculations that the expressions for effective Lagrangians up to φ^4 obtained in this realization coincide with (2.77), (2.82) on the pion mass shell. In higher orders (φ^6, φ^8, etc.) the expressions for effective Lagrangians in these two realizations are different even on mass shell. But according to the general arguments by Coleman, Wess, and Zumino [40], the physical amplitudes appear to be equal after adding one-particle reducible tree diagrams. The realization (2.91) is equivalent to one where the $O(4)$ real four-vector $U_i(x)$, which satisfies the constrain $U_i U_i^T = 1$, $i = 1, 2, 3, 4$, is used instead of the 2×2 matrix $U(x)$ [39].

The chiral effective Lagrangian (2.68) is the leading term in the expansion in pion momenta. The next term of order of p^4 consistent with Lorentz and chiral invariance, parity and G parity symmetry has the general form of [39]

$$L_{2,eff} = l_1 \left[\mathrm{Tr}\, (\partial_\mu U \partial_\mu U^+) \right]^2 + l_2\, \mathrm{Tr}\, (\partial_\mu U \partial_\nu U^+)\, \mathrm{Tr}\, (\partial_\mu U \partial_\nu U^+), \tag{2.93}$$

where l_1 and l_2 are constants. A term of the second order in quark masses should be added to (2.93). If spatial momenta of pions in the process under consideration are close to zero, $|\mathbf{p}| \ll m_\pi$, the contribution of this term is of the same order as (2.93), since $p^2 = m_\pi^2 \sim (m_u + m_d)$. Its general form is [39]

$$L'_{2,eff} = l_4\, \mathrm{Tr}\, (\partial_\mu U \partial_\mu U^+)\, \mathrm{Tr}\, [\chi(U + U^+)] + l_6\{\mathrm{Tr}\, [\chi(U + U^+)]\}^2$$
$$+ l_7\{\mathrm{Tr}\, [i\chi(U - U^+)]\}^2, \tag{2.94}$$

where

$$\chi = 2B\mathcal{M}. \tag{2.95}$$

In order to perform the next-to-leading order calculations in CET it is necessary, besides the contributions of (2.93), (2.94), to go beyond the tree approximation in the leading order Lagrangians and to calculate one-loop terms arising from (2.68), (2.81). As can be seen, the parameter of the expansion is $(1/\pi f_\pi)^2 \sim (1/500\,\mathrm{MeV})^2$ and, as a rule, small numerical coefficients also arise. Therefore, the n-loop contribution is suppressed compared with the leading order tree approximation by the factor $[p^2/(\pi f_\pi)^2]^n$. Loop integrals are divergent and require renormalization. Renormalization can be performed in any scheme which preserves the symmetry of the theory. This can be dimensional regularization or a method where finite imaginary parts of the scattering amplitudes are calculated and the whole amplitudes are reconstructed using dispersion relations (an example of such calculation is given below) or any other. The counterterms arising in loop calculations (pole contributions at $d \to 4$ in dimensional regularization or subtraction constants in the dispersion relation approach) are absorbed by the coupling constants of the next order effective Lagrangian, like l_1 and l_2 in (2.93). Theoretically unknown constants l_i are determined by comparison with experimental data.

As a result of loop calculations and of taking account of higher-order terms in the effective Lagrangian, the coupling constant f_π entering (2.68), (2.81) acquires some contributions and is no longer equal to the physical pion decay constant defined by (2.7). For this reason, the coupling constant f_π in (2.68), (2.81) should be considered as a bare one, f_π^0, which will coincide with the physical f_π after taking account of all higher-order corrections. A similar statement refers to the connection between m_π^2 and $m_u + m_d$ (2.83). If B is considered to be a constant parameter of the theory, then relation (2.83) is modified by higher-order terms. Particularly, in next-to-leading order [43]

$$m_\pi^2 = \tilde{m}_\pi^2 \left[1 + c(\mu) \frac{m_\pi^2}{f_\pi^2} + \frac{m_\pi^2}{16\pi^2 f_\pi^2} \ln \frac{m_\pi^2}{\mu^2} \right], \qquad (2.96)$$

where

$$\tilde{m}_\pi^2 = B(m_u + m_d), \qquad (2.97)$$

μ is the normalization point and $c(\mu)$ is the μ-dependent renormalized coupling constant expressed in terms of l_i. (The total correction is independent of μ.) The appearance of the term $\sim m_\pi^2 \ln m_\pi^2$ which is nonanalytic in m_π^2 (or m_q) – the so-called chiral logarithm – is a specific feature of chiral perturbation theory. The origin of its appearance are infrared singularities of the corresponding loop integrals. f_π also contains the chiral logarithm [43]:

$$f_\pi = f_\pi^0 \left[1 + c_1(\mu) \frac{m_\pi^2}{f_\pi^2} - \frac{m_\pi^2}{8\pi^2 f_\pi^2} \ln \frac{m_\pi^2}{\mu^2} \right]. \qquad (2.98)$$

Examples of loop calculations are given in Problems 2.1 and 2.2.

Straightforward is the generalization to three massless quarks, i.e. to massless u, d, and s quarks and an $SU(3)_L \times SU(3)_R$ symmetric Lagrangian. The matrix $U(x)$ is a 3×3 unitary matrix, the leading order Lagrangians are of the same form as (2.68), (2.81) with the obvious difference that the quark mass matrix $\mathcal{M}$ is now a 3×3 matrix. In the formulae for axial and vector currents (2.61), (2.71), τ_i should be replaced by Gell-Mann matrices $\lambda_n, n = 1, \ldots 8$, and the same replacement must be done in the exponential realization of $U(x)$:

$$U(x) = \exp \left(i \frac{\sqrt{2}}{f_\pi} \sum_n \lambda_n \varphi_n(x) \right), \qquad (2.99)$$

where $\varphi_n(x)$ is the octet of pseudoscalar meson fields. Because the algebra of λ_n matrices differs from that of τ_i and, particularly, the anticommutator λ_n, λ_m does not reduce to δ_{nm}, there is no linear realization as simple as (2.91) in this case.

The symmetry breaking Lagrangian (2.81) in the order of φ_n^2 – the mass term in the pseudoscalar meson Lagrangian – is nondiagonal in meson fields: the effective Lagrangian contains a term proportional to $(m_u - m_d)A\varphi_3\varphi_8$. The presence of this term means that the eigenstates of the Hamiltonian, π^0 and η mesons, are not eigenstates of the Q^3 and Q^8 generators of $SU(3)_V$: in η there is an admixture of the isospin 1 state (the pion) and in the pion there is an admixture of an isospin 0 state [44, 45].

In general we can write:

$$H = \frac{1}{2}\tilde{m}_\pi^2\varphi_3^2 + \frac{1}{3}\tilde{m}_\eta^2\varphi_8^2 + A(m_u - m_d)\varphi_3\varphi_8 + \text{kinetic terms.} \qquad (2.100)$$

The physical π and η states arise after an orthogonal transformation of the Hamiltonian (2.100):

$$|\pi\rangle = \cos\theta \, |\varphi_3\rangle - \sin\theta \, |\varphi_8\rangle$$
$$|\eta\rangle = \sin\theta \, |\varphi_3\rangle + \cos\theta \, |\varphi_8\rangle. \qquad (2.101)$$

It can be shown [44, 8] that the constant A in (2.100) is equal to

$$A = \frac{1}{\sqrt{3}}\frac{m_\pi^2}{m_u + m_d} \qquad (2.102)$$

and the mixing angle (at small θ) is given by

$$\theta = \frac{1}{\sqrt{3}}\frac{m_\pi^2}{m_\eta^2 - m_\pi^2}\frac{m_u - m_d}{m_u + m_d}. \qquad (2.103)$$

This result is used in many problems where isospin is violated, e.g. the decay rate $\psi' \to J/\psi\pi^0$ [17] and the amplitude of $\eta \to \pi^+\pi^-\pi^0$ decay. The isospin violating amplitude $\eta \to \pi^+\pi^-\pi^0$ is found to be [46, 12] (Eq. (2.103) was exploited in its derivation):

$$T_{\eta \to \pi^+\pi^-\pi^0} = \frac{\sqrt{3}}{2f_\pi^2}\frac{m_u - m_d}{m_s - (m_u + m_d)/2}\left(s - \frac{4}{3}m_\pi^2\right), \qquad (2.104)$$

where $s = (p_\eta - p_{\pi^0})^2$.

In the three-flavour case, the next-to-leading Lagrangian contains several additional terms compared with (2.93), (2.94) [8, 38]

$$L'_{2eff} = l_3 \, \text{Tr}\left(\partial_\mu U\partial_\mu U^+\partial_\nu U\partial_\nu U^+\right) + l_5 \, \text{Tr}\left[\partial_\mu U\partial_\mu U^+\chi(U + U^+)\right]$$
$$+ \quad l_8 \, \text{Tr}\left(\chi U^+\chi U^+ + U\chi^+ U\chi^+\right). \qquad (2.105)$$

In the case of three flavours, a term of different origin proportional to the totally antisymmetric tensor $\varepsilon_{\mu\nu\lambda\sigma}$ arises at the order of p^4. As was pointed out by Wess and Zumino, [47] its emergence is due to anomalous Ward identities for vector and axial nonsinglet currents. Witten [48] has presented the following heuristic argument in favour of this term. The leading and next-to-leading Lagrangians (2.68), (2.81), (2.93), (2.94), (2.105) are invariant under discrete symmetries $U(x) \to U^+(x)$, $U(x,t) \to U(-x,t)$. According to (2.72) the former transformation is equivalent to $\varphi_i(x) \to -\varphi_i(x)$. In the case of pions, this operation coincides with G-parity, but for the octet of pseudoscalar mesons this is not the case. Particularly, such symmetry forbids the processes $K^+ K^- \to \pi^+ \pi^- \pi^0$ and $\eta\pi^0 = \pi^+\pi^-\pi^0$, which are allowed in QCD. In QCD, the symmetry under change of sign of pseudoscalar meson fields is valid only if supplemented by space reflection, i.e. $\varphi_i(-x, t) \to -\varphi_i(x, t)$. Therefore, one may add to the chiral Lagrangian a term which is invariant under the latter operation, but violates separately $x \to -x$ and $\varphi_i(x) \to -\varphi_i(x)$. Evidently, such term is proportional to $\varepsilon_{\mu\nu\lambda\sigma}$. The general form of the term added to the equation of motion is unique:

$$\frac{1}{8} f_\pi^2 \mathrm{Tr}\left(-\partial_\mu^2 U^+ + U^+ \partial_\mu^2 U \cdot U^+\right) +$$
$$\lambda \varepsilon_{\mu\nu\lambda\sigma} \mathrm{Tr}\left\{U^+ \partial_\mu U \cdot U^+ \partial_\nu U \cdot U^+ \partial_\lambda U \cdot U^+ \partial_\sigma U\right\} = 0, \qquad (2.106)$$

where λ is a constant. (Other nonleading terms are omitted). Eq. (2.106) is noninvariant under $U^+ \to U$ and $x \to -x$ separately, but conserves parity. However, (2.106) cannot be derived from a local Lagrangian in four-dimensional space-time, because the trace of the second term on the left-hand side of (2.106) vanishes after cyclic permutation. Witten [48] has shown that the Lagrangian can be represented formally as an integral over some five-dimensional manifold, where the Lagrangian density is local. The integral over this manifold reduces to its boundary, which is precisely four-dimensional space-time. In the first nonvanishing order in meson fields the contribution to the Lagrangian (the so-called Wess–Zumino term [47]) is equal to: [47]–[49]

$$\Lambda_{WZ}(U) = n \frac{1}{15\pi^2 f_\pi^2} \int d^4 x \, \varepsilon_{\mu\nu\lambda\sigma} \, \mathrm{Tr}\left(\Phi \partial_\mu \Phi \partial_\nu \Phi \partial_\lambda \Phi \partial_\sigma \Phi\right), \qquad (2.107)$$

where $\Phi = \sum \lambda_m \varphi_m$. The coefficient n in (2.107) is an integer number [48]. This statement follows from the topological properties of mapping of four-dimensional space-time into $SU(3)$ manifold produced by the field U. It is clear from (2.107) that $L_{WZ} = 0$ in the case of two flavours: the only antisymmetric tensor in flavour indices is ε^{ikl} and it is impossible to construct from the derivatives of pion fields an expression antisymmetric in coordinates.

In order to find the value of n, it is instructive to consider the interaction with the electromagnetic field. In this case, the Wess–Zumino Lagrangian is supplemented by terms which form together with (2.107) a gauge-invariant Lagrangian [47]

$$L_{WZ}(U, A_\mu) = L_{WZ}(U) - en \int d^4 x A_\mu J_\mu + \frac{ie^2 n}{24\pi^2} \int d^4 x \varepsilon_{\mu\nu\lambda\sigma} (\partial_\mu A_\nu) A_\lambda$$
$$\times \mathrm{Tr}\left[e_q^2(\partial_\sigma U) U^+ + e_q^2 U^+ (\partial_\sigma U) + e_q U e_q U^+ (\partial_\sigma U) U^+\right], \qquad (2.108)$$

where

$$J_\mu = \frac{1}{48\pi^2} \varepsilon_{\mu\nu\lambda\sigma} \, \mathrm{Tr}\left[e_q(\partial_\nu U \cdot U^+)(\partial_\lambda U \cdot U^+)(\partial_\sigma U \cdot U^+)\right.$$
$$\left. + e_q(U^+ \partial_\nu U)(U^+ \partial_\lambda U)(U^+ \partial_\sigma U)\right], \qquad (2.109)$$

e_q is the matrix of quark charges, $e_q = \text{diag}(2/3, -1/3, -1/3)$ and e is the proton charge. The amplitude of $\pi^0 \to \gamma\gamma$ decay can be found from the last term in (2.108). It is given by

$$T(\pi^0 \to \gamma\gamma) = \frac{ne^2}{48\sqrt{2}\pi^2 f_\pi} \varepsilon_{\mu\nu\lambda\sigma} F_{\mu\nu}^{(1)} F_{\lambda\sigma}^{(2)}. \tag{2.110}$$

($F_{\mu\nu}^{(1)}$ and $F_{\lambda\sigma}^{(2)}$ are the fields of the first and second photon.) On the other hand, the same amplitude is defined in QCD by anomaly. Consider the anomaly condition [50]–[52] (see Chapter 3):

$$\partial_\mu j_{\mu5}^3 = \frac{\alpha}{2\pi} N_c (e_u^2 - e_d^2) F_{\mu\nu} \tilde{F}_{\mu\nu} = \frac{\alpha}{12\pi} N_c \varepsilon_{\mu\nu\lambda\sigma} F_{\mu\nu} F_{\lambda\sigma}, \tag{2.111}$$

where N_c is the number of colours and e_u, e_d are u and d quark charges. For the amplitude $T(\pi^0 \to \gamma\gamma)$ we have, using the PCAC condition (2.62):

$$T(\pi^0 \to \gamma\gamma) = \frac{e^2}{48\sqrt{2}\pi^2 f_\pi} N_c \varepsilon_{\mu\nu\lambda\sigma} F_{\mu\nu}^{(1)} F_{\lambda\sigma}^{(2)}. \tag{2.112}$$

Eq. (2.110) coincides with (2.112) if $n = N_c$ [48]. The other physically interesting object, the $\gamma\pi^+\pi^-\pi^0$ vertex, is defined by the second term on the right-hand side of (2.108) and is equal to

$$\Gamma(\gamma\pi^+\pi^-\pi^0) = -\frac{1}{3} ie \frac{n}{\pi^2\sqrt{2} f_\pi^3} \varepsilon_{\mu\nu\lambda\sigma} A_\mu \partial_\nu \pi^+ \partial_\lambda \pi^- \partial_\sigma \pi^0. \tag{2.113}$$

Again, if $n = N_c$, then this result agrees with the QCD calculation based on the VAAA anomaly or with the phenomenological approach where the anomaly was taken for granted [52]–[55].

2.6 Low-energy sum rules in CET

Using the CET technique important low energy sum rules can be derived which, of course, are valid also in QCD. The most interesting sum rules, tested by experiment, refer to the difference of the polarization operators of vector and axial currents. Let us define

$$\Pi_{\mu\nu}^U(q) = i \int d^4x e^{iqx} \langle 0 \mid T \left\{ U_\mu(x), \, U_\nu(0)^+ \right\} \mid 0 \rangle$$

$$= (q_\mu q_\nu - q^2 \delta_{\mu\nu}) \Pi_U^{(1)}(q^2) + q_\mu q_\nu \Pi_U^{(0)}(q^2), \tag{2.114}$$

where

$$U = V, A, \quad V_\mu = \bar{u}\gamma_\mu d, \quad A_\mu = \bar{u}\gamma_\mu\gamma_5 d, \tag{2.115}$$

V_μ and A_μ are vector and axial quark currents. The imaginary parts of the correlators are the spectral functions ($s = q^2$):

$$v_1(s)/a_1(s) = 2\pi \, \text{Im} \, \Pi_{V/A}^{(1)}(s), \quad a_0(s) = 2\pi \, \text{Im} \, \Pi_A^{(0)}(s), \tag{2.116}$$

which are measured in τ decay. (The spectral function which is isotopically related to v_1 is measured in e^+e^- annihilation.) The spin 0 axial spectral function $a_0(s)$ which is mainly saturated by the one pion state will not be of interest to us now.

$\Pi_V^{(1)}(s)$ and $\Pi_A^{(1)}(s)$ are analytical functions of s in the complex s plane with a cut along the right hand semiaxis, starting from the threshold of the lowest hadronic state: $4m_\pi^2$ for $\Pi_V^{(1)}$ and $9m_\pi^2$ for $\Pi_A^{(1)}$. Besides the cut, $\Pi_A^{(1)}(q^2)$ has a kinematical pole at $q^2 = 0$. This is a specific feature of QCD and CET which follows from the chiral symmetry in the limit of massless u, d quarks and its spontaneous violation. In this limit, the axial current is conserved and the pion is massless. Its contribution to the axial polarization operator is given by

$$\Pi_{\mu\nu}^A(q)_\pi = f_\pi^2\left(\delta_{\mu\nu} - \frac{q_\mu q_\nu}{q^2}\right). \tag{2.117}$$

When the quark masses are taken into account, then in the first order of quark masses, or equivalently in m_π^2, Eq. (2.117) is modified to:

$$\Pi_{\mu\nu}^A(q)_\pi = f_\pi^2\left(\delta_{\mu\nu} - \frac{q_\mu q_\nu}{q^2 - m_\pi^2}\right). \tag{2.118}$$

Decompose (2.118) in the tensor structures of (2.114)

$$\Pi_{\mu\nu}^A(q)_\pi = -\frac{f_\pi^2}{q^2}\left(q_\mu q_\nu - \delta_{\mu\nu}q^2\right) - \frac{m_\pi^2}{q^2} q_\mu q_\nu \frac{f_\pi^2}{q^2 - m_\pi^2}. \tag{2.119}$$

The pole in $\Pi_1^A(q^2)$ at $q^2 = 0$ is evident.

Let us write a dispersion relation for $\Pi_1^V(s) - \Pi_1^A(s)$. This should be an unsubtracted dispersion relation, since perturbative terms (besides the small contribution from the mass squares of the u, d quark) cancel in the difference, and the operator production expansion (OPE) terms decrease with $q^2 = s$ at least as s^{-2} (the term $\sim m_q\langle 0|\bar{q}q|0\rangle$ in OPE). We have

$$\Pi_1^V(s) - \Pi_1^A(s) = \frac{1}{2\pi^2}\int_0^\infty ds' \frac{v_1(s') - a_1(s')}{s' - s} + \frac{f_\pi^2}{s}. \tag{2.120}$$

The last term on the right-hand side of (2.120) represents the kinematical pole contribution. Let us go to $s \to \infty$ in (2.120). Since $\Pi_1^V(s) - \Pi_1^A(s) \to s^{-2}$ in this limit we get the sum rule (the first Weinberg sum rule [56]):

$$\frac{1}{4\pi^2}\int_0^\infty ds[v_1(s) - a_1(s)] = \frac{1}{2}f_\pi^2. \tag{2.121}$$

The accuracy of this sum rule is of the order of chiral symmetry violation in QCD, or next-order terms in CET, i.e. $\sim m_\pi^2/M^2$ (e.g. a subtraction term).

If the term $\sim m_q \langle 0|\bar{q}q|0\rangle \sim f_\pi^2 m_\pi^2$ in OPE can be neglected, then, performing in (2.120) the expansion up to $1/s^2$, we get the second Weinberg sum rule:

$$\int_0^\infty s\,ds[v_1(s) - a_1(s)] = O(m_\pi^2). \tag{2.122}$$

(For other derivations of these sum rules see [57].)

One more sum rule can be derived in CET (in its earlier version PCAC): the Das–Mathur–Okubo sum rule [58]:

$$\frac{1}{4\pi^2} \int_0^\infty ds\,\frac{1}{s}[v_1(s) - a_1(s)] = \frac{1}{6}\,f_\pi^2\langle r_\pi^2\rangle - F_A, \tag{2.123}$$

where $\langle r_\pi^2\rangle$ is the mean pion electromagnetic radius and F_A is the pion axial-vector formfactor in the decay $\pi^- \to e^-\nu_\mu\gamma$ (in fact, F_A is a constant to a high accuracy).

The sum rules (2.121),(2.122),(2.123) were checked by measurements of $v_1(s) - a_1(s)$ in τ decay. The upper limit of integrals in sum rules was restricted experimentally by $s_0 = m_\tau^2 \approx 3$ GeV^2. The comparison of the theory with experiment shows a good coincidence ($\sim 10\%$) of the first sum rule. No definite conclusion could be obtained in case of the second and third sum rules [59].

2.7 The nucleon and pion–nucleon interaction in CET

The general method of treatment of nucleon field in chiral symmetry was formulated by Weinberg [60] and Callan, Coleman, Wess, and Zumino [40],[41]. (More recent reviews are in [61],[62].) Since pion-nucleon interaction at soft pion fields is linear in the pion field and quadratic in nucleon fields it is expected that the nucleon field transforms like a square root of the pion field. So, let us put

$$u^2(x) = U(x), \tag{2.124}$$

where $U(x)$ is 2×2 unitary matrix, defined in Section 2.5. The chirally transformed matrix $U'(x)$ (2.66) corresponds to $u'^2 = U'$. Define now the matrix-valued function K by the following equalities:

$$V_L u = u'K, \qquad uV_R^+ = K^+u'. \tag{2.125}$$

K is a unitary 2×2 matrix, $K^{-1} = K^+$, that depends on V_L, V_R and U: $K = K(V_L, V_R, U)$. Under $SU(2)_L \times SU(2)_R$ transformations the two-component nucleon spinor

$$\psi = \begin{pmatrix} p \\ n \end{pmatrix} \tag{2.126}$$

transforms as

$$\psi \to \psi' = K(V_L, V_R, U)\psi. \tag{2.127}$$

(The explicit form of K for an infinitesimal axial transformation is presented in Problem 2.3.) The covariant derivative of the nucleon field is given by [61]:

$$D_\mu \psi = \partial_\mu \psi + \Gamma_\mu \psi, \tag{2.128}$$

where the connection Γ_μ has the form of

$$\Gamma_\mu = \frac{1}{2} \left\{ u^+ (\partial_\mu - ie A_\mu Q) u + u(\partial_\mu - ie A_\mu Q) u^+ \right\}. \tag{2.129}$$

For generality, the electromagnetic field A_μ is included, $Q = \mathrm{diag}(1, 0)$ is the nucleon charge matrix. It can be shown that in the absence of electromagnetic field, Γ_μ transforms like a gauge field under chiral transformations:

$$\Gamma'_\mu = K \Gamma_\mu K^+ + K \partial_\mu K^+. \tag{2.130}$$

Because the nucleon is massive, the expansion parameter in CET in the case of pion–nucleon interaction is the spatial nucleon momentum p. The effective πN interaction Lagrangian $L_{\pi N}^{(1)}$ valid up to the first order of p is given by:

$$L_{\pi N}^{(1)} = \bar{\psi} \left(i \gamma_\mu D_\mu - m + \frac{i}{2} g_A \gamma_\mu \gamma_5 u_\mu \right) \psi, \tag{2.131}$$

where

$$u_\mu = i u^+ \partial_\mu U \cdot u^+ = i \left(u^+ \partial_\mu u - u \partial_\mu u^+ \right). \tag{2.132}$$

The last term in (2.131) corresponds to the pion–nucleon interaction through the axial current, discussed in Section 2.4, Fig. 2.1. The constants m and g_A should be considered as bare nucleon mass and axial coupling constant, which differ from the physical ones due to CET higher-order corrections. By using $U(x)$ realizations (2.71) or (2.91), the πN interaction term in (2.131) may be reduced to

$$L_{\pi N, int}^{(1)} = -\frac{g_A}{\sqrt{2} f_\pi} \bar{\psi} \gamma_\mu \gamma_5 \tau \partial_\mu \varphi_\pi \psi - \frac{1}{2 f_\pi^2} \bar{\psi} \gamma_\mu \tau [\varphi_\pi, \partial_\mu \varphi_\pi] \psi. \tag{2.133}$$

The second term in (2.133) arises from the term proportional to Γ_μ in $D_\mu \psi$. Its existence is one of the important consequences of CET.

Problems

Problem 2.1

Find the nonanalytical correction to pion electric radius proportional to $\ln m_\pi^2$ [63],[64].

Solution

   The one-loop contribution to the pion formfactor comes from the $\pi \pi$ interaction term in the Lagrangian given by (2.77) and is equal to

$$i \frac{1}{f_\pi^2} \int \frac{d^4 k_1 d^4 k_2}{(2\pi)^4} \delta(q + k_1 - k_2)(k_1 + k_2)_\mu \frac{1}{k_1^2 - m_\pi^2} \frac{1}{k_2^2 - m_\pi^2} (p_1 + p_2)(k_1 + k_2). \tag{1}$$

Here p_1 and p_2 are the initial and final pion momenta, q is the momentum transfer, $q^2 < 0$, $p_1 + q = p_2$. The integral in (1) can be calculated in the following way. Consider the integral

$$i \int \frac{d^4k_1 d^4k_2}{(2\pi)^4} (k_1 + k_2)_\mu (k_1 + k_2)_\nu \frac{1}{k_1^2 - m_\pi^2} \frac{1}{k_2^2 - m_\pi^2} \delta^4(q + k_1 - k_2)$$

$$= A(q^2) \left(\delta_{\mu\nu} q^2 - q_\mu q_\nu \right). \tag{2}$$

The form of the right-hand side of (2) (the absence of a subtraction term) follows from gauge invariance. Calculate the imaginary part of $A(q^2)$ at $q^2 > 0$. We have

$$\text{Im } A(q^2)(\delta_{\mu\nu} q^2 - q_\mu q_\nu) = -\frac{1}{8\pi^2} \int d^4k (2k - q)_\mu (2k - q)_\nu \delta[(q - k)^2]$$

$$= \frac{1}{48\pi} \left(q^2 \delta_{\mu\nu} - q_\mu q_\nu \right). \tag{3}$$

(The pion mass can be neglected in our approximation.) $A(q^2)$ is determined by the following dispersion relation:

$$A(q^2) = \frac{1}{\pi} \int\limits_{4m_\pi^2}^{M^2} \frac{ds}{s - q^2} \text{Im } A(s) = \frac{1}{48\pi^2} \ln \frac{M^2}{4m_\pi^2 - q^2}. \tag{4}$$

(The subtraction term is omitted, M^2 is a cutoff.) Substitution of (2),(3) into (1) gives the correction to the $\gamma\pi\pi$ vertex

$$(p_1 + p_2)_\mu [F(q^2) - 1] = (p_1 + p_2)_\mu \frac{q^2}{48\pi^2 f_\pi^2} \ln \frac{M^2}{4m_\pi^2 - q^2}, \tag{5}$$

where $F(q^2)$ is the pion formfactor. The pion electric radius is defined by

$$r_\pi^2 = 6 \frac{dF(q^2)}{dq^2} \tag{6}$$

and its part nonanalytical in m_π^2 is equal to

$$r_\pi^2 = \frac{1}{8\pi^2 f_\pi^2} \ln(M^2/m_\pi^2). \tag{7}$$

Problem 2.2

Find the nonanalytical correction to the quark condensate proportional to $m_\pi^2 \ln m_\pi^2$ [65].

Solution

Using (2.84) and (2.82) we get

$$\langle 0 \mid \bar{u}u \mid 0 \rangle = -\frac{1}{2} f_\pi^2 B \left\langle 0 \mid 1 - \frac{\varphi_i^2}{f_\pi^2} \mid 0 \right\rangle. \tag{1}$$

The vacuum expectation value of φ_i^2 is given by

$$\lim_{x \to 0} \langle 0 \mid T \varphi_i(x), \varphi_i(0) \mid 0 \rangle = \frac{3i}{(2\pi)^4} \lim_{x \to 0} \int d^4 k \frac{e^{ikx}}{k^2 - m_\pi^2}$$

$$= A m_\pi^2 + C m_\pi^2 \ln m_\pi^2 + \cdots \tag{2}$$

in order to find C differentiate (2) with respect to m_π^2. We have

$$\frac{3i}{(2\pi)^4} \int d^4 k \frac{1}{(k^2 - m_\pi^2)^2} = -\frac{3\pi^2}{(2\pi)^4} \ln \frac{M^2}{m_\pi^2}. \tag{3}$$

Substitution of (3) into (1) and taking into account of (2.83) gives

$$\langle 0 \mid \bar{u}u \mid 0 \rangle = \langle 0 \mid \bar{u}u \mid 0 \rangle_0 \left(1 + \frac{3 m_\pi^2}{16 \pi^2 f_\pi^2} \ln \frac{M^2}{m_\pi^2} + A m_\pi^2 \right). \tag{4}$$

Problem 2.3

Find the $\pi - \pi$ scattering lengths in the states with isospin 0 and 2 (S. Weinberg, 1966 [42]).
Answer:

$$a_0 = \frac{7}{16\pi} \frac{m_\pi}{f_\pi^2}; \quad a_2 = -\frac{1}{8\pi} \frac{m_\pi}{f_\pi^2}.$$

Problem 2.4

Find the matrix K given by (2.125) for infinitesimal axial transformation (2.70).
Answer:

$$K = 1 - \frac{i}{2\sqrt{2} f_\pi} [\varepsilon \varphi_\pi] \tau.$$

References

[1] Vainstein, A. I. and Zakharov, V. I., *Usp. Fiz. Nauk* **100** (1970) 225.

[2] Gasser, J. and Leutwyler, H., *Nucl. Phys.* **B94** (1975) 269.

[3] Weinberg, S. in: A. Festschrift for I. I. Rabi, ed. L.Motz, Trans. New York Acad. Sci., Ser.II, **38** (1977) 185.

[4] Gasser, J. and Leutwyler, H., *Phys. Rep.* **87** (1982) 77.

[5] Appelquist, T. and Carrazone, J., *Phys. Rev.* **D11** (1975) 2856.

[6] Dashen, R., *Phys. Rev.* **183** (1969) 1245.

[7] Berestetskii, V. B., *Usp. Fiz. Nauk* **85** (1965) 393 (*Sov. Phys. Usp.* **8** (1965) 147).

[8] Gasser, J. and Leutwyler, H., *Nucl. Phys.* **B250** (1986) 465.

[9] Donoghue, J. and Perez, A., *Phys. Rev.* **D55** (1997) 7075.

[10] Cottingham, W. N., *Ann. Phys.* (N.Y.) **25** (1963) 424.

[11] Bijnens, J. and Prades, J., *Nucl. Phys.* **B490** (1997) 293.

[12] Gasser, J. and Leutwyler, H., *Nucl. Phys.* **B250** (1985) 539.

[13] Kambor, J., Wiesendanger, C. and Wyler, D., *Nucl. Phys.* **B465** (1996) 215.

[14] Anisovich, A. V. and Leutwyler, H. *Phys. Lett.* **B375** (1996) 335.

[15] Martemyanov, B. V. and Sopov, V. S., *Phys. Rev.* **D71** (2005) 017501.

[16] Leutwyler, H., *J. Moscow Phys. Soc.* **6** (1996) 1, hep-ph/9602255.

[17] Ioffe, B. L. and Shifman, M. A., *Phys. Lett.* **B95** (1980) 99.

[18] Yao, W.-M. *et al*, Review of Particle Physics, *J. Phys.* G **33** (2006) 1.

[19] Nasrallah, N., *Phys. Rev.* **D70** (2004) 116001.

[20] Aubin, C. *et al* (MILC Collab.), *Phys. Rev.* **D70** (2004) 114501.

[21] Vermaseren, J. A. M., Larin, A. A. and van Ritbergen, T., *Phys. Lett.* **B405** (1997) 327.

[22] Narison, S., *QCD as a theory of hadrons: from partons to confinement*, Cambridge Univ. Press, 2002, hep-ph/0202200.

[23] Gamiz, E. *et al*, *Nucl. Phys. Proc. Suppl.* **144** (2005) 59.

[24] Baikov, P. A., Chetyrkin, K. G. and Kühn, J. H., *Phys. Rev. Lett.* **95** (2005) 012003.

[25] Schierholz, G., *et al*, *Phys. Lett.* **B639** (2006) 307.

[26] Adami, C., Drukarev, E. G. and Ioffe, B. L., *Phys. Rev.* **D48** (1993) 1441.

[27] Eletsky, V. L. and Ioffe, B. L., *Phys. Rev.* **D48** (1993) 2304.

[28] Prades, J., *Nucl. Phys.* **B64** (Proc. Suppl.) (1998) 253.

[29] Gell-Mann, M., Oakes, R. J. and Renner, B., *Phys. Rev.* **175** (1968) 2195.

[30] Goldberger, M. L. and Treiman, S. B., *Phys. Rev.* **110** (1958) 1178.

[31] Goldstone, J., *Nuovo Cim* **19** (1961) 154.

[32] Vaks, V. G. and Larkin, A. I., ZhETF **40** (1961) 282; Nambu, Y. and Jona-Lasinio, G., *Phys. Rev.* **122** (1961) 345.

[33] Goldstone, J., Salam, A. and Weinberg, S., *Phys.Rev.* **127** (1962) 965.

[34] Coleman, S., in: *Laws of hadronic matter*, Proc. of 11 Course of the "Ettore Majorana" Intern. School of Subnuclear Physics ed. by A.Zichichi Academic Press, London and New York, 1975.

[35] Ioffe, B. L., *Nucl. Phys.* **B188** (1981) 317, E: **B192** (1982) 591.

[36] Adler, S. L., *Phys. Rev.* **137** (1965) B1022, **139** (1965) B1638.

[37] Weinberg, S., *Physica* **A96** (1979) 327.

[38] Leutwyler, H., Lectures at the XXX Internationale Universitätswochen für Kernphysik, Schladming, Austria, 1991.

[39] Gasser, J. and Leutwyler, H., *Ann. Phys.* (N.Y.) **158** (1984) 142.

[40] Coleman, S., Wess, J. and Zumino, B., *Phys. Rev.* **177** (1969) 2239.

[41] Callan, C. G., Jr., Coleman, S., Wess J. and Zumino, B., *Phys. Rev.* **177** (1969) 2247.

[42] Weinberg, S., *Phys. Rev. Lett.* **17** (1966) 616.

[43] Langacker, P. and Pagels, H., *Phys. Rev.* **D8** (1973) 4595.

[44] Ioffe, B. L., *Yad. Fiz.* **29** (1979) 1611 (*Sov. J. Nucl. Phys.* **20** (1979) 827).

[45] Gross, D. J., Treiman, S. B. and Wilczek, F., *Phys. Rev.* **D19** (1979) 2188.

[46] Osborn, H. and Wallace, D. R., *Nucl. Phys.* **B20** (1970) 23.

[47] Wess, J. and Zumino, B., *Phys. Lett.* **B37** (1971) 95.

[48] Witten, E., *Nucl. Phys.* **B223** (1983) 422.

[49] Zahed, I. and Brown, G. E., *Phys. Rep.* **142** (1986) 1.

[50] Adler, S. L., *Phys. Rev.* **177** (1969) 2426.

[51] Bell, J. S. and Jackiw, R., *Nuovo Cim* **60** (1969) 1517.

[52] Bardeen, W. A., *Phys. Rev.* **184** (1969) 1848.

[53] Terentjev, M. V., *Pisma v ZhETF* **14** (1971) 140.

[54] Adler, S., Lee, B., Treiman, S. and Zee, A., *Phys. Rev.* **D4** (1971) 3497.

[55] Terentjev, M. V., *Usp. Fiz. Nauk* **112** (1974) 37.

[56] Weinberg, S., *Phys. Rev. Lett.* **18** (1967) 507.

[57] Ioffe, B. L., Khoze, V. A., Lipatov, L. N., *Hard processes*, North Holland, Amsterdam, 1984.

[58] Das, T., Mathur, V. S., Okubo, S., *Phys. Rev. Lett.* **19** (1967) 859.

[59] Barate, R. *et al* (ALEPH Collaboration), *Eur. J. Phys.* **C4** (1998) 409.

[60] Weinberg, S., *Phys. Rev.* **166** (1968) 1568.

[61] Meissner, U.-G., *Intern. J. Mod. Phys.* **E1** (1992) 561; *Chiral nucleon dynamics*, in Lectures at the 12-th Annual Hampton Univ. Graduate Studies at CEBAF, Newport News, June 1997, hep-ph/9711365.

[62] Leutwyler, H., *PiN Newslett.* **15** (1999) 1, hep-ph/0008123; Becher, T. and Leutwyler, H., JHEP 0106 (2001) 017.

[63] Beg, M. A. and Zepeda, A., *Phys. Rev.* **D6** (1972) 2912.

[64] Volkov, M. K. and Pervushin, V. N., *Yad. Fiz.* **20** (1974) 762 (*Sov. J. Nucl. Phys.* **20** (1975) 408.

[65] Novikov, V. A., Shifman, M. A., Vainstein A. I. and Zakharov, V. I., *Nucl. Phys.* **B191** (1981) 301.

3
Anomalies

3.1 Generalities

The phenomenon of anomaly plays an important role in quantum field theory: in many cases it determines whether or not a theory is self-consistent and can be realized in the physical world and, therefore, allows one to select the acceptable physical theories. In a given theory anomalies are often related to the appearance of new quantum numbers (topological quantum numbers; see Chapter 4), result in the emergence of mass scales, determine the spectrum of physical states. So, despite its name, anomalies are a normal and significant attribute of any quantum field theory.

The term "anomaly" has the following meaning: Let the classical action of the theory obey some symmetry, i.e. let it be invariant under certain transformations. If this symmetry is violated by quantum corrections, such a phenomenon is called an "anomaly." (Reviews of anomalies are given in [1]–[4].) There are two types of anomalies – internal and external. In the first case, the gauge invariance of the classical Lagrangian is destroyed at the quantum level. The theory becomes nonrenormalizable and cannot be considered as a self-consistent theory. The standard method to solve this problem is a special choice of fields in the Lagrangian in such a way as to cancel all internal anomalies. (Such an approach is used in the standard model of electroweak interaction – the Glashow–Illiopoulos–Maiani mechanism.) An external anomaly corresponds to violation of symmetry of interaction with external sources, not related to gauge symmetry of the theory. Such anomalies arise in QCD and are considered below. There are two kinds of anomalies in QCD: axial (chiral) anomaly and scale anomaly. Both are connected with singularities of the theory at small distances (at large momenta) and with the necessity of regularization: a regularization procedure which respects the symmetry does not exist and hence the symmetry is violated by the anomaly. In QCD, evidence of anomalies came from perturbation theory but in fact their occurrence follows from general principles.

3.2 The axial anomaly

The axial anomaly in QCD is very similar to those in massless QED. For this reason let us first consider the latter. The equations of motion of QED in the external electromagnetic field $A_\mu(x)$ have the form:

$$i\gamma_\mu \frac{\partial \psi(x)}{\partial x_\mu} = m\psi(x) - e\gamma_\mu A_\mu(x)\psi(x). \tag{3.1}$$

In massless QED, classically, i.e. without account of radiative corrections, the axial current $j_{\mu5}(x)$ is conserved like the vector current,

$$\partial_\mu j_{\mu5}(x) = \partial_\mu j_\mu(x) = 0. \tag{3.2}$$

However, it appears that in quantum theory with account of radiative corrections it is impossible to keep the conservation of both currents – vector and axial. The origin for this comes from the singular character of the currents. Vector and axial currents are composite operators built from local fermion fields and the products of local operators are singular when their points coincide, as is the case with V and A currents. In order to consider the problem correctly, define the axial current by placing two fermion fields at distinct points, separated by a distance ε, and go to the limit $\varepsilon \to 0$ in the final result,

$$j_{\mu5}(x, \varepsilon) = \bar\psi\left(x + \frac{\varepsilon}{2}\right)\gamma_\mu\gamma_5 \exp\left[ie \int_{x-\varepsilon/2}^{x+\varepsilon/2} dy_\alpha A_\alpha(y)\right]\psi\left(x - \frac{\varepsilon}{2}\right). \tag{3.3}$$

The exponential factor in (3.3) is introduced in order for the operator to be locally gauge invariant. The divergence of the axial current (3.3) is equal to

$$\partial_\mu j_{\mu5}(x, \varepsilon) = 2im\bar\psi\left(x + \frac{\varepsilon}{2}\right)\gamma_5\psi\left(x - \frac{\varepsilon}{2}\right) - ie\varepsilon_\alpha\bar\psi\left(x + \frac{\varepsilon}{2}\right)\gamma_\mu\gamma_5\psi\left(x - \frac{\varepsilon}{2}\right)F_{\alpha\mu}, \tag{3.4}$$

where $F_{\alpha\mu}$ is the electromagnetic field strength. In the derivation of (3.3), the equations of motion (3.1) have been exploited and the first term of the expansion in powers of ε was retained. For simplicity assume that $F_{\mu\nu} = \text{Const}$. Take the vacuum average of Eq.(3.4). On the right-hand side of (3.4) the expression for the electron propagator in the external electromagnetic field can be used (the first term on the right-hand side of Eq.(6.273)). The vacuum averaging corresponds to taking account of corrections of first order in e^2. In massless QED, the first term on the right-hand side of (3.4) is absent and we get

$$\langle 0 \mid \partial_\mu j_{\mu5} \mid 0 \rangle = \frac{e^2}{4\pi^2} F_{\alpha\mu} F_{\lambda\sigma} \varepsilon_{\beta\mu\lambda\sigma} \frac{\varepsilon_\alpha \varepsilon_\beta}{\varepsilon^2}. \tag{3.5}$$

Since there is no specific direction in space-time the limit $\varepsilon \to 0$ should be taken symmetrically,

$$\lim_{\varepsilon \to 0} \frac{\varepsilon_\alpha \varepsilon_\beta}{\varepsilon^2} = \frac{1}{4}\delta_{\alpha\beta}. \tag{3.6}$$

Substitution of (3.6) into (3.5) gives

$$\partial_\mu j_{\mu5} = \frac{e^2}{8\pi^2} F_{\alpha\beta} \tilde{F}_{\alpha\beta}, \tag{3.7}$$

where

$$\tilde{F}_{\alpha\beta} = \frac{1}{2}\varepsilon_{\alpha\beta\lambda\sigma} F_{\lambda\sigma} \tag{3.8}$$

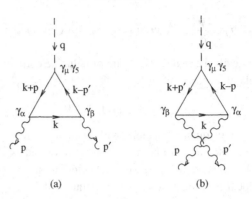

Fig. 3.1. The diagrams, representing the vacuum expectation value of axial current in the presence of external electromagnetic field in QED, (a) the direct diagram, (b) the crossing diagram.

is the dual field strength tensor. In order e^2 Eq.(3.7) can be considered as an operator equation. For this reason the symbol of vacuum averaging is omitted in (3.7). Relation (3.7) is the Adler–Bell–Jackiw anomaly [5]–[8].

In order to have a better understanding of the origin of the anomaly let us consider the same problem in momentum space. In QED, the matrix element for the transition of the axial current with momentum q into two real or virtual photons with momenta p and p' is represented by the diagrams of Fig. 3.1. The matrix element is given by

$$T_{\mu\alpha\beta}(p, p') = \Gamma_{\mu\alpha\beta}(p, p') + \Gamma_{\mu\beta\alpha}(p', p), \qquad (3.9)$$

$$\Gamma_{\mu\alpha\beta}(p, p') = -e^2 \int \frac{d^4k}{(2\pi)^4} Tr\left[\gamma_\mu\gamma_5(\not k + \not p - m)^{-1}\gamma_\alpha(\not k - m)^{-1}\gamma_\beta(\not k - \not p' - m)^{-1}\right].$$
$$(3.10)$$

Consider the divergence of the axial current $q_\mu T_{\mu\alpha\beta}(p, p'), q = p + p'$. For $q_\mu \Gamma_{\mu\alpha\beta}(p, p')$ we can write (at $m = 0$):

$$q_\mu \Gamma_{\mu\alpha\beta}(p, p') = -e^2 \int \frac{d^4k}{(2\pi)^4} \text{Tr}\left[(\not p + \not k + \not p' - \not k)\gamma_5(\not k + \not p)^{-1}\gamma_\alpha \not k^{-1}\gamma_\beta(\not k - \not p')^{-1}\right]$$

$$= -e^2 \int \frac{d^4k}{(2\pi)^4} \text{Tr}\left[-\gamma_5\gamma_\alpha \not k^{-1}\gamma_\beta(\not k - \not p')^{-1} - \gamma_5(\not k + \not p)^{-1}\gamma_\alpha \not k^{-1}\gamma_\beta\right]$$
$$(3.11)$$

Each one of the two terms in square brackets on the right-hand side of (3.11) depends after integration over k on only one 4-vector – p or p'. Each of these terms should be proportional to the totally antisymmetric unit tensor $\varepsilon_{\alpha\beta\gamma\beta}$ times the product of two different vectors. Since we have only one vector at our disposal, the result is zero. This fact seems to contradict the anomaly relation (3.7). However, we cannot trust this result. The arguments are the following: The integral (3.10) is linearly divergent. In a linearly divergent integral, it is illegitimate to shift the integration variable: such shift may result in the appearance of so-called surface terms. So, if the integration variable k in (3.10), (3.11) were changed to $k + cp + dp'$, where c and d are some numbers, then $q_\mu \Gamma_{\mu\alpha\beta}(p, p')$ would not be zero. The other argument against the calculation performed in (3.11) is that $T_{\mu\alpha\beta}(p, p')$ must satisfy the conditions of the conservation of vector current: $p_\alpha T_{\mu\alpha\beta}(p, p') = 0$, $p'_\beta T_{\mu\alpha\beta}(p, p') = 0$. The calculations performed using the same integration variable as in (3.11) show that these conditions are not fullfilled. The question arises whether it is possible to choose the integration variable in such a way that $q_\mu T_{\mu\alpha\beta} = 0$ and simultaneously $p_\alpha T_{\mu\alpha\beta} = 0$, $p'_\beta T_{\mu\alpha\beta} = 0$. Following Ref. 8, consider $\Gamma_{\mu\alpha\beta}$, defined by (3.10), where the integration variable k is shifted by an arbitrary constant vector a_λ, $k_\lambda \rightarrow k_\lambda + a_\lambda$. We can write:

$$\Gamma_{\mu\alpha\beta}(p, p'; a) = \Gamma_{\mu\alpha\beta}(p, p') + \Delta_{\mu\alpha\beta}(p, p', a) \tag{3.12}$$

$$\Delta_{\mu\alpha\beta}(p, p'; a) = \Gamma_{\mu\alpha\beta}(p, p')_{k \rightarrow k+a} - \Gamma_{\mu\alpha\beta}(p, p') \tag{3.13}$$

where $\Gamma_{\mu\alpha\beta}(p, p')$ is given by (3.10) and, therefore $q_\mu \Gamma_{\mu\alpha\beta}(p, p') = 0$ according to (3.11). $\Gamma_{\mu\alpha\beta}(p, p')_{k \rightarrow k+a}$ is obtained from (3.10) by substituting $k_\lambda \rightarrow k_\lambda + a_\lambda$. The surface term is $\Delta_{\mu\alpha\beta}(p, p'; a)$, the integral is convergent, and its calculation gives [8]:

$$\Delta_{\mu\alpha\beta} = -\frac{e^2}{8\pi^2} \varepsilon_{\mu\alpha\beta\gamma} a_\gamma. \tag{3.14}$$

Generally, a_λ is expressed in terms of two vectors involved in the problem: p and p', $a_\lambda = (a + b) p_\lambda + b p'_\lambda$. Accounting for the crossing diagram we get:

$$T_{\mu\alpha\beta}(p, p', a) = T_{\mu\alpha\beta}(p, p') - \frac{e^2}{8\pi^2} a \varepsilon_{\mu\alpha\beta\gamma} (p_\gamma - p'_\gamma). \tag{3.15}$$

The matrix element of the divergence of the axial current appears to be equal to

$$q_\mu T_{\mu\alpha\beta}(p, p'; a) = q_\mu T_{\mu\alpha\beta}(p, p') + \frac{e^2}{4\pi^2} a \varepsilon_{\alpha\beta\gamma\sigma} p_\gamma p'_\sigma. \tag{3.16}$$

As was demonstrated above, the first term on the right-hand side of (3.16) vanishes in the limit of massless quarks (the Sutherland–Veltman theorem [9], [10], see also Ref.[8]). As follows from (3.16), to ensure the conservation of the axial current it is necessary to choose $a = 0$. Such choice is just the repetition of the result already obtained in Eq. (3.11). Let us check now the conservation of the vector current. The direct calculation gives

$$p_\alpha T_{\mu\alpha\beta}(p, p'; a) = \frac{e^2}{4\pi^2} \varepsilon_{\mu\alpha\beta\gamma} p_\alpha p'_\gamma \left(1 + \frac{a}{2}\right) \tag{3.17}$$

and a similar expression for $p'_\beta T_{\mu\alpha\beta}(p, p'; a)$. As follows from (3.17), the conservation of vector current can be achieved if $a = -2$. That means that it is impossible to have simultaneously the conservation of vector and axial currents in massless QED. We are sure that the vector current is conserved, since otherwise the photon would be massive and all electrodynamics would be ruined, we must choose $a = -2$. Substitution of $a = -2$ in (3.16) gives back Eq. (3.7).

Note that the first method of deriving the anomaly – namely by using coordinate splitting in the expression of the axial current – is valid for constant external electromagnetic field since Eq.(6.273) corresponds to such a case. The second method of the derivation, based on consideration of the diagrams of Fig. 3.1 is much more general – it is valid for arbitrary varying external electromagnetic fields, including the emission of real or virtual photons. In this case, the anomaly condition has the form of

$$q_\mu T_{\mu\alpha\beta}(p, p') = \left[2mG(p, p') - \frac{e^2}{2\pi^2} \right] \varepsilon_{\alpha\beta\lambda\sigma} p_\lambda p'_\sigma. \tag{3.18}$$

In (3.18) the term proportional to the electron mass is retained and $G(p, p')$ is defined by

$$\langle p, \varepsilon_\alpha; p', \varepsilon'_\beta \mid \bar{\psi}\gamma_5\psi \mid 0\rangle = G(p, p')\varepsilon_{\alpha\beta\lambda\sigma} p_\lambda p'_\sigma, \tag{3.19}$$

where $\varepsilon_\alpha, \varepsilon'_\beta$ are photon polarizations.

The proof of the axial anomaly, Eqs. (3.7), (3.18), can be obtained also by other methods: by dimensional regularization, by Pauli–Villars regularization, or by consideration of functional integral [11],[12],[13]. In the latter, the axial anomaly arises due to noninvariance of the fermion measure in external gauge field at γ_5 transformations in the functional integral.

Although the axial current is not conserved in massless QED, there does exist a conserved, gauge-invariant axial charge [5],[7],[8]. Define

$$\tilde{j}_{\mu 5} = j_{\mu 5} - \frac{e^2}{4\pi^2} \tilde{F}_{\mu\nu} A_\nu. \tag{3.20}$$

The current $\tilde{j}_{\mu 5}$ is conserved, but is not gauge invariant. However, the axial charge

$$Q_5 = \int d^3x \, \tilde{j}_{05}(x) \tag{3.21}$$

is gauge invariant.

The axial anomaly in QED was considered till now to order of e^2. It was shown that there are no corrections to Eq. (3.7) of order e^4 [5], [6]. The argument is that at this order all radiative corrections correspond to the insertion of a photon line inside the triangle diagrams of Fig. 3.1. If the integration over the photon momentum is carried out after the integration over the fermion loop, then the fermion loop integral is convergent and there is no anomaly. [This argument was supported by direct calculation [6]]. In higher orders of perturbation theory, any insertions of photon lines and fermion loops inside the triangle diagrams of Fig. 3.1 do not give the corrections to the anomaly [5],[14]. The corrections to the Adler–Bell–Jackiw anomaly arise from higher-order diagrams like that shown in Fig. 3.2 [15]. Fig. 3.2 shows the renormalization of the anomaly term in (3.7), (3.18) to order of e^6 [5],[14],[15]. In this order [15]

$$\partial_\mu j_{\mu 5} = \frac{\alpha}{2\pi} (F_{\mu\nu}\tilde{F}_{\mu\nu})_{ext}\left(1 - \frac{3}{4} \frac{\alpha^2}{\pi^2} \ln \frac{\Lambda^2}{q^2} \right), \tag{3.22}$$

where Λ is the ultraviolet cut-off, the axial-vector vertex is renormalized and the axial-vector current, unlike the vector current, acquires an anomalous dimension.

Turn now to QCD. Here $j_{\mu 5}$ can be identified with the current of light quarks. In the case of interaction with external electromagnetic field, if $j_{\mu 5}$ corresponds to the axial current of

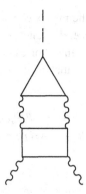

Fig. 3.2. The e^6 correction to Adler–Bell–Jackiw anomaly in QED.

one quark flavour with electric charge e_q, the anomaly has the form of Eqs. (3.7), (3.18) with the only difference that the right-hand side is multiplyed by $e_q^2 N_c$, where N_c is the number of colours. In QCD, there is also another possibility that the external fields are gluon fields. In this case, instead of (3.7) we have:

$$\partial_\mu j_{\mu 5} \doteq \frac{\alpha_s N_c}{4\pi} G^n_{\mu\nu} \tilde{G}^n_{\mu\nu}, \tag{3.23}$$

where $G^n_{\mu\nu}$ is the gluon field strength and $\tilde{G}^n_{\mu\nu}$ is its dual. Eq. (3.23) can be considered to be an operator equation and the fields $G^n_{\mu\nu}, \tilde{G}^n_{\mu\nu}$ can be represented by virtual gluons. Note, that due to the same argument as in the case of radiative corrections to the anomaly in QED, the perturbative corrections to (3.23) start from α_s^3. Evidently, the flavour octet axial current

$$j^i_{\mu 5} = \sum_q \bar{\psi}_q \gamma_\mu \gamma_5 (\lambda^i/2) \psi_q, \quad i = 1, \ldots 8 \tag{3.24}$$

is conserved in QCD. (Here λ^i are Gell-Mann $SU(3)$ matrices and the sum is performed over the flavours of light quarks, $q = u, d, s$.) Neglecting u, d, s quark masses, we have instead of (3.23):

$$\partial_\mu j^i_{\mu 5} = 0. \tag{3.25}$$

However, the anomaly persists for the singlet axial current

$$j^{(0)}_{\mu 5} = \sum_q \bar{\psi}_q \gamma_\mu \gamma_5 \psi_q, \tag{3.26}$$

$$\partial_\mu j^{(0)}_{\mu 5} = 3 \frac{\alpha_s N_c}{4\pi} G^n_{\mu\nu} \tilde{G}^n_{\mu\nu} \tag{3.27}$$

From (3.25), (3.27), it follows that because of spontaneous chiral symmetry breaking the octet of pseudoscalar mesons (π, K, η) is massless: in the approximation of $m_q \to 0$, they are Goldstown bosons, while the singlet pseudoscalar meson η' remains massive. Therefore, the occurrence of the anomaly solves the so-called $U(1)$ problem [16]. (A detailed exposition of this statement is given in [17], see also [18],[19] for reviews.)

Return now to QED and consider the matrix element of the transition of the axial current into two real or virtual photons, i.e. the function $T_{\mu\alpha\beta}(p, p')$ (3.9), described by the diagrams of Fig. 3.1, where the internal lines correspond to electron propagators. The general expression for $T_{\mu\alpha\beta}(p, p')$, which satisfyes the Bose symmetry of two photons, reads [5],[7],[20]:

$$T_{\mu\alpha\beta}(p, p') = A_1(p, p')S_{\mu\alpha\beta} - A_1(p', p)S'_{\mu\alpha\beta} + A_2(p, p')p_\beta R_{\mu\alpha}$$
$$- A_2(p', p)p'_\alpha R_{\mu\beta} + A_3(p, p')p'_\beta R_{\mu\alpha} - A_3(p', p)p_\alpha R_{\mu\beta}, \quad (3.28)$$

where

$$R_{\mu\nu} = \varepsilon_{\mu\nu\rho\sigma} p_\rho p'_\sigma, \quad S_{\mu\alpha\beta} = \varepsilon_{\mu\alpha\beta\sigma} p_\sigma, \quad S'_{\mu\alpha\beta} = \varepsilon_{\mu\alpha\beta\sigma} p'_\sigma. \quad (3.29)$$

Vector current conservation leads to

$$A_1(p, p') = (pp')A_2(p, p') + p'^2 A_3(p, p')$$

$$A_1(p', p) = (pp')A_2(p', p) + p^2 A_3(p', p) \quad (3.30)$$

Using the identity

$$\delta_{\alpha\beta}\varepsilon_{\sigma\mu\nu\tau} - \delta_{\alpha\sigma}\varepsilon_{\beta\mu\nu\tau} + \delta_{\alpha\mu}\varepsilon_{\beta\sigma\nu\tau} - \delta_{\alpha\nu}\varepsilon_{\beta\sigma\mu\nu} + \delta_{\alpha\tau}\varepsilon_{\beta\sigma\mu\nu} = 0, \quad (3.31)$$

we derive

$$p_\sigma R_{\mu\nu} - p_\mu R_{\sigma\nu} + p_\nu R_{\sigma\mu} + (pp')S_{\sigma\mu\nu} - p^2 S'_{\sigma\mu\nu} = 0$$

$$p'_\sigma R_{\mu\nu} - p'_\mu R_{\sigma\nu} + p'_\nu R_{\sigma\mu} + (pp')S'_{\sigma\mu\nu} + p'^2 S_{\sigma\mu\nu} = 0. \quad (3.32)$$

The Lorentz structures $S_{\sigma\mu\nu}$ and $S'_{\sigma\mu\nu}$ are retained in (3.28) in order to avoid kinematical singularities [20]. Let us put $p^2 = p'^2 \leq 0$. Using (3.30) and the identities (3.32), $T_{\mu\alpha\beta}(q, p, p')$ can be expressed in terms of two functions: $F_1(q^2, p^2)$ and $F_2(q^2, p^2)$ [21], [22]:

$$T_{\mu\alpha\beta}(p, p') = F_1(q^2, p^2)q_\mu \varepsilon_{\alpha\beta\rho\sigma} p_\rho p'_\sigma$$
$$- \frac{1}{2}F_2(q^2, p^2)\left[\varepsilon_{\mu\alpha\beta\sigma}(p - p')_\sigma - \frac{p_\alpha}{p^2}\varepsilon_{\mu\beta\rho\sigma} p_\rho p'_\sigma + \frac{p'_\beta}{p^2}\varepsilon_{\mu\alpha\rho\sigma} p_\rho p'_\sigma \right].$$
$$(3.33)$$

If $p^2 \neq 0$, then the formfactors $F_1(q^2, p^2) = -A_2$ and $F_2(q^2, p^2) = 2A_1$ are free of kinematical singularities [22]. Consider now the divergence

$$q_\mu T_{\mu\alpha\beta}(p, p') = [F_2(q^2, p^2) + q^2 F_1(q^2, p^2)]\varepsilon_{\alpha\beta\rho\sigma} p_\rho p'_\sigma. \quad (3.34)$$

Substitution on the left-hand side of (3.34) of the anomaly condition (3.18) gives the following sum rule [21]:

$$F_2(q^2, p^2) + q^2 F_1(q^2, p^2) = 2mG(q^2, p^2) - \frac{e^2}{2\pi^2}. \quad (3.35)$$

The functions $F_1(q^2, p^2)$, $F_2(q^2, p^2)$ and $G(q^2, p^2)$ can be represented by unsubtracted dispersion relations in q^2:

$$f_i(q^2, p^2) = \frac{1}{\pi} \int\limits_{4m^2}^{\infty} \frac{\mathrm{Im} f_i(t, p^2)}{t - q^2} dt, \quad f_i = F_1, F_2, G. \tag{3.36}$$

(The direct calculations [21], [23] show that Im $f_i(q^2, p^2)$ are decreasing at $q^2 \to \infty$. The result of the calculation of Im $F_1(q^2, p^2)$ is presented in Problem 3.1.) For the imaginary parts of F_1, F_2, G, we have the relation:

$$\mathrm{Im} F_2(q^2, p^2) + q^2 \mathrm{Im} F_1(q^2, p^2) = 2m \mathrm{Im} G(q^2, p^2). \tag{3.37}$$

From (3.35)–(3.37) we get the following sum rule:

$$\int\limits_{4m^2}^{\infty} \mathrm{Im} F_1(t, p^2) dt = \frac{e^2}{2\pi}. \tag{3.38}$$

The sum rule (3.38) has been verified explicitly by Frishman et al. [23] for $p^2 < 0$, $m = 0$, by Hořejši [21] at $p^2 = p'^2 < 0$, $m^2 \neq 0$, and by Veretin and Teryaev [24] in the general case of $p^2 \neq p'^2$, $m^2 \neq 0$.

Consider now the transition of axial current into two real photons in QCD with one flavour of unit charge. Instead of (3.38), we have

$$\int\limits_{4m_q^2}^{\infty} \mathrm{Im} F_1(t, 0) dt = 2\alpha N_c. \tag{3.39}$$

$F_2(q^2, p^2)$ should vanish at $p^2 \Rightarrow 0$ in order for $T_{\mu\alpha\beta}(p, p')$ to have no pole there, which would correspond to a massless hadronic state in channel $J^{PC} = 1^{--}$. In the limit of massless quarks, $m_q = 0$, the right-hand side of (3.35) is given by the anomaly and in QCD we have

$$F_1(q^2, 0)_{|m_q^2 = 0} = -\frac{2\alpha N_c}{\pi} \frac{1}{q^2}, \tag{3.40}$$

$$T_{\mu\alpha\beta}(p, p') = -\frac{2\alpha}{\pi} N_c \frac{q_\mu}{q^2} \varepsilon_{\alpha\beta\lambda\sigma} p_\lambda p'_\sigma. \tag{3.41}$$

The imaginary part of $F_1(q^2, 0)$ at $m_q^2 = 0$ is proportional to $\delta(q^2)$ [25]:

$$\mathrm{Im} F_1(q^2, 0)_{m_q^2 = 0} = 2\alpha N_c \delta(q^2) \tag{3.42}$$

and the sum rule (3.39) is saturated by the contribution of a zero-mass state. It is interesting to look how the limit $m_q^2 \to 0$, $q^2 \to 0$ proceeds. At $m_q \neq 0$, Im $F_1(q^2, 0)$ is equal to [25]

$$\mathrm{Im} F_1(q^2, 0) = 4\alpha N_c \frac{m^2}{q^4} \ln \frac{1 + \sqrt{1 - 4m_q^2/q^2}}{1 - \sqrt{1 - 4m_q^2/q^2}} \tag{3.43}$$

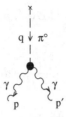

Fig. 3.3. The diagram describing the transition of isovector axial current (marked by cross) into two photons through virtual π^0.

and in the limit $m_q^2 \to 0, q^2 \to 0$ indeed we get (3.42). The most interesting case is when the current $j_{\mu 5}$ is equal to the third component of the isovector current

$$j_{\mu 5}^{(3)} = \bar{u}\gamma_\mu\gamma_5 u - \bar{d}\gamma_\mu\gamma_5 d. \qquad (3.44)$$

Then at $p^2 = p'^2 = 0, m_u^2 = m_d^2 = 0$ [25]:

$$T_{\mu\alpha\beta}(p, p') = -\frac{2\alpha}{\pi}N_c\frac{q_\mu}{q^2}(e_u^2 - e_d^2)\varepsilon_{\alpha\beta\lambda\sigma}p_\lambda p'_\sigma. \qquad (3.45)$$

The amplitude $T_{\mu\alpha\beta}$ (3.45) corresponds to the transition of isovector axial current into two photons. Eq. (3.45) is consistent with the fact that the transition proceeds through virtual π^0 and π^0 is massless at $m_q = 0$ (the pole in the amplitude at $q^2 = 0$) [25]. Then the process is described by a diagram such as Fig. 3.3. The use of relation $\langle 0 \mid j_{\mu 5}^{(3)} \mid \pi^0\rangle = \sqrt{2}if_\pi q_\mu$ determines the amplitude of $\pi^0 \to 2\gamma$ decay:

$$M(\pi^0 \to 2\gamma) = A\varepsilon_{\alpha\beta\lambda\sigma}\varepsilon_{1\alpha}\varepsilon_{2\beta}p_{1\lambda}p_{2\sigma}, \qquad (3.46)$$

where $\varepsilon_{1\alpha}$ and $\varepsilon_{2\beta}$ are the photon polarization vectors. From (3.45), the constant A is found to be

$$A = \frac{2\alpha}{\pi}\frac{1}{\sqrt{2}f_\pi} \qquad (3.47)$$

and the $\pi^0 \to 2\gamma$ decay rate is equal to

$$\Gamma(\pi^0 \to 2\gamma) = \frac{\alpha^2}{32\pi^3}\frac{m_\pi^3}{f_\pi^2}. \qquad (3.48)$$

Substitution of the π^0 mass from the Particle Data Tables [26] and $f_{\pi^0} = f_{\pi^+} = 130.7$ MeV into Eq.(3.48) gives the theoretical value of the $\pi^0 \to 2\gamma$ decay width $\Gamma(\pi^0 \to 2\gamma)_{theor} = 7.73$ eV in good agreement with the experimental value $\Gamma(\pi^0 \to 2\gamma)_{exp} = 7.8 + 0.6$ eV [26]. The accuracy of the theoretical value (3.48) is 5–7%. But one can achieve a better accuracy of the theoretical prediction. To do this it is necessary: (1) to put f_{π^0} instead of f_{π^+} in (3.48), and (2) to account for the contribution of excited states, besides π^0, to the isovector current to the sum rule (3.39). With the help of the QCD sum rules it was shown in Ref. [27] that the difference $\Delta f_\pi = f_{\pi^0} - f_{\pi^+}$ is very small: $\Delta f_\pi/f_\pi \approx -1.0 \cdot 10^{-3}$ and can be safely neglected. Among the exited states, the only significant contribution comes from the η meson because of $\eta - \pi^0$ mixing. Using the $\eta - \pi^0$ mixing angle given by (2.103) and the experimental value of $\Gamma(\eta \to 2\gamma) = 510$ eV [26] results in $\Gamma(\pi^0 \to 2\gamma)_{theor} = 7.93$ eV $\pm 1.5\%$ [27]. Theoretical predictions of $\Gamma(\pi^0 \to 2\gamma)$ which agree with this value and have the same accuracy were also obtained within the framework of chiral effective theory (CET) and $1/N_c$ expansion [28]. The recent experiment by the PriMeX Collaboration [29] gave $\Gamma(\pi^0 \to 2\gamma)_{exp} = 7.93$eV $\pm 2.1\% \pm 2.0\%$.

Despite the fact that the axial anomaly results in the appearance of a massless π^0 in the transition of isovector axial current into two photons and well predicts the π^0 decay rate, it is incorrect to say that the existence of massless pseudoscalar Goldstone bosons (at $m_q = 0$) are caused by the anomaly. The reasons are the following: Im $F_1(q^2, p^2)$ has a $\delta(q^2)$ singularity at $p^2 = 0$. According to the CET, it is expected that the same singularity will persist in the case of $p^2 \neq 0$ – the diagram of Fig. 3.3 contributes in this case as well. However, examination of Im $F_1(q^2, p^2)$ (see Problem 3.1) shows that Im$F_1(q^2, p^2)$ is a regular function of q^2 near $q^2 = 0$ at $p^2 \neq 0$ which tends to a constant at $q^2 \to 0$. The sum rules (3.38), (3.39) are satisfied by the triangle diagram contribution at $p^2 \neq 0$. The integral sum rules (3.38), (3.39) are valid, but local relation, like (3.40), is not. (It is assumed that $| p^2 |$ is less than the characteristic CET mass scale.) Therefore, by considering the case $p^2 \neq 0$, we have come to the conclusion that the existence of massless π^0 mesons cannot be attributed to the axial anomaly.

Consider now the transition of the eighth component of the octet axial current

$$j^{(8)}_{\mu 5} = \frac{1}{\sqrt{6}}(\bar{u}\gamma_\mu\gamma_5 u + \bar{d}\gamma_\mu\gamma_5 d - 2\bar{s}\gamma_\mu\gamma_5 s) \tag{3.49}$$

into two real photons at $m_u = m_d = m_s = 0$. The amplitude $F_1(q^2, 0)$ has a pole at $q^2 = 0$, which can be attributed to the η meson. The $\eta \to 2\gamma$ decay width is given by a relation analogous to (3.48):

$$\Gamma(\eta \to 2\gamma) = \frac{\alpha^2}{32\pi^3} \frac{1}{3} \frac{m_\eta^3}{f_\eta^2}. \tag{3.50}$$

However, (3.50) strongly disagrees with experiment: $\Gamma(\eta \to 2\gamma)_{theor} = 0.13$ keV (at $f_\eta = 150$ MeV) compared with $\Gamma(\eta \to 2\gamma)_{exp} = 0.510 \pm 0.026$ keV [26]. A possible explanation of this discrepancy is the effect of strong nonperturbative interactions like instantons (see Section 6.6.3), which persist in the pseudoscalar channel. The $\eta\eta'$ mixing significantly increases $\Gamma(\eta \to 2\gamma)$. Another discrepancy arises if we consider the transition $j^{(0)}_{\mu 5} \to 2\gamma$, where $j^{(0)}_{\mu 5}$ is the singlet axial current:

$$j^{(0)}_{\mu 5} = \frac{1}{\sqrt{3}}(\bar{u}\gamma_\mu\gamma_5 u + \bar{d}\gamma_\mu\gamma_5 d + \bar{s}\gamma_\mu\gamma_5 s). \tag{3.51}$$

Since at $m_u = m_d = m_s = 0$ there are poles at $q^2 = 0$ in $F_1(q^2, 0)$ for each quark flavour, the transition amplitude $T^{(0)}_{\mu\alpha\beta}(q, p, p')$ has a pole at $q^2 = 0$. The corresponding pseudoscalar meson is η'. But η' is not a Goldstone boson – it is massive! A possible explanation is the important role of instantons in $\eta' \to 2\gamma$ decay [30] and that in QCD the contribution of a diagram similar to Fig. 3.2 (with virtual gluons instead of virtual photons) and maybe a ladder (or parquet) of box diagrams is of importance here [24]. (We do not touch the theoretical determination of $\eta \to 2\gamma$ and $\eta' \to 2\gamma$ decay rates by using additional hypotheses, besides the anomaly condition; see [31] and references therein.)

Turn now back to Eqs. (3.38),(3.39). These equations are equivalent to anomaly conditions. The integrals on the left-hand side of these equations are convergent. (Im $F_1(q^2, p^2)_{q^2 \to \infty} \sim 1/q^4$; see Problem 3.1.) This means that with such interpretation the anomaly arises from a finite domain of q^2. Eq. (3.38) can be rewritten in another form [24]:

$$\lim_{q^2 \to \infty} q^2 \pi F_1(q^2, p^2) = \frac{e^2}{2\pi}. \tag{3.52}$$

This form returns us to the initial interpretation of the anomaly as corresponding to the domain of infinitely large q^2. So, it is possible to speak about the dual face of the anomaly: from one point of view, it corresponds to large q^2; from the other, its origin is connected with finite q^2. As is clear from the discussion above, both points of view are correct. These two possibilities of interpretation of the anomaly are interconnected by analyticity of the corresponding amplitudes.

't Hooft suggested the hypothesis that the singularities of amplitudes calculated in QCD on the level of quarks and gluons shall be reproduced on the level of hadrons (the so-called 't Hooft consistency condition [32]). Of course, if it were possible to prove that such singularity will not disappear after taking account of perturbative and nonperturtbative corrections, then this statement would be correct and moreover it would be trivial. But, as a rule, no such proof can be given. In examples of the realization of axial anomaly presented above (with the exception of $\pi^0 \to 2\gamma$ decay) the 't Hooft conjecture was not realized. Much better chances are for the duality conditions, such as Eq.(3.39), when the QCD amplitude integrated over some duality interval gives the same result as the corresponding hadronic amplitude integrated over the same duality interval (the so-called quark-hadron duality). Many examples of such dualities are considered in Chapter 6.

In QCD, the case when one of the photons in Fig. 3.1 is soft is of special interest [33]. If the momentum of the soft photon is p'_β and its polarization is ε'_β, then, restricting ourselves to the terms linear in p'_β, the amplitude $T_{\mu\alpha\beta}\varepsilon'_\beta$ can be represented in terms of two structure functions:

$$T_{\mu\alpha\beta}\varepsilon'_\beta = w_T(q^2)(-q^2 \tilde{f}_{\alpha\mu} + q_\alpha q_\sigma \tilde{f}_{\sigma\mu} - q_\mu q_\sigma \tilde{f}_{\sigma\alpha})$$
$$+ w_L(q^2) q_\mu q_\sigma \tilde{f}_{\sigma\alpha}, \tag{3.53}$$

where

$$\tilde{f}_{\mu\nu} = \frac{1}{2}\varepsilon_{\mu\nu\lambda\sigma}(p'_\lambda \varepsilon'_\sigma - p'_\sigma \varepsilon'_\lambda). \tag{3.54}$$

The first structure is transverse with respect to the axial current momentum q_μ, while the second one is longitudinal. From the triangle diagram, the relation [24],[34]

$$w_L(q^2) = 2w_T(q^2) \tag{3.55}$$

follows. For massless quarks, the anomaly condition gives

$$w_L(q^2) = 2\frac{\alpha}{\pi}N_c\frac{1}{q^2}. \tag{3.56}$$

Because of (3.55) this condition determines also the transverse structure function. According to the Adler–Bardeen nonrenormalization theorem, there are no perturbative corrections to the triangle diagram. But, as was demonstrated in [33], there are nonperturbative corrections which at large q^2 can be expressed through the OPE series in terms of vacuum condensates induced by external electromagnetic field.

3.3 The axial anomaly and the scattering of polarized electron (muon) on polarized gluon

Consider the scattering of longitudinally polarized electrons (muons) on longitudinally polarized gluons. The first moment of the forward scattering amplitude is proportional to the diagonal matrix element (see Chapter 7)

$$\langle g_{polar} \mid j_{\mu 5} \mid g_{polar} \rangle, \tag{3.57}$$

where the gluons are on mass shell. The corresponding Feynman diagrams are the same as in Fig. 3.1 with the only difference that the wavy lines represent now the polarized gluons and the lower vertices are the vertices of quark–gluon interaction. Put $q = 0$, $p = -p'$, $p^2 < 0$. It is convenient to use light-cone kinematics where $p_0 = p_+ + p_-/2$, $p_z = p_+ - p_-/2$, $p^2 = 2p_+p_- < 0$, and to work in an infinite momentum frame moving along the z direction. The matrix element is given by:

$$\Gamma_\mu(p) = 2ig^2 N_f \, \text{Tr}\left(\frac{\lambda^n}{2}\right)^2 \int \frac{d^4k}{(2\pi)^4} \text{Tr}\{ \not{\varepsilon}^*(\not{k}+m)\gamma_\mu\gamma_5(\not{k}+m) \not{\varepsilon}(-\not{p}+\not{k}+m) \}$$

$$\times \frac{1}{(k^2 - m^2 + i\varepsilon)^2} \frac{1}{[(p-k)^2 - m^2 + i\varepsilon]}. \tag{3.58}$$

Here, N_f is the number of flavours, λ^n, $n = 1, \ldots 8$, are the Gell-Mann $SU(3)$ matrices in colour space, m is the quark mass, assumed to be equal for all flavours, and ε_μ is the gluon polarization vector,

$$\varepsilon_\mu = \frac{1}{\sqrt{2}}(0, 1, i, 0) \tag{3.59}$$

for gluon helicity $+1$. The contribution of the crossing diagram Fig. 3.1b is equal to the direct one and is accounted for in (3.58) by the factor of 2. The evaluation of (3.58) is performed using dimensional regularization in $n \neq 4$ dimensions. According to the 't Hooft–Veltman recipe (see, e.g.[2] and Chapter 1) it is assumed that γ_5 is anticommuting with γ_μ for $\mu = 0, 1, 2, 3$, and is commuting with γ_μ for $\mu \neq 0, 1, 2, 3$. After integration over k_-, one has for the component Γ_{5+} [35]:

$$\Gamma_{5+} = -\frac{\alpha_s N_f p_+}{\pi^2} \int_0^1 dx \int \frac{d^{n-2}k_T}{[k_T^2 + m^2 + P^2 x(1-x)]^2} \left\{ k_T^2 (1-2x) - m^2 \right.$$
$$\left. -2\left(\frac{n-4}{n-2}\right) k_T^2 (1-x) \right\}, \tag{3.60}$$

where $P^2 = -p^2$. In the integration over k_- it is enough to take the residue at the pole of the last propagator in (3.58), which results in the integration domain in k_+: $0 < k_+ < p_+$, and allows $k_+ = x p_+$. The last term in (3.60) comes from the $n - 4$ regulator dimensions and is proportional to $\hat{k}^2$, where $\hat{k}$ is the projection of k into these dimensions. The azimuthal average gives $\hat{k}^2 = k_T^2(n-4)/(n-2)$.

The first term in curly brackets in (3.60) vanishes after integration over x. (In fact, the result after integration over x, which is equal to zero, is multiplyed by the divergent integral over k_T. So, strictly speaking, this term is uncertain. This problem will be discussed later.) After integration over k_T, using the rules of dimensional regularization and going to $n = 4$, we get [35]:

$$\Gamma_{5+} = -\frac{\alpha_s N_f p_+}{\pi} \left\{ 1 - \int_0^1 \frac{2m^2(1-x)dx}{m^2 + P^2 x(1-x)} \right\}. \tag{3.61}$$

The first term on the right-hand side of (3.61) arises from the last term in (3.60) and is of ultraviolet origin. As was stressed by Gribov [36] and in Ref. [35], it cannot be attributed to any definite set of quark–gluon configurations and is a result of collective effects in the QCD vacuum. In other words, this term can be considered as a local probe of gluon helicity.

The magnitude of Γ_{5+} strongly depends on the ratio m^2/P^2. At $m^2/P^2 \ll 1$

$$\Gamma_{5+} = -\frac{\alpha_s N_f p_+}{\pi}. \tag{3.62}$$

In the opposite case, $m^2/P^2 \gg 1$, the second term on the right-hand side of (3.61) almost entirely cancels the first one and we get approximately

$$\Gamma_{5+} \approx 0. \tag{3.63}$$

The real physical situation corresponds to the first case. Gluons do not exist as free particles; they are confined in hadrons and their virtualities are of the order of the inverse confinement radius squared, $P^2 \sim R_c^{-2} \gg m^2$.

Turn now to a more detailed discussion of the contribution of the first term in the curly brackets in (3.60). At fixed k_T, this contribution is zero because of the integration over x: the denominator is symmetric under the interchange $x \leftrightarrow (1-x)$, while the numerator is antisymmetric under such interchange. For the same reason, the contribution of this term vanishes under dimensional regularization at $n \neq 4$. However, the domain of low

$k_T^2 \leq P^2$ contributes to the integral over k_T^2 here. In this domain, we cannot be sure that the first term of the integrand in (3.60) has the same form as is presented there. If its form is different – we know nothing about it – and if it is not symmetric under interchange $x \to (1-x)$, then a nonvanishing infrared contribution to Γ_{5+} can arise from this term [22]. Consider the simple model with infrared cut-off in $k_\perp^2$,

$$k_T^2 > M^2(x, P^2),\qquad(3.64)$$

where $M^2(x, P^2)$ is some function of x, P^2. In the first term of (3.60), the integral over k_T^2 can be written as an integral between the limits $(0, \infty)$ which vanishes after integration over x, as before, minus the integral between the limits $(0, M^2)$. Then, neglecting the term proportional to m^2, we get instead of (3.62) the following result [22]:

$$\Gamma_{5+} = -\frac{\alpha_s N_f P_+}{\pi} \left\{ 1 - \int_0^1 dx (1-2x)[\ln r(x) - r(x)] \right\},\qquad(3.65)$$

where

$$r(x) = \frac{x(1-x)P^2}{x(1-x)P^2 + M^2(x, P^2)}.\qquad(3.66)$$

Eq. (3.65) demonstrates that the matrix element $\langle g_{polar} \mid j_{\mu 5} \mid g_{polar} \rangle$ is not entirely saturated by the ultraviolet domain connected with the anomaly, but can get a contribution from the infrared region.

Note that in the parton model Γ_{5+} is related to the part of hadron spin carried by gluons in the polarized hadron:

$$(\Delta g_1^h)_{gl} = \int_0^1 g_{1,gl}^h(x)dx = (\Gamma_{5+}/2p_+)[\, g_{gl+} - g_{gl-}\,].\qquad(3.67)$$

Here, g_{gl+} and g_{gl-} are the numbers of gluons in the hadron with helicities $+1$ and -1, respectively, $g_{1,gl}^h(x)$ is the contribution of gluons to the structure function $g_1^h(x)$. For polarized protons it was found [37],[38],[39]

$$(\Delta g_1)_{gl} = -\frac{\alpha_s N_f}{2\pi}(g_{gl+} - g_{gl-}).\qquad(3.68)$$

This relation will be exploited in Chapter 7.

3.4 The scale anomaly

The classical massless field theory with dimensionless coupling constants is scale invariant. In such theory, the energy dependence of any physical observable is determined by its dimension,

$$F(E, p_a, p_b, \ldots) = E^{d_F} f\left(\frac{p_a}{E}, \frac{p_b}{E}, \ldots \right),\qquad(3.69)$$

where d_F is the dimension of F (in energy units) and f is a dimensionless function of the ratios p_a/E, p_b/E, ... In quantum theory, scale invariance is violated because of ultraviolet divergences, which necessitates the introduction of a normalization point μ and the appearance of μ-dependent running coupling constants. Traces of the scale invariance of the classical theory still persist in quantum theory and, since the source of scale violation is known, definite predictions can be made.

Consider first the scale transformations in classical theory. Perform an infinitesimal shift of the coordinates:

$$x'_\mu = x_\mu + \xi_\mu. \tag{3.70}$$

The invariance of the theory under such transformation results in the conservation of the energy-momentum tensor $\Theta_{\mu\nu}$ (the Noether theorem). In order to derive this conservation law from a minimum action principle, it is convenient to introduce formally the metric tensor $g_{\mu\nu}(x)$ into the action, as is done in general relativity theory. This method allows one to get directly a symmetrical energy-momentum tensor. The variation of $g_{\mu\nu}(x)$ under the transformation (3.70) is given by [40]:

$$g'_{\mu\nu}(x') = g_{\mu\nu}(x) + g_{\mu\lambda}\frac{\partial \xi_\nu}{\partial x_\lambda} + g_{\nu\sigma}\frac{\partial \xi_\mu}{\partial x_\sigma} \tag{3.71}$$

(In our approach there is no need to distinguish the co- and contravariant coordinates.) In the *Classical Field Theory* by Landau and Lifshitz [40], it is shown that the variation of the action under the transformation (3.70), (3.71) is given by

$$\delta S = \int \frac{\partial \Theta_{\mu\nu}}{\partial x_\nu}\xi_\mu \sqrt{g}\, d\Omega, \tag{3.72}$$

where the integration is performed over the entire four-dimensional space-time and the energy-momentum tensor is defined by

$$\frac{1}{2}\sqrt{g}\,\Theta_{\mu\nu} = \frac{\partial}{\partial x_\lambda}\frac{\partial \sqrt{g} L}{\partial(\frac{\partial g_{\mu\nu}}{\partial x_\lambda})} - \frac{\partial \sqrt{g} L}{\partial g_{\mu\nu}}. \tag{3.73}$$

Since the shift of coordinates does not change the action, $\delta S = 0$. Because ξ_μ is arbitrary,

$$\frac{\partial \Theta_{\mu\nu}}{\partial x_\nu} = 0. \tag{3.74}$$

This is the energy-momentum conservation law. Evidently, $\Theta_{\mu\nu}$ defined by (3.73) is symmetric: $\Theta_{\mu\nu} = \Theta_{\nu\mu}$. The scale (dilatation) transformation corresponds to a special case of (3.70):

$$\xi_\mu = \alpha x_\mu. \tag{3.75}$$

Substituting (3.75) into (3.72) and integrating by parts (for $\sqrt{g}$ =Const), we have

$$\delta S = -\alpha \int \Theta_{\mu\mu} \sqrt{g}\, d\Omega. \tag{3.76}$$

Let us define the dilatation current as

$$D_\mu = \Theta_{\mu\nu} x_\nu. \tag{3.77}$$

As follows from (3.74),

$$\frac{\partial D_\mu}{\partial x_\mu} = \Theta_{\mu\mu} \tag{3.78}$$

and the dilatation current is conserved if

$$\Theta_{\mu\mu} = 0. \tag{3.79}$$

According to (3.76), this condition provides the scale invariance of the theory.

In QCD we have

$$\Theta_{\mu\nu} = -G^n_{\mu\lambda}G^n_{\nu\lambda} + \frac{1}{4}\delta_{\mu\nu}G^n_{\lambda\sigma}G^n_{\lambda\sigma} + \frac{i}{4}\sum_q \{\bar\psi_q(\gamma_\mu\nabla_\nu + \gamma_\nu\nabla_\mu)\psi_q - \bar\psi_q(\gamma_\mu\overleftarrow{\nabla}_\nu + \gamma_\nu\overleftarrow{\nabla}_\mu)\psi_q\} \tag{3.80}$$

and

$$\Theta_{\mu\mu} = \sum_q m_q\bar\psi_q\psi_q. \tag{3.81}$$

As is seen from (3.81), the massless QCD (or QED) is scale invariant on the level of classical theory.

Turn now to quantum theory and consider the properties of the dilatation current D_μ (3.77) as a quantum field operator [3],[41]. The dilatation charge is given by

$$D(t) = \int D_0(x)d^3x = tH + \tilde D(t), \quad \tilde D(t) = \int \Theta_{oi}x_i d^3x. \tag{3.82}$$

Introduce the spatial component operators of the total momentum

$$P_i = \int \Theta_{0i}(x)d^3x. \tag{3.83}$$

For any operator $A(t, x)$ we have

$$[\, P_i, A(t, x)\,] = i\frac{\partial}{\partial x_i}A(t, x); \quad A(t, x) = e^{iP_ix_i}A(t, O)e^{-iP_ix_i}, \tag{3.84}$$

and

$$[\, \tilde D, P_i\,] = iP_i. \tag{3.85}$$

Let $O(x)$ be an arbitrary colourless operator. Then, according to (3.82)

$$[\, D, O(x)\,] = -it\frac{\partial}{\partial t}O(x) + [\, \tilde D, O(x)\,] \tag{3.86}$$

Using (3.84) it can be shown [3] that

$$[\, \tilde D, O(x)\,] = -ix_i\frac{\partial}{\partial x_i}O(x) + e^{iP_ix_i}[\, \tilde D, O(t, O)\,]e^{-iP_ix_i}. \tag{3.87}$$

The commutator in the last term of (3.87) is proportional to $O(t, O)$ and we have

$$[\, D, O(x)\,] = -i\left[\, x_\mu\frac{\partial}{\partial x_\mu} + d_O\,\right]O(x). \tag{3.88}$$

(For gauge-noninvariant operators $O(x)$ additional gauge-dependent terms can appear on the right-hand side of (3.88).) The values of d_O depends on the type of the operator $O(x)$. By considering the examples of free theories it is easy to see that $d = 1$ for boson fields φ and $d = 3/2$ for fermion fields ψ.

Let us determine now the divergence of the dilatation current, equal to $\Theta_{\mu\mu}$, in massless QCD in quantum field theory [42],[43],[44]. Using (3.76) we can write

$$\partial_\mu D_\mu = \Theta_{\mu\mu} = -\frac{\partial L}{\partial \alpha} = -\frac{\partial g^2}{\partial \alpha}\frac{\partial L}{\partial g^2}, \tag{3.89}$$

where L is the QCD Lagrangian and g is the QCD running coupling constant. Redefine the gauge field in QCD: $g A_\mu = \bar{A}_\mu$. In QCD with Lagrangian expressed in terms of $\bar{A}_\mu$, g^2 appears only in front of the terms proportional to $\bar{G}^2$:

$$L = -\frac{1}{4}\frac{1}{g^2}\bar{G}^n_{\mu\nu}\bar{G}^n_{\mu\nu} + \ldots \tag{3.90}$$

The dependence of g^2 on α is given by the renormalization group equation:

$$\frac{1}{2}\frac{\partial g^2}{\partial \alpha} = \beta(\alpha_s), \quad \beta(\alpha_s) = b\alpha_s^2 + \ldots, \tag{3.91}$$

where $\beta(\alpha_s)$ is the Gell-Mann-Low β-function, $b = 11 - (2/3)N_f = 9$ for $N_f = 3$. Let us take the vacuum expectation value of (3.89), i.e. consider $G^2_{\mu\nu}$ as the mean gluon field in the QCD vacuum. Substitution of (3.90) and (3.91) in (3.89) gives:

$$\langle 0 \mid \Theta_{\mu\mu} \mid 0 \rangle = -\frac{1}{8}\beta(\alpha_s)\frac{1}{\alpha_s^2}\langle 0 \mid \bar{G}^2_{\mu\nu} \mid 0 \rangle = -\frac{\pi}{2}\beta(\alpha_s)\frac{1}{\alpha_s}\langle 0 \mid G^2_{\mu\nu} \mid 0 \rangle. \tag{3.92}$$

In one-loop approximation (3.92) reduces to:

$$\langle 0 \mid \Theta_{\mu\mu} \mid 0 \rangle = -\frac{1}{8}b\left\langle 0 \mid \frac{\alpha_s}{\pi}G^2_{\mu\nu} \mid 0 \right\rangle. \tag{3.93}$$

Since there is no preferential direction in the vacuum, the general form of $\langle 0 \mid \Theta_{\mu\nu} \mid 0 \rangle$ is:

$$\langle 0 \mid \Theta_{\mu\nu} \mid 0 \rangle = \varepsilon_v \delta_{\mu\nu}, \tag{3.94}$$

where ε_v is the energy density of the vacuum. It is reasonable to perform the renormalization and subtract the (infinite) perturbative contribution to the vacuum energy. (More precisely – see Section 6.1 of Chapter 6 – the normalization point μ in momentum space should be introduced. The perturbative contributions from vacuum fluctuations with momenta $p > \mu$ should be subtracted, the fluctuations with momenta $p < \mu$ are included into the nonperturbative part.) Then ε_v has the meaning of vacuum energy density, caused by nonperturbative fluctuations of quark and gluon fields. It follows from (3.92), (3.94) that

$$\varepsilon_v = -\frac{\pi}{8}\frac{\beta(\alpha_s)}{\alpha_s}\langle 0 \mid G^2_{\mu\nu} \mid 0 \rangle, \tag{3.95}$$

and in one-loop approximation:

$$\varepsilon_v = -\frac{9}{32}\langle 0 \mid \frac{\alpha_s}{\pi}G^2_{\mu\nu} \mid 0 \rangle. \tag{3.96}$$

So, in QCD, the vacuum energy density is expressed in terms of the gluon conden-sate [45].

The relation (3.92) is the scale (or dilatation) anomaly in QCD. The question arises: is it possible to formulate it as an operator equation, valid for any matrix element, as is the case for axial anomaly? The answer is negative for the following reasons: The axial anomaly arises from a definite type of diagram – from triangle diagrams. The scale anomaly arises from scale dependence of the QCD coupling constant g^2. If we would like to consider the matrix element of Eq. (3.89) between any hadronic states whose wave functions depend on g^2, then additional terms arising from variation of these wave functions would appear. So, Eq. (3.92) cannot be generalized to an operator equation. This can be done in one-loop only, since all other matrix elements, besides the vacuum averaging, correspond to many loops. (Note that for a special theory – supersymmetric QCD – the one-loop relation (3.93) becomes an operator equation. The origin of this is that in this theory the axial current and the energy-momentum tensor enter one supermultiplet and the axial anomaly is reproduced in the scale anomaly [3],[46],[47]).

In the presence of external electromagnetic field the anomaly condition (3.92) changes to

$$\langle 0|\Theta_{\mu\mu}|0\rangle = -\frac{\pi}{2}\frac{\beta(\alpha_s)}{\alpha_s}\langle 0|G^2_{\mu\nu}|0\rangle + \frac{\alpha}{6\pi}N_c\sum_q e_q^2 F^2_{\mu\nu}, \qquad (3.97)$$

where $F_{\mu\nu}$ is the strength of external electromagnetic field. Eq. (3.97) directly follows from (3.92) if we note that asymptotically

$$\alpha_s(Q^2) \sim \frac{1}{(b/4\pi)\ln Q^2}, \quad \alpha(Q^2) \sim -\frac{1}{(1/3\pi)\ln Q^2}. \qquad (3.98)$$

The anomaly condition (3.92) leads to useful low-energy theorems in QCD [48]. If $O(x)$ is any local operator, then the relation

$$i\int d^4x \langle 0|T\{O(x), \Theta_{\mu\mu}(0)\}|0\rangle = -d_O \langle 0|O(0)|0\rangle \qquad (3.99)$$

holds, where d_O is the canonical dimension of O. In order to prove (3.99), rescale $G_{\mu\nu}$, $gG_{\mu\nu} = \bar{G}_{\mu\nu}$. Then

$$i\int d^4x \langle 0|T\{O(x), \bar{G}^2_{\mu\nu}(0)\}|0\rangle = -\frac{\partial}{\partial(1/4g^2)}\langle 0|O(0)|0\rangle. \qquad (3.100)$$

The scale dependence of $\langle 0|O|0\rangle$ arises only from its dependence on the normalization point μ. In one-loop approximation, we have

$$\mu = M_0 \exp\left(-\frac{8\pi^2}{bg^2}\right) \qquad (3.101)$$

and on dimensional grounds the general expression for $\langle 0|O|0\rangle$ is

$$\langle 0|O(0)|0\rangle = Const\left[M_0 \exp\left(-\frac{8\pi^2}{bg^2}\right)\right]^{d_O}. \qquad (3.102)$$

Differentiation of (3.102) with respect to $1/4g^2$ and substitution into (3.100) gives (3.99). This proof was performed in 1-loop approximation, but it can be generalized to the many-loop case, where instead of b the β-function arises. Note, however, that the anomalous dimensions of the operator O were disregarded. (A more detailed proof of the low-energy theorems is presented in [48]. The delicate problem of the renormalization of divergences is also considered there.)

Two low-energy theorems for the transitions of two gluons into two photons follow from the anomaly conditions [48]:

$$\left\langle 0 \left| \frac{\alpha_s}{4\pi} G^n_{\mu\nu} \bar{G}^n_{\mu\nu} \right| \gamma(k_1)\gamma(k_2) \right\rangle = \frac{\alpha}{3\pi} N_c \sum_q e_q^2 F^{(1)}_{\mu\nu} \tilde{F}^{(2)}_{\mu\nu}, \qquad (3.103)$$

$$\left\langle 0 \left| -\pi \frac{\beta(\alpha_s)}{2\alpha_s} G^n_{\mu\nu} G^n_{\mu\nu} \right| \gamma(k_1)\gamma(k_2) \right\rangle = \frac{\alpha}{\pi} N_c \sum_q e_q^2 F^{(1)}_{\mu\nu} F^{(2)}_{\mu\nu}, \qquad (3.104)$$

where

$$F^{(i)}_{\mu\nu} = k^{(i)}_\mu e^{(i)}_\nu - k^{(i)}_\nu e^{(i)}_\mu, \quad (i = 1, 2),$$

$k^{(i)}_\mu$ and $e^{(i)}_\mu$ are the momenta and polarizations of the photons. Eq. (3.104) can be derived from (3.97) if it is noted that

$$\langle 0|\Theta_{\mu\nu}|\gamma(k_1)\gamma(k_2)\rangle = Const\left[F^{(1)}_{\mu\alpha} F^{(2)}_{\nu\alpha} - \frac{1}{4} \delta_{\mu\nu} F^{(1)}_{\alpha\beta} F(2)_{\alpha\beta} \right] \qquad (3.105)$$

in the low-energy limit. (The low-energy theorem for the matrix element $\langle 0|\Theta_{\mu\mu}|\gamma\gamma\rangle$ was considered in detail in [49]). Until now, the contributions of heavy quarks were completely ignored. The reason for this is that the vacuum expectation values of the operators involving heavy quarks fields are inversely proportional to their mass [45],[48].

3.5 The infrared power-like singularities in photon-photon, photon-gluon, and gluon-gluon scattering in massless QED and QCD. Longitudinal gluons in QCD

Longitudinal photons and gluons are not physical degrees of freedom; they cannot be emitted, absorbed, or scattered. In classical electrodynamical or chromodynamical Lagrangians, these degrees of freedom can be separated and result in Coulomb interaction between charged particles (or coloured particles in QCD). This statement holds for massive, as well as for massless electro- or chromodynamics. The virtual longitudinal photons or gluons can be, of course, scattered. As will be shown below, in massless QED the scattering cross section of nonpolarized photon on a longitudinally polarized virtual photon $\sigma_L(p^2)$ tends to a constant limit when the virtuality p^2 of the longitudinal photon tends to zero [50]. This looks like the longitudinal photon can be scattered on the photon, and, therefore, in high orders of α on any other target. This would mean that in massless QED the longitudinal photons can be emitted and absorbed, i.e. that the third degree of freedom of the electromagnetic field is revived. However, the appearance of even a tiny, but nonzero, electron mass m changes the situation cardinally: $\sigma_L(p^2) \to 0$ at $p^2 \to 0$ and such decrease starts sharply at $p^2 \sim m^2$. So, there is no soft transition from massive to massless QED. A related phenomenon is the chirality violation in processes such as $\mu^+\mu^- \to e^+e^-\gamma$ [51] or $\pi^+ \to e^+\nu\gamma$ decay [52]. If the electron mass m tends to zero, the cross section of the chirality violating process $\mu^+\mu^- \to e_R^+e_R^-\gamma$ (or $e_L^+e_L^-\gamma$) with photon production in the forward direction is going to a constant limit instead of being proportional to m^2/E^2, as should follow from chirality conservation. All these phenomena appear in quantum theory and for this reason they can be called anomalies.

Consider the scattering of nonpolarized photon on longitudinally polarized photon in more detail. The emission amplitude $T_\lambda(p)$ of the photon with momentum p satisfies the condition

$$p_\lambda T_\lambda(p) = 0 \tag{3.106}$$

for real $p^2 = 0$ and for virtual $p^2 \neq 0$ photons. This condition is a direct consequence of gauge invariance. The condition (3.106) does not mean that the emission amplitude of the longitudinal photon vanishes since the latter is proportional to $e_\lambda^L T_\lambda(p)$, where e_λ^L is the polarization of the longitudinal photon. The vector $e_\lambda^L(p)$ is by no means proportional to p_λ, but on the contrary, in the Lorentz gauge $\partial_\lambda A_\lambda(x) = 0$ it is orthogonal to it: $e^L p = 0$. Thus, in the general case $e_\lambda^L(p)T_\lambda(p) \neq 0$ although Eq. (3.106) is fulfilled. The components of the polarization vector of the virtual longitudinal photon moving along the z-axis are:

$$e_\lambda^L(p) = (e_0^L, e_x^L, e_y^L, e_z^L) = (|\boldsymbol{p}|, 0, 0, p_0)/\sqrt{-p^2} \tag{3.107}$$

(It is supposed that $p^2 < 0$ and e_λ^L is normalized by $(e_\lambda^L)^2 = 1$.) Using (3.106) and (3.107), we can write

$$e_\lambda^L(p)T_\lambda(p) = \frac{\sqrt{-p^2}}{|\boldsymbol{p}|} T_0(p). \tag{3.108}$$

As is seen from Eq.(3.108), $e_\lambda^L(p)T_\lambda(p) \to 0$ at $p^2 \to 0$, but only if $T_0(p)$ has no singularity at $p^2 \to 0$. It appears, however, that in some cases such a singularity does arise, so that $e_\lambda^L(p)T_\lambda(p)$ tends to a finite limit at $p^2 \to 0$.

Let us calculate the total annihilation cross section of a photon with momentum q and a photon with momentum p in massless QED. The value of the total cross section is determined by the sum of imaginary parts of the forward scattering amplitudes described by the graphs of Fig. 3.4.

We assume both photons to be virtual: q^2, $p^2 < 0$, the photon with momentum q being transverse and the photon with momentum p being longitudinal, $|q^2| \gg |p^2|$. The calculation is straightforward. It will be convenient to use the covariant form for the polarization of the longitudinal photon (3.107)

$$e_\lambda^L(p) = -\frac{\sqrt{-p^3}}{\sqrt{\nu^2 - p^2q^2}}\left(q_\lambda - \frac{\nu p_\lambda}{p^2}\right), \tag{3.109}$$

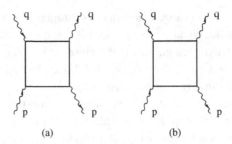

Fig. 3.4. The diagrams of photon-photon scattering: (a) direct diagram, (b) crossing diagram.

where $v = pq$. Restricting ourselves to the nonvanishing term at $p^2 \to 0$ we get for the total cross section

$$\sigma_{T(q)}^{L(p)}\bigg|_{p^2 \to 0} = \frac{16\pi\alpha^2}{Q^2}(1-x)x^2 \tag{3.110}$$

where $Q^2 = -q^2$, $x = Q^2/2v$. (The analogous calculation for the case of two longitudinal photons yields $\sigma_{L(q)}^{L(p)}\big|_{p^2 \to 0} = 0$.)

We see that despite naive expectations, the total cross section of the longitudinal photon interaction does not tend to zero when its virtuality p^2 vanishes. The reason is that the imaginary part of the forward scattering amplitude described by the graph of Fig. 3.4a has a pole at $p^2 = 0$, which compensates the factor p^2 in the denominator resulting from the products of polarization $e_\lambda^L(p)e_\sigma^L(p)$ (3.109) of the initial and final photons. Namely, the imaginary part of the forward-scattering amplitude averaged over polarizations of the hard photon q is given by

$$\frac{1}{8\pi\alpha^2}\mathrm{Im}\,T_{\lambda\sigma}(p,q) = \left(-\delta_{\lambda\sigma} + \frac{p_\lambda p_\sigma}{p^2}\right)\left\{\left[\frac{1}{2} - x(1-x)\right]\ln\left(\frac{2v}{-p^2x}\right)\right.$$
$$-1 + x(1-x)\} - \frac{p^2}{v^2}\left(q_\lambda - p_\lambda\frac{v}{p^2}\right)\left(q_\sigma - p_\sigma\frac{v}{p^2}\right)$$
$$\left. \times\left\{\left[\frac{1}{2} - x(1-x)\right]\ln\left(\frac{2v}{-p^2x}\right) - 1 + 3x(1-x)\right\}. \tag{3.111}$$

It can be easily seen that the pole $1/p^2$ in this expression which appears due to the last term in (3.111) after multiplication by the polarization vectors (3.109) gives a nonvanishing contribution in the limit $p^2 \to 0$:

$$\mathrm{Im}\,T_{\lambda\sigma}\,e_\lambda^L(p)e_\sigma^L(p) = 16\pi\alpha^2 x(1-x) \tag{3.112}$$

which being rewritten in terms of the cross section leads to (3.110).

In massive QED the situation is quite different. In the case of a massive electron, the last term in (3.111) is multiplied by

$$p^2 \bigg/ \left(p^2 - \frac{m^2}{x(1-x)}\right), \tag{3.113}$$

so that in massive QED $\sigma_T^L \to 0$ at $p^2 \to 0$.

As follows from (3.113), the sequence of limits $m^2 \to 0$, $p^2 \to 0$ or $p^2 \to 0$, $m^2 \to 0$ leads to different results. Physically, the correct order of limits is: first put p^2 to zero and afterwards $m^2 \to 0$. The reason is that the states of the electron and the photon produced in γe collisions are formed outside the formation zone, whose size is of order E/m^2, where E is the electron energy in the centre of mass system [52].

If the limit is performed in the opposite order, then the states of electron and photon cannot be separated. The origin of the appearance of the pole $1/p^2$ in(3.111) is connected with the power collinear singularity in the differential cross section

$$d\sigma \left[\gamma_T(q) + \gamma_L(p) \to e^+ e^-\right] = 8\pi\alpha^2 x^2(1-x) \frac{p^2}{v^2} \frac{d\cos\theta}{(1 - p^2 x/v - \cos\theta)^2}, \tag{3.114}$$

where θ is the angle of the emitted electron relative to the direction of the colliding photons. (The theory of the transition $m \to 0$ in QED was formulated in [53],[54] in the case of logarithmic singularities and in [51] in the case of power singularities.)

Turn now to QCD. In QCD, gluons are confined, their virtualities are of the order of the square of the inverse hadron radius: $-p^2 \sim R^{-2} \sim (500 \text{ MeV})^2$, and much larger than the square of the u and d quark masses, $(m_u + m_d)^2 \approx (12 \text{ MeV})^2$, and probably larger than the square of the s quark mass $m_s^2 = (150 \text{ MeV})^2$. Therefore, for light quarks we have $p^2/m^2 \gg 1$. This condition was already exploited in Section 3.3. As a consequence of this condition, the cross section for the scattering of a virtual photon on a longitudinal gluon is given by a formula similar to (3.110):

$$\sigma_{\gamma g L} = 8\pi\alpha\alpha_s \frac{1}{Q^2} \sum_q e_q x^2(1-x). \tag{3.115}$$

The longitudinal gluon contribution to the structure function $F_2(x)$ obtained from (3.115) is equal to (for $e_q = 1$):

$$F_2^L(x) = \frac{2}{\pi}\alpha_s x^2(1-x) \tag{3.116}$$

Since the third degree of freedom of the gluonic field – the longitudinal gluon – can be considered as hadronic constituents, besides the usually accounted for transverse gluons, the question arises as to how they will influence the evolution equations. The answer is the following [55]: The longitudinal gluons contribute to the structure functions in the next-to-leading order approximation. For this reason, accounting for them does not change the form of evolution equations and can be reduced to a change of initial conditions only. Since in practice the gluon distributions are determined from overall fits to the data, the distributions of longitudinal gluons are accounted for automatically in the analysis.

In the attempt to formulate the contribution of longitudinal gluons by using the operator product expansion (OPE) we encounter a problem. The pole $1/p^2$ in the amplitude

(3.111) cannot be described by any local operator and it is necessary to introduce nonlocal operators. A suitable nonlocal operator of the lowest dimension, corresponding to the first moment of the structure function, is

$$R_{\mu\nu} = G^n_{\alpha\nu}(D_\mu D_\nu/D^2_\lambda)G^n_{\alpha\beta}. \tag{3.117}$$

(See [55] for details.)

Generally speaking, $g_L(x)$ as a function of x differs from $g_T(x)$ and may exceed it in some cases. For instance, in pions at $x \to 1$, $g_L(x) \sim g_T(x)/(1-x)^2$, since $g_T(x)$ in pions at $x \to 1$ is suppressed by helicity conservation. In the production of particles or jets with large p_T or in the production of heavy quarks in hadronic collisions, the longitudinal gluon contribution is suppressed by the factor $\langle p^2 \rangle g_L/p_T^2 \sim (R^2 p_T^2)^{-1}$ or $(R^2 M^2)^{-1}$, where $\langle p^2 \rangle_{g_L}$ is the average virtuality of longitudinal gluons in a hadron and M is the heavy quark mass. For this reason, the contribution of longitudinal gluons to cross sections of such processes can be neglected as a rule. But there are important exceptions to this rule when the production of a given state by the transverse gluon collision is also suppressed (for instance, by chirality conservation). An example of such process may be the production in hadronic collisions of the χ_1 charmonium state with quantum number $J^{PC} = 1^{++}$. In QCD, the production of charmonium P-states occurs via two-gluon fusion $g + g \to \chi$. However, the process $g_T + g_T \to \chi_1$ is forbidden for on-shell gluons. Therefore, for transverse gluons the cross section is nonvanishing for virtual gluons only and the χ_1 production cross section by g_T and g_L fusion may be of the same order.

Problems

Problem 3.1

Calculate Im $F_1(q^2, p^2)$ according to definitions (3.9), (3.10), (3.33). Prove the sum rule (3.38). Consider the case $p^2 = 0$ and demonstrate that in this case the limit $\lim_{m^2 \to 0, q^2 \to 0}$ Im $F_1(q^2, 0) = e^2 \pi \delta(q^2)$ [25].

Answer [21]:

$$\text{Im } F_1(q^2, p^2) = -\frac{e^2}{2\pi}\frac{2p^2}{q^2}\left\{\frac{q^2+2p^2}{(q^2-4p^2)^2}\left(1-\frac{4m^2}{q^2}\right)^{1/2} + \frac{2p^2(q^2-2p^2)}{(q^2)^{1/2}(q^2-4p^2)^{5/2}}\right.$$
$$\left.\times\left[\frac{q^2-p^2}{q^2-2p^2}+m^2\frac{q^2-4p^2}{2(p^2)^2}\right]\ln\frac{q^2-2p^2-[(q^2-4m^2)(q^2-4p^2)]^{1/2}}{q^2-2p^2+[(q^2-4m^2)(q^2-4p^2)]^{1/2}}\right\}.$$

At $p^2 = 0$

$$\text{Im } F_1(q^2, 0) = \frac{e^2}{\pi}\frac{m^2}{q^4}\ln\frac{1+\sqrt{1-4m^2/q^2}}{1-\sqrt{1-4m^2/q^2}}.$$

Problem 3.2

Find the low-energy matrix element $\langle 0 \mid G^2_{\mu\nu} \mid \pi^+\pi^- \rangle$ in the limit of massless pions.

Hint: Consider $\langle 0 \mid \Theta_{\mu\nu} \mid \pi^+\pi^- \rangle$ and use CET and the anomaly condition.

Answer [56]:

$$\langle 0 \mid (-b\alpha_s/8\pi)G^n_{\mu\nu}G^n_{\mu\nu} \mid \pi^+(p_1)\pi^-(p_2)\rangle = (p_1 + p_2)^2.$$

References

[1] Treiman, S. B., Jackiw, R., Zumino, B. and Witten, E., *Current algebra and anomalies*, Princeton Series in Physics, Princeton University Press, 1985.

[2] Collins, J., *Renormalization*, Cambridge University Press, Cambridge, 1984.

[3] Shifman, M. A., Anomalies in gauge theories, *Phys. Rep.* **209** (1991) 343.

[4] Peskin, M. E. and Schroeder, D. V., *An Introduction to quantum field theories*, Addison-Wesley Publ. Company, 1995.

[5] Adler, S. L., *Phys. Rev.* **177** (1969) 2426.

[6] Adler, S. L. and Bardeen, W., *Phys. Rev.* **182** (1969) 1517.

[7] Bell, J. and Jackiw, R., *Nuovo Cim.* **A51** (1969) 47.

[8] Jackiw, R., Field theoretical investigations in current algebra, Ref. 1, p.p.81–210.

[9] Sutherland, D. G., *Nucl. Phys.* **B2** (1967) 433.

[10] Veltman, M., *Proc. Roy. Soc.* **A301** (1967) 107.

[11] Vergeles, S. N., Ph.D. Thesis, Landau Inst, 1979, published in: *Instantons, strings and conformal field theory*, Fizmatlit, Moscow, 2002 (in Russian).

[12] Migdal, A. A., *Phys. Lett.* **81B** (1979) 37.

[13] Fuijkawa, K., *Phys. Rev. Lett.* **42** (1979) 1195; *Phys. Rev.* **D21** (1980) 2848.

[14] Adler, S. L., Anomalies to all orders, in: *Fifty years of Yang-Mills theory*, ed. by G. 't Hooft, World Scientific, 2005, p.p.187–228; hep-th/0405040.

[15] Anselm, A. A. and Iogansen, A., *JETP Lett.* **49** (1989) 214.

[16] Weinberg, S., *Phys. Rev.* **D11** (1975) 3583.

[17] Dyakonov, D. I. and Eides, M. I., *Sov. Phys. JETP* 54 (1981) 2.

[18] Dyakonov, D. I. and Eides, M. I., in: *The Materials of the XVII NPI Winter School*, vol.1, pp. 55–80, Leningrad, 1982 (in Russian).

[19] Leader, E. and Predazzi, E., *An Introduction to gauge theories and modern particle physics*, Cambridge University Press, 1995.

[20] Eletsky, V. L., Ioffe, B. L. and Kogan, Ya. I., *Phys. Lett.* **122B** (1983) 423.

[21] Hořejši, J., *Phys. Rev.* **D32** (1985) 1029.

[22] Bass, S. D., Ioffe, B. L., Nikolaev, N. N. and Thomas, A. W. *J. Moscow Phys. Soc.* **1** (1991) 317.

[23] Frishman, Y., Schwimmer, A., Banks, T. and Yankelowicz, S., *Nucl. Phys.* **B117** (1981) 157.

[24] Veretin, O. L. and Teryaev, O. V., *Yad. Fiz.* **58** (1995) 2266.

[25] Dolgov, A. D. and Zakharov, V. I., *Nucl. Phys.* **B27** (1971) 525.

[26] Yao, W.-M. *et al*, Review of Particle Physics, J.Phys.G: **33** (2006) 1.

[27] Ioffe, B. L. and Oganesian, A. G., *Phys. Lett.* **B647** (2007) 389.

[28] Goity, J. L., Bernstein, A. M. and Holstein, B. R., *Phys. Rev.* **D66** (2002) 076014.

[29] De Jager, K., *et al* (PriMex Collab.), arXiv: 0801.4520.

[30] Dorokhov, A. E., *JETP Lett.* **82** (2005) 1.

[31] Shore, G. M., *Phys. Scripta* **T99** (2002) 84; hep-ph/0111165.

[32] 't Hooft, G., in: *Recent developments in gauge theories*, eds. G.'t Hooft *et al*, Plenum Press, N.Y., 1980, p.241.

[33] Vainshtein, A. I., *Phys. Lett.* **B569** (2003) 187.

[34] Achasov, N. N. *ZhETF* **101** (1992) 1713, **103** (1993) 11.

[35] Carlitz, R. D., Collins, J. C. and Mueller, A. H., *Phys. Lett.* **214B** (1988) 229.

[36] Gribov, V. N., Anomalies as a manifestation of the high momentum collective motion in the vacuum, in: V. N. Gribov, *Gauge theories and quark confinement*, Phasis, Moscow, 2002, p.p.271–288; preprint KFKI-1981-66 (1981).

[37] Efremov, A. V. and Teryaev, O. V., JINR preprint E2-88-287 (1988).

[38] Altarelli, G. and Ross, G. G., *Phys. Lett.* **212B** (1988) 391.

[39] Anselmino, M., Efremov, A. and Leader, E., *Phys. Rep.* **261** (1995) 1.

[40] Landau, L. D. and Lifshitz, E. M., *Classical field theory*, *Pergamon Press*, London, 1977.

[41] Coleman, S and Jackiw, R., *Ann. Phys.* **67** (1971) 552.

[42] Crewther, R., *Phys. Rev. Lett.* **28** (1972) 1421.

[43] Chanowitz, M. and Ellis, J., *Phys. Rev. Lett.* **40B** (1972) 397; *Phys. Rev.* **D7** (1973) 2490.

[44] Collins, J., Dunkan, L. and Joglekar, S., *Phys. Rev.* **D16** (1977) 438.

[45] Shifman, M. A., Vainshtein, A. I. and Zakharov, V. I., *Nucl. Phys.* **B147** (1979) 448.

[46] Ferrara, S. and Zumino, B., *Nucl. Phys.* **B87** (1985) 207.

[47] Shifman, M. A. and Vainshtein, A. I., *Nucl. Phys.* **B277** (1986) 456.

[48] Novikov, V. A., Shifman, M. A., Vainshtein, A. I. and Zakharov, V.I., *Nucl. Phys.* **B191** (1981) 301.

[49] Leutwyler, H. and Shifman, M., *Phys. Lett.* **B221** (1989) 384.

[50] Gorsky, A. S., Ioffe, B. L. and Khodjamirian, A. Yu., *Phys. Lett.* **227B** (1989) 474.

[51] Contopanagos, H. F. and Einhorn, M. B., *Phys. Rev.* **D45** (1992) 1291, 1322.

[52] Smilga, A. V., *Comments, Nucl. Part. Phys.* **20** (1991) 69.

[53] Kinoshita, T., *J. Math. Phys.* **3** (1962) 650.

[54] Lee, T. D. and Nauenberg, M., *Phys. Rev.* **133** (1964) B 1549.

[55] Gorsky, A. S. and Ioffe, B. L., *Part. World* **1** (1990) 114.

[56] Novikov, V. A. and Shifman, M. A., *Z. Phys.* C 8 (1981) 43.

4

Instantons and topological quantum numbers

Unlike QED, the vacuum state in QCD has nontrivial structure. In QCD vacuum there are nonperturbative fluctuations of gluon and quark fields. They are responsible for spontaneous violation of chiral symmetry and for the appearance of topological quantum numbers, which result in a complicated structure of an infinitely degenerate vacuum. The phenomenon of confinement is also attributed to these fluctuations.

Instantons were discovered in 1975 by Belavin, Polyakov, Schwarz, and Tyupkin [1]. They are the classical solutions for gluonic field in the vacuum, which indicate the nontrivial vacuum structure in QCD (papers on instantons are collected in [2]). In Euclidean gluodynamics (i.e. in QCD without quarks) at small g^2 they realize the minimum of action. The instantons carry new quantum numbers – the topological (or winding) quantum numbers n. There is an infinite set of minima of the action, labelled by the integer n and, as a consequence, an infinite number of degenerate vacuum states. In Minkowski space-time instantons represent the tunneling trajectory in the space of fields for transitions from one vacuum state to another. Therefore, the genuine vacuum wave function is a linear superposition of the wave functions of vacua of different n characterized by a parameter θ. This is analogous to the Bloch wave function of electrons in crystals – the so-called θ vacuum. θ is the analog of the electron quasimomentum in a crystal. The existence of a θ vacuum at $\theta \neq 0$ implies violation of CP-invariance in strong interactions, which is not observed until now. This problem waits for its solution.

If quarks are included in the theory, then, at least in the case of one massless quark flavour, instantons generate spontaneous breaking of chiral symmetry – the appearance of a nonvanishing value of the condensate of massless quarks.

The contributions of instantons to physical observables are proportional to $e^{-2\pi/\alpha_s}$ and can be reliably calculated if $e^{-2\pi/\alpha_s} \ll 1$. If this condition is not fulfilled, then the instanton contribution cannot be separated as a rule from other nonperturbative fluctuations. As is shown by lattice calculations, instantons are not responsible for confinement in QCD. Nevertheless, they can play a remarkable role in some physical effects, especially where chiral invariance is violated and/or the perturbative contributions are suppressed. The account of instantons is also useful for estimating nonperturbative terms not given by the operator product expansion. Such phenomena are considered in this and in the next chapters. Since instantons describe the tunneling transitions in non-Abelian gauge theories, we start from a discussion of similar phenomena in nonrelativistic quantum mechanics.

107

4.1 Tunneling in quantum mechanics

The tunneling phenomena – the penetration through potential barriers – are studied in quantum mechanics by using a quasiclassical approach: the Wentzel–Kramers–Brillouin (WKB) method. In case of a time-dependent potential, it is convenient to use the imaginary time method. This approach is a generalization of the Landau method of complex classical trajectories [3] and was introduced by Perelomov, Popov, Terentjev in 1966 [4] (a review of the modern state of the method is given in [5]). In the exposition of the method we follow the reviews [6],[7].

Consider the one-dimensional problem of a particle of mass m moving in the potential $V(x)$. The classical Lagrangian is equal to

$$L = \frac{1}{2} m \left(\frac{dx}{dt} \right)^2 - V(x). \tag{4.1}$$

(The energy scale is chosen such that $V(x) \geq 0$.) We are interested in the quantum mechanical transition amplitude, when in the initial state at $t = -t_0/2$ the wave function of the particle is equal to $\psi_i(x, -t_0/2)$ and in the final state at $t = t_0/2$ it is $\psi_f(x, t_0/2)$. The transition amplitude is given by

$$\langle f \mid e^{-iHt_0} \mid i \rangle = \sum_n e^{-iE_n t_0} \langle f \mid n \rangle \langle n \mid i \rangle, \tag{4.2}$$

where H is the Hamiltonian and the sum on the right-hand side of (4.2) is performed over the complete set of intermediate states. In order to separate the contribution of the lowest state (in field theory the vacuum state is of the main interest) put $t = -it'$ and go to the limit $t_0' \to \infty$. Then only the lowest state contribution survives on the right-hand side of (4.2):

$$\langle f \mid e^{-Ht_0'} \mid i \rangle = e^{-E_0 t_0'} \langle f \mid 0 \rangle \langle 0 \mid i \rangle. \tag{4.3}$$

It is convenient to calculate the transition amplitude by using the Feynman path integral method [8]. According to this method, the amplitude of transition of the particle from the point x_i at $t = -t_0/2$ to the point x_f at $t = t_0/2$ is equal to

$$\langle x_f \mid e^{-iHt_0} \mid x_i \rangle = N \int [\mathcal{D}x] e^{iS(x,t)}, \tag{4.4}$$

where S is the action:

$$S(x, t) = \int_{-t_0/2}^{t_0/2} L(x, t) dt \tag{4.5}$$

and the integration $[\mathcal{D}x]$ is performed over all functions $x(t)$ with boundary conditions $x(-\frac{t_0}{2}) = x_i, x(\frac{t_0}{2}) = x_f$. ($N$ is a normalization factor, inessential for us.) In order to get (4.2), Eq. (4.4) shall be multiplied by $\psi^*(x_f, t_0/2)\psi(x_i, -t_0/2)$ and integrated over x_f, x_i. Going to imaginary time in (4.5), we have:

$$iS(x, t) \rightarrow -\int_{-t_0'/2}^{t_0'/2} \left[\frac{m}{2}\left(\frac{dx}{dt'}\right)^2 + V(x)\right]dt' \equiv -S'(x, t'). \tag{4.6}$$

We will call S' the Euclidean action. (It what follows the primes in t' and S' will be omitted.) The energies and potentials will be measured in units of m and the time in units of $1/m$. So, the factor m will be omitted. The Euclidean action is positive. Suppose that it is large. This supposition corresponds to small $\hbar$, i.e. to quasiclassics in quantum mechanics and to a small coupling constant g in QCD. Then the main contribution to e^{-S} arises from the trajectory $X(t)$, where S is minimal, $S_{min} = S_0$:

$$N \int [\mathcal{D}x]e^{-S} \sim e^{-S_0}. \tag{4.7}$$

The equation for the extremal trajectory can be easily found from the requirement of the minimum of action, $\delta S = 0$:

$$\delta S = \int_{-t_0/2}^{t_0/2} dt\, \delta x(t)\left[-\frac{d^2 X}{dt^2} + V'(X)\right] = 0. \tag{4.8}$$

Since $\delta x(t)$ is arbitrary we get

$$\frac{d^2 X}{dt^2} = V'(X). \tag{4.9}$$

The boundary conditions for $X(t)$ are $X(-t_0/2) = x_i$, $X(t_0/2) = x_f$. Eq. (4.9) is the classical equation of motion in the potential $-V(x)$. In order to perform the path integration and, therefore, to find the pre-exponential factor in (4.4) we consider a deviation from the extremal trajectory

$$x(t) = X(t) + \eta(t), \tag{4.10}$$

hence

$$S[X(t) + \eta(t)] = S_0 + \int_{-t/2}^{t/2} dt\, \eta(t)\left[-\frac{1}{2}\frac{d^2\eta}{dt^2} + \frac{1}{2}V''(X)\eta\right]. \tag{4.11}$$

Let us expand $\eta(t)$ in the complete set of eigenfunctions normalized to 1 in the interval $t_0/2 > t > -t_0/2$ at $t_0 \rightarrow \infty$:

$$\eta(t) = \sum_n c_n \eta_n(t). \tag{4.12}$$

Choose $\eta_n(t)$ as the eigenfunctions of the equations

$$-\frac{d^2}{dt^2}\eta_n(t) + V''(X)\eta_n(t) = \varepsilon_n \eta_n(t) \tag{4.13}$$

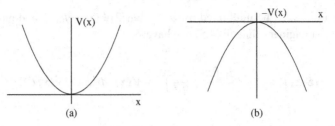

Fig. 4.1. The oscillator potential in real time (a) and in Euclidean space (b).

with eigenvalues ε_n. Then

$$S = S_0 + \frac{1}{2} \sum_n \varepsilon_n c_n^2. \tag{4.14}$$

The integration measure in (4.4) can be chosen as

$$[\mathcal{D}x] = \prod_n \frac{dc_n}{\sqrt{2\pi}}. \tag{4.15}$$

The substitution of (4.14) in (4.4) (after transition to imaginary time) gives

$$\langle x_f \mid e^{-Ht_0} \mid x_i \rangle = N e^{-S_0} \prod_n \varepsilon_n^{-1/2}. \tag{4.16}$$

In (4.16) it was tacitly assumed that all ε_n are positive, $\varepsilon_n > 0$. The case when some of the ε_n are zero will be considered below. Sometimes it is convenient to write the product of eigenvalues in the form of a determinant:

$$\prod_n \varepsilon_n^{-1/2} = \left[\det\left(-\frac{d^2}{dt^2} + V''[X(t)] \right) \right]^{-1/2}. \tag{4.17}$$

Consider the simplest example: the oscillator potential $V(x) = \omega^2 x^2/2$ in Minkowski space-time, $V(x) = -\omega^2 x^2/2$ in Euclidean space (see Fig. 4.1). Let a particle start its movement from the upper point of the hill in Fig. 4.1b at time $-t_0/2$ and return to the same point at time $+t_0/2$. The amplitude of this process calculated according to the rules presented above was found to be [7]

$$\langle x_f = 0 \mid e^{-Ht_0} \mid x_i = 0 \rangle = \left(\frac{\omega}{\pi} \right)^{1/2} (2\sinh\omega t_0)^{-1/2} \tag{4.18}$$

and at large t_0 is approximately equal to

$$\left(\frac{\omega}{\pi} \right)^{1/2} e^{-\frac{\omega t_0}{2}}. \tag{4.19}$$

As expected, the energy of the lowest state is equal to $E = \omega/2$.

Let us consider now the other example – the double well potential of Fig. 4.2. This potential can be represented as

$$V(x) = \lambda(x^2 - a^2)^2. \tag{4.20}$$

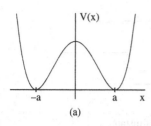 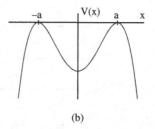

(a) (b)

Fig. 4.2. The double well potential in (a) real time (Eq.(4.1)), and (b) Euclidean expression for S (Eq. (4.6).)

Use the notation

$$8\lambda a^2 = \omega^2. \tag{4.21}$$

Classically, the lowest state of the particle in this potential is the state with zero energy at the bottom of one of the wells. The parameter ω is the energy of small oscillations around the bottom. Quantum mechanically, however, because of tunneling from one well to another, the particle simultaneously persists in both wells and the energies of the two lowest states are

$$E_0 = \frac{\omega}{2} - \sqrt{\frac{2\omega^3}{\pi\lambda}} e^{-\frac{\omega^3}{12\lambda}} \cdot \frac{\omega}{2},$$

$$E_1 = \frac{\omega}{2} + \sqrt{\frac{2\omega^3}{\pi\lambda}} \cdot e^{-\frac{\omega^3}{12\lambda}} \frac{\omega}{2}. \tag{4.22}$$

The exponentials in the second terms in (4.22) correspond to tunneling from one well to the other. According to the general theory one can expect that

$$S_0 = \omega^3/12\lambda. \tag{4.23}$$

Let us apply the general method presented above and prove that this is the case. We assume that $12\lambda/\omega^3 \ll 1$ and S_0 given by (4.23) is large. The transition amplitude from the left well to the right one is

$$\langle a \mid e^{-Ht_0} \mid -a\rangle \tag{4.24}$$

and we want to find the classical trajectory, realizing the minimum of the action S_0 in Euclidean space-time. For simplicity, we restrict ourselves to the case of zero energy, $E = 0$. In Euclidean space-time, the energy is equal to

$$E = \frac{1}{2}\left(\frac{dx}{dt}\right)^2 - V(x). \tag{4.25}$$

(See (4.6)). At $E = 0$

$$\frac{dx}{dt} = \sqrt{2V(x)} = \sqrt{2\lambda}(a^2 - x^2). \tag{4.26}$$

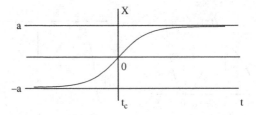

Fig. 4.3. The graph of instanton.

The situation of interest to us is when x is near the bottom of the well. So, the sign on the right-hand side of (4.26) was chosen in accord with this condition. The solution of Eq. (4.26) representing the extremal trajectory is equal to

$$X(t) = a \tanh\frac{\omega}{2}(t - t_c),\tag{4.27}$$

where t_c is an arbitrary constant. Graphically, the solution is shown in Fig. 4.3. Usually this is called the instanton (or soliton).

As is seen from (4.27) and Fig. 4.3, we have $X(t)_{|t\to-\infty} \to -a$ and $X(t)_{|t\to\infty} \to a$. At large $t \gg 1/\omega$, the instanton is a localized object:

$$X(t) - a \approx -2e^{-\omega(t-t_c)}.\tag{4.28}$$

In the limit $t_0 \to \infty$, the action is equal to

$$S_0 = \int_{-\infty}^{\infty} dt \left(\frac{dX}{dt}\right)^2 = \frac{\omega^3}{12\lambda},\tag{4.29}$$

which coincides with (4.23).

There is also a solution which corresponds to the transition from a to $-a$. It can be obtained from (4.27) by replacing t by $-t$ and is called anti-instanton. The parameter t_c in (4.27) corresponds to the position of the center of the instanton. The action S_0 (4.29) is independent of the position of the center. Therefore, one can expect that in order to get the transition amplitude (4.24), the factor e^{-S_0} should be integrated over t_c. Let us present the formal proof of this statement.

In this particular case, the general eigenvalue equations (4.13) have the form

$$-\frac{d^2}{dt^2}\eta_n(t) + \left(\omega^2 - \frac{3}{2}\frac{\omega^3}{\cosh^2\frac{\omega t}{2}}\right)\eta_n(t) = \varepsilon_n\eta_n(t).\tag{4.30}$$

(t_c was temporarily put equal to zero). The boundary conditions are $\eta_n(\pm\frac{t_0}{2}) = 0$ at $t_0 \to \infty$. These boundary conditions are fulfilled for discrete levels. There are two discrete eigenvalues of Eq. (4.30) with $\varepsilon_0 = 0$ and $\varepsilon_1 = (3/4)\omega^2$ [9]. The normalized wave function corresponding to $\varepsilon_0 = 0$ is equal to

$$\eta_0(t) = \sqrt{\frac{3\omega}{8}}\frac{1}{\cosh^2\frac{\omega t}{2}}.\tag{4.31}$$

The general formula of functional integration (4.16) cannot be applied to $\eta_0(t)$ – the zero-mode wave function. The existence of the zero mode means that the perturbations proportional to its eigenfunctions do not change the action. On the other hand, the variation of the instanton center does not change the action either. Since there is only one zero mode, one can expect that $\eta_0(t)$ is proportional to the variation of $X(t - t_c)$:

$$\eta_0(t) = A\frac{dX}{dt_c} \equiv -A\frac{dX}{dt}. \tag{4.32}$$

The substitution of (4.32) into (4.29) gives

$$A = S_0^{-1/2}. \tag{4.33}$$

It is convenient to perform the integration over t_c instead of integration over the coefficient c_0 in the functional integration $[\mathcal{D}x]$ (4.15). Using (4.12) we have:

$$\Delta\eta(t) = \eta_0(t)\Delta c_0. \tag{4.34}$$

On the other hand,

$$\Delta\eta(t) = \Delta X(t) = \frac{dX}{dt_c}\Delta t_c = S_0^{1/2}\eta_0(t)\Delta t_c. \tag{4.35}$$

Equating (4.34) with (4.35), we get:

$$dc_0 = S_0^{1/2}dt_c. \tag{4.36}$$

Finally, for the transition amplitude (4.24) we have

$$\langle a \mid e^{-Ht_0} \mid -a \rangle = \sqrt{\frac{S_0}{2\pi}}dt_c e^{-S_0} \left\{ \det'\left(-\frac{d^2}{dt^2} + V''[X(t)]\right)\right\}^{-1/2}, \tag{4.37}$$

where $\det'$ denotes that the zero-mode contribution is omitted in the determinant.

The final result with account of nonzero modes is equal to (at large t_0, $\omega t_0 \gg 1$) [7]

$$\langle +a \mid e^{-Ht_0} \mid -a \rangle = d\left(\frac{\omega}{\pi}\right)^{1/2} e^{-\frac{\omega t_0}{2}} \omega dt_c, \tag{4.38}$$

where

$$d = \left(\frac{6}{\pi}\right)^{1/2} \sqrt{S_0}e^{-S_0} \tag{4.39}$$

can be considered as the density of instantons.

The action is independent of t_c and the integration over t_c was left in (4.37), (4.38). Such variables which leave the action invariant are called collective coordinates. The integration over collective coordinates is not performed in the calculation of the determinant. Note that the factor $\sqrt{S_0}$ corresponds to the integration over t_c. This is a general statement which applies to the case of QCD instantons as well. The function $X(t)$ (4.27), which corresponds to the instanton, goes to the limits $\pm a$ at $t \to \pm\infty$. If there are n instantons with different centers t_{ci} and the positions of the centers satisfy the inequalities $\omega \mid t_{ci} - t_{cj} \mid \gg 1$, then we can sum their trajectories and evidently $X(t) \to \pm na$ at $t \to \pm\infty$. In this case, the

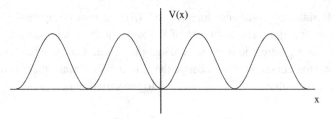

Fig. 4.4. The periodical potential in the real time.

minimal action is equal to nS_0. It is possible to introduce the topological classification of the functions, which realize the minima of the action, by their limiting values. The topological charge can be defined by

$$Q = \frac{1}{2a} \int_{-\infty}^{\infty} dy \frac{dx(t)}{dt} = \frac{x(+\infty) - x(-\infty)}{2a}. \tag{4.40}$$

Q is an integer, $Q = 0, \pm 1, \ldots$ The functions with different Q cannot be transformed into each other by any continuous deformation. The amplitude of transition from the bottom of the first well to the bottom of the second well caused by n widely separated instantons is equal to (n is odd)

$$\frac{\omega}{\pi}^{1/2} e^{-\frac{\omega t_0}{2}} d^n \int_{-t_0/2}^{t_0/2} \omega dt_{c,n} \int_{-t_0/2}^{t_{c,n}} \omega dt_{c,n-1} \ldots \int_{-t_0/2}^{t_{c,2}} \omega dt_{c,1} = \left(\frac{\omega}{\pi}\right)^{1/2} e^{-\frac{\omega t_0}{2}} d^n \frac{(\omega t_0)^n}{n!}. \tag{4.41}$$

The total amplitude is obtained by summing the contributions of all n:

$$\langle +a \mid e^{-Ht_0} \mid -a \rangle = \sum_{n=1,3\ldots}^{\infty} \left(\frac{\omega}{\pi}\right)^{1/2} e^{-\frac{\omega t_0}{2}} \frac{(\omega t_0 d)^n}{n!} = \left(\frac{\omega}{\pi}\right)^{1/2} e^{-\frac{\omega t_0}{2}} \sinh(\omega t_0 d). \tag{4.42}$$

The energies of the two lowest states, determined from (4.42) by going to the limit $t_0 \to \infty$, coincide with (4.22).

An instructive example which is useful in the consideration of the QCD vacuum is the periodical potential, Fig. 4.4.

The amplitude of transition from $-a_j$ at $-t_0/2$ to the point $+a_j$ at $t_0/2$, caused by n instantons and $\bar{n}$ anti-instantons is given by

$$\langle +a_j \mid e^{-Ht_0} \mid -a_i \rangle = \left(\frac{\omega}{\pi}\right)^{1/2} e^{-\omega t_0/2} \sum_{n,\bar{n}} d^n d^{\bar{n}} \frac{(\omega t_0)^n}{n!} \frac{(\omega t_0)^{\bar{n}}}{\bar{n}!} \delta_{n-\bar{n}-\Delta}, \tag{4.43}$$

where

$$\Delta = n(+a_j) - n(-a_i)$$

is the difference of the winding numbers of a_j and a_i states. The Kronecker symbol can be represented as

$$\delta_{n-\bar{n}-\Delta} = \int_0^{2\pi} e^{i(n-\bar{n}-\Delta)\theta} \frac{d\theta}{2\pi}.$$

(4.44)

The substitution of (4.44) in (4.43) and the summation over $n, \bar{n}$ leads to

$$\langle +a_j \mid e^{-H_0 t} \mid -a_i \rangle = \left(\frac{\omega}{\pi}\right)^{1/2} e^{-\frac{\omega t_0}{2}} \int_0^{2\pi} e^{i\Delta\theta} \frac{d\theta}{2\pi} \exp[2d\omega t_0 \cos\theta].$$

(4.45)

The energy of the lowest state is now equal to

$$E = -\frac{\omega}{2} + 2\omega d \cos\theta.$$

(4.46)

It is tacitly implied that the second term in (4.46) must be integrated over θ, according to (4.45). We found that in the periodic potential instead of discrete levels, energies are distributed over the bands and the widths of the bands are determined by the difference of winding numbers Δ. The situation here closely resembles the movement of electrons in periodic crystals, where the energies of the electrons are grouped in bands and the wave function of the electrons is the superposition of different solutions of the Schrödinger equation, characterized by one common vector – the quasimomentum (Bloch waves). Here, the role of quasimomentum is played by the parameter θ. As we shall see, a similar situation takes place in QCD. We will not go into more details of quantum mechanical models, referring the interested readers to Refs. [6],[7],[10].

4.2 Instantons and the topological current

4.2.1 Instantons in Euclidean QCD

As was demonstrated in the previous section by consideration of quantum mechanical problems, instantons correspond to the tunneling phenomena and the most suitable method of the treatment of problems is the imaginary time formalism. A similar situation takes place in QCD. Here it is convenient to work in Euclidean space-time. The rules for how to go from QCD, formulated in Minkowski space-time, to Euclidean QCD are the following: The spatial coordinates x_1, x_2, x_3 are not changed. The time coordinate x_0 is replaced by $-ix_4$,

$$x_0 \rightarrow -ix_4.$$

(4.47)

The differential operators ∂_μ are defined simply:

$$\partial_\mu = \left(\frac{\partial}{\partial x_4}, \frac{\partial}{\partial x_i}\right).$$

(4.48)

Such a definition allows one to have the standard form of the momentum operator:

$$p_\mu = -i\frac{\partial}{\partial x_\mu}.$$ (4.49)

In order to have the same form of the covariant derivative D_μ as it was in Minkowski space-time,

$$(D_\mu)^{ab} = \partial_\mu \delta^{ab} + gf^{abc}A_\mu^c,$$ (4.50)

it is necessary to put

$$A_i^{Eucl} = -A_i^{Mink}, \quad i = 1, 2, 3; \quad A_4^{Eucl} = -iA_0^{Mink}.$$ (4.51)

For the field strengths $G_{\mu\nu}^n$, we have:

$$(G_{ik}^n)^{Eucl} = (G_{ik}^n)^{Mink}, \quad (G_{4i}^n)^{Eucl} = i(G_{0i}^n)^{Mink}$$ (4.52)

and $(G_{\mu\nu}^n)^{Eucl}$ is expressed in terms of ∂_μ^{Eucl} and A_μ^{Eucl} by the same formula as in Minkowski space-time. It is convenient to define the action S as

$$S = \frac{1}{4}\int d^4x G_{\mu\nu}^2.$$ (4.53)

Then $S > 0$ in Euclidean space-time and transition probabilities are given by the matrix elements of e^{-S}.

In the case of fermions, the Dirac γ-matrices in Euclidean space-time are given by:

$$\gamma_4^E = \gamma_0, \quad \gamma_i^E = -i\gamma_i^M$$ (4.54)

and their anticommutation relations are:

$$\{\gamma_\mu, \gamma_\nu\} = 2\delta_{\mu\nu}.$$ (4.55)

The γ_5 matrix is defined as

$$\gamma_5^E = \gamma_5^M = i(\gamma_1\,\gamma_2\,\gamma_3\,\gamma_4)^M$$ (4.56)

and is Hermitian. The ψ and $\bar\psi$ operators obey the anticommutation relations

$$\{\psi(x), \psi(y)\} = \{\bar\psi(x), \bar\psi(y)\} = \{\psi(x), \bar\psi(y)\} = 0,$$ (4.57)

$\bar\psi$ is the Hermitian conjugate of ψ, $\bar\psi = \psi^+$. The last relation in (4.57) is essentially different from that in Minkowski space-time, where the anticommutator of ψ and $\bar\psi$ is proportional to $\delta(x - y)$ at $x_0 = y_0$. Because of (4.57) ψ and $\bar\psi$ must be treated as independent variables, particularly in the calculation of functional integrals. The Euclidean action for fermion fields is equal to

$$S = -\int d^4x[\bar\psi(i\gamma_\mu\nabla_\mu + im)\psi],$$ (4.58)

where

$$\nabla_\mu = \partial_\mu + ig\frac{\lambda^n}{2}A_\mu^n.$$ (4.59)

The sign in front of Eq. (4.58) is chosen in order to have Re $S > 0$ for free fermions at rest. The factor i in front of m ensures the absence of tachyon poles: the free propagator that corresponds to (4.58) is equal to

$$\frac{\not{p} + im}{p^2 + m^2} \tag{4.60}$$

and has no poles (recall that Euclidean p^2 is always positive.) Note that in Euclidean space-time the definition of the propagator as a T-product of $\psi(x)$, $\bar{\psi}(y)$ operators fails, but its definition as a Green function of the Dirac equation remains.

Consider now Euclidean gluodynamics, i.e. QCD without quarks. The action can be represented in the form of

$$S = \frac{1}{4} \int d^4x\, G^n_{\mu\nu} G^n_{\mu\nu} = \frac{1}{4} \int d^4x \left[G^n_{\mu\nu} \tilde{G}^n_{\mu\nu} + \frac{1}{2}(G^n_{\mu\nu} - \tilde{G}^n_{\mu\nu})^2 \right], \tag{4.61}$$

where $\tilde{G}^n_{\mu\nu} = (1/2)\varepsilon_{\mu\nu\lambda\sigma} G^n_{\lambda\sigma}$, $\varepsilon_{1234} = 1$. Since the last term in (4.61) is positive, the minimum of the action is realized on the self-dual gluonic fields,

$$G^n_{\mu\nu} = \tilde{G}^n_{\mu\nu}, \tag{4.62}$$

$$S_{\min} = \frac{1}{4} \int d^4x\, G^n_{\mu\nu} \tilde{G}^n_{\mu\nu}. \tag{4.63}$$

The integrand in (4.63) is the total divergence

$$G^n_{\mu\nu} \tilde{G}^n_{\mu\nu} = \partial_\mu K_\mu, \tag{4.64}$$

$$K_\mu = 2\varepsilon_{\mu\nu\gamma\delta} \left(A^n_\nu \partial_\gamma A^n_\delta - \frac{1}{3} g f^{nmp} A^n_\nu A^m_\gamma A^p_\delta \right) \tag{4.65}$$

or

$$K_\mu = \varepsilon_{\mu\nu\gamma\delta} \left(A^n_\nu G^n_{\gamma\delta} - \frac{1}{3} f^{nmp} A^n_\nu A^m_\gamma A^p_\delta \right). \tag{4.66}$$

Substitute (4.64) into (4.63). Assuming that $G^n_{\mu\nu}$ has no singularities at finite x, we can transform the four-dimensional integral in (4.63) into an integral over the infinitely remote three-dimensional sphere:

$$S_{\min} = \frac{1}{4} \int dV\, K_4. \tag{4.67}$$

Evidently, the action is finite only in the case when the field strengths $G^n_{\mu\nu}$ are decreasing faster than $1/x^2$ at infinity. Therefore, the contribution of the first term in (4.66) to (4.67) vanishes and we have

$$S_{\min} = -\frac{1}{12} g\varepsilon_{ikl} \int dV f^{nmp} A^n_i A^m_k A^p_l. \tag{4.68}$$

The potentials A_μ^n may not decrease faster than $1/x$ at infinity, but they should represent a pure gauge. Let us show that $S_{\min}$ is proportional to an integer n – the topological charge or winding number:

$$S_{\min} = \frac{8\pi^2}{g^2}n = \frac{2\pi}{\alpha_s}n. \tag{4.69}$$

This important fact was established in the Ref. [1]. (In the derivation we follow Ref. [11].) Suppose that the gluonic fields belong to $SU(2)$, the subgroup of the $SU(3)$ colour group. Then the group constants f^{nmp} reduce to the three-dimensional totally antisymmetric tensor ε^{nmp}. Use the matrix notation

$$A_i = -\frac{1}{2}g\tau^n A_i^n, \tag{4.70}$$

where τ^n are Pauli matrices,

$$\text{Tr}\,(\tau^m \tau^n) = 2\delta_{mn}. \tag{4.71}$$

Instead of (4.68) we have now:

$$S_{\min} = -\frac{i}{3}\frac{1}{g^2}\int dV\,\varepsilon_{ikl}\,\text{Tr}\,(A_i A_k A_l). \tag{4.72}$$

Since A_i is a pure gauge, only the second term survives in the general expression of the gauge transformation (see Chapter 1):

$$A_i' = U^{-1}A_i U + iU^{-1}\partial_i U, \tag{4.73}$$

where U is an x-dependent unitary unimodular matrix. The substitution of the last term of (4.73) into (4.72) gives

$$\begin{aligned}
S_{\min} &= -\frac{1}{3}\frac{1}{g^2}\int dV\,\varepsilon_{ikl}\,\text{Tr}\left[U^{-1}\partial_i U \cdot U^{-1}\partial_k U \cdot U^{-1}\partial_l U\right]\\
&= \frac{1}{3}\frac{1}{g^2}\int dV\,\varepsilon_{ikl}\,\text{Tr}\left[\partial_i U^{-1}\partial_k \cdot U^{-1}\partial_l U\right].
\end{aligned} \tag{4.74}$$

(The last equation of (4.74) follows from $\partial_i U^{-1}U = 0$.) We assume that $A_i(r)$ falls faster than $1/r$ at spatial infinity. This means that U is a constant matrix at $r \to \infty$, which without loss of generality may be taken equal to $\pm I$. Let us parametrize U by

$$U = \exp(\theta^a \tau^a /2i), \tag{4.75}$$

where $\theta^a(r)$ are functions of r. The calculation of the trace in (4.74) results in

$$S_{\min} = -\frac{1}{2}\varepsilon_{ikl}\varepsilon^{abc}\frac{1}{g^2}\int dV\,\partial_i\left[\hat{\theta}^a \partial_k\hat{\theta}^b \partial_l\hat{\theta}^c(\sin\theta - \theta)\right], \tag{4.76}$$

where

$$\theta = \sqrt{\theta^a\theta^a}, \qquad \hat{\theta}^a = \theta^a/\theta. \tag{4.77}$$

(Since $\hat{\theta}^a$ is a unit vector, $\varepsilon^{abc}\partial_i\hat{\theta}^a \partial_k\hat{\theta}^b \partial_l\hat{\theta}^c = 0$. This was exploited in the derivation of Eq. (4.76).) The integrand in (4.76) is a total derivative. Using Gauss' theorem we can represent the volume integral in (4.76) as the integral over the surface of an infinitely large sphere. The condition $U \to \pm I$ at $r \to \infty$ means that $\theta \to \pm 2\pi n$ at $r \to \infty$. We get

$$S_{\min} = \frac{\pi n}{g^2} \varepsilon_{ikl} \varepsilon^{abc} \int dS_i \left(\hat{\theta}^a \partial_k \hat{\theta}^b \partial_l \theta^c \right), \tag{4.78}$$

where the integration is performed over the surface of the sphere of infinitely large radius r. Suppose first, for simplicity, that $\hat{\theta}^a$ is the unit vector in the direction of $\mathbf{r}$, $\hat{\theta}^a = r^a/r$. Then the integral in (4.78) is calculated directly and we have

$$S_{\min} = \frac{8\pi^2 n}{g^2} = \frac{2\pi}{\alpha_s} n. \tag{4.79}$$

In a more general case, we can parametrize $\hat{\theta}^a$ as

$$\hat{\theta}^1 = \sin\psi \cos\varphi, \quad \hat{\theta}^2 = \sin\psi \sin\varphi, \quad \hat{\theta}^3 = \cos\psi, \tag{4.80}$$

where ψ and φ are functions of the polar and azimuthal angles α and β. In this case, $S_{\min}$ is given by:

$$S_{\min} = \frac{2\pi n}{g^2} \int_0^{2\pi} d\beta \int_0^\pi d\alpha \sin\psi \left(\frac{\partial\psi}{\partial\alpha} \frac{\partial\varphi}{\partial\beta} - \frac{\partial\psi}{\partial\beta} \frac{\partial\varphi}{\partial\alpha} \right). \tag{4.81}$$

The quantity in parenthesis is the Jacobian of the transformation from (α, β) to (ψ, φ). Therefore, the integral is the integer that counts the number of times the pair (ψ, φ) covers its unit sphere as (α, β) covers its unit sphere once. The integers n in (4.69), (4.79) can be positive or negative; the minima of the action correspond to positive n.

The proof of Eq. (4.69) can be based on very general arguments, as was originally done in Ref. [1]. The group space of the $SU(2)$ matrix U is S_3, a three-dimensional sphere. Hence, the S_3 set in group space is mapped onto S_3 in coordinate space. It is obvious that all continuous mappings $S_3 \to S_3$ fall into different classes corresponding to different coverings of the group manifold by the coordinates on the sphere. Evidently, a mapping that is characterized by a given winding number n cannot be transformed by continuous transformation into a mapping with another n. The $SU(2)$ subgroup considered above plays a special role in the $SU(3)$ colour group: the winding numbers and instanton solutions in $SU(2)$ are defined by separations of $SU(2)$ subgroups in various ways from $SU(3)$. Its special role is connected with the isomorphism of $SU(2)$ with the group of rotations in three-dimensional space.

Find now the self-dual gluonic fields realizing the minimum of the action. (We follow here Ref. [7].) Note that self-dual fields satisfy the equations of motion. Indeed,

$$(D_\mu G_{\mu\nu})^n = (D_\mu \tilde{G}_{\mu\nu})^n = \frac{1}{2} \varepsilon_{\mu\nu\lambda\sigma} (D_\mu G_{\mu\nu})^n$$

$$= \frac{1}{6} \varepsilon_{\mu\nu\lambda\sigma} (D_\mu G_{\lambda\sigma} + D_\lambda G_{\sigma\mu} + D_\sigma G_{\mu\lambda})^n = 0, \tag{4.82}$$

because of the Bianchi identity

$$D_\mu G_{\lambda\sigma} + D_\lambda G_{\sigma\mu} + D_\sigma G_{\mu\lambda} = 0. \tag{4.83}$$

The anti-self-dual fields $G_{\mu\nu} = -\tilde{G}_{\mu\nu}$, which also satisfy the equations of motion, correspond to negative winding numbers n and realize the minimum of the action equal to $(8\pi^2/g^2) \mid n \mid$. Consider an $SU(2)$ subgroup of the $SU(3)$ colour group and introduce 't Hooft symbols $\eta_{a\mu\nu}$ and $\bar{\eta}_{a\mu\nu}$ ($a = 1, 2, 3$ are isospin indices; $\mu, \nu = 1, 2, 3, 4$ are space-time indices) [12]:

$$\eta_{a\mu\nu} = \begin{cases} \varepsilon_{a\mu\nu}, & \mu, \nu = 1, 2, 3 \\ -\delta_{a\nu} & \mu = 4 \\ \delta_{a\mu} & \nu = 4 \\ \eta_{a44} = 0. \end{cases} \tag{4.84}$$

(The symbols $\bar{\eta}_{a\mu\nu}$ differ from $\eta_{a\mu\nu}$ by changing the sign of the δ-terms. Useful relations for $\eta_{a\mu\nu}$ and $\bar{\eta}_{a\mu\nu}$ are presented in the appendix to this chapter.) The generators of rotations in four-dimensional Euclidean space are of the following form [7] (see also [13]):

$$J_1^a = \frac{1}{4}\eta_{a\mu\nu}M_{\mu\nu},$$

$$J_2^a = \frac{1}{4}\bar{\eta}_{a\mu\nu}M_{\mu\nu}, \tag{4.85}$$

where

$$M_{\mu\nu} = -ix_\mu\frac{\partial}{\partial x_\nu} + ix_\nu\frac{\partial}{\partial x_\mu} + \text{spin terms} \tag{4.86}$$

are the operators of infinitesimal rotations in the (μ, ν) plane. The isospin τ-matrices are generalized to four dimensions as:

$$\tau_\mu^{\pm} = (\boldsymbol{\tau}, \mp i). \tag{4.87}$$

They satisfy the relations

$$\tau_\mu^+\tau_\nu^- = \delta_{\mu\nu} + i\eta_{a\mu\nu}\tau^a,$$

$$\tau_\mu^-\tau_\nu^+ = \delta_{\mu\nu} + i\bar{\eta}_{a\mu\nu}\tau^a. \tag{4.88}$$

Under action of the generators of rotations the matrix $\tau_\mu^+ x_\mu$ transforms in the following way:

$$e^{i\varphi_1^a J_1^a + i\varphi_2^a J_2^a}\tau_\mu^+ x_\mu = e^{-i\varphi_1^a \tau^a/2}(\tau_\mu^+ x_\mu)e^{i\varphi_2^a \tau^a/2}, \tag{4.89}$$

where φ_1^a and φ_2^a are the rotation parameters. As is seen from (4.89), $\tau_\mu^+ x_\mu$ is not invariant under rotations in four-dimensional Euclidean space. However, if this rotation will be supplemented by a rotation in the $SU(2) \times SU(2)$ isospin group (cf. Eq. (2.66) in Chapter 2), then $\tau_\mu^+ x_\mu$ becomes invariant. Since all observable objects in gauge theory are defined up to a gauge transformation, we can consider

$$U = i\tau_\mu^+ x_\mu / \sqrt{x^2} \tag{4.90}$$

as a suitable unitary unimodular matrix which determines the asymptotics of the gluonic field. In terms of 't Hooft symbols, the asymptotic expression of A_μ^a has the form of

$$A_\mu^a(x)_{x\to\infty} = -\frac{2}{g}\eta_{a\mu\nu}\frac{x_\nu}{x^2}. \tag{4.91}$$

Let us search for a solution of the self-duality equation (4.62) by putting

$$A_\mu^a(x) = -\frac{2}{g}\eta_{a\mu\nu}\frac{x_\nu}{x^2}f(x^2). \tag{4.92}$$

The boundary conditions for $f(x^2)$ are: $f(x^2)_{x^2\to\infty} = 1$, $f(x^2)_{x^2\to 0} \to \text{Const}\cdot x^2$. From (4.92) we have for the field strengths:

$$G_{\mu\nu}^a = \frac{4}{g}\left\{\eta_{a\mu\nu}\frac{f(1-f)}{x^2} + \frac{1}{x^4}[x_\mu\eta_{a\nu\lambda}x_\lambda - x_\nu\eta_{a\mu\lambda}x_\lambda][f(1-f) - x^2 f']\right\}, \tag{4.93}$$

$$\tilde{G}_{\mu\nu}^a = \frac{4}{g}\left\{\eta_{a\mu\nu}f' - \frac{1}{x^4}[x_\mu\eta_{a\nu\lambda}x_\lambda - x_\nu\eta_{a\mu\lambda}x_\lambda][f(1-f) - x^2 f']\right\}. \tag{4.94}$$

(The relations from the appendix were exploited.) It follows from the self-duality condition that

$$f(1-f) - x^2 f' = 0. \tag{4.95}$$

The solution of (4.95) is found to be

$$f(x^2) = \frac{x^2}{x^2 + \rho^2}. \tag{4.96}$$

The integration constant ρ is called the instanton radius. The gluonic field and the instanton strength are equal to

$$A_\mu^a(x) = -\frac{2}{g}\eta_{a\mu\nu}\frac{x_\nu}{x^2 + \rho^2}, \tag{}$$

$$G_{\mu\nu}^a(x) = \frac{4}{g}\eta_{a\mu\nu}\frac{\rho^2}{(x^2 + \rho^2)^2}. \tag{4.97}$$

The solutions (4.97) correspond to the position of the instanton center at the origin. In order to get the field with instanton center at x_c it is necessary to replace x by $x - x_c$ in (4.97). Using (4.97) and the relation $\eta_{a\mu\nu}\eta_{a\mu\nu} = 12$ it is easy to calculate the action. The result is: $S = 8\pi^2/g^2$, i.e. the instanton represents the gluonic field, corresponding to the winding number $n = 1$. The anti-instanton solution can be found from (4.97) by replacing $\eta_{a\mu\nu}$ by $\bar{\eta}_{a\mu\nu}$ and results in the same value of the action. Sometimes it is convenient to use the so-called singular gauge for the instanton field, when the singularity appears at the instanton centre. The transformation to the singular gauge can be done with the help of the formulae

$$g\frac{\tau^a}{2}(A_\mu^a)_{sing} = U^+ g\frac{\tau^a}{2}A_\mu^a U - iU^+\partial_\mu U,$$

$$g\frac{\tau^a}{2}(G_{\mu\nu}^a)_{sing} = U^+ g\frac{\tau^a}{2}G_{\mu\nu}^a U, \tag{4.98}$$

where A_μ^a, $G_{\mu\nu}^a$ are given by (4.97) and

$$U = i \frac{\tau_\mu^+ x_\mu}{\sqrt{x^2}}. \tag{4.99}$$

$(A_\mu^a)_{sing}$ and $(G_{\mu\nu}^a)_{sing}$ are found to be

$$(A_\mu^a)_{sing} = -\frac{2}{g} \bar\eta_{a\mu\nu} x_\nu \frac{\rho^2}{x^2(x^2 + \rho^2)},$$

$$(G_{\mu\nu}^a)_{sing} = \frac{8}{g} \bar\eta_{a\nu\rho} \left(\frac{x_\mu x_\lambda}{x^2} - \frac{1}{4} \delta_{\mu\lambda} \right) \frac{\rho^2}{(x^2 + \rho^2)^2} - (\mu \longleftrightarrow \nu). \tag{4.100}$$

The expression (4.100) for $(A_\mu^a)_{sing}$ can be represented as

$$(A_\mu^a)_{sing} = \frac{1}{g} \bar\eta_{a\mu\nu} \partial_\nu \ln\left(1 + \frac{\rho^2}{x^2} \right). \tag{4.101}$$

Eq. (4.101) can be generalized to the case of winding numbers $n > 1$ if

$$W = 1 + \sum_{i=1}^{n} \frac{\rho_i^2}{(x - x_{ic})^2} \tag{4.102}$$

would be substituted under the logarithm sign [12]. Here x_{ic} is the position of the center of i-th instanton.

4.2.2 The topological current

The existence of topological quantum numbers is a very specific feature of non-Abelian quantum field theories and, in particular, of QCD. Therefore, the study of properties of the topological charge density operator in QCD

$$Q_5(x) = \frac{\alpha_s}{8\pi} G_{\mu\nu}^n(x) \tilde{G}_{\mu\nu}^n(x) \tag{4.103}$$

and of the corresponding vacuum correlator

$$\zeta(q^2) = i \int d^4x e^{iqx} \langle 0|T\{Q_5(x),\ Q_5(0)\}|0\rangle \tag{4.104}$$

is of great theoretical interest. It follows from (4.63), (4.69) that

$$\int d^4x\, Q_5(x) = n \tag{4.105}$$

is the topological quantum number. Crewther [14] has derived Ward identities related to $\zeta(0)$, which allowed him to prove the theorem that $\zeta(0) = 0$ in any theory where there is at least one massless quark. An important step in the investigation of the properties of $\zeta(q^2)$ was achieved by Veneziano [15] and Di Veccia and Veneziano [16]. These authors considered the limit $N_c \to \infty$. Assuming that in the theory there are N_f light quarks with masses $m_i \ll M$, where M is the characteristic scale of strong interaction, Di Veccia and Veneziano found that

$$\zeta(0) = \langle 0 \mid \bar{q}q \mid 0 \rangle \left(\sum_i^{N_f} \frac{1}{m_i} \right)^{-1}, \qquad (4.106)$$

where $\langle 0 \mid \bar{q}q \mid 0 \rangle$ is the common value of the quark condensates for all light quarks and terms of order m_i/M are neglected. The concept of θ-term in the Lagrangian (see below Eq. (4.191)) was successfully exploited in [16] in deriving (4.106). Using the same concept and studying the properties of the Dirac operator, Leutwyler and Smilga [17] succeeded in proving (4.106) at any N_c for the case of two light quarks, u and d. The proof of Eq. (4.106) will be given here for the case of two light quarks u and d using the chiral effective theory (Chapter 2).

Consider QCD with N_f light quarks, $m_i \ll M \sim 1$ GeV, $i = 1, \ldots N_f$. Define the flavour singlet axial current by

$$j_{\mu 5}(x) = \sum_i^{N_f} \bar{q}_i(x)\gamma_\mu \gamma_5 q_i(x) \qquad (4.107)$$

and the polarization operator

$$P_{\mu\nu}(q) = i \int d^4 x e^{iqx} \langle 0 \mid T\{j_{\mu 5}(x), j_{\mu 5}(0)\} \mid 0 \rangle. \qquad (4.108)$$

The general form of the polarization operator is:

$$P_{\mu\nu}(q) = -P_L(q^2)\delta_{\mu\nu} + P_T(q^2)(-\delta_{\mu\nu}q^2 + q_\mu q_\nu). \qquad (4.109)$$

Because of the anomaly the singlet axial current is nonconserving:

$$\partial_\mu j_{\mu 5}(x) = 2N_f Q_5(x) + D(x), \qquad (4.110)$$

where $Q_5(x)$ is given by (4.103) and

$$D(x) = 2i \sum_i^{N_f} m_i \bar{q}_i(x)\gamma_5 q_i(x). \qquad (4.111)$$

It is well known that even if some light quarks are massless, the corresponding Goldstone bosons that arise from spontaneous chiral symmetry violation do not contribute to the singlet axial channel, i.e. to the polarization operator $P_{\mu\nu}(q)$. $P_L(q^2)$ also have no kinematical singularities at $q^2 = 0$. Therefore,

$$P_{\mu\nu}(q)q_\mu q_\nu = -P_L(q^2)q^2 \qquad (4.112)$$

vanishes in the limit $q^2 \to 0$. Calculate the left-hand side of (4.112) in the standard way: put $q_\mu q_\nu$ inside the integral in (4.108) and integrate by parts. (To do this, it is convenient to represent the polarization operator in coordinate space as a function of two coordinates x and y.) Going to the limit $q^2 \to 0$ we have

$$\lim_{q^2 \to 0} P_{\mu\nu}(q)q_\mu q_\nu = i \int d^4x \langle 0 \mid T\{2N_f Q_5(x),\ 2N_f Q_5(0)\}$$

$$+ T\{2N_f Q_5(x), D(0)\} + T\{D(x), 2N_f Q_5(0)\} + T\{D(x), D(0)\} \mid 0\rangle$$

$$+ 4 \sum_i^{N_f} m_i \langle\mid \bar{q}_i(0)q_i(0) \mid 0\rangle$$

$$+ \int d^4x \langle 0 \mid [j_{05}(x), 2N_f Q_5(0)] \mid 0\rangle \delta(x_0) = 0. \tag{4.113}$$

In the calculation of (4.113) the anomaly condition (4.110) was used. The terms proportional to quark condensates arise from the equal-time commutator $[j_{05}(x), D(0)]_{x_0=0}$ calculated by standard commutation relations. Relation (4.113), up to the last term, was first obtained by Crewther [14]. The last term, which is equal to zero according to standard commutation relations and is omitted in [14]–[17], is kept. The reason is that we deal with a very subtle situation related to anomaly, where nonstandard Schwinger terms in commutation relations may appear. (It can be shown that, in general, the only Schwinger term in this problem is given by the last term on the left-hand side of (4.113); no others can arise.) Consider also the correlator

$$P_\mu(q) = i \int d^4x\, e^{iqx} \langle 0 \mid T\{j_{\mu 5}(x),\ Q_5(0)\} \mid 0\rangle \tag{4.114}$$

and the product $P_\mu(q)q_\mu$ in the limit $q^2 \to 0$ (or q^2 of order of m_π^2, where m_π is the Goldstone boson mass). The general form of $P_\mu(q)$ is $P_\mu(q) = Aq_\mu$. Therefore, nonvanishing values of $P_\mu q_\mu$ in the limit $q^2 \to 0$ (or of order of quark mass m, if $\mid q^2 \mid \sim m_\pi^2$ – this limit will be also of interest to us later) can arise only from Goldstone boson intermediate states in (4.114). Let us estimate the corresponding matrix elements

$$\langle 0 \mid j_{\mu 5} \mid \pi \rangle = Fq_\mu, \tag{4.115}$$

$$\langle 0 \mid Q_5 \mid \pi \rangle = F'. \tag{4.116}$$

F is of the order of m since in the limit of massless quarks Goldstone bosons are coupled only to nonsinglet axial current. F' is of the order of $m_\pi^2 f_\pi \sim m$, where f_π is the pion decay constant (not considered to be small). These estimates give

$$P_\mu q_\mu \sim \frac{q^2}{q^2 - m_\pi^2} m^2 \tag{4.117}$$

and it is zero at $q^2 \to 0$, $m_\pi^2 \neq 0$, and of order of m^2 at $q^2 \sim m_\pi^2$. In what follows, we will restrict ourself to the terms linear in quark masses. So, we can put

$P_\mu(q)q_\mu = 0$ at $q \to 0$. The integration by parts on the right-hand side of (4.114) gives:

$$\lim_{q^2 \to 0} P_\mu(q)q_\mu = -\int d^4x \langle 0 \mid T\{2N_f Q_5(x), Q_5(0)\} + T\{D(x), Q_5(0)\} \mid 0 \rangle$$

$$-\int d^4x \langle 0 \mid [j_{05}, Q_5(0)] \mid 0 \rangle \delta(x_0) = 0. \tag{4.118}$$

After substitution of (4.118) in (4.113), we find the following low-energy theorem:

$$i \int d^4x \langle 0 \mid T\{2N_f Q_5(x), 2N_f Q_5(0)\} \mid 0 \rangle$$

$$-i \int d^4x \langle 0 \mid T\{D(x), D(0)\} \mid 0 \rangle - 4 \sum_i^{N_f} m_i \langle 0 \mid \bar{q}_i(0)q_i(0) \mid 0 \rangle$$

$$+i \int d^4x \langle 0 \mid \left[j_{05}^0(x), 2N_f Q_5(0) \right] \mid 0 \rangle \delta(x_0) = 0. \tag{4.119}$$

The low-energy theorem (4.119), with the last term on the left-hand side omitted, was found by Crewther [14].

Consider first the case of one massless quark, $N_f = 1$, $m = 0$. This case can be treated easily by introduction of the θ-term in the Lagrangian:

$$\Delta L = \theta \frac{\alpha_s}{4\pi} G_{\mu\nu}^n \tilde{G}_{\mu\nu}^n. \tag{4.120}$$

The matrix element $\langle 0 \mid Q_5 \mid n \rangle$ between any hadronic state $\mid n \rangle$ and vacuum is proportional to

$$\langle 0 \mid Q_5 \mid n \rangle \sim \langle 0 \mid \frac{\partial}{\partial\theta} \ln Z \mid n \rangle_{\theta=0}, \tag{4.121}$$

where $Z = e^{iL}$ and L is the Lagrangian. The gauge transformation of the quark field $\psi' \to e^{i\alpha\gamma_5}$ results in the appearance of the term

$$\delta L = \alpha \partial_\mu j_{\mu 5} = \alpha(\alpha_s/4\pi)G_{\mu\nu}^n \tilde{G}_{\mu\nu}^n \tag{4.122}$$

in the Lagrangian density. By the choice $\alpha = -\theta$, the θ-term (4.120) will be killed and $(\partial/\partial\theta) \ln Z = 0$. Therefore, $\zeta(0) = 0$ (Crewther theorem) – the first term in (4.119) vanishes as well as the second and third, since $m = 0$. From (4.119), we conclude that the anomalous commutator does indeed vanish:

$$\langle 0 \mid [j_{05}(x), Q_5(0)]_{x_0=0} \mid 0 \rangle = 0, \tag{4.123}$$

supporting the assumptions made in [14]–[16].

Let us turn now to the case of two light quarks, u, d, $N_f = 2$. This is the case of real QCD where the strange quark is considered to be heavy. Define the isovector axial current

$$j_{\mu 5}^{(3)} = \left(\bar{u}\gamma_\mu\gamma_5 u - \bar{d}\gamma_\mu\gamma_5 d\right) \Big/ \sqrt{2} \tag{4.124}$$

and its matrix element between the states of pion and vacuum

$$\langle 0 \mid j_{\mu 5}^{(3)} \mid \pi \rangle = f_\pi q_\mu, \tag{4.125}$$

where q_μ is the pion four-momentum, $f_\pi = 131$ MeV. Multiply (4.125) by q_μ. Using the Dirac equations for the quark fields we have

$$\frac{2i}{\sqrt{2}} \langle 0 \mid m_u \bar{u}\gamma_5 u - m_d \bar{d}\gamma_5 d \mid \pi \rangle = \frac{i}{\sqrt{2}} \langle 0 \mid (m_u + m_d)(\bar{u}\gamma_5 u - \bar{d}\gamma_5 d)$$
$$+ (m_u - m_d)(\bar{u}\gamma_5 u + \bar{d}\gamma_5 d) \mid \pi \rangle = f_\pi m_\pi^2, \tag{4.126}$$

where m_u, m_d are the u and d quark masses. The ratio of the matrix elements on the left-hand side of (4.126) is of order

$$\frac{\langle 0 \mid \bar{u}\gamma_5 u + \bar{d}\gamma_5 d \mid \pi \rangle}{\langle 0 \mid \bar{u}\gamma_5 u - \bar{d}\gamma_5 d \mid \pi \rangle} \sim \frac{m_u - m_d}{M}, \tag{4.127}$$

since the matrix element in the numerator violates isospin and such violation (in the absence of electromagnetism, which is assumed) can arise from the difference $m_u - m_d$ only. Neglecting this matrix element, we have from (4.126)

$$\frac{i}{\sqrt{2}} \langle 0 \mid \bar{u}\gamma_5 u - \bar{d}\gamma_5 d \mid \pi \rangle = \frac{f_\pi m_\pi^2}{m_u + m_d}. \tag{4.128}$$

Let us find $\zeta(0)$ from the low-energy sum rule (4.119), restricting ourselves to the terms linear in quark masses. Since $D(x) \sim m$, the only intermediate state in the matrix element

$$\int d^4x \langle 0 \mid T\{D(x), \ D(0)\} \mid 0 \rangle \tag{4.129}$$

in (4.119) is the one-pion state. Define

$$D_q = 2i \left(m_u \bar{u}\gamma_5 u + m_d \bar{d}\gamma_5 d \right). \tag{4.130}$$

Then

$$\langle 0 \mid D_q \mid \pi \rangle = i \langle 0 \mid (m_u + m_d)(\bar{u}\gamma_5 u + \bar{d}\gamma_5 d) + (m_u - m_d)(\bar{u}\gamma_5 u - \bar{d}\gamma_5 d)$$
$$= \sqrt{2}\frac{m_u - m_d}{m_u + m_d} f_\pi m_\pi^2, \tag{4.131}$$

where the matrix element of the singlet axial current was neglected and (4.128) was used. The substitution of (4.131) into (4.129) gives

$$i \int d^4x e^{iqx} \langle 0 \mid T\{D_q(x), \ D_q(0)\} \mid 0 \rangle_{q \to 0} = \lim_{q \to 0} \left\{ -\frac{1}{q^2 - m_\pi^2} 2 \left(\frac{m_u - m_d}{m_u + m_d} \right)^2 f_\pi^2 m_\pi^2 \right\}$$
$$= -4 \frac{(m_u - m_d)^2}{m_u + m_d} \langle 0 \mid \bar{q}q \mid 0 \rangle. \tag{4.132}$$

In the latter equation in (4.132), the Gell-Mann–Oakes–Renner relation (2.33),

$$\langle 0 \mid \bar{q}q \mid 0 \rangle = -\frac{1}{2} \frac{f_\pi^2 m_\pi^2}{m_u + m_d}, \tag{4.133}$$

and the $SU(2)$ equalities

$$\langle 0 \mid \bar{u}u \mid 0 \rangle = \langle 0 \mid \bar{d}d \mid 0 \rangle \equiv \langle 0 \mid \bar{q}q \mid 0 \rangle \tag{4.134}$$

were substituted. From (4.119) and (4.132) we finally get:

$$\zeta(0) = i \int d^4x \langle 0 \mid T\{Q_5(x),\ Q_5(0)\} \mid 0 \rangle = \frac{m_u m_d}{m_u + m_d} \langle 0 \mid \bar{q}q \mid 0 \rangle \tag{4.135}$$

in agreement with (4.106).

In a similar way, one can find the matrix element $\langle 0 \mid Q_5 \mid \pi \rangle$. Consider

$$\langle 0 \mid j_{\mu 5} \mid \pi \rangle = F q_\mu. \tag{4.136}$$

The estimate of F gives

$$F \sim \frac{m_u - m_d}{M} f_\pi \tag{4.137}$$

and after multiplying (4.136) by q_μ the right-hand side of (4.136) can be neglected. On the left-hand side we have

$$\langle 0 \mid D_q \mid \pi \rangle + 2N_f \langle 0 \mid Q_5 \mid \pi \rangle = 0. \tag{4.138}$$

The substitution of (4.131) into (4.138) results in

$$\langle 0 \mid Q_5 \mid \pi \rangle = -\frac{1}{2\sqrt{2}} \frac{m_u - m_d}{m_u + m_d} f_\pi m_\pi^2. \tag{4.139}$$

A relation of this type (with a wrong numerical coefficient) was found in [18], the correct formula was presented in [19]. From comparison of (4.135) and (4.139), it is clear that it would be wrong to calculate $\zeta(0)$ (4.135) by accounting only for pions as intermediate states: important are the constant terms, which reflect the necessity of subtraction terms in dispersion relations and are represented by terms proportional to quark condensate in (4.119). The cancellation of Goldstone boson pole terms and these constant terms result in the Crewther theorem – the vanishing of $\zeta(0)$, when one of the quark masses, e.g. m_u, is going to zero.

The validity of Eq. (4.106) in the case of three light quarks, u, d and s, can be shown by the same method in the limit of $m_u/m_s \ll 1$, $m_d/m_s \ll 1$ [20].

4.3 Instantons in Minkowski space-time

The fact that instantons realize the tunneling process between vacuums with different topological (winding) numbers in Minkowski space-time was first mentioned by Gribov [21] (see the remark in [22]). Independently, the tunneling phenomenon was revealed by Jackiw and Rebbi [23] and Callan, Dashen, and Gross [24]. The explicit demonstration of the tunneling mechanism caused by instantons in Minkowski space-time was presented by Bitar and Chang [25]. Below we follow Ref. [25].

Consider the $SU(2)$ subgroup of the $SU(3)$ colour group. Use the matrix notation (4.70) and

$$G_{\mu\nu} = -\frac{1}{2} g \tau^a G^a_{\mu\nu}, \tag{4.140}$$

$$G_{\mu\nu} = \partial_\mu A_\nu - \partial_\nu A_\mu - i[A_\mu, A_\nu]. \tag{4.141}$$

It is convenient to use also the chromoelectric and chromomagnetic field strengths:

$$E_i = G_{0i}, \tag{4.142}$$

$$B_i = \frac{1}{2} \varepsilon_{ijk} G_{jk}, \quad i, j, k = 1, 2, 3. \tag{4.143}$$

The Lagrangian density has the form of

$$\mathcal{L} = -\frac{1}{2g^2} \operatorname{Tr} G^2_{\mu\nu} = \frac{1}{g^2} \operatorname{Tr}(E^2 - B^2). \tag{4.144}$$

The Hamiltonian density is equal to

$$\mathcal{H} = \frac{1}{g^2}(E^2 + B^2). \tag{4.145}$$

The topological (winding) number n in Minkowski space-time is given by ($a = 1, 2, 3$):

$$n = \frac{g^2}{32\pi^2} \int d^4x \, G^a_{\mu\nu} \tilde{G}^a_{\mu\nu} = \frac{1}{4\pi^2} \int d^4x \, \operatorname{Tr}(EB). \tag{4.146}$$

Let us consider the vacuum transition from the state with $n = 0$ at $t = t_1 \to -\infty$ to the state with $n = 1$ at $t = t_2 \to \infty$. Introduce the family of field configurations characterized by the t-dependent parameter $\lambda(t)$. Then the determination of the tunneling amplitude is reduced to the quantum mechanical problem of the transition through a potential barrier. It can be shown [25] that the minimum of the action in such a transition arises in the case where $E(x, t)$ is proportional to $B(x, t)$ and the coefficient of proportionality is independent of x but dependent on t. (Note that this is different from the Euclidean case where we had $E = B$.) So put

$$A_0 = -\frac{x\tau}{x^2 + \lambda^2 + \rho^2} \dot{\lambda}(t), \quad A = -\frac{\lambda\tau + [x, \tau]}{x^2 + \lambda^2 + \rho^2}, \tag{4.147}$$

where the dot means differentiation with respect to t. The chromoelectric and chromomagnetic fields are equal to

$$E = -\frac{2\rho^2 \dot{\lambda}}{(x^2 + \lambda^2 + \rho^2)^2} \tau, \quad B = -\frac{2\rho^2}{(x^2 + \lambda^2 + \rho^2)^2} \tau \tag{4.148}$$

and proportional to each other. The effective Lagrangian that depends on $\lambda(t)$ is given by

$$L = \int d^3x \, \operatorname{Tr}(E^2 - B^2) = \frac{3\pi^2}{g^2} \frac{\rho^4}{(\lambda^2 + \rho^2)^{5/2}} (\dot{\lambda}^2 - 1). \tag{4.149}$$

The Lagrangian (4.149) can be represented in standard quantum mechanical form:

$$L = \frac{1}{2}m(\lambda)\dot{\lambda}^2 - V(\lambda),$$
(4.150)

where

$$m(\lambda) = \frac{6\pi^2\rho^4}{g^2(\lambda^2 + \rho^2)^{5/2}},$$
(4.151)

$$V(\lambda) = \frac{3\pi^2\rho^4}{g^2(\lambda^2 + \rho^2)^{5/2}}.$$
(4.152)

Eqs. (4.150)–(4.152) correspond to the following Hamiltonian:

$$H = \frac{p_\lambda^2}{2m(\lambda)} + V(\lambda),$$
(4.153)

$$p_\lambda = m(\lambda)\dot{\lambda}.$$
(4.154)

$E = B = 0$ at $t \to \pm\infty$. Assume that $\lambda(t) \to \pm\infty$ at $t \to \pm\infty$. Calculate the winding number $n(t)$:

$$n(t) = \frac{1}{4\pi^2} \int_{-\infty}^{t} dt \int d^3x \, \mathrm{Tr}(\boldsymbol{E}\boldsymbol{B}).$$
(4.155)

After substitution of (4.148) into (4.155) and integration we get:

$$n(t) = \frac{3}{4}\left[\frac{2}{3} + \frac{\lambda(t)}{\sqrt{\lambda^2(t) + \rho^2}}\left(1 - \frac{1}{3}\frac{\lambda^2(t)}{\lambda^2(t) + \rho^2}\right)\right].$$
(4.156)

Evidently, $n(t)_{t\to-\infty} \to 0$, $n(t)_{t\to+\infty} \to 1$, i.e. the fields (4.147) result in a transition of the vacuum from the state with winding number $n = 0$ to that with $n = 1$. The transition amplitude e^{-S} is given by the standard WKB formula:

$$S = \int_{-\infty}^{\infty} d\lambda[2m(\lambda)V(\lambda)]^{1/2}.$$
(4.157)

Substitution of (4.151), (4.152) in (4.157) and integration leads to:

$$S = \frac{6\pi^2\rho^4}{g^2}\int_{-\infty}^{\infty} d\lambda\frac{1}{(\lambda^2 + \rho^2)^{5/2}} = \frac{8\pi^2}{g^2}.$$
(4.158)

Eq. (4.158) coincides with (4.79) and supports the interpretation of the instanton as a minimal tunneling trajectory in the space of gluonic fields between vacuum states different by one unit of winding number.

4.4 Fermions in the instanton field. Atiyah–Singer theorem

Let us dwell on the consideration of quarks in the instanton field. It is clear that the interaction of heavy quarks with instanton field and its contribution to the action are proportional to $1/(m\rho)^2$ and are small, since $m\rho \gg 1$. (Here, m is the quark mass, ρ is the characteristic instanton size, $\rho \approx 1.5 - 2.0$ GeV^{-1}.) For the light quarks there is the opposite inequality: $m\rho \ll 1$. Therefore, we can restrict ourselves to the first nonvanishing terms in the expansion in powers of m. We will work in Euclidean space. The action of the fermion field is given by Eq. (4.58). The functional integration of the action over the anticommuting fermion fields results in the appearance of the determinant

$$- \det(i\gamma_\mu \nabla_\mu + im). \tag{4.159}$$

If the eigenvalues of the operator $-i\gamma_\mu \nabla_\mu$ are denoted by λ_n,

$$- i\gamma_\mu \nabla_\mu \psi_n = \lambda_n \psi_n, \tag{4.160}$$

then (4.159) is reduced to

$$- \det(i\gamma_\mu \nabla_\mu + im) = \prod_n (\lambda_n - im). \tag{4.161}$$

(We consider the system embedded in a box, so the spectrum is discrete.) Generally, λ_n are of the order of a characteristic hadronic scale and in the product on the right-hand side of (4.161) m can be safely neglected compared with λ_n. However, there can be a special case when some λ_n are zero. Let us show (see [6, 7, 12]), that in the instanton field one zero-mode solution of Eq. (4.160) exists – $u_0(x)$, which corresponds to $\lambda_0 = 0$,

$$- i\gamma_\mu \nabla_\mu u_0 = 0. \tag{4.162}$$

Decompose u_0 into left-hand and right-hand two-component spinors χ_L and χ_R:

$$u_0 = \frac{1}{2}(1 + \gamma_5)\chi_L + \frac{1}{2}(1 - \gamma_5)\chi_R, \quad \gamma_5 = \gamma_1 \gamma_2 \gamma_3 \gamma_4 = - \begin{pmatrix} 0 & 1 \\ 1 & 0 \end{pmatrix}. \tag{4.163}$$

Expressed in terms of χ_L, χ_R, Eq. (4.162) looks like

$$\gamma_4 \left[\sigma_\mu^+ \nabla_\mu (1 + \gamma_5)\chi_L - \sigma_\mu^- \nabla_\mu (1 - \gamma_5)\chi_R \right] = 0, \tag{4.164}$$

where

$$\sigma_\mu^\pm = (\boldsymbol{\sigma}, \mp i) \tag{4.165}$$

(cf. (4.87)). As follows from (4.165),

$$\sigma_\mu^+ \nabla_\mu \chi_L = 0, \tag{4.166}$$

$$\sigma_\mu^- \nabla_\mu \chi_R = 0. \tag{4.167}$$

Act on (4.166) with the operator $\sigma_\nu^- \nabla_\nu$ and on (4.167) with the operator $\sigma_\nu^+ \nabla_\nu$. The result is:

$$- \nabla_\mu^2 \chi_L = 0, \tag{4.168}$$

$$- \nabla_\mu^2 \chi_R = -4\sigma\tau \frac{\rho^2}{(x^2 + \rho^2)^2} \chi_R. \tag{4.169}$$

In the calculation we have exploited the relations for $\sigma_\mu^+, \sigma_\mu^-$ similar to (4.88) as well as the commutator

$$[\nabla_\nu, \nabla_\mu] = i\frac{g}{2}\tau^a G_{\nu\mu}^a, \tag{4.170}$$

the expression (4.97) for $G_{\mu\nu}^a$, and the identity $\bar{\eta}_{a\nu\mu}\eta_{a\nu\mu} = 0$. The operator $-\nabla_\mu^2 = (-i\nabla_\mu)^2$ is equal to the sum of squares of Hermitian operators, i.e. its eigenvalues are positive. Therefore $\chi_L = 0$. In (4.169), the state χ_R has spin $I = 1/2$ and isospin $T = 1/2$. The matrices σ act on spin variables, the matrices τ on isospin variables. There are two states with total spin + (colour) isospin $J = I + T$:

$$\frac{1}{4}(\sigma + \tau)^2 \chi_R = \begin{Bmatrix} 0 \\ 2 \end{Bmatrix} \chi_R \quad \begin{matrix} J = 0 \\ J = 1, \end{matrix} \tag{4.171}$$

corresponding to $\sigma\tau = -3$ and $\sigma\tau = 1$. Only the first solution is suitable in (4.169) because of the positivity of $-\nabla_\mu^2$. Its substitution into (4.169) gives:

$$\nabla_\mu^2 \chi_R + \frac{12\rho^2}{(x^2 + \rho^3)^2} \chi_R = 0. \tag{4.172}$$

From (4.59), with λ^n replaced by τ^n, and (4.97) we have:

$$\nabla_\mu^2 = \frac{\partial^2}{\partial x_\mu^2} - 3\frac{x^2}{(x^2 + \rho^2)^2} = \frac{1}{x^3}\frac{\partial}{\partial x}x^3\frac{\partial}{\partial x} - 3\frac{x^2}{(x^2 + \rho^2)^2} \tag{4.173}$$

and the equation for $\chi_R(x^2)$ takes the form of

$$\left[\frac{1}{x^3}\frac{\partial}{\partial x}x^3\frac{\partial}{\partial x} - 3\frac{x^2}{(x^2 + \rho^2)^2} + \frac{12\rho^2}{(x^2 + \rho^2)^2}\right]\chi_R(x^2) = 0. \tag{4.174}$$

The solution of Eq. (4.174) is

$$\chi_R = \frac{A\rho}{(x^2 + \rho^2)^{3/2}}, \tag{4.175}$$

where A is a normalization constant. The expression (4.175) shall be multiplyed by the colour – isospin wave function corresponding to $J = 0$:

$$\chi_0 = \left[\chi_s\left(\frac{1}{2}\right)\chi_c\left(-\frac{1}{2}\right) - \chi_s\left(-\frac{1}{2}\right)\chi_c\left(\frac{1}{2}\right)\right]/\sqrt{2}, \tag{4.176}$$

where χ_s and χ_c are spin 1/2 spinors in the coordinate and colour–isospin spaces and the arguments of χ_s are the spin projections. So, we have

$$u_0(x) = \frac{A}{2}(1 - \gamma_5)\frac{\rho}{(x^2 + \rho^2)^{3/2}}\chi_0. \tag{4.177}$$

Normalizing $u(x)$ by

$$\int d^4x u^+ u = 1, \tag{4.178}$$

we get finally

$$u_0(x) = \frac{1}{2}(1 - \gamma_5)\frac{1}{\pi}\frac{\rho}{(x^2 + \rho^2)^{3/2}}\chi_0. \tag{4.179}$$

The fact that only the right-hand zero-mode solution survives in the instanton field, while the left-hand solution vanishes, clearly indicates chirality violation.

Turn now to the proof of the important topological theorem – the Atiyah–Singer (or index) theorem [26], which was derived in [27] within the framework of the instanton approach (see also [12, 28, 29, 6, 30, 31]). Consider the anomaly condition (3.23) in QCD and take the vacuum average in the instanton field. As was shown above, only one colour state (related to spin) is contributing to the left-hand side of (3.23) and the factor N_c must be omitted from the right-hand side of this equation. We have:

$$\int d^4x \, \mathrm{Tr} \, \langle 0 \mid \partial_\mu j_{\mu 5}(x) \mid 0 \rangle = \frac{g^2}{16\pi^2}\int d^4x \langle 0 \mid G^a_{\mu\nu}\tilde{G}^a_{\mu\nu} \mid 0 \rangle. \tag{4.180}$$

In the instanton field, the right-hand side of (4.180) is equal to

$$\frac{g^2}{16\pi^2}\int d^4x \langle 0 \mid G^a_{\mu\nu}\tilde{G}^a_{\mu\nu} \mid 0 \rangle = \frac{g^2}{16\pi^2}\cdot\frac{32\pi^2}{g^2} = 2. \tag{4.181}$$

The left-hand side of (4.180) can be expressed in terms of the inverse Dirac operator:

$$\int d^4x \, \mathrm{Tr} \, \langle 0 \mid \partial_\mu j_{\mu 5}(x) \mid 0 \rangle = \int d^4x \partial_\mu \, \mathrm{Tr} \, \langle 0 \mid i \, \slashed{\nabla}^{-1}(x, x)\gamma_\mu\gamma_5 \mid 0 \rangle$$

$$= \int d^4x \nabla_\mu \, \mathrm{Tr}\left[\sum_n \frac{\psi_n(x)\psi_n^+(x)}{\lambda_n}\gamma_\mu\gamma_5\right] = \int d^4x \, \mathrm{Tr}\left[\sum_n \frac{\psi_n(x)\psi_n^+(x)}{\lambda_n}\cdot 2\lambda_n\gamma_5\right], \tag{4.182}$$

where $\psi_n(x)$ are the eigenfunctions defined by (4.160). (Quarks are assumed to be massless.) The states with nonzero λ_n do not contribute to the sum in (4.182): at $\lambda_n \neq 0$ the function $\gamma_5\psi_n$ is also the solution of (4.160) with eigenvalue $-\lambda_n$, these two states being orthogonal. So, only the zero-mode contribution survives and we have for the left-hand side of (4.180)

$$2\int d^4x \, \mathrm{Tr}\left[\gamma_5 u_0(x)u_0^+(x)\right] = -2. \tag{4.183}$$

Since the formulae for anti-instantons are obtained by the substitution $\eta_{a\mu\nu} \to \bar{\eta}_{a\mu\nu}$, it can be easily shown that in the case of anti-instantons we would have $+2$ in (4.183) instead of -2. For far-separated instantons and anti-instantons, the relations (4.181), (4.183) can be generalized to

$$n = n_L - n_R, \tag{4.184}$$

where n is the winding number defined above, n_L and n_R are the numbers of left-hand and right-hand zero modes of Dirac equation (4.160). Eq. (4.184) is the Atiyah–Singer (or index) theorem. This theorem has been proved in Ref. [26, 27], where its validity was demonstrated in a much more general situation (mult-instanton solutions, etc.) than that considered above. Eq. (4.184) does not alter for any number of massless flavours; if on the right-hand side of (4.184), n_L and n_R mean the numbers of zero modes for one flavour. Indeed, in this case, both sides of (4.180) are multiplied by N_f and this factor can be cancelled.

4.5 The vacuum structure in QCD

It has been shown that in QCD there is an infinite number of vacuums, each of which is characterized by a winding number n. The energies of all of them are equal. We will use the notation $\Omega(n)$ for the wave function of such a vacuum. We assume that vacuum wave functions are normalized, $\Omega^+(n)\Omega(n) = 1$, and form a complete set. Therefore, the arbitrariness in the wave functions reduces to the phase factors: $\Omega(n) \rightarrow e^{i\theta_n}\Omega(n)$, where θ is real. Let us divide the whole Euclidean space into two large parts and suppose that the field strengths are zero on the boundaries of each part, so that the potentials on the boundaries are of pure gauge. This is just the condition which allows us to consider the vacuum states in the two parts separately. The vacuum wave functions do not depend on the volume, the only dependence on the volume which survives is the dependence on windings numbers. If the winding numbers of the two parts of space are n_1 and n_2 and their wave functions are $e^{i\theta_{n_1}}\Omega(n_1)$ and $e^{i\theta_{n_2}}\Omega(n_2)$, then the wave function of the whole space is

$$e^{i\theta_{n_1+n_2}}\Omega(n_1+n_2) = e^{i\theta_{n_1}}\Omega(n_1)e^{i\theta_{n_2}}\Omega(n_2). \qquad (4.185)$$

Since

$$\Omega(n_1+n_2) = \Omega(n_1)\Omega(n_2). \qquad (4.186)$$

we get the equation

$$\theta_{n_1+n_1} = \theta_{n_1} + \theta_{n_2}. \qquad (4.187)$$

The solution of (4.187) is

$$\theta_n = n\theta. \qquad (4.188)$$

Therefore, the vacuum wave function in QCD, which is the superposition of all vacuum states with different winding numbers, has the form of

$$\Omega(\theta) = \sum_n e^{i\theta_n}\Omega(n). \qquad (4.189)$$

This state is often called the θ-vacuum. The θ-vacuums with different θ are orthogonal:

$$\Omega^+(\theta_1)\Omega(\theta_2) = 0. \qquad (4.190)$$

This statement directly follows from (4.189), because (4.189) can be considered as a Fourier transformation. Within the framework of QCD, the states with different θ form different worlds – transitions between them are strictly forbidden. The vacuum state $\Omega(\theta)$ is similar to the states in a periodic potential in quantum mechanics as discussed in Section 4.1 and reminiscent of the Bloch waves of electrons in a crystal. The energies of the θ-vacuums form bonds as a function of θ, like in the periodic potential in quantum mechanics but without transitions between different levels.

The vacuum state $\Omega(\theta)$ can be reproduced if (in Minkowski space-time) the term

$$L_\theta = \frac{g^2\theta}{32\pi^2} G_{\mu\nu}\tilde{G}_{\mu\nu} \qquad (4.191)$$

is added to the standard QCD Lagrangian. Eq.(4.191) explicitly demonstrates that θ is an observable quantity. It can be determined by measuring the correlator of topological currents $Q_5 \sim G_{\mu\nu}\tilde{G}_{\mu\nu}$, which enters *ep* polarized scattering. (See Section 6.9.6.) The term (4.191) violates P and T-invariance, but conserves C. (The possibility of P and T violation and C conservation was studied many years ago [32].) The strongest upper limit on θ was found from searches of the neutron electric dipole moment:

$$|\theta| \lesssim 10^{-9}. \qquad (4.192)$$

Within the framework of QCD, no arguments have been proposed to explain why θ is so small or even equal to zero.

4.6 The pre-exponential factor of the instanton action. The dilute gas instanton model

The calculation of the pre-exponential factor of the instanton action was done by 't Hooft [12] in two-loop approximation; the qualitative description of the calculation – without numerical factors – is presented in the review [7]. We skip here the details of the 't Hooft calculation and mainly follow [7]. Our goal is to present a reasonable model for instanton density in QCD vacuum based on the results of this calculation which could be used to estimate instanton contributions in various processes.

Consider first the gluonic part of the action. According to the standard procedure, expand the fields A_μ around the instanton solution:

$$A_\mu^a = A_{\mu,ins}^a + a_\mu^a \qquad (4.193)$$

and treat a_μ^a as a perturbation. The action, up to the second order terms in a_μ^a, has the form

$$S(A) = S_0 + \frac{1}{2}\int d^4x \, a_\mu^a \, L_{\mu\nu}^{ab}(A_{ins})a_\nu^b = \frac{8\pi^2}{g^2}$$

$$+ \frac{1}{2}\int d^4x a_\mu^a \left[D^2 a_\mu^a - D_\mu D_\nu a_\nu^a - 2g\varepsilon^{abc}G_{\mu\nu}^b a_\nu^c \right], \qquad (4.194)$$

where the instanton field must be substituted in D_μ and $G_{\mu\nu}^a$. Let us fix the gauge by adding to the action the term

$$\frac{1}{2}\int d^4x a_\mu^a (\Delta L)_{\mu\nu}^{ab} \, a_\nu^b = \frac{1}{2}\int d^4x (D_\mu a_\mu^a)^2 \qquad (4.195)$$

as well as the term representing the Faddeev–Popov ghosts:

$$\Delta S_{gh} = \int d^4x \bar{\phi}^a \, L_{gh}^{ab}\phi^b = \int d^4x \phi^a D^2\phi^a, \qquad (4.196)$$

where ϕ^a is a scalar anticommuting field. The instanton contribution to the vacuum–vacuum transition $\langle 0_T | 0 \rangle_{ins} = \langle 0 | e^{-HT} | 0 \rangle_{ins}$ is of the form

$$\langle 0_T | 0 \rangle_{ins} = \left[\det(L + \Delta L) \right]^{-1/2} \left[\det L_{gh} \right]^{+1} e^{-S_0}. \qquad (4.197)$$

The determinant of the ghosts fields enters (4.197) to the positive power $+1$ because these fields are anticommuting. The normalization of the transition amplitude is fixed by the condition that in perturbation theory $\langle 0_T | 0 \rangle_{pert} = 1$. The perturbative calculation corresponds to $A_{\mu\nu} = 0$ and $S_0 = 0$ and is divergent. So, the introduction of an ultraviolet cut-off M is necessary. It is convenient to perform the calculations in the Pauli–Villars regularization scheme and in the final results to go to the MS or $\overline{\text{MS}}$ schemes [12]. In this case, M is the mass of the Pauli–Villars additional vector field which enters the Lagrangian with negative metric. In order to achieve the desired normalization, let us divide (4.197) by the perturbative amplitude. We have

$$\langle 0_T | 0 \rangle_{ins} / \langle 0_T | 0 \rangle_{pert} = \left[\frac{\det(L + \Delta L)}{\det(L + \Delta L)_{M^2}^{pert}} \right]^{-1/2} \left[\frac{\det L_{gh}}{(\det L_{gh})_{M^2}} \right] e^{-S_0}, \qquad (4.198)$$

where the index M^2 means that the calculation is performed with the cut-off M^2.

Consider first the contribution of zero modes. As was shown in Sec. 4.1, each zero mode results in the appearance of an integral over the corresponding collective coordinate and contributes a factor proportional to $\sqrt{S_0}$ to the transition amplitude. In the case of the instanton in the SU(2) group, there are the following collective coordinates: four coordinates of the instanton center x, the instanton size ρ, and three isospace rotations, characterized by the Euler angles θ, φ, ψ. The spatial rotations are connected in a unique way with rotations in isospin space. So, only one of these rotations shall be accounted for. The perturbative factor $[\det(L + \Delta L)]^{1/2}$ in (4.198) contributes a factor of M for each collective coordinate. We have

$$\frac{\langle 0_T | 0 \rangle_{ins}}{\langle 0_T | 0 \rangle_{pert}} \sim \int d^4 x_c d\rho \cdot M^8 \rho^3 \left(\sqrt{S_0} \right)^8 e^{-S_0}. \qquad (4.199)$$

The factor ρ^3 arises from the Jacobian of the transition from the spatial integration to the integration over Euler angles. Eq. (4.199) can be rewritten as:

$$\frac{\langle 0_T | 0 \rangle_{ins}}{\langle 0_T | 0 \rangle_{pert}} = Const \int \frac{d^4 x_c}{\rho^5} \left(\frac{8\pi^2}{g_0^2} \right)^4 e^{-8\pi^2/g_0^2 + 8 \ln M\rho + \Phi}, \qquad (4.200)$$

where e^Φ gives the contributions of the non-zero modes. In (4.200), g_0^2 is the bare coupling constant, defined at the cut-off value M^2, $g_0^2 = g^2(M^2)$. The renormalization group requires that only quantities covariant under transformations of this group (in this case, the effective charge $g^2(\rho)$) can enter the final result. This requirement determines without any calculations the non-zero mode contribution $\Phi = (2/3) \ln M\rho$. (Of course, the explicit calculation confirms this value. Note that the account of the one-loop perturbative diagram

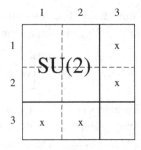

Fig. 4.5. The schematical representation of the matrix of $SU(3)$ generators. The $SU(2)$ subgroup generators are in the left-upper corner; the generators of $SU(3)$, which touch $SU(2)$ variables but are not $SU(2)$ generators, are marked by crosses.

is sufficient [12, 7]). Therefore, in the case of the $SU(2)$ colour group – which is the case studied up to now – we have the exponential factor equal to

$$-\frac{8\pi^2}{g^2(\rho)} = -\frac{8\pi^2}{g^2(M)} + 8\ln M\rho - \frac{2}{3}\ln M\rho = -\frac{8\pi^2}{g^2(M)} + \frac{22}{3}\ln M\rho. \qquad (4.201)$$

If the $SU(2)$ group is a subgroup of the $SU(3)$ colour group, then additional zero modes and the corresponding collective coordinates appear. They are related to $SU(3)$ generators which touch the $SU(2)$ subgroup variables. (They are marked by crosses in Fig. 4.5, schematically representing the matrix of the $SU(3)$ generators.) As is clear from Fig. 4.5, there are four such generators and the total number of zero modes and collective coordinates is twelve. This number is exactly equal to the coefficient in front of the "antiscreening" logarithm in the formula for $8\pi^2/g^2(\rho)$ in the $SU(3)$ colour group. After adding the nonzero-mode contribution, the effective charge for $SU(2)$ instantons in the $SU(3)$ colour group arises in the exponent, we get:

$$\frac{\langle 0_T|0\rangle_{inst}}{\langle 0_T|0\rangle_{pert}} = Const \int \frac{d^4x_c d\rho}{\rho^5} \left(\frac{8\pi^2}{g_0^2}\right)^6 e^{\frac{8\pi^2}{g^2(\rho)}}. \qquad (4.202)$$

A one-loop calculation is needed in order to find the Const in (4.202). Such calculation was performed by 't Hooft [12] (see also [7]). The result is

$$\frac{\langle 0_T|0\rangle_{ins}}{\langle 0_T|0\rangle_{pert}} = \int \frac{d^4x_c\, d\rho}{\rho^5} d_g(\rho), \qquad (4.203)$$

where the instanton density $d_d(\rho)$ caused by the gluon action is given by

$$d_g(\rho) = \frac{c_1}{(N_c-1)!\,(N_c-2)!} \left[\frac{2\pi}{\alpha_s(\rho)}\right]^{2N_c} e^{\frac{2\pi}{\alpha_s(\rho)} - c_2 N_c}, \qquad (4.204)$$

$N_c = 3$ is the number of colours, and $\alpha_s(\rho)$ must be expressed in terms of the one-loop perturbative formula. The constant $c_1 = 0.466$, the constant c_2 depends on the regularization scheme and is equal to in the Pauli–Villars regularization scheme: $c_2 = 1.679$,

in the MS dimensional regularization: $c_2 = -2.042$,

in the $\overline{MS}$ dimensional regularization: $c_2 = 1.54$.

Turn now to the account for the quark contribution to the pre-exponential factor. It is evident that the heavy quark contribution is suppressed by the factor $1/(m_q\rho)^2$, where m_q is the heavy quark mass, and can be neglected. Then light quarks give rise to the factor

$$F = \prod_f^{N_f} \frac{\det(-i\gamma_\mu\nabla_\mu - im_f)_{ins}}{\det(-i\gamma_\mu\nabla_\mu - iM)_{pert}}, \tag{4.205}$$

where the product is taken over the light quark flavours u, d, s, $N_f = 3$ is the number of light quarks. The role of light quarks in the exponential e^{-S_0} is clear: the coupling constant $\alpha_s(\rho)$ is expressed by the formula which accounts for the light quarks. Separate the factor corresponding to the zero-mode contribution in (4.205):

$$F = \frac{m_u m_d m_s}{M^3} \prod^{N_f} \frac{\det'(-i\gamma_\mu\nabla_\mu)_{ins}}{\det'(-i\gamma\nabla_\mu - iM)_{pert}}, \tag{4.206}$$

where the determinant corresponding to nonzero modes is denoted by $\det'$. Since $m_f\rho \ll 1$ and for nonzero modes $\lambda_n \sim 1/\rho$, we have neglected m_f in $\det'$. The factor $1/M^3$ cannot appear in the final result due to renormalization arguments: it must be absorbed in the definition of $\alpha_s(\rho)$ and for dimensional reasons should be replaced by ρ^3. The instanton density $d(\rho)$ which arises from gluons and light quarks is equal to

$$d(\rho) = d_g(\rho)d_q(\rho), \tag{4.207}$$

$$d_q(\rho) = m_u m_d m_s \rho^3 e^{c_f N_f}, \tag{4.208}$$

$d_g(\rho)$ is given by (4.204), where $\alpha_s(\rho)$ accounts for gluons and quarks, and $c_f = 0.292$ for the Pauli–Villars regularization, $c_f = -0.495$ for the MS scheme, and $c_f = 0.153$ for the $\overline{MS}$ scheme [12, 7].

Consider the ρ-dependence of the instanton contribution to the vacuum transition amplitude. Eqs. (4.203), (4.204), (4.207), (4.208), together with the one-loop relation

$$2\pi/\alpha_s(\rho) \sim -9\ln\rho, \tag{4.209}$$

give

$$\langle 0_T|0\rangle_{inst} \sim \rho^7. \tag{4.210}$$

The amplitude increases very steeply with ρ ! In Ref. [33] an attempt was made to account for the influence of the QCD vacuum structure – namely, the persistence of the quark condensate – on the instanton density in the case of small-size instantons. The result was (semi-quantatively) that for each quark flavour the quark mass should be replaced by

$$m_q \to m_q - \frac{2}{3}\pi^2\langle 0|\bar{q}q|0\rangle\rho^2 \tag{4.211}$$

(m_q and $\langle 0|\bar{q}q|0\rangle$ are normalized at $\mu = \rho^{-1}$). At $\rho > 1\,\mathrm{GeV}^{-1}$, the last term in (4.211) overwhelms the first one in the case of u and d quarks and is of the same order as the first term in the case of s quarks. (The numerical value of $\langle 0|\bar{q}q|0\rangle$ is given in Section 6.2).

One may guess that instantons are responsible for the appearance of quark condensates in QCD. Consider the case of one quark flavour. Differentiation with respect to m of the vacuum–vacuum amplitude gives, according to (4.58),

$$i\frac{\partial}{\partial m}\langle 0_T|0\rangle = \int d^4x\,\langle 0_T|\bar{\psi}\psi|0\rangle\,c, \qquad (4.212)$$

where the constant c is independent of m in the limit $m \to 0$. On the other hand, the differentiation of (4.206) or (4.207) results in a constant independent of m in the same limit:

$$i\frac{\partial}{\partial m}\langle 0_T|0\rangle = Const. \qquad (4.213)$$

Comparison of (4.212) with (4.213) shows that the quark condensate is nonvanishing in the instanton field in the limit of massless quarks. The case of several light quarks is less certain because additional factors, proportional to light quark masses, appear in the instanton density (4.208). But we come to the same conclusion if the substitution (4.211) holds true.

Substitution of (4.211) into (4.208) leads to an even stronger increase of $\langle 0_T|0\rangle_{inst}$ with increasing ρ than given by (4.210). Physically, it is clear that such a steep increase cannot continue in a large interval of ρ: it also must be damped very sharply. So one can expect that $\langle 0_T|0\rangle$ as a function of ρ is concentrated near some value $\rho = \rho_c$ and a reasonable model of $\langle 0_T|0\rangle$ is

$$\langle 0_T|0\rangle = \int d^4x_c\,n_0\delta(\rho - \rho_c)d\rho, \qquad (4.214)$$

where n_0 is a constant (see [30, 34] and references therein).

In such a model, it is assumed that there is a dilute gas of noninteracting instantons. Therefore, the necessary condition of the validity of the model is $n_0\rho_c^4 \ll 1$. The upper limit on the constant instanton density n_0 can be found from the value of the gluon condensate if it is supposed that the contribution to the gluon condensate of other nonperturbative fluctuations, besides instantons, is positive [30]. Then

$$\langle 0|\frac{\alpha_s}{\pi}\,G_{\mu\nu}^2|0\rangle > \int d^4x_c\,\frac{\alpha_s}{\pi}(G_{\mu\nu}^2)_{inst}n_0\delta(\rho - \rho_c) = 8\,n_0. \qquad (4.215)$$

When deriving (4.213), the expression (4.144) for $(G_{\mu\nu}^a)_{inst}$ was exploited with the substitution $x \to x - x_c$. Using the value

$$\langle 0|\frac{\alpha_s}{\pi}\,G_{\mu\nu}^2|0\rangle = (0.005 \pm 0.004)\,\mathrm{GeV}^2, \qquad (4.216)$$

(Table 1, Section 6.2), we have

$$n_0 < 0.001\,\mathrm{GeV}^4 = 0.62\,\mathrm{fm}^{-4}. \qquad (4.217)$$

The value of ρ_c has been estimated in various ways [30]. One way is the matching of the low q^2 behaviour of the correlator of topological currents $\zeta(q^2)$ given by (4.153) and (6.327) with its high q^2 dependence found in the dilute instanton gas model [35]. These estimates show that

$$\rho_c = (1.0 - 2.0) \text{ GeV}^{-1} = (0.2 - 0.4) \text{ fm}. \tag{4.218}$$

If we put for n_0 the highest value allowed by (4.217), then $n_0\rho_c^4 = (1 \pm 15) \cdot 10^{-3}$ and is small indeed. Note, that the dilute gas instanton model is a semiquantitative description of the influence of instantons on physical amplitudes. For this reason, the parameters of the model are not certain and slightly different values of the parameters can be used in the description of different processes.

4.7 Quark propagator in the instanton field

The quark propagator in the instanton field in Euclidean space satisfies the equation

$$-i(\gamma_\mu \nabla_\mu + m)S(x, y) = \delta^4(x - y), \tag{4.219}$$

where the gluon field that enters the covariant derivative (4.59) is the instanton field. The propagator can be represented as a sum over the complete set of eigenfunctions of the Dirac operator:

$$S(x, y) = \sum_n \frac{\psi_n(x)\psi_n^+(y)}{\lambda_n - im}, \tag{4.220}$$

where λ_n are eigenvalues of the equation

$$-i\gamma_\mu \nabla_\mu \psi_n = \lambda_n \psi_n \tag{4.221}$$

and it is assumed for simplicity that the system is in a box and the spectrum is discrete. By applying the operator $-i(\gamma_\mu \nabla_\mu + m)$ to (4.220) one can easily convince oneself that (4.220) satisfies Eq. (4.219).

$S(x, y)$ can be expressed in terms of the spin 0 particle propagator $(\Delta(x, y)$ in the instanton field [36]). Put

$$S = (-i\gamma_\mu \nabla_\mu + im)G = G\,(-i\gamma_\mu \overleftarrow{\nabla}_\mu + im), \tag{4.222}$$

where $\overleftarrow{\nabla}_\mu$ means the operator ∇_μ acting to the left with the opposite sign of the derivative. (The second equality in (4.222) follows from Hermiticity of S). $G(x, y)$ obeys the second-order equation

$$\left[-(\gamma_\mu \nabla_\mu)^2 + m^2\right]G = \delta^4(x - y). \tag{4.223}$$

If the field strength tensor is self-dual, $G_{\mu\nu}^n = \tilde{G}_{\mu\nu}^n$, then the following equality holds:

$$(\gamma\nabla)^2 \frac{1 + \gamma_5}{2} = \nabla^2 \frac{1 + \gamma_5}{2}. \tag{4.224}$$

The proof is straightforward. In the general relation

$$(\gamma \nabla)^2 = \nabla^2 + \frac{g}{2}\sigma_{\mu\nu}\frac{\lambda^n}{2}G^n_{\mu\nu}, \tag{4.225}$$

the last term vanishes after multiplying by $(1+\gamma_5)$ if $G_{\mu\nu} = \tilde{G}_{\mu\nu}$. Using symbolic notation we can write:

$$
\begin{aligned}
G &= \frac{1}{-(\gamma\nabla)^2 + m^2}\frac{1+\gamma_5}{2} + \frac{1}{-(\gamma\nabla)^2 + m^2}\frac{1-\gamma_5}{2} \\
&= \frac{1}{-\nabla^2 + m^2}\frac{1+\gamma_5}{2} + \frac{1}{m^2}\left[1 + (\gamma\nabla)\frac{1}{-(\gamma\nabla)^2 + m^2}(\gamma\nabla)\right]\frac{1-\gamma_5}{2} \\
&= \frac{1}{-\nabla^2 + m^2}\frac{1+\gamma_5}{2} + \frac{1}{m^2}\left[1 + (\gamma\nabla)\frac{1}{-\nabla^2 + m^2}(\gamma\nabla)\right]\frac{1-\gamma_5}{2}.
\end{aligned}
\tag{4.226}
$$

In the last equality, the anticommutativity of $(\gamma\nabla)$ with of γ_5 was exploited and (4.224) was used. S can be found from G (4.226) with the help of the identity

$$(\gamma\nabla)\left[1 + (\gamma\nabla)\frac{1}{-\nabla^2 + m^2}(\gamma\nabla)\right]\frac{1-\gamma_5}{2} = \frac{m^2}{-\nabla^2 + m^2}(\gamma\nabla)\frac{1-\gamma_5}{2}. \tag{4.227}$$

The final formula for S follows from (4.222), (4.226), (4.227) [36]:

$$
\begin{aligned}
S &= (-i\gamma\nabla + im)\,G = (-i\gamma\nabla + im)\frac{1}{-\nabla^2 + m^2}\frac{1+\gamma_5}{2} - i\frac{1}{-\nabla^2 + m^2}(\gamma\overleftarrow{\nabla})\frac{1-\gamma_5}{2} \\
&\quad + \frac{i}{m}\left[1 + (\gamma\nabla)\frac{1}{-\nabla^2 + m^2}(\gamma\overleftarrow{\nabla})\right]\frac{1-\gamma_5}{2}.
\end{aligned}
\tag{4.228}
$$

In Eq. (4.228), S expressed in terms of the propagator (Green function) of a scalar particle in the instanton field

$$\Delta = \frac{1}{-\nabla^2 + m^2}. \tag{4.229}$$

In the limit $m \to 0$, the last term in (4.228) represents the contribution of zero modes. The operator

$$P = \left[1 + \gamma\nabla\frac{1}{-\nabla^2}\gamma\overleftarrow{\nabla}\right]\frac{1-\gamma_5}{2} \tag{4.230}$$

is the projection operator on the zero-mode subspace, since

$$(\gamma\nabla)P = P(\gamma\overleftarrow{\nabla}) = 0, \quad P^2 = P. \tag{4.231}$$

At $m \to 0$, the last term in (4.228) tends to $(i/m)P$. Its singularity in m is attributed to the zero-modes contribution. But, as was demonstrated in the previous subsection, the determinant is proportional to m and the product $\det \cdot S$ tends to a finite limit at $m \to 0$. If there are several quark propagators in the amplitude under consideration, then, according to the Pauli principle, different quarks cannot be simultaneously present in the same zero-mode state and the singularity is absent in this case as well.

The quark propagator in Eq. (4.228) is expressed in terms of the propagator $\Delta(x, y)$ of a scalar particle with $SU(2)_c$ colour spin 1/2, which was defined by (4.229). $\Delta(x, y)$ was calculated in Ref. [37]. We present here the final result, where the terms of order m^2 were neglected (in the regular gauge):

$$\Delta(x, y) = \frac{1}{4\pi^2} \frac{1}{(x-y)^2} \left[\rho^2 + xy + i\eta_{a\mu\nu} x_\mu y_\nu \tau^a \right] \frac{1}{(x^2 + \rho^2)^{1/2}} \frac{1}{(y^2 + \rho^2)^{1/2}}.$$
(4.232)

The center of the instanton is chosen at the origin. In the limit $y \to x \gg \rho$, $\Delta(x, y)$ tends to the free propagator of the scalar particle:

$$\Delta(x, y) \to \frac{1}{4\pi^2} \frac{1}{(x-y)^2}.$$
(4.233)

The terms of order m^2 were accounted for in [38, 39]. The quark propagator in the anti-instanton field can be obtained from (4.228), (4.232) by the replacements $\gamma_5 \to -\gamma_5$ and $\eta_{a\mu\nu} \to \bar{\eta}_{a\mu\nu}$.

The knowledge of the quark propagator allows one to calculate polarization operators of the vector and axial quark current in the instanton field – objects which are interesting from various points of view. The polarization operator in 1-loop approximation is given by

$$\Pi_{\mu\nu}(x, y) = -i \int d^4z \, \text{Tr}[\Gamma_\mu S(x, z) \Gamma_\nu S(z, y)]$$

$$- i \int d^4z \, \text{Tr}[\Gamma_\mu S(x, z)] \, \text{Tr}[\Gamma_\nu S(z, y)],$$
(4.234)

where $\Gamma_\mu = \gamma_\mu$ stands for the vector current and $\Gamma_\mu = \gamma_\mu \gamma_5$ for the axial current. The second term on the right-hand side of (4.234) vanishes in the case of the vector current. The last term in (4.228) can contribute to (4.234) only if it is multiplied by a term proportional to m that arises from another propagator, otherwise the result will be equal to zero. Therefore, $\Pi_{\mu\nu}$ is of order m^0 at small m, the next correction is of order $m^2 \rho^2$, and, as a rule, can be disregarded.

The polarization operators of the vector and axial currents in the instanton fields were calculated by Andrei and Gross [40]. The results of these calculations are used in Chapter 6.

4.8 Appendix

Relations among $\eta_{a\mu\nu}$ and $\bar{\eta}_{a\mu\nu}$ symbols

$$\eta_{a\mu\nu} = \frac{1}{2} \epsilon_{\mu\nu\alpha\beta} \eta_{a\alpha\beta}, \quad \eta_{a\mu\nu} = -\eta_{a\nu\mu}, \quad \eta_{a\mu\nu} \eta_{b\mu\nu} = 4\delta_{ab},$$

$$\eta_{a\mu\nu} \eta_{a\mu\lambda} = 3\delta_{\nu\lambda}, \quad \eta_{a\mu\nu} \eta_{a\mu\nu} = 12,$$

$$\eta_{a\mu\nu} \eta_{a\gamma\lambda} = \delta_{\mu\gamma} \delta_{\nu\lambda} - \delta_{\mu\lambda} \delta_{\nu\gamma} + \epsilon_{\mu\nu\gamma\lambda},$$

$$\epsilon_{\mu\nu\lambda b} \eta_{a\gamma b} = \delta_{\gamma\mu} \eta_{a\nu\lambda} - \delta_{\gamma\nu} \eta_{a\mu\lambda} + \delta_{\gamma\lambda} \eta_{a\mu\nu},$$

$$\eta_{a\mu\nu}\eta_{b\mu\lambda} = \delta_{ab}\delta_{\nu\lambda} + \epsilon_{abc}\eta_{c\nu\lambda},$$

$$\epsilon_{abc}\eta_{b\mu\nu}\eta_{c\gamma\lambda} = \delta_{\mu\gamma}\eta_{a\nu\lambda} - \delta_{\mu\lambda}\eta_{a\nu\gamma} - \delta_{\nu\gamma}\eta_{a\mu\lambda} + \delta_{\nu\lambda}\eta_{a\mu\gamma},$$

$$\eta_{a\mu\nu}\bar{\eta}_{b\mu\nu} = 0, \quad \eta_{a\gamma\mu}\bar{\eta}_{b\gamma\lambda} = \eta_{a\gamma\lambda}\bar{\eta}_{b\gamma\mu}.$$

Problems

Problem 4.1

Find the relation of quark condensate to spectral function of the vacuum expectation value of quark correlator for massless quark.

Solution

Define (in Euclidean space-time):

$$\sum(x, y) = \langle 0|\psi(x)\psi^+(y)|0\rangle \tag{1}$$

and

$$\frac{1}{V}\int d^4x \sum(x, x) = \sum(0, 0) = \int_{-\infty}^{\infty} \frac{\rho(\lambda)d\lambda}{m - i\lambda}, \tag{2}$$

where V is the 4-volume of the Euclidean space, $V \to \infty$, λ – are the eigenvalues of Dirac operator given in (4.160), $\rho(\lambda)$ is the spectral function. (The continuous spectrum is considered.) $\rho(\lambda)$ is symmetrical, $\rho(\lambda) = \rho(-\lambda)$.

$$\sum(0, 0) = 2m \int_{0}^{\infty} \frac{\rho(\lambda)}{m^2 + \lambda^2}d\lambda \tag{3}$$

In the limit $m \to 0$

$$\frac{2m}{m^2 + \lambda^2} \to \pi\delta(\lambda) \tag{4}$$

and

$$\sum(0, 0) = -\langle 0|\psi^+\psi|0\rangle = \pi\rho(0) \tag{5}$$

Quark condensate is determined by zero-mode contribution of Dirac equation (4.160). (Eq.(5) is called Banks-Casher relation [41], see also [31].)

References

[1] Belavin, A. A., Polyakov, A. M., Schwarz, A. S. and Tyupkin, Yu. S., *Phys. Lett.* **59B** (1975) 85.
[2] Instantons in Gauge Theories, ed. by M.Shifman, World Scientific, 1994.

[3] Landau, L. D., *Phys. Zeit. Sowjetunion* **1** (1932) 88; **2** (1932) 46.

[4] Perelomov, A. M., Popov, V. S. and Terentjev, M. V., *ZhETF* **50** (1966) 1393; **51** (1966) 309.

[5] Popov, V. S., *Phys. Atom. Nucl.* **68** (2005) 686.

[6] Coleman, S., Aspects of Symmetry, *Cambridge University Press*, 1985, pp. 265–350, reproduced in Ref.[2].

[7] Vainshtein, A. I., Zakharov, V. I., Novikov, V. A. and Shifman, M. A., *Sov. Phys. Uspekhi* **25** (1982) 195, reproduced in Ref.[2].

[8] Feynman, R. P. and Hibbs, A. R., Quantum Mechanics and Path Integrals, McGraw-Hill, New York, 1965.

[9] Landau, L. D. and Lifshitz, E. M., Quantum Mechanics, 2nd ed., Pergamon Press, Oxford, 1965.

[10] Rajaraman, R., Solitons and Instantons, North-Holland, Amsterdam, 1982.

[11] Jackiw, R., Topological Investigations of Quantized Gauge Theories, in: Treiman, S. B., Jackiw, R., Zumino, B. and Witten, E., Current Algebra and Anomalies, Princeton Series in Physics, 1985, pp. 211–359.

[12] 't Hooft, G., *Phys. Rev.* **D14** (1976) 3432; ibid. (E) **D18** (1978) 2199.

[13] Peskin, M. E. and Schroeder, D. V., An Introduction to Quantum Field Theory, Addison-Wesley Publ. Comp., 1995, Chapter 3.

[14] Crewther, R. J., *Phys. Lett.* **70B** (1977) 349.

[15] Veneziano, G., *Nucl. Phys.* **B159** (1979) 213.

[16] Di Veccia, P. and Veneziano, G., *Nucl. Phys.* **B171** (1980) 253.

[17] Leutwyler, H. and Smilga, A., *Phys. Rev. D* **46** (1992) 5607.

[18] Gross, D. J., Treiman, S. B., and Wilczek, F., *Phys. Rev. D* **19** (1979) 2188.

[19] Ioffe, B. L. and M. A. Shifman, M. A., *Phys. Lett.* **95B** (1980) 99.

[20] Ioffe, B. L., *Phys. Atom. Nucl.* **62** (1999) 2052.

[21] Gribov, V. N., unpublished.

[22] Polyakov, A., *Nucl. Phys.* **B120** (1977) 429.

[23] Jackiw, R. and Rebbi, C., *Phys. Rev. Lett.* **37** (1976) 172.

[24] Callan, C., Dashen, R. and Gross, D., *Phys. Lett.* **63B** (1976) 334.

[25] Bitar, K. M. and Chang, S.-J., *Phys. Rev.* **D17** (1978) 486.

[26] Atiyah, M. and Singer, I., *Ann. Math.* **87** (1968) 484; **93** (1971) 119.

[27] Schwarz, A., *Phys. Lett.* **67B** (1977) 172.

[28] Brown, L., Carlitz, R., and Lee, C., *Phys. Rev.* **D16** (1977) 417.

[29] Jackiw, R. and Rebbi, C., *Phys. Rev.* **D16** (1977) 1052.

[30] Schäfer, T. and Shuryak, E. V., *Rev. Mod. Phys.* **70** (1998) 323.

[31] Smilga, A., Lectures on Quantum Chromodynamics, World Scientific, 2001.

[32] Ioffe, B. L., *ZhETF* **32** (1957) 396.

[33] Shifman, M. A., Vainshtein, A. I. and Zakharov, V. I., *Nucl. Phys.* **163** (1980) 46.

[34] Diakonov, D., *Prog. Part. Nucl. Phys.* **36** (1996) 1.

[35] Ioffe, B. L. and Samsonov, A. V., *Phys. Atom. Nucl.* **63** (2000) 1448.

[36] Brown, L. S, and Lee, C., *Phys. Rev.* **D18** (1978) 2180.

[37] Brown, L. S., Carlitz, R. D., Creamer, D. B. and Lee, C., *Phys. Rev.* **D17** (1978) 1583.
[38] Din, A. M., Finjord, F. and Zakrzewski, W. J., *Nucl. Phys.* **B153** (1979) 46.
[39] Geshkenbein, B. V. and Ioffe, B. L., *Nucl. Phys.* **B166** (1980) 340.
[40] Andrei, N. and Gross, D. J., *Phys. Rev.* **D18** (1978) 468.
[41] Banks, T. and Casher, A. *Nucl. Phys.* **B169** (1980) 103.

5

Divergence of perturbation series

It is well known that in Quantum Field Theory the perturbation series is divergent. In the case of QED a physical argument supporting this property was suggested by Dyson [1]. He performed an analytic continuation of physical quantities $O(\alpha_e)$ to negative values of the fine structure constant $\alpha_e = e^2/4\pi$. In the world with $\alpha_e < 0$ particles with the same sign of the electric charge would attract each other. In this case, the energy is not restricted from below and the vacuum state is unstable because it would decay into two clouds of electrons and positrons separated by large distances. Indeed, the attractive Coulomb potential inside the cloud with a size $\bar{r}$ consisting of k electrons would grow proportional to $\alpha_e k(k-1)/(2\bar{r})$ for $\bar{r} \sim 1/m_e$ at large k and would exceed their rest mass $k\, m_e$ at some $k = k_0$. This means that the observables $O(\alpha_e)$ should have singularities at $\alpha_e < 0$, leading to the divergence of the perturbation series $\sum (-\alpha_e)^k\, C_k$. Moreover, according to Dyson, the coefficients C_k should grow at large orders k as $b_0^{-k}\, k!$ which is proportional to the number of Feynman diagrams. Similar estimates for C_k were obtained in scalar field models [2].

A transparent physical example of the divergent perturbation series in QED is the vacuum polarization in the external magnetic field H [3]. The corresponding contribution to the Lagrangian was calculated by Schwinger [4]:

$$L = -\frac{1}{8\pi^2} \int\limits_0^\infty e^{-m^2 s} \frac{ds}{s^3} \left[e\,s\,H \operatorname{ctnh}(e\,s\,H) - 1 - \frac{1}{3}(e\,s\,H)^2 \right], \qquad (5.1)$$

where m and e are the electron mass and charge. In this case, it is easy to calculate the coefficients of the perturbative expansion for large k [3]:

$$C_k = (-1)^{k-1} \frac{2 \cdot 4^k\, (k-3)!}{\pi^k} \frac{H^{2k}}{m^{4k-4}}. \qquad (5.2)$$

Note that such a series is Borel summable. Indeed, if we perform the Borel transformation of the physical quantity $O(\alpha_e)$

$$O(\alpha_e) = \int_0^\infty \frac{db}{\alpha_e} B(b)\, e^{-b/\alpha_e}, \qquad B(b) = \sum_k (-b)^k \frac{C_k}{k!}, \qquad (5.3)$$

then the function $B(b)$ has singularities at $b = -b_0 < 0$ outside the region of integration $b > 0$ and therefore there is no ambiguity in the integration over b. A similar situation

145

related to an underbarrier transition to the new vacuum state for a negative coupling constant g is valid also in the case of the anharmonic oscillator in quantum mechanics [5]. Strictly speaking, the Dyson argument can be applied only to scalar electrodynamics (SED) because in QED the effects related to Fermi statistics lead to less rapidly growing terms $C_k \sim \sqrt{k}!$ (see the discussion in Section 5.8). Formally, the Dyson picture is related to the existence of classical solutions of the Euler–Lagrange equations in Euclidean space-time [6],[7],[8],[9]. These solutions describe underbarrier transitions between different vacuum states and the asymptotic behaviour of perturbation theory coefficients can be obtained with the use of the saddle-point approach to calculate the functional integral [6]. The possibility to use this approach was discussed initially in Ref. [10]. Another method is based on finding the dicontinuity of physical quantities considered to be analytic functions of the coupling constant [11]. In the case of non-Abelian gauge theories, there is an infinite number of vacuum states with different topological numbers and the instability of the perturbative vacuum arises already at positive α_c, which leads to the Borel nonsummability of the series $\sim \sum \alpha_c{}^k (2\pi)^{-k} k!$ (cf. (5.3)). Here, the field configuration responsible for the divergence of perturbation theory corresponds to a weakly interacting instanton–anti-instanton pair having a vanishing topological charge. The singularity of $B(b)$ in the non-Abelian models is situated on the path of integration at $b_I = 2\pi$.

But there is another reason for the divergence of perturbation theory. Namely, it turns out that at large orders k there are particular Feynman diagrams giving factorial contributions $\sim k!$ to observables [12],[13],[14],[15]. Such diagrams contain logarithmically divergent subgraphs responsible for the renormalization of the coupling constant. These contributions generate Borel plane singularities known as renormalons. A first example of the renormalon contribution was found by Lautrup [12]. He considered the electron vertex function

$$\Gamma_\nu = \gamma_\nu f(q^2) - \frac{1}{2m} \sigma_{\nu\rho} q_\rho\, g(q^2), \tag{5.4}$$

where q is the momentum transfer, $f(q^2)$ and $g(q^2)$ are the electric and magnetic formfactors. The anomalous magnetic moment can be calculated in perturbation theory $g(0) = \alpha_e/(2\pi) + \ldots$. The Feynman integral of the one loop contribution to $g(0)$ is rapidly convergent at the large photon virtuality $p^2 \to \infty$:

$$g(0) \sim \frac{\alpha_e}{\pi} \int_{m^2}^{\infty} m^4 \frac{dp^2}{p^6} d(\rho) \sim \frac{\alpha_e}{\pi} \int_0^{\infty} d\rho\, e^{-2\rho} d(\rho), \quad \rho = \ln \frac{p^2}{m^2}, \tag{5.5}$$

where $d(\rho)/p^2$ is the photon Green function. Insertion of the asymptotic expression for the Green function $d(\rho)$ in leading logarithmic approximation $\alpha_e \rho \sim 1$ gives factorially growing coefficients of the perturbation series for $g(0)$:

$$g(0) = \frac{\alpha_e}{\pi} \sum_{k=0}^{\infty} \left(\frac{\alpha_e}{3\pi} \right)^k c_k, \quad c_k \approx \int_0^{\infty} d\rho\, e^{-2\rho} \rho^k = \frac{k!}{2^{k+1}}. \tag{5.6}$$

This series is not Borel summable, because the function $B(b)$ contains a singularity at $b = 6\pi > 0$. Note that in the case of the Dirac formfactor slope $f'(0)$ the coefficients c_k

of similar multi-loop diagrams contain an extra factor 2^k and the corresponding singularity of $B(b)$ in the b-plane is situated at $b = 3\pi$.

In QCD, there are ultraviolet and infrared renormalons appearing due to the nonvanishing coefficient β_0 of the Gell-Mann–Low function. Ultraviolet renormalons are similar to the above b-plane singularity in QED; they are situated at negative values of b

$$b_u = -4\pi/\beta_0, \qquad \beta_0 = \frac{11}{3}N_c - \frac{2}{3}n_f \tag{5.7}$$

and do not lead to any problem with the Borel resummation (5.3).

Infrared renormalons exist due to logarithmic singularities of the gluon polarization operator $\Pi_g(k^2)$ at small k^2. For massless gluons and quarks, its behaviour in infrared and ultraviolet limits coincides up to a common sign. For the particular case of radiative corrections to the photon polarization operator, the pole $1/k^2$ in the propagator of the virtual gluon is cancelled at small k because of gauge invariance and the corresponding renormalon singularity is situated on the integration contour in (5.3) at $b_i = 8\pi/\beta_0 = -2b_u$. In this case, the Borel resummation contains an uncertainty [16],[17], [18] $\Delta O(\alpha_c) \sim \exp(-b_i/\alpha_c) \sim \Lambda_{QCD}^4/Q^4$, which has the same functional dependence as the gluon condensate contribution $\sim \langle 0 \mid G_{\mu\nu}^2 \mid 0 \rangle/Q^4$. Therefore, the divergence of the perturbation series is related to nonperturbative effects arising from the nontrivial structure of the physical vacuum state.

Note that renormalon singularities exist at smaller values of b than instanton–anti-instanton ones and therefore generally they are more important. But these singularities are absent in the logarithmically divergent contributions to the invariant charge and matrix elements of local operators satisfying the Callan–Symanzik equations [13]. In this section, we discuss the methods of calculating the asymptotically large orders in scalar field models, in QED, in the electroweak model, and in QCD. These results are of practical interest because at present some physical quantities in the standard model are computed in many loops and there is a need to estimate them in higher orders. In the first section, we discuss the ultraviolet renormalons from the point of view of the operator product expansion and of the renormalization group.

5.1 Renormalization group approach to renormalons

Let us consider the ultraviolet renormalon singularities in QED and QCD. The fine structure constant in both cases will be denoted by α. In the general case, the renormalons appear in the loop integrals of the type (5.5) where the integral over the momentum p is rapidly convergent at large p^2 as $\int d^4p \, / p^{2r}$ for $r = 3, 4, \ldots$. Inserting the logarithmic corrections $\sim (\alpha \ln(p^2/m^2))^k$ in the loop vertices and in the propagators we find a factorial behaviour $\sim k!$ of the large-order coefficients. Generally, the renormalon contribution can be written as an integral over the particle virtuality p^2 of the expression proportional to an effective vertex Γ. For this vertex, p^2 plays the role of an ultraviolet cut-off. The vertex has also external lines corresponding to particles of moderate virtuality. We consider a gauge invariant amplitude $F(\alpha)$ (for example, the matrix element of the product of four electromagnetic currents). The Feynman integral for this case after integration over the angles can be written as follows (cf. (5.5)):

$$A(\alpha_e) = \int_0^\infty d\rho\, e^{-(r-2)\rho} \Gamma(\alpha, \rho), \quad r = 3, 4, \ldots, \tag{5.8}$$

where $\Gamma(\alpha, \rho)$ is the corresponding vertex and $\rho = \ln \frac{p^2}{m^2}$. The parameter $r = 3, 4, \ldots$ is related to the canonical dimension of the gauge-invariant operator described by the matrix element Γ.

We consider below the ultraviolet renormalon contributions (5.8) from the point of view of the operator product expansion [13],[15],[18]. Within the framework of this approach, $\Gamma(\alpha, \rho)$ can be written as a product of the coefficient function $R(\alpha(\rho))$ (being generally a series in the running coupling constant $\alpha(\rho)$) and the matrix element $M(\alpha, \rho)$ of a local operator with ultraviolet cut-off p^2:

$$\Gamma(\alpha, \rho) = R(\alpha(\rho))\, M(\alpha, \rho), \quad R(\alpha(\rho)) = r_1 \alpha(\rho) + r_2 \alpha^2(\rho) + \ldots. \tag{5.9}$$

We have chosen the normalization point μ entering in R and M equal to $|p|$. In this case, the function $R(\alpha(\rho))$ does not contain an additional dependence on $\ln p^2/\mu^2$.

The quantity $A(\alpha_e)$ is expanded in a perturbation series

$$A(\alpha) = \sum_{n=0}^\infty a_k\, \alpha^k, \tag{5.10}$$

where the coefficients a_n can be calculated using the following contour integral representation (cf. [6]):

$$a_k = -\int_0^\infty d\rho\, e^{-(r-2)\rho} \int_L \frac{d\alpha}{2\pi i\, \alpha^{k+1}} \Gamma(\alpha, \rho). \tag{5.11}$$

The closed contour L goes around the point $\alpha = 0$ in clockwise direction. One should integrate first over α and then over ρ because generally $A(\alpha)$ contains singular contributions at $\alpha = 0$ which cannot be expanded in a series in α. In principle, $\Gamma(\alpha, \rho)$ has also nonanalytic terms related to instanton-type field configurations which will be discussed in the following sections. The positions of the renormalon and instanton singularities are opposite in sign. For large $k \gg 1$, it is possible to calculate the integrals over α and ρ by the saddle-point method. Below we construct the functional for finding this saddle-point contribution.

The renormalization group equations

$$\frac{\partial}{\partial\rho}\alpha(\rho) = -\beta(\alpha(\rho)), \quad \frac{\partial}{\partial\rho}\ln M(\alpha, \rho) = \gamma(\alpha(\rho)), \tag{5.12}$$

with the functions $\beta(\alpha)$ and $\gamma(\alpha)$ calculated in perturbation theory,

$$\beta(\alpha) = \beta_0\alpha^2 + \beta_1\alpha^3 + \ldots, \quad \gamma(\alpha) = \gamma_0\alpha + \gamma_1\alpha^2 + \ldots, \tag{5.13}$$

can be written in integral form

$$F(\alpha(\rho)) - F(\alpha) = \rho, \quad \ln M(\alpha, \rho) = \int_{F(\alpha)}^{\rho+F(\alpha)} dx\, f(x), \quad f(x) = \gamma(F^{-1}(x)), \tag{5.14}$$

where

$$F(\alpha) = -\int^{\alpha} \frac{d\alpha'}{\beta(\alpha')} = \frac{1}{\beta_0 \alpha} + \frac{\beta_1}{\beta_0^2} \ln \alpha + \dots, \quad f(x)_{x \to \infty} = \frac{\gamma_0}{\beta_0} \frac{1}{x} + \dots. \quad (5.15)$$

We have neglected above the finite renormalization effects fixing integration constants in accordance with the conditions

$$\alpha(0) = \alpha, \quad M(\alpha, 0) = 1, \quad (5.16)$$

because at the saddle point for integral (5.11) α is small for large k as will be shown below.

One can verify that $M(\alpha, \rho)$ satisfies the Callan–Symanzik equation

$$\left(-\frac{\partial}{\partial \rho} - \beta(\alpha) \frac{\partial}{\partial \alpha} + \gamma(\alpha) \right) M(\alpha, \rho) = 0.$$

Its general solution can be written in the form of

$$\ln M(\alpha, \rho) = \int^{\rho + F(\alpha)} dx \, \tilde{f}(x) - \int^{F(\alpha)} dx \, f(x),$$

where $\tilde{f}(x)$ is an arbitrary function. But from our normalization condition (5.16) for $M(\alpha, 0)$ we find that $\tilde{f}(x) = f(x)$ in accordance with the above integral representation (5.14) for $M(\alpha, \rho)$.

One can write the coefficient a_k (5.11) in the form of

$$a_k = -\int_0^{\infty} d\rho \int_{L'} \frac{\beta(\alpha) \, dy}{2\pi i \, \alpha} \chi(y) \, e^{-\int^{F(\alpha)} dx \, f(x)} \, e^{s_k(\alpha)}, \quad (5.17)$$

where the new integration variable is a function of the invariant charge

$$y = \rho + F(\alpha) = F(\alpha(\rho)). \quad (5.18)$$

We have introduced also the functions

$$s_k(\alpha) = -k \ln \alpha + (r - 2) F(\alpha), \quad (5.19)$$

$$\chi(y) = \exp\left(-(r - 2)y + \int^y dx \, \phi(x) \right), \quad (5.20)$$

$$\phi(x) = f(x) - \beta(F^{-1}(x)) \frac{\partial \ln R(F^{-1}(x))}{\partial F^{-1}(x)}, \quad (5.21)$$

where

$$\phi(x)_{x \to \infty} \approx -(1 - \frac{\gamma_0}{\beta_0}) \frac{1}{x}. \quad (5.22)$$

The contour L' goes in clockwise direction along a large circle to the right of all singularities of the integrand. This contour can be deformed in such way that the essential values of y on it are of the order of unity.

In Eq. (5.17), the integral over ρ at large k can be calculated by the saddle-point method. The position of the saddle point is given by the condition

$$\delta s_k(\tilde{\alpha}) = -\frac{k}{\tilde{\alpha}} - \frac{r-2}{\beta(\tilde{\alpha})} = 0. \tag{5.23}$$

For fixed y this equation defines the saddle point in the integral over ρ. Because $\tilde{\alpha}$ is small, we can calculate it using perturbation theory:

$$\tilde{\alpha} \approx -\frac{r-2}{\beta_0 k} - \frac{\beta_1}{\beta_0}\left(\frac{r-2}{\beta_0 k}\right)^2, \quad \ln\tilde{\alpha} \approx \ln\frac{r-2}{-\beta_0 k} + \frac{\beta_1}{\beta_0^2}\frac{r-2}{k}. \tag{5.24}$$

One can extract from the integral a smooth factor that depends on $\tilde{\alpha}$ and expand the function $s_k(\alpha)$ in a series in the small fluctuations $\Delta\rho = \rho - \tilde{\rho}$ near the saddle point:

$$s_k(\alpha) \approx k\ln\frac{-\beta_0 k}{e\,(r-2)} - (r-2)\frac{\beta_1}{\beta_0^2}\ln\frac{-\beta_0 k}{r-2} - \frac{(r-2)^2}{2k}(\Delta\rho)^2.$$

Therefore, the contour of the integration over $\Delta\rho$ goes in the right direction for $\tilde{\rho}$ to be a saddle point.

After taking the Gaussian integral over the fluctuations $\Delta\rho$ we obtain

$$a_k \approx \sqrt{\frac{2\pi}{k}}\left(\frac{-\beta_0 k}{(r-2)\,e}\right)^{k-\frac{(r-2)\beta_1}{\beta_0^2}} e^{-\frac{(r-2)\beta_1}{\beta_0^2}}\left(\frac{k}{r-2}\right)^{-\frac{\gamma_0}{\beta_0}} c(r-2), \tag{5.25}$$

where

$$c(r-2) = -\int_{L'}\frac{d\,y}{2\pi i}R(F^{-1}(y))\exp\left(-(r-2)\,y + \int^y f(x)\,d\,x\right). \tag{5.26}$$

Note that in the case of QCD, $\beta_0 > 0$ and a_k contains the factor $(-1)^k$ leading to the Borel summability of the perturbation series.

Thus, generally, the common factor $c(r-2)$ in the renormalon contribution can be found only if we can take into account nonperturbative effects. In principle, $c(r-2)$ could be zero at least for some operators [19],[20]. At $r \to 2$, large values of y are essential and it is possible to use perturbation theory to calculate $c(r-2)$:

$$c(r-2) = \int_{L'}\frac{d\,y}{2\pi i}\frac{r_1}{\beta_0}y^{-1+\frac{\gamma_0}{\beta_0}}\exp\left(-(r-2)\,y\right) = -\frac{r_1}{\beta_0}\Gamma^{-1}(1-\frac{\gamma_0}{\beta_0})(r-2)^{-\frac{\gamma_0}{\beta_0}}, \tag{5.27}$$

where r_1 was defined in Eq. (5.9). Although for all local operators $r - 2 \geq 1$, this perturbative result can be considered as an argument for nonvanishing $c(r-2)$ in the general case.

From Eq. (5.3) one can find the singularity of the Borel transformed function

$$B(b) \approx \Gamma\left(-\frac{(r-2)\beta_1}{\beta_0^2} - \frac{\gamma_0}{\beta_0}\right)(-\beta_0)^{-\frac{(r+2)\beta_1}{\beta_0^2}}(r-2+\beta_0 b)^{\frac{(r-2)\beta_1}{\beta_0^2}+\frac{\gamma_0}{\beta_0}}c(r-2). \tag{5.28}$$

Therefore, in the general case, the singular part of the Borel transformation is factorized and proportional to the renormalon "residue" $c(r-2)$. This factorised expression gives the possibility to calculate the discontinuity of $B(b)$ situated in QED on the path of integration at $b > (r-2)/|\beta_0|$.

Let us return to $A(\alpha)$ written in the form of

$$A(\alpha) = e^{(r-2)F(\alpha) - \int^{F(\alpha)} dx\, f(x)} \int_{F(\alpha)}^{\infty} dy\, e^{-(r-2)y + \int^{y} dx\, \phi(x)}. \qquad (5.29)$$

This satisfies the renormalization group equation

$$- R(\alpha) = \left(-\beta(\alpha) \frac{d}{d\alpha} - (r-2) + \gamma(\alpha) \right) A(\alpha). \qquad (5.30)$$

Therefore, there exist such functions $R(\alpha)$ for which the renormalons are absent (in particular $c(r-2) = 0$). For given $R(\alpha)$, this differential equation has a solution with the following boundary condition

$$\lim_{F(\alpha) \to \infty} \left(e^{-(r-2)F(\alpha) + \int^{F(\alpha)} dx\, f(x)} A(\alpha) \right) = 0. \qquad (5.31)$$

In the leading logarithmic approximation, the invariant charge $\alpha(\rho)$ has a Landau pole in QED at $\rho = 1/|\beta_0|$ lying on the path of integration over ρ. Taking into account that generally $\gamma_0 \neq 0$ and $\beta_1 \neq 0$, we find that this pole becomes a cut, but the integral over ρ for $A(\alpha)$ still contains an uncertainty related to different ways of integrating around this singularity. One can believe that for the complete theory the singularities $y_{\pm}$ of the integrand move into the complex plane from the real axis and are situated at symmetric points $\operatorname{Im} y_{\pm} = \pm a$. Because at large $|y|$ the asymptotics of the integrand for $c(r-2)$ can be calculated in perturbation theory due to the relation

$$\exp \int^{y} dx\, \phi(x) \approx y^{-(1 - \frac{\gamma_0}{\beta_0})},$$

these singularities should exist somewhere in the y-plane. Note that expression (5.8) in QED, written in the form

$$A(\alpha) = \int_{0}^{\infty} d\rho\, G(\rho - \Lambda)\, e^{(r-2)F(\alpha) - \int^{F(\alpha)} dx\, f(x)} c(r-2), \quad \Lambda = -F(\alpha) \approx 1/\alpha, \qquad (5.32)$$

where $G(y)$ is given by

$$G(y) = e^{-(r-2)y + \int^{y} dx\, \phi(x)} \frac{1}{c(r-2)}, \qquad (5.33)$$

can be interpreted as a dispersion relation. Indeed, at large ρ and Λ, the function $G(\rho - \Lambda)$ behaves like a smeared pole $(\rho - \Lambda)^{-1}$ because it is large at a relatively small argument and normalized due to (5.26) as

$$\int_{L'} dy\, G(y) = -2\pi i.$$

One can write an analogous dispersion-like representation also for the function $B(b)$.

In the next sections, we discuss the divergence of the perturbation theory related to the Dyson instability for various field models written in terms of functional integrals. It is helpful to begin the discussion with a similar phenomenon existing for usual integrals.

5.2 High-order estimates in zero-dimensional models

One of the reasons for the divergence of perturbation theory is related to the rapid growth of the number of the Feynman diagrams at large orders $k \gg 1$. To estimate this number it is helpful to consider a zero-dimensional scalar model with a quartic self-interaction term [8]. Its partition function is given by the integral

$$Z(g) = \int_{-\infty}^{\infty} \frac{dx}{\sqrt{2\pi}} \exp(-S(x, g)) = \sum_{k=0}^{\infty} Z_k \, g^k, \quad S(x, g) = \frac{x^2}{2} + \frac{g}{4} x^4, \qquad (5.34)$$

where

$$Z_k = \frac{(-1)^k}{4^k k!} \int_{-\infty}^{\infty} \frac{dx}{\sqrt{2\pi}} \exp\left(-\frac{x^2}{2}\right) x^{4k} = \frac{(-1)^k}{\sqrt{\pi}} \frac{\Gamma(2k + \frac{1}{2})}{\Gamma(k + 1)} \qquad (5.35)$$

and $\Gamma(x)$ is the Euler gamma-function. The asymptotic behaviour of Z_k is

$$\lim_{k \to \infty} Z_k = \frac{(-1)^k}{\sqrt{\pi k}} \left(\frac{4k}{e}\right)^k \left(1 + O\left(\frac{1}{k}\right)\right). \qquad (5.36)$$

On the other hand, the same result can be obtained by applying the saddle-point method to calculate the integral Z_k:

$$\lim_{k \to \infty} Z_k = 2 \frac{(-1)^k}{\sqrt{2\pi k}(k/e)^k} \frac{\exp\left(-S_{eff}(\tilde{x}, k)\right)}{\sqrt{\frac{\partial^2}{(\partial \tilde{x})^2} S_{eff}(\tilde{x}, k)}}. \qquad (5.37)$$

The factor 2 takes into account the contributions of two saddle points at $x = \pm \tilde{x}$,

$$S_{eff}(x, k) = \frac{x^2}{2} - k \ln \frac{x^4}{4}, \quad \frac{\partial}{\partial \tilde{x}} S_{eff}(\tilde{x}, k) = \tilde{x} - 4\frac{k}{\tilde{x}} = 0,$$

$$S_{eff}(\tilde{x}, k) = 2k - 2k \ln(2k), \quad \frac{\partial^2}{(\partial \tilde{x})^2} S_{eff}(\tilde{x}, k)) = 2, \qquad (5.38)$$

and we have integrated over the quantum fluctuations δx near the saddle point $x = \tilde{x} + \delta x$. One can represent Z_k as an integral over x and g:

$$Z_k = \int_L \frac{dg}{2\pi i g} \frac{1}{g^k} \int_{-\infty}^{\infty} \frac{dx}{\sqrt{2\pi}} \exp(-S(x, g)) \qquad (5.39)$$

with the new effective action

$$S_{eff}(x, g, k) = S(x, g) + k \ln g. \qquad (5.40)$$

The closed contour of integration L goes in counterclockwise direction around the point $g = 0$. The saddle point in the variables x, g is found from the equations

$$\frac{\partial}{\partial \tilde{x}} S(\tilde{x}, \tilde{g}) = \tilde{x} + \tilde{g} \tilde{x}^3 = 0, \quad \tilde{g} \frac{\tilde{x}^4}{4} + k = 0 \qquad (5.41)$$

and is given by

$$\tilde{x} = \pm \sqrt{k}, \quad \tilde{g} = -\frac{1}{k}. \qquad (5.42)$$

Note that the saddle point $\tilde{g}$ is situated at negative values of g. This is related to the Dyson vacuum instability. Indeed, we can easily show that the function Z has a singularity at $g = 0$ because the discontinuity on the cut $g < 0$ is nonzero:

$$
\begin{aligned}
\Delta Z(g) = Z(g + i\epsilon) - Z(g - i\epsilon) &= \int_{-(1+i)\infty}^{(-1+i)\infty} \frac{dx}{\sqrt{2\pi}} \exp(-S(x, g)) \\
&+ \int_{(1+i)\infty}^{(1-i)\infty} \frac{dx}{\sqrt{2\pi}} \exp(-S(x, g)),
\end{aligned}
\tag{5.43}
$$

where the contours of integration pass the region $x \sim 1$. At small negative values of g there are saddle points $\tilde{x} = \pm 1/\sqrt{-g}$ on these contours and we find the following result by calculating the Gaussian integrals over the fluctuations δx near these points:

$$
\lim_{g \to -0} \Delta Z(g) = -i \exp \frac{1}{4g}.
\tag{5.44}
$$

One can write the dispersion relation for $Z(g)$ in the form of

$$
Z(g) = \int_{-\infty}^{0} \frac{dg'}{2\pi i} \frac{\Delta Z(g')}{g' - g} + subtractions
\tag{5.45}
$$

and calculate the following contribution of the high-order terms of the saddle point at $g' = -1/(4k)$

$$
\lim_{k \to \infty} Z_k = (-1)^k \int_{-\infty}^{0} \frac{dg'}{2\pi i} (-g')^{-k-1} i \exp \frac{1}{4g'} = \frac{(-1)^k}{\sqrt{\pi k}} \left(\frac{4k}{e} \right)^k \left(1 + O\left(\frac{1}{k} \right) \right).
\tag{5.46}
$$

Note that in this case we have a sign-alternating behaviour of the coefficients Z_k, which leads to the possibility of summing the perturbation expansion using the Borel transformation:

$$
Z(g) = \int_{0}^{\infty} \frac{dz}{g} B(z) e^{-\frac{z}{g}}, \quad B(z) = \sum_{k=0}^{\infty} \frac{Z_k}{k!} z^k.
\tag{5.47}
$$

The series for $B(z)$ has a finite convergence radius $|z| < 1/4$ because the nearest singularity $B(z) \sim \ln(z + z_0)$ of this function is situated at $z_0 = -1/4$ beyond the region of integration.

Let us consider now another integral, which can serve as a zero-dimensional model of the partition function for the Higgs or Yang–Mills fields:

$$
Z^H(g) = \int_{-\infty}^{\infty} \frac{dx}{\sqrt{\pi}} \exp(-S^H(x, g)), \quad S^H(x, g) = x^2(1 - \lambda x)^2, \quad g = \lambda^2.
\tag{5.48}
$$

In this case, the action $S^H(x, g)$ besides $x = 0$ has another minimum at $x = 1/\lambda = 1/\sqrt{g}$ and therefore the perturbation expansion can be written as follows:

$$
Z^H(g) = 2 \sum_{k=0}^{\infty} Z_k^H g^k.
\tag{5.49}
$$

With the use of the generating function

$$\exp\left(2\lambda x^3 - \lambda^2 x^4\right) = \sum_{n=0}^{\infty} H_n(x) \frac{(\lambda x^2)^n}{n!} \tag{5.50}$$

for the Hermite polynomials

$$H_n(x) = n! \sum_{m=0}^{[n/2]} \frac{(-1)^m (2x)^{n-2m}}{m!\,(n-2m)!} = (-1)^n \, e^{x^2} \frac{d^n}{(dx)^n} e^{-x^2} \tag{5.51}$$

one can derive the following expression for the coefficients Z_k^H:

$$Z_k^H = \int_{-\infty}^{\infty} \frac{dx}{\sqrt{\pi}} e^{-x^2} H_{2k}(x) \frac{x^{4k}}{(2k)!} = \frac{(4k)!}{((2k)!)^2} \frac{\Gamma(k+1/2)}{\Gamma(1/2)}. \tag{5.52}$$

The perturbation series in this case is not Borel summable because the asymptotic behaviour of the coefficients Z_k^H,

$$\lim_{k\to\infty} Z_k^H = \frac{1}{\sqrt{\pi k}} \left(\frac{16k}{e}\right)^k \tag{5.53}$$

does not contain the factor $(-1)^k$.

Nevertheless, we can obtain for it a Borel-like representation. Indeed, let us consider the following expression defined for purely imaginary λ (negative g):

$$Y^H(g) = \frac{1}{2} \int_{-(1+i)\infty}^{(1+i)\infty} \frac{dx}{\sqrt{\pi}} \exp(-S^H(x,g)) + \frac{1}{2} \int_{-(1-i)\infty}^{(1-i)\infty} \frac{dx}{\sqrt{\pi}} \exp(-S^H(x,g)). \tag{5.54}$$

It has a singularity at $g > 0$ because its discontinuity on this cut is not zero:

$$\Delta Y^H(g) = Y^H(g+i\epsilon) - Y^H(g-i\epsilon) = \int_{-i\infty}^{i\infty} \frac{dx}{\sqrt{\pi}} \exp(-S^H(x,g)). \tag{5.55}$$

Therefore, Y^H satisfies the dispersion relation

$$Y^H(g) = \int_0^\infty \frac{dg'}{2\pi i} \frac{\Delta Y^H(g')}{g'-g} + subtractions. \tag{5.56}$$

Its real part on the cut $g > 0$ equals

$$\mathrm{Re}\, Y^H(g) = \frac{Y^H(g+i\epsilon) + Y^H(g-i\epsilon)}{2} = Z^H(g). \tag{5.57}$$

Therefore, the partition function $Z^H(g)$ is represented as a dispersion integral for $Y^H(g)$ with a principal value prescription for $g > 0$. The discontinuity $\Delta Y^H(g)$ can be calculated at $g \to +0$ by the saddle-point method. The position of the corresponding saddle point is found from the equation

$$\frac{\partial}{\partial \tilde{x}} S^H(\tilde{x}, g) = 1 - 2\lambda\tilde{x} = 0. \tag{5.58}$$

There are two saddle points since $\lambda = \pm\sqrt{g}$ but only one of them should be taken into account, for example $\tilde{x} = 1/(2\sqrt{g})$. Near this point, $S^H(x, g)$ has the form of

$$S^H(\tilde{x}, g) = \frac{1}{16g} - \frac{1}{2}(\delta x)^2, \quad x = \tilde{x} + \delta x \tag{5.59}$$

and the calculation of the Gaussian integral over the fluctuations δx gives

$$\Delta Y^H(g) = i\sqrt{2}\exp\left(-\frac{1}{16g}\right). \tag{5.60}$$

The perturbation expansion of $Y^H(g)$ for $g < 0$ is given by

$$Y^H(g) = 2\sum_{k=0}^{\infty} Z_k^H g^k \tag{5.61}$$

and the asymptotic behaviour of the coefficients Z_k^H can be calculated from the dispersion relation for $Y^H(g)$ in the form of

$$\lim_{k\to\infty} Z_k^H(g) = \lim_{k\to\infty} \int_0^{\infty} \frac{dg'}{2\pi} \left(\frac{1}{g'}\right)^{k+1} \sqrt{2}\exp\left(-\frac{1}{16g'}\right) = \frac{1}{\sqrt{\pi k}} \left(\frac{16k}{e}\right)^k \tag{5.62}$$

in agreement with the above result obtained from the explicit expression.

The function $Y^H(g)$ is Borel summable for $g < 0$

$$Y^H(g) = \int_0^{\infty} \frac{dz}{(-g)} B(z) e^{\frac{z}{g}}, \quad B(z) = 2\sum_{k=0}^{\infty} \frac{Z_k^H}{k!}(-1)^k z^k, \tag{5.63}$$

and the Borel resummation for $Z^H(g)$ for $g > 0$ is obtained by analytic continuation

$$Z^H(g) = \int_0^{\infty} \frac{dz}{g} \tilde{B}(z) e^{-\frac{z}{g}}, \quad \tilde{B}(z) = 2\sum_{k=0}^{\infty} \frac{Z_k^H}{k!} z^k, \tag{5.64}$$

where the logarithmic singularity of $\tilde{B}(z)$ at $z = 1/2$ should be avoided with the use of the principal value prescription.

5.3 Zero charge and asymptotic freedom in scalar models

In this and next sections we consider simple renormalizable scalar field theories in D-dimensional Euclidean space-time with action

$$S = \int d^D x \left(\frac{(\partial_\sigma \phi)^2}{2} + g\frac{\phi^n}{n!}\right). \tag{5.65}$$

Here the degree of nonlinearity n is assumed to be an even number and the dimension D is chosen from the condition that the coupling constant $g > 0$ is a dimensionless quantity:

$$D = \frac{2n}{n-2}. \tag{5.66}$$

Well-known examples of such theories are models with the following interactions:

$$g \int dx \frac{1}{\phi^2}, \ g \int d^3x \frac{\phi^6}{6!}, \ g \int d^4x \frac{\phi^4}{4!}. \tag{5.67}$$

The coupling constant g is assumed to be normalized at some scale $p^2 = \mu^2$ and we introduce the running coupling constant as follows:

$$g(p^2/\mu^2) = g \, \Gamma(p^2/\mu^2) \, d^{n/2}(p^2/\mu^2), \quad \Gamma(1) = d(1) = 1, \tag{5.68}$$

where $\Gamma(p^2/\mu^2)$ and $d(p^2/\mu^2)/p^2$ are the vertex function and the scalar particle Green function, respectively. For example, one can define the vertex function at the symmetric point $p_r^2 = p^2$, $p_r p_s = -p^2/(n-1)$ $(r \neq s)$. Furthermore, the action should contain the counterterms cancelling the coupling constant and mass divergences. The invariant charge $g(p^2/\mu^2)$ satisfies the equation

$$\frac{d}{d \ln(p^2/\mu^2)} g(p^2/\mu^2) = \psi(g(p^2/\mu^2)), \tag{5.69}$$

where the Gell-Mann–Low function is expanded in a perturbation series:

$$\psi(g) = \sum_{k=2}^{\infty} (-g)^k \, C_k(n). \tag{5.70}$$

The properties of the ψ-function are important for the self-consistency of the scalar field theories. The first coefficient of the perturbative expansion is positive $C_2(n) > 0$, at least for $n = 4$ and $n = 6$. In these cases, providing that the theory does not have an ultraviolet stable point $g_0 > 0$ with $\psi(g_0) = 0$ and $\psi(g)$ grows at large g more rapidly than $g \ln g$, the well-known Landau zero-charge problem exists. To go beyond the perturbation theory with the possible use of the Borel resummation (5.3), one needs to estimate high-order corrections. In the next subsections, we calculate the asymptotic behaviour of the coefficients $C_k(n)$ for $k \to \infty$.

Here we discuss an exactly solvable example of above scalar models with $n = -2$, $D = 1$. In this model, the coordinate x and field ϕ can be denoted by t and $x(t)$, respectively. In fact, this theory coincides with a nonrelativistic quantum-mechanical model with the following Hamiltonian

$$H = \frac{p^2}{2} + \frac{g}{x^2} + \frac{x^2}{2}, \ p = i\frac{d}{dx}, \tag{5.71}$$

where the "mass" term $x^2/2$ was added to have a discrete spectrum.

The normalizable solution of the Schrödinger equation $H\Psi = E\Psi$ is expressed in terms of Laguerre polynomials:

$$\Psi = x^{2s+1} \exp\left(-\frac{x^2}{2}\right) L_n^{2s+\frac{1}{2}}(x^2), \quad L_n^\alpha(z) = \Phi(-n, \alpha + 1, z), \tag{5.72}$$

$$s = \frac{-1 + \sqrt{1 + 4g}}{4} \tag{5.73}$$

and the corresponding energy is quantized as follows:

$$E_n = 2n + 2s + \frac{3}{2}. \tag{5.74}$$

Note that the ground state energy is $E_0 = 2s + \frac{3}{2}$, which does not coincide with the vacuum energy $E_0 = \frac{1}{2}$ of the harmonic oscillator at $g \to +0$. Moreover, the spacing ΔE between the neighbouring energy levels is 2 instead of 1 for the harmonic oscillator. Both properties are related to a singular character of the potential $V = g/x^2$.

At first sight, the equidistant spectrum means triviality of the interaction similar to the case of QED where we have the zero-charge phenomenon. But in fact, this quantum-mechanical model can be considered as a simple example of an asymptotically free theory of the type of QCD. Indeed, let us regularize the potential energy V_ε as follows:

$$V_{|x|>\varepsilon} = \frac{g_\varepsilon}{x^2} \tag{5.75}$$

and construct the function $g_\varepsilon = g(\varepsilon)$ for $\varepsilon \to 0$ such as to provide a nonequidistant energy spectrum for even wave functions. Generally, the energy levels of the states of even and odd parity under the substitution $x \to -x$ should be different, but for the singular potential g/x^2 they coincide with each other. After our regularization there could be a nonzero probability for a particle to penetrate through the barrier between two potential wells with $x < 0$ and $x > 0$.

To derive the functional dependence of the bare charge $g_\varepsilon \to 0$ on ε let us consider the even solution of the regularized Schrödinger equation that decreases for $x \to \pm\infty$:

$$\Psi_{x>\varepsilon} = x^{2s_\varepsilon+1} e^{-\frac{x^2}{2}} U(a, b, x^2), \quad U(a, b, z) = \int_0^\infty \frac{e^{-tz}dt}{\Gamma(a)} t^{a-1}(1+t)^{b-a-1}, \tag{5.76}$$

where $U(a, b, z)$ is the Kummer function and

$$b = 2s_\varepsilon + \frac{3}{2}, \quad E = -2a + 2s_\varepsilon + \frac{3}{2}, \quad s_\varepsilon = \frac{-1 + \sqrt{1 + 4g_\varepsilon}}{4} \approx \frac{g_\varepsilon}{2}. \tag{5.77}$$

For $x \to 0$ the function $\Psi_{x>\varepsilon}$ behaves as follows:

$$\Psi_{x\to 0} = x^{2s_\varepsilon+1} \left(\frac{\Gamma(1-b)}{\Gamma(1+a-b)} + x^{2(1-b)} \frac{\Gamma(b-1)}{\Gamma(a)} \right) (1 + O(x^2)). \tag{5.78}$$

Therefore, we obtain the following approximate result for the logarithmic derivative of the wave function at $x = \varepsilon$:

$$\frac{d}{dx} \ln \Psi_{x=\varepsilon+0} \approx -2 \frac{\Gamma(a)}{\Gamma(a-\frac{1}{2})} - 2 \frac{g_\varepsilon}{\varepsilon} = 2 \operatorname{ctn}(\pi a) \frac{\Gamma(\frac{3}{2}-a)}{\Gamma(1-a)} - 2 \frac{g_\varepsilon}{\varepsilon}. \tag{5.79}$$

On the other hand, the function $\Psi_{|x|<\varepsilon}$ can be expanded up to a quadratic term in x due to the smallness of ε. For example, if we chose $V_{|x|<\varepsilon} = g_\varepsilon/\varepsilon^2$, then the logarithmic derivative of even Ψ in the region of small x is

$$\frac{d}{dx} \ln \Psi_{0<x<\varepsilon} = 2 \frac{g_\varepsilon}{\varepsilon^2} x \tag{5.80}$$

and does not depend on a. Therefore, the quantization condition for a (and for the corresponding energy) can be written as follows:

$$2 \frac{g_\varepsilon}{\varepsilon} = \text{ctn}(\pi a) \frac{\Gamma(\frac{3}{2} - a)}{\Gamma(1 - a)}. \tag{5.81}$$

In particular, we obtain for $g_\varepsilon/\varepsilon \to \infty$

$$\lim_{g_\varepsilon/\varepsilon \to \infty} a \to -n + \frac{\varepsilon}{g_\varepsilon} \frac{1}{2\pi} \frac{\Gamma(\frac{3}{2} + n)}{\Gamma(1 + n)}, \quad n = 0, 1, 2, \ldots \tag{5.82}$$

and for $g_\varepsilon/\varepsilon \to 0$

$$\lim_{g_\varepsilon/\varepsilon \to 0} a \to -n + \frac{1}{2} - \frac{g_\varepsilon}{\varepsilon} \frac{2}{\pi} \frac{\Gamma(\frac{1}{2} + n)}{\Gamma(1 + n)}, \quad n = 0, 1, 2, \ldots \tag{5.83}$$

with some corrections to harmonic oscillator energies. Note that for negative $g_\varepsilon/\varepsilon$ the quantum number n in the large coupling limit should be shifted by unity

$$\lim_{g_\varepsilon/\varepsilon \to -\infty} a \to 1 - n + \frac{\varepsilon}{g_\varepsilon} \frac{1}{2\pi} \frac{\Gamma(\frac{1}{2} + n)}{\Gamma(n)}, \quad n = 1, 2, \ldots \tag{5.84}$$

and for $n = 0$ we obtain another asymptotic expression for a

$$\lim_{g_\varepsilon/\varepsilon \to -\infty} a = \left(\frac{2 g_\varepsilon}{\varepsilon} \right)^2. \tag{5.85}$$

This means that the ground state energy E_0 tends to $-\infty$ in this limit, which demonstrates the Dyson instability of the theory for negative $g_\varepsilon/\varepsilon$.

It is possible to give a physical interpretation of the above quantization condition for a. Indeed, consider the Schrödinger equation for the harmonic oscillator with the above Hamiltonian but without the term g/x^2. The solution that decreases for $x \to \infty$ can be expressed in terms of the parabolic cylinder function

$$\Psi^{(0)}(x) = \frac{e^{-\frac{x^2}{2}}}{\Gamma(2a - 1)} \int_0^\infty e^{-\sqrt{2}xs - \frac{s^2}{2}} s^{2a-2} \, ds, \quad E = -2a + \frac{3}{2}. \tag{5.86}$$

Its logarithmic derivative at $x = 0$ is

$$\Lambda = \frac{d}{dx} \ln \Psi^{(0)}(x)_{|x=0} = 2 \, \text{ctn}(\pi a) \frac{\Gamma(\frac{3}{2} - a)}{\Gamma(1 - a)}. \tag{5.87}$$

Thus, one can write the above quantization condition for even energy levels in the following universal form:

$$4 \frac{g_\varepsilon}{\varepsilon} = \frac{d}{dx} \ln \Psi^{(0)}(x)_{|x=0}. \tag{5.88}$$

Therefore, our theory does actually coincide with the free theory for $x > 0$ with an additional boundary condition on the logarithmic derivative of the wave function at $x = 0$.

Because the logarithmic derivative Λ has the dimension of mass similar to Λ_{QCD}, we have dimensional transmutation of the coupling constant g in this asymptotically free model:

$$\Lambda = 4\frac{g_\varepsilon}{\varepsilon}. \tag{5.89}$$

If the renormalization group equation for this model is written in the form

$$\frac{\partial g_\varepsilon}{\partial(\ln 1/\varepsilon^2)} = \Psi(g_\varepsilon), \tag{5.90}$$

then the Gell-Mann–Low function $\Psi(g)$ is given by the following expression

$$\Psi(g) = -\frac{g}{2}. \tag{5.91}$$

Presumably, all renormalizable scalar models with the interaction $V(\phi) = g\phi^{-2r}$ (r=2,3,...) also have the property of asymptotic freedom.

5.4 Renormalized strongly nonlinear scalar model

5.4.1 Strong nonlinearity and saddle-point approach

In the limit of $n \to \infty$ in (5.66), we have $D \to 2$ and the coefficients $C_k(n)$ of the perturbation theory can be calculated exactly at fixed k [22], as will be shown below. Formally, we can write the following functional integral for the n-point Green function $G(x_1, \ldots, x_n)$ in the k-th order of perturbation theory:

$$G_k = \frac{(-1)^k}{J_0 \, k!} \int \prod_x d\phi(x) \, e^{-\int d^D x \, \frac{(\partial_\sigma \phi)^2}{2}} \int \prod_{r=1}^{k} d^D x_r \, V_{int}(\phi(x_r)), \tag{5.92}$$

where

$$J_0 = \int \prod_x d\phi(x) e^{-\int d^D x \, \frac{(\partial_\sigma \phi)^2}{2}} \tag{5.93}$$

and the function $V_{int}(\phi(x))$ corresponds to the interaction term $g\phi^n/n!$ written in normal order:

$$V_{int}(\phi(x)) = g\frac{: \phi^n(x) :}{n!} = g\int_L \frac{dl}{2\pi i} \frac{\exp\left(l\phi(x) - \frac{1}{2}l^2\Delta(0)\right)}{l^{n+1}}. \tag{5.94}$$

Here the integral is performed along the contour L closed around the point $l = 0$ in counterclockwise direction, and the free Green function is defined by

$$\Delta(x) = \int \frac{d^D p}{(2\pi)^D |p|^2} e^{i\vec{p}\vec{x}} = b(n) |x|^{2-D}, \quad b(n) = \frac{1}{4}\Gamma\left(\frac{2}{n-2}\right)\pi^{-D/2}. \tag{5.95}$$

In the large-n limit, this is simplified as follows:

$$\Delta(x)_{|n\to\infty} = \frac{n}{8\pi} \exp\left(-\frac{1}{2n}(1 + \ln\pi + \gamma) - \frac{2}{n}\ln|x|\right), \tag{5.96}$$

where γ is the Euler constant.

The functional integral in the expression for G_k generally contains contributions of disconnected diagrams, but in the large-n limit they are not important. The most essential contribution arises from diagrams in which all interaction vertices are connected by an almost equal number n/k of inner lines and each vertex absorbs the same number of external lines. In this configuration we can extract the Feynman integral over the coordinates x_r in the saddle point. It turns out that this integral contains only a single logarithmic divergence in the region where all relative distances $|x_{rs}|$ tend to zero simultaneously. In fact, its integrand is equal to the product of the total number of diagrams and Green functions $\Delta(x_{rs})$.

To take account of the translational and scale invariance, one can use the Faddeev–Popov trick corresponding to the insertion of unity in the integrand,

$$1 = \int_{\epsilon^2}^{1/\mu^2} dy^2\, \delta\left(y^2 - \prod_{r=1}^{k} |x_r - x|^{2/k} \right) \int d^D x\, \delta^D \left(x - \frac{1}{k} \sum_{r=1}^{k} x_r \right) \tag{5.97}$$

with the subsequent change of the integration variables $x_r \to x_r y + x$, $l_r \to l_r y^{2/(n-2)}$. After this transformation, we can omit y everywhere leaving only the pole $1/y^2$ appearing in agreement with the Gell-Mann–Low equation for the invariant charge. The integral

$$\int d^D x \prod_{i=1}^{n} \Delta(x_i - x) \tag{5.98}$$

should be omitted, which corresponds to amputation of the external lines. In this way, the following expression for the coefficients $C_k(n)$ of the Gell-Mann–Low function is obtained after taking the Gaussian integral over $\phi(x)$:

$$C_k(n) = \frac{1}{k!} \int \prod_{r=1}^{n} d^D x_r \frac{dl_r}{l_r^{n+1} 2\pi i} \left(\sum_{i=1}^{k} l_i \right)^n \Phi(l_k; x_k), \tag{5.99}$$

where

$$\Phi(l_k; x_k) = \exp\left(-\frac{1}{2} \sum_{r \neq r'} l_r l_{r'} \Delta(x_r - x_{r'}) \right) \delta^D \left(\frac{1}{k} \sum_{r=1}^{k} x_r \right) \delta \left(\frac{1}{k} \sum_{r=1}^{k} \ln |x_r^2| \right).$$

Here x_r are the coordinates of the interaction points in the Feynman diagrams. The integrals over l_r can be calculated for $n \to \infty$ by the saddle point method. There are two saddle points: $\tilde{l}_r = \pm\sqrt{8\pi/k}$. The integration over the fluctuations δl_r around these saddle points gives the result [22]

$$C_k(n)|_{n\to\infty} = \frac{\sqrt{2}}{k!} \left(\frac{ek}{8\pi} \right)^{\frac{n(k-1)}{2}} k^n \left(2\pi n \frac{k-1}{k} \right)^{-\frac{k}{2}} e^{-(k-1)(1+\ln\pi + c_E)} z_k, \tag{5.100}$$

where the factor z_k does not depend on n and is expressed in terms of an integral over the coordinates of the interaction points in the two-dimensional space:

$$z_k = \int \prod_{r=1}^{k} d^2 x_r \prod_{s<s'} |x_s - x_{s'}|^{-\frac{4}{k}} \delta^2 \left(\frac{1}{k} \sum_{r=1}^{k} x_r \right) \delta \left(\frac{1}{k} \sum_{r=1}^{n} \ln |x_r|^2 \right). \tag{5.101}$$

This expression can be interpreted as a partition function of k particles interacting with each other through the gravitational potential $\ln |x_r - x_{r'}|$ at a temperature of $T = k/4$. Such Feynman diagrams appear if we replace our potential at large n by the Liouville interaction

$$g\frac{\phi^n}{n!} \to g\frac{\widetilde{\phi}^n}{n!} \exp\left(\frac{n\delta\phi}{\widetilde{\phi}}\right), \tag{5.102}$$

where the saddle-point field is $\widetilde{\phi} = n\sqrt{k/(16\pi)}$ and $\delta\phi$ describes the quantum fluctuations near this saddle point.

5.4.2 Holomorphic factorization of the Feynman diagrams

To calculate the integral z_k, we initially shift the variables $x_r \to x_r + x_k$ for $r < k$ and use two δ-functions to integrate over x_k. Then go to Minkowsky space by an anti-Wick rotation of the contours of integration over the second components y_r of the vectors $x_r = (z_r, y_r)$ to the imaginary axes $y_r = it_r$, and introduce the light-cone variables [22]

$$\alpha_r = t_r - z_r, \quad \beta_r = t_r + z_r, \quad d^2x_r = \frac{i}{2}d\alpha_r d\beta_r. \tag{5.103}$$

After rescaling

$$\alpha_r \to \alpha_r \alpha_1, \quad \beta_r \to \beta_r \beta_1 \tag{5.104}$$

we integrate over α_1, β_1 with the use of the third δ-function and obtain for z_r the following expression:

$$z_k = \pi \left(\frac{i}{2}\right)^{k-2} \int \prod_{r=2}^{k-1} \frac{d\alpha_r d\beta_r}{(\alpha_r\beta_r + i\epsilon)^{\frac{2}{k}}} \prod_{1\le s<s'\le k-1} ((\alpha_s - \alpha_{s'})(\beta_s - \beta_{s'}) + i\epsilon)^{-\frac{2}{k}}. \tag{5.105}$$

The additional term $i\epsilon \to 0$ in the denominators corresponds to the Feynman prescription and in the product over s and s' the equalities

$$\alpha_1 = \beta_1 = 1 \tag{5.106}$$

are implied.

If the integrals over β_r are considered to be external, then the integration over $\alpha_{r'}$ is non-zero only in the region $0 < \beta_r < 1$ for $r = 2, \dots, n - 1$ since in the opposite case all singularities in one of the variables $\alpha_{r'}$ will be on the same side of the integration contour, which would lead to the vanishing of the corresponding integral. Moreover, each region of integration $0 < \beta_{r_2} < \beta_{r_3} < \dots < \beta_{r_{k-1}} < 1$ gives the same contribution. Therefore, we can obtain for z_r the following factorized expression

$$z_k = \pi(k-2)! \, B_k \, A_k, \tag{5.107}$$

where

$$B_k = \int_0^1 d\beta_2 \beta_2^{-\frac{2}{k}} \int_0^{\beta_2} d\beta_3 . \beta_3^{-\frac{2}{k}} \dots \int_0^{\beta_{k-2}} d\beta_{k-1} \beta_{k-1}^{-\frac{2}{k}} \prod_{1\le s<s'<k} (\beta_s - \beta_{s'})^{-\frac{2}{k}} \tag{5.108}$$

and

$$A_k = \left(\frac{i}{2}\right)^{k-2} \prod_{r=2}^{k-1} \int_{-\infty}^{\infty} d\alpha_r \, (\alpha_r + i\epsilon)^{-\frac{2}{k}} \prod_{1 \leq s < s' < k} (\alpha_s - \alpha_{s'} + i\epsilon)^{-\frac{2}{k}}. \tag{5.109}$$

The integration region over α_r can also be devided in $k!/2$ subregions corresponding to different orderings of the values of α_r ($r = 2, 3, \ldots, k - 1$), $\alpha_k = 0$, and $\alpha_1 = 1$. In particular, from the region $0 < \alpha_{k-1} < \alpha_{k-2} < \ldots < \alpha_1 = 1$ we obtain the contribution $(i/2)^{k-2} B_k$. Generally, A_k can be written in factorized form

$$A_k = B_k \, D_k, \tag{5.110}$$

where D_k is proportional to the sum of the phase factors

$$D_k = \left(\frac{i}{2}\right)^{k-2} \sum_P \exp\left(-i\eta_P \frac{2}{k}\pi\right). \tag{5.111}$$

Here P denotes all possible permutations of indices $k, k - 1, \ldots, 1$, corresponding to different orderings of values of α_r. The quantity η_P is the number of the elementary pair permutations which are needed to obtain P from the normal order $k, k - 1, \ldots, 1$. The phase factors $\exp(-i\eta_P \, 2\pi/k)$ appear due to the Feynman prescription for the integration above the branch points of the integrand. Up to these factors and to the multiplier $(i/2)^{k-2}$, the value of the integral in each subregion coincides with B_k due to the conformal invariance of the integrand. Indeed, in the case of integration over all variables α_t ($t = 1, 2, \ldots k$) in the region $-\infty < \alpha_k < \alpha_{k-1} < \ldots < \alpha_1 < \infty$ the result is the same for all possible choices of indices r, s at the additional constraint $\alpha_r = 0$, $\alpha_s = 1$. Therefore [22]

$$z_k = \pi \, (k - 2)! \, B_k^2 \, D_k. \tag{5.112}$$

Let us write D_k in the form of

$$D_k = \left(\frac{i}{2}\right)^{k-2} \chi(\exp(-2\pi i/k)), \tag{5.113}$$

where the function $\chi(z)$ can be represented as follows:

$$\chi(z) = \sum_{l=0}^{\infty} N_l^k \, z^l. \tag{5.114}$$

The coefficients N_l^k are the numbers of permutations P of α_s ($s = 1, \ldots, k$ and $\alpha_k < \alpha_1$) for which $\eta_P = l$. They are expressed in terms of the number $\widetilde{N}_m^{k-2}$ of permutations of α_r ($r = 2, 3, \ldots, k - 1$), for which the number of pair permutations is m, by the simple relation

$$N_l^k = \sum_{m=max(0, n-k+2)} (n - m + 1) \, \widetilde{N}_m^{k-2}. \tag{5.115}$$

Indeed, for fixed m and n there are $n - m + 1$ possibilities to insert $\alpha_k = 0$ and $\alpha_1 = 1$ in the ordered sequence of α_r, $-\infty < \alpha_{i_1} < \alpha_{i_2} < \ldots < \alpha_{i_k} < \infty$. On the other hand, if we introduce the generating function for $\tilde{N}_m^{k-2}$ by

$$\phi^{k-2}(z) = \sum_{m=0}^{\infty} \tilde{N}_m^{k-2} z^m, \tag{5.116}$$

then we can easily find an explicit expression for it in a recursive way:

$$\phi^{k-2}(z) = \prod_{r=1}^{k-2} \frac{1 - z^r}{1 - z} = z^{\frac{k-3}{4}(k-2)} \prod_{r=1}^{k-2} \frac{z^{-r/2} - z^{r/2}}{z^{-1/2} - z^{1/2}}. \tag{5.117}$$

As a consequence of the above relation between N_l^k and $\tilde{N}_m^{k-2}$, one can express $\chi(z)$ in terms of $\phi^{k-2}(z)$

$$\chi(z) = \phi^{k-2}(z) \frac{d}{dz} \frac{1 - z^k}{1 - z} = \phi^{k-2}(z) \left(-\frac{kz^{k-1}}{1 - z} + \frac{1 - z^k}{(1 - z)^2} \right). \tag{5.118}$$

Taking into account that $z^k = 1$ for $z = \exp(-2\pi i/k)$, we obtain finally the following expression for D_k:

$$D_k = 2^{-k+1} \frac{k}{\sin(\pi \frac{1}{k})} \prod_{r=1}^{k-2} \frac{\sin(\pi \frac{r}{k})}{\sin(\pi \frac{1}{k})} = \frac{k^2 2^{-2k+2}}{(\sin(\pi/k))^k}, \tag{5.119}$$

where we have used the Gauss multiplication formula for Γ-functions:

$$\prod_{r=1}^{k-1} \sin \frac{\pi r}{k} = \pi^{k-1} \left(\prod_{m=1}^{k-1} \Gamma\left(\frac{m}{k}\right) \right)^{-2} = 2^{-k+1} k. \tag{5.120}$$

5.4.3 Gell-Mann–Low function for the strongly nonlinear model

To calculate B_k (5.108) we represent it in the form of

$$B_k = \frac{2}{k!} \lim_{\Delta \to +0} \frac{1}{2\pi} \int_0^{2\pi} \prod_{m=1}^{k} d\varphi_m \prod_{1 \le l < j \le k} |\varphi_l - \varphi_j|^{-\frac{2}{k}} \delta\left(\frac{\varphi_1 - \varphi_k}{\Delta} - 1\right), \tag{5.121}$$

where the factor $2/k!$ takes into account that $k!/2$ integration regions with different orderings of φ_m give the same contribution. Furthermore, for $\Delta \to 0$ the distances between all points on the circle are small $\sim \Delta$ and the factor $1/(2\pi)$ cancels the integral over their collective coordinate.

On the other hand, according to the Dyson hypothesis [23], proven by Wilson [24] and Hanson [25], the partition function for k electrons on the circle can be calculated analytically:

$$\Psi_k(\gamma) = (2\pi)^{-k} \int_0^{2\pi} \prod_{m=1}^{k} d\varphi_m \prod_{1 \le l < j \le k} |e^{i\varphi_l} - e^{i\varphi_j}|^{\gamma} = \frac{\Gamma(1 + \gamma k/2)}{(\Gamma(1 + \gamma/2))^k}. \tag{5.122}$$

In particular, we have

$$\Psi_k(\gamma)|_{\gamma \to -2/k} = \frac{(1 + \gamma k/2)^{-1}}{\left(\Gamma(1 - \frac{1}{k})\right)^k}. \tag{5.123}$$

The same singularity of $\Psi_k(\gamma)$ can be derived also from the integral

$$\lim_{\gamma \to -2/k} (2\pi)^{-k} \int_{|\varepsilon| \ll 1} d\varepsilon \, \delta(\varphi_1 - \varphi_k - \varepsilon) \int_0^{2\pi} \prod_{m=1}^k d\varphi_m \prod_{1 \le l < j \le k} |\varphi_l - \varphi_j|^\gamma$$

$$= \lim_{\gamma \to -2/k} \frac{B_k}{(2\pi)^{k-1}} \frac{k!}{2} \int_{|\varepsilon| \ll 1} d\varepsilon \, |\varepsilon|^{k-1+\gamma k(k-1)/2} = \frac{B_k}{(2\pi)^{k-1}} \frac{k!}{k-1} \frac{1}{1 + \gamma k/2}, \tag{5.124}$$

where we have changed the integration variables $\varphi_m \to \varphi_m |\varepsilon|$ to extract a power of $|\varepsilon|$. Thus, by comparing the last two expressions for $\Psi_k(\gamma)|_{\gamma \to -2/k}$ one can find

$$B_k = \frac{(2\pi)^{k-1}}{(k-2)! k} \left(\Gamma(1 - 1/k)\right)^{-k}. \tag{5.125}$$

The final result for z_k is given by [22]

$$z_k = \frac{\pi^{k-1}}{\Gamma(k-1)} \left(\frac{\Gamma(1/k)}{\Gamma(1 - 1/k)}\right)^k. \tag{5.126}$$

The asymptotic growth of the coefficients $C_n(k)$ at large k is very rapid:

$$C_n(k)|_{k \to \infty} = \frac{k^{n(k+1)/2}}{k^{k-1}} \left(\frac{e}{8\pi}\right)^{n(k-1)/2} (2\pi n)^{-k/2} e^{-(k+1) c_E} \frac{e^{k+3/2}}{\sqrt{2\pi}}. \tag{5.127}$$

We shall reproduce this asymptotics in the next section using a completely different approach. Therefore, the perturbation series is badly divergent especially for large values of n. We search the Gell-Mann–Low function in the form of the Mellin–Barnes representation [22] (see also [27])

$$\Psi(g) = \sum_{k=2}^\infty (-g)^k C_n(k) = -\int_{\sigma - i\infty}^{\sigma + i\infty} \frac{dk}{2i \sin(\pi k)} g^k C_n(k), \tag{5.128}$$

where $\sigma < 2$. This corresponds to one of the possibilities to sum the asymptotic series, because at small g the resummed expression reproduces it. The discontinuity of $\Psi(g)$ at $g < 0$ is given by the simple expression

$$\Delta\Psi(g)|_{g<0} = \Psi(g + i\varepsilon) - \Psi(g + i\varepsilon) = -\int_{2-\epsilon-i\infty}^{2-\epsilon+i\infty} dk \, g^k C_n(k). \tag{5.129}$$

In the limit of $n \to \infty$, the coefficients $C_n(k)$ are known analytically and have their rightmost singularity at $k \to 1$ as follows:

$$C_n(k)|_{k \to 1} = \frac{1}{\sqrt{\pi n}} (k - 1)^{3/2}. \tag{5.130}$$

As a result, one can find the strong coupling behaviour of $\Psi(g)$ [22]:

$$\lim_{g \to \infty} \Psi(g) = -\frac{g}{(\ln g)^{3/2}} \frac{1}{2\pi \sqrt{n}}. \tag{5.131}$$

Taking into account that for small g the function $\Psi(g)$ is positive,

$$\Psi(g) = 2^{-1/2} \left(\frac{e}{\pi}\right)^{n/2} \frac{1}{n\pi} e^{-1-c_E} g^2 + O(g^3), \tag{5.132}$$

we conclude that the Gell-Mann–Low function should change its sign in the interval $0 < g < \infty$ and there is a nontrivial fixed ultraviolet point where $\Psi(g) = 0$ [22]. If one can apply the above formulas which have been derived for $n \to \infty$ to the cases $n = 4$ and $n = 6$, then the positions of the ultraviolet stable points are $g = 103$ and $g = 187$, respectively.

The asymptotic behaviour of z_k for $k \to \infty$ is

$$\lim_{k \to \infty} z_k = \frac{(\pi e)^k k^{3/2}}{(2\pi)^{1/2}} \frac{\exp(-2c_E)}{\pi}. \tag{5.133}$$

On the other hand, we can attempt to apply a statistical approach to the integral (5.105), assuming that at large k the significant region of integration is fixed: $|x_{ij}| \sim 1$. In this case, one can sum over the densities of the interaction points $\rho(x)$ instead of integrating over x_i. After finding the saddle-point density $\widetilde{\rho}(x) \sim k$, it is possible to determine the asymptotics of z_k by calculating the functional integral over the quantum fluctuations $\delta\rho$ around this saddle-point solution [22]. In the next subsection, we shall find the asymptotic behaviour of the coefficients $C_n(k)$ at large k using another method.

5.5 Functional approach to the high-order estimates

5.5.1 Classical solutions, zero modes, and counterterms

As in the above zero-dimensional case, we can initially find the singularity of physical quantities on the left-hand cut $g < 0$ in the g-plane. Within a certain accuracy it is given by the expression [6, 11]

$$\Delta Z(g) \sim \exp(-S(\widetilde{\phi})), \quad S(\phi) = \int d^D x \, L(\phi), \quad L(\phi) = \frac{(\partial_\sigma \phi)^2}{2} + g\frac{\phi^n}{n!}, \tag{5.134}$$

where the saddle-point function is obtained from the Euler–Lagrange equation

$$\partial_\sigma^2 \widetilde{\phi} = g \frac{\widetilde{\phi}^{n-1}}{(n-1)!}. \tag{5.135}$$

Its spherically symmetric solutions are given by [6]

$$\widetilde{\phi}(x - x_0, y) = a(n, g) \left(\frac{y}{(x - x_0)^2 + y^2}\right)^{\frac{2}{n-2}}, \quad a^{n-2}(n, g) = -\frac{8n!}{(n-2)^2 g}, \tag{5.136}$$

where the arbitrary parameters x_0 and $y > 0$ appear as a result of the invariance of the initial action under translations and dilatations.

To justify the above choice of the classical solution, one can use the Sobolev inequality for integrals defined on arbitrary functions ϕ [21]

$$T(\phi) \geq C_S(D) \, (n! \, V(\phi))^{\frac{2}{n}}, \quad T(\phi) = \int d^D x \frac{(\partial_\sigma \phi)^2}{2}, \quad V(\phi) = \int d^D x \frac{\phi^n}{n!}, \quad (5.137)$$

where $D = 2n/(n-2)$ and

$$C_S(D) = \frac{D(D-2)}{2} \pi \left(\frac{\Gamma(D/2)}{\Gamma(D)} \right)^{2/D}. \quad (5.138)$$

The equality is realized on the above solutions $\widetilde{\phi}$ (for an arbitrary factor $a(n, g)$). It can be verified with the use of the relations

$$T(\widetilde{\phi}) = a^2(n, g) \frac{D(D-2)}{2} \pi^{D/2} \frac{\Gamma(D/2)}{\Gamma(D)}, \quad V(\widetilde{\phi}) = \frac{a^n(n, g)}{n!} \pi^{D/2} \frac{\Gamma(D/2)}{\Gamma(D)}. \quad (5.139)$$

Due to the Sobolev inequality, the maximal value of the integrand for the functional integral expressing the perturbation theorical coefficients $C_k(n)$ for $k \to \infty$ is given by above saddle-point configuration and we have [6]

$$C_k(n) \sim \frac{1}{k!} \max e^{-T(\phi)} V^k(\phi) \leq \frac{1}{k!} \max e^{-T(\widetilde{\phi})} \left(\frac{1}{n!} \right)^k \left(\frac{T(\widetilde{\phi})}{C_S(D)} \right)^{kn/2}$$

$$= \frac{1}{k!} \left(\frac{1}{n!} \right)^k \left(\frac{k \, n}{2e \, C_S(D)} \right)^{kn/2}, \quad (5.140)$$

where it was exploited the fact that at the saddle point $T(\widetilde{\phi}) = kn/2$.

One can obtain $C_k(n)$ also with the use of the dispersion approach, deriving initially the value of the action for the saddle-point function $\widetilde{\phi}$:

$$S(\widetilde{\phi}) = T(\widetilde{\phi}) + g V(\widetilde{\phi}) = (1 - \frac{2}{n}) T(\widetilde{\phi}) = \left(\frac{g_0}{g} \right)^{\frac{2}{n-2}}, \quad (5.141)$$

where

$$g_0 = -n! \frac{2}{n-2} \left(\frac{4\pi}{n-2} \right)^{\frac{n}{2}} \left(\frac{\Gamma(D/2)}{\Gamma(D)} \right)^{\frac{n-2}{2}}. \quad (5.142)$$

This means that the discontinuity for the functional integral is given by the expression [11]

$$\Delta Z(g) \sim i \, \exp \left(- (g_0/g)^{2/(n-2)} \right). \quad (5.143)$$

The calculation of the dispersion integral for $C_k(n)$ from the discontinuity $\Delta Z(g)$ gives the following asymptotic result:

$$C_k(n) \sim c_k(n),$$

$$c_k(n) \equiv \int_{-\infty}^{0} \frac{dg}{2\pi} \frac{\exp \left(- (g_0/g)^{2/(n-2)} \right)}{(-g)^{k+1}} = \left(-\frac{1}{g} \right)^k e^{-k(\frac{n}{2}-1)} \sqrt{\frac{n-2}{4\pi \, k}}, \quad (5.144)$$

where the saddle-point value $g = \tilde{g}$ is

$$\tilde{g} = k^{-\frac{n-2}{2}} \left(\frac{2}{n-2}\right)^{\frac{n-2}{2}} g_0 = -n! \left(\frac{8\pi}{(n-2)^2}\right)^{\frac{n}{2}} \left(\frac{\Gamma(D/2)}{k\,\Gamma(D)}\right)^{\frac{n-2}{2}} \tag{5.145}$$

and we have taken account of the fluctuations around this saddle point.

To find the pre-exponential factor in the asymptotic expression for the high-order coefficients γ_n of the perturbative series for the n-point vertex function,

$$(-g)\Gamma(x_1, \ldots, x_n) = \sum_{k=1}^{\infty} (-1)^k g^k \gamma_k(x_1, \ldots, x_n)), \tag{5.146}$$

one should expand the field near its saddle-point configurations:

$$\phi = \tilde{\phi}(x - x_0, y) + \delta\phi, \tag{5.147}$$

and calculate the functional integral over the quantum fluctuations $\delta\phi$ [6]:

$$\gamma_k(x_1, \ldots, x_n)$$

$$= 2\,c_k(n) \int d^D x_0 \int_0^\infty \frac{dy}{y^{D+1}} \prod_{t=1}^{n} \tilde{g}\,\frac{\tilde{\phi}^{n-1}(x_t - x_0, y)}{(n-1)!} \int \prod_x{}' d\,\delta\phi\, R, \tag{5.148}$$

where the factor 2 accounts for two equal contributions from the saddle-point fields $\pm\tilde{\phi}$. Furthermore, we have extracted from the integral the product of the fields for external particles in the saddle points and used the classical equations (5.135) to remove the propagators. The loop corrections to these propagators are not important within our accuracy, and in fact the high-order expansion of the invariant charge $g(p^2/\mu^2)$ coincides with the expansion of $g\Gamma(p^2/\mu^2)$. The symbol $\prod'$ in the expression for γ_k means that the integration over the fluctuation $\delta\phi \sim \tilde{\phi}$ should be treated in a special way because it gives the factor i in $\Delta\Gamma$, which was taken already into account in the calculation of $\Delta Z(g)$ and $c_k(n)$. The integrand $R = R(\delta\phi)$ is given by

$$R = \frac{1}{J_0} e^{-\delta\phi\,K\,\delta\phi}\,k^{D+1}\delta^D\left(\int d^D x\,\delta\rho(x)\,x\right)\delta\left(\int d^D x\,\delta\rho(x)\,\ln|x|\right) e^{-\delta S}, \tag{5.149}$$

where

$$J_0 = \int \prod_x d\,\delta\phi\, \exp\left(-\int d^D x\,\frac{(\partial_\sigma \delta\phi)^2}{2}\right) \tag{5.150}$$

and the quadratic form in the exponent is

$$\delta\phi\,K\,\delta\phi = \int d^D x\left(\frac{(\partial_\sigma \delta\phi)^2}{2} + \tilde{g}\,\frac{\tilde{\phi}^{n-2}(x, 1)}{(n-2)!}\,\frac{(\delta\phi)^2}{2}\right). \tag{5.151}$$

To integrate accurately over the zero modes related to the translational and dilatational invariance we have used the Faddeev–Popov trick by inserting in the initial functional integral the following representation of unity [6]:

$$1 = \int d^D x_0 \int_0^\infty \frac{dy}{y^{1+D}} \left(\int d^D x \, \rho(x) \right)^{D+1} \delta^D(\Delta X) \, \delta(\Delta X), \qquad (5.152)$$

where

$$\Delta X = \int d^D x \, \rho(x) \frac{x - x_0}{y}, \qquad \Delta X = \int d^D x \, \rho(x) \ln \frac{|x - x_0|}{y}, \qquad (5.153)$$

and

$$\rho(x) = -g \frac{\phi^n}{n!} \qquad (5.154)$$

can be considered as the density of the interaction points. The saddle-point solution $\widetilde{\phi}$ is assumed to satisfy these constraints. The δ-functions in the expression for R appear after the following substitution of the coordinates in the fields $\phi(x)$ entering in the definition of ρ:

$$x \to (x + x_0) \, y, \qquad \rho((x + x_0) \, y) \to y^{-D} \left(\widetilde{\rho}(x) + \delta\rho(x) \right), \qquad (5.155)$$

where the fluctuations $\delta\rho(x)$ around the saddle-point density $\widetilde{\rho}(x)$ are defined as follows:

$$\delta\rho(x) = -\widetilde{g} \frac{\widetilde{\phi}^{n-1}(x, 1)}{(n-1)!} \, \delta\phi. \qquad (5.156)$$

Finally, the term δS that appears in the expression of $R(\delta\phi)$ is the contribution of the counterterms [6]. To remove the ultraviolet divergences at $R(\delta\phi)$, it is enough to take into account for $n \geq 6$ only one counterterm for the vertex ϕ^{n-2}

$$\delta S_{n \geq 6} = -\widetilde{g} \int d^D x \frac{\widetilde{\phi}^{n-2}}{(n-2)!} \frac{\Delta(0)}{2} \qquad (5.157)$$

and two counterterms for $n = 4$

$$\delta S_{n=4} = -\widetilde{g} \int d^4 x \frac{\widetilde{\phi}^2}{2!} \frac{\Delta(0)}{2} + \widetilde{g}^2 \int d^4 x \frac{\widetilde{\phi}^4}{16} b^2(4) \int \frac{d^4 x}{x^4} \exp \left(i \frac{2 p^\mu x}{\sqrt{3}} y \right), \qquad (5.158)$$

where p^μ is the particle momentum in the normalization point μ and $\Delta(x)$ is the scalar particle Green function

$$\Delta(x) = \int \frac{d^D p}{(2\pi)^D |p|^2} e^{i p x} = b(n) |x|^{2-D}, \qquad b(n) = \frac{1}{4} \Gamma \left(\frac{2}{n-2} \right) \pi^{-D/2}. \qquad (5.159)$$

Note that the second contribution in $\delta S_{n=4}$ corresponds to three one-loop Feynman diagrams for the vertex function calculated at the pair invariants $(p_r^\mu + p_s^\mu)^2 = (2 - 2/3)\mu^2 = 4\mu^2/3$. The additional factor y in the exponent of the Fourier transformation appears as a result of the above shifting $x \to xy$ of the relative coordinates.

5.5.2 Quantum fluctuations around classical solutions

For the calculation of the integral over the fluctuations $\delta\phi$, it is helpful to take into account the symmetry of the saddle-point configuration $\widetilde{\phi}$ [6]. Recall that after the use of the Faddeev–Popov trick only the solution with $x_0 = 0$ and $y = 1$ should be considered. As a result, one is able to keep only the symmetry under the $D(D+1)/2$-parameter subgroup of the initial symmetry group including the rotational, dilatational, and conformal transformations with the number of parameters $D(D+3)/2+1$. This symmetry can be realized as a rotation group in the $D+1$-dimensional space [26]. Therefore it is natural to perform the inverse stereographic projection of the x space to the unit sphere $|Z|=1$ in the $D+1$-dimensional space with a simultaneous renormalization of the field:

$$Z_\mu = \frac{2x_\mu}{1+x^2}, \quad Z_{D+1} = \frac{x^2-1}{x^2+1}, \quad \phi(x) = \left(\frac{2}{1+x^2}\right)^{\frac{2}{n-2}} (Y_0 + \delta Y(Z)), \qquad (5.160)$$

where the classical solution $\widetilde{Y} = Y_0$ in the new variables is a constant:

$$Y_0 = a(n,g)\, 2^{-\frac{2}{n-2}}. \qquad (5.161)$$

We obtain the following helpful relations

$$\int d^D x \left(\frac{2}{x^2+1}\right)^D = S_{D+1} = 2\frac{\pi^{(D+1)/2}}{\Gamma((D+1)/2)}, \qquad (5.162)$$

$$\int d^D x\, (\partial_\sigma \delta\phi)^2 = \int dS_{D+1} \left((\nabla_t \delta Y)^2 + \frac{D(D-2)}{4}(\delta Y)^2\right), \qquad (5.163)$$

$$-\widetilde{g}\int d^D x\, \frac{\widetilde{\phi}^{n-2}}{(n-2)!}(\delta\phi)^2 = \int dS_{D+1}\frac{D(D+2)}{4}(\delta Y)^2, \qquad (5.164)$$

$$-\frac{1}{\sqrt{k}}\widetilde{g}\int d^D x\, \frac{\widetilde{\phi}^{n-1}}{(n-1)!}\delta\phi = \int dS_{D+1}\frac{D}{\sqrt{2}}\delta Y\frac{1}{\sqrt{S_{D+1}}}. \qquad (5.165)$$

Here ∇_t is the gradient tangential to the surface of the unit sphere.

It is obvious that the spherical harmonics on the $D+1$-sphere diagonalize the quadratic form because they are eigenfunctions of the spherical Laplacian operator

$$\nabla_t^2 Y_r^{(m)} = \lambda^{(m)} Y_r^{(m)}, \quad \lambda^{(m)} = m(m+D-1), \qquad (5.166)$$

where the index r denotes the components of the corresponding irreducible representation. Explicitly, the normalized spherical harmonics are given by

$$Y_r^{(m)} = S_r^{\mu_1 \cdots \mu_m} Z_{\mu_1} \cdots Z_{\mu_m}, \quad \int dS_{D+1} Y_r^{(m)} Y_{r'}^{(m')} = \delta_{m,m'}\delta_{r,r'} \qquad (5.167)$$

where the normalization coefficients $S^{\mu_1 \cdots \mu_n}$ include the subtraction of the tensors containing the Kronecker symbols to make the product of Z_{μ_r} traceless. The number of independent eigenfunctions Y_r for given m is

$$N_m(D) = \frac{\Gamma(D+m-1)}{\Gamma(D)\,\Gamma(m+1)}(D+2m-1). \qquad (5.168)$$

In particular, for the first two harmonics with $m=0$ and $m=1$ we have

$$Y^{(0)} = \frac{1}{\sqrt{S_{D+1}}}, \quad Y_\mu^{(1)} = \sqrt{\frac{D+1}{S_{D+1}}}\, Z_\mu. \qquad (5.169)$$

We expand the function $\delta Y(z)$ in a linear combination of normalized spherical harmonics:

$$\delta Y(z) = \sum_{m=0}^{\infty} \sum_{r=1}^{N_m(D)} C_{m,r} Y_r^{(m)}, \quad \int_z \prod' dY(z) = \int \prod'_{m,r} dC_{m,r}. \tag{5.170}$$

In the quadratic forms

$$\delta\varphi\, K\, \delta\phi = \sum_{m=0}^{\infty} \sum_r \frac{C_{m,r}^2}{2} \left(\left(m + \frac{D}{2}\right)\left(m + \frac{D-2}{2}\right) - \frac{D(D+2)}{4} \right), \tag{5.171}$$

$$\delta\varphi\, K_0\, \delta\phi = \sum_{m=0}^{\infty} \sum_r \frac{\tilde{C}_{m,r}^2}{2} \left(m + \frac{D}{2}\right)\left(m + \frac{D-2}{2}\right) \tag{5.172}$$

the coefficient of C_m^2 for $m = 0$ is negative and we should rotate the contour of integration over C_0^2 to the line parallel to the imaginary axis, which gives an additional factor i taken already into account in the multiplier $c_k(n)$. The coefficient of $C_{m,\mu}^2$ for $m = 1$ is zero. We should calculate the integrals over the zero modes $C_{1,\mu} = C_\mu$ with the use of the δ-functions and of the relations

$$\int d^D x\, \frac{\delta_1\rho(x)}{\sqrt{k}}\, x = \frac{D\, C_\mu}{\sqrt{2S_{D+1}}} \int dS_{D+1} x\, Y_\mu^{(1)} = \sqrt{\frac{D+1}{2}}\, C, \tag{5.173}$$

$$\int d^D x\, \frac{\delta_1\rho(x)}{\sqrt{k}}\, \ln|x| = C_{D+1}\sqrt{\frac{D+1}{2}}\, \frac{D\, S_D}{S_{D+1}} \int_0^{\pi/2} d\sin\vartheta\,(\sin\vartheta)^{D-1}\, \ln\frac{1+\cos\vartheta}{1-\cos\vartheta}$$

$$= \sqrt{\frac{D+1}{2}}\, C_{D+1}. \tag{5.174}$$

Thus, we obtain from the integration over $C_{m,r}$, taking into account the factor $1/J_0$ [6],

$$\int \prod_x d\delta\varphi\, R = k^{\frac{D+1}{2}}\, Z(D), \tag{5.175}$$

$$Z(D) = \sqrt{\frac{D-2}{4}} \left(\frac{D(D+2)}{4\pi(D+1)}\right)^{\frac{D+1}{2}} \exp\left(-\frac{1}{2}F(D) - \Delta S(D)\right), \tag{5.176}$$

where the first two factors in $Z(D)$ appear as a result of integration over the negative and zero modes C_0, $\tilde{C}_0$ and C_μ, $\tilde{C}_\mu$ ($\mu = 1, 2, \ldots, (D+1)$), respectively. The function $F(D)$ is a regularized contribution of the modes with $m \geq 2$. For $n \geq 6$, we can represent it as the following sum:

$$F(D) = \sum_{m=2}^{\infty} N_m(D)\, f_m(D), \tag{5.177}$$

$$f_m(D) = \ln\left(1 - \frac{D(D+2)}{(D+2m)(D+2m-2)}\right) + \frac{D(D+2)}{(D+2m)(D+2m-2)}. \tag{5.178}$$

The function $F(D)$ is different for $n = 4$:

$$F = \sum_{m=2}^{\infty} N_m(4) \left(\ln (1 - t(m)) + t(m) + \frac{t^2(m)}{2} \right), \tag{5.179}$$

$$t(m) = \frac{6}{(m+2)(m+1)}. \tag{5.180}$$

The terms subtracted from the logarithms provide a good convergence of the sums at large m. On the other hand, these subtractions (with the sum taken from $m = 0$) can be expressed in terms of contributions of one-loop Feynman diagrams expanded up to the second order in the external field $\sim \tilde{\varphi}^{n-2}$ (see (5.151)) and added to the counterterms δS (5.157) and (5.158). For the counterterm $\delta S(D)$ (5.157) at $n \geq 6$ modified by the first-order contribution, we obtain the following finite result:

$$\Delta S(D)_{n \geq 6} = \frac{1}{2} \sum_{m=0}^{1} \frac{N_m(D) D(D+2)}{(D+2m)(D+2m-2)} = \frac{D^2}{2(D-2)}. \tag{5.181}$$

In the case $n = 4$, the counterterm δS (5.158) contains an additional dependence on $y\mu$. After subtracting from it the first- and second-order corrections to $-F/2$ in the external field $\tilde{\varphi}^2$ we obtain for the modified counterterm [6]

$$\Delta S(y\mu) = -3 \ln(y\mu) + \Delta S, \quad \Delta S = \frac{1}{2} \sum_{m=0}^{1} (3 + 2m) \left(1 + \frac{3}{(m+2)(m+1)} \right)$$

$$+ \tilde{g}^2 \int \prod_{i=1}^{2} \frac{d^4 x_i}{16\pi^2} \frac{\tilde{\phi}^4(\frac{x_1+x_2}{2}) \exp \left(i \frac{2 p^\mu (x_1 - x_2)}{\sqrt{3}} y \right) - \tilde{\phi}^2(x_1) \tilde{\phi}^2(x_2)}{|x_1 - x_2|^4} + 3 \ln(y\mu)$$

$$+ \frac{15}{2} + \frac{9}{\pi^4} \int \frac{d^4 x_1 \, d^4 x_2}{|x_1 - x_2|^4} \left(\frac{\exp \left(i \frac{2(x_1 - x_2)_4}{\sqrt{3}} \right)}{(|\frac{x_1+x_2}{2}|^2 + 1)^4} - \frac{1}{(|x_1|^2 + 1)^2 (|x_2|^2 + 1)^2} \right) = 8 - 3\gamma, \tag{5.182}$$

where $\gamma = -\psi(1)$ is the Euler constant.

5.5.3 High-order coefficients of the Gell-Mann–Low function

As was mentioned above, the high-order behaviour of the perturbation coefficients d_k for the invariant charge

$$g(p^2/\mu^2) = \sum_{k=1}^{\infty} (-1)^k g^k d_k(p^2/\mu^2) \tag{5.183}$$

coincides up to a common sign with the Fourier transformed coefficients $\gamma_k(x_1, \ldots, x_n)$:

$$(2\pi)^D \delta^D \left(\sum_{r=1}^{n} \vec{p}_r \right) d_k(p^2/\mu^2)$$

$$= -\int \prod_{r=1}^{n} d^D x_r \left(e^{i \sum_s p_s x_s} - e^{i \sum_s p_s^\mu x_s} \right) \gamma_k(x_1, \ldots, x_n), \qquad (5.184)$$

where the product of the δ-functions appears also on the right-hand side of the equality due to the translational invariance of $\gamma_k(x_1, \ldots, x_n)$.

According to the Gell-Mann–Low equation, the large-order coefficients for the perturbative expansions of $g(p^2/\mu^2)$ and $\psi(g)$ are related as follows:

$$C_k(n) = \frac{d}{d \ln p^2} d_k(p^2/\mu^2) + (k-1) d_2(p^2/\mu^2) \frac{d}{d \ln p^2} d_{k-1}(p^2/\mu^2)$$

$$+ \frac{(k-2)(k-3)}{2} d_2^2(p^2/\mu^2) \frac{d}{d \ln p^2} d_{k-2}(p^2/\mu^2) + \ldots, \qquad (5.185)$$

where

$$d_2\left(\frac{p^2}{\mu^2}\right)\Big|_{n=4} = \frac{3}{32\pi^2} \ln \frac{p^2}{\mu^2} \qquad (5.186)$$

and the right-hand side should not depend on p^2/μ^2 due to the renormalizability of the theory. For $n \geq 6$, the coefficients d_k grow very rapidly: $d_k \geq (k!)^2$, and hence

$$C_k(n)|_{n \geq 6} = \frac{d}{d \ln p^2} d_k(p^2/\mu^2). \qquad (5.187)$$

In the case $n = 4$, they grow as $k! (16\pi^2)^{-k}$ and as a result we have at large k:

$$C_k(4) = \exp(16\pi^2 d_2(p^2/\mu^2)) \frac{d}{d \ln p^2} d_k(p^2/\mu^2)$$

$$= \left(\frac{p^2}{\mu^2}\right)^{3/2} \frac{d}{d \ln p^2} d_k(p^2/\mu^2). \qquad (5.188)$$

This means, in particular, that the second factor should compensate the p^2-dependence of the first factor.

For $n > 4$, we can use the Frullani formula to calculate the integrals over y and x_r:

$$\int_0^\infty \frac{dy}{y} \int \prod_{r=1}^{n} d^D x_r \frac{-\tilde{g}\tilde{\phi}^{n-1}(x_r)}{\sqrt{k}\,(n-1)!} \left(e^{iy \sum_s p_s x_s} - e^{iy \sum_s p_s^\mu x_s} \right)$$

$$= -s^n(D) \frac{1}{2} \ln \frac{|p|^2}{\mu^2}, \qquad (5.189)$$

where

$$s(D) = \int d^D x \left(\frac{-\tilde{g}\tilde{\phi}^{n-1}(x)}{\sqrt{k}\,(n-1)!} \right) = 2^{1+D/2} \pi^{(D-1)/4} \frac{\sqrt{\Gamma((D+1)/2)}}{\Gamma(D/2)}. \qquad (5.190)$$

But in the case $n = 4$, one should take into account an additional dependence of the modified counterterm $\Delta S(y\mu)$ on $y\mu$ with the above relation between $C_k(4)$ and $d_k(p^2/\mu^2)$. Therefore, for the factor analogous to $s''(D)$, we obtain at $n = 4$

$$s^{(4)} = -2\left(\frac{p^2}{\mu^2}\right)^{3/2}\frac{d}{d\ln p^2}\int_0^\infty\frac{dy}{y}\int\prod_{r=1}^4 d^4x_r\,\frac{\widetilde{g}\widetilde{\phi}^3(x_r)}{\sqrt{k}\,3!}\,e^{iy}\sum_s p_s x_s\,(\mu y)^3$$

$$= 3\int_0^\infty dz\,z^6\left(4\pi\sqrt{3}\,K_1(z)\right)^4,\quad z = y|p|, \tag{5.191}$$

where the function $K_1(z)$ is the Bessel function of imaginary argument:

$$K_1(z) = \frac{8}{\pi z}\int_0^\infty\cos(tz)\,dt\int_t^\infty\frac{\sqrt{|x|^2 - t^2}\,|x|\,d|x|}{(|x|^2 + 1)^3} = \frac{1}{z}\int_0^\infty\frac{\cos(tz)\,dt}{(t^2 + 1)^{3/2}}. \tag{5.192}$$

Thus, for $n \geq 6$, the high-order asymptotics of the coefficients $C_n(k)$ is given by [6]

$$\widetilde{C}_n(k) = c_k(n)\left(\sqrt{k}\,s(D)\right)^n k^{\frac{D+1}{2}}Z(D)$$

$$= (-\widetilde{g})^{-k}\exp(k(1 - n/2))\,k^{(n+D)/2}\,\widetilde{C}_n, \tag{5.193}$$

where

$$\widetilde{C}_n = \left(2^{1+D/2}\pi^{(D-1)/4}\frac{\sqrt{\Gamma((D+1)/2)}}{\Gamma(D/2)}\right)^n\frac{e^{\frac{D^2/2}{D-2}}}{\sqrt{4\pi}}\left(\frac{D(D+2)}{4\pi(D+1)}\right)^{(D+1)/2}e^{-\frac{F(D)}{2}},$$

$$F(D) = \sum_{m=2}^\infty N_m\left(\ln\left(\frac{4(D+m)(m-1)}{(D+2m)(D+2m-2)}\right) + \frac{D(D+2)}{(D+2m)(D+2m-2)}\right).$$

For $n = D = 4$, the asymptotics is [6]

$$\widetilde{C}_4(k) = c_k(4)\,k^4\,s^{(4)}\frac{1}{\sqrt{2}}\left(\frac{6}{5\pi}\right)^{5/2}e^{-\frac{1}{2}F-\Delta S} = \left(\frac{k}{16\pi^2 e}\right)^k k^4\,\widetilde{C}_4, \tag{5.194}$$

where

$$\widetilde{C}_4 = e^{-\frac{1}{2}F+3c_E-8}\frac{2^{\frac{19}{2}}3^4}{5^{5/2}}\pi\int_0^\infty dz\,z^6\,(K_1(z))^4,\quad K_1(z) = \frac{1}{z}\int_0^\infty\frac{dt\,\cos(tz)}{(t^2 + 1)^{\frac{3}{2}}},$$

$$F = \sum_{m=2}^\infty\frac{2m + 1}{\alpha_m}\left(\ln(1 - \alpha_m) + \alpha_m + \frac{\alpha_m^2}{2}\right),\quad \alpha_m = \frac{6}{(m+1)(m+2)}. \tag{5.195}$$

In particular, for the case $k \gg n \gg 1$, we obtain the asymptotic behaviour (5.127), which was found in the previous subsection in the other limit of $n \gg k \gg 1$.

As for the case $n = D = 4$, one can compare the above asymptotic expression $\widetilde{C}_4(k)$ with the known analytic results for $k = 2, 3, 4$. We find for the ratios of the asymptotic and exact expressions $\widetilde{C}_4(k)/C_4(k) \approx 0.1,\, 0.7,\, 1.1$, respectively. The relative correction to the above asymptotic formula for $n = 4$ at $k \to \infty$ was calculated by Kubyshin [27]

$$C_4(k)/\widetilde{C}_4(k) \approx 1 - 4.7/k. \tag{5.196}$$

The asymptotic estimates for the multicomponent case were performed in Ref. [7].

5.6 Series divergence in models with gauge interactions

5.6.1 Classical equations for scalar models in the presence of gauge fields

To apply the saddle-point approach to estimate the high orders of perturbation theory, one should find the solution of classical equations with the minimal value of action. We have shown in the previous section that for scalar renormalizable models such a problem is solved with the use of the Sobolev inequalities. In the case of an n-component scalar field in the Euclidean $D = 4$ space-time, this inequality is of the following form (cf. (5.138)):

$$\int \left(\sum_{i=1}^{n} \phi_i^2 \right)^2 d^4x \le \frac{3}{32\pi^2} \left(\int \sum_{i=1}^{n} (\partial_\sigma \phi_i)^2 d^4x \right)^2. \tag{5.197}$$

It is realized as an equality only for solutions of classical equations for the scalar model with interaction

$$L_{int} = \lambda \left(\sum_{i=1}^{n} \frac{(\phi_i^2)^2}{2} \right). \tag{5.198}$$

These solutions have the form

$$\widetilde{\phi}_i(x) = C_i \frac{y}{(x - x_0)^2 + y^2}, \quad \sum_{i=1}^{n} C_i^2 \sim -1/\lambda \tag{5.199}$$

and provide the minimal value of the action. The parameters x_0 and y are related to the translational and dilatational symmetries of the theory and a consistent treatment of corresponding zero modes can be performed with the use of the Faddeev–Popov trick (see previous section). These zero modes appear in any renormalizable model. We choose the simplest configuration by fixing $x_0 = 0$ and $y = 1$. This solution of classical equations has a hidden ten-dimensional symmetry, which can be realized as a group of rotations of an unit sphere in the five-dimensional space. For this purpose, we perform the inverse stereographic projection of the x-space to this sphere:

$$\mathbf{Z} = \frac{2x}{x^2 + 1}, \quad Z_5 = \frac{x^2 - 1}{x^2 + 1}, \quad |Z|^2 + Z_5^2 = 1, \quad dS = 16 \frac{d^4x}{(x^2 + 1)^4}. \tag{5.200}$$

For the scalar field $\phi(x)$, we define the corresponding field on the sphere by extracting a factor

$$\phi(x) = \frac{2}{x^2 + 1} Y(Z). \tag{5.201}$$

Then the above particular solution $\widetilde{Y}$ with $x_0 = 0$ and $y = 1$ is simply a constant invariant under rotations of the sphere.

In this subsection we consider the high-order estimates for gauge models with scalar fields [28, 29, 30]. These models are the scalar electrodynamics (SED) and the Yang–Mills theory with the $SU(2)$ gauge group interacting with the charged Higgs particle in doublet representation as in the electroweak model with a vanishing Weinberg angle. It is natural to search for the most symmetric solutions of the corresponding classical equations. For this purpose, one can use the five-dimensional formalism discussed above. The vector potential A_μ and its intensity $G_{\mu\nu}$ for the Yang–Mills model can be considered to be 2×2 matrices:

$$A_\mu = A_\mu^a \frac{\tau^a}{2}, \quad G_{\mu\nu} = \partial_\mu A_\nu - \partial_\nu A_\mu - ig\,[A_\mu, A_\nu]. \tag{5.202}$$

For SED, A_μ and $F_{\mu\nu}$ are the abelian vector and tensor fields. We can introduce the corresponding five-dimensional potentials with the definitions

$$A_i(Z) = \left(\frac{1+x^2}{2}\right)^2 A_\mu(x) \frac{\partial Z_i}{\partial x_\mu}, \quad A_\mu(x) = \frac{\partial Z_i}{\partial x_\mu} A_i(Z). \tag{5.203}$$

This is compatible with the additional constraint

$$Z_i A_i = 0. \tag{5.204}$$

Then the action for the corresponding Yang–Mills model can be written as follows:

$$S = \int dS_5\, L(Y, A), \quad L(Y, A) = L_{YM} + L_S, \tag{5.205}$$

where

$$L_{YM} = \frac{1}{6}\,tr\left(L_{ij}A_k - igZ_i[A_j, A_k] + L_{jk}A_i - igZ_j[A_k, A_i] + L_{ki}A_j - igZ_k[A_i, A_j]\right)^2,$$

$$L_S = \frac{1}{2}\left|(L_{ij} - ig(Z_iA_j - Z_jA_i))\,Y\right|^2 + 2|Y|^2 + \frac{\lambda}{2}|Y|^4. \tag{5.206}$$

In the above expressions we have

$$L_{ij} = Z_i\,\partial_j - Z_j\,\partial_i. \tag{5.207}$$

The action for SED can be obtained by omitting the commutators in the above expression. Among solutions of the Euler–Lagrange equations with different values of x_0 and y we consider the solution satisfying the constraint

$$\int dS_5\, L(Y, A)\, Z_i = 0. \tag{5.208}$$

Furthermore, the following gauge for the vector field is implied

$$(\partial_i - Z_i(Z_k\partial_k))\, A_i = 0. \tag{5.209}$$

One can represent physical quantities in the form of perturbative expansions in two coupling constants

$$G(g, \lambda) = \sum_{m=0}^{\infty} \sum_{k=0}^{\infty} g^{2m} \lambda^k G_{k,m}. \tag{5.210}$$

In the case when the order k of the expansion in λ is much larger than the order m of the expansion in g^2, one can expect that in first approximation it is possible to neglect A_i and hence get the above result for the saddle-point scalar field:

$$|\tilde{Y}^{(0)}|^2 = -\frac{2}{\lambda}, \quad \tilde{Y}^{(0)} = U \sqrt{-\frac{2}{\lambda}}, \quad U^* U = 1, \tag{5.211}$$

where U is a constant isospinor in the Yang–Mills model or a phase in SED. The corresponding saddle value of the action is

$$\tilde{S} = -\frac{16\pi^2}{3\lambda}. \tag{5.212}$$

5.6.2 Search of the saddle-point configurations

Since for the function $\tilde{Y}^{(0)}$ the Sobolev bound (5.197) turns out to be an equality, we can expect that the iteration of the Euler–Lagrange equations in A_i will give a solution with the minimal action. The linearized classical equation for A_i in the field $\tilde{Y}^{(0)}$ has the form

$$\left(-\frac{1}{2} L_{ij}^2 + 2 + \frac{g^2}{2} |\tilde{Y}^{(0)}|^2 \right) A_i^{(0)} = 0, \tag{5.213}$$

where the first two terms in the brackets appeared from the contributions of L_{YM} bilinear in A. The equation for A_i^0 is an eigenvalue equation for the ratio of two coupling constants g^2/λ. We should find the minimal value for this ratio, because for fixed λ this value is equal to the radius of convergence of the series in g^2/λ. The eigenfunctions are spherical harmonics and the corresponding eigenvalues of the operator $-\frac{1}{2} L_{ij}^2$ on the symmetric traceless tensors of rank n are $n(n + 3)$. As a result, the minimal value of g^2/λ is obtained for the function which is a constant on the sphere, but this solution does not satisfy the constraint $Z_i A_i = 0$. Therefore, for small A, we choose the solution in the form of a linear combination of the first harmonics [28, 29]:

$$\tilde{A}_i^{(0)} = \varepsilon \, \eta_{ij} Z_j, \tag{5.214}$$

where ε is a small parameter and η_{ij} is a matrix in colour space in the case of the Yang–Mills theory. The corresponding value of the ratio g^2/λ is

$$\frac{g^2}{\lambda} = 6 + O(\varepsilon). \tag{5.215}$$

Note that the critical ratio 6 of g^2 and λ corresponds to the $N = 4$ supersymmetric Yang–Mills theory where three complex scalar fields belong to the adjoint representation of the gauge group.

The saddle-point solution for the vector-potential A_i^a is an isovector:

$$\tilde{A}_i^{a(0)} = \varepsilon\, \eta_{ij}^a\, Z_j. \tag{5.216}$$

Due to the condition $Z_i A_i = 0$ and the chosen gauge condition, the tensor η_{ij} should be antisymmetric:

$$\eta_{ij}^a = -\eta_{ji}^a. \tag{5.217}$$

Other constraints on the matrix η_{ij} appear from the possibility to iterate the equation for Y and A_i in the small parameter ε:

$$[\eta^a, \eta^b] = C_1 \epsilon^{abc} \eta^c, \quad \eta^a \eta^a \eta^b = C_2 \eta^b, \quad Tr\left(\eta^a \eta^b\right) \eta^b = C_3\, \eta^a, \tag{5.218}$$

where C_1, C_2 and C_3 are some constants. The reason for these constraints is the vanishing of the contribution of the first harmonics $\delta \tilde{A}_i \sim c_{ik}^a z_k$ on the left-hand side of the Euler–Lagrange equation after applying to A_i the differential operator in Eq. (5.213). The iteration is possible only if one can compensate contributions with the first harmonics on the right-hand side of the equation by the term proportional to the correction $O(\varepsilon)$ and $O(\varepsilon^2)$ to the critical value of g^2/λ.

If the constant C_1 is nonzero, then there are three independent solutions of the above equations for η^a which are unique up to rotations in the isotopic and five-dimensional coordinate space. Indeed, one can choose the constant C_1 to be unity and, as a result, the matrices $T^a = i\eta^a$ can be interpreted as a representation of the $SU(2)$ generators. These representations should be irreducible or they should consist of irreducible representations with the same weights. Such matrices can have isospin $T = 2$ acting in the entire five-dimensional space, $T = 1$ acting in the three-dimensional subspace or $T = 1/2 \times 1/2$ acting in the four-dimensional subspace [29]:

$$1)\ \eta_{ij}^a = \gamma_{cd}^i\, \gamma_{fg}^j \left(-\epsilon^{acf} \delta^{dg} - \epsilon^{acg} \delta^{df}\right), \tag{5.219}$$

$$2)\ \eta_{ij}^a = \gamma_c^i\, \gamma_d^j \left(-\epsilon^{acd}\right), \quad c,d = 1,2,3, \tag{5.220}$$

$$3)\ \eta_{ij}^a = \gamma_\mu^i\, \gamma_\nu^j\, \hat{\eta}_{\mu\nu}^a, \quad \mu, \nu = 1,2,3,4, \tag{5.221}$$

where the γ-matrices are chosen as follows:

$$\hat{Z} = Z_i \gamma^i = \frac{1}{\sqrt{2}} \begin{pmatrix} Z_4 - \frac{Z_5}{\sqrt{3}} & Z_1 & Z_2 \\ Z_1 & -Z_4 - \frac{Z_5}{\sqrt{3}} & Z_3 \\ Z_2 & Z_3 & 2\frac{Z_5}{\sqrt{3}} \end{pmatrix}, \quad \gamma_c^i = \delta_c^i, \ \gamma_\mu^i = \delta_\mu^i. \tag{5.222}$$

The tensors $\hat{\eta}_{\mu\nu}^a$ coincide with the 't Hooft matrices $\eta_{\mu\nu}^a$ up to the common factor $-1/2$:

$$\hat{\eta}^1 = \frac{1}{2}\begin{pmatrix} 0 & 0 & 0 & -1 \\ 0 & 0 & -1 & 0 \\ 0 & 1 & 0 & 0 \\ 1 & 0 & 0 & 0 \end{pmatrix}, \quad \hat{\eta}^2 = \frac{1}{2}\begin{pmatrix} 0 & 0 & 1 & 0 \\ 0 & 0 & 0 & -1 \\ -1 & 0 & 0 & 0 \\ 0 & 1 & 0 & 0 \end{pmatrix},$$

$$\widehat{\eta}^3 = \frac{1}{2}\begin{pmatrix} 0 & -1 & 0 & 0 \\ 1 & 0 & 0 & 0 \\ 0 & 0 & 0 & -1 \\ 0 & 0 & 1 & 0 \end{pmatrix}. \tag{5.223}$$

Other solutions of the above equations for η^a correspond to the commuting case $C_1 = 0$ and can be represented in the following form [29]:

$$\eta^a = l_1^a \, \eta_1 + l_2^a \, \eta_2, \tag{5.224}$$

where the matrices $\eta_{1,2}$ are equal:

$$\eta_1 = \begin{pmatrix} 0 & 1 & 0 & 0 & 0 \\ -1 & 0 & 0 & 0 & 0 \\ 0 & 0 & 0 & 0 & 0 \\ 0 & 0 & 0 & 0 & 0 \\ 0 & 0 & 0 & 0 & 0 \end{pmatrix}, \quad \eta_2 = \begin{pmatrix} 0 & 0 & 0 & 0 & 0 \\ 0 & 0 & 0 & 0 & 0 \\ 0 & 0 & 0 & 1 & 0 \\ 0 & 0 & -1 & 0 & 0 \\ 0 & 0 & 0 & 0 & 0 \end{pmatrix}. \tag{5.225}$$

Up to appropriate rotations in isospin space the three independent solutions with $C_1 = 0$ can be chosen as follows [29]:

$$4) \ (l_1^a)^2 = (l_2^a)^2 = 1, \quad l_1^a \, l_2^a = 0, \tag{5.226}$$

$$5) \ l_1^a = l_2^a, \quad (l_1^a)^2 = 1, \tag{5.227}$$

$$6) \ l_1^2 = 1, \quad l_2^a = 0. \tag{5.228}$$

5.6.3 Solutions of the classical equations

In the first case of (5.219), due to the relations

$$\widehat{Z}^3 = -\frac{\widehat{Z}}{2} + s \begin{pmatrix} 1 & 0 & 0 \\ 0 & 1 & 0 \\ 0 & 0 & 1 \end{pmatrix}, \quad s = \frac{1}{3}Tr\,\widehat{Z}^3 = Det\,\widehat{Z} \tag{5.229}$$

one can search for the solution in the form of

$$\widetilde{A}_i^a = \varepsilon^{abc}\left(a_1(s)\,\widehat{Z}\,\gamma_i + a_2(s)\,\widehat{Z}^2\,\gamma_i + a_3(s)\,\widehat{Z}^2\,\gamma_i\,\widehat{Z}\right)_{bc}, \quad \widetilde{Y} = U\,b(s), \tag{5.230}$$

where U is an arbitrary isospinor normalized by $U^+U = 1$ and we have taken into account the property that the bracket can be antisymmetrized in the indices b, c. The Euler–Lagrange equations for the functions $a_r(s)$ and $b(s)$ can be obtained from the stationarity condition of the functional S [29]

$$S = \sqrt{6}\pi^2 \int_{-1/\sqrt{54}}^{1/\sqrt{54}} ds\, L, \quad L = W\left(\frac{(a')^2}{2} + \frac{W(a_2')^2}{12} + \frac{W(a_3')^2}{36} - 2a'a_2\right)$$

$$+ 18a^2 + 9W a_2^2 + \frac{10}{3}W a_3^2 + g\left(3a^3 - 3s\,W a_2^3 + \frac{W}{9}a_3^3 + \frac{W}{2}a_2^2(3a - a_3)\right)$$

$$+ g^2 \left(\frac{9}{8} a^4 + \frac{W^2}{32} a_2^4 + \frac{W^2}{216} a_3^4 + \frac{3W}{8} a^2 a_2^2 + \frac{W}{2} a^2 a_3 (a - 9 s a_2) - \frac{W^2}{4} a_2^2 a_3 (a - s a_2) \right)$$

$$+ \frac{2}{3} W (b')^2 + 8 b^2 + 2\lambda b^4 + g^2 b^2 \left(\frac{3}{2} a^2 + \frac{1}{4} W a_2^2 + \frac{W}{12} a_3^2 \right), \tag{5.231}$$

where we have used the following notation:

$$W = 1 - 54 s^2, \quad a = a_1 + 3 s a_2 - \frac{1}{3} a_3. \tag{5.232}$$

In cases 2, 3, 5, and 6, the iteration in ϵ leads to the following ansatz for the classical solutions [29]:

$$\tilde{A}_i^a = \eta_{ik}^a Z_k \, a(t), \quad \tilde{Y} = U \, b(t), \quad t = \eta_{ij_1}^a \eta_{ij_2}^a Z_{j_1} Z_{j_2} = \sum_{r=1}^{n_\perp} Z_r^2, \tag{5.233}$$

where $n_\perp = 2, 3, 4$ is the dimension of the subspace in which the matrices η^a act. The Euler–Lagrange equation for the functions $a(t)$ and $b(t)$ can be obtained from the stationarity condition of the functional

$$S = \frac{2\pi^{5/2}}{\Gamma(\frac{n_\perp}{2}) \Gamma(\frac{5-n_\perp}{2})} \int_0^1 dt \, t^{\frac{n_\perp}{2} - 1} (1 - t)^{\frac{3-n_\perp}{2}} L_\perp, \tag{5.234}$$

where

$$L_\perp = T(T+1) \left(-g t \, a^3 + \frac{T}{4} g^2 t^2 a^4 \right) + 4t(1-t)(b')^2 + 2b^2 + \frac{\lambda}{2} b^4$$

$$+ (T(T+3-n_\perp)+1) \left(2t^2(1-t)(a')^2 + 3ta^2 + \frac{g^2}{4} ta^2 b^2 \right). \tag{5.235}$$

For case 4 of (5.226), one can obtain the following ansatz:

$$\tilde{A}_i^a = l_1^a (\eta_1)_{ik} Z_k \, a_1 + l_2^a (\eta_2)_{ik} Z_k \, a_2, \quad \tilde{Y} = U \, b, \quad (l_1^a)^2 = (l_2^a)^2 = 1, \quad l_1^a \, l_2^a = 0, \tag{5.236}$$

where the functions a_1, a_2 and b depend on two variables,

$$t_1 = -(\eta_1^2)_{ik} Z_i Z_k, \quad t_2 = -((\eta_2)^2)_{ik} Z_i Z_k. \tag{5.237}$$

The Euler–Lagrange equations for these functions are obtained from the functional

$$S = 8\pi^2 \int_0^1 dt_1 \int_0^{1-t_1} dt_2 \, (1 - t_1 - t_2)^{-\frac{1}{2}} L_4, \tag{5.238}$$

where

$$L_4 = t_2 (1 - t_2) \left(t_1 \left(\frac{\partial a_1}{\partial t_2} \right)^2 + t_2 \left(\frac{\partial a_2}{\partial t_2} \right)^2 + 2 \left(\frac{\partial b}{\partial t_2} \right)^2 \right)$$

$$+ t_1 (1 - t_1) \left(t_1 \left(\frac{\partial a_1}{\partial t_1} \right)^2 + t_2 \left(\frac{\partial a_2}{\partial t_1} \right)^2 + 2 \left(\frac{\partial b}{\partial t_1} \right)^2 \right)$$

$$- 2t_1 t_2 \left(t_1 \frac{\partial a_1}{\partial t_1} \frac{\partial a_1}{\partial t_2} + t_2 \frac{\partial a_2}{\partial t_1} \frac{\partial a_2}{\partial t_2} + 2 \frac{\partial b}{\partial t_1} \frac{\partial b}{\partial t_2} \right)$$

$$+ \frac{3}{2}(t_1 a_1^2 + t_2 a_2^2) + \frac{g^4}{2} t_1 t_2 a_1^2 a_2^2 + b^2 + \frac{\lambda}{4} b^4 + \frac{g^2}{8} b^2 (t_1 a_1^2 + t_2 a_2^2). \tag{5.239}$$

In the case of SED, the correponding expressions for solutions similar to the above cases 5, 6 are given by

$$\tilde{A}_i = (\eta_1)_{ik} Z_k \, a(t), \quad \tilde{Y} = e^{i\varphi} b(t), \quad t = -(\eta_1^2)_{ik} Z_i Z_k \tag{5.240}$$

and

$$\tilde{A}_i = (\eta_1 - \eta_2)_{ik} Z_k \, a_1(t), \quad \tilde{Y} = e^{i\varphi} b_1(t), \quad t = -((\eta_2 - \eta_1)^2)_{ik} Z_i Z_k. \tag{5.241}$$

The corresponding effective Lagrangians can be obtained from $L_\perp$ by putting $T = 0$ in it.

5.6.4 High-order estimates for gauge model with the Higgs fields

The high-order asymptotics of the coefficients of the perturbation series is expressed in terms of the effective action $F_{k,m}$ [29]:

$$G_{k,m} \sim \text{Re} \exp(-F_{k,m}), \quad F_{k,m} = \tilde{S} + m \ln g^2 + k \ln \lambda, \tag{5.242}$$

where $\tilde{S}$ is the value of S calculated on the corresponding classical solutions. The saddle-point values $\tilde{g}^2$, $\tilde{\lambda}$ of the coupling constants are found from the equations

$$m = -\frac{\partial}{\partial \ln g^2} \tilde{S}, \quad k = -\frac{\partial}{\partial \ln \lambda} \tilde{S}, \quad m + k = \tilde{S} \tag{5.243}$$

which can be written in another form:

$$m + k = \tilde{S} = \frac{s(g^2/\lambda)}{\lambda}, \quad \frac{m}{m+k} = -\frac{g^2}{\lambda} \frac{s'(g^2/\lambda)}{s(g^2/\lambda)}. \tag{5.244}$$

Here we have taken into account that $\tilde{S}$ can be represented as $s(g^2/\lambda)/\lambda$, where s is a function depending on the ratio g^2/λ. This allows us to write the asymptotics of $G_{k,m}$ in the simple form

$$G_{k,m} = \left(\frac{k+m}{16\pi^2 e} \right)^{k+m} \text{Re} \left(C \left(\frac{m}{k+m} \right) \right)^{k+m}, \tag{5.245}$$

$$C \left(\frac{m}{k+m} \right) = \frac{16\pi^2}{s \left(\frac{\tilde{g}^2}{\tilde{\lambda}} \right)} \left(\frac{\tilde{g}^2}{\tilde{\lambda}} \right)^{-\frac{m}{k+m}}, \tag{5.246}$$

where $\tilde{g}^2$, $\tilde{\lambda}$ are the saddle-point values of the coupling constants.

The simplest way to solve the Euler–Lagrange equations approximately is to use the variational approach for the corresponding functionals by representing the trial functions as superpositions of several harmonics on the five-dimensional unit sphere [29]. This approach can be justified for small m/k. In this region, we obtain for the above six types of solutions in the Yang–Mills model the following expansion of $C(x)$

$$\ln \left| C \left(\frac{m}{m+k} \right) \right| = \ln 3 - \frac{m}{k+m} \ln 6 - B_i \left(\frac{m}{k+m} \right)^2 + \dots \quad (5.247)$$

where $B_i = 1/9, 10/63, 5/18, 5/42, 5/21, 5/28$, respectively.

This means that for small m/k, the first solution with $T = 2$ gives the largest contribution to the asymptotics of $G_{k,m}$. The phase of C for small m/k is $i\pi$, which corresponds to the sign-alternating form of $G_{k,m}$. Generally, this phase depends on m/k and is different for the above six solutions.

In the opposite limit $m/k > 1$, the leading asymptotic behaviour of $G_{k,m}$ is given by the third ansatz with $T = 1/2 \times 1/2$. For this anzatz, one can find an explicit solution directly in the four-dimensional space [29]:

$$\tilde{A}^a_\mu = \frac{4}{g} \tilde{\eta}^a_{\mu\nu} x_\nu \frac{\rho^2 - 1}{(x^2 + \rho^2)(1 + \rho^2 x^2)}, \quad \tilde{\phi} = \frac{i}{g} U \frac{4\sqrt{3}}{\sqrt{(x^2 + \rho^2)(1 + \rho^2 x^2)}}, \quad (5.248)$$

where the parameter ρ is expressed in terms of the coupling constants g and λ as follows:

$$\rho = \left(12 \frac{\lambda}{g^2} - 1 \right)^{\frac{1}{4}} = e^\xi. \quad (5.249)$$

This solution describes an instanton of size $1/\rho$ embedded inside an anti-instanton of size ρ. Recall that from the beginning we have considered the average scale y equal to unity. Such type of function for $\rho \to \infty$ will be used below for the calculation of the high-order asymptotics in the pure Yang–Mills theory. The action for this solution is given by

$$S = \frac{16\pi^2}{g^2} \left(-2 + 3 \frac{\sinh(4\xi) - 4\xi}{\cosh(4\xi) - 1} \right). \quad (5.250)$$

For the case $\xi \to 0$, we obtain $g^2/\lambda \to 6$ and $\tilde{S} \to -16\pi^2/(3\lambda)$.

On the other hand, the parameter ξ can be fixed from the saddle-point conditions

$$\frac{k}{k+m} = 12e^{-2\xi} \operatorname{ctnh}(2\xi) \frac{2\xi \cosh(2\xi) - \sinh(2\xi)}{12\xi + 2(\cosh(4\xi) - 1) - 3\sinh(4\xi)} \quad (5.251)$$

and the saddle-point value of g is found from the relation

$$m + k = S. \quad (5.252)$$

There are two different solutions $\xi_{1,2}$ of these equations. For $m/k \to 0$, we have $\xi_1 \to 0$ and $\xi_2 \to 0.65 + 1.4i$. In the opposite limit $m/k \to \infty$, one can obtain $\xi_1 \to \pi i/4$ and $\xi_2 \to \infty$. Here the second branch corresponds to the noninteracting instanton–anti-instanton configuration with $S = 16\pi^2/g^2$ in the pure Yang–Mills theory where the perurbation series in g^2 is not Borel summable.

Note that in the pure Yang–Mills model ($\phi = 0$) we can find analytically the second solution corresponding to $T = 1$ [29]:

$$a = \frac{2\sqrt{2}\sinh(2\xi)}{\cosh(2\theta) + \cosh(2\xi)}, \quad \theta = \ln t, \quad \tanh(\xi) = \sqrt{\frac{3}{2}}\,e^{-i\pi/6}. \tag{5.253}$$

The action for this solution is

$$\frac{g^2}{16\pi^2} S = \frac{4}{3}e^{-i\pi/3}\left(i\frac{\sqrt{3}}{2} - \frac{\tan^{-1}(\sinh\xi)}{\sinh(\xi)}\right) \approx 2.275\,e^{-i\pi/3} \tag{5.254}$$

and its contribution to the high order asymptotics is small in comparison with the instanton–anti-instanton configuration.

In the case of scalar electrodynamics, we obtain for the above two solutions generated by the matrices η_1 and $\eta_1 - \eta_2$ the following expressions for $\ln C$ at a small ratio m/k [28]:

$$\ln\left|C_i\left(\frac{m}{m+k}\right)\right| = \ln 3 - \frac{m}{m+k} + D_i\left(\frac{m}{m+k}\right)^2 + \ldots, \tag{5.255}$$

where $D_1 = 5/21$ and $D_2 = 5/42$, respectively. Therefore, the first solution with $n_\perp = 2$ gives the leading contribution to the asymptotic behaviour of $G_{k,m}$ at $k \gg m \gg 1$. One can verify that this conclusion is valid for an arbitrary value of the ratio m/k including the pure electrodynamic case $m \gg k \sim 1$.

To obtain a more accurate expression for $G_{k,m}$ at large k and m, one should find the quantum fluctuations around the above classical solutions. The zero-mode contribution is most important and can be calculated analytically, but the Gaussian integration over fluctuations in other directions in the functional space can be performed only numerically because it is not possible to diagonalize analytically the corresponding quadratic forms. Nevertheless, for the most interesting case of the instanton–anti-instanton configuration in the pure Yang–Mills theory the explicit result will be presented in the following section.

5.7 Asymptotic estimates in the pure Yang–Mills theory

Let us discuss the high-order behaviour of the perturbation theory coefficients in the pure Yang–Mills theory with gauge group $SU(2)$ [31]. In this model, there is a series of topologically nontrivial vacuum states characterized by an integer number n (see Chapter 4). The vector potentials describing these states are pure gauge fields which, however, cannot be obtained continuously by a gauge transformation from the field $A = 0$. These vacua are degenerate. In quantum theory, there are solutions of the Yang–Mills equations in the Minkowski space-time which are responsible for the tunnel transitions between them. The simplest solution having a topological charge equal to unity is the BPST instanton (see Chapter 4)

$$\tilde{A}_\mu^a = R_{ab}\,\eta_{\mu\nu}^b\,\frac{2}{g}\frac{(x - x_0)_\nu}{(x - x_0)^2 + y^2}, \tag{5.256}$$

where R is the orthogonal matrix describing $O(3)$-colour rotations and x_0 and y are arbitrary real parameters. The t'Hooft matrix $\eta^a_{\mu\nu}$ has the form

$$\eta^a_{\mu\nu} = \varepsilon_{0a\mu\nu} + \frac{1}{2} \varepsilon_{abc} \varepsilon_{bc\mu\nu}. \tag{5.257}$$

The analogous solution with the matrix

$$\bar{\eta}^a_{\mu\nu} = \varepsilon_{0a\mu\nu} - \frac{1}{2} \varepsilon_{abc} \varepsilon_{bc\mu\nu} \tag{5.258}$$

describes the anti-instanton. The value of the action on the instanton and anti-instanton configurations is the same

$$S^{(i)} = \frac{1}{4} \operatorname{Tr} \int \tilde{G}^2_{\mu\nu} d^4x = \frac{8\pi^2}{g^2}. \tag{5.259}$$

The amplitude of the tunnel transition between two vacuum states with $\Delta n = 1$ is known (see Section 4.6)

$$\frac{\langle 0|1\rangle_{inst}}{\langle 0|1\rangle_{pert}} = \int d^4x_0 \frac{dy}{y^5} (\mu_0 y)^\nu \exp\left(-\frac{8\pi^2}{g^2(\mu_0)}\right) B \, d\Omega, \tag{5.260}$$

where $\nu = \frac{22}{3}$, the coefficient B was calculated by 't Hooft (see Chapter 4) and μ_0 is the normalization point. The integration over the Euler angles Ω with the normalized volume $\int d\Omega = 1$ corresponds to taking into account all possible orthogonal matrices R for the instanton field.

To calculate the asymptotic behaviour of the perturbation series coefficients one should find the solution of the Euler–Lagrange equations with the minimal action and a vanishing topological charge. Strictly speaking, in the pure Yang–Mills theory such a solution does not exist, but we can obtain it as a limit of the instanton–anti-instanton solution for the theory containing a complex Higgs field in doublet representation (see (5.248)). This solution is spherically symmetric, but one can deform it using conformal transformations. Moreover, in the pure Yang–Mills theory, the value of the action is the same $S \approx 2S^{(i)}$ for all weakly interacting instanton–anti-instanton configurations which can be written as follows:

$$\tilde{A}^a_\mu = R^{(1)}_{ab} \eta^b_{\mu\nu} \frac{2}{g} \frac{y_1^2(x-x_1)_\nu}{(x-x_1)^2\left((x-x_1)^2+y_1^2\right)} + R^{(2)}_{ab} \eta^b_{\mu\nu} \frac{2}{g} \frac{(x-x_2)_\nu}{(x-x_2)^2+y_2^2}, \tag{5.261}$$

where $R^{(1)}$ and $R^{(2)}$ are arbitrary 3×3 orthogonal matrices; x_1, y_1 and x_2, y_2 are parameters of the instanton and anti-instanton, respectively. It turns out that the leading contribution is obtained from two kinematical regions in which the instanton and anti-instanton have completely different scales:

$$y_1 \ll y_2 \sim |x_1 - x_2|, \tag{5.262}$$

and

$$y_2 \ll y_1 \sim |x_1 - x_2|. \tag{5.263}$$

The calculation of the action on the above trial configurations gives [31]

$$\tilde{S} = \frac{16\pi^2}{g^2} - \frac{32\pi^2}{g^2} R_{ab}^{(1)} R_{ab}^{(2)} \eta, \tag{5.264}$$

where for $y_1 \ll y_2$ we have

$$\eta = \frac{y_1^2 \, y_2^2}{\left((x_1 - x_2)^2 + y_2^2\right)^2} \ll 1. \tag{5.265}$$

The analytic properties of physical quantities in the complex g^2-plane in the Yang–Mills theory are similar to the case of the partition function Z in the zero-dimensional model with action $S = x^2(1 - \lambda x)^2$ ($\lambda = g$) considered above. Namely, it can be represented for $g^2 > 0$ as a dispersion integral with a principal value prescription for another function analytic in $g^2 < 0$. Formally, the discontinuity of this function for Z in the Yang–Mills model for $g^2 \to 0$ is given by the expression [31]

$$\Delta Z = \int \prod_{r=1}^{2} \left(d^4 x_r \frac{dy_r}{y_r^5} (\mu_0 y_r)^\nu B \, d\Omega_r \right) \exp\left(-\tilde{S}\right). \tag{5.266}$$

Here an additional condition for the product of matrices $R^{(r)}$ is implied:

$$R_{ab}^{(1)} R_{ab}^{(2)} > 0. \tag{5.267}$$

It corresponds to attractive instanton–anti-instanton interactions, which leads to a pure imaginary value for ΔZ after integration over the parameter η in $\tilde{S}$

$$\Delta Z = i\pi^3 \left(\frac{32\pi^2}{g^2}\right)^{\frac{\nu}{2}} \frac{B^2 \, e^{-16\pi^2/g^2} \, \omega}{(\nu - 2)(\nu - 1) \, \Gamma(-1 + \frac{\nu}{2})} \int d^4 x_1 \frac{dy_1}{y_1^5} (\mu_0 y_1)^{2\nu}, \tag{5.268}$$

where we have integrated also over x_2. The coefficient ω is obtained as a result of integration over the instanton and anti-instanton orientations in the colour space [31]

$$\omega = \int_{R_{ab}^{(1)} R_{ab}^{(2)} > 0} \left(R_{ab}^{(1)} R_{ab}^{(2)}\right)^{\frac{\nu}{2}} \prod_{r=1}^{2} d\Omega_r = \frac{1}{\pi} \int_0^{2\pi/3} \frac{(1 + 2\cos\phi)^{1+\frac{\nu}{2}}}{(1 + \cos\phi)(1 + \frac{\nu}{2})} \, d\phi, \tag{5.269}$$

where ϕ is the total azimuthal angle in the Euler parametrization of the matrix $R^{(1)} R^{(2)}$.

The most interesting quantity in the Yang–Mills theory is the Gell-Mann–Low function. One can calculate it using the heavy charge interaction energy $E(R)$ defined in terms of the Wilson P-exponent [8]

$$\exp(-2E(R)T) = < P \exp\left(ig \int A_\mu dz_\mu\right) > . \tag{5.270}$$

Here the integral over z goes along a rectangle with sides T and R for $R \ll T$. Its averaging in the functional integral over A for small g^2 is performed with the use of the saddle-point method around the instanton–anti-instanton configuration $\tilde{A}$.

The invariant charge can be defined as follows:

$$g^2(p) = -p^2 \int d^3 R \, e^{i\boldsymbol{p}\boldsymbol{R}} \, E(R). \tag{5.271}$$

The Gell-Mann–Low function is found from the equation

$$\psi(g^2(p)) = \frac{\partial g^2(p)}{\partial \ln p^2/\mu^2}. \tag{5.272}$$

The final result for the coefficients ψ_k in the perturbative expansion of $\psi(g^2)$,

$$\psi(g^2) = \sum_{k=2}^{\infty} (g^2)^k \, \psi_k, \tag{5.273}$$

can be written as follows [8, 31] (cf. also Ref. [32]):

$$\lim_{k\to\infty} \psi_k = -(16\pi^2)^{1-k} \, \Gamma\left(k + \frac{\nu}{2} + 8\right) C, \quad \nu = \frac{22}{3}, \tag{5.274}$$

where the constant C is given by

$$C = \frac{2^{(\frac{\nu}{2}-11)} \, 0.412}{\pi^3(\nu-2)(\nu-1)\Gamma\left(1+\frac{\nu}{2}\right)} \, I_1 \, I_2 \,. \tag{5.275}$$

(The numerical factor in (5.275) corresponds to Pauli–Villars regularization.) In (5.275) we have introduced the functions (cf.(5.269))

$$I_1 = \int_0^{\frac{2\pi}{3}} dt \, \frac{(1+2\cos t)^{1+\frac{\nu}{2}}}{1+\cos t}, \quad I_2 = \int_0^{\infty} d\rho \, \rho^{2\nu-1} \, (f_1^2(\rho) + f_2^2(\rho)), \tag{5.276}$$

defined in terms of the integrals

$$f_1(\rho) = \int_0^{\infty} r \, dr \, \sin r \left(1 + \cos \frac{\pi r}{\sqrt{r^2 + \rho^2}}\right),$$

$$f_2(\rho) = \int_0^{\infty} dr \left(r \cos r - \frac{\sin r}{r}\right) \sin \frac{\pi r}{\sqrt{r^2 + \rho^2}}. \tag{5.277}$$

Note that the asymptotic expression for ψ_k can be represented in different forms due to the relation

$$\lim_{k\to\infty} \Gamma(k + k_0) = \lim_{k\to\infty} \Gamma(k) \, k^{k_0}.$$

Because k_0 in the Yang–Mills theory is large due to a large number of zero modes, an uncertainty in the choice of the asymptotic expression for ψ_k does not allow us to verify its accuracy for existing fixed values of k. Nevertheless, this asymptotics can be used in resummations of the perturbation theory.

5.8 Asymptotic estimates in quantum electrodymamics

Although initially the Dyson argument supporting the divergence of perturbation theory was applied to QED, it was shown later that in this case there are some peculiarities related to Fermi statistics [33]. Indeed, for the anticommuting fields the approach based on the solution of classical equations is not applicable. However, because these fields enter in the QED action only in a bilinear way, one can integrate over them to obtain a determinant being a nonlocal gauge-invariant functional of the product eA_μ, where e is the electron charge and A_μ is the photon field [34]. The logarithm of the determinant gives an additional contribution to the free electromagnetic action. But it is difficult to work with such effective theory due to its nonlocality. For constant external fields $F_{\mu\nu}$, the fermionic determinant can be calculated exactly, which leads to the Heisenberg–Euler action (5.1). But in accordance with the Dyson argument for the perturbation series divergence, classical solutions for the photon fields should be found for pure imaginary values of the electric charge e. Therefore, we cannot use the Heisenberg–Euler action for our purpose. Note, however, that for us it is enough to estimate the fermionic determinant only for large complex values of the product eA_μ [34],[35]. To find its asymptotics, one can assume that the saddle-point configuration of the photon field has the most symmetric form. Furthermore, as was argued above for the case of scalar electrodynamics, there is a simple ansatz for classical solutions. Indeed, using the inverse stereographic projection of the four-dimensional space to the unit sphere in the five-dimensional space, one can write the expression for the five-component gluon field A_i in the form

$$A_i = \eta_{ij} Z_j \, a(Z_\perp^2), \tag{5.278}$$

where the matrix η is antisymmetric

$$\eta_{ij} = -\eta_{ji} \tag{5.279}$$

and satisfies the additional constraint

$$\eta_{ij}\eta_{kj} = P_{ij}. \tag{5.280}$$

Here P is the projector onto a subspace of the five-dimensional space

$$P^2 = P. \tag{5.281}$$

There are only two possibilities (5.240) and (5.241). Namely, the dimension $n_\perp$ of this subspace can be 2 or 4. Generally the argument of the function a can be written as follows

$$Z_\perp^2 = P_{ij} Z_i Z_j = \sum_{r=1}^{n_\perp} Z_r^2. \tag{5.282}$$

One can choose the matrices η_1 and η_2 for these two cases using appropriate five-dimensional rotations in the form of

$$\eta_1 = \begin{pmatrix} 0 & 1 & 0 & 0 & 0 \\ -1 & 0 & 0 & 0 & 0 \\ 0 & 0 & 0 & 0 & 0 \\ 0 & 0 & 0 & 0 & 0 \\ 0 & 0 & 0 & 0 & 0 \end{pmatrix}, \quad n_\perp = 2 \tag{5.283}$$

and

$$\eta_2 = \begin{pmatrix} 0 & 1 & 0 & 0 & 0 \\ -1 & 0 & 0 & 0 & 0 \\ 0 & 0 & 0 & 1 & 0 \\ 0 & 0 & -1 & 0 & 0 \\ 0 & 0 & 0 & 0 & 0 \end{pmatrix}, \quad n_\perp = 4. \tag{5.284}$$

We shall obtain the large field behaviour of the fermionic determinant only for the first case, because it turns out that its asymptotics is universal for all configurations of the external field. In particular, in the large field limit one can neglect the electron mass in the stationary Dirac equation

$$\gamma_\mu \, (i\partial + e\,A)_\mu \, \Psi_n = \lambda_n \, \Psi_n, \tag{5.285}$$

where

$$A_\mu = (\eta_1)_{\mu\nu} \frac{x_\nu}{x_1^2 + x_2^2} b_1(Z_\perp^2). \tag{5.286}$$

Generally, the fermion loop contribution to the effective action is expressed in terms of the eigenvalues λ_n as follows

$$\ln \mathrm{Det} \left(\gamma_\mu \, (i\partial + e\,A)_\mu \right) = \sum_n \ln \lambda_n. \tag{5.287}$$

Furthermore, for large $e\,A$ the spin effects can be neglected. As a result, instead of the Dirac equation we can solve the Klein–Gordon equation [35]

$$(i\partial + e\,A)_\mu^2 \, \Psi = \lambda^2 \, \Psi_n, \tag{5.288}$$

taking into account the double spin degeneracy of λ_n only at the end of the calculations.

The photon field for the first anzatz has the symmetry $SO(3) \times O(2)$. Therefore, the angular momentum m in the $O(2)$-subspace is quantized ($m = 0, \pm1, \pm2, \ldots$) and the wave function has the factorized form

$$\Psi(x) = e^{im\phi} \, \psi(r, x_3, x_4), \quad r^2 = x_1^2 + x_2^2, \tag{5.289}$$

where ϕ is the asimuthal angle in the plane (x_1, x_2). The function ψ satisfies the equation

$$M\psi = -\left(\frac{1}{r} \frac{\partial}{\partial r} r \frac{\partial}{\partial r} - \frac{(m-g)^2}{r^2} + \left(\frac{\partial}{\partial x_3} \right)^2 + \left(\frac{\partial}{\partial x_4} \right)^2 \right) \psi = \lambda^2 \, \psi, \tag{5.290}$$

where the field $e\,A_\mu$ for our case has only one nonzero component $A_\phi = g$ and

$$g = e\,b_1(Z_\perp^2), \quad Z_\perp^2 = \frac{4r^2}{(r^2 + x_3^2 + x_4^2)^2}. \tag{5.291}$$

We express the determinant of the operator M in terms of the integral over the Fock–Schwinger proper time s

$$\ln \mathrm{Det} M = -\int_{s_0}^{\infty} \frac{ds}{s} \sum_{m=-\infty}^{\infty} \mathrm{Tr} \exp\left(-s\left(p_r^2 + \frac{(m-g)^2}{r^2} + p_3^2 + p_4^2\right)\right), \quad (5.292)$$

where s_0 is an ultraviolet cut-off. For large M, one can use the quasiclassical approach by replacing the trace in the above expression with integrals over the phase space for the corresponding canonical degrees of freedom [35]

$$\mathrm{Tr} \rightarrow \int \frac{dp_r \, dr}{2\pi} \frac{dp_3 \, dx_3}{2\pi} \frac{dp_4 \, dx_4}{2\pi}. \quad (5.293)$$

This leads to the following result after calculating the integrals over p_r, p_x, p_y

$$\ln \mathrm{Det} M = -\frac{1}{8\pi^2} \int \frac{d^4 x}{r^4} F(g), \quad (5.294)$$

where after rescaling the integration variable $s \rightarrow s\, r^2$ we obtain for the function $F(g)$ the simple representation [35]

$$F(g) = \frac{1}{2\sqrt{\pi}} \int_{s_0}^{\infty} \frac{ds}{s^{5/2}} \sum_{m=-\infty}^{\infty} \exp\left(-s\,(m-g)^2\right). \quad (5.295)$$

Note that the field A can be interpreted as the vector-potential of an infinitely thin and long solenoid with magnetic flux $2\pi g$. The magnetic field outside this solenoid is zero, but nevertheless for g different from an integer number, the potential A is physically observable according to Bohm and Aharonov [36].

Using the relation

$$\sum_{m=-\infty}^{\infty} \exp\left(-s\,(m-g)^2\right) = \sqrt{\frac{\pi}{s}} \sum_{k=-\infty}^{\infty} \exp\left(-\frac{\pi^2}{s} k^2 + 2\pi i\, k\, g\right), \quad (5.296)$$

one can represent $F(g)$ in another form [35]

$$F_{reg}(g) = \sum_{k=-\infty}^{\infty} \frac{1}{2(\pi k)^4} \exp\left(2\pi i\, k\, g\right), \quad (5.297)$$

where the subscript reg of F implies that in the sum over k the term with $m = 0$ is absent and the constant term at $g \rightarrow 0$ is subtracted.

The function $F_{reg}(g)$ has the following properties

$$F_{reg}(g) = F_{reg}(g+1), \quad F_{reg}(g) = F_{reg}(1-g) \quad (5.298)$$

and in the interval $0 < g < 1$ it can be expressed in terms of a Bernulli polynomial

$$F_{reg}(g) = -\frac{1}{3} g^2 (1-g)^2. \quad (5.299)$$

For large imaginary values of g it behaves as follows:

$$\lim_{g \to \pm i\infty} F_{reg}(g) = -\frac{1}{3} g^4. \tag{5.300}$$

Therefore, the contribution of the photon fields described by the first ansatz, Eq. (5.286), to the effective action of the fermionic determinant at a large imaginary charge e, can be written in the following local form [35]:

$$\Delta S = \lim_{\text{Im } e \to \infty} \ln \text{Det } M = \frac{(\text{Im } e)^4}{12\pi^2} \int (A_\mu^2)^2 d^4x, \tag{5.301}$$

where we have taken into account two spin degrees of freedom for the electron in comparison with one for a spinless particle.

It turns out that the expression derived for ΔS is valid for all external fields satisfying two additional constraints [35]:

$$\partial_\mu A_\mu = 0, \quad A_\mu \partial_\mu A^2 = 0. \tag{5.302}$$

One can easily verify that both constraints are fulfilled by our trial function (5.278) or, in four-dimensional notation, by the function

$$A_\mu = (\eta_2)_{\mu\nu} x_\nu b_2(x^2) \tag{5.303}$$

with matrices $\eta_{1,2}$ given in (5.283) and (5.284) for two cases, $n_\perp = 2$ and $n_\perp = 4$.

Inserting the above expressions for A_μ in the effective action $S_{eff} = S_0 + \Delta S$, we obtain the corresponding functionals depending on the functions $b_{1,2}$

$$S_1 = \frac{\pi^2}{2} \int_0^1 dt \, (1-t)^{1/2} \left((1-t) b_1'^2 - \frac{\lambda}{128} b_1^4 t^{-2} \right), \quad t = \frac{4(x_1^2 + x_2^2)}{(1+x)^2} \tag{5.304}$$

and

$$S_2 = 2\pi^2 \int_0^\infty dt \, t^3 \left(b_2'^2 - \frac{\lambda}{8} b_2^4 \right), \quad t = x^2, \tag{5.305}$$

where

$$\lambda = \frac{e^4}{3\pi^2}. \tag{5.306}$$

The Euler–Lagrange equations for these two functionals,

$$(1-t) b_1'' - \frac{3}{2} b_1' + \frac{\lambda}{64 t^2} b_1^3 = 0 \tag{5.307}$$

and

$$t b_2'' + 3 b_2' + \frac{\lambda}{4} b_2^3 t = 0, \tag{5.308}$$

have the simple solutions

$$b_1 = \frac{8\sqrt{2}}{\sqrt{3\lambda}} \frac{t}{\frac{4}{3} - t} \tag{5.309}$$

and

$$b_2 = \frac{4\sqrt{2}}{\sqrt{\lambda}} \frac{1}{1+t^2},$$ (5.310)

respectively.

We can calculate the actions on these solutions [35]

$$S_1 = \frac{b_1}{\alpha^2}, \quad b_1 = 4\pi^3 3^{-3/2},$$ (5.311)

$$S_2 = \frac{b_2}{\alpha^2}, \quad b_2 = 4\pi^2.$$ (5.312)

The discontinuity for $\alpha < 0$ of physical quantitities can be written as follows:

$$\Delta G_{\alpha<0} \sim i \sum_{r=1}^{2} \exp\left(-\frac{b_r}{\alpha^2}\right).$$ (5.313)

Note that the value of the action for the first solution is slightly smaller then the corresponding result for second one. This means that the asymptotic behaviour of the perturbation theory coefficients in QED is governed by the first solution [35]

$$\lim_{k\to\infty} G_k \sim (-1)^k b_1^{-\frac{k}{2}} \sqrt{k!}.$$ (5.314)

There is an additional oscillatory factor from neglected corrections to the large-field asymptotics of the fermionic determinant [8]. The modification of the factorial behaviour $G_k \sim k!$, predicted by Dyson, to a weaker growth $\sim \sqrt{k!}$ is related to a compensation of contributions of Feynman diagrams with a different number of electron loops (see [37]). Physically, this compensation is explained by the Fermi statistics of electrons.

5.9 Applications of high-order estimates

Sometimes in solid state physics and in quantum field theory it is necessary to go beyond traditional perturbation theory. A typical example is the Wilson approach to second-order phase transitions [38]. Due to the universality of critical exponents for physical quantities one can use for their calculation the simple mathematical models being in the same universality class as the corresponding physical models. The most popular method for computations in a theory with space dimension $D = 3$ is based on the ϵ-expansion of the physical quantities defined in the space with $D = 4 - 2\epsilon$ [39]. For example, in the classical Ising model each node of the three-dimensional regular lattice contains the variable $\sigma_k = \pm 1$ and the corresponding Hamiltonian is [40],[41]

$$H = -J \sum_{(i,k)} \sigma_i \sigma_k - h \sum_{k} \sigma_k,$$ (5.315)

where the sum is performed over all closest vertices i and k. The parameter J is positive for ferromagnets and negative for antiferromagnets. The quantity h can be considered to be a magnetic field. The O_n-symmetric Heisenberg spin model has the more complicated Hamiltonian [40],[41]

$$H = -J \sum_{(i,k)} s_i s_k - h \sum_k s_k, \tag{5.316}$$

where s_i are classical spins.

The corresponding mathematical model is the theory of the scalar field ϕ_k ($k = 1, 2, \ldots, n$) in the D-dimensional Euclidean space with the action

$$S = \int d^D x \left(\sum_{k=1}^n \left(\frac{(\partial_\sigma \phi_k)^2}{2} + \tau_0 \frac{(\phi_k)^2}{2} \right) + \frac{\lambda_0}{4!} \left(\sum_{k=1}^n \phi_k^2 \right)^2 \right). \tag{5.317}$$

Here τ_0 and λ_0 are the bare parameters depending on the normalization point μ (for the minimal subtraction scheme [42]). The physical quantities O_i are the powers $O_i \sim \tau^{\gamma_i}$ of the small difference $\tau = T - T_0 \to 0$, where T is the current temperature and T_0 is that of the phase transition. These anomalous dimensions γ_i are calculated as functions of the quantity $\epsilon = (4 - D)/2$ which is considered to be small (but in fact $\epsilon = 1/2$). The β-function entering in the renormalization group equation for the invariant charge

$$\frac{\partial \lambda}{\partial \ln \mu} = \beta(\lambda) \tag{5.318}$$

contains in the MS-scheme [42] an additional term linear in ϵ [40, 41]

$$\beta(\lambda) = -2\epsilon\lambda + \sum_{k=2}^\infty (-1)^k C_k \lambda^k, \tag{5.319}$$

where the coefficients C_k of the perturbative expansion are calculated in the four-dimensional theory with $\epsilon = 0$. Since $C_2 > 0$ for sufficiently small ϵ there is a nontrivial infrared stable point $\tilde{\lambda}$, where

$$\beta(\tilde{\lambda}) = 0. \tag{5.320}$$

The value $\tilde{\lambda}$ depends on ϵ and therefore the critical exponents

$$\gamma_i = \sum_{k=1}^\infty (-1)^k \lambda^k d_i(k), \tag{5.321}$$

taken at this point also depend on ϵ. The quantities β and γ_i are calculated analytically within the framework of perturbation theory up to a rather large order k_0. The result for γ_i is represented in terms of a divergent series in the parameter $\epsilon = 1/2$ and is rather unstable with increasing k_0.

To improve the convergence of the series one can perform, for example, the Pade resummation of its Borel transform $B(b)$ (5.3), using for this purpose the asymptotic expressions obtained above for the coefficients C_k and $d_i(k)$ at large k [40],[41]. The most important critical exponents γ and ν enter in the expression for the Green function of the scalar particle in momentum space

$$\lim_{p,\tau \to 0} D(p, \tau) = p^{-\gamma/\nu} f(\tau p^{-1/\nu}), \tag{5.322}$$

where $f(\infty)$ is a finite number. The resummation procedure leads to values which are in good agreement with experimental data. For example, for the critical exponents in the Ising model ($n = 1$), the experimental values $\gamma = 1,237$, $\nu = 0.630$ are reproduced by the resummation with relative accuracy $\sim 10^{-3}$ [40, 41, 43].

Another traditional application of the high-order estimates is related to attempts to verify self-consistency of local quantum field theory in the cases of the ϕ^4-model and of quantum electrodynamics. As is well known, in the leading logarithmic approximation in these models the physical charge λ_c is zero in the local limit when the ultraviolet cut-off Λ tends to infinity (so-called Moscow zero-charge problem [44], [45]). Generally, such nullification of interactions takes place when the β-function does not have an ultraviolet stable zero and grows at large coupling constants $\lambda \to \infty$ sufficiently rapidly to provide convergence of the integral on the right-hand side of the renormalization group equation [46]:

$$\ln \frac{\Lambda^2}{\mu^2} = \int^{\lambda_0} \frac{d\lambda}{\beta(\lambda)}. \tag{5.323}$$

In perturbation theory $\beta(\lambda) = C_2 \lambda^2$, and the integral is convergent, which leads to the Moscow zero. One should go beyond perturbation theory to verify whether or not the above conditions for the charge nullification are valid. For the theory with interaction Hamiltonian $\lambda \phi^4 / 4!$ the coefficients C_k in the MS-scheme are known up to $k = 6$ (these and other results for the renormalization group functions are contained in Refs. [47],[48],[49]).

$$\beta(u) = 3u^2 - \frac{17}{3}u^3 + u^4 \left(\frac{145}{8} + 12\zeta(3) \right)$$

$$+ u^5 \left(-\frac{3499}{48} - 78\zeta(3) + 18\zeta(4) - 120\zeta(5) \right) + \dots, \quad u = \frac{\lambda}{16\pi^2}. \tag{5.324}$$

The results of the Pade–Borel resummation of $\beta(u)$ based on the asymptotic estimates for the coefficients C_k can be found in Refs. [50],[51],[52]. There is an opinion that the information used up to now is not enough to prove or disprove the Moscow zero. The method of resummation based on the Mellin–Barnes representation (5.128) used in Ref. [22] was developed in Ref. [51]. It was argued [51] that the expression for the coefficients $C(k)$ at $k \to \infty$ is an analytic function and therefore one can believe that the asymptotic behaviour of $\beta(\lambda)$ for large λ is governed by the leading singularity of $C(k)$. If this singularity is situated on the real axis in the interval $1 < k < 2$, then the integral $\int d\lambda / \psi(\lambda)$ is convergent for $\lambda \to \infty$ and the theory is not self-consistent [50]. But providing that $C(k)$ is regular for $k > 1$ and $C(1) \neq 0$, the asymptotic behaviour of the β-function is $\beta(\lambda) \sim \lambda$. This would lead to the divergence of the integral $\int d\lambda / \beta(\lambda)$ at large λ and, as a result, the Moscow zero would be absent for the bare coupling constant growing as $\lambda_0 \sim \exp(c\Lambda / \mu)$ at large Λ [52]. In the case of a model with the strong nonlinearity considered above, the coefficient $C(k)$ has an additional singularity at $k = 1$ [22] but nevertheless the theory is self-consistent. It will be interesting to investigate the analytic properties of the perturbation theory coefficients in the k-plane, at least in the integrable cases, for example for $N = 4$ SUSY.

In the case of QCD, the divergence of perturbation theory for many observables is governed by the infrared and ultraviolet renormalons. As was argued above, the infrared renormalon contributions lead to nonperturbative effects similar to those arising from vacuum condensates. Therefore, they presumably contain information about the hadron mass spectrum and further investigation of the relation between renormalons and the nonperturbative physics is important.

References

[1] Dyson, F. I., *Phys. Rev.* **85** (1952) 631.

[2] Hurst, C. A., *Proc. Cambr. Phil. Soc.* **48** (1952) 625; Thirring, W., *Helv. Physica Acta* **26** (1953) 33.

[3] Ioffe, B. L., *Doklady Acad. Nauk* **44** (1954) 437.

[4] Schwinger, J., *Phys. Rev.* **82** (1951) 664.

[5] Vainstein, A. I., *Decaying systems and divergence of perturbation theory series*, Preprint of Novosibirsk NPI (1964), unpublished; Bender C. M., Wu T. T., *Phys. Rev.* **184** (1969) 1231.

[6] Lipatov, L. N., *JETP Lett.* **25** (1977) 104; Sov. *Phys. JETP* **45** (1977) 216.

[7] Brezin, E., Le Guillou J. C., Zinn-Justin J., *Phys. Rev.* **D15** (1977) 1544.

[8] Bogomolny, E. B., Fateev, V. A., Lipatov, L. N., Soviet Science Reviews: Physics, Ed. by I.M. Khalatnikov (Harwood Academic, New York, 1980), Vol.2, p.247.

[9] *Large Order Behavior of Perturbation Theory*, Ed. by Le Guillou, J. C. and Zuber, J. B. (North-Holland, Amsterdam, 1990).

[10] Lam, C. S., *Nuovo Cim.* **60A** (1968) 258; Langer, J. S., *Ann. Phys.* **41** (1967) 108.

[11] Parisi, G., *Phys. Lett.* **66B** (1977) 167; Bogomolny, E. B., *Phys. Lett.* **67B** (1977) 193.

[12] Lautrup, B., *Phys. Lett.* **69B** (1978) 109.

[13] Parisi, G., *Phys. Lett.* **76B** (1978) 65.

[14] 't Hooft, G., Lections at the "Ettore Majorana" Int. School. in Erice, (1977).

[15] Parisi, G., *Phys. Rep.* **49** (1979) 215.

[16] Mueller, A. H., *Nucl. Phys.* **250B** (1985) 327.

[17] Zakharov, V. I., *Nucl. Phys.* **385B** (1992) 452.

[18] Beneke, M. *et al*, *Phys. Rev.* **D52** (1995) 3929.

[19] Faleev, S. V., Silvestrov, P. G. *Nucl. Phys.* **463B** (1996) 489.

[20] Suslov, I. M., *J. Exp. Theor. Phys.* **99** (2000) 474.

[21] Parisi, G., *Phys. Lett.* **66B** (1977) 167, **68B** (1977) 361.

[22] Lipatov, L. N., *JETP Lett.* **24** (1976) 151, *Sov. Phys. JETP* **44** (1976) 1055.

[23] Dyson, F. J., *Journ. Math. Phys.* **3** (1962) 140, 157, 166.

[24] Wilson, K., *Journ. Math. Phys.* **3** (1962) 1040.

[25] Hanson, K., *Journ. Math. Phys.* **3** (1962) 752.

[26] Adler, S. L., *Phys. Rev.* **D6** (1972) 3445.

[27] Kubyshin, Yu. A., *Theor. Mat. Fiz.* **57** (1983) 363.

[28] Bukhvostov, A. P., Lipatov, L. N., *JETP Sov. Phys.* **73** (1977) 1658.

[29] Bukhvostov, A. P., Lipatov, L. N., Malkov, E. I., *Phys. Rev.* **D19** (1979) 2974.

[30] Itzykson, C., Parisi, G., Zuber, J. B., *Phys. Rev. Lett.* **38** (1977) 306.

[31] Bogomolny, E. B., Fateev, V. A., *Phys. Lett.* **71B** 93.

[32] Brezin, E., Parisi, G., Zinn-Justin, *Phys. Rev.* **D16** (1977) 408.

[33] Parisi, G., *Phys. Lett.* **66B** (1977) 382.

[34] Itzykson, C., Parisi, G., Zuber, J. B., *Phys. Rev.* **D16** (1977) 996.

[35] Bogomolny, E. B., Fateev, V. A., *Phys. Lett.* **79B** 210.

[36] Aharonov, Y., Bohm, D., *Phys. Rev.* **115** (1959) 485; **123** (1961) 1511.

[37] Bogomolny, E. B., Kubyshin, Yu. A., *Sov. J. Nucl. Phys.* **35** (1982) 114.

[38] Wilson, K. G., *Phys. Rev.* **B4** (1971) 3174.

[39] Wilson, K. G., Fisher, M. E., *Phys. Rev. Lett.* **28** (1972) 240.

[40] Zinn-Justin, J., *Quantum Field Theory and Critical Phenomena*, Clarendon Press, Oxford (1993).

[41] Vasil'ev, A. N. *The Field Theoretic Renormalization Group in Critical Behavior Theory and Stochastic Dynamics*, CRC Press (2004).

[42] t'Hooft, G., *Nucl. Phys.* **B61** (1973) 455.

[43] Honkonen, J., Komarova, M., Nalimov, N., *Acta Phys. Slov.* **52** (2002) 303; Nucl. Phys. **B707** (2005) 493.

[44] Landau, L. D., Abrikosov, A. A., Khalatnikov, I. M., **95** (1954) 497, 773, 1177.

[45] Landau, L. D., Pomeranchuk, I. Ya., *Dokl. Akad. Nauk USSR*, **102** (1995) 489.

[46] Bogolyubov, N. N., Shirkov, D. V., *Introduction to the Theory of Quantized Fields*, Willey, New York, 1980.

[47] Gorishny, S. S., Larin, S. A., Tkachev, F. V., *Phys.Lett.* **A101** (1984) 120.

[48] Kazakov, D. I., *Phys. Lett.* **B133** (1983) 406.

[49] Kleinert, H., Neu, J., Shulte-Frohlinde, V., Chetyrkin, K. G., Larin, S. A., *Phys. Lett.* **B272** (1991) 39; **B319**, (1993) 549.

[50] Kazakov, D. I., Tarasov, O. V., Shirkov, D. V., *Theor. Mat. Phys.* **38** (1979) 15.

[51] Kubyshin, Yu. A., *Theor. Mat. Fiz.* **58** (1984) 137.

[52] Suslov, I. M., *JETP* **100** (2005) 1188.

6

QCD sum rules

6.1 Operator product expansion

In this chapter we study the correlators of hadronic currents:

$$\Pi^{ABC\cdots}(x, y_1, y_2..y_n) = iT\left\{ j^A(x), j^B(0), ij^{C_1}(y_1)\ldots ij^{C_n}(y_n) \right\}, \qquad (6.1)$$

where $j^A(x)$, $j^B(y)$, $j_1^C(y_1)\ldots$ are colourless local currents built from the field variables of the theory, i.e. from quark and gluon fields. In (6.1), we exploited the invariance of the theory under displacement of coordinates, which allowed us to choose one of the coordinates as the origin of the reference frame. If $x^2, (x - y_i)^2, (y_i - y_k)^2, i, k = 1, \ldots n$ are negative and $x_\mu, (x - y_i)_\mu, (y_i - y_k)_\mu$ tend to zero, the matrix elements of $\Pi^{ABC\cdots}(x, y_1, y_2 \ldots y_n)$ between quark and gluon states can be calculated in perturbation theory. On the other hand, the correlators of hadronic currents are related to the amplitudes of the physical processes: e^+e^--annihilation into hadrons, τ-decays, $\tau \to \nu_\tau$+hadrons, deep inelastic lepton–hadron scattering, etc. In many cases, the study of the correlators $\Pi^{ABC\cdots}$ allows one to go beyond perturbation theory. Therefore, the study of correlators is a useful instrument in the theoretical description of hadronic properties in QCD.

Consider the simplest two-point correlator $\Pi^{AB}(x)$. In 1969, K. Wilson proposed the operator product expansion (OPE) for $\Pi^{AB}(x)$ [1]:

$$\Pi^{AB}(x) = iT\left\{ j^A(x),\ j^B(0) \right\}_{x\to 0} = \sum_n C_n^{AB}(x)O_n(0), \qquad (6.2)$$

valid at $x^2 < 0$ and $x_\mu \to 0$. Here $O_n(0)$ are the local operators built from field variables and taken at the point $x_\mu = 0$. $C_n^{AB}(x)$ are the coefficient functions that one can, in principle, calculate in perturbation theory. $C_n^{AB}(x)$ have canonical dimensions of $(\text{mass})^{d_A+d_B-d_n}$, where d_A, d_B are the current dimensions, d_n is the dimension of O_n. In momentum space, OPE looks like

$$\Pi^{AB}(q) = i \int d^4x\, e^{iqx} T\left\{ j^A(x),\ j^B(0) \right\}_{q^2<0,q\to\infty} = \sum_n C_n^{AB}(q)O_n(0) \qquad (6.3)$$

Equations (6.2) and (6.3) are operator equalities: they are valid for any matrix element, provided that the characteristic momenta of the initial and final states p_i are much smaller than q: $|p_i^2| \ll Q^2 = -q^2$. The physical meaning of OPE is that the contributions of

195

small and large distances can be factorized: the contributions of small distances determine the coefficient functions, while large distances contribute to the matrix elements of the operators.

In perturbation theory in case of any renormalizable theory, the OPE is a consequence of the following intuitive considerations: Let us go to the interaction representation in (6.2) and expand $\exp[i \int d^4 x L_{int}(x)]$ in a perturbation series. In each term of this expansion, contract some field operators according to the rules of perturbation theory and leave all others as operators. Perform the renormalization procedure and subtractions, if necessary. Then at $x^2 \to 0$ the coefficient functions are dominated by the domain of small x^2. So, all operators are local and may be taken at $x_\mu = 0$. The strict mathematical proof of the validity of OPE within perturbation theory was given by Zimmermann [2].

The intuitive proof of OPE sketched above, however, leaves aside a few important points. The first is the correct formulation of the renormalization procedure. As is well known, renormalization requires an introduction of the normalization point μ (see Chapter 1). In (6.3), $\mu = Q$ was chosen as the normalization point of the $O_n(0)$ operators. By definition, at the normalization point, all perturbative corrections to the operator are absorbed in $O_n(0)$. When we claim that the contributions of large and small momenta (or distances) are factorized in OPE, we mean that the domain of integration over vacuum fluctuations at momenta k larger than μ, $k > \mu$, contributes to the coefficient functions $C_n(q)$ while the domain of integration $k < \mu$ contributes to operators O_n. In a theory with asymptotic freedom and large nonperturbative effects, like QCD, if μ is large enough, $\alpha_s(\mu)/\pi \ll 1$, the domain $k > \mu$ can be accounted for perturbatively, but the contributions of the $k < \mu$ domain are mainly of nonperturbative origin. An equivalent (or, practically equivalent) recipe can be formulated as follows: calculate $C_n(q)$, taking into account perturbative contributions from the whole domain of integration in momentum space, and assume that operators O_n and their matrix elements are given by nonperturbative interactions and that the perturbative contributions in the domain $k < \mu$ were subtracted. Therefore, each term of OPE in (6.2), (6.3), in general, is μ dependent. If the currents $j^A(x)$, $j^B(x)$ are physical currents and the polarization operator Π^{AB} is a measurable object, the sum of the OPE series is μ-independent. In practice, when only few terms of OPE are taken into account, this statement is only approximately correct. (For a more detailed discussion of this problem see ref. [3],[4].) If we consider different matrix elements of (6.2), (6.3) or compare polarization operators of different currents $j^A(x)$, $j^B(x)$, the perturbative contributions at $k < \mu$ to the same operator O_n may be different. So, the values of the matrix element of the same operator, say $\langle 0|O_n|0 \rangle$ might be slightly different, if found from different processes. (Practically, these effects are not important, because they can be significant only for high-dimension operators or in higher orders of perturbation theory.)

The second point that introduces some uncertainty in OPE is how to account for non-perturbative vacuum fluctuations at large momenta. In QCD, we know the example of such fluctuation – the instanton. Its contribution to the effective action is suppressed by a factor $\exp[-2\pi/\alpha_s(\rho)]K_n^2(Q\rho)$, where K_n is the McDonald function, ρ is the instanton size, and $n = 0$, 1 or 2, depending on the type of the currents j_A, j_B under consideration. (See

Chapter 4.) To get the final result one has to integrate over ρ with some weight factor, $w(\rho)$, that has the physical meaning of instanton density. At fixed ρ, the instanton contribution decreases exponentially at large space-like Q. The instanton contribution to the vacuum expectation value of (6.2) can be estimated as

$$\Delta\Pi_{inst}(Q^2) \sim (Q\rho)^{n-1}e^{-2Q\rho} \tag{6.4}$$

and can be neglected at large enough Q. (It should be stressed, however, that this statement holds only for space-like $Q^2 > 0$ because for time-like $Q^2 < 0$, i.e. in the physical domain of the polarization operator, $\Delta\Pi_{inst}(Q)$ oscillates.) The integration over ρ changes the situation: the results significantly depend on the behaviour of the instanton density at large ρ. In perturbation theory, where $w(\rho) \sim \rho^m$, the integral over ρ is dominated by $\rho \sim 1/Q$. In this case, since $\alpha_s \approx 4\pi/(b \ln q^2/\lambda^2)$, the instanton contribution is proportional to

$$1/Q^{b+2} \tag{6.5}$$

($b = 9$ in QCD with three light quarks). So, the instanton contribution can generate higher-order terms of OPE. In the model of dilute instanton gas, where it is assumed that the instanton density is concentrated near some fixed $\rho = \rho_0$, the $\Delta\Pi_{inst}(Q)$ behaviour is given by (6.4). In this model, the instanton contribution can be accounted for separately. (The example of such computation will be presented in Section 6.5.1).

The third point that leads to an uncertainty in OPE is the convergence of the OPE series in $1/Q^2$. The study of the potential model in nonrelativistic quantum mechanics demonstrated that OPE series converges only asymptotically [5]. As was mentioned above, in QCD the instanton contribution leads to an exponential decrease of $\Pi(Q)$ at large Q, corresponding to the asymptotic OPE series, in the case when the instanton density is concentrated at fixed $\rho = \rho_0$. Another source of an asymptotic character of OPE in QCD are higher-order diagrams. In these diagrams, the operators, representing the nonperturbative contributions, can be chosen in various ways and the number of such possibilities factorially increases with the order of the diagram, which leads to an asymptotic series for OPE (see Chapter 5). For these reasons, in all practical calculations, where OPE is exploited, it is necessary to convince oneself that the desired accuracy of the OPE series is achieved by terms that were actually taken into account.

If instead of the variable Q a fixed normalization point μ is introduced, corrections to the operators O_n appear in perturbation theory as a series in the coupling constant α_s

$$O_i(Q) = O_i(\mu) \sum_{n=0}^{\infty} d_n^{(i)}(Q, \mu)\alpha_s^n(\mu). \tag{6.6}$$

The running coupling constant α_s at the right-hand side of (6.6) is normalized at the point μ. The coefficients $d_n^{(i)}(Q, \mu)$ generally contain logarithms in powers smaller than or equal to n. The summation of leading logarithms in (6.6) can be performed using the Callan–Symanzik equation [6], [7] (see Chapter 1):

$$\left(\mu^2\frac{\partial}{\partial\mu^2} + \beta\frac{\partial}{\partial a_s} - \gamma_i a_s b\right)O_i(Q, \mu, a_s) = 0. \tag{6.7}$$

This equation follows from the fact that the value of the operator $O_i(Q)$ is independent of the normalization point μ, apart from the overall renormalization of coupling constant and field operators (see Chapter 1). Here O_i is considered as a function of Q, μ and $a_s = \alpha_s/\pi$, $\beta = \beta(a_s)$ is the Gell-Mann–Low function for QCD. The variables a_s and μ are treated as independent. The last term in brackets in (6.7) represents the anomalous dimension of O_i (actual dimension minus canonical dimension d_{O_i}), b is the first coefficient in the expansion of $\beta(a_s)$ in a_s: $\beta(a_s) = a_s^2(b + b_1 a_s + \ldots)$. Since γ_i is a function of a_s and its expansion in powers of a_s starts from the term proportional to a_s, it is convenient to separate this factor explicitly in (6.7), as well as the factor b. In what follows, γ_i will be called the anomalous dimension of operator O_i. The Callan–Symanzik equation expresses the fact that if μ^2 is varied, then α_s is varied correspondingly, these two variations compensate one another, and the value of $O_i(Q)$ remains unchanged. Evidently, $a_s(\mu)$ satisfies (6.7) without the last term

$$\frac{da_s(\mu^2)}{d\ln\mu^2} = -\beta(a_s). \tag{6.8}$$

Eq. (6.7) can be easily solved; we notice that it is similar to the advection equation describing particles in a moving medium:

$$\left[\frac{\partial}{\partial t} + v(x)\frac{\partial}{\partial x} - \rho(x)\right]D(x, t) = 0. \tag{6.9}$$

Here $v(x)$ is the velocity of the medium, $\rho(x)$ is the rate of particle production. This allows one to write immediately the solution of Eq. (6.7):

$$O_i(Q) = O_i(\mu)\exp\left[\int_{\mu^2}^{Q^2} d\left(\ln\frac{Q'^2}{\mu^2}\right)\gamma_i\left(a_s(Q'^2)\right)a_s(Q'^2)b\right], \tag{6.10}$$

where the $a_s(Q^2)$ dependence is determined by (6.8). If γ_i is approximated by a constant and $\beta(a_s)$ by $\beta(a_s) = ba_s^2$, we get from (6.10) using (6.8)

$$O_i(Q) = O_i(\mu)\left[\frac{\alpha_s(\mu)}{\alpha_s(Q)}\right]^{\gamma_i}. \tag{6.11}$$

Accounting for higher-order terms of perturbation theory we can write

$$O_i(Q) = O_i(\mu)\left(\frac{\alpha_s(\mu)}{\alpha_s(Q)}\right)^{\gamma_i}\sum_{n=0}^{\infty} b_n^{(i)}(Q, \mu)\alpha_s^n(\mu). \tag{6.12}$$

The anomalous dimension operators can be found from the renormalization group: in order to determine γ_i, it is sufficient to calculate the divergent part of the first α_s-correction to $O_i(Q)$. The currents j^A, j^B in (6.2), (6.3) have generally anomalous dimensions γ_A, γ_B. Therefore, accounting for perturbative corrections and summing the leading logarithms we get the general form of OPE

$$\Pi^{AB}(Q) = \sum_i O_i(\mu)\left(\frac{\alpha_s(\mu)}{\alpha_s(Q)}\right)^{\gamma_A + \gamma_B + \gamma_i}\sum_{n=0}^{\infty} C_n^{(i)}(Q, \mu)\alpha_s^n(\mu). \tag{6.13}$$

$(q^2 = -Q^2)$. In (6.12), all leading logarithms related to ultraviolet divergences are absorbed in anomalous dimensions. If the quark masses are neglected, then infrared divergences may appear, resulting in infrared logarithms. They are accounted for by the coefficient functions $C_n^{(i)}(Q, \mu)$. If several operators are present with the same canonical dimensions, their mixing is possible in perturbation theory. In this case, the relations (6.6), (6.12) become matrix ones. The anomalous dimensions of conserved currents are zero. Indeed, the associated charge

$$Z_A = \int d^3x j_{A,0}(x) \qquad (6.14)$$

is a number and is independent of the normalization point. (Here $j_{A,0}$ is the time-component of $j_{A,\mu}$.) This statement refers to vector and axial currents and to their octet generalization in $SU(3)_L \times SU(3)_R$ flavour symmetry in QCD, as well as to the energy-momentum tensor $\Theta_{\mu\nu}$.

6.2 Condensates

Consider now vacuum matrix elements of (6.2), (6.3), or (6.13). The vacuum expectation values $\langle 0 \mid O_i(Q) \mid 0 \rangle$ are called vacuum condensates. One of such condensates – the quark condensate – has been already discussed in Chapter 2.

In their physical properties, condensates in QCD have much in common with those appearing in condensed matter physics: such as superfluid liquid (Bose-condensate) in liquid ^{4}He, Cooper pair condensate in superconductors, spontaneous magnetization in magnetic, etc. That is why, analogously to effects in condensed matter physics, it can be expected that if one considers QCD at a finite and increasing temperature T, then there will be a phase transition at some $T = T_c$ and condensates (or a part of them) will be destroyed. (Generally, the phase transition can proceed in some interval of temperatures near T_c – the crossover domain.) Particularly, it is expected that such a phenomenon must hold for condensates responsible for spontaneous symmetry breaking. Namely, at $T > T_c$ the condensates should vanish and the symmetry be restored. (In principle, surely, QCD may have few phase transitions.)

Condensates in QCD are divided into two types: conserving and violating chirality. As was demonstrated in Chapter 2, the masses of light quarks u, d, s in the QCD Lagrangian are small compared with the characteristic scale of hadronic masses $M \sim 1$ GeV. If light quark masses are neglected, the QCD Lagrangian becomes chiral-invariant: left-hand and right-hand (in chirality) light quarks do not transfer into one another, both vector and axial currents are conserved (except for the flavour-singlet axial current, the nonconservation of which is due to anomaly). The neglect of u and d quark masses corresponds to the accuracy of the isotopic symmetry, i.e. a few percent, and the neglect of the s quark mass to the accuracy of $SU(3)$ symmetry, i.e. 10–15%. In the case of the condensates violating the chiral symmetry, perturbative vacuum mean values are proportional to light quark masses and are zero at $m_u = m_d = m_s = 0$.

For this reason, the perturbative contributions to condensates discussed above, which violate chiral symmetry, are small. So, such condensates are defined in the theory much better than those conserving chirality and, in principle, they can be found experimentally with higher accuracy.

The lattice calculations indicate that the phase transition from the phase where chiral symmetry is violated to the phase of restored chiral symmetry is of the second order.

Quark condensate can be considered as an order parameter in QCD corresponding to spontaneous violation of chiral symmetry. At the temperature $T = T_c$ of the chiral symmetry restoration it must vanish. The investigation of the temperature dependence of quark condensate in the chiral effective theory shows that $\langle 0|\bar{q}q|0\rangle$ vanishes at $T = T_c \approx 150 - 200\,\text{MeV}$ [8]. Similar indications were obtained also in lattice calculations [9].

Thus, the quark condensate: (1) has the lowest dimension ($d = 3$) as compared with other condensates in QCD; (2) determines baryons masses (see Chapter 2); (3) is the order parameter in the phase transition between the phases of violated and restored chiral symmetry. These three facts determine its important role in low-energy hadronic physics. If u and d quark masses as well as electromagnetic interactions are neglected, then $\langle 0 \mid \bar{u}u \mid 0\rangle = \langle 0 \mid \bar{d}d \mid 0\rangle$. Since $\bar{u}u$ and $\bar{d}d$ enter the QCD Lagrangian multiplied by quark masses and the Lagrangian is renormalization group invariant, the products $m_q \langle 0 \mid \bar{q}q \mid 0\rangle$, $q = u, d, s$ are renormalization group invariant, the anomalous dimensions of $\bar{u}u$, $\bar{d}d$ and $\bar{s}s$ differ by sign from the anomalous dimension of quark masses and are equal to $\gamma_{\bar{q}q} = 4/9$.

The quark condensate of strange quarks is somewhat different from $\langle 0|\bar{u}u|0\rangle$. It was found that [10]:

$$\langle 0|\bar{s}s|0\rangle/\langle 0|\bar{u}u|0\rangle = 0.8 \pm 0.1. \tag{6.15}$$

The condensate next in dimension ($d = 5$), which violates chiral symmetry is the quark–gluon one:

$$-g\langle 0|\bar{q}\sigma_{\mu\nu}\frac{\lambda^n}{2}G_{\mu\nu}^n q|0\rangle \equiv m_0^2 \langle 0|\bar{q}q|0\rangle. \tag{6.16}$$

Here $G_{\mu\nu}^n$ is the gluonic field strength tensor, λ^n are the Gell-Mann matrices, $\sigma_{\mu\nu} = (i/2)(\gamma_\mu\gamma_\nu - \gamma_\nu\gamma_\mu)$. The value of the parameter m_0^2 was found from the sum rules for baryonic resonances [11]

$$m_0^2 = 0.8 \pm 0.2 \text{ GeV}^2 \quad \text{at} \quad \mu = 1\,\text{GeV}. \tag{6.17}$$

The same value of m_0^2 was found from the analysis of B mesons by QCD sum rules [12]. The anomalous dimension of the operator in (6.16) is equal to $-2/27$. Therefore, the anomalous dimension of m_0^2 is equal to $\gamma = -14/27$.

Consider now the condensates conserving chirality. Of fundamental significance here is the gluon condensate of the lowest dimension:

$$\langle 0|\frac{\alpha_s}{\pi}G_{\mu\nu}^n G_{\mu\nu}^n|0\rangle. \tag{6.18}$$

Since the gluon condensate is proportional to the vacuum expectation value of the trace of the energy-momentum tensor $\theta_{\mu\nu}$, its anomalous dimension is zero. The existence of the gluon condensate was first indicated by Shifman, Vainshtein, and Zakharov [13]. They also obtained its numerical value from the sum rules for charmonium:

$$\langle 0| \frac{\alpha_s}{\pi} G^n_{\mu\nu} G^n_{\mu\nu} |0\rangle = 0.012 \text{ GeV}^4. \tag{6.19}$$

As was shown by the same authors, since the vacuum energy density in QCD is given by $\varepsilon = -(9/32)\langle 0|(\alpha_s/\pi)G^2|0\rangle$, the nonzero and positive value of gluon condensate implies that the vacuum energy is negative in QCD. The persistence of quark fields in vacuum destroys (or suppresses) the condensate. Therefore, if a quark is embedded in vacuum, this leads to an increase of energy. Modern determinations of the gluon condensate numerical value from hadronic τ decay data and from the charmonium sum rules are presented in Sections 6.5.3 and 6.5.4.

The $d = 6$ gluon condensate is of the form

$$g^3 f^{abc} \langle 0|G^a_{\mu\nu} G^b_{\nu\lambda} G^c_{\lambda\mu}|0\rangle, \tag{6.20}$$

(f^{abc} are the structure constants of the $SU(3)$ group). There are no reliable methods to extract its value from experimental data. There is only an estimate which follows from the model of diluted instanton gas [14]:

$$g^3 f^{abc} \langle 0|G^a_{\mu\nu} G^b_{\nu\lambda} G^c_{\lambda\mu}|0\rangle = -\frac{4}{5}(12\pi^2)\frac{1}{\rho_c^2}\left\langle 0|\frac{\alpha_s}{\pi}G^2_{\mu\nu}|0\right\rangle, \tag{6.21}$$

where ρ_c is the effective instanton radius in the given model (for estimate, one can take $\rho_c \sim (1/3 - 1/2)$fm).

The general form of d=6 condensates built from quark fields is:

$$\alpha_s \langle 0|\bar{q}_i O_\alpha q_i \cdot \bar{q}_k O_\alpha q_k|0\rangle, \tag{6.22}$$

where q_i, q_k are quark fields of u, d, s quarks, O_α are composed from Dirac and $SU(3)$ matrices. Eq. (6.22) is usually factorized: in the sum over intermediate states in all channels (i.e. if necessary, after a Fierz transformation) only the vacuum state is taken into account. The accuracy of such approximation is $\sim 1/N_c^2$, where N_c is the number of colours i.e. $\sim 10\%$. After factorization Eq. (6.22) reduces to

$$\alpha_s \langle 0|\bar{q}q|0\rangle^2, \tag{6.23}$$

if $q = u, d$. The anomalous dimension of (6.23) is $-1/9$ and can be neglected. When α_s-corrections are taken into account, then the matrix that describes the mixing of 4-quark condensates depends on the normalization point. So, even if the factorization hypothesis is assumed to be exact at the normalization point μ^2, it will be violated at Q^2. However, as one can verify, such violation of factorization does not exceed 10% in the interval $1 \lesssim Q \lesssim 2 \text{ GeV}^2$ [13]. Within the framework of the factorization hypothesis the condensates of dimension $d = 8$ containing four quarks and one gluon fields reduce to

$$\alpha_s \langle 0|\bar{q}q|0\rangle \cdot m_0^2 \langle 0|\bar{q}q|0\rangle. \tag{6.24}$$

(The definition of m_0^2 in (6.16) is used.) Note, however, that the factorization procedure in the $d = 8$ condensate case is not quite certain. For this reason, it is necessary to demand their contribution to be small.

Table 1

Condensates	Numerical values
$\langle 0 \mid \bar{q}q \mid 0 \rangle_{1\,\text{GeV}}$, $q = u, d$	$-(1.65 \pm 0.15) \cdot 10^{-2}\,\text{GeV}^3$
$\langle 0 \mid \frac{\alpha_s}{\pi} G_{\mu\nu}^2 \mid 0 \rangle$	$0.005 \pm 0.004\,\text{GeV}^4$
$\alpha_s \langle 0 \mid \bar{q}q \mid 0 \rangle^2$	$(1.5 \pm 0.2) \cdot 10^{-4}\,\text{GeV}^6$

There are few other gluon and quark–gluon condensates of dimension 8. (The full list of them is given in [15].) As a rule, the factorization hypothesis is used for their estimate. The other way to find the values of these condensates is to use the dilute instanton gas model. However, this model gives for some condensates results (at accepted values of instanton model parameters) by an order of magnitude larger than the factorization method. There are arguments that the instanton gas model overestimates the values of $d = 8$ gluon condensates [16]. Therefore, the estimates based on the factorization hypothesis are more reliable here.

The violation of the factorization ansatz is stronger for higher-dimension condensates. So, this hypothesis can be used only for their estimate by order of magnitude. The best values of condensates, extracted by QCD sum rules from experimental data, are given in Table 1 [17]. The methods of their determination are described in Section 6.5.

6.3 Condensates, induced by external fields

The meaning of such vacuum condensates can be easily understood by comparing their appearance with similar phenomena in condensed matter physics. If the condensates considered in the previous section can be compared, for instance, with ferromagnetics, where magnetization is present even in the absence of an external magnetic field, the condensates induced by an external field are similar to dia- or paramagnetics. Consider, for example, the case of a constant external electromagnetic field $F_{\mu\nu}$. In its presence, there appears a condensate induced by the external field (in the linear approximation in $F_{\mu\nu}$) [18]

$$\langle 0|\bar{q}\sigma_{\mu\nu}q|0\rangle_F = e_q \chi F_{\mu\nu} \langle 0|\bar{q}q|0\rangle. \qquad (6.25)$$

To a good approximation $\langle 0|\bar{q}\sigma_{\mu\nu}q|0\rangle_F$ is proportional to e_q, the electric charge of the quark q [18]. (See below, Section 6.9.2.) The vacuum expectation value $\langle 0|\bar{q}\sigma_{\mu\nu}q|0\rangle_F$ induced by this field violates chiral symmetry. So, it is natural to separate $\langle 0|\bar{q}q|0\rangle$ as a factor in Eq. (6.25). The universal quark flavour independent quantity χ is called the magnetic susceptibility of the quark condensate. Its numerical value was found using a special sum rule [19],[20],[21]:

$$\chi_{1\,\text{GeV}} = -(3.15 \pm 0.3)\,\text{GeV}^{-2}. \qquad (6.26)$$

(The anomalous dimension of χ is equal to $-16/27$). Another example is the external constant axial isovector field $A_\mu^{(i)}$. The interaction of its third component $A_\mu^{(3)}$ with light quarks is described by the Lagrangian

$$L' = (\bar{u}\gamma_\mu\gamma_5 u - \bar{d}\gamma_\mu\gamma_5 d)A_\mu^{(3)}. \tag{6.27}$$

In the presence of this field, there appear the condensates induced by $A_\mu^{(3)}$

$$\langle 0|\bar{u}\gamma_\mu\gamma_5 u|0\rangle_A = -\langle 0|\bar{d}\gamma_\mu\gamma_5 d|0\rangle_A = f_\pi^2 A_\mu^{(3)}, \tag{6.28}$$

where $f_\pi = 131$ MeV is the $\pi \to \mu\nu$ decay constant. The right-hand side of Eq. (6.28) is obtained assuming $m_u, m_d \to 0$, $m_\pi^2 \to 0$ and follows directly from consideration of the polarization operator of axial currents $\Pi_{\mu\nu}^A(q)$ in the limit $q \to 0$. Indeed, because of axial current conservation the nonzero contribution to $\Pi_{\mu\nu}^A(q)_{q\to 0}$ emerges only from the one-pion intermediate state. A relation similar to (6.28) holds in the case of the octet axial field. Of special interest is the condensate induced by the flavour singlet constant axial field

$$\langle 0|j_{\mu 5}^{(0)}|0\rangle_A = 3 f_0^2 A_\mu^{(0)}, \tag{6.29}$$

$$j_{\mu 5}^{(0)} = \bar{u}\gamma_\mu\gamma_5 u + \bar{d}\gamma_\mu\gamma_5 d + \bar{s}\gamma_\mu\gamma_5 s. \tag{6.30}$$

The interaction Lagrangian with the external field has the form

$$L' = j_{\mu 5}^{(0)} A_\mu^{(0)}. \tag{6.31}$$

The constant f_0 cannot be calculated by the method used when deriving Eq. (6.28), since the singlet axial current is not conserved because of the anomaly and because the singlet pseudoscalar meson η' is not a Goldstone meson. The constant f_0^2 is proportional to the topological susceptibility of vacuum [22]

$$f_0^2 = \frac{4}{3} N_f^2 \zeta'(0), \tag{6.32}$$

where N_f is the number of light quarks, $N_f = 3$, and the topological susceptibility of the vacuum $\zeta(q^2)$ is defined as

$$\zeta(q^2) = i \int d^4 x e^{iqx} \langle 0|T\{Q_5(x),\ Q_5(0)\}|0\rangle, \tag{6.33}$$

$$Q_5(x) = \frac{\alpha_s}{8\pi} G_{\mu\nu}^n(x)\tilde{G}_{\mu\nu}^n(x). \tag{6.34}$$

Here $\tilde{G}_{\mu\nu}^n$ is dual to $G_{\mu\nu}^n$: $\tilde{G}_{\mu\nu}^n = (1/2)\varepsilon_{\mu\nu\lambda\sigma} G_{\lambda\sigma}^n$. Using the QCD sum rule, one can relate f_0^2 with the fraction of proton spin Σ carried by quarks in polarized ep (or μp) scattering [22]. The value of f_0^2 was found from the experimental value of Σ (see Section 6.9.6):

$$f_0^2 = (3.5 \pm 0.5) \cdot 10^{-2} \text{GeV}^2. \tag{6.35}$$

The value of the derivative at $q^2 = 0$ of the vacuum topological susceptibility $\zeta'(0)$, (more precisely, its nonperturbative part) is equal to:

$$\zeta'(0) = (2.9 \pm 0.4) \cdot 10^{-3} \text{ GeV}^2. \tag{6.36}$$

The quantity $\zeta'(0)$ is of interest for studying properties of vacuum in QCD.

6.4 QCD sum rules method

There are many sum rules in QCD: the sum rules for deep-inelastic lepton–hadron scattering, which were originally derived (before QCD) on the basis of current algebra (their exposition is given in the book [23]) – QCD allows one to calculate α_s-corrections to these sum rules; Weinberg sum rules (see Chapter 2), low-energy sum rules [24], etc. Traditionally, however, the term QCD sum rule applies only to those sum rules discussed in this chapter. So, we shall use this name for them. The QCD sum rules were proposed by Shifman, Vainshtein, and Zakharov and applied to calculate the masses of the light mesons and their coupling constants to corresponding quark currents [13]. Later QCD sum rules for baryon masses were suggested, QCD sum rules for meson and baryon formfactors at low Q^2 were found, a new class of QCD sum rules for hadrons in external fields (baryons magnetic moments, axial β-decay coupling constants, etc.) was invented, and finally, the sum rules for hadronic and photonic structure functions (quark distributions in hadrons and photons – real and virtual) were obtained. These topics are considered in this chapter.

The main idea of QCD sum rules for the simplest case – the calculation of hadronic masses – is the following: Consider the polarization operator (6.3) for large enough $Q^2 = -q^2 > 0$ and use OPE (6.13). On the other hand, represent $\Pi_{AB}(q^2)$ through the contributions of physical states via a dispersion relation:

$$\Pi_{AB}^{phys}(q^2) = \frac{1}{\pi} \int_0^\infty \frac{\text{Im } \Pi_{AB}(s)}{s - q^2} ds. \tag{6.37}$$

These two representations of the polarization operator are equal

$$\Pi_{AB}^{QCD}(q^2) = \Pi_{AB}^{phys}(q^2). \tag{6.38}$$

(The relations (6.37), (6.38) are understood to be valid for any structure function of polarization operators.) These equalities are the desired QCD sum rules, which, in principle, give us information about the properties of hadrons in terms of QCD variables. In Im $\Pi_{AB}(s)$, it is reasonable to separate the contribution of the lowest hadronic state in the given channel (stable particle or resonance) and use one or another model for contributions of excited states. (The first who used such an approach was Sakurai [25], who considered the polarization operator of vector currents and tried to determine the ρ-meson mass and $\rho\gamma$-coupling constant identifying the ρ-meson contribution with the contribution of the bare quark loop, cut at some q_0^2).

However, Eq. (6.38) is ill defined. First, the perturbatively calculated polarization operator is divergent and requires an ultraviolet cut-off. This corresponds to the fact that subtractions are needed in the calculation of $\Pi_{AB}^{phys}(q^2)$. So, unknown polynomials appear on both sides of (6.38). Second, in order to determine the properties of the lowest hadronic state, which is our goal, its contribution to the right-hand side of (6.37) should dominate in comparison with the contribution of excited states. In fact, this is not so. Third, the OPE must be well converging (as well as the perturbative series) in the domain of Q^2 where we would like to exploit the sum rule (6.38).

In order to get rid of these drawbacks and obtain self-consistent and informative QCD sum rules, Shifman, Vainshtein, and Zakharov [13] suggested to apply the Borel transformation of (6.38). The Borel transformation of the function $f(Q^2)$ is defined by:

$$\mathcal{B}_{M^2} f(Q^2) = \lim \frac{(Q^2)^{n+1}}{n!} \left(-\frac{d}{dQ^2}\right)^n f(Q^2) \underset{\substack{n\to\infty, \ Q^2\to\infty \\ Q^2/n\to M^2=const}}{} , \qquad (6.39)$$

where M^2 is the Borel parameter. If $f(Q^2)$ is represented by the dispersion relation

$$f(Q^2) = \frac{1}{\pi} \int_0^\infty \frac{\operatorname{Im} f(s)}{s + Q^2} ds, \qquad (6.40)$$

then

$$\mathcal{B}_{M^2} f(Q^2) = \frac{1}{\pi} \int_0^\infty e^{-s/M^2} \operatorname{Im} f(s) ds, \qquad (6.41)$$

and the Borel transformation is reduced to the Laplace transformation. Note, that

$$\mathcal{B}_{M^2} \frac{1}{(Q^2)^n} = \frac{1}{(M^2)^{n-1}(n-1)!} \qquad (6.42)$$

and

$$\mathcal{B}_{M^2}(Q^2)^k \ln Q^2 = (-1)^{k+1} \Gamma(k+1)(M^2)^{k+1}. \qquad (6.43)$$

Evidently, the Borel transformation kills the subtraction terms in (6.38) and improves the convergence of the dispersion integral in (6.37). If the lowest hadronic state is separated by a gap from the excited states, then by a suitable choice of the Borel parameter M^2 it is possible to suppress the contributions of excited states compared with that of the lowest state. As follows from (6.42), the Borel transformation factorially improves the convergence of the OPE series. In what follows, only the Borel transformed QCD sum rules will be used. In the practical applications of QCD sum rules it is necessary to ensure that the OPE converges on the accounted terms and that the contribution of the lowest hadronic state, whose characteristics are determined, dominates or is at least is comparable with contributions of excited states. (For this reason the so-called finite energy sum rules – FESR, where the integral (6.37) is cut at some chosen value s_0 – cannot be considered as reliable.) The standard model of excited state contributions is based on the assumption that, starting from some s_0, this contribution is described by the continuum, given by the nonvanishing (at $q^2 \to \infty$) terms of OPE in the given Lorentz structure. The reason for such assumption is that only such terms survive at high Q^2. Of course, the perturbation series should converge at the accounted for terms for a chosen value of M^2. Since the convergence of OPE improves with increasing M^2 but the contribution of exited states increases when M^2 grows, then there is a possibility to find a window $M_1^2 < M^2 < M_2^2$ where all above mentioned requirements are fulfilled. At $M^2 > M_1^2$, the last accounted for terms in the perturbative and OPE series are small compared with the whole contribution. So, one can believe that accounting for these terms is enough for a desired accuracy of the result. At $M^2 < M_2^2$, the contribution of excited states is small or comparable with the contribution of the lowest state (but does not overwhelm it!), therefore the latter is well separated. QCD sum rules make sense only for M^2 within this window. If the window is absent, i.e. if $M_1^2 > M_2^2$, then a QCD sum rule does not exist. To ensure the validity of a QCD sum rule it is necessary to check the M^2-dependence: the left- and right-hand sides of the Borel transformed Eq. (6.38) must have a similar M^2-dependence inside the M^2-window.

The goal of QCD sum rules is to determine the properties of the lowest hadronic states. The sizes of such states are typically of the order of 0.3 fermi and the corresponding momenta are of the order of 1 GeV. In this domain, the QCD interaction is rather strong. For this reason, QCD sum rules are approximate. The sources of the uncertainties in these relations were mentioned above. Let us summarize them:

1. The QCD coupling constant $\alpha_s(Q)$ is rather large, $\alpha_s(1\ \text{GeV}) \approx 0.5$, and in practice only few terms in perturbative series are taken into account. (Usually only one; sometimes the perturbative corrections are completely neglected.)

2. The OPE series is truncated. As a rule, it is required that the highest-order term of the OPE contributes not more than 10% to the sum rule. Nonperturbative vacuum fluctuations besides the OPE can be only estimated using the instanton model.

3. The values of condensates have some uncertainties, especially for high-dimension condensates where the factorization hypothesis is accepted.

4. The separation of perturbative and nonperturbative contributions has some arbitrariness.

5. The separation of the lowest hadronic state contribution in the given channel is achieved using the model of the hadronic spectrum "pole + continuum."

The role of all these uncertainties should be estimated in each specific calculation. The precision of QCD sum rule predictions is not high – in the best case, not better than 10%. This fact is not surprising: we have to deal with strong interaction! Since all uncertainties are of a physical origin and their estimates cannot be completely reliable, any mathematical tricks, like the χ^2 best fit, could hardly improve the situation.

Despite all these uncertainties, the QCD sum rule approach is a unique method, which allows one to calculate a broad variety of hadronic characteristics from first principles of QCD, using only a restricted number of QCD parameters – vacuum condensates. If found in one sum rule, the condensate leads to definite predictions for physical observables determined by other sum rules. Thus, new physical knowledge is obtained and at the same time the whole approach is checked.

6.5 Determination of $\alpha_s(Q^2)$ and the condensates from low-energy data

6.5.1 Determination of $\alpha_s(m_\tau^2)$ from hadronic decays of τ-lepton

Collaborations ALEPH [26],[27], OPAL [28] and CLEO [29] have measured with good accuracy the relative probability of hadronic decays of the τ lepton $R_\tau = B(\tau \to \nu_\tau + \text{hadrons})/B(\tau \to \nu_e \bar{\nu}_e)$, and the vector V and axial A spectral functions. We present below the results of the theoretical analysis of these data on the basis of the OPE in QCD [30],[31] (see also [32],[33],[27]). In the perturbation series, the terms up to α_s^4 will be taken into account, in OPE – the operators up to dimension 8. We restrict ourself to the case of final hadronic states with zero strangeness.

Consider the polarization operator of hadronic currents

$$\Pi_{\mu\nu}^J = i\int e^{iqx}\langle TJ_\mu(x)J_\nu^\dagger(0)\rangle dx = (q_\mu q_\nu - \delta_{\mu\nu}q^2)\Pi_J^{(1)}(q^2) + q_\mu q_\nu \Pi_J^{(0)}(q^2), \quad (6.44)$$

where

$$J = V, A; \quad V_\mu = \bar{u}\gamma_\mu d, \quad A_\mu = \bar{u}\gamma_\mu\gamma_5 d.$$

The spectral functions measured in τ decay are imaginary parts of $\Pi_J^{(1)}(s)$ and $\Pi_J^{(0)}(s)$, $s = q^2$

$$v_1/a_1(s) = 2\pi\,\mathrm{Im}\,\Pi_{V/A}^{(1)}(s + i0), \quad a_0(s) = 2\pi\,\mathrm{Im}\,\Pi_A^{(0)}(s + i0). \quad (6.45)$$

The functions $\Pi_V^{(1)}(q^2)$ and $\Pi_A^{(1)}(q^2)$ are analytic in the complex q^2 plane with a cut along the right-hand semiaxis starting from $4m_\pi^2$ for $\Pi_V^{(1)}(q^2)$ and from $9m_\pi^2$ for $\Pi_A^{(1)}(q^2)$. The function $\Pi_A^{(1)}(q^2)$ has a kinematical pole at $q^2 = 0$, since the singularity free physical combination is $\delta_{\mu\nu}q^2\Pi_A^{(1)}(q^2)$. Because of axial current conservation in the limit of massless quarks this kinematical pole is related to the one-pion state contribution to $\Pi_A(q)$ which has the form of (2.119):

$$\Pi_{\mu\nu}^A(q)_\pi = -\frac{f_\pi^2}{q^2}\left(q_\mu q_\nu - \delta_{\mu\nu}q^2\right) - \frac{m_\pi^2}{q^2}q_\mu q_\nu\frac{f_\pi^2}{q^2 - m_\pi^2}. \quad (6.46)$$

Chiral symmetry violation may result in corrections of order $f_\pi^2(m_\pi^2/m_\rho^2)$ in $\Pi_A^{(1)}(q^2)$ (m_ρ is the characteristic hadronic mass), i.e. to the theoretical uncertainty in the magnitude of the residue of the kinematical pole in $\Pi_A^{(1)}(q^2)$ of order $\Delta f_\pi^2/f_\pi^2 \sim m_\pi^2/m_\rho^2$.

Consider first the ratio of the total probability of hadronic decays of τ leptons into states with zero strangeness to the probability of $\tau \to \nu_\tau e\bar{\nu}_e$. This ratio is given by the expression [34]

$$R_{\tau, V+A} = \frac{B(\tau \to \nu_\tau + \text{hadrons}_{S=0})}{B(\tau \to \nu_\tau e\bar{\nu}_e)} = 6|V_{ud}|^2 S_{EW}$$

$$\times \int_0^{m_\tau^2} \frac{ds}{m_\tau^2}\left(1 - \frac{s}{m_\tau^2}\right)^2\left[\left(1 + 2\frac{s}{m_\tau^2}\right)(v_1 + a_1 + a_0)(s) - 2\frac{s}{m_\tau^2}a_0(s)\right], \quad (6.47)$$

where $|V_{ud}| = 0.9735 \pm 0.0008$ is the matrix element of the Kobayashi–Maskawa matrix and $S_{EW} = 1.0194 \pm 0.0040$ is the electroweak correction [35]. Practically only the one-pion state is contributing to the last term in (6.47) and it happens to be small:

$$\Delta R_\tau^{(0)} = -24\pi^2 \frac{f_\pi^2 m_\pi^2}{m_\tau^4} = -0.008. \tag{6.48}$$

Denote

$$\omega(s) \equiv v_1 + a_1 + a_0 = 2\pi \, \text{Im} \left[\Pi_V^{(1)}(s) + \Pi_A^{(1)}(s) + \Pi_A^{(0)}(s) \right] \equiv 2\pi \, \text{Im} \, \Pi(s). \tag{6.49}$$

As follows from Eq. (6.46), $\Pi(s)$ has no kinematical pole, but only a right-hand cut. It is convenient to transform the integral in Eq. (6.47) into an integral over the circle of radius m_τ^2 in the complex s plane [36]–[38]:

$$R_{\tau, V+A} = 6\pi i |V_{ud}|^2 S_{EW} \oint_{|s|=m_\tau^2} \frac{ds}{m_\tau^2} \left(1 - \frac{s}{m_\tau^2}\right)^2 \left(1 + 2\frac{s}{m_\tau^2}\right) \Pi(s) + \Delta R_\tau^{(0)}. \tag{6.50}$$

Eq. (6.50) allows one to express $R_{\tau, V+A}$ in terms of $\Pi(s)$ at large $|s| = m_\tau^2$ where perturbation theory and the OPE are valid.

Calculate first the perturbative contribution to Eq. (6.50). To this end, use the Adler function $D(Q^2)$:

$$D(Q^2) \equiv -2\pi^2 \frac{d\Pi(Q^2)}{d \ln Q^2} = \sum_{n \geq 0} K_n a^n, \quad a \equiv \frac{\alpha_s(Q^2)}{\pi}, \quad Q^2 \equiv -s, \tag{6.51}$$

the perturbative expansion of which is known up to terms $\sim \alpha_s^4$. In the $\overline{\text{MS}}$ regularization scheme $K_0 = K_1 = 1$, $K_2 = 1.64$ [39], $K_3 = 6.37$ [40] for three flavours and for K_4 there are the estimates $K_4 = 25 \pm 25$ [41] and $K_4 = 27 \pm 16$, $K_4 \approx 50$ [42]. The renormgroup equation yields

$$\frac{da}{d \ln Q^2} = -\beta(a) = -\sum_{n \geq 0} \beta_n a^{n+2}, \tag{6.52}$$

$$\ln \frac{Q^2}{\mu^2} = -\int_{a(\mu^2)}^{a(Q^2)} \frac{da}{\beta(a)}, \tag{6.53}$$

in the $\overline{\text{MS}}$ scheme for three flavours $\beta_0 = 9/4$, $\beta_1 = 4$, $\beta_2 = 10.06$, $\beta_3 = 47.23$ [43, 44, 45]. Integrating over Eq. (6.51) and using Eq. (6.52) we get

$$\Pi(Q^2) - \Pi(\mu^2) = \frac{1}{2\pi^2} \int_{a(\mu^2)}^{a(Q^2)} D(a) \frac{da}{\beta(a)}. \tag{6.54}$$

Put $\mu^2 = m_\tau^2$ and choose some (arbitrary) value $a(m_\tau^2)$. With the aid of Eq. (6.53), one can determine $a(Q^2)$ for any Q^2 and by analytical continuation for any s in the complex plane. Then, calculating (6.54) find $\Pi(s)$ in the whole complex plane. Substitution of $\Pi(s)$ into Eq. (6.50) determines R_τ for the given $a(m_\tau^2)$ up to power corrections. (The contribution of the constant term $\Pi(\mu^2)$ vanishes after integration over the closed contour.) Thus,

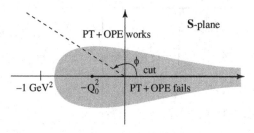

Fig. 6.1. The applicability region of PT and OPE in the complex plane s. In the shaded region, PT + OPE does not work.

knowing R_τ from experiment it is possible to find the corresponding $a(m_\tau^2)$. Note, that with such an approach there is no need to expand the denominator in Eqs. (6.53), (6.54) in inverse powers of $\ln Q^2/\mu^2$. Advantages of transforming the integral over the real axis (6.47) into a contour integral are the following: It can be expected that the applicability region of the theory presented as perturbation theory (PT) + (OPE) in the complex s-plane is off the dashed region in Fig. 6.1. It is evident that PT+OPE does not work at positive and comparatively small values of s.

As is well known in perturbation theory, in the expansion in inverse powers of $\ln Q^2$, in the first order in $1/\ln Q^2$ the running coupling constant $\alpha_s(Q^2)$ has an unphysical pole at some $Q^2 = Q_0^2$. If $\beta(a)$ is kept in the denominator in (6.53), then in n-loop approximation $(n > 1)$ a branch cut with a singularity $\sim (1 - Q^2/Q_0^2)^{-1/n}$ appears instead of the pole. The position of the singularity is given by

$$\ln \frac{Q_0^2}{\mu^2} = -\int\limits_{a(\mu^2)}^{\infty} \frac{da}{\beta(a)}. \qquad (6.55)$$

Near the singularity, the last term in the expansion of $\beta(a)$ dominates and gives the aforesaid behavior. Since the singularity has become weaker, one may expect a better convergence of the series, which would allow one to go to lower Q^2.

The real and imaginary parts of $\alpha_s(s)/\pi$, obtained as numerical solutions of Eq. (6.53) for various numbers of loops, are plotted in Fig. 6.2 as functions of $s = -Q^2$. The τ lepton mass was chosen as normalization point, $\mu^2 = m_\tau^2$ and $\alpha_s(m_\tau^2) = 0.352$ was put in. As is seen from Fig. 6.2, perturbation theory converges at negative $s < -1\,\text{GeV}^2$, and 4-loop calculations are necessary to get a good precision of the results. At positive s, especially for $\text{Im}\,(\alpha_s/\pi)$, the convergence of the series is much better. This comes from the fact that in the chosen integral form of renormalization group equation (6.53) the expansion in $\pi/\ln(Q^2/\Lambda^2)$ is avoided, the latter being not a small parameter at intermediate Q^2. (A systematic method of analytical continuation from the spacelike to time-like region with summation of π^2 terms was suggested in [46] and developed in [47].) For instance, in the next-to-leading order

$$2\pi\,\text{Im}\,\Pi(s + i0) = 1 + \frac{1}{\pi\beta_0}\left[\frac{\pi}{2} - \tan^{-1}\left(\frac{1}{\pi}\ln\frac{s}{\Lambda^2}\right)\right] \qquad (6.56)$$

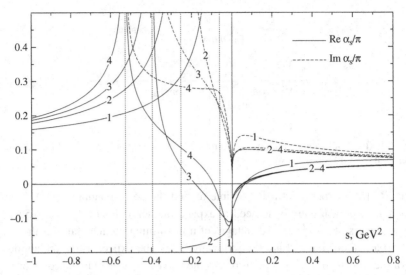

Fig. 6.2. Real and imaginary parts of $\alpha_{\overline{\mathrm{MS}}}(s)/\pi$ as exact numerical solutions of RG equation (6.53) on the real axis for different numbers of loops. The initial condition is chosen $\alpha_s = 0.352$ at $s = -m_\tau^2$, $N_f = 3$. Vertical dotted lines display the position of the unphysical singularity at $s = -Q_0^2$ for each approximation ($4 \to 1$ from left to right).

instead of

$$2\pi \, \mathrm{Im} \, \Pi(s + i0) = 1 + \frac{1}{\beta_0 \ln(s/\Lambda^2)}, \qquad (6.57)$$

which would follow in the case of small $\pi / \ln(s/\lambda^2)$.

The QCD coupling constant $\alpha_s(Q^2)$ at $Q^2 > 0$ in the low Q^2 region ($0.8 < Q^2 < 5 \, \mathrm{GeV}^2$) is plotted in Fig. 6.3. (Four loops are accounted for, $\alpha_s(m_\tau^2)$ is put to be equal to $\alpha_s(m_\tau^2) = 0.33$. As follows from the τ-decay rate $\alpha_s(m_\tau^2) = 0.352 \pm 0.020$, and the value of one standard deviation below the mean is favoured by low-energy sum rules.)

Integration over the contour allows one to obviate the dashed region in Fig. 6.1 (except for the vicinity of the positive semiaxis, the contribution of which is suppressed by the factor $(1 - s/m_\tau^2)^2$ in Eq. (6.50)), i.e. to work in the applicability region of PT + OPE.

The OPE terms, i.e. power corrections to the polarization operator, are given by the formula [13]:

$$\Pi(s)_{nonpert} = \sum_{n \geq 2} \frac{\langle O_{2n} \rangle}{(-s)^n} \left(1 + c_n \frac{\alpha_s}{\pi} \right) = \frac{\alpha_s}{6\pi \, Q^4}$$

$$\times \langle 0 \mid G_{\mu\nu}^a G_{\mu\nu}^a \mid 0 \rangle \left(1 + \frac{7}{6} \frac{\alpha_s}{\pi} \right) - \frac{2}{Q^4} (m_u + m_d) \langle 0 \mid \bar{q}q \mid 0 \rangle \frac{\alpha_s}{\pi}$$

$$+ \frac{128}{81 \, Q^6} \pi \alpha_s \langle 0 \mid \bar{q}q \mid 0 \rangle_\mu^2 \left[1 + \left(\frac{29}{24} + \frac{17}{18} \ln \frac{Q^2}{\mu^2} \right) \frac{\alpha_s}{\pi} \right] + \frac{\langle O_8 \rangle}{Q^8}, \qquad (6.58)$$

(α_s-corrections to the first and third terms in Eq. (6.58) were calculated in [48] and [49], respectively). Contributions of terms proportional to m_u^2, m_d^2 are neglected. When

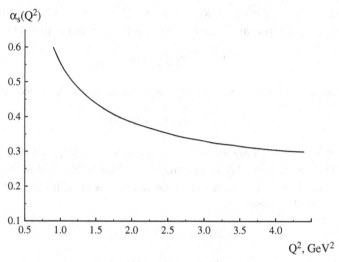

Fig. 6.3. $\alpha_s(Q^2)$ normalized at m_τ^2, $\alpha_s(m_\tau^2) = 0.33$.

calculating the $d = 6$ term, the factorization hypothesis was used. The gluon condensate of dimension $d = 6$ $g^3 \langle 0 \mid G^3 \mid 0 \rangle$ (6.20) does not contribute to the polarization operator (6.58). This is a consequence of the general theorem, proved by Dubovikov and Smilga [50], that in case of self-dual gluonic fields there are no contributions of gluon condensates of dimensions higher than $d = 4$ to vector and axial current polarization operators. Since the vacuum expectation value of the G^3 operator does not vanish for self-dual gluonic fields, this means that the coefficient in front of the condensate $g^3 \langle 0 \mid G^3 \mid 0 \rangle$ in (6.58) must vanish. The same argument refers to the dimension 8 gluon operators $g^4 G^4$ except some of them, like $g^4 [\, G^n_{\mu\alpha} G^n_{\mu\beta} - (1/4)\delta_{\alpha\beta} G^n_{\mu\nu} G^n_{\mu\nu}\,]^2$, which have zero expectation values in any self-dual field. But the latter are suppressed by a small factor $1/4\pi^2$ arising from loop integration in comparison with the tree diagram that corresponds to the $d = 8$ four-quark condensate contribution $\langle O_8 \rangle \sim \langle \bar{q} G q \cdot \bar{q} q \rangle$. The contribution from this condensate may be estimated as $\mid \langle O_8 \rangle \mid < 10^{-3}$ GeV [31] (see below, Section 6.5.2) and appears to be negligibly small. The $d = 8$ two quarks – two gluon operator $O'_8 \sim g^2 D\bar{q} G G q$ is nonfactorizable, its vacuum mean value is suppressed by $1/N_c$ and one may believe that its contribution to (6.58) is also small. It can be readily seen that $d = 4$ condensates (up to small α_s-corrections) give no contribution to the integral over the contour Eq. (6.50). $R_{\tau,V+A}$ may be represented as

$$R_{\tau,\,V+A} = 3|V_{ud}|^2 S_{EW} \left(1 + \delta'_{em} + \delta^{(0)} + \delta^{(6)}_{V+A} \right) + \Delta R^{(0)} = 3.486 \pm 0.016, \qquad (6.59)$$

where $\delta'_{em} = (5/12\pi)\alpha_{em}(m_\tau^2) = 0.001$ is the electromagnetic correction [51], $\delta^{(6)}_{A+V} = -(3.3 \pm 1.1)\cdot 10^{-3}$ is the contribution of the $d = 6$ condensate (see below) and $\delta^{(0)}$ is the PT correction. The right-hand part presents the experimental value obtained as a difference between the total probability of hadronic decays $R_\tau = 3.647 \pm 0.014$ [52] and the probability of decays into strangeness $S = -1$ states: $R_{\tau,s} = 0.161 \pm 0.007$ [53,

54, 33]. (In Ref. [27],[33] the values $R_{\tau, V+A} = 3.482 \pm 0.014, 3.478 \pm 0.016$ are given correspondingly.) For the perturbative correction it follows from Eq. (6.59) that

$$\delta^{(0)} = 0.208 \pm 0.006. \tag{6.60}$$

From (6.60) the constant $\alpha_s(m_\tau^2)$ was found employing the above described method [31]:

$$\alpha_s(m_\tau^2) = 0.352 \pm 0.020. \tag{6.61}$$

The calculation was made taking account of terms up to $\sim \alpha_\tau^4(m_\tau^2)$; the theoretical error was assumed to be equal to the last contributing term. Maybe this error is underestimated (by ~ 0.010) since the theoretical and experimental errors were added in quadratures. The value of $\alpha_s(m_\tau^2)$ (6.61) corresponds to:

$$\alpha_s(M_Z^2) = 0.121 \pm 0.002. \tag{6.62}$$

This value is in agreement with the determination [55] of $\alpha_s(m_\tau^2)$ from the complete set of available data

$$\alpha_s(M_Z^2) = 0.1189 \pm 0.0010. \tag{6.63}$$

Some nonperturbative features of QCD may be described in the instanton gas model [56] (see Chapter 4 and [57] for extensive review and the collection of related papers in [58]). Namely, one computes the correlators in the $SU(2)$-instanton field embedded in the $SU(3)$ colour group. In particular, the 2-point correlator of the vector currents had been computed long ago [59]. Apart from the usual tree-level correlator proportional to $\ln Q^2$ it has a correction which depends on the instanton position and radius ρ. In the instanton gas model these parameters are integrated out. The radius is averaged over some concentration $n(\rho)$, for which one or another model is used. Concerning the 2-point correlator of charged axial currents, the only difference from the vector case is that the term with zero modes must be taken with opposite sign. In coordinate representation, the answer can be expressed in terms of elementary functions, see [59]. An attempt to compare the instanton correlators with ALEPH data in the coordinate space was made in Ref. [60].

We shall work in momentum space. The instanton correction to the spin-J parts $\Pi^{(J)}$ of the correlator (6.44) can be written in the following form:

$$\Pi^{(1)}_{V,\,\text{inst}}(q^2) = \int_0^\infty d\rho\, n(\rho) \left[-\frac{4}{3q^4} + \sqrt{\pi}\, \rho^4 G^{30}_{13}\left(-\rho^2 q^2 \,\Big|\, {1/2 \atop 0, 0, -2} \right) \right],$$

$$\Pi^{(0)}_{A,\,\text{inst}}(q^2) = \int_0^\infty d\rho\, n(\rho) \left[-\frac{4}{q^4} - \frac{4\rho^2}{q^2} K_1^2\left(\rho\sqrt{-q^2} \right) \right],$$

$$\Pi^{(1)}_{A,\,\text{inst}}(q^2) = \Pi^{(1)}_{V,\,\text{inst}}(q^2) - \Pi^{(0)}_{A,\,\text{inst}}(q^2), \qquad \Pi^{(0)}_{V,\,\text{inst}}(q^2) = 0. \tag{6.64}$$

Here K_1 is the modified Bessel function, $G^{pq}_{mn}(z|\ldots)$ is the Meijer function. Definitions, properties, and approximations of Meijer functions can be found, for instance, in [61]. In particular, the function in (6.64) can be written as the following series:

$$\sqrt{\pi}\,G^{30}_{13}\left(z\,\middle|\,\begin{matrix}1/2\\0,0,-2\end{matrix}\right) = \frac{4}{3z^2} - \frac{2}{z} + \frac{1}{2\sqrt{\pi}}\sum_{k=0}^{\infty} z^k \frac{\Gamma(k+1/2)}{\Gamma^2(k+1)\,\Gamma(k+3)}$$

$$\times\left\{\left[\ln z + \psi(k+1/2) - 2\psi(k+1) - \psi(k+3)\right]^2\right.$$

$$\left. + \psi'(k+1/2) - 2\psi'(k+1) - \psi'(k+3)\right\}, \tag{6.65}$$

where $\psi(z) = \Gamma'(z)/\Gamma(z)$. For large $|z|$, one can obtain its approximation by the saddle-point method:

$$G^{30}_{13}\left(z\,\middle|\,\begin{matrix}1/2\\0,0,-2\end{matrix}\right) \approx \sqrt{\pi}\,z^{-3/2}e^{-2\sqrt{z}}, \qquad |z| \gg 1. \tag{6.66}$$

The formulae (6.64) should be treated in the following way. One adds Π_{inst} to the usual polarization operator with perturbative and OPE terms. But the terms $\sim 1/q^4$ must be absorbed by the operator O_4 in Eq. (6.58), since the gluonic condensate $\langle G^2\rangle$ is averaged over all field configurations, including the instanton one. Notice the negative sign of $1/q^4$ in Eq. (6.64). This arises because the negative contribution of the quark condensate $\langle m\bar{q}q\rangle$ in the instanton field exceeds the positive contribution of the gluonic condensate $\langle G^2\rangle$. In the real world, $\langle m\bar{q}q\rangle$ is negligible.

The correlators (6.64) have the appropriate analytical properties: they have cuts along the positive real axis:

$$\text{Im}\,\Pi^{(1)}_{V,\,\text{inst}}(q^2+i0) = \int_0^{\infty} d\rho\,n(\rho)\,\pi^{3/2}\rho^4 G^{20}_{13}\left(\rho^2 q^2\,\middle|\,\begin{matrix}1/2\\0,0,-2\end{matrix}\right), \tag{6.67}$$

$$\text{Im}\,\Pi^{(0)}_{A,\,\text{inst}}(q^2+i0) = -\int_0^{\infty} d\rho\,n(\rho)\,\frac{2\pi^2\rho^2}{q^2}J_1\left(\rho\sqrt{q^2}\right)N_1\left(\rho\sqrt{q^2}\right). \tag{6.68}$$

We consider below the instanton gas model. This is a model with fixed instanton radius:

$$n(\rho) = n_0\,\delta(\rho - \rho_0). \tag{6.69}$$

In Section 4.6, the following values were estimated (see also [56],[57]):

$$\rho_0 \approx 0.2 - 0.4\text{fm} \approx 1.0 - 2.0\text{GeV}^{-1}, \quad n_0 \lesssim 0.62\text{fm}^{-4} \approx (1.0) \times 10^{-3}\text{GeV}^4. \tag{6.70}$$

In fact, the instanton liquid model, with the account of the instanton self-interaction, was mainly considered in [56], but the arguments from which the estimates (6.70) follow refer also to the instanton gas model. In this case, the value of n_0 (6.70) should be considered as an upper limit (see also [57]).

Let us compute the instanton contribution to the τ-decay branching ratio. Since the instanton correlator (6.64) has a $1/q^2$ singular term in the expansion near 0 (see Eq. (6.65)), the integrals must be taken over the circle as in (6.50). In the instanton model, the function $a_0(s)$ differs from experimental δ-function, which gives a small correction. So we shall

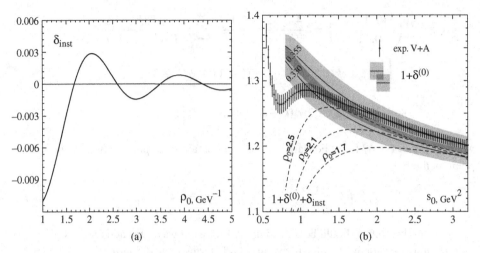

Fig. 6.4. The instanton correction to the τ decay ratio versus ρ_0 (a) and "versus τ mass" (b) for $n_0 = 1.0 \times 10^{-3}\,\mathrm{GeV^4}$. The thin solid lines in (b) are the values of $1 + \delta^{(0)}(s_0)$, where $\delta^{(0)}(s_0)$ are perturbative corrections, calculated as described in Section 6.5.1. The upper curve corresponds to $\alpha_s(m_\tau^2) = 0.355$, the lower one to $\alpha_s(m_\tau^2) = 0.330$. The shadowed regions represent the uncertainties in perturbative calculations, the dark shadowed band is their overlap. The dashed lines are $1 + \delta^{(0)}(s) + \delta_{inst}$, $\delta^{(0)}(s)$ corresponds to $\alpha_s(m_\tau^2) = 0.330$.

ignore the last term in (6.47) and consider the integral with $\Pi_{V+A}^{(1)} + \Pi_A^{(0)}$ in (6.50). The instanton correction to the τ-decay branching ratio can be brought to the following form:

$$\delta_{\text{inst}} = -48\pi^{5/2} \int_0^\infty d\rho\, n(\rho)\rho^4 G_{13}^{20}\left(\rho^2 m_\tau^2 \left|\begin{array}{c} 1/2 \\ 0, -1, -4 \end{array}\right.\right) \approx \frac{48\pi^2 n_0}{\rho_0^2 m_\tau^6} \sin\left(2\rho_0 m_\tau\right). \quad (6.71)$$

Since the parameters (6.70) are determined rather approximately, we may explore the dependence of δ_{inst} on them: δ_{inst} versus ρ_0 for fixed $n_0 = 1.0 \cdot 10^{-3}\,\mathrm{GeV^4}$ is shown in Fig. 6.4.

As seen from Fig. 6.4a the instanton correction to hadronic τ-decay is small. At the favourable value [56] $\rho_0 = 1.7\,\mathrm{GeV^{-1}}$ the instanton correction to R_τ is almost exactly zero. This confirms the calculations of $\alpha_s(m_\tau^2)$ presented above, where the instanton corrections were not taken into account.

Equation (6.71) can be used in another way. Namely, the τ mass can be considered as a free parameter s_0. The dependence of the fractional corrections $\delta^{(0)}$ and $\delta_{0.330}^{(0)} + \delta_{\text{inst}}$ on s_0 is shown in Fig. 6.4b. The result strongly depends on the instanton radius and rather significantly on the density n_0. For $\rho_0 = 1.7\,\mathrm{GeV^{-1}}$ and $n_0 = 1\,\mathrm{fm^{-4}}$, the instanton curve is outside the errors already at $s_0 \sim 2\,\mathrm{GeV^2}$ where perturbation theory is expected to work.

We have come to the conclusion that in case of a variable τ mass the instanton contribution becomes large at $s_0 < 2\,\mathrm{GeV^2}$. That means that $R_{\tau,V+A}(s_0)$ given by (6.50) cannot be represented by PT + OPE at $s_0 < 2\,\mathrm{GeV^2}$ and the results obtained in this way are not reliable.

There are many calculations of $\alpha_s(m_\tau^2)$ from the total τ-decay rate, using the same idea, which was used above – the contour improved fixed order perturbation theory [32],[34],[36],[37],[38],[52]. (For more recent ones, see [27],[33],[62].) The results of these calculations coincide with those presented above within the errors and give $\alpha_s(m_\tau^2) = 0.33 - 0.35$. From these values, by using the renormalization group one can find $\alpha_s(M_Z^2) = 0.118 - 0.121$ in agreement with $\alpha_s(M_Z^2)$ determinations from other processes (see [55],[63]).

Above, only one renormalization scheme has been considered – the $\overline{\text{MS}}$ scheme. In the BLM renormalization scheme [64], which has some advantages from the point of view of perturbative pomeron theory [65], the result is $\alpha_s(m_\tau^2) = 0.621 \pm 0.008$ [66], corresponding in the framework of the BLM scheme to the same value of $\alpha_s(M_Z^2) = 0.117 - 0.122$. At low scales, however, the $\alpha_s(Q^2)$ behaviour is significantly different from that presented in Fig. 6.3.

6.5.2 Determination of quark condensates from the $V - A$ spectral function of τ decay

In order to determine the quark condensate from τ-decay data, it is convenient to consider the difference $V - A$ of the polarization operators $\Pi_V^{(1)} - \Pi_A^{(1)}$, where the contribution of perturbative terms is absent. $\Pi_V^{(1)}(s) - \Pi_A^{(1)}(s)$ is represented by OPE:

$$\Pi_V^{(1)}(s) - \Pi_A^{(1)}(s) = \sum_{d\geq 4} \frac{O_d^{V-A}}{(-s)^{d/2}}\left(1 + c_d\frac{\alpha_s(s)}{\pi}\right). \tag{6.72}$$

The gluonic condensate contribution drops out in the $V - A$ difference and only the following condensates up to $d = 10$ remain

$$O_4^{V-A} = 2\,(m_u + m_d)\,\langle 0\,|\,\bar{q}q\,|\,0\rangle \;=\; -\,f_\pi^2 m_\pi^2,\ [13], \tag{6.73}$$

$$O_6^{V-A} = 2\pi\alpha_s\,\langle 0\,|\,(\bar{u}\gamma_\mu\lambda^a d)(\bar{d}\gamma_\mu\lambda^a u) - (\bar{u}\gamma_5\gamma_\mu\lambda^a d)(\bar{d}\gamma_5\gamma_\mu\lambda^a u)\,|\,0\rangle$$

$$= -\frac{64\pi\alpha_s}{9}\langle 0\,|\,\bar{q}q\,|\,0\rangle^2,\quad [13], \tag{6.74}$$

$$O_8^{V-A} = -8\pi\alpha_s\,m_0^2\langle 0\,|\,\bar{q}q\,|\,0\rangle^2,\quad [67],\,[30],^1 \tag{6.75}$$

$$O_{10}^{V-A} = -\frac{8}{9}\pi\alpha_s\langle 0\,|\,\bar{q}q\,|\,0\rangle^2\left[\frac{50}{9}m_0^4 + 32\pi^2\Big\langle 0\,|\,\frac{\alpha_s}{\pi}G^2\,|\,0\Big\rangle\right],\quad [68], \tag{6.76}$$

where m_0^2 is defined in Eq. (6.16). In the right-hand of (6.74), (6.75), (6.76) the factorization hypothesis was used. For the O_6 operator it is expected [13] that the accuracy of factorization hypothesis is of order $1/N_c^2 \sim 10\%$, where $N_c = 3$ is the number of colours. For operators of dimensions $d \geq 8$ the factorization procedure is not unique. (But, as a rule, the differences arising are not very large: for the $d = 8$ operator entering Eq. (6.72) it is about 20%.) The accuracy of the factorization hypothesis gets worse with increasing

---

[1] There was a sign error in the contribution of O_8 in [30].

operator dimensions: for O_8^{V-A} it is worse than for O_6^{V-A} and for O_{10}^{V-A} it is worse than for O_8^{V-A}.

Operators O_4 and O_6 have approximately zero anomalous dimensions, the O_8 anomalous dimension is equal to $-11/27$. Calculations of the coefficients in front of α_s in Eq. (6.72) gave $c_4 = 4/3$ [48] and $c_6 = 89/48$ [49]. (For O_4 the α_s^2-correction is known [48]: $(59/6)(\alpha_s/\pi)^2$.) The α_s-corrections to O_8^{V-A} are unknown; they are included into the not accurately known value of m_0^2; α_s-corrections to O_{10} are also unknown. (In this section indices $V - A$ will be omitted and O_d will mean condensates with α_s-corrections included.)

Our aim is to compare the OPE theoretical predictions with the experimental data on $V - A$ structure functions measured in τ-decay and with the help of such comparison to determine the magnitude of the most important condensate O_6. The condensate O_4 is small and is known with good accuracy:

$$O_4 = -0.5 \cdot 10^{-3} \text{ GeV}^4. \tag{6.77}$$

We put $m_0^2 = 0.8 \text{GeV}^2$ and in the analysis of the data the values of the condensates O_8 and O_{10} are taken to be equal to

$$O_8 = -2.8 \cdot 10^{-3} \text{ GeV}^8, \tag{6.78}$$

$$O_{10} = -2.6 \cdot 10^{-3} \text{ GeV}^{10}, \tag{6.79}$$

and their Q^2-dependence arising from anomalous dimensions is neglected.

In the calculation of numerical values (6.77), (6.78) it was assumed, that $a_{\bar{q}q}(1 \text{ GeV}^2) \equiv -(2\pi)^2 \langle 0 \mid \bar{q}q \mid 0 \rangle_{1\,\text{GeV}} = 0.65 \text{ GeV}^3$, $\langle 0 \mid (\alpha_s/\pi)G^2 \mid 0 \rangle = 0.005 \text{ GeV}^4$: see below, Eqs. (6.86), (6.106).

As was shown in [30], the dimension $d = 8$ four-quark operators for vector and axial currents are of opposite sign and equal in absolute values up to terms of order $1/N_c^2$: $O_8^V = -O_8^A(1 + O(N_c^{-2}))$. (The exact value of the N_c^{-2} correction is uncertain: it depends on the factorization procedure.) So, for O_8^{V+A} we have from (6.78) the estimate $\mid O_8^{V+A} \mid < 10^{-3} \text{ GeV}^8$, which was used in the calculation of $\Pi(s)_{nonpert}$, Eq.(6.58).

For $\Pi_V^{(1)}(s) - \Pi_A^{(1)}(s)$ a subtractionless dispersion relation is valid:

$$\Pi_V^{(1)}(s) - \Pi_A^{(1)}(s) = \frac{1}{2\pi^2} \int_0^\infty \frac{v_1(t) - a_1(t)}{t - s} dt + \frac{f_\pi^2}{s}. \tag{6.80}$$

(The last term on the right-hand side is the kinematical pole contribution.) The experimental data for $v_1(s) - a_1(s)$ are presented in Fig. 6.5a.

In order to improve the convergence of the OPE series as well as to suppress the contribution of the large s domain in the dispersion integral one uses the Borel transformation. Put $s = s_0 e^{i\phi}$ ($\phi = 0$ on the upper edge of the cut; see Fig. 6.1) and make the Borel transformation in s_0. As a result, we get the following sum rules for the real and imaginary parts of (6.80):

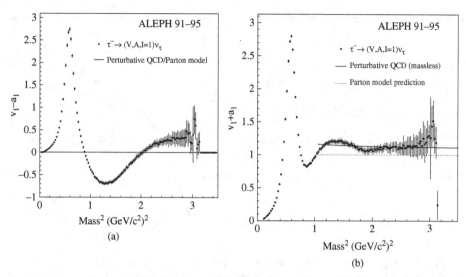

Fig. 6.5. The inclusive $(v_1 - a_1)$ (a) and $(v_1 + a_1)$ (b) spectral functions of τ-decay, measured by ALEPH [27].

$$\int_0^\infty \exp\left(\frac{s}{M^2}\cos\phi\right)\cos\left(\frac{s}{M^2}\sin\phi\right)(v_1 - a_1)(s)\frac{ds}{2\pi^2} = f_\pi^2 + \sum_{k=1}^\infty (-1)^k \frac{\cos(k\phi)O_{2k+2}}{k!\,M^{2k}},$$

(6.81)

$$\int_0^\infty \exp\left(\frac{s}{M^2}\cos\phi\right)\sin\left(\frac{s}{M^2}\sin\phi\right)(v_1 - a_1)(s)\frac{ds}{2\pi^2 M^2} = \sum_{k=1}^\infty (-1)^k \frac{\sin(k\phi)O_{2k+2}}{k!\,M^{2k+2}}.$$

(6.82)

The use of the Borel transformation along the rays in the complex plane has a number of advantages. $\cos\phi$ is negative at $\pi/2 < \phi < 3\pi/2$. Choose ϕ in the region $\pi/2 < \phi < \pi$. In this region, on the one hand, the shadowed area in Fig. 6.1 in the integrals (6.81), (6.82) is touched to a lesser degree and, on the other hand, the contribution of large s, particularly $s > m_\tau^2$ where experimental data are absent, is exponentially suppressed. At definite values of ϕ the contribution of some condensates vanishes, which may also be used. In particular, the condensate O_8 does not contribute to (6.81) at $\phi = 5\pi/6$ and to (6.82) at $\phi = 2\pi/3$, while the contribution of O_6 to (6.81) vanishes at $\phi = 3\pi/4$. Finally, a well-known advantage of the Borel sum rules is factorial suppression of higher-dimension terms of OPE. Figs. 6.6 and 6.7 present the results of the calculations of the left-hand parts of Eqs. (6.81), (6.82) on the basis of the ALEPH [26],[27] experimental data compared with OPE predictions, i.e. with the right-hand parts of these equations.

When comparing the theoretical curves with experimental data one must remember that the value of f_π, which in the figures was taken to be equal to the experimental one of $f_\pi = 130.7\,\text{MeV}$, has, in fact, a theoretical uncertainty of the order of $(\Delta f_\pi^2/f_\pi^2)_{theor} \sim m_\pi^2/m_\rho^2$, where m_ρ is a characteristic hadronic scale (say, the ρ meson mass). This uncertainty is

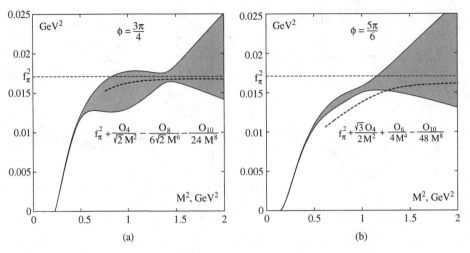

Fig. 6.6. Eq. (6.81): the left-hand part is obtained based on the experimental data, the shaded region corresponds to experimental errors; the right-hand part – the theoretical one – is represented by the dashed curve, numerical values of condensates O_4, O_8, O_{10}, O_6 are taken according to (6.77), (6.78), (6.79), (6.83); (a) $\phi = 3\pi/4$, (b) $\phi = 5\pi/6$.

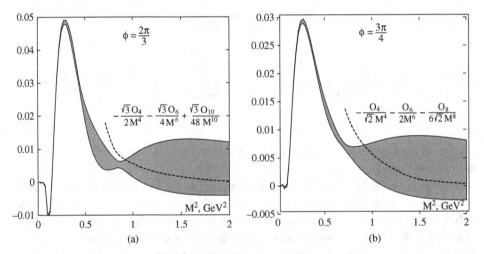

Fig. 6.7. The same as Fig.6.6, but for Eq. (6.82); (a) $\phi = 2\pi/3$, (b) $\phi = 3\pi/4$.

caused by chiral symmetry violation in QCD. Particularly, accounting for this uncertainty may lead to a better agreement of the theoretical curve with the data in Fig. 6.6b. The calculation of instanton contributions (Eq. (6.64)) shows that in all cases considered above they are less than $0.5 \cdot 10^{-3}$ at $M^2 > 0.8$ GeV^2, i.e. are well below the errors. (In some cases, they improve the agreement with the data.) The best fit of the data (the dashed curves at Figs. 6.6 and 6.7.) was achieved with the value

$$O_6 = -4.4 \cdot 10^{-3} \text{ GeV}^6. \tag{6.83}$$

From (6.83), after separating the α_s-correction $[1 + (89/48)(\alpha_s/\pi) = 1.33]$, we get

$$\alpha_s \langle 0 \mid \bar{q}q \mid 0\rangle^2 = 1.5 \cdot 10^{-4}\,\text{GeV}^6. \tag{6.84}$$

The error may be estimated as 30%. The value (6.84) agrees within the limits of errors with the previous estimate [30]. The contribution of dimension 10 is negligible in all cases at $M^2 \geq 1\,\text{GeV}^2$. It is worth mentioning that the theory, i.e. the OPE, agrees with the data at $M^2 > 0.8\,\text{GeV}^2$. The good agreement of the theoretical curves with the data confirms the chosen value of O_8 (6.78) and therefore the use of the factorization hypothesis. From (6.84), with the use of $\alpha_s(1\,\text{GeV}^2) = 0.55$ (see Fig. 6.3), we can find the value of the quark condensate at 1 GeV:

$$\langle 0|qq|0\rangle_{1\,\text{GeV}} = -1.65 \cdot 10^{-2}\,\text{GeV}^3 = -(254\,\text{MeV})^3 \tag{6.85}$$

and a convenient parameter is

$$a_{\bar{q}q}(1\,\text{GeV}^2) \equiv -(2\pi)^2\langle 0|\bar{q}q|0\rangle_{1\,\text{GeV}} = 0.65\,\text{GeV}^3. \tag{6.86}$$

The magnitude of the quark condensate (6.85) is close to that which follows from the Gell-Mann–Oakes–Renner relation (2.34).

In the past years, there were many attempts [32],[69]–[75] to determine quark condensates using $V - A$ spectral functions measured in τ decay. Unlike the approach presented above, where the analytical properties of the polarization operator were exploited in the whole complex q^2-plane which allowed one to separate the contributions of operators of different dimensions, the authors of [32],[69]–[75] considered the finite energy sum rules (FESR) (or integrals over contours) with chosen weight functions. In [69],[71], the $N_c \to \infty$ limit was used. In [70],[71],[72],[75] an attempt was made to find higher dimensional condensates (up to 18 in [75], up to 16 in [70],[71] and up to 12 in [72]). Determination of higher dimensional condensates requires fine tuning of the upper limit of integration in FESR. If the upper limit of integration s_0 in FESR is below 2 GeV^2 (such an upper limit, $s_0 = 1.47$ GeV, was chosen in [75]), then instanton-like corrections, not given by OPE, are of importance (see Section 6.5.1). The same remark refers to the case of weight factors singular at $s = 0$ like s^{-l}, $l > 0$ [32], when there is an enhancement of the contribution at low s where OPE breaks down. Keeping in mind these remarks, we have a satisfactory agreement of the values of condensate (6.83) presented above with those found in [32],[71],[75].

6.5.3 *Determination of condensates from* $V + A$ *and* V *structure functions of* τ *decay*

The perturbative corrections are calculated by integration over the whole momentum space. Therefore, in accord with the discussion in Section 6.1, the condensates, which will be extracted by comparison of the theory with experimental data, correspond to nonperturbative contributions minus the perturbative ones, integrated over the domain below the normalization point μ. As was formulated in Section 6.5.1, we treat the renormalization group equation (6.53) and the equation for polarization operator (6.54) in n-loop approximation as exact ones; the expansion in inverse logarithms is not performed. Their numerical values of condensates depend on the number of loops accounted for; that is why the condensates defined in this way are called n-loop condensates.

Consider the polarization operator $\Pi = \Pi_{V+A}^{(1)} + \Pi_A^{(0)}$, defined in (6.49) and its imaginary part

$$\omega(s) = v_1(s) + a_1(s) + a_0(s) = 2\pi \, \mathrm{Im} \, \Pi(s + i0). \tag{6.87}$$

In the parton model $\omega(s) \to 1$ at $s \to \infty$. Any sum rule can be written in the following form:

$$\int_0^{s_0} f(s) \, \omega_{\exp}(s) \, ds = i\pi \oint f(s) \, \Pi_{\text{theor}}(s) \, ds, \tag{6.88}$$

where $f(s)$ is some analytical function in the integration region. In what follows, we use $\omega_{\exp}(s)$, obtained from τ-decay invariant mass spectra published in [26],[27],[33] for $0 < s < m_\tau^2$. The experimental error of the integral (6.88) is computed as the double integral with the covariance matrix $\overline{\omega(s)\omega(s')} - \overline{\omega}(s)\overline{\omega}(s')$, which can be obtained also from the data available in Ref. [26]. In the theoretical integral in (6.88) the contour goes from $s_0 + i0$ to $s_0 - i0$ counterclockwise around all poles and cuts of the theoretical correlator $\Pi(s)$. Because of the Cauchy theorem the unphysical cut must be inside the integration contour.

The choice of the function $f(s)$ in Eq. (6.88) is actually a matter of taste. First, let us consider the usual Borel transformation:

$$B_{\exp}(M^2) = \int_0^{m_\tau^2} e^{-s/M^2} \omega_{\exp}(s) \frac{ds}{M^2} = B_{\mathrm{pt}}(M^2) + 2\pi^2 \sum_n \frac{\langle O_{2n} \rangle}{(n-1)! \, M^{2n}}. \tag{6.89}$$

We separated out the purely perturbative contribution B_{pt}, which is computed numerically according to (6.88) and Eqs. (6.51)–(6.54). Recall that the Borel transformation improves the convergence of the OPE series because of the factors $1/(n-1)!$ in front of operators and suppression of the high-energy tail contribution, where the experimental error is large. It does not suppress the unphysical perturbative cut, the main source of the error in this approach, but actually increases it since $e^{-s/M^2} > 1$ for $s < 0$. So the perturbative part $B_{\mathrm{pt}}(M^2)$ can be reliably calculated only for $M^2 \gtrsim 0.8 - 1 \, \mathrm{GeV}^2$ and higher; below this value the influence of the unphysical cut is out of control.

Both $B_{\exp}$ and B_{pt} are shown in four-loop approximation for $\alpha_s(m_\tau^2) = 0.355$ and 0.330 in Fig. 6.8. The shaded areas display the theoretical error. They are taken equal to the contribution of the last term in the perturbative Adler function expansion $K_4 a^4$ (6.51).

The contribution of the O_8 operator is of order O_8^{V-A}/N_c^2 and is negligible [30]. (In fact, it depends on the factorization procedure and is uncertain for this reason.) The contributions of gluon condensate and $D = 6$ operators are positive; the second term in (6.58) is small. So, the theoretical perturbative curve must go below the experimental points. The result shown in Fig. 6.8 favours the lower value of the QCD coupling constant $\alpha_s(m_\tau^2) = 0.330$ (or maybe $\alpha_s(m_\tau^2) = 0.340$). As is seen from Fig. 6.8, the theoretical

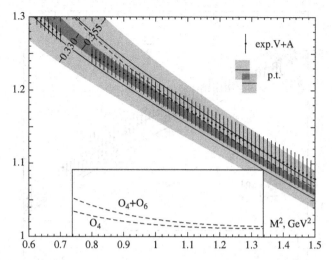

Fig. 6.8. The results of the Borel transformation of the $V + A$ correlator for two values $\alpha_s(m_\tau^2) = 0.355$ and $\alpha_s(m_\tau^2) = 0.330$. The widths of the bands correspond to PT errors; dots with dashed errors are experimental data. The dashed curve is the sum of the perturbative contribution at $\alpha_s(m_\tau^2) = 0.330$, and the condensate contributions O_4 (Eqs. (6.58), (6.91)) and O_6 (Eqs. (6.58), (6.84)).

curve (perturbative at $\alpha_s(m_\tau^2) = 0.330$ plus OPE terms) is in agreement with experiment for $M^2 \geq 0.9 \, \text{GeV}^2$.

In order to separate the contribution of the gluon condensate let us perform the Borel transformation along the rays in the complex s-plane in the same way as was done in Section 6.5.2; the real part of the Borel transform at $\phi = 5\pi/6$ does not contain the $d = 6$ operator:

$$\text{Re } B_{\exp}(M^2 e^{i5\pi/6}) = \text{Re } B_{pt}(M^2 e^{i5\pi/6}) + \pi^2 \frac{\langle O_4 \rangle}{M^4}. \tag{6.90}$$

The results are shown in Fig. 6.9. If we accept the lower value of $\alpha_s(m_\tau^2)$, we get the following constraint on the value of the gluon condensate:

$$\left\langle \frac{\alpha_s}{\pi} G_{\mu\nu}^a G_{\mu\nu}^a \right\rangle = 0.006 \pm 0.012 \text{GeV}^4, \quad \alpha_s(m_\tau^2) = 0.330, \quad M^2 > 0.8 \text{GeV}^2. \tag{6.91}$$

The theoretical and experimental errors are added in quadrature in Eq. (6.91).

Turn now to analysis of the vector correlator. (The vector spectral function was published by ALEPH in [76],[33]). In principle, this cannot give any new information in comparison with $V - A$ and $V + A$ cases, but such an analysis is an important check of the whole approach. Note that the analysis of the vector current correlator is important since it can be performed also using the experimental data on e^+e^- annihilation. The imaginary part of the electromagnetic current correlator measured there is related to the charged current correlator (6.44) by isotopic symmetry.

First, we consider the usual Borel transformation for the vector current correlator, which was originally applied in [77] to a sum rule analysis. It is defined as (6.89) with the experimental spectral function $\omega_{\exp} = 2v_1$ instead of $v_1 + a_1 + a_0$ (the normalization is

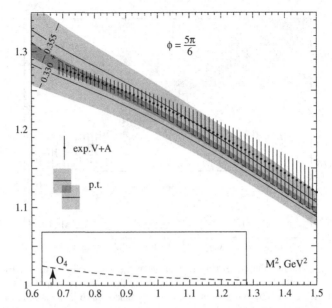

Fig. 6.9. Real part of the Borel transform (6.90) along the ray at the angle $\phi = 5\pi/6$ to the real axes. The dashed line corresponds to the gluonic condensate given by the central value of (6.91).

$v_1(s) \to 1/2$ at $s \to \infty$ in the parton model). Correspondingly, on the right-hand side one should take the vector operators $2O^V = O^{V+A} + O^{V-A}$. The numerical results are shown in Fig. 6.10. The perturbative theoretical curves are the same as in Fig. 6.8 with the $V + A$ correlator. The dashed lines display the contributions of the gluonic condensate given by Eq. (6.91), $2O_6^V = -3.5 \times 10^{-3}\,\mathrm{GeV}^6$ and $2O_8^V = O_8^{V-A} = -2.8 \times 10^{-3}\,\mathrm{GeV}^8$ added to the $\alpha_s(m_\tau^2) = 0.330$ perturbative curve. The contribution of each condensate is shown in the insert box in Fig. 6.10. Notice that for such condensate values, the total OPE contribution is small, since the positive O_4 compensates the negative O_6 and O_8. Agreement is observed for $M^2 > 0.8\,\mathrm{GeV}^2$.

The Borel transformations along the rays in the complex plane result in the same conclusion: at $M^2 > 0.8 - 0.9\,\mathrm{GeV}^2$ agreement with experiment at the 2% level is achieved with $\alpha_s(m_\tau^2) = 0.33 - 0.34$ and at the values of quark and gluon condensates given by (6.83) and (6.91).

A few words about instanton contributions: They can be calculated in the same way as in the case of $V - A$ correlators. At the chosen values of instanton gas parameters the instanton contributions are small, less than $0.5 \cdot 10^{-3}$ for $M^2 > 0.8\,\mathrm{GeV}^2$, and do not spoil the agreement of theory with experiment. The results of the presented above determinations of $\alpha_s(m_\tau^2)$ are summarized by the statement: $\alpha_s(m_\tau^2) = 0.340 \pm 0.015$ [17] (in $\overline{\mathrm{MS}}$ scheme). This value agrees well with ones obtained in [33]. The evolution to the standard normalization point M_Z^2 gives $\alpha_s(M_Z^2) = 0.1207 \pm 0.0015$. The value of $\alpha_s(M_Z^2)$ is in agreement with ones, obtained by the analysis of the whole set of data $\alpha_s(M_Z^2) = 0.1189 \pm 0.0010$ [55].

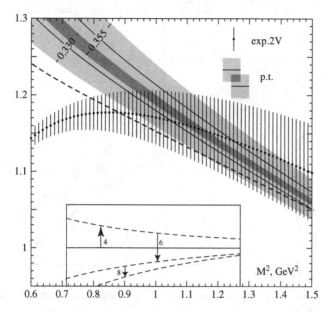

Fig. 6.10. Borel transformation for vector currents.

6.5.4 Determination of the gluon condensate and of the charmed quark mass from the charmonium spectrum

The existence of the gluon condensate was first demonstrated by Shifman, Vainshtein, and Zakharov [13]. They considered the polarization operator $\Pi_c(q^2)$ of the charmed vector current

$$\Pi_c(q^2)(q_\mu q_\nu - \delta_{\mu\nu}q^2) = i \int d^4x \, e^{iqx} \langle 0|T J_\mu(x), J_\nu(0)|0\rangle, \tag{6.92}$$

$$J_\mu(x) = \bar{c}\gamma_\mu c, \tag{6.93}$$

and calculated the moments of $\Pi_c(q^2)$

$$M_n(Q^2) = \frac{4\pi^2}{n!}\left(-\frac{d}{dQ^2}\right)^n \Pi_c(Q^2), \tag{6.94}$$

$(Q^2 = -q^2)$ at $Q^2 = 0$. The OPE for $\Pi(Q^2)$ was used and only one term in the OPE series was considered – the gluonic condensate. In the perturbative part of $\Pi(Q^2)$, only the first-order term in α_s was taken into account and a small value of α_s was chosen, $\alpha_s(m_c) \approx 0.2$. The moments were saturated by contributions of charmonium states and in this way the value of the gluon condensate (6.19) was found. The SVZ approach [13] was criticized in [78], where it was shown that the higher-order terms of OPE, namely the contributions of G^3 and G^4 operators, are of importance at $Q^2 = 0$. Reinders, Rubinstein, and Yazaki [79] have shown, however, that the SVZ results may be restored if one considers not small values of $Q^2 > 0$ instead of $Q^2 = 0$. Later there were many attempts to determine the

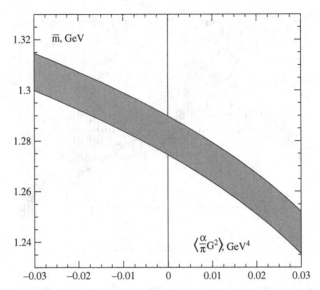

Fig. 6.11. The dependence of $\overline{m}(\overline{m})$ on $\langle 0|\alpha_s/\pi)G^2|0\rangle$ obtained at $n = 10$, $Q^2 = 0.98 \cdot 4m^2$ and $\alpha_s(Q^2 + \overline{m}^2)$.

gluon condensate by considering various processes within various approaches. In some of them, the value (6.19) (or values by a factor of 1.5 higher) was confirmed [77],[80],[81], in others it was claimed that the actual value of the gluon condensate is by a factor 2–5 higher than (6.19) [82].

From today's point of view, the calculations performed in [13] have a serious drawback. Only the first-order (NLO) perturbative correction was accounted for in [13] and a rather low value of α_s was taken, which was later not confirmed by the experimental data. The contribution of the next operator G^3 of dimension 6 was neglected, so the convergence of the operator product expansion was not tested.

There are publications [83] where the charmonium as well as bottomonium sum rules were analyzed at $Q^2 = 0$ with the account of α_s^2 perturbative corrections in order to extract the charm and bottom quark masses in various schemes. The condensate is usually taken to be zero or some another fixed value. However, the charm mass and the condensate values are entangled in the sum rules. This can be easily understood for large Q^2, where the mass and condensate corrections to the polarization operator behave as some series in negative powers of Q^2, and one may eliminate the condensate contribution to a great extent by slightly changing the quark mass. Vice versa, different condensate values may vary the charm quark mass within a few percent (see Fig. 6.11).

Therefore, in order to perform reliable calculations of the gluon condensate by studying the moments of the charmed current polarization operator, it is necessary to take account of the α_s^2 perturbative corrections to the moments, α_s-corrections to the gluon condensate contribution, the $\langle G^3 \rangle$ term in OPE, and to find the region in (n, Q^2) space where all these corrections are small. This program was realized in Ref. [84]. The basic points of this consideration are presented below.

The dispersion representation for $\Pi(q^2)$ has the form

$$R(s) = 4\pi \, \text{Im} \, \Pi_c(s + i0) \,, \quad \Pi_c(q^2) = \frac{q^2}{4\pi^2} \int_{4m_c^2}^{\infty} \frac{R(s) \, ds}{s(s - q^2)} \,, \tag{6.95}$$

where $R(\infty) = 1$ in the parton model. In the approximation of infinitely narrow widths of resonances, $R(s)$ can be written as a sum of contributions from resonances and continuum

$$R(s) = \frac{3\pi}{Q_c^2 \, \alpha_{em}^2(s)} \sum_\psi m_\psi \Gamma_{\psi \to ee} \, \delta(s - m_\psi^2) + \theta(s - s_0), \tag{6.96}$$

where $Q_c = 2/3$ is the charge of charmed quarks, s_0 is the continuum threshold (in what follows $\sqrt{s_0} = 4.6$ GeV), $\alpha_{em}(s)$ is the running electromagnetic constant, $\alpha_{em}(m_{J/\psi}^2) = 1/133.6$. The polarization operator moments are expressed through R as

$$M_n(Q^2) = \int_{4m_c^2}^{\infty} \frac{R(s) \, ds}{(s + Q^2)^{n+1}}. \tag{6.97}$$

According to (6.97), the experimental values of moments are determined by the equality

$$M_n(Q^2) = \frac{27\pi}{4\alpha_{em}^2} \sum_{\psi=1}^{6} \frac{m_\psi \Gamma_{\psi \to ee}}{(m_\psi^2 + Q^2)^{n+1}} + \frac{1}{n(s_0 + Q^2)^n}. \tag{6.98}$$

In the sum in (6.98) the following resonances were considered: $J/\psi(1S)$, $\psi(2S)$, $\psi(3770)$, $\psi(4040)$, $\psi(4160)$, $\psi(4415)$, their $\Gamma_{\psi \to ee}$ widths were taken from PDG data [63]. It is reasonable to consider the ratios of moments $M_{n1}(Q^2)/M_{n2}(Q^2)$ from which the uncertainty due to the error in $\Gamma_{J/\psi \to ee}$ largely cancels. The theoretical value of $\Pi(q^2)$ is represented as a sum of perturbative and nonperturbative contributions. It is convenient to express the perturbative contribution through $R(s)$, making use of (6.95), (6.97):

$$R(s) = \sum_{n \geq 0} R^{(n)}(s, \mu^2) \, a^n(\mu^2), \tag{6.99}$$

where $a(\mu^2) = \alpha_s(\mu^2)/\pi$. Nowadays, three terms of the expansion in (6.99) are known: $R^{(0)}$ [85] $R^{(1)}$ [86], $R^{(2)}$ [87]. They are represented as functions of the quark velocity $v = \sqrt{1 - 4m_c^2/s}$, where m_c is the quark pole mass. Since they are cumbersome we will not present them here (see [84] for details).

Nonperturbative contributions to the polarization operator have the following form (constrained by $d = 6$ operators):

$$\Pi_{nonpert}(Q^2) = \frac{1}{(4m_c^2)^2} \langle 0 \mid \frac{\alpha_s}{\pi} G^2 \mid 0 \rangle [\, f^{(0)}(z) + a f^{(1)}(z) \,]$$

$$- \frac{1}{(4m_c^2)^3} g^3 f^{abc} \langle 0 \mid G_{\mu\nu}^a G_{\nu\lambda}^b G_{\lambda\mu}^c \mid 0 \rangle F(z), \quad z = -\frac{Q^2}{4m_c^2}. \tag{6.100}$$

The functions $f^{(0)}(z)$, $f^{(1)}(z)$ and $F(z)$ were calculated in [13], [88], [89], respectively. The use of the quark pole mass is, however, not acceptable. The issue is that, in this case, the PT corrections to moments are very large in the region of interest and the perturbative series seems to diverge.

So, it is reasonable to use the $\overline{\text{MS}}$ mass $\overline{m}(\mu^2)$, taken at the point $\mu^2 = \overline{m}^2$. The calculations performed in Ref. [84] show that in the region near the diagonal in the (Q^2, n) plane, $Q^2/4m^2 = n/5-1$, all above mentioned corrections are small. For example,

$$n = 10 , \quad Q^2 = 4\overline{m}_c^2 : \quad \frac{\bar{M}^{(1)}}{\bar{M}^{(0)}} = 0.045 , \quad \frac{\bar{M}^{(2)}}{\bar{M}^{(0)}} = 1.136 , \quad \frac{\bar{M}^{(G,1)}}{\bar{M}^{(G,0)}} = -1.673, \quad (6.101)$$

(here $\bar{M}^{(k)}$ are the coefficients that multiply the contributions of terms $\sim a^k$ to the moments; $\bar{M}^{(G,k)}$ are the similar coefficients for gluonic condensate contributions).

At $a \sim 0.1$ and at the ratios of moments given by (6.101), there is good reason to believe that the PT series converges well. Such a good convergence holds (at $n > 5$) only in the case of large enough Q^2; at $Q^2 = 0$ one does not succeed in finding a value of n, such that perturbative corrections to the moments, α_s-corrections to gluonic condensates and the term $\sim \langle G^3 \rangle$ contribution would be simultaneously small.

It is also necessary to choose the scale, i.e. the normalization point μ^2 where $\alpha_s(\mu^2)$ is taken. In (6.99), $R(s)$ is a physical value and cannot depend on μ^2. Since, however, we take into account only three terms in (6.99), such μ^2 dependence can arise due to the neglected terms if an unsuitable normalization point is chosen. At large Q^2, the natural choice is $\mu^2 = Q^2$. It can be thought that at $Q^2 = 0$ the reasonable scale is $\mu^2 = \overline{m}^2$, though some numerical factor is not excluded in this equality. That is why it is reasonable to take the interpolation form

$$\mu^2 = Q^2 + \overline{m}^2, \qquad (6.102)$$

and to check the dependence of final results on a possible factor multiplying $\overline{m}^2$. Equating the theoretical value of some moment at fixed Q^2 (in the region where $M_n^{(1)}$ and $M_n^{(2)}$ are small) to its experimental value one can find the dependence of $\overline{m}$ on $\langle (\alpha_s/\pi)G^2 \rangle$ (neglecting the terms $\sim \langle G^3 \rangle$). Such a dependence for $n = 10$ and $Q^2/4m^2 = 0.98$ is presented in Fig. 6.11.

To fix both $\overline{m}$ and $\langle (\alpha_s/\pi)G^2 \rangle$, one should take not only moments but also their ratios. Fig. 6.12 shows the value of $\overline{m}$ obtained from the moment M_{10} and the ratio M_{10}/M_{12} at $Q^2 = 4m^2$ and from the moment M_{15} and the ratio M_{15}/M_{17} at $Q^2 = 8m^2$. The best values of masses of charmed quark and gluonic condensate are obtained from Fig. 6.12:

$$\overline{m}(\overline{m}^2) = 1.275 \pm 0.015 \,\text{GeV} , \quad \left\langle \frac{\alpha_s}{\pi} G^2 \right\rangle = 0.009 \pm 0.007 \,\text{GeV}^4. \qquad (6.103)$$

The calculation shows that the influence of the continuum – the last term in Eq. (6.96) – is completely negligible. Up to now, the corrections $\sim \langle G^3 \rangle$ were not taken into account. In the region of n and Q^2 used to find $\overline{m}$ and the gluonic condensate the corrections are comparatively small. They practically do not change $\overline{m}$. If the term $\sim \langle G^3 \rangle$ is estimated according to (6.21) at $\rho_c = 0.5$ fm, then the account of these corrections increases $\langle (\alpha_s/\pi)G^2 \rangle$ by $10 - 20\%$.

It should be noted that an improved accuracy of $\Gamma_{J/\psi \to ee}$ would make it possible to get a more precise value of the gluonic condensate: the widths of the horizontal bands

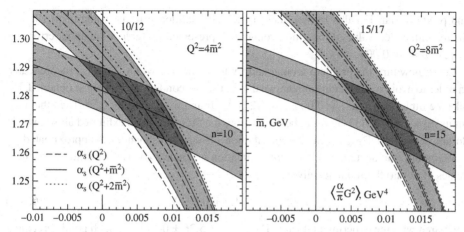

Fig. 6.12. The dependence of $\overline{m}(\overline{m})$ on $\langle 0|(\alpha_s/\pi)G^2|0\rangle$ obtained from the moments (horizontal bands) and their ratios (vertical bands) at different α_s. Left-hand figure: $Q^2 = 4\overline{m}^2$, $n = 10$, M_{10}/M_{12}; right-hand figure: $Q^2 = 8\overline{m}^2$, $n = 15$, M_{15}/M_{17}.

in Fig. 6.12 are determined mainly just by this error. In particular, this would possibly allow one to exclude the zero value of the gluonic condensate which would be extremely important. Unfortunately, Eq. (6.103) does not allow one to do this reliably. Diminution of theoretical errors which determine the width of vertical bands seems to be less real.

In order to check the result (6.103) for the gluon condensate, the pseudoscalar and axial-vector channels in charmonia were considered. The same method of moments was used and the regions in the space (n, Q^2) were found where higher-order perturbative and OPE terms are small. In the pseudoscalar case, it was shown [90] that if for $\overline{m}$ the value (6.103) is accepted and the contribution of $\langle 0|G^4|0\rangle$ condensate may be neglected, then there follows an upper limit for the gluon condensate

$$\langle 0|\frac{\alpha_s}{\pi} G^2|0\rangle \ < \ 0.008 \text{ GeV}^4. \tag{6.104}$$

The contribution of the $d = 6$ condensate $\langle 0|G^3|0\rangle$ is shown to be small. If the $\langle G^4\rangle$ condensate is accounted for and its value is estimated by the factorization hypothesis, then the upper limit of the gluon condensate increases to

$$\langle 0|\frac{\alpha_s}{\pi} G^2|0\rangle \ < \ 0.015 \text{ GeV}^4. \tag{6.105}$$

The case of the axial-vector channel in charmonia was investigated in [91] and very strong limits on the gluon condensate were found:

$$\langle 0|\frac{\alpha_s}{\pi} G^2|0\rangle = 0.005^{+0.001}_{-0.004} \text{ GeV}^4. \tag{6.106}$$

Unfortunately, (6.106) does not allow one to exclude the zero value for the gluon condensate. It should be mentioned that the allowed region in (n, Q^2) space, where all corrections are small, is very narrow in this case. This implies that one cannot check the result (6.106) by studying some other regions in (n, Q^2) as was done in the two previous cases – vector and pseudoscalar. For this reason, there is no full confidence in the value of (6.106). In the recommended value of the gluon condensate presented in Table 1 the upper error was increased up to 0.004.

Let us now turn the problem around and try to predict the width $\Gamma_{J/\psi \to ee}$ theoretically. In order to avoid a circular argument, we do not use the condensate value just obtained, but take the limits $\langle \alpha_s/\pi \, G^2 \rangle = 0.006 \pm 0.012 \, \text{GeV}^4$ found from τ-decay data. Then the mass limits $\bar{m} = 1.28 - 1.33 \, \text{GeV}$ can be found from the moment ratios exhibited above, which do not depend on $\Gamma_{J/\psi \to ee}$ if the contribution of higher resonances is approximated by a continuum (the accuracy of such approximation is about 3%). The substitution of these values of $\bar{m}$ into the moments gives

$$\Gamma_{J/\psi \to ee}^{\text{theor}} = 4.9 \pm 0.8 \, \text{keV}, \tag{6.107}$$

compared with the experimental value $\Gamma_{J/\psi \to ee} = 5.26 \pm 0.37 \, \text{keV}$. Such good agreement of the theoretical prediction with experimental data is a very impressive demonstration of the QCD sum rule effectiveness. It must be stressed that while obtaining (6.107) no additional input was used besides the condensate constraint taken from Eq. (6.91) and the value of $\alpha_s(m_\tau^2)$.

Sometimes when considering the heavy quarkonia sum rules the Coulomb-like corrections are summed up [92]–[96]. The basic argument for such summation is that at $Q^2 = 0$ and high n only small quark velocities $v \lesssim 1/\sqrt{n}$ are essential and the problem becomes nonrelativistic. So it is possible to perform the summation with the help of well-known formulae of nonrelativistic quantum mechanics for $|\psi(0)|^2$ in the case of the Coulomb interaction (see [97]).

This method was not used here for the following reasons:

1. The basic idea of our approach is to calculate the moments of the polarization operator in QCD by applying perturbation theory and OPE (left-hand sides of the sum rules) and to compare it with the right-hand sides of the sum rules, represented by the contribution of charmonium states (mainly by J/ψ in the vector channel). Therefore, it is assumed that the theoretical side of the sum rule is dual to the experimental one, i.e. the same domains of coordinate and momentum spaces are of importance on both sides. But the charmonium states (particularly J/ψ) are by no means Coulomb systems. A particular argument in favor of this statement is the ratio $\Gamma_{J/\psi \to ee}/\Gamma_{\psi' \to ee} = 2.4$. If charmonia were nonrelativistic Coulomb systems, $\Gamma_{\psi \to ee}$ would be proportional to $|\psi(0)|^2 \sim 1/(n_r + 1)^3$, and since ψ' is the first radial excitation with $n_r = 1$ this ratio would be equal to 8 (see also [97]).

2. The heavy quark-antiquark Coulomb interaction at large distances $r > r_{\text{conf}} \sim 1 \, \text{GeV}^{-1}$ is screened by gluon and light quark-antiquark clouds, resulting in string formation. Therefore the summation of the Coulombic series makes sense only when the Coulomb radius r_{Coul} is below r_{conf}. (It must be borne in mind that higher-order terms in the Coulombic series represent the contributions of large distances, $r \gg r_{\text{Coul}}$.) For charmonia we have

$$r_{\text{Coul}} \approx \frac{2}{m_c C_F \alpha_s} \approx 4 \, \text{GeV}^{-1}. \tag{6.108}$$

It is clear that the necessary condition $R_{Coul} < R_{conf}$ is badly violated for charmonia. This means that the summation of the Coulomb series in case of charmonium would be a wrong step.

3. The analysis is performed at $Q^2/4\bar{m}^2 \geq 1$. At large Q^2, the Coulomb corrections are suppressed in comparison with $Q^2 = 0$. It is easy to estimate the characteristic values of the quark velocities. At large n they are $v \approx \sqrt{(1 + Q^2/4m^2)/n}$. In the region (n, Q^2) exploited above, the quark velocity $v \sim 1/\sqrt{5} \approx 0.45$ is not small and is not in the nonrelativistic domain where the Coulomb corrections are large and legitimate.

Nevertheless, let us look at the expression of R_c, obtained after summation of the Coulomb corrections in the nonrelativistic theory [98]. It reads (to go from QED to QCD one has to replace $\alpha \to C_F \alpha_s$, $C_F = 4/3$):

$$R_{c,\,Coul} = \frac{3}{2} \frac{\pi C_F \alpha_s}{1 - e^{-x}} = \frac{3}{2} v \left(1 + \frac{x}{2} + \frac{x^2}{12} - \frac{x^4}{720} + \dots \right), \tag{6.109}$$

where $x = \pi C_F \alpha_s/v$. At $v = 0.45$ and $\alpha_s \approx 0.26$, the first three terms in the expansion (6.109) accounted for in our calculations reproduce the exact value of $R_{c,\,Coul}$ with an accuracy of 1.6%. Such deviation leads to an error of the mass $\bar{m}$ of order $(1 - 2) \times 10^{-3}$ GeV, which is completely negligible. In order to avoid misunderstanding, it must be mentioned that the value of $R_{c,Coul}$, computed by summing the Coulomb corrections in nonrelativistic theory, has not too much in common with the real physical situation. Numerically, at the chosen values of the parameters, $R_{c,Coul} \approx 1.8$, while the real value (both experimental and in perturbative QCD) is about 1.1. The goal of the arguments presented above was to demonstrate that even in the case of Coulombic systems our approach would have a good accuracy of calculation.

At $v = 0.45$, the momentum transfer from quark to antiquark is $\Delta p \sim 1$ GeV. (This is a typical domain for QCD sum rule validity.) In coordinate space it corresponds to $\Delta r_{q\bar{q}} \sim 1\,\text{GeV}^{-1}$. Comparison with potential models [98] shows that in this region the effective potential strongly differs from the Coulombic one.

4. Large compensation of various terms in the expression for the moments in the $\overline{\text{MS}}$ scheme is not achieved if only the Coulomb terms are taken into account. This means that the terms of non-Coulombic origin are more important here than the Coulombic ones.

For all these reasons, the summation of nonrelativistic Coulomb corrections is inadequate to the problem considered: it will not improve the accuracy of calculations but would be misleading.

6.6 Calculations of light meson masses and coupling constants

6.6.1 Vector mesons

Vector meson $(\rho, \omega, \varphi, K^*)$ masses and their coupling constants with quark vector currents were first calculated by QCD sum rules by Shifman, Vainshtein, and Zakharov in their pioneering paper [13]. Their calculations are repeated below with the account of higher-order terms of the perturbation series and operators of higher dimension in OPE.

Consider the case of the ρ meson. The quark current corresponding to ρ^- is equal to

$$j_\mu^\rho = \bar{u}\gamma_\mu d. \tag{6.110}$$

The expression for the vector current (6.110) polarization operator

$$\Pi_V(Q^2) = \Pi_V^{pert}(Q^2) + \Pi_V^{nonpert}(Q^2), \tag{6.111}$$

follows directly from (6.54), (6.58), (6.72)–(6.76). For Π_V^{pert}, we have up to α_s^3 terms:

$$\Pi_V^{pert}(Q^2) = \frac{1}{4\pi^2} \int\limits_{a(m_\tau^2)}^{a(Q^2)} D(a)\frac{da}{\beta(a)}, \tag{6.112}$$

$$D(a) = 1 + a + 1.64a^2 + 6.37a^3 + 50a^3, \tag{6.113}$$

$$\beta(a) = \beta_0 a^2(1 + 1.78a + 4.47a^2 + 20.9a^3), \tag{6.114}$$

where $a = \alpha_s/\pi$, $\beta_0 = 9/4$. In the OPE series, condensates up to dimension 10 are accounted for. For the condensates of dimensions 6, 8, and 10, the factorization hypothesis is assumed to hold. Several operators of dimensions 8 and 10 are omitted for the reasons mentioned in the text after Eq. (6.58). The nonperturbative part of the polarization operator has the form of

$$\Pi_V^{nonpert}(Q^2) = \frac{1}{12Q^4}\left\langle 0\left|\frac{\alpha_s}{\pi}G^2\right|0\right\rangle\left(1 + \frac{7}{6}a\right) + (m_u + m_d)\langle 0|\bar{q}q|0\rangle$$

$$\times \frac{1}{Q^4}\left[1 + \frac{1}{3}a + \left(11.6 - 4.4.\ln\frac{Q^2}{\mu^2}\right)a^2\right]$$

$$\times -\frac{32}{81}\frac{\pi}{Q^6}\alpha_s\langle 0\mid\bar{q}q\mid 0\rangle^2\left[7 + \left(\frac{685}{48} - \frac{17}{9}\ln\frac{Q^2}{\mu^2}\right)a\right] - \frac{4\pi}{Q^8}\alpha_s m_0^2\langle 0\mid\bar{q}q\mid 0\rangle^2$$

$$\times -\frac{4}{9}\pi\alpha_s\langle 0\mid\bar{q}q\mid 0\rangle^2\frac{1}{Q^{10}}\left[\frac{50}{9}m_0^4 + 32\pi^2\left\langle 0\left|\frac{\alpha_s}{\pi}G^2\right|0\right\rangle\right], \tag{6.115}$$

where μ is the normalization point. (Terms of order m_u^2, m_d^2 are neglected.) In what follows, $\mu = 1\,\text{GeV}$ is chosen. The α_s-corrections are taken into account for the contributions of the operators up to dimension 6. The α_s-corrections to higher-order operators are unknown, their possible contributions, as well as deviations from the factorization hypothesis, will be included in the errors. The phenomenological side of the sum rule is represented by the contributions of ρ meson and continuum, the imaginary part of which is equal to $\text{Im}\,\Pi_V^{pert}(s), s = q^2$ and starts at some threshold value $s = s_0$. The matrix element $\langle\rho^- \mid j_\mu^\rho \mid 0\rangle$ is known from the vector dominance model (VDM). (For a review, see e.g. [23].) Thus,

$$\langle\rho^- \mid j_\mu^\rho \mid 0\rangle = \frac{\sqrt{2}m_\rho}{g_\rho}e_\mu, \tag{6.116}$$

where m_ρ is the ρ meson mass, e_μ is the ρ meson polarization vector, and g_ρ is the $\rho - \gamma$ coupling constant. The factor $\sqrt{2}$ appears in (6.116) because in VDM g_ρ is defined as the matrix element between ρ^0 and the third component of the isovector quark current

$$j_\mu^{(3)} = \frac{1}{2}(\bar{u}\gamma_\mu u - \bar{d}\gamma_\mu d). \tag{6.117}$$

The current (6.110) is the isotopically minus component of the same isovector current times $\sqrt{2}$. So, Im $\Pi_V^{phys}(s)$ is given by

$$\text{Im } \Pi_V^{phys}(s) = \pi \frac{2m_\rho^2}{g_\rho^2}\delta(s - m_\rho^2) + \text{Im } \Pi_V^{pert}(s)\theta(s - s_0). \tag{6.118}$$

(The ρ meson width is disregarded.) It is convenient to represent $\Pi_V^{pert}(Q^2)$ by the following dispersion relation:

$$\Pi_V^{pert}(Q^2) = \frac{1}{\pi} \int \frac{\text{Im } \Pi_V^{pert}(s)}{s + Q^2} ds \tag{6.119}$$

Substituting (6.115), (6.118), (6.119) into (6.38) and performing a Borel transformation, we get the sum rule

$$\frac{1}{\pi} \int_0^{s_0} ds \text{ Im } \Pi_V^{pert}(s)e^{-s/M^2} + \frac{1}{12M^2}\left\langle 0 \left| \frac{\alpha_s}{\pi}G^2 \right| 0 \right\rangle\left(1 + \frac{7}{6}a\right)$$

$$+ \frac{1}{M^2}(m_u + m_d)\langle 0 | \bar{q}q | 0\rangle\left[1 + \frac{1}{3}a + \left(11.6 - 4.4.\ln\frac{M^2}{\mu^2}\right)a^2\right]$$

$$- \frac{16}{81} \cdot \frac{7\pi}{M^4}\alpha_s\langle 0 | \bar{q}q | 0\rangle^2\left[1 + \frac{1}{7}\left(\frac{685}{48} - \frac{17}{9}\ln\frac{M^2}{\mu^2}\right)a\right] - \frac{2\pi}{3}\frac{m_0^2}{M^6}\alpha_s\langle 0 | \bar{q}q | 0\rangle^2$$

$$- \frac{\pi}{54}\alpha_s\langle 0 | \bar{q}q | 0\rangle^2\frac{1}{M^8}\left[\frac{50}{9}m_0^4 + 32\pi^2\left\langle 0 \left| \frac{\alpha_s}{\pi}G^2 \right| 0 \right\rangle\right] = \frac{2m_\rho^2}{g_\rho^2}e^{-m_\rho^2/M^2}. \tag{6.120}$$

In (6.120), $a = \alpha_s(Q^2)/\pi$ is taken at $Q^2 = M^2$. The corrections to this approximation are beyond the accuracy of (6.120), since they are of the next order of α_s. In the term $\ln Q^2/\mu^2$, which appears in Eq. (6.115), the substitution $Q^2 \to M^2$ was done as well at the Borel transformation. The deviations from this procedure are very small, with less than 0.1% correction to the sum rule. It is important to calculate the perturbative correction to the sum rule – the first term on the left-hand side of (6.120) – with sufficient precision because this is the main correction to the parton model result. Using (6.112)–(6.114) and renormalization group equation (6.52), we get:

$$\Pi_V^{pert}(Q^2) = -\frac{1}{4\pi^2}\ln\frac{Q^2}{m_\tau^2} + \frac{1}{4\pi^2}\int_{a(m_\tau^2)}^{a(Q^2)}\frac{da}{\beta_0 a}\frac{1 + 1.64a + 6.37a^2 + 50a^3}{1 + 1.78a + 4.47a^2 + 20.9a^3}. \tag{6.121}$$

As can be seen from (6.121), the coefficients of equal powers of a in the numerator and the denominator of the second term in (6.121) are of the same magnitude. Therefore, we can write the numerator of (6.121) as the sum of the polynomial in the denomerator plus the rest terms and expand in powers of the latter. The estimate of these terms shows that their

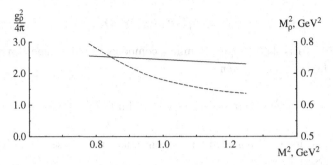

Fig. 6.13. The dependence of $g_\rho^2/4\pi$ (solid line) and m_ρ^2 (dashed line, right-hand scale) on the Borel parameter M^2.

contribution to Im Π_V^{pert} is about 10% of the total. So, with such accuracy Im $\Pi_V^{pert}(s)$ reduces to

$$4\pi \text{ Im } \Pi_V^{pert}(s) = 1 + \frac{1}{\pi\beta_0} \text{ Im ln } \frac{a(s)}{a(m_\tau^2)}. \qquad (6.122)$$

As follows from Fig. 6.2, at $s > 0.8$ GeV^2 $\alpha_s(s)$ can be calculated at the same accuracy in one-loop approximation with the result

$$4\pi \text{ Im } \Pi_V^{pert} = 1 + \frac{1}{\pi\beta_0} \left[\frac{\pi}{2} - \tan^{-1}\frac{1 + \beta_0 a(m_\tau^2) \ln\frac{s}{m_\tau^2}}{\pi\beta_0 a(m_\tau^2)} \right]. \qquad (6.123)$$

At $s < 0.8$ GeV^2, the values of Re $\alpha_s(s)$ and Im $\alpha_s(s)$ are presented in Fig. 6.2. Im $\Pi_V^{pert}(s)$ is then found using (6.122). These two ways of determining Im $\Pi_V^{pert}(s)$ match well at $s = 0.8$ GeV^2.

The values of $g_\rho^2/4\pi$ as a function of the Borel parameter M^2 found from the sum rule (6.120) are plotted in Fig. 6.13. The numerical values of condensates, presented in Section 6.2 (Table 1), were used, $\alpha_s(Q^2)$ was taken from Fig. 6.3, $s_0 = 1.5$ GeV^2 was chosen, the experimental value of $m_\rho^2 = 0.602$ GeV^2 was substituted. The interval in M^2 where the sum is reliable can be estimated as $0.8 < M^2 < 1.2$ GeV^2. At $M^2 < 0.8$ GeV^2, there is no confidence in the convergence of the perturbative series; at $M^2 = 1.2$ GeV^2, the continuum comprises about 35% of the total contribution. Fig. 6.13 demonstrates that $g_\rho^2/4\pi$ is very stable in this interval of M^2 and the mean value of $g^2/4\pi$ is

$$\frac{g_\rho^2}{4\pi} = 2.48, \qquad (6.124)$$

in comparison with the experimental value (see [23]):

$$\left(\frac{g_\rho^2}{4\pi}\right)_{exp} = 2.54 \pm 0.23. \qquad (6.125)$$

Let us estimate the errors in the theoretical value (6.124). The main contribution to the sum rule (6.120), besides the parton model term, arises from perturbative α_s-corrections. They comprise about 12% of the total (at $M^2 = 1$ GeV^2). Among nonperturbative terms, the main role belongs to the term, proportional to $\alpha_s \langle 0 \mid \bar{q}q \mid 0 \rangle^2$, the contribution of which is negative and about 4%. The convergence of OPE is good: the last terms of OPE contribute about 1% of the total. The main uncertainty arises from the model of the hadronic spectrum. The continuum contribution is about 25% of the total (at $M^2 = 1$ GeV^2) and its possible uncertainty may be estimated as 20–30%. Therefore, the accuracy of (6.124) is about 5–7%. It should be stressed that such accuracy of theoretical prediction is very high considering that we are dealing with the theory of strong interactions and a large coupling constant! Note also that the perturbative corrections play a modest role in the final result and the role of nonperturbative terms is even less.

In order to find the theoretical value of m_ρ^2, let us apply the differential operator $(-1/M^2)^{-1}\partial(-1/M^2)$ to (6.120). The obtained result as a function of M^2 is plotted in Fig. 6.13. The confidence interval in M^2 significantly shrinks in the differentiated sum rule in comparison with the sum rule (6.120). The reason is that even at $M^2 = 1$ GeV^2 the continuum contribution, subtracted from the parton model term, is practically equal to the total left-hand side of the differentiated sum rule. Therefore, the optimistic estimate of the confidence interval is: $0.8 < M^2 < 1$ GeV^2. The remarkable dependence of m_ρ^2 on M^2 in Fig. 6.13 and its significant variation with the variation of s_0 indicate that the error in the determination of m_ρ^2 is much larger than in case of g_ρ^2 and, probably, comprises about 20%.

From Fig. 6.13 and taking into account this error, we have

$$(m_\rho^2)_{th} = 0.72 \pm 0.15 \text{ GeV}^2 \tag{6.126}$$

in comparison with the experimental value

$$(m_\rho^2)_{exp} = 0.602 \pm 0.001 \text{ GeV}^2. \tag{6.127}$$

This example clearly demonstrates that differentiation of the QCD sum rule is a dangerous operation: the confidence interval shrinks significantly and the accuracy of the sum rule is reduced.

The coupling constants and masses of the ω, φ and K^* mesons can be calculated in a similar way. The sum rule for the ω meson in the accepted approximation coincides exactly with (6.120) – ρ and ω mesons are degenerate. For φ and K^*, the strange quark mass m_s must be accounted for, particularly the m_s^2 term for φ (see [13]).

6.6.2 Axial mesons

Let us now consider the QCD sum rules for the axial meson a_1 with isospin $T = 1$ ($J^{PC} = 1^{++}$). The corresponding quark current is equal to

$$j_\mu^A(x) = \bar{u}\gamma_\mu\gamma_5 d \tag{6.128}$$

and the polarization operator is

$$\Pi_{\mu\nu}^A(q) = i \int d^4x e^{iqx} \langle 0 \mid T \left\{ j_\mu^A(x), j_\nu^A(0)^+ \right\} \mid 0 \rangle. \tag{6.129}$$

Separate in the expression for the polarization operator the transverse structure

$$(q_\mu q_\nu - q^2 \delta_{\mu\nu}) \Pi^A(q^2) \tag{6.130}$$

to which axial mesons are contributing. Neglect the a_1 width although experimentally it is large – about 300–400 MeV. The sum rule for the a_1 meson after Borel transformation has the form of

$$\frac{1}{\pi} \int^{s_0} \mathrm{Im}\, \Pi^A_{pert}(s) e^{-s/M^2} + \frac{1}{12M^2} \left\langle 0 \left| \frac{\alpha_s}{\pi} G^2 \right| 0 \right\rangle \left(1 + \frac{7}{6} a\right) - \frac{1}{M^2}(m_u + m_d)$$

$$\times \langle 0 | \bar{q}q | 0 \rangle \left[1 + \frac{7}{3}a + \left(7 - \ln \frac{\mu^2}{M^2}\right) a^2\right] + \frac{\pi}{M^4} \frac{176}{81} \alpha_s \langle 0 | \bar{q}q | 0 \rangle^2$$

$$\times \left[1 + a\left(\frac{917}{528} + \frac{149}{326} \ln \frac{\mu^2}{M^2}\right)\right] + \frac{2\pi}{3} \frac{m_0^2}{M^6} \alpha_s \langle 0 | \bar{q}q | 0 \rangle^2 + \frac{\pi}{54} \frac{1}{M^6}$$

$$\times \langle 0 | \bar{q}q | 0 \rangle^2 \left[\frac{50}{9} m_0^4 + 32\pi^2 \left\langle 0 \left| \frac{\alpha_s}{\pi} G^2 \right| 0 \right\rangle\right] = 2 \frac{m_{a_1}^2}{g_{a_1}^2} e^{-\frac{m_{a_1}^2}{M^2}} + f_\pi^2. \tag{6.131}$$

The OPE leading terms of $d = 0, 4, 6$ were calculated in [13], the term of $d = 8$ was found in [30], the term of $d = 10$ was calculated in [68]. The magnitude of the last two terms are equal to the corresponding ones in the case of vector currents, but the signs are opposite. The α_s-corrections to OPE terms were calculated in [48] and [49]. The factorization procedure was employed in the calculation of $d \geq 6$ terms. The anomalous dimensions of $d = 8$ and $d = 10$ terms (as well as the unknown α_s-corrections) were disregarded because the error, arising from the factorization procedure, probably overwhelms their magnitude. At Borel transformation, the α_s dependence on Q^2 was disregarded and Q^2 was replaced by M^2; the same was done with $\ln \mu^2/Q^2$. The f_π^2 term on the right-hand side of (6.131) arises from the kinematical pole in $\Pi^A_{\mu\nu}$ (see Eq. 2.119). Note that the contributions of all OPE corrections are positive in (6.131), while in (6.120) they were negative (except for the gluon condensate contribution). This circumstance results in the necessity to choose larger values of the Borel parameter M^2 in (6.131) in comparison with (6.120).

The values of $g_{a_1}^2/4\pi$ and $m_{a_1}^2$ calculated from the sum rule (6.131) are plotted in Fig. 6.14 ($m_{a_1}^2$ was found by differentiation of the sum rule with respect to $(1/M^2)$). s_0 was chosen equal to 2 GeV^2. The confidence interval in M^2 can be estimated as $1.0 < M^2 < 1.4$ GeV^2: at $M^2 = 1.4$ GeV^2 the continuum contribution in the differentiated sum rule is about 60%, at $M^2 = 1.0$ GeV^2 the perturbative and nonperturbative corrections are each about 10%. The mean theoretical value of $m_{a_1}^2$ is equal to

$$(m_{a_1}^2)_{theor} = 1.32 \pm 0.18 \text{ GeV}^2 \tag{6.132}$$

in comparison with experimental data

$$(m_{a_1}^2)_{exp} = 1.51 \pm 0.09 \text{ GeV}^2. \tag{6.133}$$

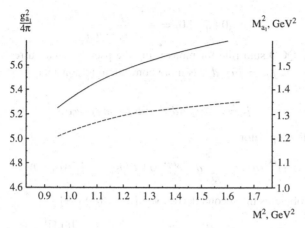

Fig. 6.14. The values $g^2_{a_1}/4\pi$ (solid line, left scale) and $m^2_{a_1}$ (dashed line, right scale) found from the sum rule (6.131).

The main theoretical error arises from the uncertainty in the continuum threshold s_0. The theoretical value of the a_1 meson coupling constant with the axial current is given by

$$\left(\frac{g^2_{a_1}}{4\pi}\right)_{theor} = 5.60 \pm 0.15. \tag{6.134}$$

This value agrees with the data determined from experiment. However, because of the large a_1 width the latter is model dependent.

6.6.3 Pseudoscalar mesons

Consider first the longitudinal part of the polarization operator $q_\mu q_\nu \Pi^{(0)}_A$ defined by (6.44). The OPE for it in the limit $m_u, m_d \to 0$ reads:

$$q_\mu q_\nu \Pi^{(0)}_A(Q^2) = 2q_\mu q_\nu \frac{1}{Q^4}(m_u + m_d)\langle 0 \mid \bar{q}q \mid 0 \rangle. \tag{6.135}$$

No perturbative terms, nor any higher-order terms of OPE, contribute to $q_\mu q_\nu \Pi^{(0)}_A(Q^2)$ in this limit. The proof of this statement can be obtained by considering $q_\mu q_\nu \Pi_{\mu\nu,A}$, where $\Pi_{\mu\nu,A}$ is given by (6.44) and using current algebra, as was done in Section 2.3 of Chapter 2. The corrections to Eq. (6.135) are proportional to $(m_u + m_d)^2$. The physical part of $q_\mu q_\nu \Pi^{(0)}_A(Q^2)$ looks like

$$q_\mu q_\nu \left(-f^2_\pi \frac{1}{q^2 - m^2_\pi} + \text{contributions of excited states}\right). \tag{6.136}$$

Comparison of (6.135) with (6.136) shows that the coupling constants of all excited states of nonzero mass are proportional to $m_u + m_d$ in the limit of massless u and d quarks. The pion mass is also vanishing in this limit. Expanding the denominator in (6.136) in m^2_π/q^2 we get by comparing (6.135) and (6.136) the relation derived in Chapter 2:

$$\langle 0 \mid \bar{q}q \mid 0 \rangle = -\frac{1}{2} \frac{f_\pi^2 m_\pi^2}{m_u + m_d}. \tag{6.137}$$

Write now the QCD sum rule for pions using the pseudoscalar current. Instead of the pseudoscalar current $j_{ps} = i\bar{u}\gamma_5 d$ it is more convenient to consider the divergence of the axial current

$$j_5 = \partial_\mu \bar{u}\gamma_\mu \gamma_5 d = i(m_u + m_d)\bar{u}\gamma_5 d, \tag{6.138}$$

and the polarization operator

$$\Pi^{(5)}(q^2) = i \int d^4 x e^{iqx} \langle 0 \mid T\{j_5(x), \; j_5^+(0)\} \mid 0 \rangle. \tag{6.139}$$

A more suitable object is the second derivative of $\Pi^{(5)}(q^2)$ [99]

$$\Pi^{(5)}(q^2)'' = \frac{d^2}{d(q^2)^2} \Pi^{(5)}(q^2) = \frac{2}{\pi} \int ds \frac{\mathrm{Im}\, \Pi^{(5)}(s)}{(s - q^2)^3}. \tag{6.140}$$

The QCD sum rule after Borel transformation has the form [99] ($l_M = \ln(M^2/\mu^2)$):

$$\frac{3}{8\pi^2}\Bigg\{1 + \left(\frac{11}{3} + \gamma_E - 2l_M\right)a + \left(21.5 - \frac{139}{6}l_M - \frac{17}{2}\gamma_E l_M + \frac{17}{4}l_M^2\right)a^2$$

$$+ \left(52.7 - 223.5l_M + 10.28l_M^2 - 9.21l_M^3\right)a^3 + \frac{\pi^2}{3}\left\langle 0 \left| \frac{\alpha_s}{\pi} G^2 \right| 0 \right\rangle \frac{1}{M^4} + \frac{896}{81}\pi^3 \alpha_s$$

$$\times \langle 0 \mid \bar{q}q \mid 0 \rangle^2 \frac{1}{M^6}\Bigg\} = 4\frac{\langle 0 \mid \bar{q}q \mid 0 \rangle^2}{f_\pi^2 M^4} + \text{excited states contribution.} \tag{6.141}$$

The terms $\sim m_q \langle 0 \mid qq \mid 0 \rangle$ are omitted. On the right-hand side of (6.141), the GMOR relation (6.137) was exploited. (In [99], the α_s^4-corrections as well as α_s-corrections to $d = 4$ and $d = 6$ power terms were also calculated.) It is easy to see that the sum rule (6.141) cannot be fulfilled. (The statement that the QCD sum rule does not work in the PS channel was made in [100],[14],[24].) Indeed, the pion contribution – the first on the right-hand side of (6.141) – comprises 60% of the perturbative terms of the left-hand side at $M^2 = 1$ GeV^2. Even if we assume that all the rest on the right-hand side is represented by excited state contributions, then the equality achieved at $M^2 = 1$ GeV^2 will be destroyed with increasing M^2, because the M^2-dependence of left and right sides of (6.141) are quite different. There are, in principle, two ways to get rid of the problem:

1. To say that the sum rule (6.141) is valid at high M^2 only, i.e. that PT + OPE is dual to excited state contributions. This statement is trivial and gives us no new information.

2. To suppose that at $M^2 \sim 1 - 2$ GeV^2 the vacuum fluctuations not given by OPE are important in PS sum rules.

One can present an argument in favour of the second possibility. It is well known that isospin zero physical states in pseudoscalar nonets are the eighth component of the $SU(3)$ octet η and $SU(3)$ singlet η' with quark contents $\mid \eta \rangle = \mid \bar{u}u + \bar{d}d - 2\bar{s}s\rangle/\sqrt{6}$ and $\mid \eta' \rangle = \mid \bar{u}u + \bar{d}d + \bar{s}s\rangle/\sqrt{3}$. In vector nonets the situation is different: $\mid \omega \rangle = \mid \bar{u}u + \bar{d}d\rangle/\sqrt{2}$ and

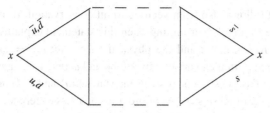

Fig. 6.15. The diagram, representing $\bar{u}u + \bar{d}d \Leftrightarrow \bar{s}s$ mixing in the polarization operator. The solid lines are quarks, the dashed lines are gluons.

$| \varphi \rangle = | \bar{s}s \rangle$. This means that in the attempt to describe pseudoscalar mesons in QCD by QCD sum rules, the mixing transitions $\bar{u}u + \bar{d}d \to \bar{s}s$ are as important as the diagonal ones $\bar{u}u \to \bar{u}u$. In perturbation theory (and in OPE), the term in the polarization operator, corresponding to the quark pair mixing transition, is given in the first nonvanishing order of α_s by the diagram of Fig. 6.15 (for PS-current).

Evidently, this diagram is proportional to α_s^2, contains a small numerical factor due to three-loop integration and is much smaller than the bare-loop diagram corresponding to the diagonal $\bar{u}u \to \bar{u}u$ transition. Therefore, the large mixing in the PS nonet can be attributed to nonperturbative vacuum fluctuations not given by OPE. In QCD, one example is known of such a nonperturbative solution for gluon field – the instanton. So, let us consider the polarization operator corresponding to the mixing $\bar{u}u \to \bar{s}s$ of axial currents:

$$\Pi_{\mu\nu}^{su}(q) = i \int d^4x e^{iqx} \left\langle 0 \left| T \left\{ j_\mu^{A,s}(x), \; j_\nu^{A,u}(0) \right\} \right| 0 \right\rangle. \tag{6.142}$$

Suppose that quarks are moving in the field on instanton and neglect their interaction with gluons. In the dilute instanton gas approximation, we get the factorized expression for $\Pi_{\mu\nu}^{su}(Q)$ by integrating over the instanton position [100]:

$$\Pi_{\mu\nu}^{su}(Q) = -B_\mu^{A,s}(Q) B_\nu^{A,u}(-Q), \tag{6.143}$$

$$B_\mu^{A,i}(Q) = \int e^{iQx} \, \mathrm{Tr} \left\{ \gamma_\mu \gamma_5 G^i(x, x) \right\} d^4x, \tag{6.144}$$

where $G^i(x, y)$, $i = s, u$ is the quark Green function in the instanton field. (The center of the instanton is at the origin.)

It is instructive to show that if instead of the axial current, a vector current is considered, then there is no mixing induced by instantons: $B_\mu^{V,i} \equiv 0$. Indeed, the expression $\mathrm{Tr} \{\gamma_\mu G^i(x, x)\}$ is invariant under the transformations, which does not change the Lagrangian of the quark in the instanton field. The instanton is a vector in the $SU(2)$ subgroup of the colour group $SU(3)$. Let us consider a G-parity transformation in the colour space, i.e. the product of charge conjugation and a 180° rotation around the second axis in the $SU(2)$ subgroup of the colour group. The Lagrangian of the quark in the instanton field is invariant under such transformation. Since $\mathrm{Tr} \{\gamma_\mu G^i(x, x)\}$ is odd under charge conjugation, it changes the sign under such transformation and therefore is identically zero:

$$\mathrm{Tr} \left\{ \gamma_\mu G^i(x, x) \right\} = 0. \tag{6.145}$$

The same statement follows also from vector current conservation. If the instanton mechanism is responsible for quark pair mixing, then this statement explains why such mixing is absent in vector meson nonet and the physical states are $|\omega\rangle = |\bar{u}u + \bar{d}d\rangle/\sqrt{2}$ and $|\varphi = |\bar{s}s\rangle$ [100]. The question can arise: why are ρ_0 and ω physical states but not $|\bar{u}u\rangle$ and $|\bar{d}d\rangle$? The answer to this question is evident: the states $|\bar{u}u\rangle$ and $|\bar{d}d\rangle$ are almost degenerate: the energy splitting of these states is of order of the mass difference of u and d quarks, i.e. few MeV. So one can expect that even very small mixing (e.g. by exchange of perturbative gluons) is considerably larger (in the energy scale) than $m_u - m_d$ and will result in the generation of ρ_0 and ω as physical states.

Turn back to Eqs. (6.143), (6.144). The quark Green function in the field of instanton was calculated in Refs. [59],[101],[102] and is presented in Eqs. (4.228), (4.232). For massless quarks, its substitution in (6.144) results in:

$$B_\mu^A(Q) = -i\rho^2 Q_\mu K_0(Q\rho) \tag{6.146}$$

and

$$\Pi_{\mu\nu}^{su}(Q) = -Q_\mu Q_\nu \Pi_A^{(0)su}(Q) = -Q_\mu Q_\nu K_0^2(Q\rho)\rho^4. \tag{6.147}$$

In (6.146), (6.147), $K_0(x)$ is the McDonald function and ρ is the instanton radius. $\Pi_{\mu\nu}^{su}(Q)$ (6.147) is longitudinal and therefore contributes to the formation of pseudoscalar η and η' mesons. Let us estimate its magnitude in the instanton gas model with fixed instanton radius $\rho = \rho_c$, where the instanton density is given by (6.69). The diagonal longitudinal polarization operator in the instanton field $\Pi_A^{(0)uu}(Q) = \Pi_A^{(0)ss}(Q)$ is given by the second equation (6.64):

$$\Pi_a^{(0)uu}(Q) = \frac{4\rho_c^2}{Q^2} K_1^2(Q\rho)n_0. \tag{6.148}$$

The first term in this equation, proportional to $1/Q^4$, is omitted because it corresponds to the OPE term (6.135) and is small in the real world. At the instanton gas parameter $\rho_c = 1.5\,\text{GeV}^{-1}$ and $Q = 1\,\text{GeV}$ the ratio $\Pi_A^{(0)qq}/\Pi_A^{(0)(qs)} \approx 2$, $(\bar{q}q = (\bar{u}u+\bar{d}d)/\sqrt{2})$, i.e. the diagonal and nondiagonal polarization operators are of the same order. Therefore, in the limit of massless u, d, s quarks, the physical states in the isospin zero pseudoscalar channel belong to an $SU(3)$ octet or to an $SU(3)$ singlet and the lowest states are η and η'. The same statement follows from the persistence of the anomaly in the divergence of the $SU(3)$ singlet axial current, which shows, that η' is not a Goldstone boson. (See Chapter 3.) Let us check that the nondiagonal polarization operator $\Pi_A^{(0)su}(Q)$ is approximately saturated by contributions of η and η' mesons. Define their coupling constants to axial currents by:

$$\frac{1}{\sqrt{6}}\langle 0 \mid \bar{u}\gamma_\mu\gamma_5 u + \bar{d}\gamma_\mu\gamma_5 d - 2\bar{s}\gamma_\mu\gamma_5 s \mid \eta\rangle = if_\eta q_\mu,$$

$$\frac{1}{\sqrt{3}}\langle 0 \mid \bar{u}\gamma_\mu\gamma_5 u + \bar{d}\gamma_\mu\gamma_5 d + \bar{s}\gamma_\mu\gamma_5 s \mid \eta'\rangle = if_{\eta'} q_\mu. \tag{6.149}$$

Write the sum rule for $\Pi_A^{(0)su}(Q)$ (6.147) and perform the Borel transformation. The asymptotic expansion of $K_0(Q\rho)$

$$K_0(Q\rho) \approx \left(\frac{\pi}{2Q\rho}\right)^{1/2} e^{-Q\rho} \tag{6.150}$$

can be used. ($Q\rho_c \approx 1.5 - 1.7$ and the correction term to (6.150) is less than 10%.) The formula for Borel transformation [24] is

$$\mathcal{B}\frac{1}{Q}e^{-2Q\rho} = \frac{1}{\sqrt{\pi}}M\exp(-M^2\rho^2). \tag{6.151}$$

Neglecting the contributions of excited states, we get the sum rule

$$\frac{1}{3}f_{\eta'}^2 e^{-m_{\eta'}^2/M^2} - \frac{1}{3}f_\eta^2 e^{-m_\eta^2/M^2} = \frac{\sqrt{\pi}}{2}M e^{-M^2\rho_c^2} n_0\rho_c^3. \tag{6.152}$$

The coupling constant f_η can be estimated as $f_\eta \approx 150\,\text{MeV}$. At $M^2 = 1\,\text{GeV}^2$ and $n_0 = 0.5 \cdot 10^{-3}\,\text{GeV}^4$ the right-hand side of (6.152) is about to $1.5 \cdot 10^{-4}\,\text{GeV}^2$. Then from (6.152) we get: $f_{\eta'} \approx 200\,\text{MeV}$. This value is 30% higher than the result of phenomenological analysis [103],[104], but bearing in mind the crudeness of the model the agreement can be considered as satisfactory. (Note that the s quark mass was taken into account in [103],[104] but not in the present discussion.)

At least one can say that qualitatively the instanton contribution can explain the formation of meson states in pseudoscalar and vector nonets – the presence of mixing of quark pairs $\bar{u}u + \bar{d}d \leftrightarrow \bar{s}s$ in the first case and its absence in the second. Taking account of instanton contributions in the sum rules for pseudoscalar mesons is necessary. Probably the same statement is appropriate for scalar mesons.

6.7 Sum rules for baryon masses

6.7.1 Determination of the proton mass

The calculation of the proton mass is evidently a first-rate problem of nonperturbative QCD [10],[11],[105],[106]. At the same time, such calculation is an important check of the QCD sum rule approach and OPE since, as was already stressed, the proton mass is of nonperturbative origin, arises due to chirality violation and should be expressed in the QCD sum rule method through chirality violating condensates. At first sight, a direct relationship of the proton mass m and quark condensate seems unlikely since from dimensional considerations it must have the form of

$$m^3 = -c\langle 0 \mid \bar{q}q \mid 0\rangle, \tag{6.153}$$

where c is a numerical coefficient. In order for (6.153) to be valid, c should be about 50 – a value which seems to be improbably large. However, as will be shown, such a relation does approximately take place indeed.

Consider the polarization operator

$$\Pi(p) = i \int d^4x e^{ipx} \langle 0 \mid T\{\eta(x), \bar{\eta}(0)\} \mid 0\rangle, \tag{6.154}$$

where $\eta(x)$ is the three-quark current with proton quantum numbers built from u and d quark fields. We restrict ourselves to consideration of currents without derivatives. The reason is that the presence of derivatives shifts the effective momenta in (6.154) to higher values where the convergence of OPE is worse and the role of excited states is higher. Generally, $\Pi(p)$ has the form

$$\Pi(p) = \not{p}\,\Pi_1(p^2) + \Pi_2(p^2). \tag{6.155}$$

$\Pi_1(p^2)$ is the chirality conserving structure function, $\Pi_2(p^2)$ violates chirality. In what follows, u and d quark masses will be neglected. Then, in OPE, $\Pi_2(p^2)$ is proportional to chirality violating condensates. There are two three-quark currents with proton quantum numbers:

$$\eta_1(x) = \left(u^a(x)C\gamma_\mu u^b(x) \right) \gamma_5\gamma_\mu d^c(x)\varepsilon^{abc}, \tag{6.156}$$

$$\eta_2(x) = \left(u^a(x)C\sigma_{\mu\nu} u^b(x) \right) \sigma_{\mu\nu}\gamma_5 d^c(x)\varepsilon^{abc}, \tag{6.157}$$

where $u^a(x)$, $d^c(x)$ – are the four-component spinor fields of u and d quarks, a, b, c – are colour indices, C is the charge conjugation matrix, and $C^T = -C$, $C^+C = 1$ (T stands for transposition). Any of the currents η_1 or η_2, as well as their linear combination, can be used as the current η in (6.154), but the choice of η_1 is much preferable for our goal – the proton mass calculation [10],[107]. In order to see this, perform the Fierz transformation and introduce right- and left-hand spinors. We have

$$\eta_1(x) = 2[\,(u^a C d^b)\gamma_5 u^c - (u^a C\gamma_5 d^b)u^c\,]\varepsilon^{abc}$$

$$= 4[\,(u_R^a C d_R^b)u_L^c - (u_L^a C d_L^b)u_R^c\,]\varepsilon^{abc}, \tag{6.158}$$

$$\eta_2(x) = 2[\,(u^a C d^b)\gamma_5 u^c + (u^a C\gamma_5 d^b)u^c\,]\varepsilon^{abc}$$

$$= -4[\,(u_R^a C d_R^b)u_R^c - (u_L^a C d_L^b)u_L^c\,]\varepsilon^{abc}. \tag{6.159}$$

Comparison of (6.158) and (6.159) shows that the properties of the polarization operator (6.154) constructed from the currents η_1 or η_2 are sharply different. Namely, in the case of the η_1 current, the OPE for the chirality violating structure $\Pi_2(p^2)$ starts from the vacuum expectation value $\langle 0 \mid \bar{q}q \mid 0 \rangle$, while in the case of the η_2 current, the first term of OPE in $\Pi_2(p^2)$ is proportional to $\langle 0 \mid \bar{q}q \mid 0 \rangle^3$. Indeed, the chirality violating condensates appear only if the right-hand spinor is paired with a left-hand spinor. In the case of the η_2 current this is possible only if three spinors are paired at once. Therefore, in the case of the η_2 the current $\Pi_2(p^2)$ is strongly suppressed compared with $\Pi_2(p^2)$ in the case of the η_1 current and also with $\Pi_1(p^2)$ at $\mid p^2 \mid \sim 1\,\mathrm{GeV}^2$. ($\Pi_1(p^2)$ is of the same order in both cases.) If the structure functions are represented by contributions of hadronic states via dispersion relation (6.37), then at $p^2 < 0$, $\Pi_1(p^2) > 0$ all resonances contribute to Im $\Pi_1(p^2)$ with the same sign. But the signs of contributions of positive and negative parity resonances to Im $\Pi_2(p^2)$ are opposite. The suppression of $\Pi_2(p^2)$ in the case of the η_2 current indicates strong compensation of contributions of the lowest state and excited states with the opposite parity. For this reason, the current

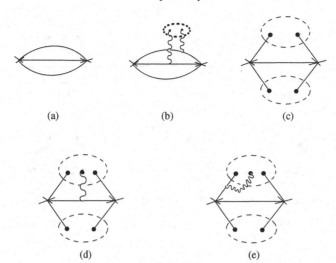

Fig. 6.16. Feynman diagrams for $\Pi_1(p^2)$ – the chirality conserving part of the polarization operator. The solid lines correspond to quarks, wavy lines to gluons, dots outlined by dashed lines stand for the mean vacuum values of the field operators, and crosses stand for interactions with the external current.

$\eta_2(x)$ is not suitable for QCD sum rules and even its admixture to η_1 is not acceptable [10],[107]. In what follows, when considering QCD sum rules for proton only the current $\eta_1(x)$ (Ioffe current) will be used (the index 1 will be omitted). The currents $\eta_1(x)$, $\eta_2(x)$ are renormcovariant, i.e. they transform through themselves under renormalization group transformations. The anomalous dimension of the current $\eta_1(x)$ is equal to $-2/9$ [108].

The operators up to dimension 8 are accounted for when calculating the chirality conserving part $\Pi_1(p^2)$ of the polarization operator (6.154). They are: the unit operator corresponding to the bare loop diagram (Fig. 6.16a) (the α_s-corrections will be discussed later); the $d = 4$ operator – gluon condensate $\langle 0 \mid (\alpha_s/\pi)G_{\mu\nu}^n G_{\mu\nu}^n \mid 0 \rangle$ (Fig. 6.16b); the $d = 6$ four-quark condensate $\langle 0 \mid \bar{\psi}\Gamma\psi \cdot \bar{\psi}\Gamma\psi \mid 0 \rangle$ (Fig. 6.16c); and the $d = 8$ four-quark + gluon condensate $\langle 0 \mid \bar{\psi}\Gamma\psi\bar{\psi}\Gamma\psi G \mid 0 \rangle$ (Figs. 6.16d and 6.16e). In the diagram of Fig. 6.16d, the soft gluon entering the quark–gluon condensate is emitted by a hard quark. In the diagram of Fig. 6.16e, the gluon is emitted by the soft quark and enters the quark–gluon condensate. The contribution of the diagram of Fig. 6.16e is found by expansion of vacuum expectation value $\langle 0 \mid q_\alpha^a(x)\bar{q}_\beta^a(0) \mid 0 \rangle$ in powers of x and using the equations of motion. The factorization hypothesis is used in the calculation of $d = 6$ and $d = 8$ operator contributions. The operators of dimensions $d = 3 - 9$ are accounted for in the calculation of the chirality violating structure $\Pi_2(p^2)$. The first term of OPE ($d = 3$) is proportional to the quark condensate (Fig. 6.17a). Because of the structure of the quark current $\eta_1(x)$ (6.156) only the d quark condensate contributes here. The next term of OPE ($d = 5$) represents the contribution of the quark–gluon condensate (6.16), Figs. 6.17b, c. The calculation shows, however, that the diagrams where gluons are emitted by hard (Fig. 6.17b) and soft (Fig. 6.17c) quarks exactly compensate one another. So, the contribution of the

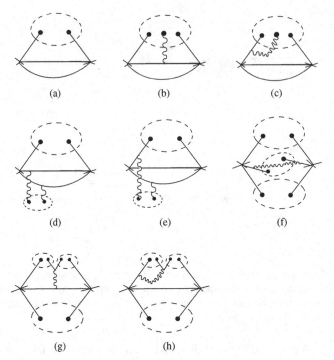

Fig. 6.17. Feynman diagrams for $\Pi_2(p^2)$. The notation is the same as in Fig. 6.16.

quark–gluon condensate to $\Pi_2(p^2)$ appears to be zero. The diagrams corresponding to the $d = 7$ operator $\langle 0 \mid \bar{d}d \mid 0 \rangle \langle 0 \mid (\alpha_s/\pi)G^2 \mid 0 \rangle$ (factorization is assumed) are shown in Figs. 6.17d, e. Finally, the diagrams of Figs. 6.17f, g, h represent the contributions of the $d = 9$ operator $\alpha_s \langle 0 \mid \bar{q}q \mid 0 \rangle^3$. Note that this contribution is enhanced in comparison with contributions of other $d = 9$ operators (e.g. $\bar{q}qG^3$) because there are no loop integrations in Figs. 6.17f, g, h. (Each loop integration results in appearance of a factor of $(2\pi)^2$ in the denominator.) A similar situation arises in the case of $d = 6$ operators (Figs. 6.17, d, e). For this reason we neglect the contributions of all other operators of dimension 6 and 9 leading to loop integrations.

The calculation technique is described in Section 6.8. We present the results – the sum rules for the proton polarization operator after the Borel transformation [10],[11],[18] (the early calculations, where not all terms in (6.160), (6.161) were accounted for, were done also in [105],[106],[109]):

$$M^6 E_2(M) L^{-4/9} c_0(M) + \frac{1}{4} b M^2 E_0(M) L^{-4/9} + \frac{4}{3} a_{\bar{q}q}^2 c_1(M) - \frac{1}{3} a_{\bar{q}q}^2 \frac{m_0^2}{M^2}$$

$$= \tilde{\lambda}_N^2 \exp\left(-\frac{m^2}{M^2}\right), \tag{6.160}$$

$$2 a_{\bar{q}q} M^4 E_1(M) c_2(M) + \frac{272}{81} \frac{\alpha_s(M)}{\pi} \frac{a_{\bar{q}q}^3}{M^2} - \frac{1}{12} a_{\bar{q}q} b = m \tilde{\lambda}_N^2 \exp\left(-\frac{m^2}{M^2}\right). \tag{6.161}$$

Here m is the proton mass, furthermore

$$a_{\bar{q}q} = -(2\pi)^2 \langle 0 \mid \bar{q}q \mid 0 \rangle, \tag{6.162}$$

$$b = (2\pi)^2 \langle 0 \mid \frac{\alpha_s}{\pi} G_{\mu\nu}^2 \mid 0 \rangle, \tag{6.163}$$

$$L = \frac{\alpha_s(\mu^2)}{\alpha_s(M^2)}, \tag{6.164}$$

$$E_n(M) = \frac{1}{n!} \int\limits_0^{s_0/M^2} z^n e^{-z} dz. \tag{6.165}$$

The matrix element of the current η between the proton state with momentum p and polarization r and the vacuum is defined as

$$\langle 0 \mid \eta \mid p \rangle = \lambda_N \upsilon^{(r)}(p), \tag{6.166}$$

where $\upsilon^{(r)}(p)$ is the proton spinor normalized as $\bar{\upsilon}^{(r)}(p)\upsilon^{(r)}(p) = 2m$. The constant $\tilde{\lambda}_N$ is related to λ_N by

$$\tilde{\lambda}_N^2 = 2(2\pi)^4 \lambda_N^2. \tag{6.167}$$

The contribution of excited states to the physical (right-hand) sides of the sum rules is approximated by a continuum starting from s_0 and given by terms of OPE not vanishing at $p^2 \rightarrow \infty$. These contributions were transferred to the left-hand sides of the sum rules and result in the appearance of the factors $E_n(M)$ there. Strictly speaking, the continuum thresholds in (6.160) and (6.161) can be different since the contributions of opposite parity states enter (6.160) with the same sign but appear in (6.161) with opposite signs, which may lead to different effective values of continuum thresholds. For simplicity, equal values of s_0 were chosen in (6.160) and (6.161). The factors L^γ reflect the anomalous dimensions of the currents and operators. The factors $c_i(M)$, $i = 0, 1, 2$ are α_s-corrections which were accounted for only for the main terms in the sum rules. The anomalous dimensions of higher-order terms of OPE are neglected. Let us first estimate crudely the proton mass. Neglect high-order terms of OPE in (6.160), (6.161) as well as continuum contributions and α_s-corrections. Divide (6.161) by (6.160) and put $M = m$. We get Ioffe formula [10]

$$m^3 = -2(2\pi)^2 \langle 0 \mid \bar{q}q \mid 0 \rangle \tag{6.168}$$

i.e. a relation of the form of (6.153). Numerically, at the value of the quark condensate given in Table 1, we have

$$m = 1.09 \, \text{GeV} \tag{6.169}$$

in surprisingly good agreement with the experimental value $m = 0.94 \, \text{GeV}$.

The α_s-corrections to the sum rules were calculated in [110]–[113]. The calculations of α_s-corrections to the first term in (6.160) are performed in the standard way using dimensional regularization. The terms proportional to $\ln Q^2/\mu^2$ are absorbed by the anomalous dimension factor. When calculating the α_s-corrections to the first term of (6.161), there appear not only ultraviolet but also infrared singularities. The latter are absorbed by the

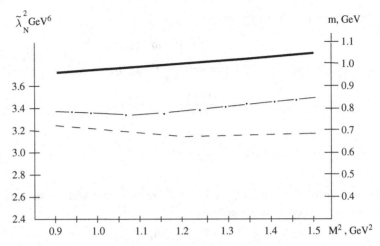

Fig. 6.18. The proton mass sum rules, Eqs. (6.160) and (6.161). The dashed and dash-dotted curves give $\tilde{\lambda}_N^2$, determined from (6.160) and (6.161), respectively, using the experimental value of m (left scale). The solid line gives m as the ratio of (6.161) to (6.160).

counterterm renormalizing the quark condensate value [111]. The problem of validity of the factorization hypothesis arises when α_s-corrections to four-quark condensate contributions are calculated (the third term in (6.160)). The most suitable way to treat this problem is to renormalize each operator separately, take into account their mixing, and use the factorization hypothesis in the final answer. As a result, the α_s-corrections are found to be (after Borelization) [111]–[113]

$$c_0(M) = 1 + \left(\frac{53}{12} + \gamma_E\right)\frac{\alpha_s(M^2)}{\pi}, \qquad (6.170)$$

$$c_1(M) = 1 + \left(-\frac{1}{6} + \frac{1}{3}\gamma_E\right)\frac{\alpha_s(M^2)}{\pi}, \qquad (6.171)$$

$$c_2(M) = 1 + \frac{3}{2}\frac{\alpha_s(M^2)}{\pi}, \qquad (6.172)$$

where γ_E is the Euler constant, $\gamma_E = 0.577$. During the Borelization, $\alpha_s(Q^2)$ was taken to be a constant at the value $Q^2 = M^2$ since the deviations from the constant are of order α_s^2.

The values of m, found as the ratio of (6.161) to (6.160), and $\tilde{\lambda}_N^2$ from (6.160) and (6.161) at $m = m_{\text{exp}} = 0.94\,\text{GeV}$ are plotted in Fig. 6.18 as functions of M^2. The following values of the parameters were chosen: The values of quark and gluon condensates were taken from Table 1 and correspond to $a_{\bar{q}q} = 0.65\,\text{GeV}^3$ and $b = 0.20\,\text{GeV}^4$, the continuum threshold $s_0 = 2.5\,\text{GeV}^2$, the normalization point $\mu^2 = 1\,\text{GeV}^2$, $\alpha_s(\mu^2) = 0.55$, the M^2 dependence of $\alpha_s(M^2)$ from Fig. 6.3. As follows from Fig. 6.18, the proton mass m is equal to $m = 0.98\,\text{GeV}$ in good agreement with experiment and in accord with the crude estimate (6.168), (6.169). The mean value of $\tilde{\lambda}_N^2$ is $\tilde{\lambda}_N^2 = 3.2\,\text{GeV}^6$. Let us determine the window $M_1^2 < M^2 < M_2^2$, where we can believe in the sum rules (6.160), (6.161). At low M^2, the most restrictive is the growth of perturbative corrections: as is demonstrated in Section 6.5, the perturbative series breaks down at $M^2 \approx 0.9 - 1.0\,\text{GeV}^2$. So, we can put $M_1^2 = 0.9\,\text{GeV}^2$. (The α_s-correction to the first term in (6.160) is very large; this problem is discussed below.) The last terms of OPE accounted for in (6.160), (6.161) are still small here in comparison with the total sum: the OPE series is probably still converging at $M^2 = 0.9\,\text{GeV}^2$. The third term in (6.160) is comparable with the first one. This is caused by the absence of loop integration in this term and does not mean any violation of OPE expansion. The upper value of M^2 is determined by an increasing role of the continuum. At central values of $M^2 \approx 1.0 - 1.1\,\text{GeV}^2$ the continuum contribution in (6.160) is about 30% less than the total sum. (In (6.161) the situation is even better.) At $M^2 = 1.3 - 1.4\,\text{GeV}^2$, the continuum contribution in (6.160) is 1.5–2.0 times greater than the total sum. Therefore, the reasonable choice is $M_2^2 = 1.3 - 1.4\,\text{GeV}^2$.

Let us estimate the possible uncertainties of the calculations. The value of the gluon condensate influences the results weakly: a 100% uncertainty in $\langle 0 \mid (\alpha_s/\pi)G^2 \mid 0 \rangle$ gives less than 4% error in m and $\tilde{\lambda}_N^2$. The error in the quark condensate quoted in Table 1 results in a 5% error in the proton mass and a 10–15% error in $\tilde{\lambda}_N^2$. The influence of the continuum threshold value can be estimated if we put $s_0 = 2.2\,\text{GeV}^2$ instead of the $s_0 = 2.5\,\text{GeV}^2$ used in the calculation of the curves presented in Fig. 6.18. Such variation results in a 3% increase of m and about 10% decrease of $\tilde{\lambda}_N^2$. The main uncertainty comes from α_s-corrections. The reason is that the α_s-correction to the first term in (6.161), $c_0(M)$ – Eq. (6.170) – is enormously large and about 80% at $M^2 \approx 1\,\text{GeV}^2$. The first term in (6.160) gives about half of the total contribution to the left-hand side of this equation. In the ratio of (6.161) to (6.160), i.e. in the value of m, this effect is partly compensated by the α_s-correction to (6.161). So, if we completely omit the α_s-corrections, then the mass would increase by 12% and $\tilde{\lambda}_N^2$ decrease by 20–25%. The more optimistic supposition is that α_s-corrections are given by the first-order term with accuracy about 50%. Then they result to the error in the proton mass about 6% and in $\tilde{\lambda}_N^2$ about 10–12%.

It is difficult to sum all these errors. If we consider them as independent and sum them in quadrature, then we would get an error in m of about 10% and an error in $\tilde{\lambda}_N^2$ of about 20%. These estimates look reasonable. So, the final result of the proton mass calculation by QCD sum rules is (see also [17],[114]):

$$m = 0.98 \pm 0.10\,\text{GeV}\,, \tag{6.173}$$

$$\tilde{\lambda}_N^2 = 3.2 \pm 0.6\,\text{GeV}^6. \tag{6.174}$$

The ratio of Eqs. (6.161) to (6.160) was used above for the determination of the proton mass. Sometimes [13],[105],[106],[114], another method is exploited: to take the logarithmic derivative of the sum rule with respect to $(1/M^2)$. This method directly gives the value of m^2 and is the only one possible in the case where there is only one sum rule – the case of spinless bosons. In the case of the sum rules (6.160), (6.161) considered above the differentiation of (6.160) is useless because it would result in an enormously large continuum contribution – about three times larger than the total sum of all accounted for terms at the central values of $M^2 \approx 1.0 - 1.1\,\text{GeV}^2$. The differentiation of Eq. (6.161) also leads to an increase of the role of the continuum, but here its contribution appears to be not unacceptably large – about 20% greater than the total sum of all terms. The calculation of the proton mass by differentiation of (6.161) gives $m = 1.04\,\text{GeV}$ in agreement with (6.173). This calculation has the advantage that the problem of large α_s-corrections in (6.160) is avoided. The method of differentiation of the sum rules has, however, a disadvantage: the results significantly depend on the value of the continuum threshold s_0.

The instanton contribution was disregarded in the calculation presented above. An attempt to take it into account was done in [115],[116]. The instanton gas model was considered where the instantons are noninteracting, the distribution of instantons over their radius ρ is concentrated at $\rho = \rho_c$, and the instanton density is given by (6.69). The quark propagator and the polarization operator in the instanton field were calculated. It must be stressed that, in order to avoid double counting, the power-like terms $\sim (Q^2)^{-n}$ must be omitted from $\Pi(Q)^{inst}$ since they are already accounted for in OPE. Such procedure was performed in [116]. It was found that instantons do not contribute to $\Pi_1(Q)$ in the case of the quark current η (6.156). The instanton contribution to $\Pi_2(Q)$ essentially depends on the effective quark mass m_q^* in the instanton field $\Pi_2^{inst} \sim 1/m_q^{*2}$ which introduces a serious uncertainty into results of the calculation. The formulae presented in [116] show that at $n_0 = 1 \cdot 10^{-3}\,\text{GeV}^4$, $\rho_c = (1/3)\text{fm} = 1.7\,\text{GeV}^{-1}$ and m_q^*=300 MeV (this is the standard choice of the instanton gas model parameters; see [57],[117]), $\Pi_2^{inst}(M)$ amounts to about 5–7% of $\Pi_2(M)$ at $M^2 \approx 1.0 - 1.2\,\text{GeV}^2$ and results in an increase of the proton mass. It is possible that the instanton gas model overestimates the instanton contribution since one may expect that the interaction of instantons results in its suppression. In any case, the estimate presented above shows that the instanton contribution is within the limits of the errors of the sum rule calculation. The good coincidence of the results obtained in different ways as well as their agreement with experiment is also an argument in favour that instanton contribution is negligible in this problem.

6.7.2 The masses of octet baryons

In order to find the masses of octet baryons it is enough to calculate the masses of Σ and Ξ baryons. Then the Λ mass will be found using the Gell-Mann–Okubo mass formula

$$m_N + m_\Xi = \frac{3}{2}m_\Lambda + \frac{1}{2}m_\Sigma \qquad (6.175)$$

valid in the first order of $SU(3)$ breaking. The quark currents corresponding to Σ^+ and Ξ^- can be easily found from the proton current (6.156) by replacing $d \to s$ in case of Σ^+ and $u \to s$ in case of Ξ^-. We have

$$\eta_{\Sigma^+} = (u^a C\gamma_\mu u^b)\gamma_5\gamma_\mu s^c \cdot \varepsilon^{abc}, \qquad (6.176)$$

$$\eta_{\Xi^-} = (s^a C\gamma_\mu s^b)\gamma_5\gamma_\mu d^c \cdot \varepsilon^{abc}. \qquad (6.177)$$

In our calculation the strange quark mass m_s is accounted for in the first order and the deviation of the strange quark condensate from the u quark condensate:

$$f = \frac{\langle 0 \mid \bar{s}s \mid 0 \rangle}{\langle 0 \mid \bar{u}u \mid 0 \rangle} - 1. \qquad (6.178)$$

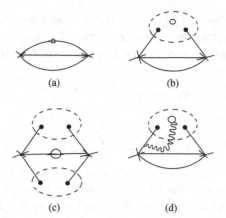

Fig. 6.19. The diagrams, corresponding to quark mass corrections to Π_Σ. The notation is the same as in Fig. 6.16. The small circles mean the mass term insertion.

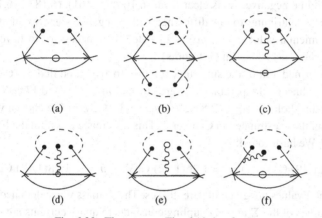

Fig. 6.20. The same as Fig. 6.19, but for Π_Ξ.

The diagrams corresponding to contribution of the strange quark mass in the polarization operators are shown in Fig. 6.19 (for Σ^+) and in Fig. 6.20 for Ξ^-. Because of the structure of the quark current η_Σ (6.176), the diagrams in Fig. 6.19b and d contribute to the chirality conserving polarization operator $\Pi_{1\Sigma}$ and the diagrams in Fig. 6.16a and c to the chirality violating operator $\Pi_{2\Sigma}$. Similarly, the diagrams in Fig. 6.20a and c contribute to $\Pi_{1\Xi}$ (in fact, the first one vanishes) and all other diagrams in Fig. 6.20 to $\Pi_{2\Xi}$. After Borelezation, the terms proportional to m_s in the polarization operators for Σ and Ξ are found to be [10],[118]:

$$\Delta\Pi_{1\Sigma} = -2m_s a_{\bar{q}q} E_0(M)L^{-4/9}M^2 - \frac{1}{3}m_0^2 a_{\bar{q}q}m_s L^{-14/27}, \qquad (6.179)$$

$$\Delta\Pi_{2\Sigma} = 2m_s M^6 E_2(M)L^{-8/9} + \frac{8}{3}m_s a_{\bar{q}q}^2, \qquad (6.180)$$

$$\Delta\Pi_{1\Xi} = -\frac{2}{3}m_s m_0^2 a_{\bar{q}q} L^{-1}, \qquad (6.181)$$

$$\Delta\Pi_{2\Xi} = 4m_s a_{\bar{q}q}^2. \qquad (6.182)$$

(In the higher-order terms of OPE ($d > 6$), the corrections proportional to m_s are neglected.) In order to get the sum rules for Σ and Ξ besides the contributions given by (6.179)–(6.182) the following changes must be made in (6.160), (6.161): in the sum rules for Σ, in (6.161)

$$a_{\bar{q}q} \rightarrow a_{\bar{s}s} = (1 + f)a_{\bar{q}q}, \tag{6.183}$$

in the sum rules for Ξ, in (6.160)

$$a_{\bar{q}q}^2 \rightarrow a_{\bar{s}s}^2 = (1 + f)^2 a_{\bar{q}q}^2. \tag{6.184}$$

As is seen from (6.179)–(6.182), the baryon mass is increased when terms proportional to m_s are taken into account. Indeed, the baryon mass is determined as the ratio Π_2/Π_1. The contribution of m_s terms is positive for Π_2 and negative for Π_1. Therefore, we have come to the conclusion that strange baryons are heavier than the nucleon if $f = 0$. However, if only the terms proportional to m_s would be accounted for, then the mass difference $m_\Xi - m_\Sigma$ would be negative. As is clear from (6.160), (6.161), (6.183), (6.184) the term proportional to f contributes to this difference with a negative sign. So, in order to reproduce the experimental value $m_\Xi - m_\Sigma \approx 130\,\text{MeV}$ it is necessary to have $f < 0$ and numerically of the order of m_s (in GeV units).

We have performed a fit of the sum rules (6.160), (6.161) modified by (6.179)–(6.184). The following values of the parameters were chosen: $m_s(1\,\text{GeV})=0.17\,\text{GeV}$, $f = -0.2$, the continuum thresholds – $s_{0\Sigma} = 2.8\,\text{GeV}^2$, $s_{0\Xi} = 3.0\,\text{GeV}^2$. (The chosen value of m_s is 20% higher than those estimated in Chapter 2. This difference is within the limits of errors allowed there.) We have found:

$$m_\Sigma = 1.16(1.19)\,\text{GeV}, \quad m_\Xi = 1.26(1.32)\,\text{GeV}, \quad m_\Lambda = 1.08(1.12)\,\text{GeV}. \tag{6.185}$$

The experimental values are given in parentheses. The Λ mass was calculated according to (6.175). The values of the Σ and Ξ coupling constants to quark currents are given by

$$\tilde{\lambda}_\Sigma^2 = 5.0\,\text{GeV}^6, \quad \tilde{\lambda}_\Xi^2 = 7.0\,\text{GeV}^6. \tag{6.186}$$

The values of the Σ and Ξ-masses are about 30–50 MeV lower than the experimental ones. It must be stressed that the values of m_Σ and m_Ξ cannot be considered as theoretical predictions, because two new parameters – m_s and f (besides $s_{0\Sigma}$ and $s_{0\Xi}$) – were used in their determination. It can be said merely that the good agreement of m_Σ and m_Ξ with the data shows that the chosen values of m_s and f are reasonable. Note that α_s-corrections to m_s terms were not accounted for. This may lead to a 20–30% error in m_s and f and, correspondingly, about 50 MeV error in m_Σ, m_Ξ.

6.7.3 Masses of baryon resonances

The masses of baryon resonances can also be predicted by QCD sum rules, but it is hard to expect high accuracy of such a prediction. The reason is that the model of the hadronic spectrum used in this approach – an infinitely narrow lowest-state separated by a gap from the continuum – is far from a perfect representation of the spectrum in this case:

the resonances are situated on backgrounds. Nevertheless, the QCD sum rules for the first excited states can predict the basic feature of the hadronic spectrum: the sequence of the levels and, although with less accuracy than in the case of stable particles, the resonance masses. For highly excited states, the QCD sum rule method does not work. The physical reason is evident. These states are orbital or radial excitations of the bound quark system; their sizes are large. The treatment of such systems are outside the scope of perturbation theory and OPE.

Let us start with the case of the isobar Δ (isospin $T = 3/2$, $J^P = 3/2^+$). It is convenient to consider Δ^{++}, built from three u quarks. There is only one three-quark current without derivatives with the quantum numbers of Δ^{++}

$$\eta_\mu^\Delta(x) = (u^a(x) C \gamma_\mu u^b(x)) u^c(x) \varepsilon^{abc}. \tag{6.187}$$

The polarization operator

$$\Pi_{\mu\nu}^\Delta(x) = i \int d^4x \, \langle 0 \, | T \, \{\eta_\mu^\Delta(x), \bar\eta_\nu^\Delta(0)\} | \, 0 \rangle \tag{6.188}$$

has few kinematical structures. Below, we present the results of the calculation of these structures where only the most important terms of OPE are accounted for [10],[11],[105],[106]: the unit operator, $\langle 0|\bar u u|0\rangle$, $\langle 0|\bar u u|0\rangle^2$ and $m_0^2\langle 0|\bar u u|0\rangle$, the last ones within the framework of the factorization hypothesis (the perturbative corrections are neglected):

$$\Pi_{\mu\nu}^\Delta(p) = \Pi_{\mu\nu}^{\Delta,1} + \Pi_{\mu\nu}^{\Delta,2}, \tag{6.189}$$

where $\Pi_{\mu\nu}^{\Delta,1}$ and $\Pi_{\mu\nu}^{\Delta,2}$ correspond to chirality conserving and chirality violating structures.

$$\begin{aligned}
\Pi_{\mu\nu}^{\Delta,1}(p) = {} & \frac{p^4}{10(2\pi)^4} \ln \frac{-p^2}{\Lambda^2} \left\{ \delta_{\mu\nu} \not{p} - \frac{5}{16} \gamma_\mu \gamma_\nu \not{p} + \frac{5}{16}(p_\nu \gamma_\mu - p_\mu \gamma_\nu) \right. \\
& \left. + \frac{1}{16}(p_\nu \gamma_\mu + p_\mu \gamma_\nu) - \frac{p_\mu p_\nu}{p^2} \not{p} \right\} + \frac{4}{3} \frac{\langle 0 \, | \, \bar u u \, | \, 0\rangle^2}{p^2} \left(1 + \frac{7}{12} \frac{m_0^2}{p^2}\right) \\
& \times \left\{ \delta_{\mu\nu} \not{p} - \frac{3}{8} \gamma_\mu \gamma_\nu \not{p} + \frac{3}{8}(p_\nu \gamma_\mu - p_\mu \gamma_\nu) - \frac{1}{8}(p_\nu \gamma_\mu + p_\mu \gamma_\nu) \right\},
\end{aligned} \tag{6.190}$$

$$\begin{aligned}
\Pi_{\mu\nu}^{\Delta,2}(p) = {} & -\frac{4}{3} \frac{\langle 0 \, | \, \bar u u \, | \, 0\rangle}{(2\pi)^2} p^2 \ln \frac{-p^2}{\Lambda^2} \left\{ \delta_{\mu\nu} - \frac{5}{16} \gamma_\mu \gamma_\nu + \frac{1}{4}(p \gamma_\mu - p_\mu \gamma_\nu) \frac{\not{p}}{p^2} \right. \\
& \left. - \frac{1}{2} \frac{p_\mu p_\nu}{p^2} \right\} + \frac{2}{3} m_0^2 \frac{\langle 0 \, | \, \bar u u \, | \, 0\rangle}{(2\pi)^2} \left\{ \ln\left(\frac{-q^2}{\Lambda^2}\right) \left(\delta_{\mu\nu} - \frac{1}{4} \gamma_\mu \gamma_\nu\right) \right. \\
& \left. + \frac{1}{2}(p_\nu \gamma_\mu - p_\mu \gamma_\nu) \frac{\not{p}}{p^2} - \frac{p_\mu p_\nu}{p^2} \right\}.
\end{aligned} \tag{6.191}$$

On the phenomenological side of the sum rule represented by dispersion relations, the Δ^{++} contribution is proportional to the matrix element

$$\sum_\tau \langle 0|\eta_\mu(0)|\Delta(p,r)\rangle \langle \Delta(p,r)|\bar\eta_\nu(0)|0\rangle, \tag{6.192}$$

where the summation goes over the spin projection r of the isobar Δ with momentum p, $p^2 = m_\Delta^2$. The matrix element entering (6.192) is equal to

$$\langle 0|\eta_\mu(0)|\Delta(p,r)\rangle = \lambda_\Delta v_\mu^{(r)}(p), \qquad (6.193)$$

where $v_\mu^{(r)}(p)$ is an isospin-vector of the isobar in the Rarita–Schwinger formalism. $v_\mu^{(r)}$ satisfies the equalities $\gamma_\mu v_\mu^{(r)}(p) = 0$, $p_\mu v_\mu^{(r)}(p) = 0$, and the Dirac equation $(\not{p} - m_\Delta)v_\mu^{(r)}(p) = 0$. The summation over r in (6.192) can be performed using (6.193) and taking into account these subsidiary conditions. The result is:

$$\sum_r \langle 0|\eta_\mu|\Delta(p,r)\rangle\langle\Delta(p,r)|\bar{\eta}_\nu(0)|0\rangle = -\lambda_\Delta^2 \left\{ \left[\delta_{\mu\nu}\hat{p} - \frac{1}{3}\gamma_\mu\gamma_\nu \not{p} + \frac{1}{3}(\gamma_\mu p_\nu - \gamma_\nu p_\mu) \right.\right.$$

$$\left.\left. - \frac{2}{3}\frac{p_\mu p_\nu \not{p}}{m_\Delta^2} \right] + \left[\delta_{\mu\nu} - \frac{1}{3}\gamma_\mu\gamma_\nu + \frac{1}{3}(\gamma_\mu p_\nu - \gamma_\nu p_\mu)\frac{\not{p}}{m_\Delta^2} - \frac{2}{3}\frac{p_\mu p_\nu}{m_\Delta^2} \right] m_\Delta \right\}. \qquad (6.194)$$

On the physical side of the sum rule, the polarization operator $\Pi_{\mu\nu}^\Delta$ has contributions not only from resonances with spin $3/2$ but also from those with spin $1/2$. The relations between various Lorentz structures in this case are quite different from (6.194). Comparison of (6.194) with (6.190), (6.191) shows that the relations between the Lorentz structures in (6.190), (6.191), and (6.194) are very close to one another, i.e. (6.190), (6.191) reproduces well the dominance of the Δ isobar. Note that the sign in front of the second square bracket in (6.194) is appropriate for the positive parity of Δ. In the case of negative parity the sign would be opposite. The sign of the corresponding first term in (6.191) – the main term of OPE in $\Pi_{\mu\nu}^{\Delta,2}$ – is also positive after Borel transformation. Therefore QCD predicts that the lowest baryonic state with quantum numbers $T = 3/2$, $J = 3/2$ has positive parity.

The spin $1/2$ baryons do not contribute to the Lorentz structures $\delta_{\mu\nu}\hat{p}$ and $\delta_{\mu\nu}$ in (6.190), (6.191). The sum rules for the corresponding structure functions, where one may expect dominance of the Δ^{++} isobar, are [10],[11]:

$$M^6 E_2(M)L^{4/27} - \frac{25}{72}bM^2 E_0(M)L^{4/27} + \frac{20}{3}a_{\bar{q}q}^2 L^{28/27} - \frac{35}{9}a_{\bar{q}q}^2 \frac{m_0^2}{M^2}L^{14/27}$$

$$= \tilde{\lambda}_\Delta^2 e^{-m_\Delta^2/M^2}, \qquad (6.195)$$

$$\frac{20}{3}M^4 a_{\bar{q}q}E_1(M)L^{16/27} - \frac{10}{3}a_{\bar{q}q}m_0^2 M^2 E_0(M)L^{2/27} - \frac{20}{27}\frac{\alpha_s}{\pi}\frac{a_{\bar{q}q}^3}{M^2}L^{40/27} - \frac{5}{18}a_{\bar{q}q}b L^{16/27}$$

$$= m_\Delta \tilde{\lambda}_\Delta^2 e^{-m_\Delta^2/M^2}, \qquad (6.196)$$

where $\tilde{\lambda}_\Delta^2 = 5(2\pi)^2\lambda_\Delta^2$. α_s-corrections are not accounted for in (6.195), (6.196) except for anomalous dimension factors. (The anomalous dimension of the current η_Δ (6.187) is equal to $2/27$). The sum rules (6.195), (6.196) are fitted in the interval of M^2 from 1.2 to 1.6 GeV^2 at $s_0 = 3.6$ GeV^2. The best fit is achieved at $m_\Delta = 1.30$ GeV and $\tilde{\lambda}_\Delta^2 = 12$ GeV^2. However, the value of $\tilde{\lambda}_\Delta^2$ determined from (6.196) demonstrates a remarkable

M^2 dependence in this interval. This means that the accuracy of the sum rules (6.195), (6.196) is worse than in the case of the nucleon. Probably, m_Δ and $\tilde{\lambda}_\Delta^2$ are determined with 15% and 30% accuracy, respectively.

The mass splitting in the decuplet can be found in the same way as was done in the case of the baryon octet. It is enough to calculate $m_{\Sigma^*} - m_\Delta$, since all mass differences in the decuplet are equal because of the Gell-Mann–Okubo relation:

$$m_{\Sigma^*} - m_\Delta = m_{\Xi^*} - m_{\Sigma^*} = m_\Omega - m_{\Xi^*} \tag{6.197}$$

The Σ^{*+} quark current is given by:

$$\eta_{\Sigma^*} = \sqrt{\frac{1}{3}} \left[2(u^a C\gamma_\mu s^b)u^c + (u^a C\gamma_\mu u^b)s^c) \right] \varepsilon^{abc}. \tag{6.198}$$

The left-hand sides of the sum rules for Σ^* follow from the sum rules for Δ by adding the terms

$$\Delta \Pi^{\Sigma^*,1} = 5m_s a_{\bar{q}q} M^2 L^{4/27} \left(E_0 - \frac{1}{2}\frac{m_0^2}{M^2} L^{-14/27} \right) + \frac{40}{9} f a_{\bar{q}q}^2 L^{28/27}, \tag{6.199}$$

$$\Delta \Pi^{\Sigma^*,2} = \frac{5}{2}m_s M^6 E_2 L^{-8/27} + \frac{10}{3}m_s a_{\bar{q}q}^2 L^{16/27} + \frac{20}{9} f a_{\bar{q}q} M^4 E_1 L^{16/27}. \tag{6.200}$$

These sum rules were fitted at $m_s(1\text{ GeV}) = 170$ MeV, $f = -0.2\,\text{GeV}^2$ in the interval $1.3 < M^2 < 1.7\,\text{GeV}^2$ of the Borel mass parameters with the result

$$m_{\Sigma^*} - m_\Delta = 0.12 \pm 0.05(0.15)\text{ GeV}. \tag{6.201}$$

The experimental data are in parenthesis.

Turn now to considering nonstrange baryon resonances with isospin $T = 1/2$ and spin $3/2$. The three-quark current can be chosen as

$$\eta_{1\mu} = \left[(u^a C\sigma_{\rho\lambda} d^b)\sigma_{\rho\lambda}\gamma_\mu u^c - (u^a C\sigma_{\rho\lambda} u^b)\sigma_{\rho\lambda}\gamma_\mu d^c \right] \varepsilon^{abc}. \tag{6.202}$$

(This is not a unique choice: there is one more local three-quark current with the same quantum numbers; see below.) The current $\eta_{1\mu}$ is renormcovariant – it transforms through itself under renormalization group transformations. The anomalous dimension of $\eta_{1\mu}$ is equal to $2/27$ [108]. As before, in order to separate the contributions of spin $3/2$ states it is convenient to consider the structure functions that multiply the tensor structures $\delta_{\mu\nu}\hat{p}$ and $\delta_{\mu\nu}$. The sum rules for these structure functions are: ($\tilde{\lambda}_R^2 = (5/24)(2\pi)^4\lambda_R^2$) [11]

$$M^6 E_2(M)L^{4/27} - \frac{5}{36}bM^2 E_0(M)L^{4/27} - \frac{10}{3}a_{\bar{q}q}^2 L^{28/27} + \frac{35}{18} a_{\bar{q}q}^2 \frac{m_0^2}{M^2} L^{14/27}$$

$$= \tilde{\lambda}_R^2 e^{-m_R^2/M^2}, \tag{6.203}$$

$$-\left[\frac{10}{3}a_{\bar{q}q}M^4 E_1(M)L^{16/27} - \frac{5}{3}a_{\bar{q}q}M^2 m_0^2 E_0(M)L^{2/27} - \frac{220}{81}\frac{\alpha_s}{\pi}\frac{a_{\bar{q}q}^3}{M^2}L^{40/27} - \frac{5}{36}a_{\bar{q}q}b\,L^{16/27}\right]$$

$$= (\pm)m_R\tilde{\lambda}_R^2\,e^{-m_R^2/M^2}. \tag{6.204}$$

The + or − signs on the right-hand side of (6.204) correspond to positive and negative parities of the resonance R. Since the first – the main term – on the left-hand side of (6.204) is negative, we have come to the conclusion that in QCD the parity of the lowest baryonic resonance with $T = 1/2$ and $J = 3/2$ is negative. This nontrivial prediction is in full accord with experiment (the $N(1520)$ resonance [63]).

The sum rules (6.203), (6.204) are appropriate for large $M^2 > 1.6\,\text{GeV}^2$ only, because the contributions of the third and fourth terms in (6.203) – higher-order terms of OPE – overwhelm the main term at lower M^2. But at such a large M^2, the continuum contribution dominates the sum rule. Therefore, only a semi-quantitative estimate of the resonance mass can be obtained from the sum rules (6.203), (6.204). If we put $s_0 = 5\,\text{GeV}^2$, then

$$m_R\left(T = 1/2,\, J^P = \frac{3}{2}^-\right) = 1.7 \pm 0.3\,\text{GeV}. \tag{6.205}$$

Consideration of the second renormcovariant quark current with quantum numbers $T = 1/2$, $J = 3/2$ [10]

$$\eta_{2\mu} = \left[(u^a C d^b)\gamma_\mu u^c - (u^a C\gamma_5 d^b)\gamma_\mu\gamma_5 u^c\right]\varepsilon^{abc} \tag{6.206}$$

results in the same qualitative conclusion: the parity of the lowest baryonic resonance with $T = 1/2$ and $J = 3/2$ is negative. However, this consideration does not allow one to improve the estimate (6.205) of the resonance mass.

6.8 Calculation technique

In this section, the basic formulae for QCD sum rules calculations are presented. (For QCD sum rule technique see also the review [119].) The calculation of the nonperturbative parts of polarization operators, i.e. two-point functions, is convenient to perform in coordinate space. The terms of OPE calculated in coordinate space can be transformed into momentum space using the formula (for $n \geq 2$):

$$\int d^4x\, e^{ipx}\frac{1}{(x^2)^n} = i(-1)^n\frac{2^{4-2n}\pi^2}{\Gamma(n-1)\Gamma(n)}(p^2)^{n-2}\ln(-p^2) + P_{n-2}(p^2), \tag{6.207}$$

where $P_{n-2}(p^2)$ is an inessential polynomial in p^2 which will be killed by the Borel transformation. For $n = 1$ and the case of $n = 2$ with factors x_μ, $x_\mu x_\nu$, $x_\mu x_\nu x_\lambda$ in the numerator there are well-known formulae:

$$\int d^4x e^{ipx} \frac{1}{x^2} = -i\frac{4\pi^2}{p^2}; \quad \int d^4x e^{ipx} \frac{x_\mu}{x^2} = 8\pi^2 \frac{p_\mu}{p^4}; \tag{6.208}$$

$$\int d^4x e^{ipx} \frac{x_\mu}{x^4} = 2\pi^2 \frac{p_\mu}{p^2}; \quad \int d^4x e^{ipx} \frac{x_\mu x_\nu}{x^4} = \frac{2\pi^2 i}{p^4}(2p_\mu p_\nu - \delta_{\mu\nu} p^2); \tag{6.209}$$

$$\int d^4x e^{ipx} \frac{x_\mu x_\nu x_\lambda}{x^4} = \frac{4\pi^2}{p^6}[\, p^2(\delta_{\mu\nu} p_\lambda + \delta_{\mu\lambda} p_\nu + \delta_{\nu\lambda} p_\mu) - 4p_\mu p_\nu p_\lambda]. \tag{6.210}$$

For the reader's convenience, we present here once more the equations of motion in coordinate space:

$$(\nabla_\mu \gamma_\mu + im_q)\psi_q = 0, \tag{6.211}$$

$$D_\mu^{nm} G_{\mu\nu}^m = g \sum_q \bar{\psi}_q \gamma_\nu \frac{\lambda^n}{2} \psi_q, \tag{6.212}$$

where ∇_μ and D_μ^{nm} are covariant derivatives in fundamental and adjoint representations, respectively:

$$\nabla_\mu = \partial_\mu + ig\frac{\lambda^n}{2} A_\mu^n, \tag{6.213}$$

$$D_\mu^{nm} = \partial_\mu \delta^{nm} - gf^{nlm} A_\mu^l. \tag{6.214}$$

To calculate nonperturbative corrections it is convenient to represent the gluonic field as a sum of perturbative and nonperturbative parts

$$A_\mu = a_\mu + A_\mu^{nonpert} \tag{6.215}$$

and consider $A_\mu^{nonpert}$ as a background field in which the perturbative gluons are moving. In this section only $A_\mu^{nonpert}$ is accounted for and the index "nonpert" will be omitted. The most convenient way to calculate the OPE terms is to use the Fock–Schwinger [120],[121] (or fixed point) gauge. In application to QCD sum rules such a method was developed by Smilga [122]. The gauge condition is:

$$(x - x_0)_\mu A_\mu^n(x) = 0, \tag{6.216}$$

where x_0 is some fixed point. The gauge condition (6.216) breaks the translational symmetry. Evidently, this symmetry should be restored in all gauge-invariant physical observables. This property can be used as a check of the calculations. When considering the polarization operators $\Pi(x, y)$, it is possible to put $y = x_0 = 0$. In case of more complicated objects, such as 3-point functions, it is necessary to keep $x_0 \neq 0$. Let us temporarily make $x_0 = 0$. Condition (6.216) allows one to express $A_\mu^n(x)$ at small x as a series in x through field strength $G_{\mu\nu}^n(0)$

$$A_\mu^n(x) = \frac{1}{2}x_\nu G_{\nu\mu}^n(0) + \frac{1}{3}x_\alpha x_\nu (D_\alpha G_{\nu\mu}(0))^n + \frac{1}{4 \cdot 2!}x_\alpha x_\beta x_\nu (D_\alpha D_\beta G_{\nu\mu}(0))^n + \dots \tag{6.217}$$

The proof of (6.217) follows from the relation $x_\mu D_\mu = x_\mu \partial_\mu$, which is a consequence of (6.214) and gauge condition (6.216) at $x_0 = 0$. (The detailed proof was given in [122],[123].) A representation similar to (6.217) holds for the quark field:

$$\psi(x) = \psi(0) + x_\alpha \nabla_\alpha \psi(0) + \frac{1}{2} x_\alpha x_\beta \nabla_\alpha \nabla_\beta \psi(0) + \dots \qquad (6.218)$$

The representations (6.217) and (6.218) are very useful in the QCD sum rules approach, because the interactions of hard quarks and gluons with vacuum fields is treated as interaction with soft quark and gluon fields, the wavelengths of which are much larger than the wavelengths of hard quarks and gluons. Therefore, the vacuum fields entering the condensates can be considered as slowly varying background fields for which the expansions (6.217), (6.218) are valid.

Consider the equation for the quark propagator

$$S_{\alpha\beta}(x, z) = \langle 0 \mid T\{\psi_\alpha(x), \bar{\psi}_\beta(z)\} \mid 0 \rangle, \qquad (6.219)$$

$$\left[i\gamma_\mu \left(\partial_\mu + ig \frac{\lambda^n}{2} A_\mu^n(x) \right) - m \right] S(x, z) = i\delta^{(4)}(x - z), \qquad (6.220)$$

in the weak slowly varying external gluon field $A_\mu^n(x)$. We use the gauge condition (6.216) at $x_0 = z$ and restrict ourselves to the first term in the expansion (6.217). Eq.(6.220) can be solved iteratively. The solution, valid up to terms $\sim G_{\mu\nu}^2$ is given by [124]:

$$S(x, z) = \frac{i}{2\pi^2} \left\{ \frac{\slashed{u}}{u^4} + \frac{i}{2} \frac{m}{u^2} - \frac{1}{16} g\lambda^a G_{\alpha\beta}^a \varepsilon_{\alpha\beta\sigma\rho} \gamma_5 \gamma_\rho \frac{u_\sigma}{u^2} + \frac{i}{4} g\lambda^a G_{\lambda\beta}^a z_\beta \frac{u_\lambda \slashed{u}}{u^4} \right.$$
$$\left. - \frac{1}{384} g^2 \lambda^a \lambda^b G_{\mu\nu}^a G_{\mu\nu}^b \frac{\slashed{u}}{u^4} \left[z^2 u^2 - (zu)^2 \right] \right\}, \qquad (6.221)$$

where $u_\mu = x_\mu - z_\mu$, $\varepsilon_{0123} = 1$. (The quark mass is accounted for to the first order.) In practical applications, particularly in calculations of 3- and 4-point functions, it is convenient to use (6.221) expressed as an integral over momenta [125]:

$$S(x, z) = \frac{i}{(2\pi)^4} \int d^4 p\, e^{-ip(x-z)} \left\{ \frac{\slashed{p} + m}{p^2} - \frac{1}{4} g\lambda^a G_{\alpha\beta}^a \varepsilon_{\alpha\beta\sigma\rho} \gamma_5 \gamma_\rho \frac{p_\sigma}{p^4} \right.$$
$$+ \frac{1}{4} g\lambda^a G_{\alpha\beta}^a z_\beta (\gamma_\alpha p^2 - 2p_\alpha \slashed{p}) \frac{1}{p^4} + \frac{1}{192} g^2 \lambda^a \lambda^b G_{\mu\nu}^a G_{\mu\nu}^b$$
$$\left. \left[-z^2 p^2 \slashed{p} - 4 \slashed{p}(pz)^2 + 2 \slashed{z} p^2 (pz) \right] \frac{1}{p^6} \right\}. \qquad (6.222)$$

Taking account of the next term in the expansion (6.217) results in the addition of the term $S^{(1)}(p)$ to the quark propagator. It is convenient to represent this term in momentum space and for our further purposes it is enough to put $z = x_0 = 0$. (The reader can easily get the general expression, corresponding to $z \neq 0$.) In the limit of massless quarks, $S^{(1)}(p)$ is given by

$$S^{(1)}(p) = -\frac{i}{3} g \frac{1}{p^8} \lambda^n (D_\alpha G_{\nu\mu})^n \left\{ p^2 [\delta_{\alpha\nu} p^2 + \slashed{p}(\gamma_\alpha p_\nu + \gamma_\nu p_\alpha) - 4 p_\alpha p_\nu p^2 \right\} \gamma_\mu.$$
$$(6.223)$$

The contributions of the next order term to the quark propagator in the expansion (6.217) was calculated in [126] and reproduced in [119]. In a similar way, the gluon Green function (gluon propagator) $S_{\mu\nu}^{kp}(x, y)$ in the external gluon field can be found. The equation for $S_{\mu\nu}^{kp}(x, y)$ is given by

$$D_\lambda^{nm}(x) D_\lambda^{mk}(x) S_{\mu\nu}^{kp}(x, z) = i\delta^{np}\delta^4(x - z), \tag{6.224}$$

where D_λ^{nm} is defined by (6.214). The solution of (6.224) up to dimension $d = 3$ gluonic operators has the form $(x - z = u)$:

$$\begin{aligned}
S_{\mu\nu}^{np}(u, z) = -\frac{i}{(2\pi)^4} \int \frac{d^4k}{k^2} e^{-iku} \Big\{ \delta_{\mu\nu}\delta^{np} - g\delta_{\mu\nu} f^{npl} \frac{1}{k^2} \Big[-ik_\lambda z_\alpha G_{\alpha\lambda}^l \\
- \frac{2}{3}\Big(z_\beta z_\alpha k_\lambda - i\frac{z_\beta}{k^2}(k^2\delta_{\alpha\lambda} - 2k_\alpha k_\lambda)\Big)(D_\alpha G_{\beta\lambda})^l + \frac{1}{3}\Big(z_\alpha + 2i\frac{k_\alpha}{k^2}\Big)(D_\lambda G_{\alpha\lambda})^l \Big] \\
- 2g f^{npl} \frac{1}{k^2}\Big[G_{\mu\nu}^l + \Big(z_\alpha + 2i\frac{k_\alpha}{k^2}\Big)(D_\alpha G_{\mu\nu})^l \Big] \Big\}.
\end{aligned} \tag{6.225}$$

Equations (6.221)–(6.223) allow one to find the OPE terms which contain vacuum gluon field strengths up to the second order. The vacuum averaging of gluon field bilinears is performed according to

$$\langle 0 \mid G_{\mu\nu}^a G_{\lambda\sigma}^b \mid 0 \rangle = \frac{1}{96}\delta^{ab}(\delta_{\mu\lambda}\delta_{\nu\sigma} - \delta_{\mu\sigma}\delta_{\nu\lambda})\langle 0 \mid G_{\mu\nu}^a G_{\mu\nu}^a \mid 0 \rangle. \tag{6.226}$$

In order to calculate of the contributions of the $d \geq 5$ operators it is necessary to take account of the x dependence of the vacuum averages $\langle 0 \mid q_\alpha^i(x)\bar{q}\,_\beta^k(0) \mid 0 \rangle$, $q = u, d, s$. Up to dimension 7 we have $(i, k = 1, 2, 3)$:

$$\begin{aligned}
\langle 0 \mid q_\alpha^i(x)\bar{q}\,_\beta^k(0) \mid 0 \rangle = -\frac{1}{12}\delta^{ik}(\delta_{\alpha\beta} - \frac{i}{4}m_q \not{x}_{\alpha\beta}\langle 0 \mid \bar{q}(0)q(0) \mid 0 \rangle \\
+ \frac{1}{2^6 \cdot 3}\delta^{ik}\delta_{\alpha\beta}x^2 g \Big\langle 0 \Big| \bar{q}\sigma_{\mu\nu}\frac{\lambda^n}{2} G_{\mu\nu}^n q \Big| 0 \Big\rangle \\
- i\frac{x^2 \not{x}_{\alpha\beta}\delta^{ik}}{2^3 \cdot 3^5} g^2\langle 0 \mid \bar{q}q \mid 0 \rangle^2 - \frac{x^4}{2^9 \cdot 3^3}\delta_{\alpha\beta}\delta^{ik}\langle 0 \mid g^2G^2 \mid 0 \rangle\langle 0 \mid \bar{q}q \mid 0 \rangle. \tag{6.227}
\end{aligned}$$

The term linear in the quark mass was retained in the first term in (6.227). In the last two terms of (6.227) the factorization hypothesis was exploited. Within the framework of the factorization hypothesis one allows to reduce the vacuum expectation values of four-quark operators to the quark condensate:

$$\langle 0 \mid \bar{q}\,_\alpha^i\bar{q}\,_\beta^k q_\gamma^l q_\delta^m \mid 0 \rangle = \frac{1}{144}\langle 0 \mid \bar{q}q \mid 0 \rangle^2(\delta^{im}\delta^{kl}\delta_{\alpha\delta}\delta_{\beta\gamma} - \delta^{il}\delta^{km}\delta_{\alpha\gamma}\delta_{\beta\delta}),$$

$$i, k, l = 1, 2, 3; \quad \alpha, \beta, \gamma, \delta = 0, 1, 2, 3. \tag{6.228}$$

This equation is widely exploited in the derivation of sum rules. Below are several relations useful for calculations of the OPE terms:

$$\langle 0 \mid q_\alpha^i\bar{q}\,_\beta^k G_{\lambda\sigma}^n \mid 0 \rangle = -\frac{1}{2^6 \cdot 3}(\sigma_{\lambda\sigma})_{\alpha\beta}\frac{(\lambda^n)^{ik}}{2}\Big\langle 0 \Big| \bar{q}\sigma_{\mu\nu}\frac{\lambda^m}{2} q G_{\mu\nu}^m \Big| 0 \Big\rangle, \tag{6.229}$$

$$\langle 0 \mid q_\alpha^i \bar{q}_\beta^{\ k} (D_\rho G_{\mu\nu})^n \mid 0 \rangle = \frac{-g}{2^5 \cdot 3^3} \langle 0 \mid \bar{q}q \mid 0 \rangle^2 (\lambda^n)^{ik} (\delta_{\rho\nu}\gamma_\mu - \delta_{\rho\mu}\gamma_\nu)_{\alpha\beta}, \qquad (6.230)$$

$$\langle 0 \mid \bar{q}_\alpha^{\ k} (\nabla_\sigma \nabla_\rho \nabla_\tau q_\beta)^i \mid 0 \rangle = -i \frac{g^2}{2^4 \cdot 3^5} \delta^{ik} \langle 0 \mid \bar{q}q \mid 0 \rangle^2 (\gamma_\sigma \delta_{\rho\tau} + \gamma_\tau \delta_{\rho\sigma} - 5\gamma_\rho \delta_{\sigma\tau}), \tag{6.231}$$

$$\langle 0 \mid \bar{q}_\alpha^{\ k} (\nabla_\sigma q_\beta)^i G_{\rho\mu}^n \mid 0 \rangle = \frac{g}{2^6 \cdot 3^3} \langle 0 \mid \bar{q}q \mid 0 \rangle^2 (\lambda^n)^{ik} (\delta_{\sigma\rho}\gamma_\mu - \delta_{\sigma\mu}\gamma_\rho - i\varepsilon_{\sigma\rho\mu\xi}\gamma_5\gamma_\xi)_{\beta\alpha},$$

$$\varepsilon_{0123} = 1. \tag{6.232}$$

The derivation of (6.229) is evident: it is enough to multiply (6.229) by $(\sigma_{\lambda\sigma})_{\beta\alpha}(\lambda^n)^{ki}$ and sum over α, β, k, i. In order to derive (6.230), note that the general expression of the left-hand side following from Lorentz and colour invariance is

$$\langle 0 \mid q_\alpha^i \bar{q}_\beta^{\ k} (D_\rho G_{\mu\nu})^n \mid 0 \rangle = A(\lambda^n)^{ik} (\delta_{\rho\nu}\gamma_\mu - \delta_{\rho\mu}\gamma_\nu)_{\alpha\beta}. \tag{6.233}$$

(The term proportional to $\varepsilon_{\rho\mu\nu\sigma}\gamma_5\gamma_\sigma$ cannot appear in (6.233) because of the Bianchi identity $D_\rho G_{\mu\nu} + D_\nu G_{\rho\mu} + D_\nu G_{\rho\mu} + D_\mu G_{\nu\rho} = 0$.) Multiply (6.233) by $\delta_{\rho\mu}(\gamma_\nu)_{\beta\alpha}(\lambda^n)^{ki}$, sum over all repeated indices, and take into account the equation of motion (6.212). As a result we obtain

$$A = \frac{g}{2^9 \cdot 3} \left\langle 0 \left| (\bar{q}\gamma_\mu \lambda^n q) \cdot \left(\sum_{q'} \bar{\psi}_{q'} \gamma_\mu \lambda^n \psi_{q'} \right) \right| 0 \right\rangle = -\frac{g \langle 0 \mid \bar{q}q \mid 0 \rangle^2}{2^5 \cdot 3^3} \tag{6.234}$$

because of (6.228). The substitution of (6.234) into (6.233) gives (6.230). Equations (6.231) and (6.232) are derived in Problems 6.1 and 6.2.

To demonstrate how this technique really works, let us calculate the OPE series for the polarization operator of vector current (6.110) $\Pi_{\mu\nu}(q)$, used in Section 6.5.3 for checking QCD at low energies and in Section 6.6.1 for constructing QCD sum rules for vector mesons. The general form of $\Pi_{\mu\nu}(q)$ is

$$\Pi_{\mu\nu}(q) = \int d^4x e^{iqx} \Pi_{\mu\nu}(x) = (q_\mu q_\nu - q^2 \delta_{\mu\nu}) \Pi(q^2), \tag{6.235}$$

$$\Pi_{\mu\nu}(x) = i \langle 0 \mid T\{j_\mu(x), j_\nu^+(0)\} \mid 0 \rangle, \tag{6.236}$$

where $j_\mu(x)$ is given by (6.110). We ignore here the perturbative corrections and calculate OPE contributions up to operators of dimension 6. The first terms of OPE correspond to the bare-loop contribution, $d = 0$ (Fig. 6.21), and to the gluon condensate contribution, $d = 4$ (Fig. 6.22).

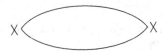

Fig. 6.21. The bare-loop contribution to $\Pi_{\mu\nu}(q)$. The solid lines correspond to quark propagators, the crosses to the external vector currents.

Fig. 6.22. The diagram describes the gluon condensate contribution to $\Pi_{\mu\nu}(q)$. The wavy lines correspond to interaction of quarks with the background gluon fields $G_{\mu\nu}$, the dots, surrounded by the dashed line, mean the vacuum averaging of gluon fields. All other notations are as in Fig. 6.21.

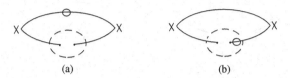

Fig. 6.23. The contribution of the operator $m\bar{\psi}\psi$ to $\Pi_{\mu\nu}$. The circle corresponds to a mass insertion, the dots surrounded by a dashed line mean the quark in the condensate phase. (a) Mass insertion in the free propagator; (b) accounting for the quark mass in the expansion of $\langle 0 \mid q(x), \bar{q}(0) \mid 0 \rangle$ in x.

We restrict ourselves to the terms linear in the quark mass m. In coordinate space, the part of the polarization operator $\Pi_{\mu\nu}(x)$ corresponding to Figs. 6.21 and 6.22 is given by

$$\Pi^{(1)}_{\mu\nu}(x) = -i \operatorname{Tr} \left\{ S(x)\gamma_\nu S(-x)\gamma_\mu \right\}, \tag{6.237}$$

where the quark propagator $S(x)$ is given by the first and third terms in (6.221). (We can put $z = 0$). The simple calculation gives

$$\Pi^{(1)}_{\mu\nu}(x) = \frac{i}{\pi^4} \left[-\frac{3}{x^8}(2x_\mu x_\nu - x^2)\delta_{\mu\nu} - \frac{g^2}{384}\frac{1}{x^4}(2x_\mu x_\nu + x^2\delta_{\mu\nu})\langle 0 \mid G^2 \mid 0 \rangle \right], \tag{6.238}$$

or in momentum space:

$$\Pi^{(1)}_{\mu\nu}(q) = (q_\mu q_\nu - q^2\delta_{\mu\nu})\frac{1}{4\pi^2} \left[-\ln(-q^2) + \frac{\pi^2}{3}\frac{1}{q^4}\left\langle 0 \left| \frac{\alpha_s}{\pi}G^2_{\mu\nu} \right| 0 \right\rangle \right]. \tag{6.239}$$

The other dimension-4 operator is $m\bar{\psi}\psi$ ($m = m_u, m_d$; $\psi = u, d$). There are two types of diagrams corresponding to its contributions to $\Pi_{\mu\nu}$. In the first (Fig. 6.23a), the quark mass insertion is accounted for in one of the quark propagators (the second term in (6.221)), while for the other propagator, the expression in terms of the quark condensate (the first term in (6.227)) is substituted. In the diagrams of the second type, the term proportional to m in the expansion of $\langle 0 \mid q(x), \bar{q}(0) \mid 0 \rangle$ in x (the second term in (6.227)) is taken into account in one of the quark propagators, while the other quark propagator is free.

In momentum space, the complete expression is

$$\Pi^{(2)}_{\mu\nu}(q) = (q_\mu q_\nu - q^2\delta_{\mu\nu})\frac{1}{q^4}(m_u + m_d)\langle 0 \mid \bar{u}u \mid 0 \rangle. \tag{6.240}$$

We have put $\langle 0 \mid \bar{u}u \mid 0 \rangle = \langle 0 \mid \bar{d}d \mid 0 \rangle$).

Turn now to the calculation of $d = 6$ operator contributions proportional to the vacuum averages of four-quark fields. There are three types of these. In the first one, all quarks

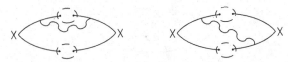

Fig. 6.24. The contribution of the first type of the operator $\alpha_s(\bar{\psi}\psi)^2$ to $\Pi_{\mu\nu}$ (see text). The wavy lines correspond to perturbative gluon propagators, the other notations are the same as in Figs. 6.21–6.23.

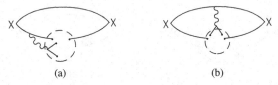

(a)   (b)

Fig. 6.25. The contributions of the operator $\alpha_s(\bar{\psi}\psi)^2$ of the second (a) and third types (b) to $\Pi_{\mu\nu}$. The quark lines enclosed in dashed circles mean the quark pairs produced in vacuum gluon background fields.

are in the condensate phase and the high momentum exchange between the quark pairs is realized by a hard gluon (Fig. 6.24). The calculation of the Fig. 6.24 contribution is a standard perturbative calculation with only one difference, that the vacuum average of quark fields is performed according to (6.227). Within the framework of the factorization hypothesis the result is:

$$\Pi_{\mu\nu}^{(3.1)}(q) = -\frac{32\pi}{9}\frac{\alpha_s}{Q^6}(q_\mu q_\nu - q^2\delta_{\mu\nu})\langle 0 \mid \bar{u}u \mid 0\rangle^2. \tag{6.241}$$

The second contribution arises from the expansion in x of $\langle 0 \mid q(x)q(0) \mid 0\rangle$ and is represented by the diagram Fig. 6.25a. Its calculation is straightforward: the vacuum average of one quark propagator is replaced by the term proportional to $\langle 0 \mid \bar{q}q \mid 0\rangle^2$ in (6.227), the other one by the free-quark propagator. In coordinate space, we get:

$$\Pi_{\mu\nu}^{(3.2)}(x) = i\frac{g^2}{\pi^2}\frac{1}{2\cdot 3^4}\frac{1}{x^2}\langle 0 \mid \bar{q}q \mid 0\rangle^2(2x_\mu x_\nu - x^2\delta_{\mu\nu}), \tag{6.242}$$

and in momentum space:

$$\Pi_{\mu\nu}^{(3.2)}(q) = g^2\frac{8}{3^4 Q^6}(4q_\mu q_\nu - q^2\delta_{\mu\nu})\langle 0 \mid \bar{q}q \mid 0\rangle^2. \tag{6.243}$$

In order to find the third contribution, see Fig. 6.25b. One of the quark lines in the diagram should be replaced by (6.223) and the vacuum averaging should be performed according to Eq. (6.230). The result is

$$\Pi_{\mu\nu}^{(3.3)}(q) = -\frac{8g^2}{3^4 Q^6}(2q_\mu q_\nu + q^2\delta_{\mu\nu})\langle 0 \mid \bar{q}q \mid 0\rangle^2. \tag{6.244}$$

The sum of terms proportional to $\alpha_s\langle 0 \mid \bar{q}q \mid 0\rangle^2$ is given by

$$\Pi_{\mu\nu}^{(3)}(q) = -\frac{32\pi}{9}\left(1 - \frac{2}{9}\right)(q_\mu q_\nu - q^2\delta_{\mu\nu})\frac{1}{Q^6}\alpha_s\langle 0 \mid \bar{q}q \mid 0\rangle^2. \tag{6.245}$$

Equations (6.239), (6.240), (6.245) give the terms of OPE up to $d = 6$ for the vector current polarization operator used in Sections 6.5.3 and 6.6.1.

6.9 Static properties of hadrons

6.9.1 The general approach. Spectral representations of the polarization operators in constant external fields

Consider hadrons in a constant external field F. F can be the constant electromagnetic field $F_{\mu\nu}$ (the electric charge $e = \sqrt{4\pi\alpha_{em}}$ is included into $F_{\mu\nu}$), the constant axial potential A_μ, etc. In this case, the term

$$L' = \int d^4x \, j^{ext}(x) F \qquad (6.246)$$

is added to the QCD Lagrangian. Here j^{ext} is the quark (or gluon) current corresponding to interaction with the external field. As before, calculate the polarization operator $\Pi(p^2)$ using perturbation theory and OPE. Retain only the terms linear in F. Then

$$\Pi(p^2) = i^2 \int d^4x \, e^{ipx} \left\langle 0 \left| T \left\{ \eta(x), \int d^4z \, j^{ext}(z), \bar{\eta}(0) \right\} \right| 0 \right\rangle F, \qquad (6.247)$$

where $\eta, \bar{\eta}$ are currents with the quantum numbers of the hadron whose interaction vertex with the external field F, $\Gamma(p^2) \equiv \Gamma(p^2, p^2; 0)$, $\Pi = \Gamma F$, we would like to determine. On the other hand, using a dispersion relation, $\Pi(p^2)$ is represented through contributions of the hadronic states. Among these, the contribution of the lowest state in the given channel is of interest and must be separated. The desirable hadron interaction vertex with constant external field, which is of our interest, can be found if the contributions of excited states are reliably estimated and are small.

When carrying out this program it is essential to represent correctly hadronic contributions to $\Pi(p^2)$, or, equivalently, to $\Gamma(p^2)$, accounting for all possible terms.

In order to get the dispersion representation of the polarization operator in the external field $\Pi(p^2)$ or the vertex function at zero momentum transfer $\Gamma(p^2, p^2, 0)$, it is convenient to start from the case when the momentum transfer $q = p_2 - p_1$ is nonzero, but q^2 is small and negative. (All the following refers to the coefficient function at any Lorentz tensor structure.) The general double-dispersion representation of $\Gamma(p_1^2, p_2^2, q^2)$ in variables p_1^2, p_2^2 at fixed $q^2 < 0$ has the form

$$\Gamma(p_1^2, p_2^2, q^2) = \int\limits_0^\infty \int\limits_0^\infty \frac{\rho(s_1, s_2, q^2)}{(s_1 - p_1^2)(s_2 - p_2^2)} ds_1 ds_2 + P(p_1^2) f_1(p_2^2, q^2) + P(p_2^2) f_2(p_1^2, q^2), \qquad (6.248)$$

where $P(p^2)$ is a polynomial. (It can be shown that at $q^2 < 0$ there are no anomalous thresholds.) For simplicity, assume that the currents η and $\bar{\eta}$ are Hermitian conjugate of each other corresponding to the diagonal matrix element between hadronic states. In this case, $\rho(s_1, s_2, q^2)$ is symmetric in s_1, s_2 and the functions f are the same in the second and third terms on the right-hand side of (6.248). The second and third terms on the right-hand side of (6.248) play the role of subtraction functions in the double-dispersion relation. The function $f(p^2, q^2)$ can be represented by a single-variable dispersion relation in p^2.

It is clear that the dispersion representation (6.248) holds also in the limit $q \to 0$, $p_2^2 \to p_1^2 \equiv p^2$, where $\Gamma(p^2) \equiv \Gamma(p^2, p^2, 0)$ is a function of one variable p^2. At first sight, it seems that a single-variable dispersion relation can be written for $\Gamma(p^2)$. Indeed, decomposing the denominator in (6.248), we can write

$$\int_0^\infty ds_1 \int_0^\infty ds_2 \frac{\rho(s_1, s_2)}{(s_1 - p^2)(s_2 - p^2)} = \int \int \frac{\rho(s_1, s_2)}{s_1 - s_2} ds_1 ds_2 \left(\frac{1}{s_2 - p^2} - \frac{1}{s_1 - p^2} \right).$$

$$(6.249)$$

In the first (second) term on the right-hand side of (6.249), the integration over $s_1(s_2)$ can be performed and the result has the form of single-variable dispersion relation. Such transformation is, however, misleading because, in general, the integrals

$$\int ds_1 \frac{\rho(s_1, s_2)}{s_1 - s_2} = - \int ds_2 \frac{\rho(s_1, s_2)}{s_1 - s_2} \tag{6.250}$$

are ultraviolet divergent. This ultraviolet divergence cannot be cured by subtractions in the single-variable dispersion relation: only the subtractions in the double-dispersion representation (6.248) may be used. It is evident that the procedures, eliminating the subtraction terms and leading to fast-converging dispersion integrals in standard single-variable dispersion representations, such as the Borel transformation in p^2, do not help here.

Let us consider two examples. The first one corresponds to the determination of the proton magnetic moment. In this case, $\eta(x)$ is given by (6.156) and the current j^{ext} is equal to

$$\frac{1}{2} j_\mu^{em}(x) x_\nu, \tag{6.251}$$

where j_μ^{em} is the electromagnetic quark current. In (6.251), the fixed-point gauge $x_\mu A_\mu^{em}(x){=}0$ was chosen for the electromagnetic potential and the equation

$$A_\mu^{em}(x) = \frac{1}{2} x_\nu F_{\nu\mu} \tag{6.252}$$

was used. The simple loop diagram – the contribution for the lowest-dimension operator $F_{\mu\nu}$ is shown in Fig. 6.26. It is clear that the spectral function $\rho(s_1, s_2)$ in (6.248) corresponding to the diagram of Fig. 6.26 is proportional to $\delta(s_1 - s_2)$.

Fig. 6.26. The bare-loop diagram, corresponding to the determination of nucleon magnetic moments. The solid lines correspond to quark propagators, the crosses mean the action of currents $\eta, \bar\eta$, and the bubble corresponds to quark interaction with the external field $F_{\mu\nu}$.

The separation of the chirality conserving structure (proportional to $\not{p}$) in $\Pi(p^2)$ results in the statement that the dimension of ρ is equal to 2 (see Sec. 6.9.2). So the general form of $\rho(s_1, s_2)$ in (6.248) is

$$\rho(s_1, s_2) = a s_1 s_2 \delta(s_1 - s_2), \tag{6.253}$$

where a is a constant. The substitution of (6.253) into (6.248) gives for the first term on the right-hand side of (6.248) at $p_1^2 = p^2 \equiv p^2$

$$\Gamma(p^2) = a \int_0^\infty \frac{s_1^2 ds_1}{(s_1 - p^2)^2}. \tag{6.254}$$

In this simple example, the dispersion representation is reduced to a single-variable dispersion relation, but with the square of $(s_1 - p^2)$ in the denominator. Of course, by integrating by parts, (6.254) may be transformed to the standard dispersion representation. However, the boundary term arising at such transformation must be accounted for; it does not vanish even after the Borel transformation was applied. This means that, even in this simplest case, the representation (6.248) is not equivalent to a single-variable dispersion relation.

The second example corresponds to the determination of the twist 4 correction to the Bjorken sum rule for polarized deep inelastic electron–nucleon scattering [127]. Here the external current j^{ext} in (6.246) is given by

$$U_\mu = -\frac{1}{2} \bar{q} g \varepsilon_{\mu\nu\lambda\sigma} G^a_{\lambda\sigma} \lambda^a q, \tag{6.255}$$

where $G^a_{\mu\nu}$ is the gluonic field strength tensor. An example of the bare-loop diagram is shown in Fig. 6.27. In this case, unlike the previous one, $\rho(s_1, s_2)$ is not proportional to $\delta(s_1 - s_2)$. This stems from the fact that in the discontinuity over p_1^2 at $q \neq 0$ and $p_1^2 \neq p_2^2$ only the left-hand part of diagram Fig. 6.27 is touched and the loop integration on the right-hand part still persists. For the tensor structure selected in Ref.[127], $\rho(s_1, s_2)$ has dimension 4 and is proportional to $s_1 s_2$. We see that in this example the general form of the dispersion representation (6.248) must be used in the limit

$$q^2 \to 0, \quad p_1^2 \to p_2^2 = p^2. \tag{6.256}$$

Let us represent $\Gamma(p^2, p^2; 0)$ in terms of contributions of hadronic states, using (6.248) and separating the contribution of the lowest hadronic state in the channels with momentum p. Consider the first term on the right-hand side of (6.248) at $q^2 = 0$, $p_1^2 = p_2^2 = p^2$. As is seen from Fig. 6.28, it is convenient to divide the whole integration region in s_1, s_2 into three domains: (I) $0 < s_1 < s_0, 0 < s_2 < s_0$; (II) $0 < s_1 < s_0, s_0 < s_2 < \infty$; $s_0 < s_1 < \infty$,

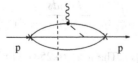

Fig. 6.27. The bare-loop diagram for the twist 4 correction to the Bjorken sum rule for deep inelastic electron–nucleon scattering. The dashed vertical line corresponds to the discontinuity over p_1^2 at $p_1^2 \neq p_2^2$.

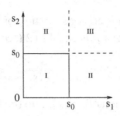

Fig. 6.28. The integration domains in s_1, s_2 plane.

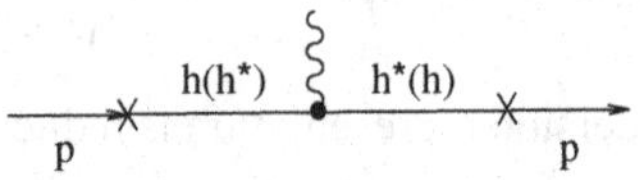

$$
\xrightarrow[\;p\;]{\times} \quad \overset{h(h^*)}{\text{---}} \; \bullet \; \overset{h^*(h)}{\text{---}} \quad \xrightarrow[\;p\;]{\times}
$$

Fig. 6.29. The schematical representation of $h \longrightarrow h^*$ ($h^* \longrightarrow h$) transitions in the external field.

$0 < s_2 < s_0$; and (III) $s_0 < s_1 < \infty$, $s_0 < s_2 < \infty$. Adopt the standard in QCD sum rule model of hadronic spectra: the lowest hadronic state plus continuum, starting from some threshold s_0. Then in domain I only the lowest hadronic state h contributes and

$$
\rho(s_1, s_2) = G\lambda^2 \delta(s_1 - m^2)\delta(s_2 - m^2), \tag{6.257}
$$

where m is the mass of this state, λ is the transition constant of the hadron in the current η. (For a baryon $\langle B|\bar{\eta}|0\rangle = \lambda_B \bar{v}_B$ where v_B is the baryon spinor.) G is the coupling constant of the hadron with external field, which we would like to determine from the sum rule. In domain III the higher-order terms in OPE may be neglected and the contribution of hadronic states is with good accuracy equal to the contribution of the bare-quark loop (like Figs. 6.26 or 6.27) with perturbative corrections. The further application of the Borel transformation in p^2 essentially suppresses this contribution.

The consideration of the domain II contribution is the most troublesome and requires an additional hypothesis. Assume, using duality arguments, that in this domain also the contribution of hadronic states is approximately equal to the contribution of the bare-quark loop. The accuracy of this approximation may be improved by subtracting the lowest hadronic state contributions proportional to $\delta(s_1 - m^2)$ or $\delta(s_2 - m^2)$ from each strip of domain II. The terms of the latter type also persist in the functions $f(p_1^2)$, $f(p_2^2)$ in (6.248). They correspond to the process when the current $\bar{\eta}$ produces hadron h from the vacuum, and under the action of the external current j the transition $h \to h^*$ to an excited state occurs or vice versa (Fig. 6.29). At $p_1^2 = p_2^2 = p^2$ these contributions have the form

$$
\int_{s_0}^{\infty} \frac{b(s)\,ds}{(p^2 - m^2)(s - p^2)} \tag{6.258}
$$

with some unknown function $b(s)$. The term (6.258) will be accounted for separately on the right-hand side of (6.248). The term (6.258) must be added to the right-hand side of (6.248) independently of the form of the bare-loop contribution $\rho(s_1, s_2)$. The term (6.258) may persist even if $\rho(s_1, s_2) = 0$, when the OPE for the vertex function $\Gamma(p^2, p^2, 0)$ with zero momentum transfer starts from the condensate terms. (6.258) may be written as

$$\int_{s_0}^{\infty} ds\, b(s) \left(\frac{1}{p^2 - m^2} + \frac{1}{s - p^2} \right) \frac{1}{s - m^2}. \tag{6.259}$$

The functions $f(p^2)$ in (6.248) can be represented by a dispersion relation as

$$f(p^2) = \int_0^{\infty} \frac{d(s)}{s - p^2} ds. \tag{6.260}$$

The integration domain in (6.260) can be also divided into two parts: $0 < s < s_0$ and $s_0 < s < \infty$. According to our model, the contribution of the first part is approximated by the h-state contribution, the second one by the continuum. These two parts look like the contributions of the first and the second terms in the bracket in (6.259).

Now we can formulate the recipe how the sum rule can be written [128]. On the phenomenological side – the right-hand side of the sum rule – there is the contribution of the lowest hadronic state h and the unknown term (6.259), corresponding to the nondiagonal transition $h \to h^*$ in the presence of an external field:

$$\underbrace{\frac{\lambda^2 G}{(p^2 - m^2)^2}}_{W^2} + \int_{}^{\infty} ds\, b(s) \frac{1}{s - m^2} \left(\frac{1}{p^2 - m^2} + \frac{\alpha(s)}{s - p^2} \right). \tag{6.261}$$

(The coefficient α reflects the possibility that in the function f the ratio of terms proportional to $(p^2 - m^2)^{-1}$ and $(s - p^2)^{-1}$ may differ from 1 as is the case in (6.259).) The continuum contribution, corresponding to the bare loop (or also to the higher-order terms in OPE if their discontinuity does not vanish at $s \to \infty$) is transferred to the left-hand side of the sum rule. Here it is cancelled by the bare-loop contribution from the same domain of integration. As a result, in the double-dispersion representation of the bare loop the domain of integration over s_1, s_2 is restricted to $0 < s_1, s_2 < s_0$. Finally, apply the Borel transformation in p^2 to both sides of the sum rule. On the left-hand side – the QCD side – the contribution of the bare loop has the form

$$\int_0^{s_0} ds_1 \int_0^{s_0} ds_2 \rho(s_1, s_2) \frac{1}{s_1 - s_2} \left[e^{-s_2/M^2} - e^{-s_1/M^2} \right]$$

$$= 2P \int_0^{s_0} ds_2 \int_0^{s_0} ds_1 \frac{\rho(s_1, s_2)}{s_1 - s_2} e^{-s_2/M^2}, \tag{6.262}$$

where P means the principal value and the symmetry of $\rho(s_1, s_2)$ was used. The right-hand side of the sum rule is equal to

$$G \frac{\lambda^2}{M^2} e^{-m^2/M^2} - A e^{-m^2/M^2} + e^{-m^2/M^2} \int_{s_0}^{\infty} ds\, b(s) \frac{\alpha(s)}{s - m^2} \exp\left[-(s - m^2)/M^2 \right], \tag{6.263}$$

where

$$A = \int_{S_0}^{\infty} ds \, \frac{b(s)}{s - m^2}. \tag{6.264}$$

Two remarks in connection with Eqs. (6.262), (6.263) are necessary. If the discontinuity $\rho(s_1, s_2)$ of the bare loop is proportional to $\delta(s_1 - s_2)$, $\rho(s_1, s_2) = \rho(s_1)\delta(s_1 - s_2)$, as in the diagram Fig. 6.26, then Eq. (6.262) reduces to

$$\frac{1}{M^2} \int_0^{s_0} ds_1 e^{-s_1/M^2} \rho(s_1). \tag{6.265}$$

In this case, at $s_0 \gg M^2$ the continuum contribution is suppressed exponentially and the dependence on the value of the continuum threshold s_0 is weak. If, however, $\rho(s_1, s_2)$ has no such form and is a polynomial in s_1, s_2, as in the diagram Fig. 6.27, then, as can be seen from (6.262), the bare-loop diagram contribution has a power-like dependence on s_0. In this case, the QCD sum rule calculation lost a part of its advantage in comparison with finite energy sum rules.

In the case when the double discontinuity of the bare-loop diagram $\rho(s_1, s_2)$ is proportional to $\delta(s_1 - s_2)$, this form will be absent in the radiation correction terms. Here (6.265) is invalid and the more general expression (6.262) must be used. This will result in the appearance of $\ln\left(s_0/(-p^2)\right)$ in the final answer; s_0 plays the role of the ultraviolet cut-off. The third term in (6.263) is suppressed in comparison with the second term by a factor smaller than $\exp[-(s_0 - m^2)/M^2] \leq 1/4$. (It is supposed that α is of order 1.) There-fore, if $A \ll G\lambda^2/M^2$, then this term can be safely neglected. The first term in (6.263) of interest to us can be determined by applying to the sum rule the differential operator $(\partial/\partial(1/M^2))e^{m^2/M^2}$, which kills the second term in (6.263). After the constant G is found in this way, the unknown parameter A can be determined by fitting the sum rule. If it will be found that $A \sim G\lambda^2/M^2$, then the contribution of the last term in (6.263), which has been up to now neglected, must be accounted for as an additional error in the final result.

6.9.2 Nucleon magnetic moments

Let us dwell now on the calculation of proton and neutron magnetic moments [18],[129],[130]. In accord with (6.247), the polarization operator of currents bearing nucleon quantum numbers in the weak constant electromagnetic field $F_{\mu\nu}$ has the form

$$i \int d^4x e^{ipx} \langle 0 \mid T\{\eta(x), \, \bar{\eta}(0)\} \mid 0 \rangle_F = \Pi^{(0)}(p) + \Pi_{\mu\nu}(p) F_{\mu\nu}, \tag{6.266}$$

where $\Pi^{(0)}(p)$ is the polarization operator when the field is absent. The current η_p with proton quantum numbers is given by (6.156); for the neutron the substitution $u \leftrightarrow d$ must be done. Now our interest is concentrated on the second term in (6.266).

Use OPE and classify the operator vacuum expectation values according to their dimen-sions d. Evidently, the operator of the lowest dimension $d = 2$ is $F_{\mu\nu}$ itself. The field-induced vacuum expectation value

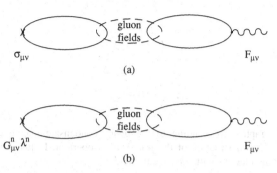

Fig. 6.30. The diagrams corresponding to quark pair mixing, $\bar{q}q \to \bar{q}'q'$, and resulting in deviations from Eq.(6.272): (a) the diagrams corresponding to Eq.(6.267); (b) the diagrams corresponding to Eqs. (6.269), (6.270).

$$\langle 0 \mid \bar{q}\sigma_{\mu\nu}q \mid 0\rangle_F = \chi_q \langle 0 \mid \bar{q}q \mid 0\rangle F_{\mu\nu} \qquad (6.267)$$

is of the next dimension, $d = 3$, and this is the reason for its importance in the problem under consideration. It can be shown that four-dimensional vacuum expectation values are absent [18]. There are three vacuum expectation values of dimension 5:

$$\langle 0 \mid \bar{q}q \mid 0\rangle F_{\mu\nu}, \qquad (6.268)$$

$$g\langle 0 \mid \bar{q}\frac{1}{2}\lambda^n G^n_{\mu\nu}q \mid 0\rangle_F = \kappa_q F_{\mu\nu} \langle 0 \mid \bar{q}q \mid 0\rangle, \qquad (6.269)$$

$$-ig\varepsilon_{\mu\nu\lambda\sigma}\left\langle 0 \left| \bar{q}\gamma_5\frac{1}{2}\lambda^n G^n_{\lambda\sigma}q \right| 0\right\rangle_F = \xi_q F_{\mu\nu}\langle 0 \mid \bar{q}q \mid 0\rangle. \qquad (6.270)$$

Among six-dimensional operators the vacuum average

$$\langle 0 \mid \bar{q}q \mid 0\rangle\langle 0 \mid \bar{q}\sigma_{\mu\nu}q \mid 0\rangle_F \qquad (6.271)$$

is accounted for (within the framework of the factorization hypothesis). All other $d = 6$ operators contain the gluon field strength tensor; their contributions are small, because they result in loop integrations and in the appearance of additional factors of $(2\pi)^{-2}$ or $(2\pi)^{-4}$. For this reason, the contributions of these operators are neglected. The contributions of higher-order operators will be discussed below.

The factors χ_q, κ_q and ξ_q ($q = u, d$) are proportional to quark charges with good accuracy:

$$\chi_q = e_q\chi, \quad \kappa_q = e_q\kappa, \quad \xi_q = e_q\xi, \qquad (6.272)$$

where χ, κ and ξ are flavour independent. (χ is called the quark condensate magnetic susceptibility.) The deviations from (6.272) are described by the diagrams of Fig. 6.30. Evidently, for massless quarks the diagrams of Fig. 6.30 with gluon exchange are zero in any order of perturbation theory because of chirality conservation. Chirality violation might appear due to instantons, but, as was shown in Section 6.6.3 (see also [100]) in the case of the vector current which acts in (6.267), (6.269), (6.270), the instantons (in the dilute-gas approximation) do not contribute to the diagrams of Fig. 6.30. The amplitude of Fig. 6.30 has a certain resemblance to that of $\varphi - \omega$ mixing. Therefore, the experimental smallness of $\varphi - \omega$ mixing is also an argument in favour of (6.272).

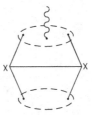

Fig. 6.31. The graph corresponding to the quark condensate magnetic susceptibility contribution to the odd structure of the polarization operator. Dotted lines encircle the nonperturbative interactions with the vacuum fields.

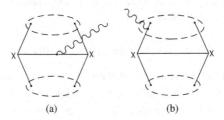

(a)   (b)

Fig. 6.32. The diagrams describing the contributions of vacuum expectation value $F_{\mu\nu}\langle 0 \mid \bar{q}q \mid 0 \rangle^2$ to the odd structure of polarization operator.

Turn now to the description of the terms in the operator expansion which will be taken into account in the calculation of $\Pi_{\mu\nu}(p)$ in QCD (the left-hand sides of the desired sum rules). Three different tensor structures contribute to $\Pi_{\mu\nu}(p)$: $\not{p}\sigma_{\mu\nu} + \sigma_{\mu\nu}\not{p}$, $i(p_{\mu}\gamma_{\nu} - p_{\nu}\gamma_{\mu})\not{p}$ and $\sigma_{\mu\nu}$. The first structure includes an odd number of γ-matrices and conserves chirality. The second and the third include an even number of γ-matrices and violate chirality.

Let us consider first the left-hand side of the sum rule at the odd structure $\not{p}\sigma_{\mu\nu} + \sigma_{\mu\nu}\not{p}$. The lowest-dimension operator contributing to this structure is $F_{\mu\nu}(d = 2)$ and the corresponding Feynman diagram is depicted in Fig. 6.26. Because of chirality conservation, the $d = 3$ and $d = 5$ operators do not contribute to the odd structure, so the next in dimension are six-dimensional operators. The contribution of the field-induced vacuum expectation value $\langle 0 \mid \bar{q}q \mid 0 \rangle \times \langle 0 \mid \bar{q}\sigma_{\mu\nu}q \mid 0 \rangle_F (d = 6)$ corresponds to diagram of Fig. 6.31.

In the sum rule at the odd structure $\not{p}\sigma_{\mu\nu} + \sigma_{\mu\nu}\not{p}$, we take into account the contributions of vacuum expectation values of eight-dimensional operators assuming the factorization hypothesis. There are four such vacuum expectation values, $F_{\mu\nu}\langle 0|\bar{q}q|0\rangle^2$, $-g\langle 0|\bar{q}\sigma_{\alpha\beta}(\frac{1}{2}\lambda^n)G^n_{\alpha\beta}q|0\rangle\langle 0|\bar{q}\sigma_{\mu\nu}q|0\rangle_F = e_q\chi m_0^2 F_{\mu\nu}\langle 0|\bar{q}q|0\rangle^2$, $e_q\kappa F_{\mu\nu}\langle 0|\bar{q}q|0\rangle^2$, and $e_q\xi F_{\mu\nu}\langle 0|\bar{q}q|0\rangle^2$. The corresponding diagrams are depicted in Figs. 6.32–6.34. The estimate of eight-dimensional terms is important. Indeed, as will be seen, the contribution of a six-dimensional term to the sum rules is rather large, even larger than the Fig. 6.26 contribution since the magnitude of the quark condensate magnetic susceptibility χ is large. The eight-dimensional terms (especially the term $\sim \chi m_0^2 \langle 0|\bar{q}q|0\rangle^2$) can be considered as the first correction to the six-dimensional term. Therefore, their account is necessary since it

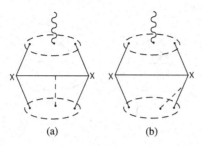

Fig. 6.33. The diagrams describing the contribution of vacuum expectation value $e_q \chi m_0^2 F_{\mu\nu} \langle 0 \mid \bar{q}q \mid 0 \rangle^2$.

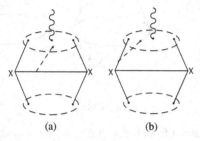

Fig. 6.34. The diagrams describing the contributions of vacuum expectation values $e_q \kappa F_{\mu\nu} \langle 0 \mid \bar{q}q \mid 0 \rangle^2$ and $e_q \xi F_{\mu\nu} \langle 0 \mid \bar{q}q \mid 0 \rangle^2$ to the odd structure of $\Pi_{\mu\nu}$.

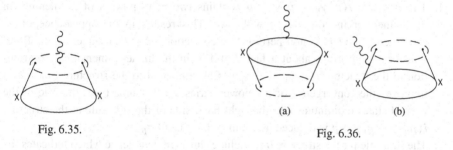

Fig. 6.35.   Fig. 6.36.

permits checking the convergence of the operator expansion series. Of course, the results will be convincing (and really they are) if contributions of eight-dimensional terms are much smaller than those of $d = 6$.

The lowest-dimension operator which contributes to chirality violating structures in $\Pi_{\mu\nu}(p)$ is $\langle 0|\bar{q}\sigma_{\mu\nu}q|0\rangle_F$ ($d = 3$); see Fig. 6.35. The next in dimension with $d = 5$ are the operator vacuum expectation values $\langle 0|\bar{q}q|0\rangle F_{\mu\nu}$, Fig. 6.36, and those of Eqs. (6.269), (6.270) corresponding to Fig. 6.37. Assuming factorization again we are left in even structures with two seven-dimensional operator vacuum expectation values $\langle 0|\bar{q}\sigma_{\mu\nu}q|0\rangle_F \langle 0|G_{\alpha\beta}^n G_{\alpha\beta}^n|0\rangle$ and $-g\langle 0|\bar{q}\sigma_{\alpha\beta}(\frac{1}{2}\lambda^n)G_{\alpha\beta}^n q|0\rangle F_{\mu\nu}$. The diagram corresponding to the first one is shown in Fig. 6.38. The diagram corresponding to the second one is evident.

As is seen, the interval of dimensions $d = 2 - 8$ for the odd structure is larger than the interval of dimensions $d = 3 - 7$ for even structures. Therefore, one may expect a better

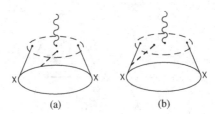

Fig. 6.37.

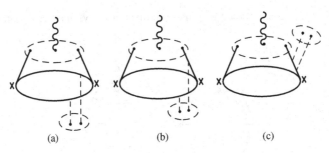

Fig. 6.38.

accuracy of the results obtained from the sum rule at odd structure. As for even structure, we shall use only the sum rule at the structure $i(p_\mu \gamma_\nu - p_\nu \gamma_\mu)\ \not{p}$. The reasons are the following:

1. The structure $i(p_\mu \gamma_\nu - p_\nu \gamma_\mu)\ \not{p}$ contains two extra powers of momentum in the numerator in comparison with $\sigma_{\mu\nu}$. This results in the appearance of an extra factor of $1/p^2$ in non-perturbative corrections. As a consequence, the Borel transformation brings about a factor of $1/n$ in the higher-dimensional contribution to the structure $i(p_\mu \gamma_\nu - p_\nu \gamma_\mu)\ \not{p}$ as compared to the structure $\sigma_{\mu\nu}$ which improves the convergence of the power series. At the same time, the role of the excited states (continuum) on the right-hand side of the sum rule at the structure $i(p_\mu \gamma_\nu - p_\nu \gamma_\mu)\ \not{p}$ is reduced as compared to that of $\sigma_{\mu\nu}$.
2. The sum rule for the structure $\sigma_{\mu\nu}$ includes infrared divergence which indicates the appearance of unknown infrared nonfactorizable vacuum expectation values.
3. Instantons contribute significantly to the sum rule for this structure, do not contribute to the chiral odd structure $\not{p}\sigma_{\mu\nu} + \sigma_{\mu\nu}\ \not{p}$, and give only a small contribution to the structure $i(p_\mu \gamma_\nu - p_\nu \gamma_\mu)\ \not{p}$ [131].

The presence of electromagnetic field results in the appearance of additional terms in Eq. (6.227) proportional to $F_{\mu\nu}$:

$$\langle 0 \mid T\{q_\alpha^i(x), \bar{q}_\beta^k(0)\} \mid 0 \rangle_F = i\delta^{ik} \frac{1}{32\pi^2 x^2} e_q F_{\mu\nu} (\not{x}\sigma_{\mu\nu} + \sigma_{\mu\nu}\ \not{x})_{\alpha\beta}$$

$$- \frac{1}{24}\delta^{ik}(\sigma_{\mu\nu})_{\alpha\beta}\langle 0 \mid \bar{q}\sigma_{\mu\nu}q \mid 0\rangle_F + \frac{1}{288}\delta^{ik}(\sigma_{\mu\nu})_{\alpha\beta}e_q(F_{\mu\nu}x^2 + 2F_{\mu\rho}x_\rho x_\nu)$$

$$+ \frac{1}{586}\delta^{ik}\langle 0 \mid \bar{q}q \mid 0\rangle e_q F_{\mu\nu}\left[\sigma_{\mu\nu}x^2(\kappa + \xi) - 2x_\rho x_\nu \sigma_{\rho\mu}\left(\kappa - \frac{1}{2}\xi\right)\right]. \tag{6.273}$$

The calculations give for the odd structure:

$$\Pi_{\mu\nu}(p)_{\text{odd}} = -\frac{1}{16\pi^4}(\sigma_{\mu\nu}\,\not{p}+\not{p}\sigma_{\mu\nu})\left\{\frac{1}{2}e_u p^2 \ln\frac{\Lambda^2}{-p^2} + \frac{1}{3}e_u\chi\frac{a_{\bar{q}q}^2}{p^2}\left(1+\frac{m_0^2}{8p^2}\right)\right.$$

$$\left. - \frac{a_{\bar{q}q}^2}{6p^4}\left[e_d + \frac{2}{3}e_u - \frac{1}{3}e_u(\kappa-2\xi)\right]\right\}$$ (6.274)

and for the even structure $i(p_\mu\gamma_\nu - p_\nu\gamma_\mu)\,\not{p}$:

$$\Pi_{\mu\nu}(p)_{\text{even}} = i\frac{a_{\bar{q}q}}{16\pi^4}(p_\mu\gamma_\nu - p_\nu\gamma_\mu)\,\not{p}\left\{\left(e_u + \frac{1}{2}e_d\right)\frac{1}{p^2}\right.$$

$$\left. - \frac{1}{3}e_d\chi\left[\ln\frac{\Lambda^2}{-p^2} + \frac{\pi^2}{6p^4}\left\langle 0\left|\frac{\alpha_s}{\pi}G_{\mu\nu}^2\right|0\right\rangle\right]\right\}.$$ (6.275)

Here Λ^2 is the ultraviolet cut-off. (The details of the calculation are presented in [18].)

The one-proton contribution to $\Pi_{\mu\nu}(p)$ is given by

$$\Pi_{\mu\nu}(p) = -\frac{1}{4}\frac{\lambda_N^2}{(p^2-m^2)^2}\left\{\mu_p(\sigma_{\mu\nu}\,\not{p}+\not{p}\sigma_{\mu\nu}) + 2\sigma_{\mu\nu}m\mu_p\right.$$

$$\left. + \sigma_{\mu\nu}\mu_p^a(p^2-m^2)/m + 2i\mu_p^a(p_\mu\gamma_\nu - p_\nu\gamma_\mu)\,\not{p}/m\right\},$$ (6.276)

where μ_p and μ_p^a are the proton total and anomalous magnetic moments in nuclear magneton units, λ_N is the proton transition amplitude in the current η_p as determined in Section 6.7.1. The term resulting from the proton $\to$ excited state transition corresponding to the second term in (6.263) is equal to

$$\Pi_{\mu\nu}(p)_{p\to N^*} = \frac{1}{4}\lambda_N^2\frac{1}{p^2-m^2}\left\{A_p(\sigma_{\mu\nu}\,\not{p}+\not{p}\sigma_{\mu\nu})\right.$$

$$\left. + 2A_p\sigma_{\mu\nu}m + 2iB_p(p_\mu\gamma_\nu - p_\nu\gamma_\mu)\,\not{p}/m\right\}.$$ (6.277)

Perform the Borel transformation of the sum rules and neglect the last term in the phenomenological part of the sum rules (6.263). The sum rules for the invariant functions at the structures $\sigma_{\mu\nu}\,\not{p}+\not{p}\sigma_{\mu\nu}$ and $i(p_\mu\gamma_\nu - p_\nu\gamma_\mu)\,\not{p}$ taking account of anomalous dimensions are:

$$e_u M^4 E_2(M)L^{-4/9} + \frac{a_{\bar{q}q}^2}{3M^2}L^{4/9}\left[-\left(e_d + \frac{2}{3}e_u\right) + \frac{1}{3}e_u(\kappa-2\xi)\right.$$

$$\left. - 2e_u\chi\left(M^2 L^{-16/27} - \frac{1}{8}m_0^2 L^{-10/9}\right)\right] = \frac{1}{4}\tilde{\lambda}_N^2 e^{-m^2/M^2}\left(\frac{\mu_p}{M^2} + A_p\right),$$ (6.278)

$$ma_{\bar{q}q}\left\{e_u + \frac{1}{2}e_d + \frac{1}{3}e_d\chi M^2\left[E_1(M) + \frac{b}{24M^4}\right]L^{-16/27}\right\}$$

$$= \frac{1}{4}\tilde{\lambda}_N^2 e^{-m^2/M^2}\left(\frac{\mu_p^a}{M^2} + B_p\right).$$ (6.279)

(We use the notations of Section 6.7.1.) The anomalous dimensions of κ and ξ are unknown. The continuum contribution is represented by the double-dispersion relation (6.248) which in this case reduces to the form given by (6.254). The sum rules for the neutron are obtained from Eqs. (6.278), (6.279) by substitution: $e_u \leftrightarrow e_d, \mu_p, \mu_p^a \to \mu_n$, $A_p, B_p \to A_n, B_n$.

In order to get rid of the constants χ, κ and ξ, let us multiply the sum rule (6.278) for the proton by e_d and for the neutron by e_u and subtract one from the other. Similarly, multiply the sum rule (6.279) for the proton by e_u and for the neutron by e_d and subtract one from the other. The resulting expressions can be represented in the form of

$$\mu_p e_d - \mu_n e_u + M^2 (A_p e_d - A_n e_u) = \frac{4a_{\bar{q}q}^2}{3\tilde{\lambda}_N^2} e^{m^2/M^2} (e_u^2 - e_d^2) L^{4/9}, \tag{6.280}$$

$$\mu_p^a e_u - \mu_n e_d + M^2 (B_p e_u - B_n e_d) = \frac{4a_{\bar{q}q} m M^2}{\tilde{\lambda}_N^2} e^{m^2/M^2} (e_u^2 - e_d^2). \tag{6.281}$$

In order to eliminate the unknown single-pole contributions still remaining on the left-hand side of Eqs. (6.280), (6.281) we apply the differential operator $1 - M^2 \partial/\partial M^2$ to these equations and obtain

$$\mu_p e_d - \mu_n e_u = \frac{4a_{\bar{q}q}^2}{3\tilde{\lambda}_N^2} (e_u^2 - e_d^2) \left(1 - M^2 \frac{\partial}{\partial M^2}\right) e^{m^2/M^2} L^{4/9}, \tag{6.282}$$

$$\mu_p e_u - \mu_n e_d = e_u + \frac{4a_{\bar{q}q} m}{\tilde{\lambda}_N^2} (e_u^2 - e_d^2) \left(1 - M^2 \frac{\partial}{\partial M^2}\right) M^2 e^{m^2/M^2}. \tag{6.283}$$

The magnetic moments μ_p and μ_n can be approximately determined by setting $M = m$, disregarding anomalous dimensions and substituting for the residue $\tilde{\lambda}_N^2$ the value

$$\tilde{\lambda}_N^2 = \frac{2a_{\bar{q}q} M^4}{m} e^{m^2/M^2} \bigg|_{M^2 = m^2}, \tag{6.284}$$

which follows from the mass sum rules (6.161) neglecting both α-corrections and continuum contributions. Solving in this approximation Eqs. (6.282), (6.283) we arrive at the elegant results:

$$\mu_p = \frac{8}{3}\left(1 + \frac{1}{6}\frac{a_{\bar{q}q}}{m^3}\right), \tag{6.285}$$

$$\mu_n = -\frac{4}{3}\left(1 + \frac{2}{3}\frac{a_{\bar{q}q}}{m^3}\right). \tag{6.286}$$

Numerically, at $a_{\bar{q}q} = 0.65 \,\text{GeV}^3$ (see Table 1 and Section 6.7.1) we get from (6.285), (6.286) $\mu_p = 3.01, \mu_n = -2.03$ in comparison with the experimental values $\mu_p = 2.79, \mu_n = -1.91$. In a more rigorous treatment, the study of the M^2-dependence of Eqs.

(6.282), (6.283) in the confidence interval $0.9 < M^2 < 1.3\,\mathrm{GeV}^2$ gives as the best fit the values

$$\mu_p = 2.7, \quad \mu_n = -1.7 \tag{6.287}$$

with an estimated error of about 10%. In the fit we have used the value $\tilde{\lambda}_N^2 = 3.2 \pm 0.6\,\mathrm{GeV}^6$ found in Section 6.7.1. When performing the differentiation in (6.282), (6.283) the anomalous dimension factor $L^{4/9}$ was considered to be a constant. The reason is that the deviations of $L^{4/9}$ from a constant are of order of α_s-corrections, which were disregarded. The main sources of errors are the neglected α_s-corrections and the error in the magnitude of $\tilde{\lambda}_N^2$. In (6.274), (6.275), the gluon condensate contribution was accounted for only in the case when it was multiplied by the numerically large factor χ. The calculations of gluon condensate contributions, in all other cases, show that its influence on the values of magnetic moments is completely negligible, less than 1% [132].

The constants A_p, A_n, B_p, B_n can be determined from (6.278)–(6.281) by substituting into these equations the values of μ_p and μ_n shown in (6.287). Eq. (6.280) determines the combination $A_p + 2A_n = 0.27\,\mathrm{GeV}^{-2}$, which is to be compared with $(\mu_p + 2\mu_n)/M^2 \approx 0.7\,\mathrm{GeV}^{-2}$. As noted in Section 6.9.1, the last neglected term in (6.263) is in this case less than $(A_p + 2A_n)e^{-(s_0-m^2)/M^2} \sim 0.027\,\mathrm{GeV}^{-2}$, i.e. it is negligible compared with $(\mu_p + 2\mu_n)/M^2$. From (6.281), we get $2B_p + B_n = 0.15\,\mathrm{GeV}^{-2}$ in comparison with $(2\mu_p^a + \mu_n)/M^2 \approx 1.7\,\mathrm{GeV}^{-2}$. The last term in (6.263) is also negligible. The values of κ and ξ were calculated by studying special sum rules and found to be $\kappa = -0.34 \pm 0.1$, $\xi = -0.74 \pm 0.2$ [133]. However, this determination is less certain than the determination of χ because of higher dimensions of the corresponding operators.

A remark about the role of instantons in the sum rules for nucleon magnetic moments. In Ref. [131], it was demonstrated that in the dilute instanton gas approximation, instantons do not contribute to the sum rules at the structures $\not{p}\sigma_{\mu\nu}+\sigma_{\mu\nu}\not{p}$ and $i(p_\mu\gamma_\nu-p_\nu\gamma_\mu)\not{p}$, considered above. However, their contributions to the sum rule at the structure $\sigma_{\mu\nu}$ are large for the standard instanton gas parameters. The consideration of this sum rule with account of instantons, performed in [131], leads to the values of nucleon magnetic moments, which are in agreement with those found above.

6.9.3 Hyperon magnetic moments

The same technique is applied to calculate the hyperon magnetic moments. The only difference with the case of the nucleon is the necessity to account for terms proportional to the strange quark mass m_s. The contribution of these terms is of order of 15–20% in the values of the Σ and Ξ magnetic moments. As in the case of the nucleon, it is possible to find such combinations of the sum rules, in which the unknown susceptibilities χ, κ and ξ are excluded. These sum rules have the following form [134]:

For Σ^+, Σ^- hyperons:

$$2\mu_{\Sigma^-} + \mu_{\Sigma^+} = \frac{m}{m_\Sigma}\frac{4a_{qq}^2}{3\cdot\tilde{\lambda}_\Sigma^2}\left(1 - M^2\frac{\partial}{\partial M^2}\right)L^{4/9}e^{m_\Sigma^2/M^2}, \tag{6.288}$$

$$\mu_{\Sigma^+} - \mu_{\Sigma^-} = 2 + \frac{4m}{\tilde{\lambda}_\Sigma^2}\left(1 - M^2\frac{\partial}{\partial M^2}\right)\left[a_{\bar{q}q}M^2(1+f)\right.$$

$$\left. + m_s M^2 E_1\left(\frac{s_0}{M^2}\right)L^{-8/9} + \frac{4}{9}\frac{a_{\bar{q}q}^2 m_s}{M^2}\right]e^{m_\Sigma^2/M^2}; \quad (6.289)$$

For Ξ^0, Ξ^- hyperons:

$$2\mu_{\Xi^-} + \mu_{\Xi^0} = -2 - \frac{4a_{\bar{q}q}m}{\tilde{\lambda}_\Xi^2}\left(1 - M^2\frac{\partial}{\partial M^2}\right)\left(M^2 + \frac{17}{36}\frac{a_{\bar{q}q}m_s}{M^2}\right)e^{m_\Xi^2/M^2}, \quad (6.290)$$

$$\mu_{\Xi^0} - \mu_{\Xi^-} = -\frac{m}{m_\Xi}\frac{4a_{\bar{q}q}}{\tilde{\lambda}_\Xi^2}\left(1 - M^2\frac{\partial}{\partial M^2}\right)\left[2m_s M^2 L^{-4/9} + \frac{1}{3}a_{\bar{q}q}(1+f)^2 L^{4/9}\right]e^{m_\Xi^2/M^2}.$$

$$(6.291)$$

(The notation of Section 6.7.1 and 6.7.2 is used.) The factors m/m_Σ and m/m_Ξ correspond to measuring the hyperon magnetic moments in nuclear magnetons.

The Λ hyperon magnetic moment cannot be obtained by this method. The sum rule for μ_Λ has nothing to combine with in order to exclude the χ, κ, and ξ susceptibilities. Moreover, the terms proportional to $\chi_s \sim \langle 0 \mid \bar{s}\sigma_{\mu\nu}s \mid 0\rangle_F$ appear in the sum rule for the Λ hyperon magnetic moment, which means the appearance of a new parameter. The approximate calculation of μ_Λ can be done by using the $SU(3)$ relation $\mu_\lambda = \mu_n/2$ and accounting for a kinematical factor m/m_Λ, corresponding to the μ_Λ measurement in nuclear magnetons. In this way we get $\mu_\Lambda = -0.72$. Although with less precision, the magnetic moment of Λ can be also calculated by using the sum rule, where the values found in separate sum rules [135] are substituted for χ, χ_s, κ, ξ. The Σ^0 magnetic moment is determined by isospin symmetry:

$$\mu_{\Sigma^0} = \frac{1}{2}(\mu_{\Sigma^+} + \mu_{\Sigma^-}). \quad (6.292)$$

Finally, the magnetic moment of the $\Sigma^0\lambda$ transition is given by the relation

$$\mu_{\Sigma\lambda} = \frac{1}{2\sqrt{3}}(3\mu_\lambda + \mu_{\Sigma^0} - 2\mu_n - 2\mu_{\Xi^0}), \quad (6.293)$$

which is valid in the linear approximation in the $SU(3)$ violating parameters [136].

The results of the calculations of baryon octet magnetic moments are presented in Table 2. In calculations of hyperon magnetic moments according to the sum rules (6.288)–(6.291) the numerical values of parameters m_s, f and s_0 were taken from Section 6.7.2, and the experimental values for nucleon and hyperon masses were substituted. For comparison, the experimental data [63] and the quark model predictions are also presented. In the latter, the experimental values of proton, neutron, and λ magnetic moments are used as input.

Within the limits of the expected theoretical errors (10–15%) the results of the sum rule calculations are in agreement with the data. The exceptional case is the Ξ hyperon, where the difference between theory and experiment is larger. The latter can be addressed to a significant M^2 dependence of the sum rules for the Ξ-mass and magnetic moments,

Table 2 Magnetic moments of the baryon octet.

	p	n	Σ^+	Σ^0	Σ^-	Ξ^0	Ξ^-	λ	$\Sigma\lambda$
sum rules	2.70	-1.70	2.70	0.79	-1.12	-1.65	-1.05	-0.72^b	1.54^b
quark model	2.79^a	-1.91^a	2.67	0.78	-1.09	-1.44	-0.49	-0.61^a	1.63
experiment	2.79	-1.91	2.46	—	-1.16	-1.25	-0.65	-0.61	1.61

[a] Input data.
[b] Approximate value, calculated on the basis of $SU(3)$ relations.

which in turn may be related to a larger role of m_s-corrections. In conclusion, it must be emphasized that no new parameters, besides those found in the calculations of baryon masses, enter the above used sum rules for baryon magnetic moments.

6.9.4 Magnetic moments of Δ^{++}, Ω^- baryons, and ρ meson

The magnetic moments of Δ^{++} and Ω^- members of the baryon decuplet were also calculated by the QCD sum rules [137]. In the calculation the generalized method was used, valid not only for a constant, but for a variable external field [138]. Using this method, Δ^{++} and Ω^- magnetic formfactors were calculated at $0 < Q^2 < 0.8\,\text{GeV}^2$. For the Δ^{++} and Ω^- magnetic moments the following results were found [137]:

$$\mu_{\Delta^{++}} = 6.2(3.7 - 7.5)\ [63]; \quad \mu_{\Omega^-} = -3.1\ (-2.024 \pm 0.056)\ [139]. \quad (6.294)$$

(The experimental values are presented in parentheses.) In calculating the Ω magnetic formfactor, the terms proportional to the strange quark mass were disregarded, resulting in worse accuracy of the results. The baryon octet magnetic moments were also calculated by using the sum rules, where some chosen values of vacuum susceptibilities were substituted and a strong violation of factorization hypothesis for vacuum expectation value of four-quark operators was assumed (the value of the gluon condensate was taken about six times greater than given in Table 1, Section 6.2) [140]. The results for Δ^{++} and Ω^- are:

$$\mu_{\Delta^{++}} = 4.1 \pm 1.3, \quad \mu_{\Omega^-} = -1.5 \pm 0.5. \quad (6.295)$$

Because of many arbitrary assumptions it is difficult to estimate the reliability of these calculations.

The ρ meson magnetic moment was calculated by the method presented above taking account of α_s-corrections, [141]. It was obtained:

$$\mu_\rho = (1.8 \pm 0.3)/2m_\rho. \quad (6.296)$$

The fact that the gyromagnetic ratio of the ρ meson appears close to 2 ($\mu_\rho \approx 1/m_\rho$) is very interesting: it demonstrates that at low momentum transfer the ρ meson can be considered as a Yang–Mills vector meson in close correspondence with the vector dominance hypothesis.

6.9.5 Nucleon- and hyperon-coupling constants with axial currents

The knowledge of nucleon- and hyperon-coupling constants with axial currents is important for several reasons: (1) these constants determine the β-decay of neutrons and
hyperons; (2) due to the Goldberger–Treiman relation (2.42), the knowledge of the
nucleon-coupling constant with axial current g_A allows one to find theoretically the value
of the pion–nucleon coupling constant; (3) the nucleon-coupling constants with isovector
octet in flavour $SU(3)$ and singlet axial current determine the values of the Bjorken sum
rules for polarized deep inelastic $e(\mu)N$ scattering, which allows one to study the spin
content of the nucleon.

Let us start with the calculation of the nucleon-coupling constant g_A [142, 143]. For this
goal, add the term

$$L' = \int d^4x j^3_{\mu 5}(x) A^3_\mu \tag{6.297}$$

to the QCD Lagrangian. Here

$$j^3_{\mu 5} = \left(\bar{u}\gamma_\mu\gamma_5 u - \bar{d}\gamma_\mu\gamma_5 d\right)\Big/\sqrt{2} \tag{6.298}$$

is the third component of the axial current and A_μ is the external axial potential, constant in space-time. The presence of the external field A^3_μ results in the appearance of
additional terms linear in A^3_μ in the expression (6.227) for the vacuum average of quark
fields:

$$
\left\langle 0 \left| u^a_\alpha(x), \bar{u}^b_\beta(0) \right| 0 \right\rangle_A = \Bigg[\frac{1}{2\pi^2 x^4}\delta^{ab}(A^3 x)(\gamma_5 \,\slashed{x})_{\alpha\beta} + \frac{1}{72}\delta^{ab}\langle 0 \mid \bar{q}q \mid 0 \rangle
$$

$$
\times (\slashed{x}\,\slashed{A}^3\gamma_5 - \slashed{A}^3\,\slashed{x}\gamma_5)_{\alpha\beta} + \frac{1}{12}f^2_\pi\delta^{ab}(\slashed{A}^3\gamma_5)_{\alpha\beta}
$$

$$
+ \frac{1}{216}\delta^{ab} f^2_\pi m^2_1\left(\frac{5}{2}x^2\,\slashed{A}^3\gamma_5 - (A^3 x)\,\slashed{x}\gamma_5\right)_{\alpha\beta}\Bigg]\Big/\sqrt{2}. \tag{6.299}
$$

In the case of the d quark correlator, the right-hand side of (6.299) changes its sign. The
appearance of the third term on the right-hand side of (6.299) was elucidated in Section 6.3.
The parameter m^2_1, introduced in [144] (in [144] it was denoted by δ^2) is defined by the
equality:

$$\frac{1}{2}g\varepsilon_{\alpha\beta\mu\nu}\left\langle 0 \left| \bar{u}\gamma_\alpha G^n_{\mu\nu}\frac{1}{2}\lambda^n u \right| 0 \right\rangle_A = \frac{1}{\sqrt{2}}f^2_\pi m^2_1 A_\beta. \tag{6.300}$$

The numerical value of m^2_1 was found in [144] by considering special sum rules. Reeval-
uation of these sum rules at the numerical values of α_s and condensates, presented in this
chapter, gives $m^2_1 = 0.15\,\text{GeV}^2$.

Among three tensor structures of the polarization operator, $\slashed{A}^3\gamma_5$, $(\slashed{p}\,\slashed{A}^3 - \slashed{A}^3\,\slashed{p})\gamma_5$, and
$2(A^3 p)\,\slashed{p}\gamma_5$, the most suitable is the third one, because it contains the highest power of p.

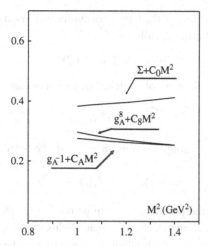

Fig. 6.39. The M^2-dependence of $g_A - 1 + C_A M^2$, Eq. (6.301), $g_A^8 + C_8 M^2$, Eq. (6.307) and $\Sigma^{fit} + C_0^{fit} M^2$ at $f_0^2 = 4 \cdot 10^{-2}$ GeV^2, Eq. (6.324).

So, the sum rule for the coefficient function at this structure is considered. The sum rule for the determination of g_A was found to be [142],[143],[145]–[147]:

$$1 + \frac{8}{9\tilde{\lambda}_N^2} e^{m^2/M^2} \left[a_{\bar{q}q}^2 L^{4/9} + 2\pi^2 f_\pi^2 m_1^2 M^2 - \frac{1}{4} a_{\bar{q}q}^2 \frac{m_0^2}{M^2} + \frac{5}{3} \pi \alpha_s f_\pi^2 \frac{a_{\bar{q}q}^2}{M^2} \right] = g_A + C_A M^2.$$
(6.301)

In the derivation of (6.301), the fact was exploited that the bare loop and gluon condensate contributions are equal to those in the sum rule for the nucleon mass. The account of this fact leads to appearance of the first term, equal to 1, on the left-hand side of (6.301). Note that the main term of OPE of dimension 3 proportional to f_π^2 is absent in (6.301) – it cancelled. Therefore, the deviation of g_A from 1 is entirely connected with condensates and one can expect that it is relatively small. The dependence of $g_A + C_A M^2 - 1$ on M^2 is plotted in Fig. 6.39.

The fitted value of g_A is given by

$$g_A = 1.24 \pm 0.05$$
(6.302)

in comparison with the experimental value $g_A = 1.270 \pm 0.003$ [63]. The constant C_A is small: $C_A = -0.04$ GeV^{-2}, which confirms the legitimacy of neglecting the last term in (6.253). The pion–nucleon interaction coupling constant $g_{\pi NN}^2/4\pi$ can be determined from (6.302) with the help of the Goldberger–Treiman relation (2.42):

$$\frac{g_{\pi NN}^2}{4\pi} = \frac{m^2 g_A^2}{2\pi f_\pi^2} = 12.6 \pm 1.0$$
(6.303)

in good agreement with the experimental value $(g_{\pi NN}^2/4\pi)_{exp} = 13.5 \pm 0.05$ (see, e.g. [148]).

The axial-coupling constants of the baryon decuplet were also determined by the same method [149]. The result for the Δ isobar is:

$$g_{A\Delta\Delta} = 1.40 \pm 0.07, \tag{6.304}$$

which corresponds to the following pion–delta coupling constant:

$$\frac{g_{\pi\Delta\Delta}^2}{4\pi} = 27.6 \pm 2.8. \tag{6.305}$$

Consider now the nucleon-coupling constant g_A^8 with the octet axial current

$$j_{\mu 5}^8 = \frac{1}{\sqrt{6}} \left(\bar{u}\gamma_\mu\gamma_5 u + \bar{d}\gamma_\mu\gamma_5 d - 2\bar{s}\gamma_\mu\gamma_5 s \right). \tag{6.306}$$

The same technique as in the previous case can be used here. The only difference is that, in the expression for the vacuum average of quark fields (6.297), the overall factor $1/\sqrt{2}$ should be replaced by $1/\sqrt{6}$ and f_π should be changed to f_8 – the mean meson octet decay constant. In the limit of massless u, d, s quarks $f_8 = f_\pi$. However, $SU(3)$ symmetry is violated and experimentally $f_K = 0.160\,\text{GeV} = 1.23 f_\pi$. The approximation linear in the s quark mass m_s gives $f_\eta = 1.28 f_\pi$. We put below $f_8^2 = 2.6 \cdot 10^{-2}\,\text{GeV}^2$. The sum rule for g_A^8 has the form:

$$-1 + \frac{8}{9\tilde{\lambda}_N^2} e^{m^2/M^2} \left[6\pi^2 f_8^2 M^4 E_1 L^{-4/9} + 14\pi^2 f_8^2 m_1^2 M^2 E_0 L^{-8/9} + a_{\bar{q}q}^2 L^{4/9} \right.$$

$$\left. - \frac{1}{4} a_{\bar{q}q}^2 \frac{m_0^2}{M^2} - \frac{1}{9}\pi\alpha_s f_8^2 \frac{a_{\bar{q}q}^2}{M^2} \right] = g_A^8 + C_8 M^2. \tag{6.307}$$

The sum rule for the nucleon mass was used in the derivation of (6.307), but unlike the sum rule for g_A, it results to appearance of the item -1 appears in the left-hand side of (6.307), but not $+1$, as in case of g_A. This negative term is compensated by the large, proportional to f_8^2 first term in the square bracket in (6.307). The values of $g_A^8 + C_8 M^2$ as a function of M^2 are plotted in Fig. 6.39. The following result was found:

$$g_A^8 = 0.45 \pm 0.15, \quad C_8 = -0.15 \pm 0.05\,\text{GeV}^{-2}. \tag{6.308}$$

The large errors in (6.308) are caused by strong compensation of negative and positive terms in (6.307) and by uncertainties in $\tilde{\lambda}_N^2$ and f_8^2. The value of the parameter C_8 is comparable with g_A^8. Therefore, the neglected contribution of the last term in (6.253), estimated as $C_8 M^2 e^{-(s_0 - m^2)/M^2}$, was also included in the errors in g_A^8. The theoretical value of g_A^8 (6.308) can be compared with the experimental value $(g_A^8)_{\text{exp}} = 0.59 \pm 0.02$ found from the data on baryon octet β-decay under the assumption of strict $SU(3)$ flavour symmetry [150]. The F and D β-decay axial-coupling constants in the baryon octet are determined from (6.302), (6.308) using the relations (assuming exact $SU(3)$ symmetry):

$$g_A = F + D,$$
$$g_A^8 = 3F - D. \tag{6.309}$$

The results are:

$$F = 0.42 \pm 0.04, \quad D = 0.82 \pm 0.08. \tag{6.310}$$

6.9.6 The nucleon-coupling constant with singlet axial current, the proton spin content, and QCD vacuum topological susceptibility

The method used in the previous subsection cannot be directly applied to the calculation of the nucleon-coupling constant with the singlet axial current

$$j_{\mu 5}^0 = \bar{u}\gamma_5\gamma_\mu u + \bar{d}\gamma_\mu\gamma_5 d + \bar{s}\gamma_\mu\gamma_5 s, \tag{6.311}$$

since $j_{\mu 5}^0$ is not conserved due to the anomaly

$$\partial_\mu j_{\mu 5}^0(x) = 2N_f Q_5(x), \tag{6.312}$$

$$Q_5(x) = \frac{\alpha_s}{8\pi} G_{\mu\nu}^n(x)\tilde{G}_{\mu\nu}^n(x). \tag{6.313}$$

Here $N_f = 3$, $\tilde{G}_{\mu\nu}^n = (1/2)\varepsilon_{\mu\nu\lambda\sigma}G_{\lambda\sigma}^n$, and it is assumed that u, d, s quarks are massless. $Q_5(x)$ is called topological charge density operator (see Chapters 3, 4). As before, add to the QCD Lagrangian the term

$$L' = \int d^4x j_{\mu 5}^0(x)A_\mu^0, \tag{6.314}$$

where A_μ^0 is the constant external axial potential interacting with the singlet axial current $j_{\mu 5}^0$. However, because of nonconservation of $j_{\mu 5}^0$ and the fact that the η' meson is not a Goldstone boson, the polarization operator $\Pi_{\mu\nu}(x) = iT\{j_{\mu 5}^0(x), j_{\mu 5}^0(0)\}$ cannot be saturated by the η' contribution and a relation like (6.28) does not take place. So, we put

$$\langle 0 \mid j_{\mu 5}^0 \mid 0 \rangle_{A^0} = 3f_0^2 A_\mu^0 \tag{6.315}$$

and consider f_0^2 as a new parameter to be determined. The constant f_0^2 is related to the topological susceptibility. Using (6.314) we can write

$$\langle 0 \mid j_{\mu 5}^0 \mid 0 \rangle_{A_0} = \lim_{q\to 0} i \int d^4x\, e^{iqx} \left\langle 0 \left| T\left\{ j_{\nu 5}^0(x),\ j_{\mu 5}^0(0) \right\} \right| 0 \right\rangle A_\nu^0 \equiv \lim_{q\to 0} P_{\nu\mu}(q)A_\nu^0. \tag{6.316}$$

The general structure of $P_{\mu\nu}(q)$ is:

$$P_{\mu\nu}(q) = -P_L(q^2)\delta_{\mu\nu} + P_T(q^2)(-\delta_{\mu\nu}q^2 + q_\mu q_\nu). \tag{6.317}$$

There are no massless states in the spectrum of the singlet polarization operator $P_{\mu\nu}(q)$ even for massless quarks. $P_{T,L}(q^2)$ also have no kinematical singularities at $q^2 = 0$. Therefore, the nonvanishing value $P_{\mu\nu}(0)$ comes entirely from $P_L(q^2)$. Multiplying $P_{\mu\nu}(q)$ by $q_\mu q_\nu$ in the limit of massless u, d, s quarks, we get

$$q_\mu q_\nu P_{\mu\nu} = -P_L(q^2)q^2 = 4N_f^2 i \int d^4x e^{iqx} \langle 0 \mid T\{Q_5(x), Q_5(0)\} \mid 0 \rangle. \tag{6.318}$$

Going to the limit $q^2 \to 0$ we have

$$f_0^2 = -\frac{1}{3}P_L(0) = \frac{4}{3}N_f^2\zeta'(0), \tag{6.319}$$

where $\zeta(q^2)$ is the topological susceptibility

$$\zeta(q^2) = i \int d^4x e^{iqx} \langle 0 \mid T\{Q_5(x), Q_5(0)\} \mid 0 \rangle. \tag{6.320}$$

Crewther derived Ward identities related to $\zeta(0)$ which allowed him to prove the theorem that $\zeta(0) = 0$ in any theory where there is at least one massless quark [151]. So, $\zeta(q^2) \approx \zeta'(0)q^2$ at small q^2 in the limit of massless quarks. Our goal is to find the proton coupling constant with singlet axial current, i.e. the matrix element $\langle p \mid j^0_{\mu5} \mid p \rangle$. According to current algebra, this matrix element is proportional to the part Σ of the spin projection of the completely longitudinally (along the momentum) polarized proton carried away by quarks:

$$2ms_\mu \Sigma = \langle p \mid j^0_{\mu5} \mid p \rangle, \tag{6.321}$$

where s_μ is the proton spin vector [23] (see Chapter 7). Within the parton model, Σ is defined by

$$\Sigma = \Delta u + \Delta d + \Delta s, \tag{6.322}$$

where

$$\Delta q = \int_0^1 dx[\, q_+(x) - q_-(x)\,] \tag{6.323}$$

and $q_+(x)$ [$q_-(x)$] are distributions of quarks q carrying the fraction x of the proton momentum with spin parallel (antiparallel) to the proton spin in longitudinally polarized protons.

In the same way as in Section 6.9.5, consider the polarization operator $\Pi(p)$ in the external singlet axial potential A^0_μ and separate the coefficient function at the tensor structure $2(A^0 p) \not p \gamma_5$. The sum rule has the same form as Eq.(6.307) with substitutions: $f_8^2 \to f_0^2$, $f_8^2 m_1^2 \to h_0$, $g_A^8 \to \Sigma$, $C_8 \to C_0$ [146]:

$$-1 + \frac{8}{9\tilde\lambda_N^2} e^{m^2/M^2} \left[6\pi^2 f_0^2 M^4 E_1 L^{-4/9} + 14\pi^2 h_0 M^2 E_0 L^{-8/9} + a_{\bar q q}^2 L^{4/9} \right.$$
$$\left. - \frac{1}{4} a_{\bar q q}^2 \frac{m_0^2}{M^2} - \frac{1}{9}\pi\alpha_s f_0^2 \frac{a_{\bar q q}^2}{M^2} \right] = \Sigma + C_0 M^2. \tag{6.324}$$

The parameter h_0 is defined by

$$g \left\langle 0 \left| \bar u \gamma_\alpha \frac{\lambda^n}{2} \tilde G^n_{\alpha\beta} u \right| 0 \right\rangle_{A^0} = h_0 A^0_\beta. \tag{6.325}$$

A special sum rule was suggested for its determination [152]. The calculation gives: $h_0 = 3.5 \cdot 10^{-4}$ GeV^4. The sum rule (6.324) represents Σ as a function of f_0^2. The left-hand side of (6.324), which looks like a complicated function of M^2, should be in fact well fitted by a linear function of M^2 in the confidence interval $0.9 < M^2 < 1.4$ GeV^2. The best values of $\Sigma = \Sigma^{fit}$ and $C_0 = C_0^{fit}$ are found from the following χ^2 fitting procedure:

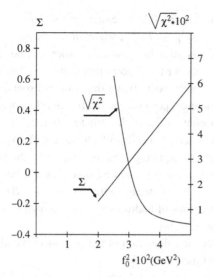

Fig. 6.40. Σ (solid line, left ordinate) and $\sqrt{\chi^2}$, Eq. (6.326) (solid line, right ordinate) as a function of f_0^2.

$$\sqrt{\chi^2} = \left\{ \frac{1}{n} \sum_{i=1}^{n} \left[\Sigma_i^{fit} + C_0^{fit} M_i^2 - R(M_i^2) \right]^2 \right\}^{1/2} \Big/ \left(\Sigma^{fit} + C_0^{fit} \bar{M}^2 \right) = \min,$$

(6.326)

where $R(M^2)$ is the left-hand side of Eq.(6.324) and $\bar{M}^2 \approx 1.0\,\text{GeV}^2$. The values of Σ^{fit} as a function of f_0^2 are plotted in Fig. 6.40 together with $\sqrt{\chi^2}$.

The dependence of $\sqrt{\chi^2}$ on f_0^2 demonstrates that theoretically the lower limit on f_0^2 and Σ^{fit} can be established: $f_0^2 \geq 3.2 \cdot 10^{-2}\,\text{GeV}^2$, $\Sigma^{fit} \geq 0.15$. In order to find f_0^2, we can fix Σ at the experimental value $\Sigma_{\exp} = 0.20 \pm 0.07$ in the $\overline{\text{MS}}$ regularization scheme [153] (see also [154] and [147] for a discussion). Then $f_0^2 = (3.5 \pm 0.5) \cdot 10^{-2}\,\text{GeV}^2$ in agreement with the theoretical lower limit. The substitution of $f_0^2 = 3.5 \cdot 10^{-2}\,\text{GeV}^2$ in (6.319) results in

$$\zeta'(0) = (2.9 \pm 0.4) \cdot 10^{-3}\,\text{GeV}^2.$$

(6.327)

The corresponding value of C_0^{fit} is equal to $0.06\,\text{GeV}^2$. The errors mentioned above include the error arising from the neglected last term in (6.253).

6.10 Three-point functions and formfactors at intermediate momentum transfers

This section is devoted to a consideration of the next in complexity problem in QCD – the determination of dynamical characteristics of hadrons such as their electromagnetic formfactors and three-hadron vertices. The 3-point functions $\Gamma_{A,B;C}(p^2, p'^2; q^2)$ are considered. Let us assume that all variables p^2, p'^2, and q^2 of the 3-point function

$\Gamma_{A,B;C}(p^2, p'^2; q^2)$ are in the Euclidean region, $p^2, p'^2, q^2 < 0$, and large enough in absolute values, $|p^2|, |p'^2|, |q^2| \gg R_{conf}^{-2}$, where R_{conf} is the confinement radius. Then the idea of the calculation is straightforward: in order to find the transition amplitude $\Gamma_{A \to B}(Q^2)$, $Q^2 = -q^2$, consider the 3-point function for suitably chosen quark currents and use OPE. On the other side, write down the double-dispersion relations in channels A and B for invariant structures, perform the double-Borel transformation in p^2 and p'^2, and saturate the double discontinuities of the structure function by the contributions of the low-lying hadronic states A and B. The formfactor of the $A \to B$ transition $F_{A \to B}(Q^2)$ at intermediate Q^2 is found by equating these two representations. If the third channel is also saturated by the contribution of the lowest state C, then the coupling constant g_{ABC} can be found and, consequently, the decay width $\Gamma(C \to A + \bar{B})$. The calculation method is the same as in the calculations of hadron masses presented in Sections 6.6–6.8, but the amount of work grows considerably.

As an example, consider the calculation of the pion electromagnetic formfactor [155]–[157]. The suitable vertex function is:

$$\Gamma_{\mu\nu,\lambda}(p, p'; q) = -\int d^4x \, d^4y \, e^{i(p'x - qy)} \langle 0 | T \{ j_{\nu5}^+(x), \, j_\lambda^{em}(y), \, j_{\mu5}(0) \} | 0 \rangle, \quad (6.328)$$

where $q = p' - p$, $j_{\mu5}(x) = \bar{u}(x)\gamma_\mu\gamma_5 d(x)$ is the axial current, j_λ^{em} is the electromagnetic current,

$$j_\mu^{em} = \frac{2}{3}\bar{u}\gamma_\mu u - \frac{1}{3}\bar{d}\gamma_\mu d. \quad (6.329)$$

The bare-loop diagrams, corresponding to the contribution of the dimension-0 unit operator are shown in Fig. 6.41a, b. The perturbative corrections are disregarded, u and d quarks are considered as massless. In OPE, we restrict ourselves to the contributions of the operators of dimensions $d \leq 6$: the gluon condensate $\langle 0|0(\alpha_s/\pi)G_{\mu\nu}^2|0\rangle$ and four-quark condensates. For the latter, the factorization hypothesis is accepted and therefore the four-quark condensates reduce to $\alpha_s \langle 0 | \bar{q}q | 0\rangle^2$. The contribution of the three-gluon condensate $\langle 0|G^3|0\rangle$ is neglected.

The vertex function $\Gamma_{\mu\nu;\lambda}(p, p'; q)$ includes a number of different tensor structures: $P_\mu P_\nu P_\lambda$, $P_\lambda P_\mu q_\nu$, $P_\lambda \delta_{\mu\nu}$ etc. (The notation $P = (p + p')/2$ is introduced here.) The most suitable structure for our purposes is $P_\mu P_\nu P_\lambda$. The reasons are: (1) the one-pion state contributes to this structure; (2) the terms singular in Q^2, like $1/Q^2$, $1/Q^4$, which do not allow extension to low Q^2 the applicability domain of the approach, are absent in the coefficient function at this structure, whereas they persist at other structures.

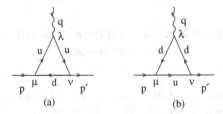

Fig. 6.41. The bare-loop diagrams, corresponding to $\Gamma_{\mu\nu,\lambda}(p, p'; q)$, Eq. (6.328).

The double-dispersion representation of the structure function has the form of (6.248).[2] The last two terms in (6.248) will be eliminated by the double-Borel transformation. So, our interest is concentrated on the first term. The contribution of the lowest state in the channels, corresponding to axial currents, i.e. the pion contribution to the spectral function $\rho_{\mu\nu,\lambda}(p, p'; q)$, is given by:

$$\rho_{\mu\nu,\lambda}(p, p'; q) = \langle 0 \mid j_{\nu 5}^+(0) \mid \pi^+(p)\rangle\langle\pi^+(p) \mid j_{\lambda}^{em}(0) \mid \pi^+(p)\rangle\langle\pi^+(p) \mid j_{\mu 5}(0) \mid 0\rangle$$

$$\times \delta(s)\delta(s') = f_\pi^2 p_\nu' p_\mu (p_\lambda + p_\lambda') F_\pi(Q^2)\delta(s)\delta(s')$$

$$= 2f_\pi^2 F_\pi(Q^2)\left[P_\mu P_\nu P_\lambda + \frac{1}{2}P_\lambda(P_\mu q_\nu - P_\nu q_\mu) - \frac{1}{4}q_\mu q_\nu P_\lambda\right]\delta(s)\delta(s'),$$

$$(6.330)$$

where $Q^2 = -q^2$, $f_\pi = 131\,\text{MeV}$, $F_\pi(Q^2)$ is the pion electromagnetic formfactor and pions are assumed to be massless. The hadronic spectrum is described in the same way, as in Section 6.9 (Fig. 6.28). In domain I, the one-pion state contributes, and domains II and III refer to the continuum, represented by the bare-loop diagrams of Fig. 6.41. The lowest-order diagram Fig. 6.41a or b is equal to

$$\Gamma_{\mu\nu\lambda}^{(0)}(p, p'; q) = \frac{i}{(2\pi)^4}\int\frac{d^4k}{k^2(p'-k)^2(p-k)^2}\,\text{Tr}\left\{\gamma_\nu\gamma_5\,\not{k}\gamma_\mu\gamma_5(\not{p}-\not{k})\gamma_\lambda(\not{p}'-\not{k})\right\}.$$

$$(6.331)$$

We are interested only in the part of the amplitude expressible in the double-dispersive form Eq. (6.248). The most direct way to obtain it is to calculate the double discontinuity of the amplitude, i.e. the spectral functions $\rho^{(0)}(s, s', Q^2)$. To perform this, one has to put the quark lines in the graphs of Fig. 6.41 on mass-shell and substitute the denominators of the quark propagators for the δ functions according to Cutkosky's rule $k^{-2} \to -2\pi i\delta(k^2)$. In such a way, one is left with expressions including the following integrals:

$$I = \int d^4k\delta(k^2)\delta[(p-k)^2]\delta[(p'-k)^2] = \pi/2\lambda^{1/2},\qquad(6.332)$$

$$I_\mu = \int d^4k k_\mu\delta(k^2)\delta[(p-k)^2]\delta[(p'-k)^2]$$

$$= -\frac{\pi}{2\lambda^{3/2}}\left[s'(s-s'-Q^2)p_\mu + s(s'-s-sQ^2)p_\mu'\right],\qquad(6.333)$$

where

$$\lambda(s, s', Q^2) = (s + s' + Q^2)^2 - 4ss',\qquad(6.334)$$

as well as $I_{\mu\nu}$ and $I_{\mu\nu\lambda}$ with factors $k_\mu k_\nu$ and $k_\mu k_\nu k_\lambda$, respectively. The calculation of the I integral is most easily performed in a particular reference system, say, in the Breit system. The integrals I_μ, $I_{\mu\nu}$, and $I_{\mu\nu\lambda}$ can be found by a standard recursive procedure. The double-spectral density $\rho_{\mu\nu,\lambda}(s, s', Q^2)$ is equal to the double discontinuity divided

---

[2] The QCD sum rules for the vertex function using double-dispersion relations were originally considered by Khodjamirian, who investigated the radiative transitions in charmonium [158].

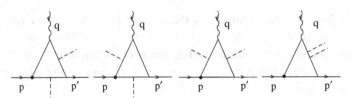

Fig. 6.42. The diagrams describing the terms of OPE proportional to $\langle 0 \mid G_{\mu\nu}^2 \mid 0 \rangle$.

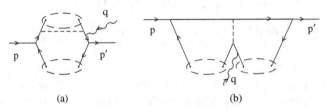

(a)   (b)

Fig. 6.43.

by $(2\pi i)^2 = -4\pi^2$. After a rather cumbersome calculation, the spectral function at the structure $P_\mu P_\nu P_\lambda$ was found to be [155]–[157]:

$$
\rho^{(0)}(s, s', Q^2) = \frac{3Q^4}{2\pi^2}\left[\left(\frac{\partial}{\partial Q^2}\right)^2 + \frac{1}{3}Q^2\left(\frac{\partial}{\partial Q^2}\right)^3\right]\lambda^{-1/2}(s, s', Q^2)
$$

$$
= \frac{3Q^4}{2\pi^2}\lambda^{-7/2}\left[3\lambda(x+Q^2)(x+2Q^2) - \lambda^2 - 5Q^2(x+Q^2)^3\right], \quad (6.335)
$$

where $x = s + s'$ and λ is given by (6.334).

The power corrections proportional to the gluon condensate are described by the diagrams of Fig. 6.42. In the calculations one uses the expressions for the quark propagator in the soft gluon field given by Eqs. (6.221), (6.222). After double-Borel transformation in p^2, p'^2 the term proportional to $\langle 0 \mid G^2 \mid 0 \rangle$ at the structure $\sim P_\mu P_\nu P_\lambda$ is:

$$
\mathcal{B}_{M^2}\mathcal{B}'_{M'^2}\Gamma_{\mu\nu,\lambda}^{\langle G^2 \rangle}(p^2, p'^2, Q^2) = \frac{\alpha_s}{\pi}\frac{\langle 0 \mid G_{\rho\sigma}^n G_{\rho\sigma}^n \mid 0 \rangle}{6M^4}P_\mu P_\nu P_\lambda. \quad (6.336)
$$

(The Borel parameters M^2, M'^2 were taken to be equal.) There are a few types of contributions which generate the term of OPE proportional to the square of the quark condensate. The graphs in which the virtual gluon is hard with virtuality p^2, p'^2 (Fig. 6.43a) or q^2 (Fig. 6.43b) are calculated in an elementary way using Feynman rules.

For the calculation of diagram Fig. 6.44, where only one soft quark pair is present explicitly, it is necessary to expand the quark field $\psi(x)$ in x up to the third order in accord with (6.218) and exploit the relation (6.231). Finally, there are graphs comprising a soft quark pair and a soft gluon (see Fig. 6.45). If substituting the expansions for $A_\mu^n(x)$ and $\psi(x)$, Eqs. (6.217), (6.218), one sees that the six-dimensional vacuum average $\langle 0|\bar{\psi}\psi|0\rangle^2$ can be generated in two ways. First, the quark field can be taken at the origin where the second term of the expansion (6.217) $A_\mu(x)$ is relevant. The calculation of this term is performed using Eq. (6.230). The second way to obtain the required vacuum average $\langle 0|\bar{\psi}\psi|0\rangle^2$ is

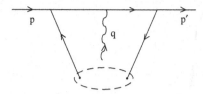

Fig. 6.44.

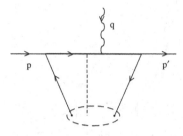

Fig. 6.45.

to take the third term in the expansion (6.218) and the first term in (6.217). In this case, the matrix element (6.232) arises. After summing all terms together the $\alpha_s \langle 0 \mid \bar{q}q \mid 0 \rangle^2$ contribution to the $P_\mu P_\nu P_\lambda$ structure appears to be (see [156] for details):

$$\mathcal{B}_{M^2} \mathcal{B}'_{M'^2} \Gamma^{\langle \bar{q}q \rangle^2}_{\mu\nu,\lambda} = \frac{416\pi}{81 M^4} \alpha_s \langle 0 \mid \bar{q}q \mid 0 \rangle^2 \left(1 + \frac{2Q^2}{13 M^2}\right) P_\mu P_\nu P_\lambda. \qquad (6.337)$$

Eqs. (6.248), (6.330), (6.336), (6.337) allows one to find the sum rule for the pion electromagnetic formfactor $F_\pi(Q^2)$

$$\frac{4}{f_\pi^2} \left\{ \int_0^{s_0} ds \int_0^{s_0} ds' e^{-(s+s')/M^2} \rho^{(0)}(s, s', Q^2) + \frac{\alpha_s}{48\pi M^2} \langle 0 \mid G^a_{\mu\nu} G^a_{\mu\nu} \mid 0 \rangle \right.$$

$$\left. + \frac{52\pi}{81 M^4} \alpha_s \langle 0 \mid \bar{\psi}\psi \mid 0 \rangle^2 \left(1 + \frac{2Q^2}{13 M^2}\right) \right\} = F_\pi(Q^2). \qquad (6.338)$$

As follows from (6.335), (6.338), the main term on the left-hand side of (6.338) decreases as Q^{-4} as $Q^2 \to \infty$, while the power corrections grow with Q^2. This means that sum rule (6.338) is inapplicable at large Q^2 (practically, at $Q^2 > 4\,\text{GeV}^2$) where power corrections become large and uncontrollable. At small Q^2, the main term has an unphysical singularity $\sim Q^4 \ln Q^2$ resulting from massless quark propagators and testifying to the inapplicability of our formulae at small Q^2. The fact that this method is inapplicable at small Q^2 could be foreseen a priori since when approaching the physical region the operator expansion stops working. However, the singularity at $Q^2 \to 0$ is weak. So, one may believe that the present approach is valid up to a rather small Q^2 and the extrapolation from the values of Q^2, where the results are legitimate, to $Q^2 = 0$ can have good accuracy. A reasonable choice of s_0 is $s_0 = 1.2\,\text{GeV}^2$. At this energy, $\sqrt{s_0} \approx 1.1\,\text{GeV}$, the axial meson a_1 starts to contribute to the selected structure in $\Gamma_{\mu\nu,\lambda}$ (6.328). The confidence interval

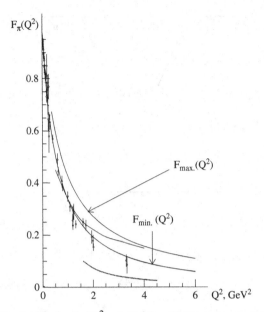

Fig. 6.46. The pion formfactor $F_\pi(Q^2)$: the thick solid line is the result of the calculation according to Eq. (6.338); the thin solid lines are the upper and lower limits, obtained from the data in the time-like region [163]; the dotted line is the asymptotic formula (6.339). The points are the experimental data [159]–[162].

in Borel mass parameter is $0.9 < M^2 < 1.3\,\text{GeV}^2$. At $M^2 = 1.3\,\text{GeV}^2$, $s_0 = 1.2\,\text{GeV}^2$, $Q^2 = 1\,\text{GeV}^2$ the continuum contribution is about 60%. The study of the M^2 dependence of the left-hand side of (6.338) shows that at $Q^2 = 1\,\text{GeV}^2$, $s_0 = 1.2\,\text{GeV}^2$, the left-hand side of (6.338) is M^2 independent within $0.9 < M^2 < 1.3\,\text{GeV}^2$ with accuracy of about 10%. At $Q^2 = 1\,\text{GeV}^2$ and $M^2 = 1\,\text{GeV}^2$, the continuum contribution comprises 30% of the total and gluon and quark condensate corrections are about 10% each. With Q^2 increasing the confidence interval in M^2 decreases and at $Q^2 = 4\,\text{GeV}^2$ shrinks almost to zero: at $M^2 = 1\,\text{GeV}^2$ the continuum contribution is almost equal to the main term and the quark condensate contribution comprises about 40% of it. The lowest value of Q^2 where this approach has good accuracy is perhaps $Q^2 = 0.5\,\text{GeV}^2$. Therefore, the results are reliable for $0.5 \lesssim Q^2 \lesssim 3\,\text{GeV}^2$. The best choice of the parameters is $s_0 = 1.2\,\text{GeV}^2$, $M^2 = 1\,\text{GeV}^2$. The Q^2 dependence of the pion formfactor $F_\pi(Q^2)$ calculated according to Eq. (6.338) is plotted in Fig. 6.46 in comparison with experimental data. The presented data were obtained by measuring $F_\pi(Q^2)$ at $Q^2 > 0$, i.e. in the space-like region [159]–[162]. The upper and lower limits on $F_\pi(Q^2)$ found from the data for $Q^2 < 0$ (in the time-like region) using the analyticity conditions [163] are also shown. For comparison, the asymptotic dependence curve [164]–[169]:

$$F_\pi(Q^2) = 8\pi f_\pi^2 \frac{\alpha_s(Q^2)}{Q^2} \qquad (6.339)$$

is also shown. The agreement with experiment for $0.5 < Q^2 < 3\,\text{GeV}^2$ is good enough. From Fig. 6.46, it is clear that for $Q^2 < 3\,\text{GeV}^2$ the pion formfactor calculated by the QCD sum rule method is in much better agreement with experiment than the asymptotic formula (6.339): the so-called soft mechanism in exclusive processes, described by diagrams such as Fig. 6.41, is dominating over the so-called hard mechanism, when the struck quark is exchanged with the spectator quark by a hard gluon.

Let us try to saturate $F_\pi(Q^2)$ by the ρ meson contribution. Within the framework of the vector dominance model we have

$$F_\pi(Q^2) = \frac{g_{\rho\pi\pi}}{g_\rho} \frac{m_\rho^2}{Q^2 + m_\rho^2}, \qquad (6.340)$$

where $g_{\rho\pi\pi}$ is the $\rho\pi\pi$ coupling constant and g_ρ is the $\rho - \gamma$ coupling constant defined by (6.116). Comparison of (6.340) with the curve of $F_\pi(Q^2)$ in Fig. 6.46 shows that (6.340) well describes $F_\pi(Q^2)$ at intermediate Q^2 and approximately $g_{\rho\pi\pi}/g_\rho \approx 1.0$. The numerical value of g_ρ was found in Section 6.6.1: $g_\rho^2/4\pi = 2.48$ (6.124). Using this value and the known ratio $g_{\rho\pi\pi}/g_\rho$, it is possible to calculate the ρ meson width:

$$\Gamma_{\rho\pi\pi} = \frac{g_{\rho\pi\pi}^2}{48\pi} m_\rho (1 - 4m_\pi^2/m_\rho^2)^{3/2} \approx 131\,\text{MeV} \qquad (6.341)$$

in reasonable agreement with the experimental value $\Gamma_{\rho\pi\pi} = 150\,\text{MeV}$.

The ρ meson electromagnetic formfactors [156], the electromagnetic formfactors for $\rho\pi$ and $\omega\pi$ transitions [171, 172], the axial formfactor of the $\omega\rho$ transition [170] and many others were calculated by the same method. The calculation of ρ meson formfactors demonstrates that in the region of $Q^2 \sim 1 - 3\,\text{GeV}^2$ the behaviour of the formfactors $F_{TT}^\rho(Q^2)$, $F_{LT}^\rho(Q^2)$, $F_{LL}^\rho(Q^2)$ (T means the transverse and L the longitudinal ρ meson polarizations) have nothing in common with their asymptotical behaviour. Quark counting and chirality conservation lead to the following asymptotics: $F_{LL}^\rho(Q^2) \sim 1/Q^2$, $F_{LT}^\rho(Q^2) \sim 1/Q^3$, and $F_{TT}^\rho(Q^2) \sim 1/Q^4$. The QCD sum rule calculation shows that at $Q^2 \sim 1 - 3\,\text{GeV}^2$, $F_{LT}^\rho(Q^2)$ is much larger than $F_{LL}^\rho(Q^2)$ and decreases more slowly than is expected asymptotically, F_{LL}^ρ decreases very fast. The calculations of the axial formfactor of the $\omega \to \rho$ transition allowed one to determine the $g_{\omega\rho\pi}$ coupling constant, which was found to be in good agreement with experiment [170, 172]. The attempt to calculate the nucleon formfactor by this method failed: there were no windows in the space of Borel parameters, where QCD sum rules could be applied. A new method was invented for the calculation of nucleon magnetic formfactors for $0 \le Q^2 \le 1\,\text{GeV}^2$ [138]. Since this method is rather special it will not be presented here.

6.11 Valence quark distributions in hadrons

6.11.1 The method

Quark and gluon distributions in hadrons are not calculated in QCD starting from first principles. It is only possible to calculate their evolution with the square of momentum transfer Q^2, which characterizes the scale where these distributions are measured (see Chapter 7).

In the case of the nucleon, the standard procedure is the following: At some fixed $Q^2 = Q_0^2$, a form of the quark and gluon distributions is assumed which is described by a number of free parameters. Then the evolution with Q^2 of the distributions is calculated in perturbative QCD and by comparing with data of deep inelastic lepton–nucleon scattering, prompt photon production and other hard processes, the best fit for these parameters is found. The quark and gluon distributions in pions are determined in a similar way from data on lepton pair production in pion-nucleon collisions (the Drell–Yan process). But here one needs an additional hypothesis about the connection of the fragmentation function at time-like Q^2, which is measured in Drell–Yan processes, with parton distributions defined at space-like Q^2. For other mesons and baryons, as well as for some distributions in polarized nucleon unmeasured up to now (such as $h_1(x, Q^2)$), we have no information about parton distributions from experiment and have to rely on models. In order to check the form of quark distributions, used as input in evolution equations, as well as to have the possibility to find them in case of unstable hadrons, the direct QCD calculation of these distributions would be of great value.

The principal difficulty in the problem of finding the structure functions (parton distributions) in QCD is that one should calculate the amplitude of deep inelastic forward scattering, the space-time description of which is characterized by large distances in the t channel. Since in QCD the dynamics of processes at large distances is determined by the yet unclear confinement mechanism, this circumstance prevents the calculation of the initial parton distributions, used as input in evolution equations. In OPE, the essential role of large distances in the t channel manifests itself in the fact that when trying to extrapolate the amplitudes to the point $t = 0$, there appear unphysical singularities and, as a consequence, it turns out to be necessary to take into account new unknown vacuum averages in OPE, the number of which is infinite for the case of the forward-scattering amplitude. The idea how to overcome this difficulty is the following [173, 174]. Consider the imaginary part (in the s channel) of the 4-point correlator, corresponding to the forward scattering of two quark currents, one of which has the quantum numbers of the hadron of interest and the other is electromagnetic (or weak). Suppose that the virtualities of the photon and hadronic current q^2 and p^2 are large and negative $|q^2| \gg |p^2| \gg R_c^{-2}$, where R_c is the confinement radius (q is the virtual photon momentum, p is the momentum carried by the hadronic current). In this case, the imaginary part in the s channel $[s = (p + q)^2]$ of forward-scattering amplitude is dominated by small distance contributions at intermediate x. (The standard notation is used: x is the Bjorken scaling variable, $x = -q^2/2\nu$, $\nu = pq$). The proof of this statement is based on the fact that for the imaginary part of the forward-scattering amplitude the position of the singularity in momentum transfer closest to zero is determined by the boundary of the Mandelstam spectral function and is given by the equation

$$t_0 = -4 \frac{x}{1 - x} p^2. \tag{6.342}$$

Therefore, if $|p^2|$ is large and x is not small, then even at $t = 0$ (the forward amplitude) the virtualities of intermediate states in the t channel are large enough for OPE to be applicable.

As follows from (6.342), the approach is invalid at small x. (Since in real calculations $-p^2$ are of order $1 - 2\,\text{GeV}^2$, small x means in fact $x \lesssim 0.1 - 0.2$.) This statement is evident a priori, because at small x Regge behaviour is expected which cannot be described within the framework of OPE. The approach is also invalid at x close to 1. This is the domain of resonances, also outside the scope of OPE. The fact that this method of calculation of quark distributions in hadrons is invalid at $x \ll 1$ and at $1 - x \ll 1$ follows from the theory itself: the OPE diverges in these two domains. Therefore, calculating higher-order terms of OPE makes it possible to estimate up to which numerical values of x, in the small and large x domain, the theory is reliable in each particular case.

The consideration of the 4-point function with equal momenta of hadronic currents in initial and final states has, however, a serious drawback. The origin of this drawback comes from the fact that in the case of the 4-point function, which corresponds to the forward-scattering amplitude with equal initial and final hadron momenta, the Borel transformation does not provide suppression of all excited state contributions: the nondiagonal matrix elements, like

$$\langle 0|j^h|h^* \rangle \langle h^*|j^{em}(x)j^{em}(0)|h \rangle \langle h|j^h|0 \rangle \tag{6.343}$$

are not suppressed in comparison with the matrix element of interest

$$\langle 0|j^h|h \rangle \langle h|j^{em}(x)j^{em}(0)|h \rangle \langle h|j^h|0 \rangle, \tag{6.344}$$

proportional to the hadron h structure function. (Here h is the hadron whose structure function we would like to calculate, h^* is the excited state with the same quantum numbers as h, j^h is the quark current with quantum numbers of hadron h, j^{em} is the electromagnetic current.) This effect is similar to those which arise in the case of QCD sum rules in a constant external field (see Section 6.9, Fig. 6.29). In order to avoid the additional differentiation with respect to the Borel parameter used in Section 6.9, which reduces the accuracy of the results, let us consider the nonforward 4-point correlator [175, 176]:

$$\Pi(p_1, p_2; q, q') = -i \int d^4x d^4y d^4z\, e^{ip_1x + iqy - ip_2z}$$

$$\times \langle 0|T \left\{ j^h(x),\ j^{em}(y),\ j^{em}(0),\ j^h(z) \right\} |0 \rangle. \tag{6.345}$$

Here p_1 and p_2 are the initial and final momenta carried by hadronic current j^h, q and $q' = q + p_1 - p_2$ are the initial and final momenta carried by virtual photons. (Lorentz indices are omitted). It will be essential to have nonequal p_1, p_2 and treat p_1^2, p_2^2 as two independent variables. However, we may put $q'^2 = q^2$ and $t = (p_1 - p_2)^2 = 0$. We are interested in the imaginary part of $\Pi(p_1^2, p_2^2, q^2, s)$ in the s channel:

$$\text{Im}\,\Pi(p_1^2, p_2^2, q^2, s) = \frac{1}{2i}\left[\Pi(p_1^2, p_2^2, q^2, s + i\varepsilon) - \Pi(p_1^2, p_2^2, q^2, s - i\varepsilon) \right]. \tag{6.346}$$

The double-dispersion relation in p_1^2, p_2^2 takes place for Im $\Pi(p_1^2, p_2^2, q^2, s)$:

$$\text{Im } \Pi(p_1^2, p_2^2, q^2, s) = a(q^2, s) + \int_0^\infty \frac{\varphi(q^2, s, u)}{u - p_1^2} du + \int_0^\infty \frac{\varphi(q^2, s, u)}{u - p_2^2} du$$

$$+ \int_0^\infty du_1 \int_0^\infty du_2 \frac{\rho(q^2, s, u_1, u_2)}{(u_1 - p_1^2)(u_2 - p_2^2)}. \quad (6.347)$$

The nondiagonal matrix elements contribute to the first three terms in (6.347). Apply the double-Borel transformation in p_1^2, p_2^2 to (6.347). This transformation kills the first three terms on the right-hand side of (6.347) and we have

$$\mathcal{B}_{M_1^2} \mathcal{B}_{M_2^2} \text{ Im } \Pi(p_1^2, p_2^2, q^2, s) = \int_0^\infty du_1 \int_0^\infty du_2 \rho(q^2, s, u_1, u_2) \exp\left[-\frac{u_1}{M_1^2} - \frac{u_2}{M_2^2} \right]. \quad (6.348)$$

The integration region in the u_1, u_2 plane can be divided in four areas in the same manner, as was done in Section 6.9.1 (Fig. 6.28). The contribution of area I to ρ corresponds to the lowest state contribution and is equal to

$$\rho(u_1, u_2, x, Q^2) = g_h^2 \cdot 2\pi F_2(x, Q^2) \delta(u_1 - m_h^2) \delta(u_2 - m_h^2), \quad (6.349)$$

where g_h is defined by

$$\langle 0 | j_h | h \rangle = g_h. \quad (6.350)$$

(For simplicity, we consider the case of a Lorentz scalar hadronic current.) If in Im $\Pi(p_1, p_2, q, q')$ the structure proportional to $P_\mu P_\nu$ $[P_\mu = (p_1 + p_2)_\mu/2]$ is considered, then in the lowest twist approximation $F_2(x, Q^2)$ is the structure function depending on the Bjorken scaling variable x and weakly on $Q^2 = -q^2$. The same model of the hadronic spectrum as in Section 6.9.1 is assumed: the contributions of areas II and III are referred to as continuum, represented by the bare-loop diagram (see Figs. 6.47 and 6.51 below). So, we are left with the sum rule:

$$\text{Im } \Pi_{QCD}^0 + \text{Power corrections} = 2\pi F_2(x, Q^2) g_h^2 \exp\left(-m_h^2(1/M_1^2 + 1/M_2^2) \right),$$

$$\text{Im } \Pi_{QCD}^0 = \int_0^{s_0} \int_0^{s_0} \rho^0(u_1, u_2, x, Q^2) \exp\left(u_1/M^2 + u_2/M_2^2 \right), \quad (6.351)$$

where $\rho^0(u_1, u_2, x, Q^2)$ is the bare-loop spectral function. One advantage of this method is that after double-Borel transformation unknown contributions of the areas II are exponentially suppressed (actually, the contribution of area II to the bare loop is zero, since $\rho^0 \sim \delta(u_1 - u_2)$).

In what follows, only the terms of the first nonvanishing order in the expansion in p_1^2/q^2, p_2^2/q^2 are retained, which corresponds to taking account of the lowest-twist contributions (twist 2) to the structure functions. Since no additional quark loops, besides the

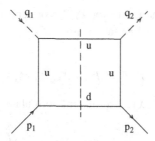

Fig. 6.47. The bare-loop diagram describing the u quark distribution in pion. Dashed lines with arrows correspond to virtual photons, solid lines to hadron currents. The vertical dashed line indicates the cut propagators in Im $\Pi_{\mu\nu\lambda\sigma}$.

bare-quark loops shown e.g. in Figs. 6.47 and 6.51, are accounted for, this approach allows one to find only valence quark distributions in hadrons at moderate $Q_0^2 \sim 3 - 5 \, \text{GeV}^2$. The evolution equations must be exploited in order to go to higher Q^2. The symmetrical Borel transformation will be performed and we put $M_1^2 = M_2^2 = 2M^2$. The factor of 2 is introduced in order that the values of M^2 can be identified with the values of M^2 used in the sum rules for hadron masses. The hadronic currents, which enter Eq. (6.343), are composed by the fields of valence quarks forming the hadron h.

6.11.2 Valence-quark distributions in pion and ρ meson

It is enough to find the distribution of the valence u quark in π^+, since $\bar{d}(x) = u(x)$. The most suitable hadronic current in this case is the axial current

$$j_{\mu 5} = \bar{u}\gamma_\mu\gamma_5 d. \tag{6.352}$$

In order to find the u quark distribution, the electromagnetic current is chosen as u quark current with unit charge

$$j_\mu^{em} = \bar{u}\gamma_\mu u. \tag{6.353}$$

The bare loop of the Fig. 6.47 contribution is given by

$$\text{Im } \Pi_{\mu\nu\lambda\sigma} = -\frac{3}{(2\pi)^2} \frac{1}{2} \int \frac{d^4k}{k^2} \frac{1}{(k + p_2 - p_1)^2} \delta[(q + k)^2]\delta[p_1 - k)^2]$$
$$\times \text{Tr}\left\{\gamma_\lambda \not{k}\gamma_\mu(\not{k} + \not{q})\gamma_\nu(\not{k} + \not{p}_2 - \not{p}_1)\gamma_\sigma(\not{k} - \not{p}_1)\right\}. \tag{6.354}$$

The tensor structure chosen to construct the sum rule is a structure proportional to $P_\mu P_\nu P_\lambda P_\sigma / \nu$, $[P = (p_1 + p_2)/2]$. The reasons are the following: As is known, the results of the QCD sum rule calculations are more reliable if invariant amplitudes at kinematical structures with maximal dimension are used. Different $p_1 \neq p_2$ are important for us only in denominators, where they allow one to separate the terms in dispersion relations. In numerators, one may confine oneself to consideration of terms proportional to 4-vector P_μ and ignore the terms $\sim r_\mu = (p_1 - p_2)_\mu$. Then the factor $P_\mu P_\nu$ provides the contribution of the $F_2(x)$ structure function and the factor $P_\lambda P_\sigma$ corresponds to the contribution of spin zero states (the factor $1/v$ is a scaling factor: $w_2 = F_2/v$).

Let us use the notation

$$\Pi_{\mu\nu\lambda\sigma} = (P_\mu P_\nu P_\lambda P_\sigma/v)\Pi(p_1^2, p_2^2, x) + \ldots . \tag{6.355}$$

For the calculation of $\Pi(p_1^2, p_2^2, x)$ the following formula is exploited:

$$\int d^4 k \frac{1}{(p_1 - k)^2} \frac{1}{(p_2 - k)^2}\delta[\,(p_1+q-k)^2]\delta(k^2) = \frac{\pi}{4vx}\int\limits_0^\infty \frac{1}{u - p_1^2}\frac{1}{u - p_2^2}du. \tag{6.356}$$

(The higher-twist terms are omitted; see [175] for details.) The result is:

$$\text{Im }\Pi(p_1^2, p_2^2, x) = \frac{3}{\pi}x^2(1-x)\int\limits_0^\infty du_1\int\limits_0^\infty du_2 \frac{\delta(u_1 - u_2)}{(u_1 - p_1^2)(u_2 - p_2^2)}. \tag{6.357}$$

The use of Eqs. (6.351), (6.357) and the expression (2.7) for $\langle 0 \mid j_{\mu 5} \mid \pi \rangle$, gives the u quark distribution in the pion in bare-loop approximation

$$u_\pi(x) = \frac{3}{2\pi^2}\frac{M^2}{f_\pi^2}x(1 - x)\left(1 - e^{-s_0/M^2}\right)e^{m_\pi^2/M^2}. \tag{6.358}$$

Before going to more accurate consideration taking account of higher-dimension operators and leading order perturbative corrections, let us discuss in more detail the unit operator contribution in order to estimate whether it is reasonable. The calculation of the pion-decay constant f_π in the same approximation results in [13]:

$$f_\pi^2 = \frac{1}{4\pi^2}M^2(1 - e^{-s_0/M^2}). \tag{6.359}$$

The relation (6.359) follows from the sum rule correlator of the axial currents (6.129), if another separation of the tensor structures than in Section 6.6.2 is performed: the structures at $q_\mu q_\nu$ and $\delta_{\mu\nu}$ are separated. The relation (6.359) arises from the bare-loop contribution to the structure at $q_\mu q_\nu$. Substitution (6.359) into (6.358) gives

$$u_\pi(x) = 6x(1 - x). \tag{6.360}$$

One can note that

$$\int\limits_0^1 u_\pi(x)dx = 1, \tag{6.361}$$

in agreement with the fact that in the quark–parton model there is one valence quark in the pion. Also, it can be easily verified, that

$$\int\limits_0^1 xu_\pi(x)dx = 1/2, \tag{6.362}$$

corresponding to naive quark model where no sea quarks exist. So one can say that formally the unit operator contribution corresponds to the naive parton model. The relations (6.361), (6.362) demonstrate that the Borel parameter M^2 in the Borelized double-dispersion relation is correctly identified with those in the sum rule for masses.

The perturbative and nonperturbative corrections must be added to Eq. (6.358). Taking account of perturbative corrections in leading order, i.e. proportional to $\ln Q^2/\mu^2$, can be easily done using the DGLAP evolution equations (see Chapter 7). In nonperturbative corrections the terms of OPE up to dimension $d = 6$ were considered. In the calculation, one should use the quark propagator expansion (6.221) up to the third-order term (not presented in (6.221)) and the expansion of the gluon propagator (6.225). The calculations of OPE terms are complicated and were performed in Ref. [175] by using the REDUCE program for analytical calculations. It was found that the sum of all diagrams proportional to the gluon condensate is zero after double Borelization. Similarly, all terms proportional to the dimension $d = 6$ gluon condensate $\langle 0|G^3|0\rangle$ are exactly cancelled. (These statements refer only to the structure of Im $\Pi_{\mu\nu\lambda\sigma}$ proportional to $P_\mu P_\nu P_\lambda P_\sigma$). The origin of such cancellations is unclear until now.

Among the contributions of $d = 6$ vacuum expectation value of the four-quark operator there are diagrams with no loops, like those shown in Fig. 6.48. Since this approach is valid at intermediate x only, such diagrams, which are proportional to $\delta(1 - x)$, are out the domain of applicability of the method and must be omitted. The diagrams, arising from perturbative evolution of the latter, are also omitted. The calculations of the $\alpha_s \langle 0 | \bar{q}q | 0\rangle^2$ contribution are cumbersome, some details are given in [175]. The final result for the valence u quark distribution is given by:

$$
xu_\pi(x) = \frac{3}{2\pi^2} \frac{M^2}{f_\pi^2} x^2(1 - x) \left\{ \left[1 + \left(\frac{\alpha_s(2M^2)\ln(Q_0^2/2M^2)}{3\pi} \right) \right. \right.
$$
$$
\left. \times \left(\frac{1 + 4x\ln(1 - x)}{x} - \frac{2(1 - 2x)\ln x}{1 - x} \right) \right]
$$
$$
\left. \times \left(1 - e^{-s_0/M^2} \right) - \frac{\alpha_s(2M^2)\alpha_s a_{\bar{q}q}^2}{(2\pi)^2 \cdot 3^7 \cdot 2^4 \cdot M^6} \frac{\omega(x)}{x^3(1 - x)^3} \right\}, \qquad (6.363)
$$

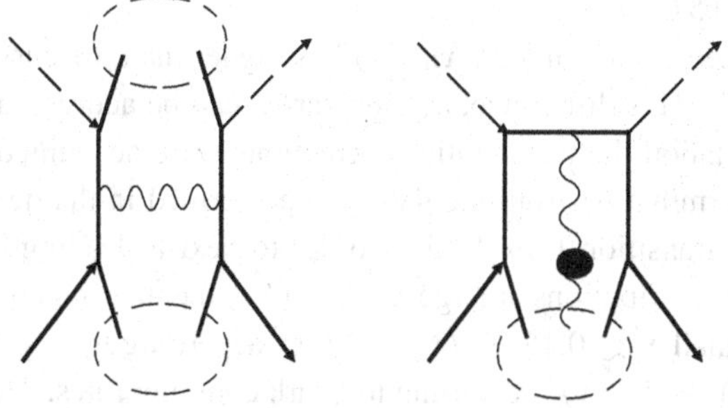

Fig. 6.48. Examples of the nonloop diagrams of dimension-6. Wavy lines correspond to gluons, a dot means a derivative.

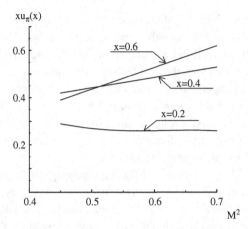

Fig. 6.49. The M^2-dependence of $xu_\pi(x)$ for various x.

where $\omega(x)$ is a fourth-order polynomial in x,

$$\omega(x) = (-5784x^4 - 1140x^3 - 20196x^2 + 20628x - 8292)\ln 2$$

$$+ 4740x^4 + 8847x^3 + 2066x^2 - 2553x + 1416. \tag{6.364}$$

In (6.363), the normalization point $\mu^2 = M_1^2 = M_2^2 = 2M^2$ was chosen. The parameter $\alpha_s a_{\bar{q}q}^2$ is approximately renormalization group invariant and can be taken at $M^2 = 1\,\text{GeV}^2$, $\alpha_s a_{\bar{q}q}^2 = 0.234\,\text{GeV}^6$. The continuum threshold is chosen as $s_0 = 1.2\,\text{GeV}^2$ to be equal to the mass square in the axial channel, where the influence of the axial a_1 resonance starts. The analysis of the sum rule (6.363) shows that it is valid in the region $0.20 < x < 0.70$; for $M^2 > 0.5\,\text{GeV}^2$, the power corrections are less than 30% and the continuum contribution is small ($< 20\%$). However, due to the factors $1/x$ and $1/(1-x)$ in (6.363) there is a rapid increase of the perturbative corrections at the borders of this interval, up to 40–50%, which shows that this method is inapplicable at $x < 0.2$ and $x > 0.7$. There is good stability in the Borel mass parameter in the domain $0.4 < M^2 < 0.7\,\text{GeV}^2$, see Fig. 6.49. (Note, that the effective parameter, characterizing the magnitude of perturbative corrections is $2M^2$.) The pion valence u quark distribution $xu_\pi(x, Q_0^2)$ at $Q_0^2 = 5\,\text{GeV}^2$ is shown in Fig. 6.50 for $M^2 = 0.5\,\text{GeV}^2$.

At $x > 0.3$, the theoretical curve of $xu_\pi(x)$ lies higher than the curve determined from the data in Ref. [177]. This difference, at least partly, can be addressed by the fact that in the theoretical calculation the perturbative corrections were accounted for in the leading order, while the determination from the data was performed in the next-to-leading order. It is known that the transition from leading order to next-to-leading order results in the suppression of quark distributions at large x, $x > 0.3$ and enhancement at low x.

Suppose that at small $x \lesssim 0.15$ $u_\pi(x) \sim 1/\sqrt{x}$ according to Regge behaviour and at large $x \gtrsim 0.7$ $u_\pi(x) \sim (1 - x)^2$ according to quark counting rules. Then, matching these functions with (6.363), one can find the numerical values of the first and second moments of the u quark distribution

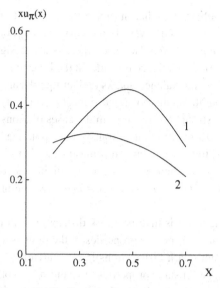

Fig. 6.50. Pion valence quark distribution $xu_\pi(x)$ as a function of x (line 1). The line 2 shows $xu_\pi(x)$ extracted from pion Drell–Yan process and prompt photon production in pion–nucleon collisions, Ref. [177].

$$\mathcal{M}_1 = \int_0^1 u_\pi(x)dx \approx 1.05, \tag{6.365}$$

$$\mathcal{M}_2 = \int_0^1 xu_\pi(x)dx \approx 0.26. \tag{6.366}$$

The results only slightly depend on the points of matching. $\mathcal{M}_2$ is the fraction of the pion momentum carried by the valence u quark. $\mathcal{M}_1$ is the number of u quarks in π^+ and should be $\mathcal{M} = 1$. The error in (6.365) may be estimated as 20%. This comes from the fact that about a half of the contribution to $\mathcal{M}_1$ arises from the region $x < 0.2$, where the extrapolation formula was used. In (6.366), the contributions of regions $x < 0.2$ and $x > 0.7$ comprise 20% of the result. Therefore, one can expect that the accuracy of (6.366) is, at least, not worse than 20%.

The fact that $\mathcal{M}_1$ was found close to 1 confirms the validity of the method. Since due to charge invariance the distribution $\bar{d}_\pi(x)$ is equal to $u_\pi(x)$, the numerical value of $\mathcal{M}_2$ (6.366) means that valence quarks and antiquarks in the pion are carrying about half of the pion momentum. The other half must be attributed to the momentum carried by gluons and sea quarks. This important conclusion, known from experiment, is strongly supported by the calculation presented above.

Valence quark distributions in longitudinally – along the direction of flight – and transversely – perpendicular to this direction – polarized ρ mesons were calculated by the same method [176]. It was found that the u quark distribution in longitudinally polarized

ρ mesons has some resemblance to that in pions: the contributions of vacuum averages $\langle 0 \mid G^2 \mid 0 \rangle$ and $\langle 0 \mid G^3 \mid 0 \rangle$ in OPE vanish, the curve of $xu_\rho^L(x)$ has a shape similar to $xu_\pi(x)$ with its maximum at $x \approx 0.5$. However, there is also a significant difference: the fraction of momentum carried by valence u quarks in the longitudinally polarized ρ meson is $\mathcal{M}_2^L \approx 0.40$. This means that valence quarks and antiquarks are carrying about 80% of the total momentum in longitudinally polarized ρ mesons and only about 20% are left for gluons and sea quarks – a strong difference from the cases of pions and nucleons.

The shape of the u quark distribution in transversely polarized ρ mesons is different from that of pions or longitudinally polarized ρ mesons: $xu_\rho^T(x)$ increases monotonically in the domain $0.2 < x < 0.6$, where $xu_\rho^T(x)$ can be reliably calculated. There is even an indication of a plateau in $xu_\rho^T(x)$ at $0.2 < x < 0.4$ and a two-hump structure in $xu_\rho^T(x)$ is not excluded.

The main physical conclusion is that the quark distributions in pions and ρ mesons do not have much in common. The specific properties of the pion as a Goldstone boson manifest themselves in different quark distributions in comparison with ρ. $SU(6)$ symmetry, probably, may take the place for static properties of π and ρ but not for their internal structure. This fact is not surprising. Quark distributions make sense in fast-moving hadrons. However, $SU(6)$ symmetry cannot be generalized self-consistently to the relativistic case.

6.11.3 Valence-quark distributions in nucleons

The method, developed in Section 6.11, can be applied to the calculation of valence-quark distributions in nucleons [174, 178]. The object of investigation is the 4-current correlator

$$T_{\mu\nu}(p_1, p_2; q, q') = -i \int d^4x d^4y d^4z e^{i(p_1 x + q y - p_2 z)}$$
$$\times \left\langle 0 \left| T \left\{ \eta(x), j_\mu^{u,d}(y), j_\nu^{u,d}(0), \bar\eta(z) \right\} \right| 0 \right\rangle, \quad (6.367)$$

where $\eta(x)$ is the three-quark current with proton quantum numbers (Ioffe current), defined by (6.156). Choose the currents $j_\mu^u = \bar u \gamma_\mu u$, $j_\mu^d = \bar d \gamma_\mu d$, i.e. as electromagnetic currents, which interact only with u(d) quarks with unit charge. Such a choice allows one to get the sum rules separately for the distribution functions of u and d quarks in protons. Quark distributions in neutrons u_n, d_n are related to them by $u_n = d_p$, $d_n = u_p$. We will follow the general method outlined in Section 6.11.1.

Let us select the most suitable invariant amplitude. Consider the contribution of the intermediate state of a proton to the imaginary part of $T_{\mu\nu}$ in the s-channel:

$$\text{Im } T_{\mu\nu}^{(p)} = \lambda_N^2 \frac{1}{p_1^2 - m^2} \sum_{r,r'} v^r(p_1)$$
$$\times \text{Im } \left\{ -i \int d^4x e^{-iqy} \cdot \langle p_1, r | T\{j_\mu(y), j_\nu(0)\} | p_2, r' \rangle \right\} \cdot \bar v^{r'}(p_2) \frac{1}{p_2^2 - m^2},$$
$$(6.368)$$

where $v^r(p)$ is the proton spinor with polarization r and momentum p, λ_N is the coupling constant of the proton with the current $\langle 0|\eta|p, r\rangle = \lambda_N v^r(p)$.

In order to choose the most suitable invariant amplitude, rewrite Eq. (6.368) in the form of

$$\text{Im } T_{\mu\nu} = \lambda_N^2 \frac{1}{(p_1^2 - m^2)(p_2^2 - m^2)} \sum_{r,r'} v^r(p_1)$$

$$\times \left(\bar{v}^r(p_1) \text{ Im } \tilde{T}_{\mu\nu}^{(p)} v^{r'}(p_2) \right) \bar{v}^{r'}(p_2), \tag{6.369}$$

where $\text{Im } \tilde{T}_{\mu\nu}^{(p)}$ is the amplitude before averaging in proton spin, m is the proton mass.

The general form of this amplitude is

$$\text{Im } \tilde{T}_{\mu\nu}^{(p)}(p_1, p_2) = C_1 P_\mu P_\nu + C_2 (P_\mu \gamma_\nu + P_\mu \gamma_\nu) + C_3 \not{P} P_\mu P_\nu + \dots$$

$$+ \quad \text{(terms with } t), \tag{6.370}$$

where $P = (p_1 + p_2)/2; t = p_1 - p_2$.

Let us now take into account that we are interested in such a combination of invariant amplitudes that appears in the spin-averaged matrix element

$$\sum_r \bar{v}^r(p_1) \text{ Im } \tilde{T}_{\mu\nu}^{(p)} v^r(p_2) \tag{6.371}$$

at the kinematical structure $P^\mu P^\nu$ (since this structure in the limit $p_1 \rightarrow p_2 \equiv p$ transforms into $p^\mu p^\nu$, the coefficient at which is just $F_2(x)$).

Using the equation of motion

$$\not{p}_{1,2} v^r(p_{1,2}) = \frac{1}{2} (\not{p}_{1,2} + m) v^r(p_{1,2}) \tag{6.372}$$

and

$$\sum_r v_\alpha^r(p_{1,2}) \bar{v}_\beta^r(p_{1,2}) = (\not{p}_{1,2} + m)_{\alpha\beta} \tag{6.373}$$

one can see that the combination of invariant amplitudes in Eq. (6.370) that arise in Eq. (6.369) at the kinematical structure $\not{P} P^\mu P^\nu$ coincides (up to a numerical factor) with the combination of invariant functions at the structure $P_\mu P_\nu$ in (6.371).

Thus we come to the conclusion that the sum rules should be written for an invariant amplitude at the kinematical structure $\not{P} P^\mu P^\nu$ (in what follows we will denote it $\text{Im } T/\nu$).

So, the sum rules for nucleons have the form

$$\frac{2\pi}{4M^4} \frac{\bar{\lambda}_N^2}{32\pi^4} x q^{u,d}(x) e^{-m^2/M^2} = \text{Im } T_{u,d}^0 + \text{Power corrections} \tag{6.374}$$

where $\bar{\lambda}_N^2 = 32\pi^4 \lambda_N^2$; $q^{u,d}(x)$ are distribution functions of u(d) quarks in the proton, $\text{Im } T^0$ is a perturbative contribution, i.e. of a bare loop with perturbative corrections. (The continuum contribution, i.e., of regions II, III of Fig. 6.28 should be subtracted from Im T. Note that, in fact, the contribution of regions II to the bare loop is zero, since $\rho^0 \sim \delta(u_1 - u_2)$). As before, we put the Borel parameters M_1^2, M_2^2 in the double-Borel transformation

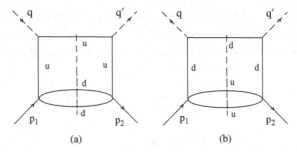

Fig. 6.51. The bare-loop diagrams corresponding to determination of u quark (a) and d quark (b) distributions in the proton. The vertical dashed lines mean that the imaginary parts of the diagrams are taken.

to be equal: $M_1^2 = M_2^2 = 2M^2$. The bare-loop diagrams are shown in Fig. 6.51. In the calculation of the bare-loop diagrams, the following formulae are used

$$\int \frac{d^4k[1; k^2; (p_1 - k)p_2]}{(p_1 - k)^2(p_2 - k)^2} \delta[(p_1 + q - k)^2]\theta(k^2)$$
$$= \frac{\pi}{4vx}(1 - x) \int du \frac{u}{(p_1^2 - u)(p_2^2 - u)} \left[1; \frac{1}{2}u(1 - x); \frac{1}{4}u(1 + x)\right]. \quad (6.375)$$

(The terms which vanish at the double-Borel transformation are omitted.) The results after the double-Borel transformation are the same as in the case for equal $p_1 = p_2$ [174]:

$$\text{Im } T_{u(d)}^0 = \varphi_0^{u(d)}(x) \frac{M^2}{32\pi^3} E_2(M), \quad (6.376)$$

where $E_2(M)$ is given by (6.165) and

$$\varphi_0^u(x) = x(1 - x)^2(1 + 8x), \quad \varphi_0^d(x) = x(1 - x)^2(1 + 2x). \quad (6.377)$$

The substitution of Eq. (6.376) into the sum rules (6.374) results in

$$xq(x)_0^{u(d)} = \frac{2M^6 e^{m^2/M^2}}{\bar{\lambda}_N^2} \varphi_0^{u(d)}(x) \cdot E_2(M). \quad (6.378)$$

In the bare-loop approximation the moments of the quark structure functions are equal to

$$\int_0^1 q_0^d(x)dx = \frac{M^6 e^{m^2/M^2}}{\bar{\lambda}_N^2} E_2, \quad \int_0^1 q_0^u(x)dx = 2\frac{M^6 e^{m^2/M^2}}{\bar{\lambda}_N^2} E_2. \quad (6.379)$$

Making use of relation $\bar{\lambda}_N^2 e^{-m^2/M^2} = M^6 E_2$ which follows from the sum rule for the nucleon mass (see (6.160)) in the same approximation, we get

$$\int_0^1 q_0^d(x)dx = 1,$$

$$\int_0^1 q_0^u(x)dx = 2. \tag{6.380}$$

In the bare-loop approximation, there also appears the sum rule for the second moment:

$$\int_0^1 x(q_0^u(x) + q_0^d(x))dx = 1. \tag{6.381}$$

One can show that relations (6.380), (6.381) hold also when taking into account power corrections proportional to the quark condensate square in the sum rules for the 4-point correlator and in the sum rules for the nucleon mass. Relations (6.380) reflect the fact that the proton has two u quarks and one d quark. Relation (6.381) expresses the momentum conservation law – in the bare-loop approximation the entire momentum is carried by the valence quarks. Therefore, the sum rules (6.380), (6.381) demonstrate that the zero-order approximation is reasonable. In the real physical theory, the regions $x \ll 1$ and $1 - x \ll 1$ are outside the frame of our consideration. However, in the noninteracting quark model which corresponds to the bare-loop approximation, the whole region $0 \le x \le 1$ should be considered and relations (6.381), (6.381) should hold.

The perturbative corrections to the bare loop are calculated using the evolution equations (DGLAP equations) in the leading order. If the quark distributions $q(x, Q_0^2)$ are determined at the point Q_0^2, then in leading order the corrections are proportional to $\alpha_s(\mu^2) \ln Q_0^2/\mu^2$, where μ^2 is the normalization point. It is natural to choose μ^2 to be equal to the Borel parameter $M_1^2 = M_2^2 = 2M^2$. The results take the form:

$$d^{LO}(x) = d_0(x)\left\{1 + \frac{4}{3}\frac{\alpha_s(2M^2)}{2\pi}\ln(Q_0^2/2M^2)\left[\frac{1}{2} + x + \ln\frac{(1-x)^2}{x}\right.\right.$$
$$\left.\left. + \frac{-5 - 17x + 16x^2 + 12x^3}{6(1-x)(1+2x)} - \frac{(3-2x)x^2\ln(1/x)}{(1-x)^2(1+2x)}\right]\right\}, \tag{6.382}$$

$$u^{LO}(x) = u_0(x)\left\{1 + \frac{4}{3}\frac{\alpha_s(2M^2)}{2\pi}\ln(Q_0^2/2M^2)\left[1/2 + x + \ln((1-x)^2/x)\right.\right.$$
$$\left.\left. + \frac{7 - 59x + 46x^2 + 48x^3}{6(1-x)(1+8x)} - \frac{(15-8x)x^2\ln(1/x)}{(1-x)^2(1+8x)}\right]\right\}, \tag{6.383}$$

where $u_0(x)$ and $d_0(x)$ are the bare-loop contributions given by (6.378). In the sum rules for quark distributions, the terms of OPE up to dimension $d = 6$ are accounted for: the contributions of gluon condensate ($d = 4$) and quark condensate square ($d = 6$). The contribution of the dimension-6 operator $g^3 f^{abc} G_{\mu\nu}^a G_{\nu\lambda}^b G_{\lambda\mu}^6$ is neglected since it arises through two loop diagrams and is suppressed by the factor $1/2\pi$. The sum rules for valence u and d quark distributions are [178]:

$$xu(x) = \frac{M^6 e^{m^2/M^2}}{\tilde{\lambda}_N^2} \left\{ 2x(1-x)^2(1+8x)E_2\left(\frac{s_0}{M^2}\right)\left[1 + \frac{u^{LO}(x, Q_0^2)}{u_0(x)}\right] \right.$$

$$+ \frac{\pi^2}{6}(11 + 4x - 31x^2)\frac{1}{M^4}\left\langle 0\left|\frac{\alpha_s}{\pi}G^2\right|0\right\rangle E_0\left(\frac{s_0}{M^2}\right)$$

$$\left. + \frac{1}{324\pi}\frac{1}{1-x}(215 - 867x + 172x^2 + 288(1-x)\ln 2)\frac{1}{M^6}\alpha_s a_{\bar{q}q}^2 \right\}, \quad (6.384)$$

$$xd(x) = \frac{M^6 e^{m^2/M^2}}{\tilde{\lambda}_N^2} \left\{ 2x(1-x)^2(1+2x)E_2\left(\frac{s_0}{M^2}\right)\left[1 + \frac{d^{LO}(x, Q_0^2)}{d_0(x)}\right] \right.$$

$$\left. - \frac{\pi^2}{3x}(1 - 2x^2)\frac{1}{M^4}\left\langle 0\left|\frac{\alpha_s}{\pi}G^2\right|0\right\rangle E_0\left(\frac{s_0}{M^2}\right) - \frac{2}{81\pi}\frac{19 - 43x + 36x^2}{1-x}\frac{1}{M^6}\alpha_s a_{\bar{q}q}^2 \right\}.$$

$$(6.385)$$

For u quarks, the interval of x where the sum rules are valid is $0.2 < x < 0.7$. For d quarks, the interval is narrower: $0.3 < x < 0.6$. The reason is that at $x = 0.2$ the gluon condensate, due to the factor $1/x$ in the second term in (6.385), contributes about 50% to the total $d(x)$ and, therefore, invalidates the result. The typical M^2 dependence of $xu(x)$ and $xd(x)$ are plotted in Fig. 6.52 for $x = 0.4$. As can be seen, the M^2 dependence is not negligible, which indicates a not very good accuracy of the calculation. For d quarks at $x = 0.7$, the M^2 dependence is even stronger: $xd(x)$ changes by a factor 2 for the variation of M^2 in the interval $1 < M^2 < 1.4\,\text{GeV}^2$. The valence quark distributions $xu(x)$ and $xd(x)$ determined by (6.384) and (6.385) are shown in Fig. 6.53 at $Q_0^2 = 5\,\text{GeV}^2$ and $M^2 = 1.1\,\text{GeV}^2$. The value of $\tilde{\lambda}_N^2$ was chosen to be $\tilde{\lambda}_N^2 = 2.6\,\text{GeV}^2$, corresponding to (6.174) at the lower border of error. The magnitudes of condensates are the same as used in the previous calculations: $\langle 0|\frac{\alpha_s}{\pi}G^2|0\rangle = 0.005\,\text{GeV}^4$, $\alpha_s a_{\bar{q}q}^2 = 0.234$, $s_0 = 2.5\,\text{GeV}^2$. The calculated valence quark distributions are compared with those found from the whole set of the data using evolution

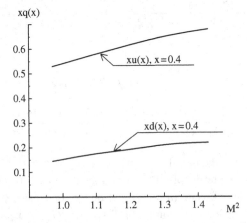

Fig. 6.52. The M^2 dependence of $xu(x)$ and $xd(x)$ at $x = 0.4$, $Q_0^2 = 5\,\text{GeV}^2$, which follows from (6.384), (6.385).

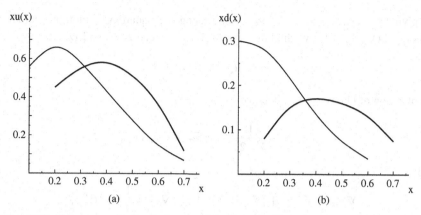

Fig. 6.53. Valence-quark distributions $xu(x)$ (a) and $xd(x)$ (b) (thick solid lines) in comparison with those found from the full set of the data by using evolution equations Ref. [179] (thin solid lines).

equations in next-to-leading order approximation by the MRST collaboration [179]. (The data presented by the CTEQ collaboration [180] only slightly differ from [179].) There is some difference between the data presented in Ref. [181] and Refs. [179] and [180], which may be attributed to the lower normalization point $\mu^2 = 0.3\,\text{GeV}^2$ used in [181].

The valence quark distributions found from the sum rule describe the data only semi-quantitatively. The disagreement is more noticeable in the case of the d quark. A part of this disagreement may be attributed to the fact that the sum rule calculation was performed accounting for perturbative terms in leading order, while the data analysis was done in next-to-leading order. The transition from leading order to next-to-leading order would result in suppression of the high-x domain and enhancement of the low-x domain in quark distributions, i.e. to shrinking of the disagreement. The other source of disagreement is more serious – the inherent default of QCD sum rules to treat the domain of x close to 1 in the calculations of valence quark distributions. Note that in both cases considered above – the cases of pion and nucleon – the perturbative and power corrections are working in the right direction in order to bring the parton model results more close to experiment. Despite the drawbacks of the method, one can believe that in cases of unknown quark distributions in hadrons the QCD sum rule approach would allow one to find them at least semiquantitatively.

Problems

Problem 6.1

Derive Eq. (6.231).

Solution

The general expression for the left hand side of (6.231) is:

$$\langle 0 \mid \bar{q}_\alpha^k (\nabla_\sigma \nabla_\rho \nabla_\tau q_\beta)^i \mid 0 \rangle = \delta^{ik} (B \gamma_\sigma \delta_{\rho\tau} + C \gamma_\rho \delta_{\sigma\tau} + D \gamma_\tau \delta_{\rho\sigma})_{\beta\alpha}. \qquad (1)$$

Multiplying (1) by $(\gamma_\tau)_{\alpha\beta}$ or by $(\gamma_\sigma)_{\alpha\beta}$ and using the equation of motion for massless quarks (6.211), $(m_q = 0)$, we find the relationships between B, C, and D:

$$D = B, \quad C = -5B. \tag{2}$$

Multiplication of (1) by $(\gamma_\rho)_{\alpha\beta}\delta^{ik}\delta_{\sigma\tau}$ gives:

$$B = -\frac{1}{3^3 \cdot 2^5}\langle 0 \mid \bar{q}\nabla_\sigma \not{\nabla}\nabla_\sigma q \mid 0\rangle. \tag{3}$$

Since

$$\not{\nabla}\nabla_\sigma = \nabla_\sigma \not{\nabla} + [\not{\nabla}, \nabla_\sigma], \quad [\nabla_\rho, \nabla_\sigma] = \frac{1}{2}ig\lambda^n G^n_{\rho\sigma}, \tag{4}$$

then, accounting for the equations of motion,

$$\langle 0 \mid \bar{q}\nabla_\sigma \not{\nabla}\nabla_\sigma q \mid 0\rangle = ig\left\langle 0 \left| \bar{q}\nabla_\sigma \gamma_\rho \frac{\lambda^n}{2}G^n_{\rho\sigma} q \right| 0 \right\rangle$$

$$= ig\left\langle 0 \left| \bar{q}\frac{\lambda^n}{2}\gamma_\rho q (D_\sigma G_{\rho\sigma})^n \right| 0 \right\rangle + ig\left\langle 0 \left| \bar{q}\frac{\lambda^n}{2}\gamma_\rho G^n_{\rho\sigma} \nabla_\sigma q \right| 0 \right\rangle. \tag{5}$$

Deriving (5) we used the relation:

$$\left[\nabla_\sigma, \frac{1}{2}\lambda^n G^n_{\rho\sigma}\right] = \frac{1}{2}\lambda^n (D_\sigma G_{\rho\sigma})^n. \tag{6}$$

On the other hand, acting in the left-hand side of (5) with the operator $\nabla_\sigma \not{\nabla}$ to the left and using (4), we have

$$\langle 0 \mid \bar{q}\nabla_\sigma \not{\nabla}\nabla_\sigma q \mid 0\rangle = -ig\left\langle 0 \left| \bar{q}\frac{\lambda^n}{2}\gamma_\rho G^n_{\rho\sigma} \nabla_\sigma q \right| 0 \right\rangle. \tag{7}$$

The comparison of (5) and (7) gives

$$\left\langle 0 \left| \bar{q}\frac{\lambda^n}{2}\gamma_\rho G^n_{\rho\sigma} \nabla_\sigma q \right| 0 \right\rangle = -\frac{1}{2}\left\langle 0 \left| \bar{q}\frac{\lambda^n}{2}\gamma_\rho q (D_\sigma G_{\rho\sigma})^n \right| 0 \right\rangle \tag{8}$$

and

$$\langle 0 \mid \bar{q}\nabla_\sigma \not{\nabla}\nabla_\sigma q \mid 0\rangle = \frac{1}{2}ig\left\langle 0 \left| \bar{q}\frac{\lambda^n}{2}\gamma_\rho q (D_\sigma G_{\rho\sigma})^n \right| 0 \right\rangle. \tag{9}$$

Substituting Eq. (6.212) (for one-quark flavour) into (9) and the result into (3), we have

$$B = i\frac{g^2}{3^3 \cdot 2^8}\langle 0 \mid \bar{q}\gamma_\rho\lambda^n q \cdot \bar{q}\gamma_\rho\lambda^n q \mid 0\rangle = -i\frac{g^2}{3^5 \cdot 2^4}\langle 0 \mid \bar{q}q \mid 0\rangle^2, \tag{10}$$

because of the factorization hypothesis. Formulae (1), (2), and (10) give (6.231).

Problem 6.2

Derive Eq. (6.232)

Solution

The general expression for left hand side of (6.232) has the form:

$$\langle 0 \mid \bar{q}^i_\alpha \nabla_\sigma q^k_\beta G^n_{\mu\nu} \mid 0 \rangle = [\, E(\delta_{\sigma\mu}\gamma_\nu - \delta_{\sigma\nu}\gamma_\mu) + iF\varepsilon_{\sigma\mu\nu\lambda}\gamma_5\gamma_\lambda](\lambda^n)^{ki}. \tag{1}$$

Multiplying (1) by $(\gamma_\sigma)_{\alpha\beta}$, using Eq. (6.211) ($m_q = 0$) and the relation

$$\varepsilon_{\sigma\mu\nu\lambda}\gamma_\sigma\gamma_\lambda\gamma_5 = i(\gamma_\mu\gamma_\nu - \gamma_\nu\gamma_\mu), \quad \varepsilon_{0123} = 1 \tag{2}$$

we find:

$$E = -F. \tag{3}$$

Multiply (1) by $\delta_{\sigma\mu}(\gamma_\nu)_{\alpha\beta}(\lambda^n)^{ik}$. Then

$$E = \frac{1}{3 \cdot 2^8} \langle 0 \mid \bar{q}\lambda^n \gamma_\nu \nabla_\sigma q G^n_{\sigma\nu} \mid 0 \rangle = \frac{1}{3 \cdot 2^8} \left\langle 0 \left| \bar{q}\frac{\lambda^n}{2}\gamma_\rho q (D_\sigma G_{\rho\sigma})^n \right| 0 \right\rangle. \tag{4}$$

The substitution of (6.212) in (4) and the use of the factorization hypothesis gives

$$E = \frac{-g}{3 \cdot 2^{10}} \langle 0 \mid \bar{q}\lambda^n\gamma_\rho q \cdot \bar{q}\lambda^n\gamma_\rho q \mid 0 \rangle = \frac{g}{3^3 \cdot 2^6} \langle 0 \mid \bar{q}q \mid 0 \rangle^2. \tag{5}$$

Eq. (6.232) follows directly from (3) and (5).

Problem 6.3

Write the sum rule to determine the quark condensate magnetic susceptibility. Find its numerical value [19],[20].

Solution

Define the function $\chi(q^2)$ by the equality

$$(\delta_{\mu\nu}q_\nu - \delta_{\nu\lambda}q_\mu)\langle 0 \mid \bar{u}u \mid 0 \rangle \chi(q^2) = \int d^4x e^{iqx} \langle 0 \mid T\{\bar{u}(x)\gamma_\lambda u(x), \bar{u}(0)\sigma_{\mu\nu}u(0)\} \mid 0 \rangle. \tag{1}$$

It is easy to see that if the contribution of the diagram Fig. 6.30a is neglected, then $\chi(0)$ coincides with the quark condensate magnetic susceptibility, $\chi(0) = \chi$. Put $\mu = \lambda$ in (1). Then

$$\langle 0 \mid \bar{u}u \mid 0 \rangle \chi(q^2) = \Pi(q^2), \tag{2}$$

where $\Pi(q^2)$ is defined by:

$$3q_\nu\Pi(q^2) = \int d^4x e^{iqx} \langle 0 \mid T\{\bar{u}(x)\gamma_\mu u(x), \bar{u}(0)\sigma_{\mu\nu}u(0)\} \mid 0 \rangle. \tag{3}$$

Write the OPE for $\Pi(q^2)$. Since the right-hand side of (3) violates chirality, only chirality violating operators contribute to $\Pi(q^2)$. Restricting ourselves to the contributions of the operators up to dimension $d = 5$, we have

$$-\chi(Q^2) = \frac{2}{Q^2} L^{-16/27} - \frac{2}{3} \frac{m_0^2}{Q^4} L^{-2/27}, \tag{4}$$

where $Q^2 = -q^2$ and $L = \alpha_s(Q^2)/\alpha_s(\mu)$ reflect the anomalous dimensions of the operators. The following subtractionless dispersion relation is valid for $\chi(Q^2)$:

$$-\chi(Q^2) = \frac{1}{\pi} \int\limits_0^\infty \frac{\rho(s)ds}{s + Q^2} = \frac{2}{Q^2} L^{-16/27} - \frac{2}{3} \frac{m_0^2}{Q^4} L^{-2/27}. \tag{5}$$

Since the integral in (5) is rapidly converging, $\rho(s)$ can be saturated by contributions of the lowest resonances to a good approximation. The further treatment of Eq. (5) can be performed in various ways. The simplest is to saturate the integral by contributions of ρ and ρ' mesons ($m_\rho^2 = 0.6\,\text{GeV}^2$, $m_{\rho'}^2 = 2.1\,\text{GeV}^2$), neglect the anomalous dimension factors, expand the left-hand side of (5) in $1/Q^2$ and compare the coefficients at $1/Q^2$ and $1/Q^4$. In this way, we find [19]:

$$-\chi(Q^2) \approx \frac{4}{Q^2 + m_\rho^2} - \frac{2}{Q^2 + m_{\rho'}^2}. \tag{6}$$

(The value m_0^2 accidentally disappears from (6) because approximately $(2/3)m_0^2 \approx m_\rho^2$). Eq. (6) can be extrapolated to $Q^2 = 0$ and we get:

$$\chi(0) \approx -5.7\,\text{GeV}^{-2}. \tag{7}$$

The more refined treatment taking account of anomalous dimensions and continuum (besides the resonances) result in the value, is given in Eq. (6.26) (see [21]).

References

[1] Wilson, K. G., *Phys. Rev.* **179** (1969) 1499.

[2] Zimmermann, W. in: *Lectures on elementary particles and quantum field theories at high energies*, ed. by S. Deser, M. Grisaru, H. Pendelton, MIT Press, Cambridge, Mass, 1971, v.1.

[3] Novikov, V. A., Shifman, M. A., Vainshtein, A. I. and Zakharov, V. I., *Nucl. Phys.* **B249** (1985) 445.

[4] Shifman, M. A., Lecture given at the 1997 Yukawa International Seminar *Non-perturbative QCD-structure of QCD vacuum*, Kyoto, Dec. 2–12, 1997, hep-ph/9802214.

[5] Ioffe, B. L., *JETP Lett.* **58** (1993) 876.

[6] Callan, C. G., *Phys. Rev.* **D2** (1970) 1541.

[7] Symanzik, K., *Comm. Math. Phys.* **18** (1970) 227.

[8] Gerber, P. and Leutwyler, H., *Nucl. Phys.* **B321** (1989) 387.

[9] Chen, P. *et al*, *Phys. Rev.* **D64** (2001) 014503.

[10] Ioffe, B. L., *Nucl. Phys.* **B188** (1981) 317; Errata **192** (1982) 591.

[11] Belyaev, V. M. and Ioffe, B. L., *Sov. Phys. JETP* **56** (1982) 493.

[12] Dosch, H. G. and Narison, S., *Phys. Lett.* **417B** (1998) 173.

[13] Shifman, M. A., Vainshtein, A. I. and Zakharov, V.I., *Nucl. Phys.* **B147** (1979) 385, 448.

[14] Novikov, V. A., Shifman, M. A., Vainstein, A. I. and Zakharov, V. I., *Phys. Lett.* **86B** (1979) 347.

[15] Nikolaev, S. N. and Radyushkin, A. V., *Sov. J. Nucl. Phys.* **39** (1984) 91.

[16] Ioffe, B. L. and Samsonov, A. V., *Phys. Atom. Nucl.* **63** (2000) 1448.

[17] Ioffe, B. L., *Progr. Part. Nucl. Phys.* **56** (2006) 232.

[18] Ioffe, B. L. and Smilga, A. V., *Nucl. Phys.* **B232** (1984) 109.

[19] Belyaev, V. M. and Kogan, I. I., *Yad. Fiz.* **40** (1984) 1035.

[20] Balitsky, I. I., Kolesnichenko, A. V. and Yung, A. V., *Sov. J. Nucl. Phys.* **41** (1985) 138.

[21] Ball, P., Braun, V. M. and Kivel, N., *Nucl. Phys.* **B649**(2003) 263.

[22] Ioffe, B. L. and Oganesian, A. G., *Phys. Rev.* **D57** (1998) R6590.

[23] Ioffe, B. L., Khoze, V. A. and Lipatov, L. N., *Hard processes*, North Holland, Amsterdam, 1984.

[24] Novikov, V. A., Shifman, M. A., Vainshtein, A. I. and Zakharov, V. I., *Nucl. Phys.* **B 191** (1981) 301.

[25] Sakurai, J. J., *Phys. Lett.* **46 B** (1973) 207.

[26] Barate, R. *et al*, ALEPH Collaboration, *Eur. Phys. J.* **C 4** (1998) 409, **C 11** (1999) 599, The data files are taken from http://alephwww.cern.ch/ALPUB/paper/ paper.html.

[27] Barate, R. *et al*, ALEPH Collaboration, *Phys. Rep.* **421** (2005) 191.

[28] Ackerstaff, K. *et al*, OPAL Collaboration, *Eur. Phys. J.* **C7** (1999) 571; Abbiendi, G. *et al*, ibid, **13** (2002) 197.

[29] Richichi, S. J. *et al*, CLEO Collaboration, *Phys. Rev.* **D 60** (1999) 112002.

[30] Ioffe, B. L. and Zyablyuk, K. N. *Nucl. Phys.* **A 687** (2001) 437.

[31] Geshkenbein, B. V., Ioffe, B. L. and Zyablyuk, K. N., *Phys. Rev.* D **64** (2001) 093009.

[32] Davier, M., Höcker, A., Girlanda, R. and Stern, J., *Phys. Rev.* D **58** (1998) 96014.

[33] Davier, M., *et al*, arXiv:0803.0979.

[34] Pich, A., *Proc. of QCD 94 Workshop*, Monpellier, 1994; *Nucl. Phys. Proc.Suppl.* 39 BC (1995) 326.

[35] Marciano, W. J. and Sirlin, A., *Phys. Rev. Lett.* **61** (1998) 1815.

[36] Braaten, E., *Phys. Rev. Lett.* **60** (1988) 1606; *Phys. Rev.* D **39** (1989) 1458.

[37] Narison S., and Pich, A., *Phys. Lett.* **211B** (1988) 183.

[38] Le Diberder, F. and Pich, A., *Phys. Lett.* **286B** (1992) 147.

[39] Chetyrkin, K. G., Kataev, A. L. and Tkachov, F. V., *Phys. Lett.* **85B** (1979) 277; Dine, M. and Sapirshtein, J., *Phys. Rev. Lett.* **43** (1979) 668; Celmaster, W. and Gonsalves, R., ibid, **44** (1980) 560.

[40] Surgaladze, L. R. and Samuel, M. A., *Phys. Rev. Lett.* **66** (1991) 560; Goryshny, S. G., Kataev, A. L. and Larin, S. A., *Phys. Lett.* **259B** (1991) 144.

[41] Kataev, A. L. and Starshenko, V. V., *Mod. Phys. Lett.* **A 10** (1995) 235.

[42] Baikov, P. A., Chetyrkin, K. G. and Kühn, J. P., *Phys. Rev.* D **67** (2003) 074026.

[43] Tarasov, O. V., Vladimirov, A. A. and Zharkov, A. Yu., *Phys. Lett.* **93B** (1980) 429; Larin, S. A. and Vermaseren, J. A. M., ibid, **303B** (1993) 334.

[44] van Ritbergen, T., Vermaseren, J. A. M. and Larin, S. A., *Phys. Lett.* **400B** (1997) 379; ibid **404B** (1997) 153.

[45] Czakon, M., *Nucl. Phys.* **B710** (2005) 485.

[46] Radyushkin, A. V., JINR E2-82-159, hep-ph/9907228.

[47] Pivovarov, A., *Nuovo Cimento* A **105** (1992) 813.

[48] Chetyrkin, K. G., Gorishny, S. G. and Spiridonov, V. P., *Phys. Lett.* **160B** (1985) 149.

[49] Adam, L.-E. and Chetyrkin, K. G., *Phys. Lett.* **329B** (1994) 129.

[50] Dubovikov, M. S. and Smilga, A. V., *Nucl. Phys.* **185B** (1981) 109.

[51] Braaten, E. and Lee, C. S., *Phys. Rev.* D **42** (1990) 3888.

[52] Davier, M., 8 Intern. Symposium on Heavy Flavour Physics, Southampton, England, 1999, hep-ex/9912094.

[53] Barate, R. *et al*, ALEPH Collaboration, *Eur. Phys. J.* **C 11** (1999) 599.

[54] Abbiendi, G. *et al*, OPAL Collaboration, *Eur. Phys. J.* **C 19** (2001) 653.

[55] Bethke, S., *Prog. Part. Nucl Phys.* **58** (2007) 351.

[56] Shuryak, E. V., *Nucl. Phys.* B **198** (1982) 83.

[57] Schäfer, T. and Shuryak, E. V., *Rev. Mod. Phys.* **70** (1998) 323.

[58] *"Instantons in Gauge Theories,"* ed. by M. Shifman, World Scientific, 1994.

[59] Andrei, N. and Gross, D. J., *Phys. Rev.* D **18** (1978) 468.

[60] Shafer, T. and Shuryak, E. V., *Phys. Rev. Lett.* **86** (2001) 3973.

[61] Luke, Y. L., *Mathematical functions and their approximations*, NY, Academic Press 1975; Bateman, H. and Erdelyi, A., *Higher transcendental functions*, Vol. I, Krieger, Melbourne, FL, 1953.

[62] Baikov, P. A., Chetyrkin, K. G. and Kühn, J. H., *Phys. Rev. Lett.* **95** (2005) 012003.

[63] Yao, W.-M. *et al*, Particle Data Group, Review of Particle Physics, *J. Phys.* G **33** (2006) 1.

[64] Brodsky, S. J., Lepage, G. P. and Mackenzie, P. B., *Phys. Rev.* D **28** (1983) 228.

[65] Brodsky, S. J., Fadin, V. S., Kim, V. T., Lipatov, L. N. and Pivovarov, G. B., *JETP Lett.* **70** (1999) 155.

[66] Brodsky, S. J., Menke, S., Merino, C. and Rathsman, J., *Phys. Rev.* D **67** (2003) 055008.

[67] Dubovikov, M. S. and Smilga, A. V. , *Yad. Fiz.* **37** (1983) 984; Grozin, A. and Pinelis, Y., *Phys. Lett.* **166B** (1986) 429.

[68] Zyablyuk, K. N., *Eur. Phys. J.* **C38** (2004) 215.

[69] Bijnens, J., Gamiz, E. and Prades, J., *JHEP* **10** (2001) 009.

[70] Cirigliano, V., Golowich, E. and Maltman, K., *Phys. Rev.* D **68** (2003) 054013.

[71] Friot, S., Greynat, D. and de Rafael, E., *JHEP* **10** (2004) 043.

[72] Rojo, J. and Lattore, J. I., *JHEP* 0401 (2004) 055.

[73] Dominguez, C. A. and Schilcher, A., *Phys. Lett.* **581B** (2004) 193; Bordes, J., Dominguez, C. A. Penarrocha, J., Schilcher, K., *JHEP* 0602:037 (2006).

[74] Chiulli, S., Sebu, G., Schilcher, K. and Speiberger, G. *Phys. Lett.* **595B** (2004) 359.

[75] Narison, S., *Phys. Lett.* **B624** (2005) 223.

[76] Barate, R. *et al*, ALEPH Coollaboration, *Z. Phys.* C **76** (1997) 15.

[77] Eidelman, S. I., Kurdadze, L. M. and Vainstein, A. I. *Phys. Lett.* **82B** (1979) 278.

[78] Nikolaev, S. N. and Radyushkin, A. V., *JETP Lett.* **37** (1982) 526.

[79] Reinders, L. J., Rubinstein, H. R. and Yazaki, S. *Phys. Lett.* **138B** (1984) 340; Phys. Reports, **127** (1985) 1.

[80] Narison, S., *QCD spectral sum rules*, World Scientific, 1989, *Phys. Lett.* **387B** (1996) 162; Novikov, V. A., Shifman, M. A., Vainstein, A. I., Voloshin, M. B. and Zakharov, V. I. *Nucl. Phys.* B **237** (1984) 525; Miller, K. J. and Olsson, M. G., *Phys. Rev.* D **25** (1982) 1247.

[81] Colangelo, P. and Khodjamirian, A., *At frontier of Particle Pysics, Handbook of QCD, Boris Ioffe Festschrift*, v.3, p.1495, World Scientific, 2001.

[82] Bertlmann, R. A., *Nucl. Phys.* B **204** (1982) 387; Baier, V. N. and Pinelis, Yu. F., *Phys. Lett.* **116B** (1982) 179, *Nucl. Phys.* B 229 (1983) 29; Narison, G. S. and Tarrach, R., *Z. Phys.* C **26** (1984) 433; Bertlmann, R. A., Dominguez, C. A., Loewe, M., Perrottet, M. and de Rafael, E., *Z. Phys.* C **39** (1988) 231; Baikov, P. A., Ilyin, V. A. and Smirnov, V. A., *Phys. Atom. Nucl.* **56** (1993) 1527; Broadhurst, D. J., Baikov, P. A., Ilyin, V. A., Fleischer, J., Tarasov, O. V. and Smirnov, V. A., *Phys. Lett.* **329B** (1994) 103; Geshkenbein, B. V., *Phys. Atom. Nucl.* **59** (1996) 289.

[83] Jamin, M. and Pich, A., *Nucl. Phys.* B **507** (1997) 334; Eidemüller, M. and Jamin, M., *Phys. Lett.* **498B** (2001) 203; Kühn, J. N. and Steinhauser, M., *Nucl. Phys.* B **619** (2001) 588.

[84] Ioffe, B. L. and Zyablyuk, K. N., *Eur. Phys. J.* C **27** (2003) 229.

[85] Berestetsky, V. B. and Pomeranchuk, I. Ya., *JETP* **29** (1955) 864.

[86] Schwinger, J., *Particles, Sources, Fields*, Addison-Wesley Publ., 1973, V.2.

[87] Hoang, A. H., Kühn, J.H. and Teubner, T., *Nucl. Phys.* B **452** (1995) 173; Chetyrkin, K. G., Kühn, J. H. and Steinhauser, M., *Nucl. Phys.* B **482** (1996) 231; Chetyrkin, K. G. *et al*, *Nucl. Phys.* B **503** (1997) 339; Chetyrkin, K. G. *et al*, *Eur. Phys. J.* C **2** (1998) 137.

[88] Broadhurst, D. J. *et al*, *Phys. Lett.* **329B** (1994) 103.

[89] Nikolaev, S. N. and Radyushkin, A. V., *Sov. J. Nucl. Phys.* **39** (1984) 91.

[90] Zyablyuk, K. N., *JHEP* 0301:081 (2003).

[91] Samsonov, A. V., hep-ph/0407199.

[92] Khoze, V. A. and Shifman, M. A. *Sov. Phys. Usp.* **26** (1983) 387.

[93] Voloshin, M. B., *Int. J. Mod. Phys.* A **10** (1995) 2865.

[94] Jamin, M., Pich, A., *Nucl. Phys.* B **507** (1997) 334.

[95] Chetyrkin, K. G., Hoang, A. H., Kühn, J. H., Steinhauser, M. and Teubner, T., *Eur. Phys. J.* C **2** (1998) 137.

[96] Kühn, J. H., Penin, A. A. and Pivovarov, A. A., *Nucl. Phys.* B **534** (1998) 356.

[97] Landau, L. and Lifshitz, E., *Quantum mechanics: nonrelativistic theory*, Pergamon Press, 1977.

[98] Eichten, E. *et al*, *Phys. Rev.* D **21** (1980) 203.

[99] Chetyrkin, K. G. and Khodjamirian, A., *Eur. Phys. J.* **C46** (2006) 721.

[100] Geshkenbein, B. V. and Ioffe, B. L., *Nucl. Phys.* B **166** (1980) 340.

[101] Brown, L. S., Carlitz, R. D., Cremer, D. B. and Lee, C., *Phys. Rev.* D **17** (1978) 1583.

[102] Brown, L. S. and Lee, C., *Phys. Rev.* D **18** (1978) 2180.

[103] Feldmann, Th., Kroll, P. and Stech, B., *Phys. Rev.* D **58** (1998) 114006.

[104] Feldmann, Th., Kroll, P. and Stech, B., *Phys. Lett.* **B449** (1999) 339.

[105] Chung, Y., Dosch, H.G., Kremer, M. and Schall, D., *Phys. Lett.* **102B** (1981) 175.

[106] Chung, Y., Dosch, H. G., Kremer, M. and Schall, D., *Nucl. Phys.* **B197** (1982) 55.

[107] Ioffe, B. L., *Z. Phys.* **C 18** (1983) 67.

[108] Peskin, M. E., *Phys. Lett.* **88B** (1979) 128.

[109] Espriu, D., Pascual, P. and Tarrach, R., *Nucl. Phys,* **B214** (1983) 285.

[110] Jamin, M., *Z. Phys.* **C37** (1988) 635.

[111] Jamin, M., Dissertation thesis, Heidelberg preprint HD-THEP-88-19, 1988.

[112] Ovchinnikov, A. A., Pivovarov, A. A. and Surguladze, L. R., *Sov. J. Nucl. Phys.* **48** (1988) 358.

[113] Oganesian, A. G., hep-ph/0308289.

[114] Sadovnikova, V. A., Drukarev, E. G. and Ryskin, M. G., *Phys. Rev.* **D72** (2005) 114015.

[115] Dorokhov, A. E. and Kochelev, N. I., *Z. Phys.* **C47** (1990) 281.

[116] Forkel, H. and Banerjee, M., *Phys. Rev. Lett.* **71** (1993) 484.

[117] Diakonov, D., *Prog. Part. Nucl. Phys.* **36** (1996) 1.

[118] Belyaev, V. M. and Ioffe, B. L., *Sov. Phys. JETP* **57** (1983) 716.

[119] Novikov, V. A., Shifman, M. A., Vainshtein, A. I. and Zakharov, V. I., *Fortschr. Phys.* **32** (1984) 11, 585.

[120] Fock, V. A., *Sov. J. Phys.* **12** (1937) 44.

[121] Schwinger, J., *Phys. Rev.* **82** (1951) 664; Schwinger, J., *Particles, Sources and Fields*, vols 1 and 2, Addison-Wesley (1963).

[122] Smilga, A. V., *Yad. Fiz.* **35** (1982) 473.

[123] Shifman, M., *Nucl. Phys.* **B173** (1980) 13.

[124] Shifman, M., *Yad. Fiz.* **36** (1982) 1290.

[125] Ioffe, B. L. and Smilga, A. V., *Nucl. Phys.* **B216** (1983) 373.

[126] Shuryak, E. V. and Vainshtein, A. I., *Nucl. Phys.* **B201** (1982) 141.

[127] Balitsky, I. I., Braun, V. M. and Kolesnichenko, A. V. *Phys. Lett.* **242B** (1990) 245; Errata **318B** (1993) 648.

[128] Ioffe, B. L. *Phys. At. Nucl.* **58** (1995) 1408.

[129] Ioffe, B. L. and Smilga A. V., *Pis'ma v ZhETF* **37** (1983) 250.

[130] Balitsky, I. I. and Yung, A. V., *Phys. Lett.* **129B** (1983) 384.

[131] Aw, M., Banerjee, M. K. and Forkel, H., *Phys. Lett.* **454B** (1999) 147.

[132] Wilson, S. L., Pasuparthy, J. and Chiu, C. B., *Phys. Rev.* **D36** (1987) 1451.

[133] Kogan, I. I. and Wyler, D., *Phys. Lett.* **274B** (1992) 100.

[134] Ioffe, B. L. and Smilga, A. V., *Phys. Lett.* **133B** (1983) 436.

[135] Pasuparthy, J., Singh, J. P., Wilson, S. L. and Chiu, C. B., *Phys. Rev.* **36** (1987) 1553.

[136] Okubo, S., *Phys. Lett.* **4** (1963) 14.

[137] Belyaev, V. M., CEBAF-TH-93-02, hep-ph/9301257.

[138] Belyaev, V. M. and Kogan Ya. I., *Int. J. Mod. Phys.* **A8** (1993) 153.

[139] Wallace, N. B. *et al*, *Phys. Rev. Lett.* **74** (1995) 3732.

[140] Lee, X., *Phys. Rev.* **D57** (1998) 1801.

[141] Samsonov, A. V., JHEP, 0312: 061 (2003), hep-ph/0308065.

[142] Belyaev, V. M. and Kogan, Ya. I., Pis'ma v ZhETF **37** (1983) 611.

[143] Belyaev, V. M. and Kogan, Ya. I., *Phys. Lett.* **136B** (1984) 273.

[144] Novikov, V. A. *et al*, *Nucl. Phys.* **B237** (1984) 525.

[145] Belyaev, V. M., Ioffe, B. L. and Kogan, Ya. I., *Phys. Lett.* **151B** (1985) 290.

[146] Ioffe, B. L. and Oganesian, A. G., *Phys. Rev.* **D 57** (1998) R6590.

[147] Ioffe, B. L., *Surveys in High Energy Physics* **14** (1999) 89.

[148] Machleidt, R. and Li, C. G., *Proc. of the 5-th International Symposium on Meson-Nucleon Physics and the Structure of the Nucleon*, vol. II, πN Newsletter **9** (1993) 37.

[149] Belyaev, V. M., Blok, B. Yu. and Kogan, Ya. I., *Sov. J. Nucl. Phys.* **41** (1985) 280.

[150] Hsueh, S. Y. *et al*, *Phys. Rev.* **D38** (1988) 2056.

[151] Crewther, R. J., *Phys. Lett.* **70B** (1977) 349.

[152] Ioffe, B. L. and Khodjamiryan, A. Yu., *Sov. J. Nucl. Phys.* **55** 1992 1701.

[153] Adeva, B. *et al*, SMC Collaboration, *Phys. Rev.* **D58** (1998) 112002.

[154] Abe, K. *et al*, E154 Collaboration, *Phys. Lett.* **405B** (1997) 180.

[155] Ioffe, B. L. and Smilga, A. V., *Phys. Lett.* **114B** (1982) 353.

[156] Ioffe, B. L. and Smilga, A. V., *Nucl. Phys.* **B216** (1983) 373.

[157] Nesterenko, A. V. and Radyushkin, A. V., *Phys. Lett.* **115B** (1982) 410.

[158] Khodjamirian, A. Yu., *Phys. Lett.* **90B** (1980) 460.

[159] Bebek, C. J. *et al*, *Phys. Rev.* **D9** (1974) 1229.

[160] Bebek, C. J. *et al*, *Phys. Rev.* **D13** (1976) 25.

[161] Bebek, C. J. *et al*, *Phys. Rev.* **D17** (1978) 1693.

[162] Volmer, J. *et al*, *Phys. Rev. Lett.* **86** (2001) 1713.

[163] Geshkenbein, B. V., *Phys. Rev.* **D61** (2000) 033009.

[164] Chernyak, V. L. and Zhitnitski, A. R., Pis'ma v ZhETF **25** (1977) 544.

[165] Chernyak, V. L. and Zhitnitski, A. R., *Phys. Rep.* **112** (1984) 175.

[166] Lepage, G. P. and Brodsky, S. J., *Phys. Lett.* **87B** (1979) 359.

[167] Farrar, G. P. and Jackson, D. R., *Phys. Rev. Lett.* **43** (1979) 246.

[168] Efremov, A. V. and Radyushkin, A. V., *Phys. Lett.* **94B** (1980) 245.

[169] Dunkan, A. and Mueller, A. H., *Phys. Rev.* **D21** (1980) 1636.

[170] Eletsky, V. L., Ioffe, B. L. and Kogan, Ya. I, *Phys. Lett* **122B** (1983) 423.

[171] Eletsky, V. L. and Kogan, Ya. I, *Z. Phys.* C **20** (1983) 357.

[172] Eletsky, V. L. and Kogan, Ya. I, *Yad. Fiz.* **39** (1984) 138.

[173] Ioffe, B. L., *JETP Lett.* **42** (1985) 327.

[174] Belyaev, V. M. and Ioffe, B. L., *Nucl. Phys.* **B310** (1988) 548.

[175] Ioffe, B. L. and Oganesian, A. G., *Eur. Phys. J.* **C13** (2000) 485.

[176] Ioffe, B. L. and Oganesian, A. G., *Phys. Rev.* **D63** (2001) 096006.

[177] Martin A. D., Roberts, R. G., Stirling, W. J. and Sutton, P. J., *Phys. Rev.* **D45** (1992) 2349.

[178] Ioffe, B. L. and Oganesian, A. G., *Nucl. Phys.* **A714** (2003) 145.

[179] Martin, A. D., Roberts, R. G., Stirling, W. J. and Thorn, R. S., MRST Collaboration, *Eur. Phys. J.* **C23** (2002) 73.

[180] Lai, H. L. *et al*, CTEQ Collaboration, *Eur. Phys. J.* **C12** (2000) 375.

[181] Glučk, M., Reya, E., Vogt, A., *Eur. Phys. J.* **C5** (1998) 461.

7
Evolution equations

7.1 Introduction

The theory of the deep inelastic lepton–hadron scattering, as well as of other hard processes in QCD [1], is based on the parton model suggested by Bjorken [2] and Feynman [3]. (The phenomenology of hard processes and their description in the framework of the parton model is presented in the book [4].) The parton model was motivated by the scaling law in the deep inelastic lepton–nucleon scattering suggested by Bjorken [5]. Generally, the structure functions of the deep inelastic lepton–nucleon scattering $W_i(v, q^2)$, are the functions of two variables: q^2, the square of momentum transfer, and $v = pq$, where p is the nucleon momentum. According to the scaling law, at high $| q^2 |$ and v, the structure functions have the form:

$$W_i(v, q^2) = (Q^2)^n F_i(x_{Bj}), \tag{7.1}$$

where $Q^2 = -q^2$, $x_{Bj} = Q^2/2v$ and the power n is given by the canonical dimension of W_i. The scaling law in the deep inelastic lepton–nucleon scattering can be justified by considering this process in the coordinate space. The cross section of the inclusive reaction $lN \rightarrow l'+$ everything (the imaginary part of the forward lN scattering amplitude) is proportional to the integral

$$\int d^4x e^{iqx} \Big\langle p \mid [j_\mu(x), j_\nu(0)] \mid p \Big\rangle, \tag{7.2}$$

where $| p \rangle$ means the nucleon state with momentum p and $j_\mu(x)$ is the electromagnetic current in the case of eN or μN scattering and the weak current in the case of vN scattering. It was proved [6], that at high $| q^2 | \rightarrow \infty$ the amplitude (7.2) is dominated by the domain near the light cone, $x^2 \sim 1/q^2 \rightarrow 0$ in the coordinate space. Therefore, the amplitude in the coordinate space is a function of one variable – the longitudinal distance along the light cone (Ioffe time) times the factor depending on x^2 and fixed by the behavior of the commutator near the light cone. The canonical dimension in (7.1) corresponds to the free-particle commutator. So, the conclusion follows that the deep inelastic lepton–nucleon cross section is described as a superposition of cross sections for the lepton scattering off the free nucleon constituents – partons. In QCD, this phenomenon is slightly modified due to the asymptotic freedom – the decrease of the coupling constant $\alpha_s(Q^2)$ at large Q^2. Since the decrease is only logarithmic, the structure functions F_i depend weakly on Q^2, $F_i = F_i(x_{Bj}, \alpha_s(Q^2))$.

The Q^2-dependence of the structure functions is derived in QCD by two methods. The results of these approaches are equivalent. It is supposed that at some fixed $Q^2 = Q_0^2$ the functions $F_i(x_{Bj}, Q_0^2)$ are known. Then $F_i(x_{Bj}, Q^2))$ are calculated through $F_k(x_{Bj}, Q_0^2)$ using the evolution equations.

In the first approach, the light-cone dominance of the deep inelastic scattering amplitudes is exploited and the expansion of the matrix element in (7.2) over the distances x along the light cone is performed [7]. The main contribution to $W_i(v, q^2)$ arises from the operators of the lowest twist, $t = 2$. The twist of an operator is defined as the difference between its dimension d and Lorentz spin j. The operators of higher twists result in the terms suppressed by powers of Q^2 and can be omitted at large Q^2. Contributions of individual operators can be obtained by taking moments of $F_i(x, Q^2)$:

$$M_k^{(i)}(Q^2) = \int\limits_0^1 x^k dx F_i\left(x, Q^2\right). \tag{7.3}$$

Then, the renormalization group equations are used for $M_k^{(i)}(Q^2)$, which allows to calculate them from their initial values $M_k^{(i)}(Q_0^2)$. The structure function are reconstructed from the known values of their moments. This method is exposed in details, e.g. in the books [8],[9], where the references to original papers are given.

The second method, presented in this chapter, uses the leading logarithm expansion. As was explained in Chapter 1, in QCD at large s or Q^2 the effective parameter of the perturbation theory apart from α_s can contain an additional large factor being a certain power of $\ln s$ or $\ln Q^2$. (The similar situation takes place in QED.) The leading-order terms in the deep inelastic lepton–hadron scattering and other hard processes are proportional to $(\alpha_s \ln Q^2)^n$. The sum of such terms results in the leading-order approximation, which is valid under conditions:

$$\alpha_s(Q^2) \ll 1, \quad \alpha_s(\mu^2) \ll 1, \quad \alpha_s(Q^2) \ln Q^2/\mu^2 \sim 1. \tag{7.4}$$

The leading-order approximation for the inelastic lepton–hadron scattering reproduces the parton model results.

For the deep inelastic lepton–hadron scattering at large energies $\sqrt{s}$ and large momentum transfers Q the transverse momenta of produced particles are strongly ordered. In this case one can use the probabilistic picture to express the lepton–hadron cross section in terms of the cross sections for the collision of point-like partons. We review below the parton model (see also [4]) and discuss its modification in QCD where the parton transverse momenta grow logarithmicly with Q. Later, the evolution equations for parton distributions and for fragmentation functions are derived. We calculate their splitting kernels and find their solutions. The relation with the renormalization group in the framework of the Wilson operator product expansion is outlined. The evolution equations are generalized to the case of the parton correlators related to matrix elements of the so-called quasipartonic operators. The case of the twist-3 operators describing the large-Q^2 behaviour of the structure function $g_2(x)$ is studied in more detail. We discuss also the double-logarithmic asymptotics of structure functions at small x. The next-to-leading corrections to the splitting kernels in QCD are reviewed.

For exclusive processes, such as the backward e^+e^--scattering in QED the effective parameter is $\alpha_{em} \ln^2 s$ [10],[11]. The corresponding physical quantities in the double-logarithmic approximation (DLA) are obtained by calculating and summing the asymptotic contributions $\sim \alpha_{em}^n \ln^{2n} s$ in all orders of the perturbation theory. Their region of applicability is

$$\alpha_{em} \ll 1, \quad \alpha_{em} \ln^2 s \sim 1. \tag{7.5}$$

In DLA, the transverse momenta $|k_r^\perp|$ of the virtual and real particles are large and implied to be strongly ordered because the first $\ln s$ in a loop is obtained as a result of integration over the energy of a relatively soft particle and another logarithm – over its transverse momentum (or emission angle). Instead of calculating the asymptotics of each individual diagram, one can initially divide the integration region in several subregions depending on the ordering of the particle transverse momenta and sum subsequently their contributions with the same orderings over all diagrams of the perturbation theory. It gives a possibility to write an evolution equation with respect to the logarithm of the infrared cut-off λ [12]. In the Regge kinematics for the forward-scattering processes, the integrals over transverse momenta are convergent, which leads in these cases to the effective parameter of the perturbation theory $\alpha \ln s$ [13] or even $\alpha^2 \ln s$ [14].

7.2 Parton model in QCD

As was argued in Section 7.1, in hard processes hadrons can be considered as superpositions of the point-like bare particles – partons having quantum numbers of quarks and gluons [2], [3]. For example, in the framework of the parton model the cross section for the inclusive Z production in hadron–hadron collisions $h+h \to Z+anything$ is expressed in terms of the product of the inclusive probabilities $D_{h_1}^q(x_1)$, $D_{h_2}^{\bar{q}}(x_2)$ to find the quark and antiquark with the energies $x_{1,2}\sqrt{s/4}$ inside the colliding hadrons h_1, h_2 and the Born cross section $\sigma_{q\bar{q}\to Z}(s\,x_1 x_2)$ for the Z-boson production in the quark-antiquark collisions [15], [16]. This expression is integrated over the Feynman components $x_{1,2}$ of the quark and antiquark momenta (see Fig. 7.1)

$$\sigma_{h_1 h_2 \to Z} = \frac{1}{3} \sum_q \int dx_1 dx_2 \left(D_{h_1}^q(x_1)\, D_{h_2}^{\bar{q}}(x_2) + D_{h_1}^{\bar{q}}(x_1)\, D_{h_2}^q(x_2) \right) \sigma_{q\bar{q}\to Z}(s x_1 x_2).$$

$$\tag{7.6}$$

Here the factor $\frac{1}{3}$ appears because in the quark and antiquark distributions $D_h^{q,\bar{q}}(x)$ the sum over three colour states of the quark is implied, but the Z-boson is produced in annihilation of quark and antiquark with an opposite colour.

Initially, the parton model was applied to the description of the deep inelastic scattering of leptons off hadrons [2]. For example, the differential cross section of the inclusive

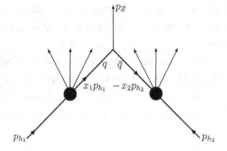

Fig. 7.1. Partonic description of the Z-production in $h_1 h_2$ collisions.

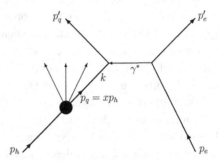

Fig. 7.2. Partonic description of the deep inelastic eh scattering.

scattering of an electron with the momentum p_e off a hadron with the momentum $p = p_h$ can be written as follows [4] (see Fig.7.2)

$$d\sigma_\gamma = \frac{\alpha^2}{\pi} \frac{1}{q^4} L_{\mu\nu} W_{\mu\nu} \frac{d^3 p'_e}{(pp_e)E'}, \tag{7.7}$$

where p'_e and E' are the momentum and energy of scattered electron and only the exchange of one photon with the momentum $q = p_e - p'_e$ was taken into account (for large q^2, one should also add the Z-boson exchange, which leads in particular to the parity nonconservation effects proportional to the axial charge $g_a \sim T_3$). The electron tensor $L_{\mu\nu}$ is calculated explicitly

$$L_{\mu\nu} = \frac{1}{2} tr \left(\hat{p}'_e \gamma_\mu \hat{p}_e \gamma_\nu \right) = 2 \left(p'_{e\mu} p_{e\nu} + p'_{e\nu} p_{e\mu} - \delta_{\nu\mu} (p'_e p_e) \right). \tag{7.8}$$

The hadronic tensor $W_{\mu\nu}$ is expressed in terms of matrix elements of the electromagnetic current J_μ^{el}

$$W_{\mu\nu} = \frac{1}{4} \sum_n \langle p | J_\mu^{el}(0) | n \rangle \langle n | J_\nu^{el}(0) | p \rangle (2\pi)^4 \delta^4(p + q - p_n). \tag{7.9}$$

For nonpolarized electron and target it can be written as a sum of two contributions proportional to structure functions $F_{1,2}(x, Q^2)$ using the properties of the gauge invariance and the parity conservation as follows

$$\frac{1}{\pi} W_{\mu\nu} = -\left(\delta_{\mu\nu} - \frac{q_\mu q_\nu}{q^2}\right) F_1(x, Q^2) + \left(p_\mu - \frac{q_\mu(pq)}{q^2}\right)\left(p_\nu - \frac{q_\nu(pq)}{q^2}\right) \frac{F_2(x, Q^2)}{pq}.$$

$$(7.10)$$

Here Q^2 and x are the Bjorken variables

$$Q^2 = -q^2, \quad x = \frac{Q^2}{2pq} \quad 0 \le x \le 1. \tag{7.11}$$

The structure functions $F_{1,2}(x, Q^2)$ do not depend on Q^2 for fixed x and large Q^2 in the framework of the Bjorken–Feynman parton model

$$\lim_{Q \to \infty} F_{1,2}(x, Q^2) = F_{1,2}(x), \tag{7.12}$$

which corresponds to the Bjorken scaling [2]. In this model, the structure functions can be calculated in the impulse approximation as a sum of the structure functions for charged partons averaged with the partonic distributions (see Fig.7.2). From the point of view of the Wilson operator product expansion [7], the Bjorken scaling means that the corresponding twist-2 operators have canonical dimensions, as in the free theory.

The charged partons are assumed to be fermions (quarks). The hadronic tensor for the quarks can be calculated using the relation

$$|n\rangle \langle n| = \int \frac{d^4 p_n}{(2\pi)^3} \, \delta(p_n^2 - m^2) \, \theta(E_n) \tag{7.13}$$

in the form

$$\frac{1}{\pi} W^f_{\mu\nu} = \frac{e_q^2}{2}\delta\left((k+q)^2\right)\frac{1}{2}\mathrm{Tr}\,(\slashed{k}\,\gamma_\mu(\slashed{q}+\slashed{k})\gamma_\nu) = -e_q^2\frac{1}{2}\delta^\perp_{\mu\nu}\,\delta(\beta - x), \tag{7.14}$$

where k is the quark momentum and $\beta = \frac{kq}{pq}$ is its Feynman parameter. (The quark mass is neglected in (7.14).)

In (7.14), $\delta^\perp_{\mu\nu}$ is the projector to the transverse subspace orthogonal to p and q

$$-\delta^\perp_{\mu\nu} = -\left(\delta_{\mu\nu} - \frac{q_\mu q_\nu}{q^2}\right) + \left(p_\mu - \frac{q_\mu(pq)}{q^2}\right)\left(p_\nu - \frac{q_\nu(pq)}{q^2}\right)\frac{2x}{pq}. \tag{7.15}$$

As follows from (7.10) and (7.15) in the framework of the quark–parton model the Callan–Gross relation between F_1 and F_2 is valid [17]

$$F_2(x) = 2x F_1(x), \tag{7.16}$$

where the expression

$$F_2(x) = x \sum_{i=q,\bar{q}} e_i^2 \, n^i(x) \tag{7.17}$$

corresponds to the impulse approximation for the cross section. The quantity $n^q(x)$ is the quark distribution in the hadron, normalized in such a way that the electric charge conservation for the proton takes the form

$$1 = \sum_{i=q,\bar{q}} e_i \int_0^1 dx\, n^i(x). \tag{7.18}$$

The structure functions can be expressed in terms of the cross sections σ_t and σ_l for the scattering of the virtual photons off protons with the transverse (t) and longitudinal (l) polarization [4]. In the quark–parton model, we have $\sigma_l = 0$ in accordance with the Callan–Gross relation between F_1 and F_2.

Another important deep inelastic process is the scattering of neutrinos and antineutrinos off hadrons. It is induced by the Z-boson exchange (a neutral current contribution). In this case, the expressions for inclusive cross sections are similar to the case of the ep scattering (7.10), but one should take into account the parity-violating contribution (see, for example, [4]).

The process of the neutrino–hadron scattering with a charged lepton in the final state is related to the t-channel exchange of the W-boson. The cross section of the corresponding inclusive process, in which one measures only a produced electron or positron with the momentum p'_e, is given by [4]

$$d\sigma^{\nu,\bar{\nu}} = \frac{g^4}{16\pi^3} \left(\frac{1}{2\sqrt{2}}\right)^2 \left(\frac{1}{q^2 - M_W^2}\right)^2 L^{\nu,\bar{\nu}}_{\mu\rho} W^{\nu,\bar{\nu}}_{\mu\rho} \frac{d^3 p'_e}{\left(p p'_{\nu,\bar{\nu}}\right) E'}. \tag{7.19}$$

Taking into account that the initial neutrino or antineutrino have fixed helicities, we obtain

$$L^{\nu,\bar{\nu}}_{\mu\rho} = \mathrm{Tr}\left(\not{p}'_e \gamma_\mu (1 \pm \gamma_5)\, \not{p}_{\nu,\bar{\nu}} \gamma_\rho (1 \pm \gamma_5)\right)$$
$$= 8\left(p'_{e\mu} p^{\nu,\bar{\nu}}_\rho + p'_{e\rho} p^{\nu,\bar{\nu}}_\mu - \delta_{\rho\mu}(p'_e p_{\nu,\bar{\nu}}) \mp i\epsilon_{\mu\rho\lambda\delta} p^{\nu,\bar{\nu}}_\lambda p'_{e\delta}\right). \tag{7.20}$$

For the hadronic tensor, one obtains a more complicated spin structure in comparison with the deep inelastic electron–hadron scattering where the parity nonconservation effects were absent

$$\frac{1}{\pi} W^{\nu,\bar{\nu}}_{\mu\rho} = -\left(\delta_{\mu\rho} - \frac{q_\mu q_\rho}{q^2}\right) F^{\nu,\bar{\nu}}_1 + \left(p_\mu - \frac{q_\mu(pq)}{q^2}\right)\left(p_\rho - \frac{q_\rho(pq)}{q^2}\right) \frac{F^{\nu,\bar{\nu}}_2}{pq}$$
$$- i\epsilon_{\mu\rho\lambda\delta}\, p_\lambda\, q_\delta \frac{F^{\nu,\bar{\nu}}_3}{2pq}. \tag{7.21}$$

In the framework of the parton model for the neutrino and antineutrino scattering off proton we can derive [4]

$$F^\nu_1 = \frac{1}{2x} F^\nu_2 = d(x)\cos^2\theta_c + \bar{u}(x) + s(x)\sin^2\theta_c, \tag{7.22}$$

$$\frac{1}{2} F^\nu_3 = d(x)\cos^2\theta_c - \bar{u}(x) + s(x)\sin^2\theta_c, \tag{7.23}$$

$$F^{\bar{\nu}}_1 = \frac{1}{2x} F^{\bar{\nu}}_2 = u(x) + \bar{d}(x)\cos^2\theta_c + \bar{s}(x)\sin^2\theta_c, \tag{7.24}$$

$$\frac{1}{2} F^{\bar{\nu}}_3 = u(x) - \bar{d}(x)\cos^2\theta_c - \bar{s}(x)\sin^2\theta_c. \tag{7.25}$$

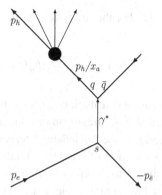

Fig. 7.3. Inclusive $e\bar{e}$-annihilation to hadrons in the parton model.

Here $u(x)$, $d(x)$, $s(x)$, $\bar{u}(x)$, $\bar{d}(x)$, $\bar{s}(x)$ are the parton distributions of the corresponding quarks in the proton and θ_c is the Cabbibo angle – the parameter of the Cabbibo–Kobayashi–Maskawa (CKM) matrix (we neglect the presence of heavier quarks in the proton). The elements of the CKM matrix appear in the quark–W-boson interaction vertex.

The simplest process, in which one can measure the distribution $n_q^h(z)$ of the hadron h with the relative momentum z inside the quark q is the inclusive annihilation of the e^+e^--pair into hadrons. In this process, only one hadron h having the momentum p is detected. Its differential cross section corresponding to the intermediate photon production (neglecting the Z-boson contribution) is [4] (see Fig.7.3)

$$d\sigma_\gamma = \frac{4\alpha^2}{\pi} \frac{1}{s^3} \bar{L}_{\mu\nu} \overline{W}_{\mu\nu} \frac{d^3 p}{E}, \quad E = |p|, \quad s = q^2, \tag{7.26}$$

where

$$\bar{L}_{\mu\nu} = \frac{1}{4} \text{Tr} \left(\not{p}_e \gamma_\mu \not{p}_{\bar{e}} \gamma_\nu \right) = p_{e\mu} p_{\bar{e}\nu} + p_{e\nu} p_{\bar{e}\mu} - \delta_{\mu\nu} (p_e p_{\bar{e}}) \tag{7.27}$$

and

$$\overline{W}_{\mu\nu} = \frac{1}{8} \sum_n \langle 0 | J_\mu^{el}(0) | n, h \rangle \langle n, h | J_\nu^{el}(0) | 0 \rangle (2\pi)^4 \delta^4 (q - p_n - p_h). \tag{7.28}$$

We have from the gauge invariance and the parity conservation

$$\frac{1}{\pi} \overline{W}_{\mu\nu} = -\left(\delta_{\mu\nu} - \frac{q_\mu q_\nu}{q^2} \right) \overline{F}_1(x_a) + \left(p_\mu - \frac{q_\mu (pq)}{q^2} \right) \left(p_\nu - \frac{q_\nu (pq)}{q^2} \right) \frac{\overline{F}_2(x_a)}{pq}, \tag{7.29}$$

where

$$x_a = \frac{2pq}{q^2} < 1, \quad q^2 = s. \tag{7.30}$$

In the parton model, the inclusive annihilation $e^+ e^- \to h + \dots$ is described as the process where initially e^+ and e^- produce the pair $q\bar{q}$ and later q or $\bar{q}$ transform into the hadron system with a measured particle h. For the structure functions $\overline{F}_1(x_a, q^2)$ and $\overline{F}_2(x_a, q^2)$, we obtain in this model:

$$\overline{F}_1(z) = -\frac{z}{2}\overline{F}_2(z) = \frac{3}{z}\sum_q e_q^2(h_q(z) + h_{\overline{q}}(z)), \tag{7.31}$$

where $h_q(z)$ and $h_{\overline{q}}(z)$ are inclusive distributions of hadrons h inside the corresponding partons q and $\overline{q}$ (they are called also the fragmentation functions). The factor 3 is related to the number of coloured quarks in the fundamental representation of the gauge group $SU(3)$. Note that the total cross section of the e^+e^--annihilation in hadrons in the parton model behaves at large s similar to the cross section for the e^+e^--annihilation in the $\mu^+\mu^-$-pair, which can be proved using more general arguments [18].

In QCD, the Bjorken scaling for the structure functions is violated [19], [20], [21], [22] and the quark and hadron distributions depend logarithmically on Q^2 in accordance with the evolution equations [23]–[26].

7.3 Evolution equations for parton distributions

In the framework of the parton model, one can introduce the wave functions of the hadron in its infinite momentum frame $|p| \to \infty$ with the following normalization condition:

$$1 = \sum_n \int \prod_{i=1}^n \frac{d\beta_i d^2 k_{\perp i}}{(2\pi)^2} \, |\Psi(\beta_1, k_{\perp 1}; \beta_2, k_{\perp 2}; \ldots; \beta_n, k_{\perp n})|^2 \, \delta(1 - \sum_{i=1}^n \beta_i) \, \delta^2\left(\sum_{i=1}^n k_{\perp i}\right), \tag{7.32}$$

where k_i are the parton momenta, $k_{\perp i}$ are their components transverse to the hadron momentum p and β_i are the components along p, $(k_i p) = \beta_i p^2$. The wave function $\Psi(\beta_1, k_{\perp 1}; \beta_2, k_{\perp 2}; \ldots; \beta_n, k_{\perp n})$ satisfies the Schrödinger equation $H\Psi = E(p)\Psi$ and contains in the perturbation theory the energy propagators simplified in the infinite momentum frame as follows

$$\left(E(p) - \sum_{i=1}^n E(k_i)\right)^{-1} = 2|p|\left(m^2 - \sum_{i=1}^n \frac{m_i^2 + k_{i\perp}^2}{\beta_i}\right)^{-1}. \tag{7.33}$$

In the field theories, the integrals in the right-hand side of the normalization condition for Ψ are divergent at large momenta. We regularize them by introducing the ultraviolet cut-off Λ^2 over the transverse momenta $k_{i\perp}$:

$$\left|k_{i\perp}^2\right| < \Lambda^2. \tag{7.34}$$

Note that in gauge theories it is more natural to use the dimensional regularization of the ultraviolet divergencies to keep the gauge invariance. In this case, instead of Λ one introduces another dimensional parameter – the normalization point μ.

The amplitudes of the physical processes do not depend on Λ due to the property of the renormalizability, corresponding to a possibility to compensate the ultraviolet divergences by an appropriate choice of the bare-coupling constant depending on Λ. In particular, the renormalized-coupling constant does not depend on Λ. In a hard process of the type of the deep inelastic ep scattering it is convenient to fix Λ as follows

$$\Lambda^2 \sim Q^2. \tag{7.35}$$

In this case, the strong interaction does not modify significantly the hard subprocess and we obtain the usual formulae of the parton model for physical amplitudes. Note that to preserve the partonic picture in the gauge theories of the types of QED and QCD, one should use a physical gauge in which the virtual vector particles contain only the states with the positive norm. For the deep inelastic ep scattering it is convenient to choose the light-cone gauge

$$A_\mu q'_\mu = 0, \quad q' = q - \frac{q^2}{2pq} p, \quad q'^2 = 0, \tag{7.36}$$

where q is the virtual photon momentum and p is the proton momentum. In this gauge, the propagator of the gauge boson is

$$D_{\mu\nu}(k) = -\frac{\Lambda_{\mu\nu}}{k^2} = \frac{\sum_{i=1,2} e^i_\mu(k') e^{i*}_\nu(k')}{k^2} + \frac{q'_\mu q'_\nu}{(kq')^2}, \quad \Lambda_{\mu\nu} = \delta_{\mu\nu} - \frac{k_\mu q'_\nu + k_\nu q'_\mu}{kq'}. \tag{7.37}$$

On the mass-shell $k'^2 = 0$ $(k' = k - \frac{k^2}{xs} q')$ it contains only the physical polarization vectors $e^i_\mu(k')$ $(i = 1, 2)$ satisfying the constraints

$$e^i_\mu(k') k'_\mu = e^i_\mu(k') q'_\mu = 0. \tag{7.38}$$

Note that the last contribution $q'_\mu q'_\nu / (kq')^2$ to $D_{\mu\nu}(k)$ is usually unimportant because it does not contain the pole $1/k^2$ or the propagator is multiplied by q'_μ or q'_ν.

The polarization vectors have the following Sudakov decomposition

$$e^i = e^i_\perp - \frac{ke^i_\perp}{kq'} q' \tag{7.39}$$

and therefore they are parametrized by their transverse components.

The Λ-dependence of partonic wave functions expressed in terms of the physical-coupling constant is determined by the renormalization group:

$$|\Psi(\beta_1, k_{\perp 1}; \beta_2, k_{\perp 2}; \ldots; \beta_n, k_{\perp n}|^2 \sim \prod_r Z_r^{n_r}. \tag{7.40}$$

In the right-hand side of this relation we omitted the factor depending on β_r and $k_{\perp r}$. Further, $\sqrt{Z_r} < 1$ are the renormalization constants for the wave functions of the corresponding partons r and Z_r is the probability to find a physical particle r in the corresponding one-particle bare state. The quantity n_r is the number of the bare particles r for their total number n.

To find the Λ-dependence of the right-hand side of the normalization condition (7.32) from the integration limits Λ for $k_{\perp i}$, one should take into account that the largest integral contribution occurs from the momentum configuration of the type of the Russian doll, when the constituent particles in the initial hadron consist of two partons, each of these partons

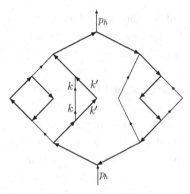

Fig. 7.4. Normalization condition for the partonic wave function.

again consists of two other partons, and so on. In each step of this parton branching, the transverse momenta of the particles significantly grow and only one of the last partons in this chain of decays reaches the largest possible value $|k_\perp| = \Lambda$. It is related to the fact that only for such configuration the number of the energy propagators (7.33) with large denominators is minimal. Moreover, in each of the denominators we can leave only two terms containing $\boldsymbol{k}_{\perp r} \approx -\boldsymbol{k}_{\perp \bar{r}}$ maximal on the corresponding branching level

$$\left(E(p) - \sum_i E(k_i) \right)^{-1} \approx -2\,|\boldsymbol{p}|\left(k_{\perp r}^2 \left(\frac{1}{\beta_r} + \frac{1}{\beta_{\bar{r}}} \right) \right)^{-1}.$$

Further, the quantum-mechanical interference of amplitudes with different decay schemes is not important in the normalization condition (7.32) within the leading logarithmic accuracy (see Fig.7.4), because in an opposite case the large transverse momentum in a loop diagram would enter in $k \geq 3$ energy denominators, which would lead to the loss of $\ln \Lambda$ in the integral over this momentum.

Therefore, the most essential contribution in the derivative of (7.32) over Λ appears from the upper limits of the integrals over $k_\perp^2 \simeq k_\perp'^2$ for two partons p, p' produced at the end of the decay chain (see Fig.7.4). This largest transverse momenta enter only in two energy denominators corresponding to the square $|\Psi|_{r \to pp'}^2$ of the wave function of the decaying parton r. Really $|\Psi|_n^2$ for n partons is factorized in the product of $|\Psi|_{r \to pp'}^2$ and $|\Psi|_{n-1}^2$ in which Λ^2 is substituted by $\boldsymbol{k}_\perp^2$

$$|\Psi|_n^2 = |\Psi|_{r \to pp'}^2 \, |\Psi|_{n-1}^2, \quad \Lambda^2 \to \boldsymbol{k}_\perp^2. \tag{7.41}$$

In accordance with this factorization property, after the differentiation of limits of integration over $k_\perp$ it is convenient to shift the summation variable $n - 1 \to n$.

Thus, the differentiation of the normalization condition (7.32), taking into account Eq. (7.40), gives the relation

$$0 = \sum_r \bar{n}_r \left(\frac{d \ln Z_r}{d \ln(\Lambda^2)} + \gamma_r \right), \quad \gamma_r = \sum_{p,p'} \gamma_{r \to pp'}, \quad \gamma_{r \to p\,p'} = \frac{d\,\|\Psi_{r \to pp'}\|^2}{d \ln(\Lambda)^2}. \tag{7.42}$$

Here $\|\Psi_{r\to pp'}\|^2 \sim g^2$ is the one-loop contribution to the norm of the wave function of the parton r related to its transition to two more virtual partons p and p'

$$\|\Psi_{r\to pp'}\|^2 = \int_{|k_\perp|<\Lambda} \frac{d^2k_\perp}{(2\pi)^2} \int_0^{\beta_r} |\Psi(\beta_p, k_\perp; \beta_{p'}, -k_\perp)|^2_{r\to pp'} d\beta_p, \quad \beta_r = \beta_p + \beta_{p'}.$$
(7.43)

With the use of dimensional considerations, we obtain

$$|\Psi(x, k_\perp; y-x, -k_\perp)|^2_{r\to pp'} = \frac{g^2(2\pi k_\perp^2)}{|k_\perp|^2} \frac{1}{y} f_{r\to pp'}(x/y).$$
(7.44)

The anomalous dimension γ_r of the field r also depends on the coupling constant (here, in comparison with that of Chapters 1 and 6, a different normalization for γ_r is used)

$$\gamma_{r\to pp'} = \frac{g^2(\Lambda^2)}{8\pi^2} \int_0^1 f_{r\to pp'}(x)dx.$$
(7.45)

The partial anomalous dimension $\gamma_{r\to pp'}$ describes the probability for the parton r to be in the state (p, p'). In Eq. (7.42) the quantity

$$\bar{n}_r = \sum_{\{n_s\}} n_r \int \prod_{i=1}^n \frac{d\beta_i d^2k_{\perp i}}{(2\pi)^2} |\Psi(\beta_1, k_{\perp 1}; \beta_2, k_{\perp 2}; \ldots; \beta_n, k_{\perp n}|^2 \delta\left(1 - \sum_{i=1}^n \beta_i\right) \delta^2\left(\sum_{i=1}^n k_{\perp i}\right)$$
(7.46)

is an averaged number of partons r in the hadron. Because this number is different for different hadrons, we derive the relation

$$\frac{dZ_r}{d\ln(\Lambda^2)} = -\gamma_r Z_r,$$
(7.47)

which coincides with the Callan–Simanzik equation for the renormalization constants. All γ_r are positive, since for $\gamma_r < 0$ the probability Z_r to find a physical particle in the bare state would tend to infinity at $\Lambda^2 \to \infty$.

Let us introduce now a more general quantity – the density of the number of partons l in the hadron as a function of the Feynman parameter x of this parton

$$n_l(x) = \sum_{\{n_s\}} \int \prod_{i=1}^n \frac{d\beta_i d^2k_{\perp i}}{(2\pi)^2} |\Psi(\beta_1, k_{\perp 1}; \quad \ldots; \beta_n, k_{\perp n}|^2 \delta$$

$$\times \left(1 - \sum_{i=1}^n \beta_i\right) \delta^2\left(\sum_{i=1}^n k_{\perp i}\right) \sum_{i\in l} \delta(\beta_i - x).$$
(7.48)

Note that

$$\bar{n}_r = \int_0^1 dx\, n_r(x).$$
(7.49)

The differentiation of $|\Psi|^2$ in expression (7.48) over $\ln \Lambda^2$ according to (7.40) and (7.47) gives the extra factor

$$C_\Psi = -\sum_t n_t \gamma_t.$$
(7.50)

As for the differentiation of the limits of integration over the momenta $k_{\perp i}^2$, one should take into account that with a leading logarithmic accuracy only two partons from their total number n can reach the largest transverse momentum $|k_\perp| = \Lambda$. There are two possibilities. In the first case, the parton $i \in l$ and another parton $\widetilde{i}$ have the maximal transverse momentum $k_{\perp i} = -k_{\perp \widetilde{i}}$. The parton of the type r decaying into the pair $i, \widetilde{i}$ has the Sudakov variable $y = \beta_i + \beta_{\widetilde{i}} > \beta_i$. The integration over β_i is performed with the use of the δ-function $\delta(\beta_i - x)$ or their sum $\delta(\beta_i - x) + \delta(\beta_{\widetilde{i}} - x)$ for $\widetilde{i} \in l$. Instead of summing over $i \in l$ in (7.48), we can perform summation over all partons $j \in r$ summing later over their different types r. The summation over $\widetilde{i}$ for fixed j is automatically performed. By substituting the summation index $n - 1$ in (7.48) by n after the differentiation in $\ln \Lambda^2$, one can obtain for the first case the following result

$$
\left(\frac{d}{d \ln \Lambda^2} n_l(x) \right)_1 = \frac{g^2(\Lambda)}{8\pi^2} \sum_r \int_x^1 \frac{dy}{y} \, n_r(y) \sum_{\widetilde{i}} \phi_{r \to l \widetilde{i}} \left(\frac{x}{y} \right)
$$

$$
= \frac{g^2(\Lambda)}{8\pi^2} \sum_r \int_x^1 \frac{dy}{y} \, n_r(y) \, \phi_{r \to l} \left(\frac{x}{y} \right), \tag{7.51}
$$

where we introduced the elementary inclusive probabilities

$$
\phi_{r \to l \widetilde{i}} \left(\frac{x}{y} \right) = f_{r \to l \widetilde{i}} \left(\frac{x}{y} \right), \quad (\widetilde{i} \neq l); \quad \phi_{r \to l \widetilde{i}} \left(\frac{x}{y} \right) = 2 f_{r \to l \widetilde{i}} \left(\frac{x}{y} \right), \quad (\widetilde{i} = l); \tag{7.52}
$$

$$
\phi_{r \to l} \left(\frac{x}{y} \right) = \sum_{\widetilde{i}} \phi_{r \to l \widetilde{i}} \left(\frac{x}{y} \right). \tag{7.53}
$$

with the use of relations (7.43) and (7.44). We obtain also the following expression for the anomalous dimension of the field r

$$
\gamma_{r \to l \widetilde{i}} = \frac{g^2(\Lambda)}{8\pi^2} \int_0^y \frac{dx}{y} \frac{x}{y} \phi_{r \to l \widetilde{i}} \left(\frac{x}{y} \right). \tag{7.54}
$$

In the second case, the pair of partons with the maximal transverse momentum $|k_\perp|$ does not include the extracted parton $i \in l$ with its Sudakov variable β_i in Eq.(7.48) and the differentiation of the integration limit over $\ln \Lambda^2$ leads to the factor in (7.48) being the sum of $n - 2$ anomalous dimensions γ_r for $n - 1$ partons at the previous branching level. This contribution after the substitution $n - 1 \to n$ would cancel $n - 1$ terms from n terms related to the differentiation of the factor $|\Psi|^2$ (see (7.50)). The sum of these two contributions is simplified as follows

$$
\left(\frac{d}{d \ln \Lambda^2} n_l(x) \right)_\Psi + \left(\frac{d}{d \ln \Lambda^2} n_l(x) \right)_2 = -\gamma_l \, n_l(x). \tag{7.55}
$$

Summing (7.51) and (7.55), we obtain the equation of Dokshitzer, Gribov, Lipatov, Altarelli and Parisi (DGLAP) [23]

$$
\frac{d}{d \ln \Lambda^2} n_l(x) = -\gamma_l \, n_l(x) + \frac{g^2(\Lambda)}{8\pi^2} \sum_r \int_x^1 \frac{dy}{y} \, n_r(y) \, \phi_{r \to l} \left(\frac{x}{y} \right). \tag{7.56}
$$

In QCD, this equation can be written as

$$\frac{d}{d\xi} n_l(x,\xi) = -w_l\, n_l(x,\xi) + \sum_r \int_x^1 \frac{dy}{y}\, w_{r\to l}\left(\frac{x}{y}\right) n_r(y,\xi), \qquad (7.57)$$

where $w_{r\to l}(z)$ is an elementary inclusive probability and the evolution variable ξ is defined as

$$\xi = -\frac{2N_c}{\beta_0} \ln \frac{\alpha(Q^2)}{\alpha_\mu}. \qquad (7.58)$$

It is expressed in terms of the QCD running coupling constant in one-loop approximation as follows

$$d\xi = \frac{\alpha(Q^2)\, N_c}{2\pi}\, d\ln Q^2, \quad \alpha(Q^2) = \frac{\alpha_\mu}{1 + \beta_0 \frac{\alpha_\mu}{4\pi} \ln \frac{Q^2}{\mu^2}}, \quad \beta_0 = \frac{11}{3}N_c - n_f \frac{2}{3}. \quad (7.59)$$

The quantity w_r is given below

$$w_r = \sum_k \int_0^1 dx\, x\, w_{r\to k}(x). \qquad (7.60)$$

The last relation corresponds to the energy-conservation constraint

$$\sum_k \int_0^1 dx\, x\, n_k(x,\xi) = 1. \qquad (7.61)$$

The DGLAP equation has a simple probabalistic interpretation and it is analogous to the balance equation for the densities of various gases being in chemical equilibrium. Indeed, the first term in its right-hand side describes the decrease of the number of partons l as a result of their decay to other partons in the opening phase space $d\xi$. On the other hand, the second term is responsible for its increase due to the possibility that the other partons r can decay into the states containing the parton l.

Traditionally, the evolution equation is written in other form

$$\frac{d}{d\xi} n_l(x,\xi) = \sum_r \int_x^1 \frac{dy}{y}\, P_{r\to l}\left(\frac{x}{y}\right) n_r(y,\xi), \qquad (7.62)$$

where

$$P_{r\to l}(z) = w_{r\to l}(z) - \delta_{r,l}\, \delta(1-z)\, w_r. \qquad (7.63)$$

In QCD, the kernels $P_{r\to r}(z)$ contain the singularities at $z = 1$ because the diagonal transition probabilities $w_{r\to r}(z)$ for the gluon $(r = g)$ and quark $(r = q)$ have the poles $\sim 1/(1-z)$. We can extract these poles explicitely

$$w_{g\to g}(z) = \frac{2z}{1-z} + w_{g\to g}^{reg}(z), \quad w_{q\to q}(z) = \frac{C_F}{N_c}\frac{1+z^2}{1-z} + w_{q\to q}^{reg}(z), \quad C_F = \frac{N_c^2 - 1}{2N_c}, \tag{7.64}$$

where $w_{g \to g}^{reg}(z)$ and $w_{q \to q}^{reg}(z)$ are analytic functions at $z = 1$. The regular contributions $w_{g \to g}^{reg}(z)$ and $w_{q \to q}^{reg}(z)$ will be calculated in the next section. Analogously, from the total decay probabilities we can extract divergent integrals in the form

$$w_g = \int_0^1 dz \, \frac{2}{1-z} + w_g^{reg}, \qquad w_q = \frac{C_F}{N_c} \int_0^1 dz \, \frac{2}{1-z} + w_q^{reg}. \qquad (7.65)$$

Then the splitting kernels $P_{r \to r}(z)$ are written as follows

$$P_{g \to g}(z) = P_{g \to g}^{sing}(z) + P_{g \to g}^{reg}(z), \qquad P_{g \to g}^{sing}(z) = \frac{2z}{1-z} - \delta(1-z) \int_0^1 dz' \frac{2}{1-z'},$$

$$P_{q \to q}(z) = P_{q \to q}^{sing}(z) + P_{q \to q}^{reg}(z), \qquad P_{q \to q}^{sing}(z) = \frac{C_F}{N_c} \left(\frac{1+z^2}{1-z} - \delta(1-z) \int_0^1 dz' \frac{2}{1-z'} \right). \qquad (7.66)$$

Here $P_{g \to g}^{reg}(z)$ and $P_{q \to q}^{reg}(z)$ do not contain divergencies and will be presented later. Indeed providing these kernels act on smooth parton distributions $n_g(x)$ and $n_q(x)$ we obtain the following results for the integral from the singular parts of the kernels

$$\int_0^1 dz \, P_{g \to g}^{sing}(z) \, n_g(z) = \int_0^1 dz \, \frac{2z}{(1-z)_+} \, n_g(z) \qquad (7.67)$$

and

$$\int_0^1 dz \, P_{q \to q}^{sing}(z) \, n_q(z) = \frac{C_F}{N_c} \int_0^1 dz \frac{1+z^2}{(1-z)_+} \, n_q(z), \qquad (7.68)$$

where we used the Altarelli–Parisi notation [23]

$$\int_0^1 dz \, \frac{1}{(1-z)_+} \, f(z) \equiv \int_0^1 dz \, \frac{1}{1-z} \, (f(z) - f(1)). \qquad (7.69)$$

It means that the result of integration of the splitting kernels with smooth functions is finite.

By integrating Eq. (7.57) over x one can obtain a simpler evolution equation for the average number of partons

$$n_k = \int_0^1 d x \, n_k(x, \xi). \qquad (7.70)$$

Namely,

$$\frac{d}{d\xi} n_k = -w_k n_k + \sum_r w_{r \to k} \, n_r, \qquad w_{r \to k} = \int_0^1 dx \, w_{r \to k}(x). \qquad (7.71)$$

Note that generally in this case the divergencies in the right-hand side are not compensated, which is related with the known double-logarithmic Sudakov-type behaviour of the parton multiplicities.

For the electric charge and other additive quantum numbers Q_k, we have the conservation law

$$\frac{d}{d\xi} \sum_k Q_k n_k = 0 \qquad (7.72)$$

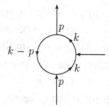

Fig. 7.5. Matrix element of the conserved current or energy-momentum tensor.

fulfilled due to the following property

$$\sum_k Q_k \, w_{r \to k} = w_r \, Q_r \tag{7.73}$$

of the elementary inclusive probability $w_{r \to k}$. The splitting kernels, respectively, satisfy the relations

$$\sum_k Q_k \int_0^1 dz \, P_{r \to k}(z) = 0. \tag{7.74}$$

The energy conservation

$$\frac{d}{d\xi} \sum_k \int_0^1 dx \, x \, n_k(x, \xi) = 0 \tag{7.75}$$

is valid due to another important property of the splitting kernels

$$\sum_k \int_0^1 dz \, z \, P_{r \to k}(z) = 0. \tag{7.76}$$

In turn, these sum rules can be used for finding the kernels $w_{r \to i}(x)$. For this purpose, on one hand one should calculate the Feynman diagram Fig.7.5 for the matrix elements of the conserved currents $j_\mu(z)$ or energy-momentum stress tensor $T_{\mu\nu}(z)$ between the states with the same momentum p with the Sudakov parametrization of the momentum k

$$k = \alpha q' + x p + k_\perp. \tag{7.77}$$

On the other hand, we can write the partonic expressions for the matrix element of the current $j_\mu q'_\mu$

$$\xi s' \int_0^1 dx \quad w_{r \to i}(x) = i \int \frac{|s'| d\alpha dx \, d^2 k_\perp}{2(2\pi)^4} \frac{g^2(k_\perp^2) \sum_t |\gamma_{r \to i,t}|^2}{-s(1-x)\alpha - k_\perp^2 + i\varepsilon} \frac{s'x}{\left(sx\alpha - k_\perp^2 + i\varepsilon\right)^2}, \tag{7.78}$$

in the case when the particles r and i have the same conserved quantum number Q, and

$$\xi s'^2 \int_0^1 dx \, x \, w_{r \to i}(x) = i \int \frac{|s'| d\alpha \, dx \, d^2 k_\perp}{2(2\pi)^4} \frac{g^2(k_\perp^2) \sum_t |\gamma_{r \to i,t}|^2}{-s(1-x)\alpha - k_\perp^2 + i\varepsilon} \frac{s'^2 x^2}{\left(sx\alpha - k_\perp^2 + i\varepsilon\right)^2} \tag{7.79}$$

for the energy-momentum tensor $T_{\mu\nu}q'_\mu q'_\nu$. Here the transition amplitude $\gamma_{r\to i,t}$ is calculated in the helicity basis. This basis is convenient because the helicity is conserved for the matrix elements of j_μ and $T_{\mu\nu}$. We find below the amplitudes $\gamma_{r\to i,t}$ for all possible parton transitions in QCD. In the above expressions, the integrals over α are nonzero only if the Sudakov variable x belongs to the interval $0 < x < 1$. They can be calculated by residues leading to the following expressions for splitting kernels

$$w_{r\to i}(x) = \sum_t \frac{|\gamma_{r\to i,t}|^2}{N_c} \frac{x(1-x)}{2|k_\perp|^2}, \tag{7.80}$$

which do not depend on $k_\perp$ because $\gamma_{r\to i,t} \sim k_\perp$.

7.4 Splitting kernels in the Born approximation

7.4.1 Transition from a gluon to two gluons

We start with the discussion of the splitting kernels $w_{r\to k}\left(\frac{x}{y}\right)$ for the transition from gluon to gluon. The gluon Yang–Mills vertex for the transition $(p,\sigma) \to (k,\mu) + (p-k,\nu)$ can be written as follows (see Fig. 7.6)

$$\gamma_{\sigma\mu\nu} = (p-2k)_\sigma \delta_{\mu\nu} + (p+k)_\nu \delta_{\sigma\mu} + (k-2p)_\mu \delta_{\nu\sigma} \tag{7.81}$$

up to the colour factor f_{abc} being the gauge-group structure constant which enters in the commutation relations for the generators:

$$[T_a, T_b] = i f_{abc} T_c. \tag{7.82}$$

On the mass-shell, the numerators of the gluon propagators in the light-cone gauge coincide with the projectors to physical states. The corresponding polarization vectors satisfying also the Lorentz condition $ek' = 0$ have the following Sudakov representation

$$e_\sigma(p) = e_\sigma^\perp, \quad e_\mu(k') = e_\mu^\perp - \frac{ke^\perp}{kq'}q'_\mu, \quad e_\nu(p-k') = e_\nu^\perp - \frac{(p-k)e^\perp}{(p-k)q'}q'_\nu,$$

$$k' = k - \frac{k^2}{2kq'}q'. \tag{7.83}$$

After multiplication of the Yang–Mills vertex $\gamma_{\sigma\mu\nu}$ with these polarization vectors one can introduce the tensor $\gamma_{\sigma\mu\nu}^\perp$ with transverse components according to the definition

$$\gamma_{\sigma\mu\nu}e^\sigma(p)e^\mu(k')e^\nu(p-k') \equiv \gamma_{\sigma\mu\nu}^\perp e_\perp^\sigma(p)e_\perp^\mu(k')e_\perp^\nu(p-k'), \tag{7.84}$$

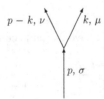

Fig. 7.6. Yang–Mills vertex.

where

$$\gamma^\perp_{\sigma\mu\nu} = -2k^\perp_\sigma \, \delta^\perp_{\mu\nu} + \frac{2}{1-x} k^\perp_\nu \, \delta^\perp_{\sigma\mu} + \frac{2}{x} k^\perp_\mu \, \delta^\perp_{\nu\sigma}, \quad x = \frac{kp}{q'p}. \tag{7.85}$$

The gluons moving in the z-direction with the helicities $\lambda = \pm 1$ are described by the polarization vectors

$$e^\pm_\perp = \frac{1}{\sqrt{2}}(e^1 \pm ie^2). \tag{7.86}$$

We put the helicity λ of the initial gluon with the momentum p equal to $+1$. For $\lambda = -1$, the results can be found from the obtained expressions by changing the signs of helicities of the final particles. The nonzero matrix elements

$$\gamma_{\lambda_1\lambda_2} = \gamma^\perp_{\sigma\mu\nu} e_{+\sigma}(p) e^*_{\lambda_1\mu}(k') e^*_{\lambda_2\nu}(p-k') \tag{7.87}$$

are

$$\gamma_{++} = \sqrt{2}\frac{k^*}{x(1-x)}, \quad \gamma_{+-} = \sqrt{2}\frac{x}{1-x}k, \quad \gamma_{-+} = \sqrt{2}\frac{1-x}{x}k, \tag{7.88}$$

where $k = k_1 + ik_2$, $k^* = k_1 - ik_2$.

The dimensionless quantities

$$w_{1+\to i}(x) = \frac{x(1-x)}{2\,|k|^2} \sum_{t=\pm} |\gamma_{it}|^2, \tag{7.89}$$

are the the elementary inclusive probabilities (we do not write the colour factor $f_{abc}f_{a'bc} = N_c\delta_{aa'}$, because it is included in the definition of ξ)

$$w_{1+\to 1+}(x) = \frac{1+x^4}{x(1-x)}, \quad w_{1+\to 1-}(x) = \frac{(1-x)^4}{x(1-x)}. \tag{7.90}$$

For the total probability of the gluon transition to gluons we get:

$$w_{g\to gg} = \int_0^1 \frac{1+x^4+(1-x)^4}{x(1-x)} x \, dx = \frac{1}{2}\int_0^1 \frac{1+x^4+(1-x)^4}{x(1-x)} dx, \tag{7.91}$$

where the factor x in the integrand is substituted by $\frac{1}{2}$ due to its symmetry under the substitution $x \to 1-x$. The divergency of $w_{g\to gg}$ at $x = 0, 1$ is cancelled in the evolution equations. There is also the contribution to w_g from the quark-antiquark state (see below):

$$w_{g\to q\bar q} = \frac{n_f}{2N_c} \int_0^1 \left(x^2 + (1-x)^2\right) dx = \frac{n_f}{3N_c}. \tag{7.92}$$

To present the evolution equation in the traditional form (see, e.g. ([8],[9])) we write the splitting kernels for the gluon-gluon transitions as follows

$$\begin{aligned} P_{1+\to 1+}(x) &= w_{1+\to 1+}(x) + w_g \, \delta(x-1) \\ &= \frac{2x}{(1-x)_+} + \frac{(1+x)(1-x^2)}{x} + \left(\frac{11}{6} - \frac{n_f}{3N_c}\right)\delta(x-1), \end{aligned} \tag{7.93}$$

$$P_{1^+ \to 1^-}(x) = w_{1^+ \to 1^-}(x) + w_g = \frac{(1-x)^3}{x}, \tag{7.94}$$

where the notations (7.69) were used. For the splitting kernel describing unpolarised gluon-gluon transitions, we obtain

$$P_{g \to g}(x) = \frac{2x}{(1-x)_+} + 2\frac{(1-x)(1+x^2)}{x} + \left(\frac{11}{6} - \frac{n_f}{3N_c}\right)\delta(x-1). \tag{7.95}$$

The matrix elements for the anomalous dimension matrix describing gluon-gluon transitions can be written as follows

$$w_{r \to k}^j = \int_0^1 dx \ w_{r \to k}(x)\left(x^{j-1} - x\,\delta_{rk}\right) - \frac{n_f}{3N_c}\,\delta_{rk}. \tag{7.96}$$

Thus, we obtain

$$w_{1^+ \to 1^+}^j = 2\psi(1) - 2\psi(j-1) - \frac{1}{j+2} - \frac{1}{j+1} - \frac{1}{j} - \frac{1}{j-1} + \frac{11}{6} - \frac{n_f}{3N_c}. \tag{7.97}$$

where $\psi(z)$ is the Euler function,

$$\psi(z) = \frac{d}{dz}\ln\Gamma(z), \quad \psi(1) = -C \approx -0.577 \tag{7.98}$$

and

$$w_{1^+ \to 1^-}^j = -\frac{1}{j+2} + \frac{3}{j+1} - \frac{3}{j} + \frac{1}{j-1}. \tag{7.99}$$

The anomalous dimension of tensors $G_{\mu_1\sigma}D_{\mu_2}\ldots G_{\mu_j\sigma}$, corresponding to the vector (electric) gluon field is:

$$w_{1 \to 1}^{jv} = 2\psi(1) - 2\psi(j-1) - \frac{2}{j+2} + \frac{2}{j+1} - \frac{4}{j} + \frac{11}{6} - \frac{n_f}{3N_c}. \tag{7.100}$$

Note that the regular term $\frac{11}{6} - \frac{n_f}{3N_c}$ is proportional to the function β_0 entering the expression for the running of the QCD-coupling constant. It seems to be related to a supersymmetric generalization of QCD.

Further, the anomalous dimension for the axial tensors $G_{\mu_1\sigma}D_{\mu_2}\ldots \widetilde{G}_{\mu_j\sigma}$ in the gluodynamics is

$$w_{1 \to 1}^{ja} = 2\psi(1) - 2\psi(j) - \frac{4}{j+1} + \frac{2}{j} + \frac{11}{6} - \frac{n_f}{3N_c}. \tag{7.101}$$

The energy conservation sum rule for $j = 2$

$$w_{1 \to 1}^{2v} + w_{1 \to 1/2}^{2v} = 0. \tag{7.102}$$

is fulfilled as it can be verified from the expression for $w_{1 \to 1/2}^{jv}$ obtained in the next subsection. Because the contribution $w_{1 \to 1/2}^{jv}$ is proportional to n_f, one can verify that $w_{1 \to 1}^{2v} = 0$ at $n_f = 0$.

7.4.2 Transition from a gluon to a quark pair

The propagator of the massless fermion can be written in the form

$$G(k) = \frac{\not{k}}{k^2}, \quad \not{k} = \sum_{\lambda=\pm} u^\lambda(k') \overline{u^\lambda(k')} + \frac{k^2}{2kq'} \not{q}', \quad k' = k - \frac{k^2}{xs} q', \quad (7.103)$$

where $k'^2 = 0$. The last contribution in $\not{k}$ is absent on the mass shell or providing that the vertex neighboring to the propagator G is $\not{q}'$. The massless fermion with the momentum k and the helicity $\lambda/2$ is described by the spinor

$$u^\lambda(k) = \sqrt{k_0} \begin{pmatrix} \varphi^\lambda \\ \lambda\varphi^\lambda \end{pmatrix}, \quad \overline{u}\gamma_\mu u = 2k_\mu, \quad \not{k} = \sum_\lambda u^\lambda \overline{u}^\lambda, \quad (7.104)$$

where the Pauli spinor φ satisfies the equation

$$\sigma k \varphi^\lambda - \lambda k_0 \varphi^\lambda = \begin{pmatrix} k_3 - \lambda k_0 & k_1 - ik_2 \\ k_1 + ik_2 & -k_3 - \lambda k_0 \end{pmatrix} \begin{pmatrix} \varphi_1^\lambda \\ \varphi_2^\lambda \end{pmatrix} = 0, \quad k_0 = |\boldsymbol{k}|. \quad (7.105)$$

In the light-cone frame $p = p_3 \to \infty$, we have

$$\varphi^+ \simeq \begin{pmatrix} 1 \\ \frac{k}{2xp} \end{pmatrix}, \quad \varphi^- \simeq \begin{pmatrix} \frac{-k^*}{2xp} \\ 1 \end{pmatrix}. \quad (7.106)$$

The massless antifermion with the momentum $p - k$ and the helicity $\lambda'/2$ is described by the spinor

$$v^{\lambda'}(-p + k) = \sqrt{p_0 - k_0} \begin{pmatrix} \chi^{\lambda'} \\ -\lambda'\chi^{\lambda'} \end{pmatrix}, \quad (7.107)$$

where the Pauli spinor χ satisfies the equation

$$\sigma(p - k)\chi^{\lambda'} + \lambda'(p_0 - k_0)\chi^{\lambda'} = 0. \quad (7.108)$$

In the light-cone frame, we have

$$\chi^- \simeq \begin{pmatrix} 1 \\ \frac{-k}{2(1-x)p} \end{pmatrix}, \quad \chi^+ \simeq \begin{pmatrix} \frac{k^*}{2(1-x)p} \\ 1 \end{pmatrix}, \quad (7.109)$$

and v^λ satisfies the equation $\gamma_5 v^\lambda = -\lambda v^\lambda$. Note that in the case of the left Pauli neutrino we have only $\lambda = -1$ for v and $\lambda' = 1$ for $\overline{v}$ in correspondence with the eigenvalues of the matrix γ_5 for its eigenfunctions u^λ and $v^{\lambda'}$, respectively.

Thus, for the matrix element of the vertex describing the transition of the gluon with momentum $\boldsymbol{p}$ and helicity 1 into a pair of fermions (see Fig. 7.7), we have

$$\gamma^{\lambda\lambda'} = \overline{u}^\lambda(\overrightarrow{k}) \frac{\gamma^1 + i\gamma^2}{\sqrt{2}} v^{\lambda'}(-p + k) = 2\sqrt{(p_0 - k_0)k_0}\lambda\,\delta_{\lambda,-\lambda'}\,\sqrt{2}\varphi^{\lambda*} \begin{pmatrix} 0 & 1 \\ 0 & 0 \end{pmatrix} \chi^{\lambda'}.$$

$$(7.110)$$

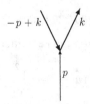

Fig. 7.7. Gluon-quark-antiquark vertex

The nonzero matrix elements $\gamma^{\lambda\lambda'}$ are given by:

$$\gamma^{+-} = -\sqrt{2}\,k\,\sqrt{\frac{x}{1-x}}, \qquad \gamma^{-+} = -\sqrt{2}\,k\,\sqrt{\frac{1-x}{x}}. \tag{7.111}$$

Now, let us use the following representation for the elementary inclusive probabilities, which was obtained above,

$$w_{1^+\to q}(x) = \frac{n_f}{2N_c}\,\frac{x(1-x)}{2\,|k|^2}\,\sum_t |\gamma^{qt}|^2. \tag{7.112}$$

Here $\frac{1}{2}$ is the colour factor, n_f is the number of different types of quarks, and the factor $1/N_c$ takes into account that N_c is included in the definition of ξ. Thus, we get

$$w_{1^+\to q^+}(x) = w_{1^+\to\bar{q}^+}(x) = \frac{n_f}{2N_c}x^2, \qquad w_{1^+\to q^-}(x) = w_{1^+\to\bar{q}^-}(x) = \frac{n_f}{2N_c}(1-x)^2. \tag{7.113}$$

We write also these splitting kernels in the traditional notations, corresponding to the evolution equation (7.62)

$$P_{1^+\to q^+}(x) = \frac{n_f}{2N_c}x^2, \qquad P_{1^+\to q^-}(x) = \frac{n_f}{2N_c}(1-x)^2,$$

$$P_{g\to q}(x) = P_{g\to\bar{q}} = \frac{n_f}{2N_c}(x^2 + (1-x)^2). \tag{7.114}$$

The total probability for the gluon transition to quarks is

$$w_{1\to q,\bar{q}} = \frac{n_f}{2N_c}\int_0^1 \left(x^2 + (1-x)^2\right)dx = \frac{n_f}{3N_c}. \tag{7.115}$$

For QED, the colour factor $\frac{n_f}{2N_c}$ is absent and we obtain $w_{1\to\frac{1}{2}\frac{1}{2}} = \frac{2}{3}$. Taking into account that in the perturbation theory $\xi = \frac{\alpha}{2\pi}\ln\Lambda^2$ for QED, the probability for the photon to be in the e^+e^--pair state is $\frac{\alpha}{3\pi}\ln\Lambda^2$ in agreement with the known result for the charge renormalization in this theory.

The nonvanishing matrix elements for the anomalous dimension matrix are

$$w^j_{1^+\to q^+} = \frac{n_f}{2N_c}\frac{1}{j+2}, \qquad w^j_{1^+\to q^-}(x) = \frac{n_f}{2N_c}\left(\frac{1}{j} - \frac{2}{j+1} + \frac{1}{j+2}\right) \tag{7.116}$$

and these quantities for the vector and axial current are

$$w^{jv}_{1\to 1/2} = \frac{n_f}{N_c}\left(\frac{1}{j} - \frac{2}{j+1} + \frac{2}{j+2}\right), \tag{7.117}$$

$$w_{1\to 1/2}^{ja} = \frac{n_f}{N_c}\left(-\frac{1}{j} + \frac{2}{j+1}\right),$$ (7.118)

where we added the gluon transitions to the quark and antiquark.

7.4.3 Transition from a quark to the quark–gluon system

The amplitude for the transition of a quark with the helicity $+$ to quark and gluon (see Fig. 7.8) can be written as follows in the light-cone gauge for the gluon-polarization vector:

$$\bar{u}^+(k)\left(e_\perp^{*\lambda} + \frac{k_\perp e^{*\lambda}}{(p-k,q')}\right)\slashed{q}'\,u^+(p) = \sqrt{k_0 p_0}\,2\varphi^{+*}\sigma e^{*\lambda}\varphi^+ + 2k_\perp e^{*\lambda}\frac{\sqrt{x}}{1-x}.$$ (7.119)

Therefore, we have

$$\bar{u}^+(k)\left(\slashed{e}_\perp^{+*} + \frac{k_\perp e^{+*}}{(p-k,q')}\right)\slashed{q}'\,u^+(p) = \sqrt{2}k^*\frac{1}{\sqrt{x}(1-x)},$$ (7.120)

$$\bar{u}^+(k)\left(\slashed{e}_\perp^{-*} + \frac{k_\perp e^{-*}}{(p-k,q')}\right)\slashed{q}'\,u^+(p) = \sqrt{2}k\frac{\sqrt{x}}{(1-x)}.$$ (7.121)

Thus, the splitting kernels are

$$w_{1/2+\to 1+}(x) = \frac{c}{x}, \quad w_{1/2+\to 1-}(x) = c\frac{(1-x)^2}{x}, \quad w_{1/2+\to 1/2+}(x) = c\frac{1+x^2}{1-x},$$ (7.122)

where

$$c = \frac{N_c^2 - 1}{2N_c^2} = \frac{C_F}{N_c}$$ (7.123)

is the colour factor for the corresponding loop (note that N_c is included in ξ)

The total contribution to $w_{1/2}$ is

$$w_{1/2} = \frac{C_F}{2N_c}\int_0^1 \left(\frac{1+x^2}{1-x} + \frac{1+(1-x)^2}{x}\right)dx.$$ (7.124)

The splitting kernels for the quark–gluon transitions in the evolution equation (7.62) are

$$P_{1/2+\to 1+}(x) = \frac{C_F}{N_c}\frac{1}{x}, \quad P_{1/2+\to 1-}(x) = \frac{C_F}{N_c}\frac{(1-x)^2}{x},$$

$$P_{q\to g}(x) = \frac{C_F}{N_c}\frac{1+(1-x)^2}{x}.$$ (7.125)

Fig. 7.8. Quark–gluon–quark vertex

Analogously, the splitting kernels for the quark-quark transition in (7.62)

$$P_{1/2^+ \to 1/2^+}(x) = \frac{C_F}{N_c} \left(\frac{1+x^2}{(1-x)_+} + \frac{3}{2} \delta(1-x) \right), \tag{7.126}$$

where we used the notations (7.69).

The corresponding anomalous dimensions are

$$w^j_{1/2^+ \to 1^+} = \frac{C_F}{N_c} \frac{1}{(j-1)}, \quad w^j_{1/2^+ \to 1^-} = \frac{C_F}{N_c} \left(\frac{1}{j-1} - \frac{2}{j} + \frac{1}{j+1} \right), \tag{7.127}$$

$$\begin{aligned} w^j_{1/2^+ \to 1/2^+} &= \frac{C_F}{N_c} \int_0^1 \frac{1+x^2}{1-x} (x^{j-1} - 1) dx \\ &= \frac{C_F}{N_c} \left(2\psi(1) - 2\psi(j) - \frac{1}{j+1} - \frac{1}{j} + \frac{3}{2} \right). \end{aligned} \tag{7.128}$$

The vector and axial contributions are

$$w^{jv}_{1/2 \to 1} = \frac{C_F}{N_c} \left(\frac{2}{j-1} - \frac{2}{j} + \frac{1}{j+1} \right), \tag{7.129}$$

$$w^{ja}_{1/2 \to 1} = \frac{C_F}{N_c} \left(\frac{2}{j} - \frac{1}{j+1} \right). \tag{7.130}$$

We have the sum rules

$$w^{1v}_{1/2 \to 1/2} = 0, \quad w^{2v}_{1/2 \to 1} + w^{2v}_{1/2 \to 1/2} = 0, \tag{7.131}$$

expressing the conservation of the baryon charge and the energy, respectively, in the quark decay.

As it is seen from the above formulae, there are two Dokshitzer relations among the matrix elements of the anomalous dimension matrix for $n_f = N_c$

$$\frac{N_c}{C_F} \left(w^{jv}_{1/2 \to 1} + w^{jv}_{1/2 \to 1/2} \right) = w^{jv}_{1 \to 1} + w^{jv}_{1 \to 1/2} = 2\psi(1) - 2\psi(j-1) - \frac{3}{j} + \frac{3}{2}, \tag{7.132}$$

$$\frac{N_c}{C_F} \left(w^{ja}_{1/2 \to 1} + w^{ja}_{1/2 \to 1/2} \right) = w^{ja}_{1 \to 1} + w^{ja}_{1 \to 1/2} = 2\psi(1) - 2\psi(j) - \frac{2}{j+1} + \frac{1}{j} + \frac{3}{2}. \tag{7.133}$$

The relations (7.132), (7.133) can be derived using arguments based on the supersymmetry (SUSY). In the supersymmetric generalization of the Yang–Mills theory, the gluon and its partner, gluino, are unified in one multiplet. The gluino is a Majorana fermion which coincides with the corresponding antiparticle and belongs to an adjoint representation of the gauge group. In this model, the total probability of finding both gluon and gluino with the parameter x should not depend on the spin $s = 1, 1/2$ of the initial particle, since they are components of the supermultiplet.

On the other hand, for the probability of gluon transition to the $q\bar{q}$ pair in QCD we have the extra factor n_f/N_c in comparison with its transition to two gluinos, because the ratio of the number of the fermion states for these theories is $2n_f$ and the ratio of the corresponding colour factors is $1/(2N_c)$. Analogously, for the probability of the transition of the quark to the gluon and quark we have the additional colour factor $\frac{C_F}{N_c}$ – the ratio of two Casimir operators in comparison with the transition to the gluon and gluino.

Collecting the results obtained above, we can write the evolution equation for parton distributions in leading order approximation, i.e. where all terms, proportional to $[\alpha_s(Q^2)\ln(Q^2/\mu^2)]^n$ are accounted, but terms $\sim \alpha_s^k(Q^2)[\alpha_s(Q^2)\ln Q^2/\mu^2]^n, k \geq 1$ are disregarded (DGLAP equations):

$$\frac{d}{d\ln Q^2}n_g(x, Q^2) = \frac{\alpha_s(Q^2)}{2\pi}\int_x^1 \frac{dy}{y}\left\{P_{q\to g}\left(\frac{x}{y}\right)\sum_f[n_f(y, Q^2) + n_{\bar f}(y, Q^2)]\right.$$

$$\left. + P_{g\to g}\left(\frac{x}{y}\right)n_g(y, Q^2)\right\}$$

$$\frac{d}{d\ln Q^2}n_f(x, Q^2) = \frac{\alpha_s(Q^2)}{2\pi}\int_x^1 \frac{dy}{y}\left\{P_{q\to q}\left(\frac{x}{y}\right)n_f(y, Q^2) + P_{g\to q}\left(\frac{x}{y}\right)n_g(y, Q^2)\right\}$$

$$\frac{d}{d\ln Q^2}n_{\bar f}(x, Q^2) = \frac{\alpha_s(Q^2)}{2\pi}\int_x^1 \frac{dy}{y}\left\{P_{q\to q}\left(\frac{x}{y}\right)n_{\bar f}(y, Q^2) + P_{g\to q}\left(\frac{x}{y}\right)n_g(y, Q^2)\right\},$$

$$(7.134)$$

where

$$P_{q\to q}(z) = \frac{4}{3}\left[\frac{1+z^2}{(1-z)_+} + \frac{3}{2}\delta(1-z)\right]$$

$$P_{g\to q}(z) = \frac{1}{2}\left[z^2 + (1-z)^2\right]$$

$$P_{q\to g}(z) = \frac{4}{3}\frac{1+(1-z)^2}{z}$$

$$P_{g\to g}(z) = 6\left[\frac{1-z}{z} + \frac{z}{(1-z)_+} + z(1-z) + \left(\frac{11}{12} - \frac{n_f}{18}\right)\delta(1-z)\right]. \quad (7.135)$$

To solve the evolution equation it is necessary to fix the initial conditions: the values of $n_g, n_f, n_{\bar f}$ at some $Q^2 = Q_0^2$. Usually, it is chosen the low $Q_0^2 \sim 2 - 5$ GeV^2. The x-dependence of $n_i(x, Q_0^2)$ is taken from intuitive considerations – from the expected behaviour at $x \to 0$ and $x \to 1$ (see [4]). A few numerical parameters are introduced in $n_i(x, Q_0^2)$. These parameters are determined by the best fit of the solution of evolution equations to the whole set of experimental data. The parton densities are devided into a singlet:

$$N^{sing} = n_u + n_{\bar u} + n_d + n_{\bar d} + n_s + n_{\bar s} + n_g \quad (7.136)$$

and a nonsinglet part:

$$N_{T=1}^{nonsing} = n_u - n_d + n_{\bar{d}} - n_{\bar{u}}, \tag{7.137}$$

$$N_{T=0}^{nonsingl} = \sum_i n_i^{octet} + n_{\bar{i}}^{octet}. \tag{7.138}$$

Each part is described by separate equations. Only quark and antiquark distributions contribute to nonsinglet evolution equations.

7.5 OPE on light cone and parton model

As already was mentioned, the cross section of the inclusive reaction $lN \rightarrow l'$+everything is proportional to

$$W_{\mu\nu}(q, p) = \int d^4 z e^{iqz} \langle p \mid [j_\mu(z), j_\nu(0)] \mid p \rangle. \tag{7.139}$$

(See [2] and the books [4],[9], the second term in the commutator $\sim j_\nu(0) j_\mu(z)$ can be omitted, since its contribution vanishes for the matrix elements between the proton states $\mid p \rangle$.) It was proved in [6] (see also [4]), that at high $\mid q^2 \mid \rightarrow \infty$ the amplitude (7.139) is dominated by the domain near the light cone, $z^2 = 0$. Since we expect that the parton model is valid at high $\mid q^2 \mid$, we can put $z^2 = 0$ in (7.139) and consider the integrand as a function of z_λ along light cone. For simplicity, we consider one flavour and the symmetrical part of the tensor $W_{\mu,\nu}$. Further,

$$j_\mu(z) = \bar{q}(z) \gamma_\mu q(z), \tag{7.140}$$

where $q(z)$ is the quark field. In the leading order in α_s, the product $j_\mu(z) j_\nu(0)$ is given by

$$\bar{q}(z)\gamma_\mu q(z)\bar{q}(0)\gamma_\nu q(0) + \bar{q}(z)\gamma_\mu q(z)\bar{q}(0)\gamma_\nu q(0) \tag{7.141}$$

and the contracted terms shall be substituted by free propagators:

$$S(z) = \frac{1}{\pi} \, \not{z} \delta'(z^2). \tag{7.142}$$

After substitution of (7.142) in (7.141) and simple algebra, we get:

$$W_{\mu\nu}^{sym}(q, p) = \frac{1}{\pi} \int d^4 z e^{iqz} S_{\mu\alpha\nu\beta} \delta'(z^2) z_\alpha \langle p \mid \bar{q}(z) \gamma_\beta q(0) \mid p \rangle, \tag{7.143}$$

where

$$S_{\mu\alpha\nu\beta} = \delta_{\mu\alpha}\delta_{\nu\beta} + \delta_{\mu\beta}\delta_{\alpha\nu} - \delta_{\mu\nu}\delta_{\alpha\beta}. \tag{7.144}$$

Expand now the matrix element in (7.143) in powers of z. (Because of $\delta'(z^2)$ factor in (7.143) it is the expansion along the light cone.) In the Fock–Schwinger gauge, $z_\lambda \partial_\lambda$ can be replaced by $z_\lambda \nabla_\lambda$ where ∇_λ is the covariant derivative. After transfer of the derivatives to the right, we have:

$$W_{\mu\nu}^{sym} = \frac{1}{\pi} \int d^4 z e^{iqz} S_{\mu\alpha\nu\beta} \delta'(z^2) z_\alpha \sum_{\substack{n-odd \\ n=1,3}} \frac{1}{n!} z_{\mu_1} \ldots z_{\mu_n} \langle p \mid \bar{q}(0) \gamma_\beta \nabla_{\mu_1} \ldots \nabla_{\mu_n} q(0) \mid p \rangle.$$

(7.145)

The summation in (7.145) is going over odd $n = 1, 3 \ldots$. The reason is that $W_{\mu\nu}^{sym}(q, p)$ is an even function of q_λ, $W^{sym}(q, p) = W^{sym}(-q, p)$. Let us keep only the terms proportional to $\delta_{\mu\nu}$ and $p_\mu p_\nu$. All others can be restored from the requirement of gauge invariance. The substitution of (7.144) into (7.145) results in:

$$W_{\mu\nu}^{sym} = -\frac{1}{2\pi} \int d^4 z e^{iqz} \partial_\alpha \delta(z^2) \delta_{\mu\nu} \sum_{n=1,3,\ldots} \frac{1}{n!} z_{\mu_1} \ldots z_{\mu_n} \langle p \mid \bar{q}(0) \gamma_\alpha \nabla_{\mu_1} \ldots \nabla_{\mu_n} q(0) \mid p \rangle$$

$$-\left[\frac{1}{2\pi} \int d^4 z e^{iqz} \delta(z^2) \sum_{\substack{n-even \\ n\geq 0}} \frac{1}{n!} z_{\mu_1} \ldots z_{\mu_n} \langle p \mid \bar{q}(0) \gamma_\mu \nabla_\nu \nabla_{\mu_1} \ldots \nabla_{\mu_n} q(0) \mid p \rangle + (\mu \to \nu) \right].$$

(7.146)

The first term in (7.146) reduces to:

$$W_{\mu\nu}^{sym(1)} = \frac{i}{2\pi} \delta_{\mu\nu} \int d^4 z e^{iqz} q_\alpha \delta(z^2) \sum_{n=1,3 \ldots} z_{\mu_1} \ldots z_{\mu_n} \langle p \mid \bar{q} \gamma_\alpha \nabla_{\mu_1} \ldots \nabla_{\mu_n} q \mid p \rangle.$$

(7.147)

The term, proportional to:

$$\langle p \mid \bar{q} \gamma_\alpha \nabla_{\mu_1} \ldots \nabla_\alpha \ldots \nabla_{\mu_n} q \mid p \rangle$$

(7.148)

is dropped, because its spin is lower than the spin of the corresponding term in (7.147), which results to additional suppression factor $1/Q^2$ (see below). The general form of the matrix element in (7.147) is:

$$\langle p \mid \bar{q}(0) \gamma_\alpha \nabla_{\mu_1} \ldots \nabla_{\mu_n} q(0) \mid p \rangle = i^{n+1} p_\alpha p_{\mu_1} \ldots p_{\mu_n} a_n,$$

(7.149)

where a_n are some numbers. Using the equalities

$$\int d^4 z e^{iqz} \delta(z^2) z_{\mu_1} \ldots z_{\mu_n} = \frac{\partial^n}{\partial q_{\mu_1} \ldots \partial q_{\mu_n}} \frac{1}{i^n} \int d^4 z e^{iqz} \delta(z^2),$$

(7.150)

$$\int d^4 z e^{iqz} \delta(z^2) = -\frac{2\pi}{q^2},$$

(7.151)

$$\frac{\partial^n}{\partial q_{\mu_1} \ldots \partial q_{\mu_n}} = 2^n q_{\mu_1} \ldots q_{\mu_n} \left(\frac{\partial}{\partial q^2} \right)^n + \text{trace terms},$$

(7.152)

the first term in (7.146) can be calculated and appears to be equal ($\nu = pq$):

$$W_{\mu\nu}^{sym(1)} = -\frac{1}{2} \delta_{\mu\nu} \sum_{n=1,3\ldots} \left(\frac{2\nu}{-q^2} \right)^{n+1} a_n = -\frac{1}{2} \delta_{\mu\nu} \sum_n \frac{a_n}{x^{n+1}}.$$

(7.153)

The second term in (7.146) is calculated in similar way. We have:

$$W_{\mu\nu}^{sym(2)} = -\frac{1}{2\pi} \int d^4z \, e^{iqz} \delta(z^2) \sum_{n=0,2...} \frac{1}{n!} \langle p \mid \bar{q}\gamma_\mu \nabla_\nu \nabla_{\mu_1} \cdots \nabla_{\mu_n} \cdots \nabla_{\mu_n} q \mid p \rangle$$

$$= \sum_{n=1,3...} \left(\frac{2\nu}{-q^2}\right)^n \frac{a_n}{\nu} p_\mu p_\nu = \frac{1}{\nu} \sum_n \frac{1}{x^n} p_\mu p_\nu a_n. \tag{7.154}$$

Finally,

$$W_{\mu\nu}^{sym} = \left(-\delta_{\mu\nu}\frac{1}{2x} + \frac{1}{\nu}p_\mu p_\nu\right) \sum_{n=1,3...} \frac{a_n}{x^n}. \tag{7.155}$$

From comparison with (7.10) follows the Gallan–Gross relation (7.16). According to (7.17) the infinite sum

$$\sum_{n=1,3}^{\infty} \frac{a_n}{x^{n+1}} = n(x) \tag{7.156}$$

can be interpreted as the quark density. Note that if there are two conciding indices in (7.149), as happened in (7.148), i.e. the spin of the operator is lower, by two units, then in the right hand-side appears the factor p^2. For dimensional grounds, such a term is suppressed by $1/Q^2$. Therefore, in OPE the contributions of various operators to the cross section of deep inelastic scattering is determined not by their dimensions d, but by their twist

$$t = d - j, \tag{7.157}$$

where j is the spin. The lowest possible twist is 2. Note also, that the conservation low (7.61) in terms of OPE corresponds to the contribution of the operator of the energy-momentum tensor.

7.6 Evolution equations for fragmentation functions

Let us consider now the evolution equations for the fragmentation functions $\bar{D}_i^h(z) = h_i(z)$, which are the distributions of the hadron h carrying the longitudinal fraction z of the momentum of the parton i. The quark-fragmentation function $h_q(z)$ enters, for example, the cross section for the inclusive production of hadrons (see (7.26)). The initial parton is assumed to be a highly virtual particle. The virtualities of particles decrease in the process of their decay to other partons (cf. Fig.(7.4)) in such way that the partons forming the hadron have virtualities of the order of the value of the characteristic QCD scale Λ_{QCD}.

Let us find the normalization condition for the wave function $\psi_i(k_1, .., k_n)$ of the virtual parton i with momentum k which is a superposition of the near-mass-shell partons with momenta $k_1, .., k_n$. Consider the nonrenormalized Green functions in the light-cone gauge. One can present such Green functions for the quark and gluon in the form (cf. (7.37) and (7.103))

$$G(k) = \frac{\not{k}}{k^2} g(k^2, kq') + \frac{\not{q}'}{kq'} f(k^2, kq'),$$

$$D_{\mu\nu}(k) = \frac{\Lambda_{\mu\nu}}{k^2} d(k^2, kq') + \frac{q'_\mu q'_\nu}{(kq')^2} c(k^2, kq'), \tag{7.158}$$

where we used the light-cone vector q' (7.36). In our kinematics, we have for the parton virtuality (see (7.34) and (7.35)) the restriction from above

$$k^2 \ll 2kq' \sim \Lambda^2 \sim Q^2, \tag{7.159}$$

where Q^2 is a characteristic large scale in the corresponding hard process. For example, in the e^+e^--annihilation to hadrons this scale is the virtuality of the intermediate photon. The functions g, f, d, and c depend in the leading logarithmic approximation only on the ratio k^2/Λ^2.

Because at $k^2 = \Lambda^2$ the nonrenormalized Green functions coincide with the free propagators, we have the following normalization conditions

$$g(k^2/\Lambda^2)|_{k^2=\Lambda^2} = 1, \quad f(k^2/\Lambda^2)|_{k^2=\Lambda^2} = 0,$$

$$d(k^2/\Lambda^2)|_{k^2=\Lambda^2} = 1, \quad c(k^2/\Lambda^2)|_{k^2=\Lambda^2} = 0. \tag{7.160}$$

Let us write the dispersion relations for the functions g and d in the form

$$\frac{1}{k^2} g(k^2/\Lambda^2) = \frac{Z_q}{k^2} + \frac{1}{\pi} \int_{m^2}^{\Lambda^2} \frac{d\tilde{k}^2}{k^2 - \tilde{k}^2 + i\epsilon} \, \mathrm{Im}\left(-\frac{1}{\tilde{k}^2} g(\tilde{k}^2/\Lambda^2)\right),$$

$$\frac{1}{k^2} d(k^2/\Lambda^2) = \frac{Z_g}{k^2} + \frac{1}{\pi} \int_{m^2}^{\Lambda^2} \frac{d\tilde{k}^2}{k^2 - \tilde{k}^2 + i\epsilon} \, \mathrm{Im}\left(-\frac{1}{\tilde{k}^2} d(\tilde{k}^2/\Lambda^2)\right), \tag{7.161}$$

where Z_q and Z_g are squares of the renormalization constants for corresponding fields.

Using the constraints (7.160), we obtain the following sum rules

$$1 = Z_q + \frac{1}{\pi} \int_{m^2}^{\Lambda^2} d\tilde{k}^2 \, \mathrm{Im}\left(-\frac{1}{\tilde{k}^2} g(\tilde{k}^2/\Lambda^2)\right),$$

$$1 = Z_g + \frac{1}{\pi} \int_{m^2}^{\Lambda^2} d\tilde{k}^2 \, \mathrm{Im}\left(-\frac{1}{\tilde{k}^2} d(\tilde{k}^2/\Lambda^2)\right). \tag{7.162}$$

They correspond to the normalization conditions for the wave functions $\psi_i(k_1, \ldots, k_n)$ which will be derived below.

For this purpose, let us find g and d from Eqs. (7.158)

$$g(k^2/\Lambda^2) = \frac{k^2}{2kq'} \frac{1}{2} \mathrm{Tr} \, \slashed{q}' G(k), \quad d(k^2/\Lambda^2) = \frac{1}{2} \Lambda_{\mu\nu} D_{\mu\nu}(k). \tag{7.163}$$

Since the nonphysical contributions cancel in the propagators, one can express in these relations the propagators through the projectors to the physical states (see (7.37) and (7.103))

$$D_{\mu\nu}(k) \sim \frac{\sum_{i=1,2} e^i_\mu(k') e^{i*}_\nu(k')}{k^2}, \quad G(k) \sim \frac{\sum_{\lambda=\pm} u^\lambda(k') \bar{u}^\lambda(k')}{k^2}, \quad k' = k - \frac{k^2}{xs} q', \tag{7.164}$$

Note that these representations are in an agreement with the normalization conditions for the Dirac spinors and polarization vectors

$$\frac{1}{2kq'} \overline{u^\lambda(k')} \hat{q}' u^{\lambda'}(k') = \delta^{\lambda\lambda'}, \quad e^\lambda_\mu(k') e^{\lambda'*}_\nu(k') \Lambda_{\mu\nu} = \delta^{\lambda\lambda'}. \tag{7.165}$$

In relations (7.161) and (7.162), one can use the following expressions for the imaginary parts of the corresponding Green functions in terms of the amplitudes $M^\lambda_{i \to h_n}$ which describe transitions of the initial parton i to the hadron states h_n

$$2\text{Im}\left(-\frac{1}{\tilde{k}^2} g(\tilde{k}^2/\Lambda^2)\right) = \frac{1}{\tilde{k}^4} \frac{1}{2} \sum_{\lambda=\pm 1/2} \sum_{h_n} \int d\Omega_n |M^\lambda_{q \to h_n}|^2,$$

$$2\text{Im}\left(-\frac{1}{\tilde{k}^2} d(\tilde{k}^2/\Lambda^2)\right) = \frac{1}{\tilde{k}^4} \frac{1}{2} \sum_{\lambda=\pm 1} \sum_{h_n} \int d\Omega_n |M^\lambda_{g \to h_n}|^2, \tag{7.166}$$

where we averaged over the helicities $\pm s$ of the initial parton. Note that one can omit this averaging due to the parity conservation in strong interactions. The phase space $d\Omega_n$ for momenta of the final hadrons in the Sudakov variables defined in (7.32) can be written as follows

$$d\Omega_n = \prod_{i=1}^{n} \frac{d^3 k_i}{2|k_i|(2\pi)^3} (2\pi)^4 \delta^4(\tilde{k} - \sum_{i=1}^{n} k_i)$$

$$= \prod_{i=1}^{n-1} \frac{d\beta_i \, d^2 k_{i\perp}}{2\beta_i (2\pi)^3} \frac{2\pi}{1 - \sum_{i=1}^{n-1} \beta_i} \delta(\tilde{k}^2 - \sum_i \frac{k_{i\perp}^2}{\beta_i}), \tag{7.167}$$

where

$$1 > \beta_i > 0, \quad \tilde{\beta} = 1 - \sum_{i=1}^{n} \beta_i, \quad \tilde{\alpha} s = \tilde{k}^2 < \Lambda^2. \tag{7.168}$$

Using Eqs. (7.162) and (7.166), one can get the following normalization condition for the wave function of the parton in the space of hadron states

$$1 - Z_q = \frac{1}{2} \sum_{\lambda=\pm 1/2} \sum_{h_n} \int \prod_{i=1}^{n-1} \frac{d\beta_i \, d^2 k_{i\perp}}{2\beta_i (2\pi)^3} \frac{2\pi}{1 - \sum_{i=1}^{n-1} \beta_i} \frac{|M^\lambda_{q \to h_n}|^2}{\left(\sum_i \frac{k_{i\perp}^2}{\beta_i}\right)^2},$$

$$1 - Z_g = \frac{1}{2} \sum_{\lambda=\pm 1/2} \sum_{h_n} \int \prod_{i=1}^{n-1} \frac{d\beta_i \, d^2 k_{i\perp}}{2\beta_i (2\pi)^3} \frac{2\pi}{1 - \sum_{i=1}^{n-1} \beta_i} \frac{|M^\lambda_{g \to h_n}|^2}{\left(\sum_i \frac{k_{i\perp}^2}{\beta_i}\right)^2}. \tag{7.169}$$

The above expressions are applied, strictly speaking, only to renormalized field theories of the type of QED, for which bare particles (partons) coincide in their quantum numbers with physical particles (hadrons). In QCD, quarks and gluons do not exist in the free state and therefore the amplitude $M^\lambda_{q \to h_n}$ does not have any physical meaning. Nevertheless, we shall use the above expressions also in the QCD case, assuming that the final hadron states include necessarily also soft quarks and gluons, which are annihilated with other soft particles appearing from the decay of other hard partons produced in the initial state.

The kinematics in which one can obtain the leading logarithmic terms for the fragmentation functions is different from the kinematics, which gives the main contribution to the normalization of the partonic wave function of a hadron (see Fig.(7.4)). We shall discuss it below.

In a one-loop approximation ($n = 2$) for the transitions $i \to (j, k)$, the normalization condition for the parton wave function takes the form

$$1 - Z_i = \sum_{\lambda_j, \lambda_k} \int_0^1 \frac{d\beta}{\beta(1 - \beta)} \int \frac{d^2 k_\perp}{\left(\frac{k_\perp^2}{\beta} + \frac{k_\perp^2}{1-\beta} \right)^2} |M_{i \to j,k}^{\lambda_i; \lambda_j, \lambda_k}|^2, \tag{7.170}$$

where $k, k' - k$ are momenta of two final partons and λ_j, λ_k are their polarizations, respectively. It is obvious that this expression coincides formally with the analogous one-loop contribution to the normalization condition of the hadron wave function (see, for example, (7.32)). Nevertheless, there is an essential difference in the interpretation of two results. Namely, in (7.32) the wave function Ψ has the small energy propagator $1/\Delta E$ for the final partons, whereas in (7.169) we have such propagator for the initial parton. The difference is especially obvious if one considers the wave functions for the state containing several ($n > 2$) final particles with their momenta $k_1, k_2, \ldots, k_n$ and the Sudakov components $x_1, x_2, \ldots, x_n$. In this case, for the intermediate state of r partons with momenta $q_1, q_2, \ldots, q_r$ the hadron wave function has the energy propagator

$$\left(m^2 - \sum_{k=1}^{r} \frac{q_k^2}{\beta_k} \right)^{-1}, \tag{7.171}$$

but for the parton wave function the corresponding propagator is

$$\left(\sum_{i=1}^{n} \frac{k_i^2}{\beta_i} - \sum_{k=1}^{r} \frac{q_k^2}{\beta_k} \right)^{-1}. \tag{7.172}$$

As was mentioned above, the second important peculiarity in the case of fragmentation functions in comparison with parton distributions is related to the kinematics for the flow of particle transverse momenta. In the parton shower drawn in Fig.(7.4) for the hadron wave function, each pair of partons is produced with significantly larger transverse momenta than the transverse momentum of the initial particle, but in the case of the transition of a parton to hadrons each produced a pair of partons (j, l) flies almost in the same direction as the initial parton (i) and has a much smaller relative transverse momentum. To show it, let us consider the contribution of this elementary transition $i \to j, l$ to the denominator of the energy propagator (7.172) and impose on momenta the condition of its relative smallness

$$\frac{k_i^2}{\beta_i} - \frac{k_j^2}{\beta_j} - \frac{k_l^2}{\beta_l} \ll k_i^2. \tag{7.173}$$

It is obvious that the left-hand side is zero for the planar kinematics when the emission angles $\theta \sim k_\perp/\beta$ for their momenta are the same

$$\frac{k_i}{\beta_i} = \frac{k_j}{\beta_j} = \frac{k_l}{\beta_l}. \tag{7.174}$$

We obtain the logarithmic contribution over the transverse momenta of produced particles i, j providing that

$$\left(\frac{k_j}{\beta_j} - \frac{k_i}{\beta_i}\right)^2 \ll \left(\frac{k_i}{\beta_i}\right)^2. \tag{7.175}$$

It means that the relative angles $\Delta\theta_i$ of produced pairs decrease in the course of the subsequent decays

$$1 \gg \Delta\theta_1 \gg \Delta\theta_2 \gg \ldots \gg \Delta\theta_r. \tag{7.176}$$

Thus, the hadron and parton wave functions are different even on the partonic level, when instead of final hadrons h we consider the partons with a fixed virtuality $k^2 \sim \Lambda_{QCD}^2$. Nevertheless, for the last case, in the region of fixed β the evolution equations for the parton distributions $D_h^k(x)$ and for the fragmentation functions $\bar{D}_h^k(x)$ in the leading logarithmic approximation coincide and one can get the Gribov–Lipatov relation [23]–[25]

$$\bar{D}_h^k(x) = D_h^k(x), \tag{7.177}$$

which illustrates a duality between the theoretical descriptions of hard processes in terms of hadrons and partons. In the next-to-leading approximation, this relation is violated [27] because in the fragmentation function at small x there appear double-logarithmic contributions $\sim \alpha^2 \ln^3 x \ln Q^2$, which are absent in the parton distributions.

Another interesting equality valid in the Born approximation is the so-called Drell–Levy–Yan relation [15]

$$\bar{D}_k^h(x) = (-1)^{2(s_k - s_h) + 1} x D_h^k\left(\frac{1}{x}\right), \tag{7.178}$$

where s_h and s_k are the spins of the corresponding particles. This relation is violated in QED and QCD already in the leading logarithmic approximation, because the point $x = 1$ turns out to be a singular point $D_h^k(x) \sim (1 - x)^a$ with a noninteger value of a. In each order of the perturbation theory we have a polynomial of logarithms $\ln(1 - x)$ with the coefficients which are analytic functions of x. The receipt for the analytic continuation of $D_h^k(x)$ around the point $x = 1$ compatible with the relation (7.178) is simple: the arguments of logarithms should be taken as the modules $|1 - x|$ at $x > 1$ with an analytic continuation of the coefficients of the polynomials [24]. (For the connection of $D_h^k(x)$ and $\bar{D}_k^h(x)$ near $x = 1$ see [4]). The fragmentation functions are discussed in more detail in Chapter 8.

7.7 Parton distributions in QCD in LLA

To solve the evolution equations (7.57) for the parton distributions $n_k(x)$ in the leading logarithmic approximation (LLA) of QCD one can use the scale invariance of their integral kernels to the dilatations $x \to \lambda x$ in the Bjorken variable. This invariance is related to conservation of the total angular momentum j in the crossing channel t. Thus, we search for solution $n_k(x)$ of the DGLAP equations for QCD in the form of the Mellin representation in the variable $\ln(1/x)$

$$n_k(x) = \int_{\sigma-i\infty}^{\sigma+i\infty} \frac{dj}{2\pi i} \left(\frac{1}{x}\right)^j n_k^j, \tag{7.179}$$

where n_k^j are the analytically continued moments n_k^j of the parton distributions

$$n_k^j = \int_0^1 dx \, x^{j-1} n_k(x). \tag{7.180}$$

These quantities have simple physical meaning. Namely, for integer j they coincide with the matrix elements of the so-called twist-2 operators O_k^j

$$n_k^j(s) = < h|O_k^j|h >, \qquad O_k^j = q_{\mu_1}' q_{\mu_1}' \cdots q_{\mu_j}' O_k^{\mu_1\mu_2\ldots\mu_j}. \tag{7.181}$$

For the nonpolarized case, the corresponding gauge-invariant tensors are

$$O_q^{\mu_1\mu_2\ldots\mu_j} = i^{j-1} S_{\{\mu_1\ldots\mu_j\}} \bar\psi \, \gamma_{\mu_1} \nabla_{\mu_2} \ldots \nabla_{\mu_j} \psi, \tag{7.182}$$

$$O_g^{\mu_1\mu_2\ldots\mu_j} = i^{j-2} S_{\{\mu_1\ldots\mu_j\}} G_{\sigma\,\mu_1} D_{\mu_2} \ldots D_{\mu_{j-1}} G_{\sigma\,\mu_j}. \tag{7.183}$$

The corresponding pseudotensor operators appeared in the polarized case are

$$\widetilde{O}_q^{\mu_1\mu_2\ldots\mu_j} = i^{j-1} S_{\{\mu_1\ldots\mu_j\}} \bar\psi \gamma_5 \gamma_{\mu_1} \nabla_{\mu_2} \ldots \nabla_{\mu_j} \psi, \tag{7.184}$$

$$\widetilde{O}_g^{\mu_1\mu_2\ldots\mu_j} = i^{j-2} S_{\{\mu_1\ldots\mu_j\}} G_{\sigma\,\mu_1} D_{\mu_2} \ldots D_{\mu_{j-1}} \widetilde{G}_{\sigma\,\mu_j}. \tag{7.185}$$

In the above relations, $S_{\{\mu_1\ldots\mu_j\}}$ means the symmetrization and subtraction of traces in the corresponding Lorentz indices, D_μ are the covariant derivatives, and $\widetilde{G}_{\sigma\,\mu}$ is the dual tensor of the gluon field. The above operators have the dimension $d = j + 2$ in the mass units. Therefore, all of them have the twist $t = d - j$ equal to 2. There are also four extra operators

$$O_{q,\sigma}^{\mu_1\mu_2\ldots\mu_j} = i^{j-1} S_{\{\mu_1\ldots\mu_j\}} \bar\psi \gamma_\sigma^\perp \gamma_{\mu_1} \nabla_{\mu_2} \ldots \nabla_{\mu_j} \psi, \tag{7.186}$$

$$O_{g,\sigma_1,\sigma_2}^{\mu_1\mu_2\ldots\mu_j} = i^{j-2} S_{\{\mu_1\ldots\mu_j\}} G_{\sigma_1\,\mu_1}^\perp D_{\mu_2} \ldots D_{\mu_{j-1}} G_{\sigma_2\,\mu_j}^\perp \tag{7.187}$$

$$\widetilde{O}_{q,\sigma}^{\mu_1\mu_2\ldots\mu_j} = i^{j-1} S_{\{\mu_1\ldots\mu_j\}} \bar\psi \gamma_5 \gamma_\sigma^\perp \gamma_{\mu_1} \nabla_{\mu_2} \ldots \nabla_{\mu_j} \psi, \tag{7.188}$$

$$\widetilde{O}_{g,\sigma_1,\sigma_2}^{\mu_1\mu_2\ldots\mu_j} = i^{j-2} S_{\{\mu_1\ldots\mu_j\}} G_{\sigma_1\,\mu_1}^\perp D_{\mu_2} \ldots D_{\mu_{j-1}} \widetilde{G}_{\sigma_2\,\mu_j}^\perp, \tag{7.189}$$

with the twist 2, but they do not enter the description of deep inelastic processes.

In LLA, the momenta n_k^j obey the system of the ordinary differential equations

$$\frac{\partial}{\partial \xi} n_k^j = \sum_r w_{r\to k}^j n_r^j, \tag{7.190}$$

which correspond to the renormalization group equations for the operators O_k^j. Here the variable ξ is defined in eq. (7.58) and (7.59) as follows

$$\xi(Q^2) = \frac{2N_c}{\beta_s} \ln \frac{\alpha_\mu}{\alpha(Q^2)} \tag{7.191}$$

and the elements of the anomalous dimension matrix $w_{r \to k}^j$ are calculated above. This matrix has the block-diagonal form in an appropriate basis and can be easily diagonalized

$$\sum_r w_{r \to k}^j n_r^j(s) = w^j(s) n_k^j(s). \tag{7.192}$$

Its eigenvalues $w^j(s)$ and eigenfunctions $n_k^j(s)$ describe the asymptotic behaviour of the multiplicatively renormalized matrix elements of twist 2 operators O_s^j

$$n_k^j(s) =< h|O_s^j|h > \tag{7.193}$$

at large ultraviolet cut-off $\Lambda^2 = Q^2$

$$n_k^j(s, Q^2) = n_k^j(s, Q_0^2) \exp\left(\Delta\xi\, w^j(s)\right), \quad \Delta\xi = \xi(Q^2) - \xi(Q_0^2). \tag{7.194}$$

Note that the asymptotic freedom, corresponding to logarithmic vanishing of $\alpha(Q^2)$ at large Q^2, leads to a rather weak dependence of $n_k^j(s)$ from Q^2. The value of the matrix element $n_k^j(s)$ at $Q^2 = Q_0^2$ is an initial condition for the evolution equation and should be extracted from the experimental data.

The moments of the parton distributions n_k^j are linear combinations of the eigenfunctions $n_k^j(s, Q^2)$ (7.194). Partly, the coefficients in these combinations are fixed by the quantum numbers of the corresponding twist-2 operators. But there are operators with the same quantum numbers and constructed from gluon or quark fields. They are mixed with others in the process of renormalization. The linear combinations of the parton distributions which are multiplicatively renormalizable are obtained from diagonalization of the anomalous dimension matrix $w_{r \to k}^j$ calculated above.

We write below the moments n_k^j entering expression (7.179) for the parton distributions $n_k(x)$ being solutions of the DGLAP equations (7.57) in LLA. The helicity of the initial hadron with its spin s is assumed to be $+s$. The simplest result is obtained for momenta of the flavour nonsinglet distributions

$$\frac{1}{2}\left(n_q(j) - n_{\bar{q}}(j)\right)_{Q^2} = \frac{1}{2}\left(n_q(j) - n_{\bar{q}}(j)\right)_{Q_0^2} \exp\left(\Delta\xi\, w_{1/2^+ \to 1/2^+}^j\right), \tag{7.195}$$

where the helicity is conserved $\lambda_h = \lambda_q$ and due to (7.128) the anomalous dimension $w_{1/2^+ \to 1/2^+}^j$ is

$$w_{1/2^+ \to 1/2^+}^j = \frac{N_c^2 - 1}{2N_c^2}\left(2\psi(1) - 2\psi(j) - \frac{1}{j+1} - \frac{1}{j} + \frac{3}{2}\right). \tag{7.196}$$

For the flavour singlet quantum numbers in the crossing channel the corresponding expressions are more complicated. We begin with the distributions of nonpolarized quarks and gluons ($i = 1/2, 1$)

$$\frac{1}{2}\left(n_{i+}(j) + n_{i-}(j)\right) = \sum_{s=1,2} V_s^i(j) \exp\left(\Delta\xi\, w_s^{jv}\right). \tag{7.197}$$

The anomalous dimension matrix for the nonpolarized case was calculated above (see expressions (7.128), (7.129), (7.117), and (7.100))

$$w_{1/2\to 1/2}^{jv} = \frac{N_c^2 - 1}{2N_c^2}\left(2\psi(1) - 2\psi(j) - \frac{1}{j+1} - \frac{1}{j} + \frac{3}{2}\right),$$

$$w_{1/2\to 1}^{jv} = \frac{N_c^2 - 1}{2N_c^2}\left(\frac{2}{j-1} - \frac{2}{j} + \frac{1}{j+1}\right),$$

$$w_{1\to 1/2}^{jv} = \frac{n_f}{N_c}\left(\frac{1}{j} - \frac{2}{j+1} + \frac{2}{j+2}\right),$$

$$w_{1\to 1}^{jv} = 2\psi(1) - 2\psi(j-1) - \frac{2}{j+2} + \frac{2}{j+1} - \frac{4}{j} + \frac{11}{6} - \frac{n_f}{3N_c}. \tag{7.198}$$

The coefficients $V_s^i(j)$ and quantities w_s^{jv} for $s = 1, 2$ are respectively two eigenfunctions and eigenvalues of the anomalous dimension matrix $w_{k\to i}^{ja}$

$$w^{jv} V_i(j) = \sum_{k=1,1/2} w_{k\to i}^{jv} V_k(j) \tag{7.199}$$

with the initial condition

$$\sum_{r=1,2} V_i^r(j) = \frac{1}{2}\left(n_{i+}(j) + n_{i-}(j)\right)\big|_{Q^2=Q_0^2} \tag{7.200}$$

fixed by the experimental data at $Q^2 = Q_0^2$.

In a similar way, one can construct the solution of the evolution equations (7.57) which describe the Q^2-dependence of the difference of distributions for the partons with the same and opposite helicities with respect to that of the initial hadron

$$\frac{1}{2}\left(n_{i+}(j) - n_{i-}(j)\right) = \sum_{s=1,2} A_i^s(j) \exp\left(\Delta\xi\, w_s^{ja}\right) \tag{7.201}$$

using the anomalous dimension matrix for polarized particles (see (7.128), (7.130)), (7.118), and (7.101)

$$w_{1/2+\to 1/2+}^{ja} = \frac{N_c^2 - 1}{2N_c^2}\left(2\psi(1) - 2\psi(j) - \frac{1}{j+1} - \frac{1}{j} + \frac{3}{2}\right),$$

$$w_{1/2\to 1}^{ja} = \frac{N_c^2 - 1}{2N_c^2}\left(\frac{2}{j} - \frac{1}{j+1}\right);$$

$$w_{1\to 1/2}^{ja}(x) = \frac{n_f}{N_c}\left(-\frac{1}{j} + \frac{2}{j+1}\right),$$

$$w_{1\to 1}^{ja} = 2\psi(1) - 2\psi(j) - \frac{4}{j+1} + \frac{2}{j} + \frac{11}{6} - \frac{n_f}{3N_c}. \tag{7.202}$$

In this case, the coefficients $A_i^s(j)$ and quantities w_s^{ja} are obtained from the eigenvalue equation for the above matrix $w_{k \to i}^j$

$$w^{ja} A_i(j) = \sum_{k=1,1/2} w_{k \to i}^{ja} A_k(j) \tag{7.203}$$

with the additional condition

$$\sum_{s=1,2} A_i^s(j) = \frac{1}{2} (n_{i+}(j) - n_{i-}(j))|_{Q^2=Q_0^2} \tag{7.204}$$

fixed by the experimental data at $Q^2 = Q_0^2$.

One can assume, for example, that already for $Q^2 = Q_0^2$ the formulae of the parton model are correct. Then for $Q^2 \gg Q_0^2$, we have

$$n_k(x, Q^2) = \sum_r \int_x^1 \frac{dy}{y} n_r(y, Q_0^2) W_{r \to k}(x/y), \tag{7.205}$$

where $W_{r \to k}(x)$ is an inclusive probability to find the hard parton k inside a comparatively soft parton r and $n_r(y, Q_0^2)$ is a parton distribution function at the scale Q_0^2.

The inclusive probabilities $W_{r \to k}(x)$ are normalized as follows

$$W_{r \to k}(x)|_{Q^2=Q_0^2} = \delta_{r,k}\, \delta(x - 1). \tag{7.206}$$

Their momenta

$$W_{r \to k}(j) = \int_0^1 dx\, x^{j-1}\, W_{r \to k}(x) \tag{7.207}$$

can be calculated in LLA

$$\frac{1}{2} \left(W_{q \to q}(j) - W_{q \to \bar{q}}(j) \right) = \frac{1}{2} \left(W_{\bar{q} \to \bar{q}}(j) - W_{q \to \bar{q}}(j) \right) = \exp\left(\Delta\xi\, w_{1/2+ \to 1/2+}^j \right) \tag{7.208}$$

and

$$\frac{1}{2} \left(W_{r+ \to i+}(j) + W_{r+ \to i-}(j) \right) = \sum_{s=\pm} V_{r \to i}^s(j) \exp\left(\Delta\xi\, w_s^{jv} \right), \tag{7.209}$$

$$\frac{1}{2} \left(W_{r+ \to i+}(j) - W_{r+ \to i-}(j) \right) = \sum_{s=\pm} A_{r \to i}^s(j) \exp\left(\Delta\xi\, w_s^{ja} \right), \tag{7.210}$$

where

$$w_{\pm}^{jv} = \frac{w_{1/2 \to 1/2}^{jv} + w_{1 \to 1}^{jv}}{2} \pm \sqrt{\frac{1}{4} \left(w_{1/2 \to 1/2}^{jv} - w_{1 \to 1}^{jv} \right)^2 + w_{1/2 \to 1}^{jv} w_{1 \to 1/2}^{jv}},$$

$$w_{\pm}^{ja} = \frac{w_{1/2 \to 1/2}^{ja} + w_{1 \to 1}^{ja}}{2} \pm \sqrt{\frac{1}{4} \left(w_{1/2 \to 1/2}^{ja} - w_{1 \to 1}^{ja} \right)^2 + w_{1/2 \to 1}^{ja} w_{1 \to 1/2}^{ja}}, \tag{7.211}$$

$$V_{q\to q}^{\pm}(j) = -\frac{w_{\mp}^{jv} - w_{1/2\to 1/2}^{jv}}{w_{\pm}^{jv} - w_{\mp}^{jv}}, \quad V_{g\to q}^{\pm}(j) = \frac{w_{1\to 1/2}^{jv}}{w_{\pm}^{jv} - w_{\mp}^{jv}},$$

$$V_{q\to g}^{\pm}(j) = \frac{w_{1/2\to 1}^{jv}}{w_{\pm}^{jv} - w_{\mp}^{jv}}, \quad V_{g\to g}^{\pm}(j) = \frac{w_{\pm}^{jv} - w_{1/2\to 1/2}^{jv}}{w_{\pm}^{jv} - w_{\mp}^{jv}}, \tag{7.212}$$

$$A_{q\to q}^{\pm}(j) = -\frac{w_{\mp}^{ja} - w_{1/2\to 1/2}^{ja}}{w_{\pm}^{ja} - w_{\mp}^{ja}}, \quad A_{g\to q}^{\pm}(j) = \frac{w_{1\to 1/2}^{ja}}{w_{\pm}^{ja} - w_{\mp}^{ja}},$$

$$A_{q\to g}^{\pm}(j) = \frac{w_{1/2\to 1}^{ja}}{w_{\pm}^{ja} - w_{\mp}^{ja}}, \quad A_{g\to g}^{\pm}(j) = \frac{w_{\pm}^{ja} - w_{1/2\to 1/2}^{ja}}{w_{\pm}^{ja} - w_{\mp}^{ja}}. \tag{7.213}$$

In particular, we have

$$\frac{1}{2}\left(W_{q\to q}(1) - W_{q\to \bar q}(1)\right) = \frac{1}{2}\left(W_{\bar q\to \bar q}(1) - W_{q\to \bar q}(1)\right) = 1, \tag{7.214}$$

which corresponds to the baryon number conservation. Further, for momenta of the nonpolarized parton distributions with $j = 2$ one can get

$$W_{q\to q}(2) = \frac{w_{+}^{2v} - w_{1/2\to 1/2}^{2v}}{w_{+}^{2v}} + \frac{w_{1/2\to 1/2}^{2v}}{w_{+}^{2v}} \exp\left(\Delta\xi\, w_{+}^{2v}\right),$$

$$W_{q\to g}(2) = \frac{w_{1/2\to 1/2}^{2v}}{w_{+}^{2v}}\left(1 - \exp\left(\Delta\xi\, w_{+}^{2v}\right)\right),$$

$$W_{g\to q}(2) = \frac{w_{+}^{2v} - w_{1/2\to 1/2}^{2v}}{w_{+}^{2v}}\left(1 + \exp\left(\Delta\xi\, w_{+}^{2v}\right)\right),$$

$$W_{g\to g}(2) = \frac{w_{1/2\to 1/2}^{2v}}{w_{+}^{2v}} + \frac{w_{+}^{2v} - w_{1/2\to 1/2}^{2v}}{w_{+}^{2v}} \exp\left(\Delta\xi\, w_{+}^{2v}\right), \tag{7.215}$$

where $w_{-}^{2v} = 0$, because the energy-momentum tensor $\theta_{\mu\nu}(x)$ is not renormalized. From these expressions, we obtain the sum rule

$$W_{h\to g}(2) + W_{h\to q}(2) = 1 \tag{7.216}$$

valid for any initial particle h, which corresponds to the parton energy conservation in the hadron infinite momentum frame $p \to \infty$. Moreover, at $\Delta\xi \to \infty$, where the contribution $\exp\left(\Delta\xi\, w_{+}^{2v}\right)$ tends to zero, the average energy taken by gluon or quark is independent of the type of the initial particle h

$$W_{h\to g}(2) = \frac{w_{1/2\to 1/2}^{2v}}{w_{+}^{2v}}, \quad W_{h\to q}(2) = \frac{w_{+}^{2v} - w_{1/2\to 1/2}^{2v}}{w_{+}^{2v}}. \tag{7.217}$$

This property is a consequence of factorization of the parton distribution momenta $D_k^l(j)$ for each multiplicatively renormalized operator

$$D_k^l(j) = \sum_{r=1,2} b_k(r)b^l(r) \exp\left(\Delta\xi\, w^j\right).$$

For the quasi-elastic region $x \to 1$, the large values of the Lorentz spin $j \sim 1/(1-x)$ are essential in the Mellin representation for $n_k(x)$. In the limit $j \to \infty$, the anomalous dimensions are simplified

$$w_+^j \approx w_{1/2 \to 1/2}^j \approx w_+^{jv} \approx w_+^{ja} \approx \frac{N_c^2 - 1}{2N_c^2} \left(-2 \ln j + \frac{3}{2} \right),$$

$$w_-^j \approx w_-^{jv} \approx w_-^{ja} \approx -2 \ln j + \frac{11}{6} - \frac{n_f}{3N_c}. \tag{7.218}$$

Using also Eqs. (7.212) and (7.213), one can verify that in the quasi-elastic regime the inclusive probability $W_{q \to g}(x)$ is small and in other transitions the helicity sign is conserved leading to the relation $A \approx V$. The momenta of nonvanishing inclusive probabilities at large j are

$$W_{q+ \to g+}(j) \approx \frac{1}{2} \frac{N_c^2 - 1}{N_c^2 + 1} \frac{1}{j \ln j} \left(e^{\Delta\xi \, w_+^j} - e^{\Delta\xi \, w_-^j} \right),$$

$$W_{g+ \to q+}(j) \approx \frac{2 n_f N_c}{N_c^2 - 1} W_{q+ \to g+}(j),$$

$$V_{q+ \to q+}(j) \approx e^{\Delta\xi \, w_+^j}, \quad V_{g+ \to g+}(j) \approx e^{\Delta\xi \, w_-^j}. \tag{7.219}$$

Similar results are valid for transitions between partons with negative helicities.

Inserting the above expressions in the Mellin integral representation (7.179), we can obtain the following behaviour of inclusive probabilities

$$W_{g+ \to q+}(j) \approx \frac{2 n_f N_c}{N_c^2 - 1} V_{q+ \to g+}(j), \quad W_{q+ \to g+}(x)|_{x \to 1}$$

$$\approx \frac{1}{2} \frac{N_c^2 - 1}{N_c^2 + 1} \left(\frac{\exp\left(\frac{3}{4} \frac{N_c^2 - 1}{N_c^2} \Delta\xi \right)}{\Gamma(1 + \frac{N_c^2 - 1}{N_c^2} \Delta\xi)} \frac{(1 - x)^{\frac{N_c^2 - 1}{N_c^2} \Delta\xi}}{\ln \frac{1}{1-x}} - \frac{\exp\left((\frac{11}{6} - \frac{n_f}{3N_c}) \Delta\xi \right)}{\Gamma(1 + 2\Delta\xi)} \frac{(1-x)^{2\Delta\xi}}{\ln \frac{1}{1-x}} \right),$$

$$W_{q+ \to q+}(x)|_{x \to 1} \approx \frac{\exp\left(\frac{3}{4} \frac{N_c^2 - 1}{N_c^2} \Delta\xi \right)}{\Gamma(\frac{N_c^2 - 1}{N_c^2} \Delta\xi)} (1 - x)^{-1 + \frac{N_c^2 - 1}{N_c^2} \Delta\xi},$$

$$W_{g+ \to g+}(x)|_{x \to 1} \approx \frac{\exp\left((\frac{11}{6} - \frac{n_f}{3N_c}) \Delta\xi \right)}{\Gamma(2\Delta\xi)} (1 - x)^{-1 + 2\Delta\xi}. \tag{7.220}$$

Note that the expressions for $W_{q+ \to q+}(x)|_{x \to 1}$ and $W_{g+ \to g+}(x)|_{x \to 1}$ contain the sum of the Sudakov double-logarithmic terms.

Let us consider now the small-x behavior of inclusive probabilities $W_{g+ \to g+}(x)$. It is related to the singularities of the anomalous dimension matrix w (7.197), (7.198), and (7.202) at $j \to 1$. Only two elements of this matrix are singular

$$\lim_{j \to 1} w_{1/2 \to 1}^{jv}(x) = \frac{N_c^2 - 1}{2N_c^2} \left(\frac{2}{j - 1} - \frac{3}{2} \right), \quad \lim_{j \to 1} w_{1 \to 1}^{jv} = \frac{2}{j - 1} - \frac{11}{6} - \frac{n_f}{3N_c}, \tag{7.221}$$

which leads to the singular eigenvalue

$$w_+^{jv} \approx \frac{2}{j-1} - \frac{11}{12} + \frac{n_f}{6N_c} \frac{N_c^2 - 2}{N_c^2} \tag{7.222}$$

entering the following nonvanishing momenta $W_{r \to k}(j)$ (7.209)

$$\lim_{j \to 1} W_{q \to g}(j) \approx \frac{N_c^2 - 1}{2N_c} \exp(\Delta\xi \, w_+^{jv}), \qquad \lim_{j \to 1} W_{g \to g}(x) \approx \exp(\Delta\xi \, w_+^{jv}). \tag{7.223}$$

For $x \to 0$, the Mellin integral (7.179) has a saddle point at

$$j - 1 = \sqrt{\frac{2\Delta\xi}{\ln\frac{1}{x}}}. \tag{7.224}$$

The calculation of the integral by the saddle-point method gives the following result for the inclusive probabilities

$$W_{q \to g}(x) \approx \frac{N_c^2 - 1}{2N_c} W_{g \to g}(x),$$

$$W_{g \to g}(x) \approx \frac{1}{x} \exp\left(2\sqrt{2\Delta\xi \ln\frac{1}{x}}\right) \left(\frac{\Delta\xi}{8\pi^2 \ln^3\frac{1}{x}}\right)^{1/4} \exp\left(\left(-\frac{11}{12} + \frac{n_f}{6N_c} \frac{N_c^2 - 2}{N_c^2}\right)\Delta\xi\right). \tag{7.225}$$

Thus, the gluon distributions grow rapidly with increasing $\Delta\xi$ and decreasing x. It turns out, that by gathering the powers of $g^2 \ln 1/x$ in in all orders of perturbation theory in the framework of the BFKL approach, one can obtain an even more significant increase of these distributions [13] (see Chapter 9).

The leading singularity of the anomalous dimension $w_{1/2 \to 1/2}^j$ (7.197) of the nonsinglet quark distribution is at $j = 0$ and

$$\lim_{j \to 0} w_{1/2 \to 1/2}^j \approx \frac{N_c^2 - 1}{2N_c^2} \left(\frac{1}{j} - \frac{1}{2}\right). \tag{7.226}$$

The small-x behaviour of this distribution is less singular

$$\lim_{x \to 0} \frac{1}{2} \left(W_{q \to q}(x) - W_{q \to \bar{q}}(x)\right) \sim \exp\sqrt{\frac{2\Delta\xi \ln\frac{1}{x} (N_c^2 - 1)}{N_c^2}}. \tag{7.227}$$

7.8 Hard processes beyond the LLA

The results of higher-order term calculations in perturbative QCD are now available. The first task is to find parton distributions in hadrons with better precision. The starting point is the factorization theorem. (The exhaustive review of the factorization theorem is given in [26].) In the case of deep inelastic scattering structure functions, the factorization theorem has the form:

$$F_b(x, Q, \alpha_s(\mu)) = f_b^a\left(x, \frac{Q}{\mu}, \alpha_s(\mu)\right) \otimes P_a\left(x, \frac{Q}{\mu}, \alpha_s(\mu)\right). \tag{7.228}$$

Here F_b is the original partonic structure function of hadron, f_b^a is the distribution function of parton a in hadron, P_a is the universal splitting function, and μ is the normalization point. The factor $f_b^a(x, \frac{Q}{\mu}, \alpha_s(\mu))$ absorbs all collinear and soft singularities of the original partonic structure function $F_b(x, Q, \alpha_s(\mu))$. The right-hand side of (7.228) depends on μ, while $F_b(x, Q, \alpha_s)$ is μ-independent (besides of the running couple $\alpha_s(\mu)$). However, this statement is correct when all perturbative series are summed. When only few terms of the series are accounted, then the result can be μ-dependent. Therefore, the μ-dependence of the result can be used as a check of the convergency of the series.

7.8.1 Anomalous dimensions of twist-2 operators in two loops

Soon after the discovery of the asymptotic freedom in QCD, several groups of physisists calculated the splitting kernels of the DGLAP equations and the related anomalous dimension matrices for the twist-2 operators in the two-loop approximation [27], [28], [29]. The calculations of anomalous dimensions were performed in the $\overline{\text{MS}}$-renormalization scheme. In the same scheme, the one-loop corrections to the coefficient functions were obtained. Because the obtained results are rather cumbersome, we consider here only the nonsinglet flavour quantum numbers, where the corrections are comparatively simple. The knowledge of the anomalous dimensions $\gamma_j(\alpha_s)$ and $\widetilde{\gamma}_j(\alpha_s)$ in this case allows one to find the Q^2-dependence of the matrix elements n_q^j, $\widetilde{n}_q^j$ of the tensor and pseudotensor operators O, $\widetilde{O}$ constructed only from the quark fields

$$\frac{\partial n_q^j}{\partial \ln Q^2} = \gamma_j(\alpha_s(Q^2)) \, n_q^j, \qquad \frac{\partial \widetilde{n}_q^j}{\partial \ln Q^2} = \widetilde{\gamma}_j(\alpha_s(Q^2)) \, \widetilde{n}_q^j, \qquad (7.229)$$

where $\alpha_s(Q^2)$ is the QCD running coupling constant and the functions $\gamma_j(\alpha_s)$, $\widetilde{\gamma}_j(\alpha_s)$ are expanded in the series over α_s

$$\gamma_j(\alpha_s) = \sum_{n=1}^{\infty} \left(\frac{\alpha_s}{2\pi}\right)^n a_n(j), \qquad \widetilde{\gamma}_j(\alpha_s) = \sum_{n=1}^{\infty} \left(\frac{\alpha_s}{2\pi}\right)^n \widetilde{a}_n(j), \qquad (7.230)$$

where in the Born approximation we have

$$a_1(j) = \widetilde{a}_1(j) = C_F \left(2\psi(1) - 2\psi(j) - \frac{1}{j+1} - \frac{1}{j} + \frac{3}{2} \right), \qquad (7.231)$$

where

$$C_F = \frac{N_c^2 - 1}{2N_c} \qquad (7.232)$$

is the Casimir operator for the quark. Note that our anomalous dimension is normalized in such a way that it differs from conventional γ by the factor $-1/2$. Taking into account this difference, we obtain in two loops [27]

$$a_2(j) = C_F C_A \, a_2^F(j) + \left(C_F^2 - \frac{C_F C_A}{2} \right) a_2^{FF}(j) + C_F T_F \, N_F \, a_2^{FFF}(j), \qquad (7.233)$$

where

$$C_A = N_c, \quad T_F = \frac{1}{2} \tag{7.234}$$

are the colour factors appearing, respectively, in one gluon and quark loops and N_F is the number of light quarks. The coefficients $a_2^i(j)$ are given below [27]

$$a_2^F(j) = S_1(j) \left(\frac{134}{9} + \frac{2(2j+1)}{j^2(j+1)^2} \right) + S_2(j) \left(-\frac{13}{3} + \frac{2}{j(j+1)} \right)$$
$$- 4S_1(j)S_2(j) - \frac{43}{24} - \frac{1}{9} \frac{151j^4 + 263j^3 + 97j^2 + 3j + 9}{j^3(j+1)^2}. \tag{7.235}$$

$$a_2^{FF}(j) = \frac{4(2j+1)}{j^2(j+1)^2} S_1(j) - 4 \left(2S_1(j) - \frac{1}{j(j+1)} \right) (S_2(j) + 2S_{-2}(j)) + 16S_{-2,1}(j)$$
$$- 8S_3(j) - 8S_{-3}(j) + 6S_2(j) - \frac{3}{4} - 2\frac{3j^3 + j^2 - 1}{j^3(j+1)^2} + 4(-1)^j \frac{2j^2 + 2j + 1}{j^3(j+1)^2}, \tag{7.236}$$

$$a_2^{FFF}(j) = -\frac{40}{9} S_1(j) + \frac{8}{3} S_2(j) + \frac{1}{3} + \frac{4}{9} \frac{11j^2 + 5j - 3}{j^2(j+1)^2}. \tag{7.237}$$

Actually the tensor anomalous dimension $\gamma_j(\alpha_s)$ is obtained from even values of j and $\tilde{\gamma}_j(\alpha_s)$ corresponds to odd values of j. In the above expressions, we used the following definitions of harmonic sums

$$S_a(j) = \sum_{k-1}^{j} \frac{1}{k^a}, \quad S_{a,b,c\ldots}(j) = \sum_{k-1}^{j} \frac{1}{k^a} S_{b,c\ldots}(k), \quad .S_{-a,b,c\ldots}(j) = \sum_{k-1}^{j} \frac{(-1)^k}{k^a} S_{b,c\ldots}(k). \tag{7.238}$$

Their analytic continuation to the complex values of j is performed with the use of the relations

$$S_a(j) = \psi(j+1) - \psi(1), \quad S_{a,b,c\ldots}(j) = \sum_{k-1}^{\infty} \left(\frac{S_{b,c\ldots}(k)}{k^a} - \frac{S_{b,c\ldots}(k+j)}{(k+j)^a} \right),$$

$$.S_{-a,b,c\ldots}(j) = \sum_{k-1}^{\infty} \frac{(-1)^k}{k^a} S_{b,c\ldots}(k) + (-1)^j \sum_{k-1}^{\infty} \frac{(-1)^k}{(k+j)^a} S_{b,c\ldots}(k+j). \tag{7.239}$$

In such a way, we obtain two different functions $a_2(j)$ and $\tilde{a}_n(j)$, corresponding to the analytic continuation from even and odd values of j. They correspond to the positive and negative signatures, respectively. It then is possible to verify the double-logarithmic prediction of Section 7.14 concerning the singularity of the anomalous dimensions at $j = 0$

$$\gamma_j(\alpha_s) = \frac{j}{2} - \sqrt{\frac{j^2}{4} - \frac{\alpha_s}{2\pi} C_F} = \frac{\alpha_s}{2\pi j} C_F \cdot + \left(\frac{\alpha_s}{2\pi j} C_F \right)^2 + \cdots \tag{7.240}$$

$$\tilde{\gamma}_j(\alpha_s) = \gamma_j(\alpha_s) - 4 \left(\frac{\alpha_s}{2\pi} \right)^2 \left(C_F^2 - \frac{C_F C_A}{2} \right) + \cdots. \tag{7.241}$$

In the next-to-leading approximation the Gribov–Lipatov relation (7.177)

$$D_{rs}(x) = \overline{D}_{rs}(x) \tag{7.242}$$

between the parton distributions (PD) and the fragmentation functions (FF) is violated [27], [28]. For the singlet case at small $j - 1$ an important physical effect of this violation is the appearence of two different effective parameters of the perturbative expansion $\alpha_s/(j-1)$ and $\alpha_s/(j-1)^2$ for the deep inelastic ep scattering and the e^+e^--annihilation, respectively. Nevertheless, the difference of the corresponding elements of the anomalous dimension matrices for PD and FF is simple [27], [28] and this fact deserves an explanation. Such an explanation was presented by Dokshitzer with collaborators [30]. From the physical arguments of ordering the parton processes in their "fluctuation lifetimes" they suggested the use of two different evolution parameters in the scattering and annihilation channels. In their approach, the two evolution equations can be written in an universal form [30]

$$\frac{\partial}{\partial \ln Q^2} D(x, Q^2) = \int_0^1 \frac{dz}{z} \, P(z, \alpha_s(z^{-1}Q^2)) \, D(\frac{x}{z}, z^\sigma Q^2), \tag{7.243}$$

where $\sigma = -1\,(+1)$ for the space-like (time-like) region. Here it is assumed that the new splitting kernel is identical in two considered channels, which corresponds to the Gribov–Lipatov relation. In this case, by returning to the evolution equations written in the traditional form, we can obtain for the difference of the splitting kernels in the space-like (S) and time-like (T) regions the following result for the nonsinglet case in two loops [30]

$$\frac{1}{2}\left(P_{ns}^{(2),T}(x) - P_{ns}^{(2),S}(x)\right) = \int_0^1 dz \int_0^1 dy \delta(x - yz) \, P_{qq}^{(1)}(z) \ln z \left(P_{qq}^{(1)}(y)\right)_+. \tag{7.244}$$

This result coincides with that obtained by direct calculations in Ref. [27]. There exists another relation between the structure functions of the deep inelastic ep scattering and e^+e^--annihilation. This relation obtained by Drell, Levy, and Yan [31]. In terms of the nonsinglet splitting function, it can be written as follows

$$P_{ns}^T(x) = -x^{-1} P_{ns}^S(x^{-1}), \tag{7.245}$$

which corresponds to an analytic continuation of $P_{ns}^S(z)$ in the region $z = x^{-1} > 1$. Together with the Gribov–Lipatov relation it could impose a constraint on $P_{ns}^S(x)$. In the j-representation this constraint means that the anomalous dimension should be a function of the total momentum square $J^2 = j(j+1)$. But in QCD, the point $x = 1$ is a singular point for the partonic distributions even in LLA. It is possible, however, to give a certain meaning to this continuation, if in the polynomials over $\ln(1-x)$ with coefficients being analytic functions of x at $x = 1$ one substitutes the logarithm argument by $(x-1)$ for $x > 1$ [24]. In terms of the splitting functions in the Born approximation, we have only the pole at $x = 1$ and therefore the analytic continuation around this point is possible. In higher orders of the perturbation theory, the splitting kernels have more complicated logarithmic singularities at $x = 1$. Nevertheless, it is possible to generalize Gribov–Lipatov and Drell–Levy–Yan relations also to this case [31].

7.8.2 Partonic distributions in next-to-next-to leading order from the fit of the data

Partonic splitting functions in next-to-next-to leading order were calculated in [32]. The fit to the data in next-to-leading order and next-to-next-to leading order was performed by MRST ([33] and references therein), CTEQ collaboration [34], Jimenez–Delgado and Reya [35] and Alekhin [36]. The data on $e(\mu)N \rightarrow e(\mu)+$ all, $\nu(\bar{\nu})N \rightarrow \nu(\bar{\nu})+$ all and the Drell–Yan process were exploited. We present here the partonic distributions in proton, obtained by MRST (Figs. 7.9a, b, c).

The initial parton distributions – the input of evolution equations – were taken in the form:

$$xf_i(x, Q_0^2) = A_i x^{-\lambda_i}(1-x)^{\beta_i}(1 + \gamma_i \sqrt{x} + \delta_i x), \quad Q_0^2 = 4 \, \text{GeV}^2. \tag{7.246}$$

This form of initial distributions was chosen from the following considerations:

1. At small x (and fixed Q^2) the Regge behaviour is reproduced.
2. At $x \rightarrow 1$ $xf_i(x)$ looks like counting rules formulae.

As is seen from Figs. 7.9a–7.9c, the most remarkable variation with Q^2 appears in gluon distributions. But, on the other hand, they are among the least well determined parton

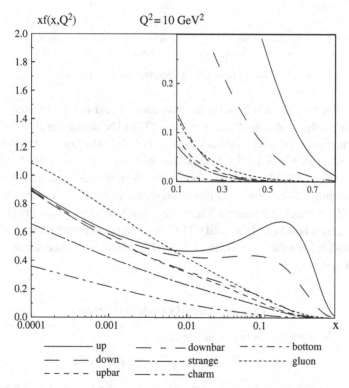

Fig. 7.9. The parton distributions in proton at $Q^2 = 10 \, \text{GeV}^2$ by MRST [33], courtesy of A. V. Sidorov. On the main figure, the gluon distribution is multiplied by 0.1. On the insert diagram, the parton distributions at $x > 0.1$ are shown in more detail.

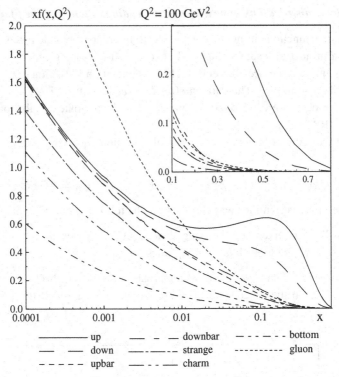

Fig. 7.9. The same as in Fig.7.9a, but at $Q^2 = 100\ \text{GeV}^2$.

distributions. For u-quark distributions, the difference of next-to-next-to leading order and next-to-leading order is of order 5% at $x > 10^{-2}$ [33]. (The detailed analysis of the uncertainties of parton-distribution functions was done in [37],[38].) Note that below $x = 10^{-3}$ (or perhaps even below $x = 10^{-2}$), there is no confidence in the analysis of deep inelastic scattering based on DGLAP equations, since in this domain large logarithms $\ln(1/x)$ arise, which are not properly accounted in these equations. The account of large $\ln(1/x)$ was done by BFKL approach (Chapter 9). The results of different groups are in agreement, but some disagreement between MRST and CTEQ arises in gluon distribution at $x \sim 0.3$ [37]. Also, Alekhin [36] got the value $\alpha_s(m_Z) = 0.114$, which contradicts the world average $\alpha_s(m_Z) = 0.119$.

7.8.3 Longitudinal structure function

The longitudinal structure function is defined by:

$$F_L(x, Q^2) = F_2(x, Q^2) - 2x F_1(x, Q^2) \qquad (7.247)$$

In the parton model, $F_L(x, Q^2)$ vanishes (Callan–Gross relation; Eq.(7.17)). The vanishing of F_L persists, if the target and quark masses are disregarded, as well as nonperturbative corrections (i.e. only the lowest-twist contributions are accounted) and the quark transverse

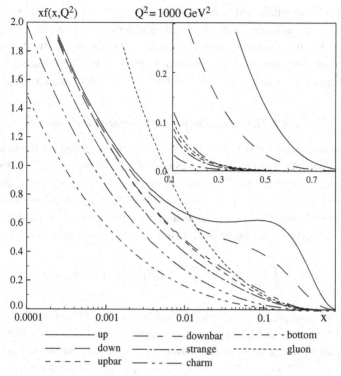

Fig. 7.9. The same as in Fig.7.9a, but at $Q^2 = 1000$ GeV^2.

momentum $k_\perp$ is neglected. If the quark-transverse momentum and the quark mass are accounted, then the simple relation can be derived in the parton model [4]:

$$F_L = 2x\frac{\sigma_L}{\sigma_T} = 8x\frac{\overline{k_\perp^2} + m_q^2}{Q^2}, \qquad (7.248)$$

where $\overline{k_\perp^2}$ is the mean square of quark-transverse momentum. In QCD, the nonvanishing transverse momenta of quarks arise due to interaction, i.e. as α_s-correction. The calculation of α_s-corrections to the structure functions results in the following formula for F_L [39]:

$$F_L(x, Q^2) = \frac{\alpha_s(Q^2)}{4\pi}\int_x^1\frac{dy}{y}\frac{x^2}{y^2}\left[\frac{16}{3}F_2(y, Q^2) + 8\delta_f N_f\left(1 - \frac{x}{y}\right)yg(y, Q^2)\right], \quad (7.249)$$

where $g(y, Q^2)$ is the gluon distribution and δ_f is a number, depending on quark charges. For ep scattering and $N_f = 4$, $\delta_f = 5/18$; for $\nu(\overline{\nu})p$ scattering $\delta_f = 1$. $F_L(x, Q^2)$ increases at small x. In this case, the main contribution arises from the first term in the right-hand side of (7.249). The estimation gives that at $x \sim 10^{-3} - 10^{-4}$ $F_L(x, Q^2) \sim 0.3 - 0.5$ at moderate $Q^2 \sim 10 - 100$ GeV^2. This expectation qualitatively agrees with experimental data [40]. Eq.(7.249) represents the leading-order contribution to F_L. The next-to-leading order and next-to-next-to leading order contributions were also calculated

[41],[42],[43]. The qualitative estimations of F_L presented above are valid with account of next-to-leading order and next-to-next-to leading order corrections. Besides perturbative contribution to F_L, there are nonperturbative terms, given by higher-twist operators [44]. The crude estimations show that they are much smaller, than perturbative contributions (~ 0.05) and are more or less randomly distributed over all domain of x [43],[44].

7.8.4 The sum rules for the structure functions

There are few sum rules for structure functions, which are called by the names of their discoverers: Adler (A), Gross–Llewellyn–Smith (GLS), Bjorken (Bj^{pol} – for polarized $e(\mu)N$-scattering), Bjorken (Bj^{unpol} – for unpolarized $\nu(\nu)p$ scattering), and Gottfried (G). The derivation of these sum rules was given in [4], where there are also references to original papers. Generally, in QCD these sum rules got perturbative and nonperturbative (higher-twist) corrections. An exceptional case is the Adler sum rule:

$$A = \int\limits_0^1 \frac{dx}{x}[F_2^{\bar{\nu}p}(x, Q^2) - F_2^{\nu p}(x, Q^2)] = 4T_3 = 2, \tag{7.250}$$

where T_3 is the third projection of target (proton) isospin. (The Cabibbo angle θ_c was neglected, i.e. the strange and charmed particles were not accounted in F_2. Note that $F_2^{\bar{\nu}p} = F_2^{\nu n}$.) The Adler sum rule was derived, basing on $SU(2)$ current algebra and supposing that the difference $F_2^{\bar{\nu}p}(x) - F_2^{\nu p}(x)$ satisfies unsubtracted dispersion relation. Since these statements are very general, the Adler sum rule has no perturbative or nonperturbative corrections. The experimental check [45] shows the absence of significant Q^2 variation in the range $2 < Q^2 < 200$ GeV and gives

$$A = 2.02 \pm 0.40. \tag{7.251}$$

The Gottfrid sum rule is the charged-lepton analog of (7.250) and has the form ($a = \alpha_s/\pi$)

$$G = \int\limits_0^1 \frac{dx}{x}[F_2^{ep}(x, Q^2) - F_2^{en}(x, Q^2)] = \frac{1}{3}\left[1 + 0.035a + 3.7C_F(C_F - 0.5C_A)a^2\right]$$

$$- \frac{2}{3}\int\limits_0^1 dx(\bar{d}(x) - \bar{u}(x)), \tag{7.252}$$

$C_F = (N_c^2 - 1)/2N_c$, $C_A = N_c$. The α_s- and α_s^2-corrections are very small. (The summary of these corrections is presented in [46].) If the nucleon sea would be flavour symmetric, $\bar{u}(x) = \bar{d}(x)$, we would have

$$G = \frac{1}{3}. \tag{7.253}$$

However, the experimental data of NMC collaboration [47] give

$$G(Q^2 = 4\text{GeV}^2) = 0.235 \pm 0.026 \tag{7.254}$$

There are arguments that nonperturbative corrections are also small [46]. So, the most plausible conclusion, following from the comparison of the Gottfried sum rule with the data, is the flavour asymmetry of the sea in the nucleon. This statement was conclusively confirmed by Fermilab E866/Nusea collaboration [48].

GLS, Bj^{pol}, and Bj^{unpol} sum rules are:

$$GLS = \frac{1}{2} \int_0^1 dx [F_3^{\bar{\nu}p}(x, Q^2) + F_3^{\nu p}(x, Q^2] = 3[1 + a(1 + k_1 a + k_2 a^2)], \quad (7.255)$$

$$Bj^{pol} = \int_0^1 dx [g_1^{ep}(x, Q^2) - g_1^{en}(x, Q^2)] = \frac{1}{6} g_A [1 + a(1 + k_1 a + k_2 a^2 + ca^3)], \quad (7.256)$$

$$Bj^{unpol} = \int_0^1 dx [F_1^{\bar{\nu}p}(x, Q^2) - F_1^{\nu p}(x, Q^2)] = 1 - \frac{a}{2} C_F (1 + u_1 a + u_2 a^2), \quad (7.257)$$

$g_A = 1.257 \pm 0.003$. The perturbative corrections k_1, k_2, u_1, u_2 were calculated in [49],[50]:

$$k_1 = -0.333 N_f + 5.58, \quad k_2 = 0.177 N_f^2 - 7.61 N_f + 41.4, \quad (7.258)$$

$$u_1 = -0.444 N_f + 5.75, \quad u_2 = 0.239 N_f^2 - 9.5 N_f + 54.2, \quad (7.259)$$

the coefficient c was estimated in [51], $c \approx 130$. There are nonperturbative corrections of twist 4 to the sum rules (7.255)–(7.257). The estimation of this correction to Bj^{pol} sum rule, performed in [52] by the QCD sum rule method indicated that it is small at $Q^2 > 5$ GeV^2. However, this estimation was critized in [53]. The comparison with experiment shows the agreement with perturbative contributions at the level of 10–15% in the case of GLS sum rule (at $Q^2 \approx 2 - 10$ GeV^2, see e.g. [54] and references therein). The agreement with experiment of the Bjorken polarized sum rule will be discussed in more detail in Section 7.12.2.

7.9 Parton-number correlators

The knowledge of the parton-wave function makes it possible to calculate more complicated quantities – the parton correlators. The simplest generalization of the parton-number distribution is the parton-number correlator having the representation

$$< h | n_{l_1} n_{l_2} \ldots n_{l_k} | h > = Z_h \prod_{r=1}^k \delta_{l_r h}$$

$$+ \sum_{n=2}^{\infty} \sum_{i_1 \ldots i_n} n_{l_1} n_{l_2} \ldots n_{l_k} \int \prod_{i=1}^{n} \frac{d^2 k_{\perp i}}{(2\pi)^3} \frac{d\beta_i}{\beta_i} \theta(\beta_i) Z_g^{n_g} Z_q^{n_q} Z_{\bar{q}}^{n_{\bar{q}}} |\psi|^2 (2\pi)^3 \delta^2$$

$$\times \left(\sum_i k_{\perp i} \right) \delta \left(1 - \sum_i \beta_i \right), \tag{7.260}$$

where $Z_s^{1/2}$ ($s = g, q, \bar{q}$) are the renormalization constants for the corresponding wave functions.

By differentiating this expression over the variable ξ (7.58) related to the ultraviolet cut-off for the transverse momenta of partons and taking into account that in the essential integration region parton-transverse momenta rapidly grow after each partonic decay, we obtain the equation for the parton-number correlators

$$\frac{d}{d\xi} < h| \prod_{r=1}^{k} n_{l_r} |h> = - \sum_s W_s < h|n_s \prod_{r=1}^{k} n_{l_r} |h$$

$$+ \sum_s \sum_{r=1}^{k} \sum_{i_1, i_2} W_{s \to (i_1, i_2)} < h|n_s \prod_{r=1}^{k} \left(n_{l_r} - \delta_{l_r s} + \delta_{l_r i_1} + \delta_{l_r i_2} \right) |h> \tag{7.261}$$

with the initial conditions

$$< h| \prod_{r=1}^{k} n_{l_r} |h>_{|\xi=0} = \prod_{r=1}^{k} \delta_{l_r h}. \tag{7.262}$$

In the right-hand side of this equation, the terms containing $(k + 1)$ factors of n_l are cancelled due to the relation

$$W_s = \sum_{i_1, i_2} W_{s \to (i_1, i_2)} \tag{7.263}$$

and therefore it relates the parton-number correlators of the k-th order with low-order correlators which can be considered as nonhomogeneous contributions.

To illustrate the essential features of the solution of such an equation we neglect in it the quark and antiquark contributions considering the pure Yang–Mills model. Further, we consider only the total number of gluons with two helicities

$$n = n_+ + n_-. \tag{7.264}$$

Introducing the following function of the gluon numbers

$$T^{(k)}(n) = n(n+1) \ldots (n+k-1) \tag{7.265}$$

one can verify from (7.261) that its matrix elements satisfy the linear evolution equation

$$\frac{d}{d\xi} < h|T^{(k)}(n)|h> = W k < h|T^{(k)}(n)|h>, \quad < h|T^{(k)}(n)|h>_{|\xi=0} = k!. \tag{7.266}$$

Its solution is simple

$$< h|T^{(k)}(n)|h> = k! \, e^{k W \xi}. \tag{7.267}$$

Further, the probability to find a hadron in the state with the n partons is

$$P_n = \sum_{i_1...i_n} \int \prod_{i=1}^{n} \frac{d^2k_{\perp i}}{(2\pi)^3} \frac{d\beta_i}{\beta_i} \theta(\beta_i) Z_g^{n_g} Z_q^{n_q} Z_{\bar{q}}^{n_{\bar{q}}} |\psi|^2 (2\pi)^3 \delta^2 (\sum_i k_{\perp i}) \delta(1 - \sum_i \beta_i),$$

(7.268)

where, in particular (for h coinciding with one of partons)

$$P_1 = Z_h.$$

(7.269)

It is obvious that due to the normalization condition for the wave function together with Eq. (7.266) W_n for the pure gluonic case satisfies the sum rules

$$\sum_{n=1}^{\infty} P_n = 1, \quad \sum_{n=1}^{\infty} T^{(k)}(n) P_n = \bar{n}^k = e^{kW\xi}$$

(7.270)

and the evolution equation

$$\frac{d}{d\xi} P_n = W(-n P_n + (n-1) P_{n-1}), \quad (P_n)|_{\xi=0} = \delta_{n,1}.$$

(7.271)

In an explicit form, its solution written below

$$P_n = \frac{1}{\bar{n}} \left(\frac{\bar{n} - 1}{\bar{n}} \right)^{n-1}, \quad \bar{n} = e^{W\xi}$$

(7.272)

is different from the Poisson distribution

$$P_n^P = \frac{1}{n!} (\bar{n})^n e^{-\bar{n}}.$$

(7.273)

Note that P_n satisfies the sum rules (7.270) and for $\bar{n} \to \infty$ it has the property of the KNO-scaling [55]

$$\lim_{\bar{n}\to\infty} P_n = \frac{1}{\bar{n}} f\left(\frac{n}{\bar{n}}\right), \quad f(x) = e^{-x}.$$

(7.274)

In the real QCD case, we can generalize the above discussion by introducing the probability $P(n_g^+, n_g^-, n_q^+, n_q^-, n_{\bar{q}}^+, n_{\bar{q}}^-)$ to find a hadron in the states with a certain number n_t of partons $(t = g^\pm, q^\pm, \bar{q}^\pm)$ with definite helicities $\lambda = \pm s$. This probability satisfies the equation

$$\frac{d}{d\xi} P(n_t) = - \sum_r n_r W_r P(n_t) + \sum_r \sum_{i,j} (n_r + 1 - \delta_{r,i} - \delta_{r,j})$$

$$\times W_{r\to(i,j)} P\left(n_t + \delta_{t,r} - \delta_{t,i} - \delta_{t,j}\right).$$

(7.275)

In the second term of the right-hand side of this equation $\sum_t n_t$ is smaller by one than that in the left-hand side. Therefore, after the Mellin transformation

$$P(n_t) = \int_{-i\infty}^{\infty} \frac{dv}{2\pi i} e^{v\xi} P_v(n_t)$$

(7.276)

one can obtain a recurrent relation for $P_v(n_t)$.

It is natural to introduce the more informative quantity for the description of the hadron structure – the exclusive distributions of partons as functions of their Feynman parameters β_r^t $(r = 1, 2, \ldots n_t)$, where t is the sort of a parton (including its helicity)

$$f_{n_t}(\beta_r^t) = \frac{1}{\prod_t n_t!} \sum_{i_1 \ldots i_n} \int \prod_{t,l} \frac{d^2 k_{\perp l}^t}{(2\pi)^3 \beta_l^t} \prod_t Z_t^{n_t} |\psi_h(k_{\perp l}^t, \beta_l^t)|^2 (2\pi)^3$$

$$\times \delta \left(\sum k_{\perp l}^t \right) \delta \left(1 - \sum \beta_l^t \right). \tag{7.277}$$

This quantity satisfies the following evolution equation

$$\frac{d}{d\xi} f_{n_t}(\beta_r^t) = -\sum_s n_s W_s f(\beta_r^s)$$

$$+ \sum_{s,m} \sum_{i,j} W_{s \to (i,j)} \left(\beta_m^s, \beta_{l_1}^i, \beta_{l_2}^j \right) \delta \left(\beta_m^s - \beta_{l_1}^i - \beta_{l_2}^j \right) f \left(n_t + \delta_{ts} - \delta_{ti} - \delta_{tj} \right) (\beta_l^t) \, d\beta_m^s \tag{7.278}$$

with the initial conditions

$$f_{n_t}(\beta_r^t)|_{\xi=0} = \delta_{n_t 1} \prod_{i \neq 1} \delta_{n_i 0} \delta(\beta^n - 1), \tag{7.279}$$

One can define the generating functional for these exclusive distributions

$$I(\phi_t) = \sum_{n_t} \int \prod_t \prod_{l=1}^{n_t} \phi_t \left(\beta_l^t \right) d\beta_l^t \, f_{n_t} \left(\beta_r^t \right), \tag{7.280}$$

where $\phi_t(\beta)$ $(t = g_\pm, q_\pm, \bar{q}_\pm)$ are some auxiliary fields. The evolution equation for $f_{n_t}(\beta_r^t)$ is equivalent to the following equation in the variational derivatives $\delta/(\delta\phi_s(x))$ for the functional $I(\phi_t)$

$$\frac{d}{d\xi} I = -\sum_s W_s \int \phi_s(x) \frac{\delta I}{\delta\phi_s(x)} dx$$

$$+ \sum_s \sum_i \sum_j \int dx_1 \, dx_2 \, W_{s \to (i,j)}(x_1 + x_2, x_1, x_2) \, \phi_i(x_1)\phi_j(x_2) \frac{\delta I}{\delta\phi_s(x_1 + x_2)} \tag{7.281}$$

with the initial condition

$$I|_{\xi=0} = \phi_h(1). \tag{7.282}$$

Exclusive distributions can be calculated by taking the functional derivative from I

$$f_{n_t}(\beta_l^t) = \frac{1}{\prod_t n_t!} \prod_t \prod_l \frac{\delta I}{\delta\phi_t(\beta_l^t)} \Big|_{\phi_t(\beta)=0}. \tag{7.283}$$

Moreover, we can calculate from this functional also inclusive correlators of partons having the Feynman parameters β_l^t

$$D_{n_t}(\beta_l^t) = \frac{1}{\prod_t n_t!} \prod_t \prod_l \frac{\delta I}{\delta\phi_t(\beta_l^t)} \Big|_{\phi_t(\beta)=1}. \tag{7.284}$$

It is related with the fact that the generating functional J for the inclusive correlators is expressed in terms of I

$$J(\phi_t) = \sum_{n_t} \int \prod_t \prod_{l=1}^{n_t} d\beta_l^t D_{n_t}(\beta_l^t), \tag{7.285}$$

related to the functional I as follows

$$J(\phi_t) = I(1 + \phi_t). \tag{7.286}$$

Due to the normalization condition for the wave function, the functionals I and J are normalized as follows

$$I(1) = J(0) = 1. \tag{7.287}$$

The energy-momentum conservation $\sum_r \beta_r = 1$ imposes the following constraint for the functional I

$$I(e^{\lambda\beta}\phi(\beta)) = e^\lambda I(\phi(\beta)). \tag{7.288}$$

The generating functional for the inclusive probabilities satisfies the following equation

$$\frac{d}{d\xi}J = -\sum_s W_s \int \phi_s(x) \frac{\delta J}{\delta\phi_s(x)} dx$$

$$+ \sum_s \sum_i \sum_j \int dx_1 \, dx_2 \, W_{s\to(i,j)}(x_1 + x_2, x_1, x_2) \, (\phi_i(x_1)$$

$$+ \phi_j(x_2) + \phi_i(x_1)\phi_j(x_2)) \frac{\delta J}{\delta\phi_s(x_1 + x_2)} \tag{7.289}$$

with the initial condition

$$J(\phi_t)|_{\xi=0} = 1 + \phi_h(1). \tag{7.290}$$

The inclusive distributions can be obtained from J by the functional differentiation

$$D_{n_t}(\beta_l^t) = \frac{1}{\prod_t n_t!} \frac{\delta J}{\delta\phi_t(\beta_l^t)}\Big|_{\phi_t(\beta)=0}. \tag{7.291}$$

From the above expressions, we can get the evolution equation for the functions $D_{n_t}(\beta_l^t)$. It has a recurrence form and allows one to calculate it these functions in terms of the simple inclusive probabilities $D_h^i(\beta)$

To solve of the evolution equation for the functional I, we can use a simple mathematical trick and consider the so-called characteristic equation

$$\frac{d}{d\xi}\phi_s(x, \xi) = -W_s \phi_s(x, \xi)$$

$$+ \sum_{k,r} \int dx_1 \, dx_2 \, \phi_k(x_1, \xi) \, \phi_r(x_2, \xi) \, \delta(x_1 + x_2 - x) \, W_{s\to(k,r)}(x, x_1, x_2). \tag{7.292}$$

Let us assume that for any function $\phi_k(x, \xi)$ one can calculate the initial condition

$$\phi_k(x, 0) = \chi_k \left(\phi_r(x, \xi), x \right), \tag{7.293}$$

where χ_s are functions of x and functionals from $\phi_r(x, \xi)$. Then it can be easily verified that the solution of the evolution equation for I can be constructed as follows

$$I = \chi_h \left(\phi_r(x, \xi), 1 \right). \tag{7.294}$$

Note that this expression satisfies the initial condition.

To illustrate this mathematical method, we consider the pure Yang–Mills theory and only the most singular terms in the splitting kernels

$$W_{s \to (i,k)}(x, x_1, x_2) = 2 \frac{x}{x_1 x_2}, \qquad W = \int_0^1 \frac{d x}{x(1 - x)}, \tag{7.295}$$

which corresponds to the $N = 4$ extended supersymmetric theory where the anomalous dimension is universal for all twist-2 operators

$$w_j = 2 \int_0^1 \frac{d x}{x(1 - x)} \left(x^{j-1} - x \right) = 2\psi(1) - 2\psi(j - 1). \tag{7.296}$$

In this case, the characteristic equation takes the form

$$\frac{d}{d\xi} \phi(x, \xi) = -W \phi(x, \xi) + 2 \int \frac{x \, dx_1 \, dx_2}{x_1 x_2} \delta(x_1 + x_2 - x) \phi(x_1, \xi) \phi(x_2, \xi). \tag{7.297}$$

For simplicity, we disregard complications related to infrared divergences of integrals and consider W as a constant. Then the solution can be searched for in the form of the Mellin transformation

$$\chi(p, \xi) = \int_0^\infty \frac{d x}{x} e^{-px} \phi(x, \xi), \tag{7.298}$$

where $\phi(x, \xi)$ satisfies the equation

$$\frac{d}{d\xi} \chi(p, \xi) = -W \chi(p, \xi) + 2 \chi^2(p, \xi). \tag{7.299}$$

Therefore, we can find $\chi(p, \xi)$ for any initial condition $\chi(p, 0)$

$$\chi(p, \xi) = \frac{W \chi(p, 0) e^{-W \xi}}{W - 2 \chi(p, 0) \left(1 - e^{-W \xi} \right)}, \tag{7.300}$$

which allows us to calculate the generating functional I

$$I(\phi) = \int_{-i\infty}^{i\infty} \frac{d p}{2\pi i} e^p \frac{W \int_0^\infty e^{-xp} \frac{d x}{x} \phi(x, \xi)}{2 \left(1 - \exp(-W\xi) \right) \int_0^\infty e^{-xp} \frac{d x}{x} \phi(x, \xi) + W \exp(-W\xi)}. \tag{7.301}$$

Note, however, that for self-consistency of the approach one should regularize the splitting kernel in the infrared region, for example, by introducing the infinitesimal parameter $\epsilon \to 0$ in such way to conserve its scale invariance

$$W_{s \to (i,k)}^{(\epsilon)}(x, x_1, x_2) = 2 \frac{x^{1-2\epsilon}}{(x_1 x_2)^{1-\epsilon}}, \qquad W^{(\epsilon)} = \int_0^x \frac{x^{1-2\epsilon} \, d x_1}{x_1^{1-\epsilon} (x - x_1)^{1-\epsilon}} \approx \frac{2}{\epsilon}. \tag{7.302}$$

In this case, the characteristic equation is more complicated

$$\frac{d}{d\xi} \phi^{(\epsilon)}(x, \xi) = -W^{(\epsilon)} \phi^{(\epsilon)}(x, \xi) + 2 \int_0^x \frac{x^{1-2\epsilon} \, dx_1}{x_1^{1-\epsilon} (x - x_1)^{1-\epsilon}} \phi^{(\epsilon)}(x_1, \xi) \phi^{(\epsilon)}(x - x_1, \xi).$$

$$(7.303)$$

We are not going to solve it. Note, that for the considered model, the inclusive parton distribution can be easily calculated

$$D(x) = \int_{-i\infty}^{i\infty} \frac{d\,j}{2\pi i} \left(\frac{1}{x}\right)^j e^{w_j \xi}. \tag{7.304}$$

7.10 Deep inelastic electron scattering off the polarized proton

One of the most interesting processes is the deep inelastic electron scattering off the polarized target. We consider the scattering off the proton. The differential cross section for finding an electron with a definite momentum in the final state is expressed in terms of the antisymmetric tensor describing the imaginary part of the forward-scattering amplitude for the virtual photons with momenta q and polarization indices μ and ν

$$W_{\mu\nu}^A = \frac{1}{2} \left(W_{\mu\nu} - W_{\nu\mu}\right) = \frac{i}{pq} \epsilon_{\mu\nu\lambda\sigma} q_\lambda \left(s_\sigma \, g_1\left(x, Q^2\right) + s_{\sigma\perp} \, g_2\left(x, Q^2\right)\right), \quad (7.305)$$

where $\epsilon_{\mu\nu\lambda\sigma}$ is the completely antisymmetric tensor ($\epsilon_{0123} = 1$). The quantities $g_i(x, Q^2)$ ($i = 1, 2$) are the corresponding structure functions. (The kinematics and phenomenological description are given in [4].) The four-dimensional pseudovector s^σ describes the spin state of the completely polarized proton with the wave function $U(p)$

$$s_\sigma = \bar{U}(p)\gamma_\sigma \gamma_5 U(p) = -2m \, a_\sigma, \quad \bar{U}(p)U(p) = 2m. \tag{7.306}$$

The vector a^σ is a parameter of the proton-density matrix

$$\Lambda(p) = \frac{1}{2} (\not{p} + m)(1 - \gamma_5 \not{a}), \quad \gamma_5 = -i \, \gamma_0\gamma_1\gamma_2\gamma_3 = \begin{pmatrix} 0 & -1 \\ -1 & 0 \end{pmatrix} \tag{7.307}$$

and has the properties

$$a \, p = 0, \quad a^2 = -1, \quad a_\perp = a - \frac{a \, q}{p \, q} \, p. \tag{7.308}$$

The tensor $W_{\mu\nu}^A$ is proportional to the imaginary part of the photon-proton scattering amplitude $T_{\mu\nu}^A$

$$W_{\mu\nu}^A = \frac{1}{\pi} \, \text{Im} \, T_{\mu\nu}^A \tag{7.309}$$

with the representation

$$T_{\mu\nu}^A = \frac{1}{2} \left(T_{\mu\nu} - T_{\nu\mu}\right) = \frac{i}{pq} \epsilon_{\mu\nu\lambda\sigma} q_\lambda \left(s_\sigma \, \bar{g}_1\left(x, Q^2\right) + s_{\sigma\perp} \, \bar{g}_2\left(x, Q^2\right)\right). \tag{7.310}$$

Use the operator product expansion for the product of the electromagnetic currents

$$J_\mu(x) = \sum_q e_q \, \bar{\psi}_q(x) \, \gamma_\mu \, \psi_q(x) \tag{7.311}$$

entering the expression for $T_{\mu\nu}$

$$T_{\mu\nu} = i \int d^4x \, e^{iqx} \, < p|T(J_\mu(x) J_\nu(0))|p > . \tag{7.312}$$

In the Bjorken limit $q^2 \sim 2pq \gg m^2$ on the light cone $x^2 \sim q^{-2}$ we obtain [56],[57]

$$T_{\mu\nu}^A = -i \, \epsilon_{\mu\nu\rho\sigma} \frac{q_\rho}{Q^2} \sum_{n=0}^\infty (1 + (-1)^n) \left(\frac{2}{Q}\right)^n q_{\mu_1} \cdots q_{\mu_n} \sum_k \varphi_{1,n}^k(Q^2) < h|R_{1\sigma\mu_1\ldots\mu_n}^k|h >$$

$$+ \frac{i}{Q^2} \left(\epsilon_{\mu\rho\lambda\sigma} \, q_\nu q^\rho - \epsilon_{\nu\rho\lambda\sigma} \, q_\mu \, q^\rho - q^2 \, \epsilon_{\mu\nu\lambda\sigma} \right) \sum_{n=0}^\infty (1 + (-1)^n)$$

$$\times \, q_{\mu_1} \cdots q_{\mu_{n-1}} \sum_k \varphi_{2,n}^k(Q^2) < h|R_{2\sigma\lambda\mu_1\ldots\mu_{n-1}}^k|h > . \tag{7.313}$$

Here the index k enumerates various pseudotensor local operators R_i^k ($i=1,2$) with the positive charge parity. They are constructed from the quark and gluon fields in a gauge-invariant way. The quantities $\varphi_{i,n}^k(Q^2)$ are the corresponding coefficient functions. We should take into account such operators R_i^k which would lead to the largest asymptotic contribution to $T_{\mu\nu}^A$ in the Bjorken limit. In particular, it means that the tensors $R_{1\sigma\mu_1\ldots\mu_n}^k$ and $R_{2\sigma\lambda\mu_1\ldots\mu_{n-1}}^k$ should be symmetric and traceless in indices σ, $\mu_1, \ldots, \mu_n$ and λ, $\mu_1, \ldots \mu_{n-1}$, respectively. Further, they should have the minimal dimensions d in the mass units for a given number of indices. Examples of such operators are given below

$$R_{1\sigma\mu_1\ldots\mu_n}^q = i^n \, S_{\{\sigma\,\mu_1\ldots\mu_n\}} \, \bar{\psi}_q \gamma_5 \, \gamma_\sigma \, \nabla_{\mu_1} \cdots \nabla_{\mu_n} \psi_q, \tag{7.314}$$

$$R_{2\sigma\lambda\mu_1\ldots\mu_{n-1}}^q = i^n \, S_{\{\mu_1\ldots\mu_{n-1}\}} \, A_{[\sigma\,\mu_1]} \, \bar{\psi}_q \gamma_5 \, \gamma_\sigma \, \nabla_{\mu_1} \cdots \nabla_{\mu_{n-1}} \psi_q, \tag{7.315}$$

where the sign S means the symmetrization of the tensor in the corresponding indices and subtraction of traces. The sign A implies its antisymmetrization. In the above expression, ∇_μ is the covariant derivative

$$\nabla_\mu = \partial_\mu + ig \, A_\mu, \tag{7.316}$$

where $A = t_a \, A^a$ and t_a are the colour-group generators in the fundamental representation. In principle, each derivative can be applied to the left-quark field with an opposite sign without changing the value of the matrix element between the initial and final proton states with the same momentum p. For the deep inelastic ep scattering, only the operators with even values of n are taken into account because they have the positive charge parity. The tensor $R_{1\sigma\mu_1\ldots\mu_n}^q$ symmetrized over all indices has the Lorentz spin $j = n + 1$ and the canonical dimension $d = 3 + n$. Further, due to one antisymmetrization the tensor

$R^q_{2\sigma\lambda\mu_1...\mu_{n-1}}$ has the Lorentz spin $j = n$ and the canonical dimension $d = 3 + n$. The difference between the canonical dimension and the Lorentz spin of an operator is its twist

$$t = d - j. \tag{7.317}$$

It is obvious that the twists of operators R^q_1 and R^q_2 are $t = 2$ and $t = 3$, respectively. Increasing the value of the twist leads usually to diminishing the power of Q for the corresponding contribution in the cross section. However, for the deep inelastic scattering off the polarized target the two above operators give a comparable contribution to the structure function $g_2\left(x, Q^2\right)$.

The matrix elements of the operators $R^{q'}_i$ between the free-quark states in the Born approximation can be easily calculated

$$< q|R^{q'}_{1\sigma\mu_1...\mu_n}|q> = -\delta_{qq'} S_{\{\mu_1...\mu_n\}} \left(\frac{s_\sigma}{n+1} p_{\mu_1} \cdots p_{\mu_n} + \frac{n}{n+1} s_{\mu_1} p_\sigma p_{\mu_2} \cdots p_{\mu_n} \right) \tag{7.318}$$

and

$$< q|R^{q'}_{2\sigma\mu_1...\mu_n}|q> = -\frac{1}{2} \delta_{qq'} S_{\{\mu_1...\mu_n\}} \left(s_\sigma p_{\mu_1} \cdots p_{\mu_n} - s_{\mu_1} p_\sigma p_{\mu_2} \cdots p_{\mu_n} \right). \tag{7.319}$$

In the Born approximation, the structure functions are

$$g_1(x) = \frac{1}{2} e_q^2 \delta(x-1), \quad g_2(x) = 0,$$

$$\bar{g}_1(x) = e_q^2 \frac{x}{x^2 - 1 - i0}, \quad \bar{g}_2(x) = 0, \tag{7.320}$$

The formulas of the operator product expansion give the same result for $\bar{g}_i(x)$, providing that the coefficient functions in the same approximation are

$$\varphi^q_{1,n} = e_q^2, \quad \varphi^q_{2,n} = 2 \left(\frac{n}{n+1} \right)^2 e_q^2. \tag{7.321}$$

We used here the identity

$$q_\nu q_\rho \epsilon_{\mu\rho\lambda\sigma} - q_\mu q_\rho \epsilon_{\nu\rho\lambda\sigma} - q^2 \epsilon_{\mu\nu\lambda\sigma} = q_\rho \left(q_\sigma \epsilon_{\mu\nu\rho\lambda} - q_\lambda \epsilon_{\mu\nu\rho\sigma} \right). \tag{7.322}$$

Because the form of the operator product expansion does not depend on the target, one can construct the photon–hadron scattering amplitude $T^A_{\mu\nu}$ using the same coefficient functions $\varphi^q_{i,n}$ as follows

$$T^A_{\mu\nu} = -i \, \epsilon_{\mu\nu\rho\sigma} \frac{q_\rho}{Q^2} \sum_{n=0}^{\infty} \sum_q e_q^2 \left(1 + (-1)^n\right) \left(\frac{1}{x}\right)^n <h \left| R^q_{1\sigma...} + \frac{2n}{n+1} R^q_{2\sigma...} \right| h>. \tag{7.323}$$

Here we introduced the light-cone projections of the Lorentz tensors defined below

$$O_{...} = \frac{q'_{\mu_1}}{pq} \cdots \frac{q'_{\mu_n}}{pq} O_{\mu_1...\mu_n}, \quad q' = q - xp, \quad q'^2 = 0+ \tag{7.324}$$

and used the relation

$$\frac{n}{n+1}\frac{q_\rho}{pq}\left(q_\sigma\epsilon_{\mu\nu\rho\lambda}-q_\lambda\epsilon_{\mu\nu\rho\sigma}\right)\frac{q_{\mu_1}}{pq}\cdots\frac{q_{\mu_{n-1}}}{pq}R^q_{2\sigma\lambda\mu_1\ldots\mu_{n-1}}=-\epsilon_{\mu\nu\rho\sigma}q_\rho\,R^q_{2\sigma\ldots} \quad (7.325)$$

valid for the matrix elements between the states with the same momentum p.

Note, that for the case of the large quark mass, there is another twist-3 operator

$$R^q_{3\sigma\mu_1\ldots\mu_n}=m_q\,i^{n-1}\,S_{\{\mu_1\ldots\mu_n\}}A_{\sigma\mu_1}\,\bar\psi\gamma_5\,\gamma_\sigma\,\gamma_{\mu_1}\nabla_{\mu_2}\cdots\nabla_{\mu_{n-1}}\,\psi_q. \quad (7.326)$$

Moreover, in the Born approximation one can obtain the relation for the matrix elements of these twist-3 operators between the quark states:

$$< q|R^{q'}_{3\sigma\mu_1\ldots\mu_n}|q>=2\;<q|R^{q'}_{3\sigma\mu_1\ldots\mu_n}|q>\,. \quad (7.327)$$

However, the above expression for the amplitude $T^A_{\mu\nu}$ written in terms of the Wilson operator expansion remains valid also in the case of massive quarks. To verify it, one should calculate also its matrix elements between the states containing gluons.

For this purpose, using the Heisenberg equation of motion for the quark fields

$$(i\,\nabla\!\!\!\!/-m_q)\,\psi_q=0,\quad\bar\psi(i\,\nabla\!\!\!\!/-m_q)=0 \quad (7.328)$$

we can obtain the following operator identity

$$R^\perp_{2\sigma\ldots}-\frac{1}{2}\,R^\perp_{3\sigma\ldots}=\frac{1}{2n}\sum_{l=1}^{n-1}(n-l)\,Y^{\perp(l)}_{\sigma\ldots}, \quad (7.329)$$

where the operators are constructed from the fields $\bar\psi$, ψ, $G_{\rho\lambda}$ and their covariant derivatives. For the deep inelastic scattering, where only operators with the positive charge parity are essential, n is even and, in fact, only the following linear combinations of $Y^{\perp(l)}_{\sigma\ldots}$ appear [57], [58], [59]

$$Y^{\perp(n-l)}_{\sigma\ldots}-Y^{\perp(l)}_{\sigma\ldots}=R^{\perp(l)}_{4\sigma\ldots}-R^{\perp(n-l)}_{4\sigma\ldots},$$

$$R^{\perp(l)}_{4\sigma\ldots}=\bar\psi\gamma_5\gamma.(i\nabla.)^l\,g\,G^\perp_{\sigma.}\,(i\nabla.)^{n-l-1}\psi \quad (7.330)$$

and

$$Y^{\perp(n-l)}_{\sigma\ldots}+Y^{\perp(l)}_{\sigma\ldots}=R^{\perp(l)}_{5\sigma\ldots}+R^{\perp(n-l)}_{5\sigma\ldots},$$

$$R^{\perp(l)}_{5\sigma\ldots}=\bar\psi\gamma.(i\nabla.)^l\,g\,\tilde G^\perp_{\sigma.}\,(i\nabla.)^{n-l-1}\psi, \quad (7.331)$$

where

$$G_{\rho\lambda}=t_a G^a_{\rho,\lambda},\quad\tilde G_{\rho\lambda}=\frac{1}{2}\,\epsilon_{\rho\lambda\sigma\eta}\,G_{\sigma\eta},\quad G^a_{\rho,\lambda}=\partial_\rho A^a_\lambda-\partial_\lambda A^a_\rho+igf^{abc}A^c_\rho A^b_\lambda. \quad (7.332)$$

The calculation of the matrix elements of the operators $R_{2\sigma\ldots}$ between the quark and quark–gluon states using the operator identity (7.329) shows that indeed the operator product expansion in the Born approximation is given by the above expression. We confirm this result below also basing on the parton ideas [58], [59].

To begin with, recall that the numerator of the gluon propagator $D_{\mu\nu}(k)$ in the axial gauge $q'_\mu A^\mu = 0$ and the numerator of the quark propagator $G(k)$ for positive energies $kq'/pq' = \beta > 0$ can be expressed in terms of projectors on physical states with two helicities $\lambda = \pm s$

$$-\delta_{\mu\nu} + \frac{k_\mu q'_\nu + k_\nu q'_\mu}{kq'} = \sum_{\lambda=\pm 1} e_\mu^\lambda(k')\, e_\nu^{\lambda*}(k') + k^2 \frac{q'_\mu q'_\nu}{(kq')^2}, \tag{7.333}$$

$$\not{k} + m_q = \sum_{\lambda=\pm 1/2} u_\lambda(k')\, \bar{u}_\lambda(k') + \frac{k^2 - m_q^2}{s\beta}\, \not{q}' \tag{7.334}$$

providing that the last contributions proportional to the particle virtuality are small. In the above relations, we have for the gluon and quark momenta

$$k' = k - \frac{k^2}{2kq'}\, q' = \beta p + k_\perp - \frac{k_\perp^2}{\beta s}\, q', \quad k'^2 = 0 \tag{7.335}$$

and

$$k' = k - \frac{k^2 - m_q^2}{2kq'}\, q' = \beta p + k_\perp - \frac{k_\perp^2 - m_q^2}{\beta s}\, q', \quad k'^2 = m_q^2, \tag{7.336}$$

respectively.

For the longitudinally polarized hadron, where

$$a_L = \frac{1}{m}\, \lambda_h \left(p - \frac{m^2}{pq'}\, q' \right), \quad a_L^2 = -1, \quad \lambda_= \pm 1, \tag{7.337}$$

the virtual gluon propagators in the light-cone gauge are not attached to the quark line between the photon vertices γ_μ and γ_ν, because such contribution would lead to an additional large denominator in the quark propagators, which is not compensated by a nominator. As for the quark lines, the numerators in their propagators can be simplified as follows

$$\gamma_\mu(\not{k} + \not{q} + m)\gamma_\nu \to -i\epsilon_{\mu\nu\lambda\sigma}\, q_\lambda \gamma_5 \gamma_\sigma$$

$$\gamma_\nu(\not{k} - \not{q} + m)\gamma_\mu \to -i\epsilon_{\mu\nu\lambda\sigma}\, q_\lambda \gamma_5 \gamma_\sigma. \tag{7.338}$$

This leads to the following expression for the cross section

$$W_{\mu\nu}^A = -\frac{i}{pq}\, \epsilon_{\mu\nu\lambda\sigma} q_\lambda \sum_q \frac{1}{2} e_q^2 \int\, <\gamma_5 \gamma_\sigma> (\delta(\beta - x) + \delta(\beta + x))\, d\beta, \tag{7.339}$$

where $<\gamma_5 \gamma_\sigma>$ implies the corresponding vertex, and the integration over the Sudakov parameter β for the neighbouring virtual quark is not performed. The second term in the bracket describes the antiquark contribution in the Dirac picture, where $\beta < 0$. Of course, the physical antiquark energy is positive. Because for the longitudinal polarization s_h, the indices μ, ν are transversal, γ_σ is effectively multiplied by q'_σ, which allows us to leave in the neighbouring quark propagators only projectors on the physical states with two helicities leading to the result

$$\frac{1}{s} \int < \gamma_5 \not{q}' > (\delta(\beta - x) + \delta(\beta + x)) \, d\beta = \sum_{r=q,\bar{q}} \left(n_r^+(x) - n_r^-(x) \right). \qquad (7.340)$$

Thus, we get the parton expression for the structure function $g_1(x)$ in terms of quark and antiquark distributions with helicities $\lambda = \pm 1/2$

$$g_1(x) = \sum_{r=q,\bar{q}} \frac{e_r^2}{2} \lambda_h \left(n_r^+(x) - n_r^-(x) \right). \qquad (7.341)$$

Let us consider now a transversely polarized hadron. Taking into account that, in this case, one of two indices μ and ν is transversal and other is longitudinal, corresponding to the substitution of one of two photon vertices by $\not{q}'$, we can calculate the intermediate quark propagator obtaining the relation

$$(g_1(x) + g_2(x)) \, S_\sigma^\perp = -\sum_q \frac{1}{2} e_q^2 \int d\beta \, < \gamma_5 \gamma_\sigma^\perp > (\delta(\beta - x) + \delta(\beta + x)). \qquad (7.342)$$

It coincides in fact with the result, which can be derived from the expression for $T_{\mu\nu}$ written in terms of the operator product expansion. To show it in the light-cone gauge $A_\mu q'_\mu = 0$ we apply the operator identity

$$A_{\sigma\ldots}^q \equiv R_{1\sigma\ldots}^q + \frac{2n}{n+1} R_{2\sigma\ldots}^q = \bar{\psi}_q \gamma_5 \gamma_\sigma (i\partial.)^n \psi_q. \qquad (7.343)$$

Further, using the relation

$$< h|A_{\sigma\ldots}^q|h > = \int d\beta \, \beta^n \, < \gamma_5 \gamma_\sigma > \qquad (7.344)$$

one can calculate $W_{\mu\nu}^A$

$$W_{\mu\nu}^A = -\frac{i}{pq} \epsilon_{\mu\nu\lambda\sigma} q_\lambda \sum_q \frac{1}{2} e_q^2 \frac{1}{\pi} \mathrm{Im} \sum_{n=0}^{\infty} (1 + (-1)^n) \left(\frac{1}{x}\right)^n < h|A_{\sigma\ldots}^q|h >$$

$$= -\frac{i}{pq} \epsilon_{\mu\nu\lambda\sigma} q_\lambda \sum_q \frac{1}{2} e_q^2 \int < \gamma_5 \gamma_\sigma > (\delta(\beta - x) + \delta(\beta + x)) \, d\beta. \qquad (7.345)$$

Note that the operator $A_{\sigma\ldots}^q$ does not have a definite twist and therefore it is not renormalized in a multiplicative way. Therefore, it is more consistent to use the representation of $W_{\mu\nu}$ in terms of the operators $R_{1\sigma\ldots}$ and $R_{2\sigma\ldots}$. Due to the relativistic invariance, the different components of the vector $R_{1\sigma\ldots}$ are proportional. For example,

$$< h|R_{1\sigma\ldots}^\perp|h > = s_\sigma^\perp \frac{pq'}{sq'} \frac{1}{n+1} < h|R_{1\ldots}|h > . \qquad (7.346)$$

Analogously, for the operator $A_{\sigma...}^{q\perp}$ using the equations of motion we obtain

$$< h|A_{\sigma...}^{q\perp}|h > = s_\sigma^\perp \frac{pq'}{sq'} \frac{1}{n+1} < h|R_{1...}|h >$$

$$+ \frac{n}{n+1} < h|R_{3\sigma...}^\perp|h > + \frac{1}{n+1} \sum_{l=1}^{n-1}(n-l) < h|Y_{\sigma...}^{l\perp}|h > . \tag{7.347}$$

One can introduce the generating functions $E(\beta)$, $A(\beta)$, $C(\beta)$, and $Y(\beta_1, \beta_2)$, which describe the matrix elements of the corresponding operators

$$\lambda_h \int d\beta \beta^n < \frac{\gamma_5 \bar{q}'}{2pq'} > = \int d\beta \beta^n E(\beta),$$

$$< h|A_{\sigma n}^\perp|h > = \int d\beta \, \beta^n < \gamma_5 \gamma_\sigma^\perp > \equiv s_\sigma^\perp \int d\beta \, \beta^{n-1} A(\beta),$$

$$< h|R_{3\sigma...}^\perp|h > = \int d\beta \, \beta^{n-1} < \gamma_5 \gamma_\sigma^\perp \gamma. > \equiv s_\sigma^\perp \int d\beta \, \beta^{n-1} C(\beta),$$

$$< h|Y_{\sigma...}^{l\perp}|h > \equiv \int d\beta_1 d\beta_2 \, \beta_1^{l-1} \beta_2^{n-l-1} Y(\beta_1, \beta_2). \tag{7.348}$$

The notation $\gamma.$ was explained in (7.324). We have in particular

$$g_1(x) = \sum_q \frac{e_q^2}{2x} \left(E_q(x) - E_q(-x) \right),$$

$$g_1(x) + g_2(x) = -\sum_q \frac{e_q^2}{2x} \left(A_q(x) - A_q(-x) \right). \tag{7.349}$$

Further, due to the equation of motion one obtains the relation among these functions [58],[59]

$$\left(1 - \beta \frac{d}{d\beta}\right) A(\beta) = E(\beta) - \beta \frac{d}{d\beta} C(\beta) + \beta \int \frac{d\beta_1}{\beta_1 - \beta} \left(\frac{\partial}{\partial \beta} Y(\beta_1, \beta) + \frac{\partial}{\partial \beta_1} Y(\beta, \beta_1) \right). \tag{7.350}$$

To solve this differential equation concerning $A(\beta)$, the integration constant should be chosen from the Burkhardt–Cottingham sum rule

$$\int_0^1 g_2(x) dx = 0 \tag{7.351}$$

equivalent to the equality

$$\int_{-1}^1 \frac{d\beta}{\beta} A(\beta) = -\int_{-1}^1 \frac{d\beta}{\beta} E(\beta). \tag{7.352}$$

Thus, we obtain the relation

$$g_1(x) + g_2(x) = \int_x^1 \frac{dx'}{x'} g_1\left(x'\right) - \sum_q \frac{e_q^2}{2} \int_x^1 \frac{dx'}{x'} K\left(x'\right),$$

$$K\left(x'\right) = -\frac{d}{dx'} C_q\left(x'\right) + \int \frac{dx_1}{x_1 - x'} \left(\frac{\partial}{\partial x'} Y_q\left(x_1, x'\right) + \frac{\partial}{\partial x_1} Y_q\left(x', x_1\right) \right) + \left(x' \to -x'\right).$$

(7.353)

Providing that the current quark masses are small and the gluon contributions are not large, one can neglect the term containing $K\left(x'\right)$ in the above expression. In this case, it is reduced to the Wandzura–Wilczek relation

$$g_1(x) + g_2(x) \approx \int_x^1 \frac{dx'}{x'} g_1\left(x'\right),$$

(7.354)

which appears if we take into account only the contribution of the twist-2 operator.

The matrix element of the current $\gamma_5 \gamma_\sigma^\perp$ between hadron states contains also the contribution from the off-mass shell quarks, because we can not neglect the additional term $\sim \hat{q}'$ in their propagators. To solve this problem, let us use the following identity for this current

$$\gamma_5 \gamma_\sigma^\perp = \gamma_5 \frac{\not{q}' k_\sigma^\perp}{kq'} + m_q \gamma_5 \gamma_\sigma^\perp \frac{\not{q}'}{kq'} - \frac{(\not{k} - m_q) \gamma_5 \not{q}' \gamma_\sigma^\perp}{2kq'} - \frac{\gamma_5 \not{q}' \gamma_\sigma^\perp (\not{k} - m_q)}{2kq'}.$$

(7.355)

In the last two terms, the factors $(\not{k} - m_q)$ cancel the nearest quark propagator and the corresponding gluon–quark vertex $g\, t^a\, \gamma_\rho$ turns out to be at the same space-time point as the current producing the new vertices

$$D_1(\beta_1, \beta) s_\sigma^\perp = \left\langle g\, \not{A}^\perp \frac{\gamma_5 \not{q}' \gamma_\sigma^\perp}{pq'} \right\rangle, \quad D_2(\beta, \beta_1) s_\sigma^\perp = \left\langle g\, t^a \frac{\gamma_5 \not{q}' \gamma_\sigma^\perp}{pq'} \not{A}^\perp \right\rangle,$$

(7.356)

where the integration over the Sudakov parameters β and β_1 respectively for ingoing and outgoing quarks is not performed and $A_\rho^\perp$ means the transverse component of the gluon field in the momentum space. Such a procedure corresponds to perturbation theory and leads to the relation

$$A(\beta) = B(\beta) + C(\beta) - \int d\beta_1 D(\beta_1, \beta),$$

(7.357)

where

$$D(\beta_1, \beta) = \frac{1}{2} \left(D_1(\beta_1, \beta) + D_2(\beta, \beta_1) \right).$$

(7.358)

Here we introduced the functions $B(\beta)$ and $C(\beta)$ according to the following definitions

$$B(\beta) s_\sigma^\perp = \left\langle \gamma_5 \frac{\not{q}' k_\sigma}{pq'} \right\rangle, \quad C(\beta) s_\sigma^\perp = \left\langle m_q \gamma_5 \gamma_\sigma^\perp \frac{\not{q}'}{pq'} \right\rangle.$$

(7.359)

It is important that in the vertices B, C, D_1, D_2 the nearest quarks can be considered as real particles with momenta k' and two physical helicities, since the extra term $\sim \not{q}'$ in numerators of their propagators does not give any contribution due to the relation $\not{q}'\, \not{q}' = 0$. Similarly, the gluon absorbed in the vertices D_1, D_2 in the axial gauge also can be considered as a real particle with two possible polarizations, because the extra term $\sim q'_\mu q'_\nu$ in the numerator of its propagator gives the vanishing contribution. The corresponding operators are simple examples of a large class of quasipartonic operators of arbitrary twists which will be considered below.

7.11 Parton distributions in polarized nucleon

7.11.1 Evolution equations for parton distributions in longitudinally polarized nucleon

Let us use the notation:

$$\Delta n_r\left(x, Q^2\right) = n_r^+\left(x, Q^2\right) - n_r^-\left(x, Q^2\right), \quad r = u, d, s, g, \tag{7.360}$$

where $n_r^+(x, Q^2)$ and $n_r^-(x, Q^2)$ are correspondingly the densities of partons, polarized along or opposite to the longitudinally polarized nucleon. Using the splitting kernels presented in Section 7.4, the evolution equations for $\Delta n_r(x)$ can be written in the form analogous to Eqs.(7.134). It is convenient to consider separately the equations for singlet $\Delta \sum(x, Q^2) = \sum_r \Delta n_r(x, Q^2)$ and nonsinglet $\Delta n_{NS} = \Delta n_3, \Delta n_8$ parts of quark densities. The evolution equations are (see [60],[61] for reviews) for singlet:

$$\frac{d}{d \ln Q^2} \Delta \sum\left(x, Q^2\right) = \frac{\alpha_s\left(Q^2\right)}{2\pi} \left[\int_x^1 \frac{dy}{y} \Delta P_{qq}\left(\frac{x}{y}\right) \Delta \sum\left(y, Q^2\right) \right.$$

$$\left. + 2N_f \int_x^1 \frac{dy}{y} \Delta P_{g \to q}\left(\frac{x}{y}\right) \Delta g\left(y, Q^2\right) \right], \tag{7.361}$$

$$\frac{d}{d \ln Q^2} \Delta g\left(x, Q^2\right) = \frac{\alpha_s\left(Q^2\right)}{2\pi} \left[\int_x^1 \frac{dy}{y} \Delta P_{q \to g}\left(\frac{x}{y}\right) \Delta \sum\left(y, Q^2\right) \right.$$

$$\left. + \int_x^1 \frac{dy}{y} \Delta P_{g \to g}\left(\frac{x}{y}\right) \Delta g\left(y, Q^2\right) \right] \tag{7.362}$$

for nonsinglet:

$$\frac{d}{d \ln Q^2} \Delta n\left(x, Q^2\right)_{NS} = \frac{\alpha_s\left(Q^2\right)}{2\pi} \int_x^1 \frac{dy}{y} \Delta P_{q \to q}\left(\frac{x}{y}\right) \Delta n_{NS}\left(y, Q^2\right), \tag{7.363}$$

where in leading order:

$$\Delta P_{q \to q}^{(z)} = \frac{4}{3}\left[\frac{1+z^2}{(1-z)_+} + \frac{3}{2}\delta(z-1)\right],$$

$$\Delta P_{g \to q}(z) = \frac{1}{2}\left[z^2 - (1-z)^2\right] = \frac{1}{2}(2z-1),$$

$$\Delta P_{q \to g}(z) = \frac{4}{3} \frac{1 - (1-z)^2}{z},$$

$$\Delta P_{g \to g}(z) = 3\left[(1+z)^4 \left(\frac{1}{z} + \frac{1}{(1-z)_+} \right) - \frac{(1-z)^3}{z} + \left(\frac{11}{6} - \frac{1}{9} \right) \delta(1-z) \right]. \quad (7.364)$$

The splitting kernels $\Delta P_{q \to q}$ and $\Delta P_{g \to q}$ satisfy the equalities:

$$\int_0^1 dz \Delta P_{q \to q}(z) = \int_0^1 dz \Delta P_{g \to q}(z) = 0,$$

corresponding to chirality conservation. Note that Δg contribution is infrared unstable and depends on the factorization scheme [62]. Particularly, in the $\overline{\text{MS}}$-renormalization scheme $\int_0^1 dx \Delta g(x, Q^2)$ does not contribute to the first moment of $g_1(x)$. In the renormalization scheme proposed in [63], where it was accepted that the gluon virtuality $| p^2 |$ is much larger than the square of quark mass m^2, $| p^2 | \gg m^2$ and dimensional regularization was used:

$$\Delta_g \int_0^1 dx g_1(x) = -\frac{N_f}{18\pi} \alpha_s(Q^2) \int_0^1 dx \Delta g(x, Q^2) \quad (7.365)$$

(See also [64].) The spin dependent splitting kernels ΔP_{ij} in NLO were calculated in [65], [66], [67]. The analysis of experimental data in next-to-leading order (in $\overline{\text{MS}}$ scheme) gave the polarized parton distribution presented in Fig.(7.10) (see the reviews in [60],[61],[68]).

7.11.2 Sum rules

The Bjorken sum rule for polarized DIS was discussed in Section 7.8.4, where the results for perturbative corrections were presented. E155 collaboration at SLAC found experimentally $Bj^{pol} = 0.176 \pm 0.003 \pm 0.007$ [69] and SMS collaboration got [70] $Bj^{pol} = 0.174^{+0.024}_{-0.012}$ (both at $Q^2 = 5 \text{ GeV}^2$) in comparison with the theoretical value 0.185 ± 0.005. The good agreement of theoretical and experimental values indicates that nonperturbative corrections to B_j^{pol} are small. Note, however, that about 10% of experimental contributions to the integral in (7.256) comes from the region $x < 0.01$, where the data points are absent and the extrapolation of experimental curve is needed. The Burkhardt–Cottingham (BC) sum rule [71] looks simple

$$BC = \int_0^1 g_2(x) dx = 0. \quad (7.366)$$

In [4] it was argued that this sum rule is valid if only the lowest twist terms are accounted. In SLAC experiments, it was found [72] $(BC)_p = -0.042 \pm 0.008$ for proton and $(BC)_d = -0.006 \pm 0.011$ for deuterium at $Q^2 = 5 \text{ GeV}^2$. (The measurements were performed at $0.02 < x < 0.8$). These numbers cannot be considered as a trustworthy confirmation of

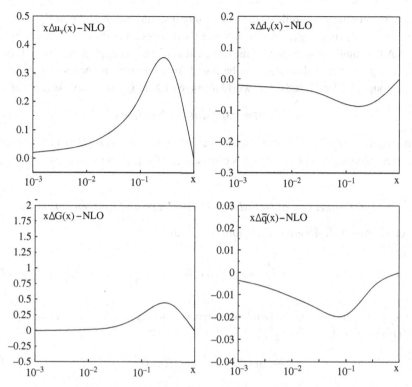

Fig. 7.10. Polarized parton distribution in next-to-leading order fits at $Q^2 = 4\,\text{GeV}^2$.

BC sum rule, because $g_2(x)$ is small and consistent with zero in the whole domain of x except $x \approx 0.5$, where $g_{2,p}(0.5) \approx -0.06$ and $g_{2,d}(0.5) \approx -0.04$ [72]. JLab experiments [73] on transversally polarized ^3He at low Q^2 showed the agreement BC sum rule for neutron with zero value at $0 < Q^2 < 0.9\,\text{GeV}^2$ at the accuracy about 0.005.

An intriguing problem arises, if the integrals

$$\Gamma_{p,n}(Q^2) = \int\limits_0^1 dx g_{1,p,n}(x) \tag{7.367}$$

are considered. Neglecting higher-twist contributions, we have

$$\Gamma_{p,n}\left(Q^2\right) = \frac{1}{12}\left\{\left[1 - a - 3.58a^2 - 20.2a^3 - ca^4\right]\left[\pm g_A + \frac{1}{3}g_A^8\right]\right.$$
$$\left. + \frac{4}{3}\left[1 - \frac{1}{3}a - 0.55a^2 - 4.45a^3\right]\Sigma\right\} - \gamma\frac{\alpha_s\left(Q^2\right)}{6\pi}\Delta g\left(Q^2\right) \tag{7.368}$$

(The higher-order perturbative coefficients were calculated in [74], where the references on previous works can be found.) In (7.368), $g_A^8 = 3F - D = 0.59 \pm 0.02$ [75] is the axial-octet-coupling constant with hyperons, $\Sigma = \Delta u + \Delta d + \Delta s$ is equal to the part of parton (neutron) spin projection, carried by quarks in completely longitudinally polarized nucleon. The coefficient γ reflects the uncertainty in contribution of Δg to $\Gamma_{p,n}$, discussed above. Since Δg is not large, $|\Delta g| < 0.5$ (see the discussion in [61]) and the coefficient in front of Δg is small, contribution of the last term in (7.368) is negligible. The values of Σ, which is of great physical interest can be found from the measurements of $\Gamma_{p,n}$, done by various groups and Eq.(7.368). The world average of Σ at $Q^2 = 3$ Gev^2 is (see e.g. [61])

$$\Sigma = 0.30 \pm 0.01(stat.) \pm 0.02(evol.) \tag{7.369}$$

This means that only 30% of proton spin is carried by quarks. This phenomenon until now has no clear physical explanation and is often called "the proton spin crises."

7.11.3 Connection of Gerasimov–Drell–Hearn and Bjorken sum rules

The Gerasimov–Drell–Hearn (GDH) sum rule reads:

$$\int\limits_0^\infty \frac{d\nu}{\nu} \mathrm{Im} G_{1;p,n}(\nu, 0) = -\frac{1}{4}\kappa_{p,n}^2, \tag{7.370}$$

where $\kappa_{p,n}$ are the nucleon anomalous magnetic moments, $\kappa_p = 1.79$, $\kappa_n = -1.91$. $G_1(\nu, Q^2)$ is related to $g_1(x, Q^2)$ by

$$\frac{\nu}{m^2} \mathrm{Im} G_1(\nu, Q^2) = g_1(x, Q^2). \tag{7.371}$$

The GDH sum rule follows from the Low theorem and unsubtracted dispersion relation representation in ν for $G_1(\nu, Q^2)$ (see [4]). Since $g_{1,p}(x, Q^2)$ is positive at large Q^2 and the integral (7.370) is negative, this means that $g_{1,p}(x, Q^2)$ change sign, when Q^2 decreases. That, in turn, means the presence of large nonperturbative contributions to $g_{1,p}(x, Q^2)$. The phenomenological model realizing the smooth transition from large Q^2 to $Q^2 = 0$ was suggested in [76],[77],[78]. The model is based on vector dominance, accounts for the low nucleon resonance contributions, and is in good agreement with experiment.

7.11.4 Transversity structure function

Besides the structure functions considered above, there is a very specific structure function $h_1(x, Q^2)$ of twist 2 [79]. Unlike F_1, F_2, g_1, q_2, which conserve chirality, $h_1(x, Q^2)$ violates chirality. For this reason, $h_1(x, Q^2)$ does not contribute to electroproduction or $\nu(\bar{\nu})N$ scattering, where chirality is conserving. The most convenient definition of $h_1(x, Q^2)$ is the light-cone representation [80]:

$$i \int\limits_{-\infty}^\infty \frac{d\lambda}{2\pi} e^{i\lambda x} \langle p, s \mid \overline{\psi}(0)\sigma_{\mu\nu}\gamma_5\psi(\lambda n) \mid p, s \rangle = 2h_1\left(x, Q^2\right)(s_{\perp\mu}p_\nu - s_{\perp\nu}p_\mu) \tag{7.372}$$

Here n is the light-cone vector of dimension of (mass)$^{-1}$, $n^2 = 0$, $n^+ = n_0 + n_z = 0$ $np = 1$, p and s are proton momentum and spin vectors, $s^2 = -1$, $ps = 0$. (We restrict

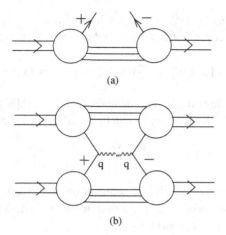

Fig. 7.11. (a) The diagrammatic representation of the transversity; $+, -$ means the quark helicities; (b) the polarized Drell–Yan process.

ourselves by the terms of twist 2 in (7.372).) As is seen from (7.372), $h_1(x, Q^2)$ corresponds to transversally polarized nucleon and can be called transversity distribution. In parton language, $h_1(x, Q^2)$ can be described in following way [79],[80],[81],[61]. Suppose that in infinite momentum frame the proton is moving in z-direction and is polarized in x-direction, perpendicular to z. Then

$$h_1(x) = \sum_r n_{r,+}(x) - n_{r,-}(x), \quad r = u, d, s \tag{7.373}$$

where $n_{r,+}(x)$ and $n_{r,-}(x)$ are correspondingly the quark densities, polarized along and opposite to proton spin. In helicity basis, the transversity corresponds to helicity-flip process and can be represented by the diagram of Fig.7.11a. It can be measured in the polarized Drell–Yan process (Fig.7.11b).

The first moment of the transversity distribution

$$H = \int_0^1 dx h_1(x, Q^2) \tag{7.374}$$

is proportional to the proton-tensor charge [80]

$$\langle p, s \mid \bar{\psi} i \sigma_{\mu\nu} \gamma_5 \psi \mid p, s \rangle = (s_\mu + p_\nu - s_{\nu} + p_\mu) H \tag{7.375}$$

It can be shown [82], that $h_1(x, Q^2)$ can be also expressed through the correlator of axial and scalar currents:

$$T_\mu(p, q) = i \int d^4 x e^{iqx} \langle p, s \mid \frac{1}{2} T \{j_{\mu5}(x), j(0) + j(x), j_{\mu5}(0)\} \mid p, s \rangle, \tag{7.376}$$

$$\mathrm{Im} T_\mu(p, q) = -\pi \left(s_\mu - \frac{qs}{q^2} q_\mu \right) h_1(x, q^2) + \left(p_\mu - \frac{vq_\mu}{q^2} \right) (qs) l_1(s, q^2)$$

$$+ \varepsilon_{\mu\nu\lambda\sigma} p_\nu q_\lambda s_\sigma (qs) l_2 \left(x, q^2 \right). \tag{7.377}$$

As was proved in [80], $h_1\left(x, Q^2\right)_2$ satisfy the inequality $\mid h_1\left(x, Q^2\right)_r \mid < n_r\left(x, Q^2\right)$ for each quark flavour. Moreover, Soffer [83] proved the more strict inequalities

$$\mid h_1\left(x, Q^2\right)_r \mid < \frac{1}{2}[n_r\left(x, Q^2\right) + \Delta n_r(x, Q^2)] \tag{7.378}$$

valid in leading order. However, as was demonstrated in [84],[85] (7.378) is violated in next-to-leading order. The transversity h_1 can be found by measurement of the asymmetry

$$A_{s_A s_B} = \frac{\sigma\left(s_A, s_B\right) - \sigma\left(s_A, -s_B\right)}{\sigma\left(s_A, s_B\right) + \sigma\left(s_A, s_B\right)} \tag{7.379}$$

in production of $e^+ e^- (\mu^+ \mu^-)$ pairs in collision of two polarized nucleons A and B (or nucleon A and antinucleon B). In the case when both A, B or $A, \overline{B}$ are polarized transversally, the asymmetry is equal [79]

$$A_{TT}^{(x,y)} = \frac{\sin^2\theta \cos\varphi}{1 + \cos^2\theta} \frac{\Sigma e_r h_{1,r}^A(x) h_{1,r}^B(y)}{\Sigma e_r^2 n_r^A(x) n_r^B(y)}, \tag{7.380}$$

where θ and φ are the polar and azimuthal angles of $e^+ e^- (\mu^+ \mu^-)$ pair relative to collision axes and transverse polarization direction. Some other effects for measuring the transversity were also suggested (see [61] for review). Up to now, no definite results were obtained.

7.12 Evolution equations for quasipartonic operators

As it was demonstrated in the Section 7.10, the equations of motion allow us to reduce the set of all twist-3 operators appearing in the operator product expansion of two electromagnetic currents in the light-cone gauge

$$A_\mu q'^\mu = A_. = 0, \quad q' = q + xp \tag{7.381}$$

to the class of operators whose matrix elements can be calculated between on-mass shell parton states (quarks and gluons) with two helicities. It turns out, that almost all twist-4 operators appearing in the power corrections $\sim 1/Q^2$ to structure functions of the deep inelastic lepton–hadron scattering also can be reduced to this class of quasipartonic operators (QPO) [59],[86]. Apart from the on-mass shell requirement it is natural also to impose the additional constraint on QPO: they should not contain explicitly the strong coupling constant [86]. Indeed, in the parton model the form of strong interactions is essential only when one calculates parton distributions whereas the matrix elements of the operators between parton states do not depend on the QCD dynamics. Below we shall neglect the quark mass. It will give us an opportunity to use the conformal symmetry of the theory.

The mass-shell condition will be fulfilled in the light-cone gauge providing that corresponding QPO are invariant under the following field transformations

$$\psi \to \psi + \gamma_. \chi, \quad \bar{\psi} \to \bar{\psi} + \bar{\chi} \gamma_., \quad A_\mu \to A_\mu + q'_\mu \phi, \tag{7.382}$$

where χ, $\bar{\chi}$, and ϕ are arbitrary spinor and scalar functions. Therefore, these quantities can be constructed as a product of the structures (we also take into account the independence from the QCD coupling constant g)

$$\gamma.\psi, \quad \bar{\psi}\gamma., \quad G^{\perp}_{.\mu} = \partial.A^{\perp}_{\mu}, \quad D. = \partial., \quad \gamma.. \tag{7.383}$$

Of course, the composite operators should be colourless, but on an intermediate step we consider the operators with an open colour as it takes place for above structures. In principle, the fields ψ, $\bar{\psi}$ can belong to an arbitrary representation of the colour group. For example, in the supersymmetric models gluinos are transformed according to the adjoint representation of the gauge group. But, as a rule, the fields ψ and $\bar{\psi}$ are assumed to be usual massless quark fields.

Below we list examples of QPO for the twist 2:

$$\bar{\psi}\gamma.(i\partial.)^n\psi, \quad \bar{\psi}\gamma_5\gamma.(i\partial.)^n\psi, \quad \bar{\psi}\gamma.\gamma^{\perp}_{\rho}(i\partial.)^n\psi,$$

$$\left(-i\partial.A^{\perp}_{\rho}\right)(i\partial.)^{n+1}A^{\perp}_{\rho}, \quad \epsilon^{\perp\perp}_{\rho_1\rho_2}\left(-i\partial.A^{\perp}_{\rho_1}\right)(i\partial.)^{n+1}A^{\perp}_{\rho_2}, \quad S_{\rho_1\rho_2}\left(-i\partial.A^{\perp}_{\rho_1}\right)(i\partial.)^{n+1}A^{\perp}_{\rho_2}, \tag{7.384}$$

twist 3:

$$\left((-i\partial.)^{n_1}\bar{\psi}\right)\gamma.\gamma_5(i\partial.A^{\perp}_{\sigma})(i\partial.)^{n_2}\psi, \quad f_{abc}\epsilon^{\perp\perp}_{\rho_1\rho_2}\left((-i\partial.)^{n_1}A^{a\perp}_{\rho_1}\right)(i\partial.A^{b\perp}_{\sigma})(i\partial.)^{n_2}A^{c\perp}_{\rho_2}, \tag{7.385}$$

and twist 4:

$$\left((-i\partial.)^{n_1}\bar{\psi}\right)\gamma.\left((i\partial.)^{n_2}\psi\right)\left((-i\partial.)^{n_3}\bar{\psi}\right)\gamma.(i\partial.)^{n_4}\psi,$$

$$\left((-i\partial.)^{n_1}A^{\perp}_{\rho}\right)\left((i\partial.)^{n_2}A^{\perp}_{\rho}\right)\left((-i\partial.)^{n_3}A^{\perp}_{\sigma}\right)(i\partial.)^{n_4}A^{\perp}_{\sigma}, \tag{7.386}$$

where $\epsilon^{\perp\perp}_{\rho_1\rho_2}$ is an antisymmetric tensor in the two-dimensional space ($\epsilon^{\perp\perp}_{12} = 1$).

One can easily verify that the twist of QPO coincides with the number k of constituent fields ψ, $\bar{\psi}$, $A^{\perp}_{\sigma}$

$$t = d - j = k. \tag{7.387}$$

They can be written in the form [86]

$$O^{\{r\}} = \Gamma^{\{r\}}_{\rho_1...\rho_l} \prod_{t=1}^{l}(i\partial.)^{n_t+s_t-1} \varphi^{r_t}_{\rho_t}, \tag{7.388}$$

where $\varphi^{r_t}_{\rho_t}$ is one of these fields with the spin index ρ_t and the colour index r_t. The factor $\Gamma^{\{r\}}$ is a numerical matrix satisfying the constraints

$$q'_{\rho_t} \Gamma^{\{r\}}_{\rho_1...\rho_l} = 0, \quad \hat{q}'_{\bar{\rho}\rho_t} \Gamma^{\{r\}}_{\rho_1...\rho_l} = 0 \tag{7.389}$$

in vector and spinor indices. Generally, we consider the matrix elements of such operators between the hadron states h, h' with the different momenta p_h, $p_{h'}$ [86]

$$< h'|O|h >, \tag{7.390}$$

They appear, for example, in the theoretical description of the electro- and photoproduction of photons [87]. The simple case of the matrix element between the vacuum and the hadron state

$$< 0|O|h > \tag{7.391}$$

corresponds to the calculation of the hadron-wave functions at small parton distances $x^2 \sim 1/Q^2 \to 0$ [88],[89],[90],[91] appearing in the theory of hadron formfactors and other exclusive processes at large momentum transfers.

To have the universal formulas for particles and antiparticles it is convenient to change the normalization of the quark spinors and gluon polarization vectors by introducing the new wave functions [86]

$$\xi^\lambda(k) = \frac{\not{k} \not{q}'}{s} u^\lambda(p'), \quad s = 2p'q', \quad k^2 = s\alpha\beta + k_\perp^2 = 0,$$

$$\eta_\rho^\lambda(k) = \beta e_\rho^\lambda(p') - \frac{2}{s} q_\rho' k_\perp^\sigma e_\sigma^\lambda, \quad \lambda = \pm 1, \quad p = p' + \frac{m_h^2}{s} q', \tag{7.392}$$

where the basis spinors $u^\lambda(p')$ and vectors $e^\lambda(p')$ satisfy the following eigenvalue equations

$$\gamma_5 u^\lambda(p') = -\lambda u^\lambda(p'), \quad \not{p}' u^\lambda(p') = 0, \quad p_0' > 0,$$

$$i\epsilon_{\rho\tau}^{\perp\perp} e_\tau^\lambda(p') = \lambda e_\tau^\lambda(p'), \quad p'e_\tau^\lambda(p') = q'e_\tau^\lambda(p') = 0, \quad \epsilon_{\rho\tau}^{\perp\perp} = \frac{2}{s}\epsilon_{\rho\tau\alpha\beta} p_\alpha' q_\beta' \tag{7.393}$$

and the normalization conditions

$$\bar\xi^{\lambda'}(k)\gamma_\mu\xi^\lambda(k) = 2\beta k_\mu \delta_{\lambda\lambda'}, \quad \eta_\rho^{\lambda'}(k)\eta_\rho^\lambda(k) = -\beta^2 \delta_{\lambda\lambda'}. \tag{7.394}$$

Inverse transformations for spinors with positive and negative energies and for the polarization vectors is

$$u^\lambda(k) = \beta^{-1/2} \xi^\lambda(k), \quad v^\lambda(-k) = \beta^{-1/2} \gamma_2 \xi^{-\lambda*}(k), \quad e^\lambda(k) = \beta^{-1}\eta^\lambda(k). \tag{7.395}$$

It is helpful also to impose on $u^\lambda(p')$ and $e^\lambda(p')$ additional constraints by fixing their phases to satisfy the equalities

$$v^\lambda(-k) = |\beta|^{-1/2} \xi^{-\lambda}(-k), \quad \eta^{\lambda*}(k) = \eta^{-\lambda}(-k). \tag{7.396}$$

The above relations allow us to relate analytically the amplitude for the particle production to that for the annihilation of the antiparticle with an opposite helicity. In such normalization, the numerators of the propagators for the partons entering the vertex can be written as follows

$$\not{k} = \beta^{-1} \sum_{\lambda=\pm 1} \xi^\lambda(k')\bar\xi^\lambda(k') + \frac{k^2}{s\beta}\not{q}', \quad k' = k - \frac{k^2}{\beta s} q',$$

$$-\delta_{\mu\nu} + \frac{k_\mu q_\nu' + k_\nu q_\mu'}{kq'} = \beta^{-2} \sum_{\lambda=\pm 1} \eta_\mu^\lambda(k')\eta_\nu^{\lambda*}(k') + \frac{4k^2}{s^2\beta^2} q_\mu' q_\nu'. \tag{7.397}$$

Because for QPO the last terms in the right-hand sides of the above equalities give vanishing contributions one can express the hadron-matrix elements of these operators as the sum of products of the matrix elements $O_{\lambda_1...\lambda_l}$ between the on-mass shell parton states with definite helicities $\lambda_1 ... \lambda_l$ integrated with the inclusive parton correlation functions $N_{\lambda_1...\lambda_l}(\beta_1, ... \beta_l)$ [86]

$$< h'|O^{\{r\}}|h >= \sum_{\lambda_1...\lambda_l} \int d\beta_1 ... d\beta_l \; O^{\{r\}}_{\lambda_1...\lambda_l} \prod_{t=1}^{l} \beta_t^{n_t} \; N^{\{r\}}_{\lambda_1...\lambda_l}(\beta_1, ... \beta_l), \qquad (7.398)$$

where

$$O^{\{r\}}_{\lambda_1...\lambda_l} = \Gamma^{\{r\}}_{\rho_1...\rho_l} \prod_{t=1}^{l} \beta_t^{-1} \xi^{\lambda_t}_{\rho_t}(k'_t), \qquad (7.399)$$

$$N^{\{r\}}_{\lambda_1...\lambda_l}(\beta_1, ... \beta_l) = \int d^4k_1 ... d^4k_l M^{\{r\}}_{\rho_1...\rho_l}(k_1, ..., k_l) \prod_{t=1}^{l} \frac{\bar{\xi}^{\lambda_t}_{\rho_t}(k'_t)}{k_t^2} \delta\left(\beta_t - \frac{k_t q'}{p' q'}\right). \qquad (7.400)$$

In the above expressions, for simplicity, we denote by $\xi^{\lambda_t}_{\rho_t}(k'_t)$ the parton-wave functions which for various t could be $\xi^{\lambda}(k')$, $\bar{\xi}^{-\lambda}(-k')$ or $\eta^{\lambda}_{\rho}(k')$. The product $O^{\{r\}}_{\lambda_1...\lambda_l} \prod_{t=1}^{l} \beta_t^{n_t}$ corresponds to the parton-matrix element of the operator $O^{\{r\}}_{\underset{\cdots}{}}$ which includes the derivatives $(i\partial)^{n_t}$ acting on the quark or gluon fields.

The parton-correlation functions (PCF) $N^{\{r\}}_{\lambda_1...\lambda_l}(\beta_1, ... \beta_l)$ contain the δ-function, corresponding to the energy-momentum conservation

$$\beta = \sum_{t=1}^{l} \beta_t, \qquad (7.401)$$

where β is the difference of the Sudakov variables for ingoing and outgoing hadrons. In the case of the usual deep inelastic scattering β is zero, but it is not zero for the case of different momenta of initial and final hadrons. PCF can be calculated in terms of scalar products of parton-wave functions for initial and final hadrons integrated over the momenta and summed over quantum numbers of nonobserved partons. Generally, the function $N^{\{r\}}_{\lambda_1...\lambda_l}(\beta_1, ... \beta_l)$ describes several products of different wave functions which correspond to two possible signs for each β_t. Namely, for a positive sign, the corresponding parton belongs to the initial state and, for an opposite sign, it is in the final state. Note that PCF are similar to the density matrix in the quantum mechanics.

Another important property of QPO is that they are closed under the renormalization in the leading logarithmic approximation [86]. Namely, if one calculates the one-loop correction to their matrix elements between parton states, the resulting operators after using the equations of motion can be reduced again to QPO. Moreover, the integral kernel in the evolution equation for the matrix elements of QPO is expressed as a sum of pair-splitting kernels $\Phi^{r_i r_k}_{r_{i'} r_{k'}}$ [86]

$$\frac{\partial}{\partial \tilde{\xi}} N^{\{r\}}(\beta_1, \ldots \beta_l)$$

$$= \sum_{i<k} \sum_{r_{i'}r_{k'}} \int d\beta_{i'} \, d\beta_{k'} \, \Phi_{r_{i'}r_{k'}}^{r_i r_k}(\beta_i, \beta_k | \beta_{i'}, \beta_{k'}) \, \delta\left(\beta_{i'} + \beta_{k'} - \beta_i - \beta_k\right)$$

$$N^{\{r'\}}(\beta_1, \ldots, \beta_{i'} \ldots \beta_{k'} \ldots \beta_l). \tag{7.402}$$

Here indices r include the types of particles (quark or gluon) and their colour, flavour, and helicity. In accordance with Ref.[86], we changed here the normalization of the "time" variable $\tilde{\xi}$ in comparison with expression (7.58)

$$\tilde{\xi} = \frac{1}{b} \ln\left(1 + b \, \frac{g^2}{16\pi^2} \ln \frac{Q^2}{\Lambda_{QCD}^2}\right), \qquad b = \frac{11}{3} N_c - \frac{2}{3} n_f. \tag{7.403}$$

All pair-splitting kernels $\Phi_{r_{i'}r_{k'}}^{r_i r_k}$ are calculated for both signs of the Sudakov variables β_r [86]. They describe interactions of quarks and gluons in all possible helicity, colour and flavour states and do not contain infrared divergencies. The pair kernels with quantum numbers of quark or gluon have contributions from the corresponding virtual particles in the t-channel. The evolution equations for PCF in the coordinate representation have many interesting properties, including their conformal invariance under the Möbius transformations

$$z \rightarrow \frac{az + b}{cz + d}, \tag{7.404}$$

where $z = x^\mu q'_\mu$ is the light-cone variable. This symmetry makes it possible to find the eigenfunctions of the pair-splitting kernels in terms of the Gegenbauer polynomials [92]

$$R_{n+1}^{\bar{q}q}(\beta_1, \beta_2) = \sum_{k=0}^{n} (-1)^k \, \frac{n! \, (n+2)! \, \beta_1^k \, \beta_2^{n-k}}{k! \, (k+1)! \, (n-k)! \, (n-k+1)!}$$

$$R_{n+2}^{gg}(\beta_1, \beta_2) = \sum_{k=0}^{n} (-1)^{k+1} \, \frac{n! \, (n+4)! \, \beta_1^k \, \beta_2^{n-k}}{k! \, (k+2)! \, (n-k)! \, (n-k+2)!},$$

$$R_{n+\frac{3}{2}}^{qg}(\beta_1, \beta_2) = \sum_{k=0}^{n} (-1)^k \, \frac{n! \, (n+3)! \, \beta_1^k \, \beta_2^{n-k}}{k! \, (k+1)! \, (n-k)! \, (n-k+2)!}, \tag{7.405}$$

where β_1 and β_2 are the Sudakov variables of two corresponding partons with spins s_1 and s_2, respectively. The degree n of the polynomials is related with the Lorentz spin j of the twist-2 operators by the relation

$$j = n + s_1 + s_2. \tag{7.406}$$

It is convenient, however, to pass to the momenta representation using the relation

$$N^{r_1 \cdots r_l}(n_1, \ldots, n_l) = \int \prod_{t=1}^{l} \left(\frac{d\beta_t}{(n_t + 1)!} \, \beta_t^{n_t}\right) N^{\{r\}}(\beta_1, \ldots, \beta_l). \tag{7.407}$$

For these functions, the evolution equation takes the form [86]

$$\frac{\partial}{\partial \xi} N^{r_1 \cdots r_l}(n_1, \ldots n_l)$$

$$= \sum_{i < k} \sum_{r_{i'}, r_{k'}} \sum_{n'_i, n'_k} \left[\Phi^{r_i r_k}_{r_{i'} r_{k'}} \right]^{n_i n_k}_{n'_i n'_k} \delta_{n_{i'} + n_{k'} + \Delta S_{ik}, \, n_i + n_k} N^{r_1 \cdots r_l}(n_1, \ldots, n_{i'} \ldots n_{k'} \ldots n_l),$$

$$(7.408)$$

where

$$\Delta S_{ik} = s_{i'} + s_{k'} - s_i - s_k. \tag{7.409}$$

The new pair kernels are related to those in the β-representation by the relations [86]

$$\sum_{n_{1'} n_{2'}} \left[\Phi^{r_1 r_2}_{r_{1'} r_{2'}} \right]^{n_1 n_2}_{n_{1'} n_{2'}} \frac{\beta_{1'}^{n_{1'}}}{(n_{1'} + 1)!} \frac{\beta_{2'}^{n_{2'}}}{(n_{2'} + 1)!}$$

$$= \int d\beta_1 \, d\beta_2 \, \frac{\beta_1^{n_1}}{(n_1+1)!} \frac{\beta_2^{n_2}}{(n_2+1)!} \, \delta(\beta_1 + \beta_2 - \beta_{1'} - \beta_{2'}) \, \Phi^{r_i r_k}_{r_{i'} r_{k'}}(\beta_i, \beta_k | \beta_{i'}, \beta_{k'}). \tag{7.410}$$

Due to the conformal invariance the kernels $\left[\Phi^{r_1 r_2}_{r_{1'} r_{2'}} \right]^{n_1 n_2}_{n_{1'} n_{2'}}$ are completely determined by their eigenfunctions and eigenvalues [86]

$$\left[\Phi^{r_1 r_2}_{r_{1'} r_{2'}} \right]^{n_1 n_2}_{n_{1'} n_{2'}} = \frac{a_{r_{1'}}(n_{1'}) \, a_{r_{2'}}(n_{2'})}{a_{r_1}(n_1) \, a_{r_2}(n_2)} \sum_j \left(V^{r_1 r_2} \right)^j_{n_1 n_2} \left(V^{r_{1'} r_{2'}} \right)^j_{n_{1'} n_{2'}} \Lambda^{r_1 r_2}_{r_{1'} r_{2'}}(j)$$

$$\times \left(\frac{j+2}{j-1} \right)^{\frac{1}{2} \Delta S_{12}}, \tag{7.411}$$

where ΔS_{12} was defined in Eq.(7.409),

$$a_q(n) = \sqrt{n+1}, \qquad a_g(n) = \sqrt{\frac{n+1}{n+2}}, \tag{7.412}$$

and

$$\left(V^{r_1 r_2} \right)^j_{n_1 n_2} = (-1)^{n_1 + 2s_1 - 1} \, C^{j, \Delta S}_{\frac{n_1 + n_2}{2}, \frac{n_2 - n_1}{2}; \, \frac{n_1 + n_2}{2} + S, \, \frac{n_2 - n_1}{2} + \Delta S},$$

$$S = s_1 + s_2, \qquad \Delta S = s_1 - s_2. \tag{7.413}$$

The factor $C^{j,m}_{j_1, m_1; j_2, m_2}$ is the Clebsch–Gordon coefficient which is independent of the colour, flavour, and helicity of the partons. Such dependence takes place only in the matrices $\Lambda^{r_1 r_2}_{r_{1'} r_{2'}}(j)$ acting on the corresponding indices of PCF. They are presented below (see [86]):

$$\Lambda^{qq}_{qq}(j) = \frac{1}{2} (Q_6 + Q_3) \left(\frac{N_c - 1}{N_c} P_6 - \frac{N_c + 1}{N_c} P_3 \right)$$

$$\times \left[(P_V + P_A + P_T) \left(2 S_j - \frac{3}{2} \right) - \frac{P_V + P_A}{j(j+1)} \right]$$

$$+ (-1)^j (Q_3 \to -Q_3, \quad P_3 \to -P_3, \quad P_A \to -P_A), \tag{7.414}$$

$$\Lambda_{\bar{q}q}^{\bar{q}q}(j) = -(Q_1 + Q_8)\left(\frac{N_c^2 - 1}{N_c}P_1 - \frac{1}{N_c}P_8\right)$$

$$\times \left[(P_V + P_A + P_T)\left(2S_j - \frac{3}{2}\right) - \frac{P_V + P_A}{j(j+1)}\right] - \frac{2}{3}n_f Q_1 P_8 P_V \delta_{j,1}, \quad (7.415)$$

$$\Lambda_{gg}^{gg}(j) = \left(N_c P_1 + P_{10} + P_{\bar{10}} - P_{27} + \frac{N_c}{2}P_{8d} + \frac{N_c}{2}P_{8f}\right)$$

$$\times \left[(P_V + P_A + P_T)\left(\frac{11}{6} - 2S_j - \frac{n_f}{3N_c}\right)\right.$$

$$\left. + \frac{4}{j(j+1)}\left(P_V + P_A + \frac{3P_V}{(j-1)(j+2)}\right)\right]$$

$$+ (-1)^j \left(P_{8f} \to -P_{8f}, \quad P_A \to -P_A\right), \quad (7.416)$$

$$\Lambda_{\bar{q}q}^{gg}(j) = \sqrt{n_f}\, Q_1 \left[\left(\sqrt{\frac{N_c^2 - 1}{N_c}}P_1 + \sqrt{\frac{N_c^2 - 4}{2N_c}}P_{8d} + i\sqrt{\frac{1}{2}N_c}\,P_{8f}\right)\right.$$

$$\left. \times \frac{(j+2)(P_V + P_A) + \frac{4P_V}{j-1}}{j(j+1)} + (-1)^j \left(P_{8f} \to -P_{8f}, \quad P_A \to -P_A\right)\right], \quad (7.417)$$

$$\Lambda_{gg}^{\bar{q}q}(j) = \frac{1}{2}\sqrt{n_f}\, Q_1 \left[\left(\sqrt{\frac{N_c^2 - 1}{N_c}}P_1 + \sqrt{\frac{N_c^2 - 4}{2N_c}}P_{8d} - i\sqrt{\frac{1}{2}N_c}\,P_{8f}\right)\right.$$

$$\left. \times \frac{(j-1)(P_V + P_A) + \frac{4P_V}{j+2}}{j(j+1)} + (-1)^j \left(P_{8f} \to -P_{8f}, \quad P_A \to -P_A\right)\right], \quad (7.418)$$

$$\Lambda_{qg}^{qg}(j) = -(N_c P_3 + P_6 - P_{15})$$

$$\times \left[(P_V + P_A + P_T)\left(2S_{j-\frac{1}{2}} + \frac{2}{2j+1} - \frac{5}{3} + \frac{n_f}{6N_c}\right) - \frac{2(P_V + P_A)}{\left(j - \frac{1}{2}\right)\left(j + \frac{3}{2}\right)}\right]$$

$$+ \left(\frac{1}{N_c}P_3 + P_6 - P_{15}\right)\frac{(-1)^{j-\frac{1}{2}}}{j + \frac{1}{2}}\left[P_T - \frac{2(P_V + P_A)}{\left(j - \frac{1}{2}\right)\left(j + \frac{3}{2}\right)}\right]. \quad (7.419)$$

Here P_r $(r = V, A, T)$ are projectors for vector (V), axial (A), and tensor (T) helicity states of two partons

$$(P_V)_{\lambda_1'\lambda_2'}^{\lambda_1\lambda_2} = \frac{1}{2}\delta_{\lambda,-\lambda_2}\delta_{\lambda_1',-\lambda_2'}, \quad (P_A)_{\lambda_1'\lambda_2'}^{\lambda_1\lambda_2} = \frac{1}{2}\lambda_1\lambda_1'\delta_{\lambda,-\lambda_2}\delta_{\lambda_1',-\lambda_2'},$$

$$(P_T)^{\lambda_1 \lambda_2}_{\lambda'_1 \lambda'_2} = \frac{1}{2} \left(1 + \lambda_1 \lambda'_1\right) \delta_{\lambda,\lambda_2} \delta_{\lambda'_1,\lambda'_2}, \quad \lambda_r = \pm 1. \tag{7.420}$$

The projectors P_r on the colour states $r = 1, 8f, 8d, 10, \bar{10}, 27$ (two gluons), $r = 3, 6$ (two quarks or antiquarks), $r = 1, 8$ (quark-antiquark pair) and $r = 3, 6, 15$ (gluon and quark or antiquark) can be easily constructed (see, for example, Ref. [93]). Similar expressions are valid also for the projectors Q_r on the various flavour states $r = 1, 8$ (quark-antiquark pair) and $r = 3, 6$ (two quarks or two antiquarks) for the $SU(n_f)$-flavour group corresponding to massless u, d, s quarks. We write below only projections for the transitions between two gluons and a quark-antiquark pair

$$(P_1)^{ab}_{ij} = \left(N_c \left(N_c^2 - 1\right)\right)^{-\frac{1}{2}} \delta_{ab} \delta_{ij}, \quad (P_{8d})^{ab}_{ij} = \left(\frac{2N_c}{N_c^2 - 4}\right)^{\frac{1}{2}} d_{abc} t^c_{ij},$$

$$(P_{8f})^{ab}_{ij} = f_{abc} t^c_{ij}, \quad (Q_1)^{f_1 f_2} = (n_f)^{-\frac{1}{2}} \delta_{f_1 f_2}, \quad [t_a, t_b] = i f_{abc} t_c. \tag{7.421}$$

The above relations give a possibility to write the evolution equation for any QPO with the Lorentz spin j as a system of linear equations. The rank of this system grows rapidly with increasing j. The diagonalisation of the anomalous dimension matrix allows us to find the anomalous dimensions of the multiplicative renormalized operators. As an example, we write below this system for twist-3 operators responsible for the violation of the Bjorken scaling in the structure function $g_2(x)$.

7.13 Q^2-dependence of the twist-3 structure functions for the polarized target

The twist-3 operators appearing in the operator product expansion of two electromagnetic currents j_μ and j_ν in the antisymmetric tensor $T_{\mu\nu}$ (7.313) which describe the electon scattering off the polarized target can be reduced to two quasipartonic operators (7.385) using equations of motion [58, 86]. To calculate their matrix elements (7.398) between hadron states one should introduce, respectively, two parton correlation functions

$$N(\beta_1, \beta_2, \beta_3) \equiv N^{\bar{q}qg}(\beta_1, \beta_2, \beta_3), \quad M(\beta_1, \beta_2, \beta_3) \equiv N^{ggg}(\beta_1, \beta_2, \beta_3). \tag{7.422}$$

The evolution equations (7.402) are simplified in this case, as it will be shown below.

To begin, we note that the matrix elements of the operators (7.385) vanish if the signs of helicities of all three partons are the same. Due to parity conservation, it is enough to consider only the case when two helicities are positive and one is negative. For the operators containing fermion fields the helicities of quark and antiquark are opposite and therefore we can choose the positive sign for the gluon helicity. Moreover, PCF $N_8(\beta_1, \beta_2, \beta_3)$ with the flavour octet quantum number (for $SU(3)$-flavour group) do not mix with pure gluonic operators. Further, for the flavour-singlet case the combination $N_1^{-++}(\beta_1, \beta_2, \beta_3) - N_1^{+-+}(\beta_2, \beta_1, \beta_3)$ has the negative charge parity and therefore it does not give any contribution to $T_{\mu\nu}$. Due to its positive charge parity, the colour structure of the gluonic operator is proportional to the antisymmetric structure constant f_{abc}. This means that the Bose statistics make it possible to express the corresponding correlation functions with various gluon helicities only in terms of $M^{-++}(\beta_1, \beta_2, \beta_3)$

$$M^{-++}(\beta_1, \beta_2, \beta_3) = -M^{-++}(\beta_1, \beta_3, \beta_3) = -M^{+-+}(\beta_2, \beta_1, \beta_3)$$

$$= -M^{++-}(\beta_3, \beta_2, \beta_1) = M^{++-}(\beta_2, \beta_3, \beta_1) = M^{+-+}(\beta_3, \beta_1, \beta_2). \quad (7.423)$$

Thus, one should consider the following correlation functions

$$N_8(\beta_1, \beta_2, \beta_3) \equiv N_8^{-++}(\beta_1, \beta_2, \beta_3),$$

$$N_1(\beta_1, \beta_2, \beta_3) \equiv N_1^{-++}(\beta_1, \beta_2, \beta_3) + N_1^{+-+}(\beta_2, \beta_1, \beta_3),$$

$$M(\beta_1, \beta_2, \beta_3) \equiv M^{-++}(\beta_1, \beta_2, \beta_3). \quad (7.424)$$

The evolution equations for N_8 in the momenta basis have the form (7.408) [58],[86]

$$\frac{\partial}{\partial \tilde{\xi}} N_8(n_1, n_2, n_3) = \left[\left(\Phi_{\bar{q}q}^{\bar{q}q} \right)_{V+A} \right]_{n_{1'} n_{2'}}^{n_1 n_2} N_8(n_{1'}, n_{2'}, n_3) + \left[\left(\Phi_{\bar{q}q}^{\bar{q}q} \right)_{V-A} \right]_{n_{1'} n_{2'}}^{n_1 n_2} N_8(n_{1'}, n_{2'}, n_3)$$

$$+ \left[\left(\Phi_{\bar{q}g}^{\bar{q}g} \right)_{V+A} \right]_{n_{1'} n_{3'}}^{n_1 n_3} N_8(n_{1'}, n_2, n_{3'}) + \left[\left(\Phi_{qg}^{qg} \right)_T \right]_{n_{2'} n_{3'}}^{n_2 n_3} N_8(n_1, n_{2'}, n_{3'}), \quad (7.425)$$

where

$$\left(\Phi_{r_{1'} r_{2'}}^{r_1 r_2} \right)_{V \pm A} \equiv \frac{1}{2} \left(\left(\Phi_{r_{1'} r_{2'}}^{r_1 r_2} \right)_V \pm \left(\Phi_{r_{1'} r_{2'}}^{r_1 r_2} \right)_A \right). \quad (7.426)$$

The evolution equations for N_1 and M are more complicated [58],[86]

$$\frac{\partial}{\partial \tilde{\xi}} N_1(n_1, n_2, n_3) = \left[\left(\Phi_{\bar{q}q}^{\bar{q}q} \right)_{V+A} \right]_{n_{1'} n_{2'}}^{n_1 n_2} N_1(n_{1'}, n_{2'}, n_3)$$

$$+ \left[\left(\Phi_{\bar{q}q}^{\bar{q}q} \right)_{V-A} \right]_{n_{1'} n_{2'}}^{n_1 n_2} N_1(n_{1'}, n_{2'}, n_3) + \left[\left(\Phi_{\bar{q}g}^{\bar{q}g} \right)_{V+A} \right]_{n_{1'} n_{3'}}^{n_1 n_3} N_1(n_{1'}, n_2, n_{3'})$$

$$+ \left[\left(\Phi_{qg}^{qg} \right)_T \right]_{n_{2'} n_{3'}}^{n_2 n_3} N_1(n_1, n_{2'}, n_{3'}) + 4 \left[\left(\Phi_{gg}^{\bar{q}q} \right)_{V+A} \right]_{n_{1'} n_{2'}}^{n_1 n_2} M(n_{1'}, n_{2'}, n_3) \quad (7.427)$$

and

$$\frac{\partial}{\partial \tilde{\xi}} M(n_1, n_2, n_3) = 2 \left[\left(\Phi_{gg}^{gg} \right)_{V+A} \right]_{n_{1'} n_{2'}}^{n_1 n_2} M(n_{1'}, n_{2'}, n_3)$$

$$+ 2 \left[\left(\Phi_{gg}^{gg} \right)_{V+A} \right]_{n_{1'} n_{3'}}^{n_1 n_3} M(n_{1'}, n_2, n_{3'}) + \left[\left(\Phi_{gg}^{gg} \right)_T \right]_{n_{2'} n_{3'}}^{n_2 n_3} M(n_1, n_{2'}, n_{3'})$$

$$+ \left[\left(\Phi_{\bar{q}q}^{gg} \right)_{V+A} \right]_{n_{1'} n_{2'}}^{n_1 n_2} N_1(n_{1'}, n_{2'}, n_3) + \left[\left(\Phi_{\bar{q}q}^{gg} \right)_{V+A} \right]_{n_{1'} n_{3'}}^{n_1 n_3} N_1(n_{1'}, n_2, n_{3'}), \quad (7.428)$$

where the sum over $n_{i'}, n_{k'}$ is implied with the additional constraint

$$n_{i'} + n_{k'} + \Delta S_{ik} = n_i + n_k. \quad (7.429)$$

Here S_{ik} is defined in Eq. (7.409). The pair kernels $\left(\Phi^{r_1 r_2}_{r_1' r_2'}\right)_s$, $(s = V, A, T)$ in the above equations are enlisted in Eq. (7.411), and in the expressions for $\Lambda^{r_1 r_2}_{r_1' r_2'}$ one should leave only the contributions containing the projectors (7.420) to the corresponding spin states V, A, or T. In particular, in the kernels describing the transition between partons $\bar{q}q$ and gg we have only the $V + A$ intermediate state in accordance with the fact that the helicity of the third gluon is positive. The coefficients 4 and 2 in equations for N_1 and M in front of some kernels appear due to our definition of N_1 and the symmetry relations for M (7.423). The colour and flavour structures are not shown, since for each pair kernel only one colour and flavour projector gives a nonzero contribution.

Let us consider, for example, the evolution equation for the singlet-flavour state with j=3. In this case, the independent components of momenta $N_1(n_1, n_2, n_3)$ and $M(n_1, n_2, n_3)$ are

$$x \equiv N_1(0, 0, 2), \quad y \equiv N_1(0, 1, 1), \quad z \equiv N_1(1, 0, 1), \quad u \equiv N_1(2, 0, 0), \qquad (7.430)$$

and

$$v \equiv N_1(2, 0, 0), \quad w \equiv N_1(1, 1, 0), \quad m \equiv iM(0, 1, 0) = -iM(0, 0, 1). \qquad (7.431)$$

Using expressions (7.414)–(7.419), we can write the matrix of anomalous dimensions for the corresponding operators [86]. It turns out that it contains information about the anomalous dimension for lower spins $j = 1$ and $j = 2$. Indeed, because the total derivative from the local operator does not change its anomalous dimension, we obtain, using relation (7.407), that the linear combination

$$\sum_{k=1}^{l} (n_k + 2) N(n_1, \ldots, n_{k-1}, n_k + 1, n_{k+1}, \ldots, n_l) \qquad (7.432)$$

has the same anomalous dimension matrix as PCF $N(n_1, \ldots, n_l)$ with lower j.

Therefore the components of $j = 2$ state

$$r = 2y + 3u + 2w, \quad s = 2z + 3v + 2w, \quad t = 3x + 2y + 2z \qquad (7.433)$$

are closed under the Q^2-evolution [58],[86]:

$$\frac{d}{d\bar{\xi}} r = \left(-\frac{3}{2}N_c + \frac{1}{N_c} - \frac{2}{3}n_f\right) r + \left(-\frac{2}{3N_c} - \frac{1}{3}n_f\right) s + \left(\frac{1}{3}N_c - \frac{1}{3N_c}\right) r,$$

$$\frac{d}{d\bar{\xi}} s = \left(-\frac{2}{3N_c} - \frac{1}{3}n_f\right) r + \left(-\frac{7}{4}N_c + \frac{1}{4N_c} - \frac{2}{3}n_f\right) s + \left(\frac{1}{2}N_c + \frac{1}{6N_c}\right) t,$$

$$\frac{d}{d\bar{\xi}} t = \left(-\frac{1}{2}N_c - \frac{1}{2N_c}\right) r + \left(\frac{3}{4}N_c + \frac{1}{4N_c}\right) s + \left(-\frac{11}{6}N_c - n_f\right) r. \qquad (7.434)$$

In turn, one can separate from this system the evolution equation for the operator with j=1

$$\frac{d}{d\bar{\xi}} (r + s + t) = \left(-N_c - \frac{1}{6N_c} - n_f\right) (r + s + t) \qquad (7.435)$$

and its large-q^2 behaviour is

$$r + s + t \sim \exp\left(\left(-N_c - \frac{1}{6N_c} - n_f\right)\bar{\xi}\right). \tag{7.436}$$

The components of the operators having really $j = 2$ are related as follows

$$t = -r - s, \tag{7.437}$$

where r and s satisfy the evolution equation

$$\frac{d}{d\bar{\xi}} r = \left(-\frac{11}{6}N_c - \frac{2}{3}n_f\right)r + \left(-\frac{1}{3}N_c - \frac{1}{3N_c} - \frac{1}{3}n_f\right)s,$$

$$\frac{d}{d\bar{\xi}} s = \left(-\frac{1}{2}N_c - \frac{5}{6N_c} - \frac{1}{3}n_f\right)r + \left(-\frac{9}{4}N_c + \frac{1}{12N_c} - \frac{2}{3}n_f\right)s. \tag{7.438}$$

The multiplicatively renormalized PCF with $j = 2$ depend on Q^2 as follows

$$\left(-\frac{1}{3}N_c - \frac{1}{3N_c} - \frac{1}{3}n_f\right)s + \left(-\frac{11}{6}N_c - \frac{2}{3}n_f - \lambda_1\right)r \sim e^{\lambda_2\bar{\xi}},$$

$$\left(-\frac{1}{3}N_c - \frac{1}{3N_c} - \frac{1}{3}n_f\right)s + \left(-\frac{11}{6}N_c - \frac{2}{3}n_f - \lambda_2\right) \sim e^{\lambda_1\bar{\xi}}, \tag{7.439}$$

where $\lambda_{1,2}$ are two solutions of the secular equation

$$\left(-\frac{11}{6}N_c - \frac{2}{3}n_f - \lambda\right)\left(-\frac{9}{4}N_c + \frac{1}{12N_c} - \frac{2}{3}n_f - \lambda\right)$$

$$= \left(-\frac{1}{3}N_c - \frac{1}{3N_c} - \frac{1}{3}n_f\right)\left(-\frac{1}{2}N_c - \frac{5}{6N_c} - \frac{1}{3}n_f\right). \tag{7.440}$$

In turn, the three components of PCF with $j = 3$ are expressed through independent ones

$$y = -\frac{3}{2}u - w, \quad z = -\frac{3}{2}v - w, \quad x = -\frac{2}{3}y - \frac{2}{3}z \tag{7.441}$$

which satisfy the evolution equations

$$\frac{d}{d\bar{\xi}} u = \left(-\frac{29}{12}N_c + \frac{7}{6N_c} - \frac{8}{15}n_f\right)u + \left(\frac{1}{12}N_c - \frac{1}{2N_c} - \frac{1}{5}n_f\right)v$$

$$+ \left(-\frac{1}{3}N_c - \frac{7}{6N_c} - \frac{4}{15}n_f\right) + \frac{17}{90}\sqrt{\frac{1}{2}N_c n_f}\, m,$$

$$\frac{d}{d\bar{\xi}} v = \left(\frac{3}{20}N_c - \frac{3}{20N_c} - \frac{1}{5}n_f\right)u + \left(-\frac{59}{20}N_c + \frac{37}{60N_c} - \frac{8}{15}n_f\right)v$$

$$+ \left(-\frac{2}{5}N_c - \frac{23}{30N_c} - \frac{4}{15}n_f\right) - \frac{1}{30}\sqrt{\frac{1}{2}N_c n_f}\, m,$$

$$\frac{d}{d\xi}w = \left(-\frac{3}{4}N_c - \frac{7}{8N_c} - \frac{1}{5}n_f\right)u + \left(-\frac{1}{2}N_c - \frac{1}{8N_c} - \frac{1}{5}n_f\right)v$$

$$+ \left(-\frac{37}{12}N_c + \frac{3}{2N_c} - \frac{3}{5}n_f\right) - \frac{7}{60}\sqrt{\frac{1}{2}N_c n_f}\, m,$$

$$\frac{d}{d\xi}m = \sqrt{\frac{1}{2}N_c n_f}\left(\frac{37}{20}u - \frac{23}{20}v - \frac{7}{10}w\right) - \left(\frac{197}{60}N_c + n_f\right)m. \qquad (7.442)$$

Again, one can construct the multiplicatively renormalized PCF and find their anomalous dimensions by solving the corresponding secular equation.

For larger j, the derivation of similar evolution equations is straightforward, but the construction of the multiplicatively normalized operators and their anomalous dimensions can be done only numerically, because the secular equation contains the polynomials $P_n(\lambda)$ of the rank $n > 4$.

7.14 Infrared evolution equations at small x

7.14.1 Double logarithms and the DGLAP equation

For the nonsinglet structure functions, the pure gluonic intermediate states in the crossing channel are absent and as a result their behaviour at small x is not related to the pomeron exchange. The main contribution in this case is obtained from the ladder diagrams having the quark-antiquark intermediate states in the t-channel [12]. In the Born approximation, the amplitude for the forward scattering of the virtual photon off a massless quark has the form

$$F_B = -e^2\left(\frac{\gamma_\mu(\not{p}+\not{q})\gamma_\nu}{(p+q)^2} + \frac{\gamma_\nu(\not{p}-\not{q})\gamma_\mu}{(p-q)^2}\right), \qquad (7.443)$$

where e is the quark electric charge, p is the momentum of the quark ($p^2 = 0$), and q is the photon momentum

$$q = q' + xp, \quad (q')^2 = 0, \quad x = \frac{Q^2}{2pq}, \quad Q^2 = -q^2. \qquad (7.444)$$

Taking into account that the γ-matrices in F_B are calculated between the quark spinors, we can write F_B in another form

$$F_B = F_B^+ + F_B^-, \qquad (7.445)$$

where

$$F_B^+ = e^2\,\delta_{\mu\nu}^\perp\,\not{q}'\left(\frac{1}{2pq' - Q^2} - \frac{1}{-2pq' - Q^2}\right),$$

$$F_B^- = e^2\,\varepsilon_{\mu\nu}^\perp\,\gamma_1\gamma_2\,\not{q}'\left(\frac{1}{2pq' - Q^2} + \frac{1}{-2pq' - Q^2}\right). \qquad (7.446)$$

Here $\delta_{\mu\nu}^\perp$ is the Kronecker symbol ($\delta_{11}^\perp = \delta_{22}^\perp = -1$) and $\varepsilon_{\mu\nu}^\perp$ is the unit antisymmetric tensor ($\varepsilon_{12}^\perp = 1$) in the two-dimensional transverse subspace. Because the matrix elements of $F_B^\pm$ between the spinors with fixed helicities $\lambda = \pm 1$ contain the additional factor $2pq'$,

$$\bar{u}_\lambda(p) \, \rlap{/}q' u_\lambda(p) = 2pq', \quad \bar{u}_\lambda(p)\gamma_1\gamma_2 \, \rlap{/}q' u_\lambda(p) = -\lambda 2pq', \tag{7.447}$$

the expressions F_B^+ and F_B^- describe contributions with the positive and negative signatures, respectively.

In double-logarithmic approximation (DLA), the spin and colour structure of the Born term is not changed. As a result, the amplitude for the photon-quark forward scattering with the infrared cut-off $|k_\perp|^2 > \mu^2$ can be written as follows

$$F(s, Q^2; \mu^2) = F^+(s, Q^2; \mu^2) + F^-(s, Q^2; \mu^2), \tag{7.448}$$

$$F^+(s, Q^2; \mu^2) = F_B^+ \int_{\sigma-i\infty}^{\sigma+i\infty} \frac{dj}{2\pi i} \left(\frac{1}{x}\right)^j \int_{-i\infty}^{+i\infty} \frac{d\gamma}{2\pi i} \left(\frac{Q^2}{\mu^2}\right)^\gamma \xi_+(j) \, f_+(j, \gamma),$$

$$F^-(s, Q^2; \mu^2) = F_B^- \int_{\sigma-i\infty}^{\sigma+i\infty} \frac{dj}{2\pi i} \left(\frac{1}{x}\right)^j \int_{-i\infty}^{+i\infty} \frac{d\gamma}{2\pi i} \left(\frac{Q^2}{\mu^2}\right)^\gamma \xi_-(j) \, f_-(j, \gamma),$$

where $f_p(j, \gamma)$ are the t-channel partial waves with the corresponding signatures $p = \pm 1$ and $\xi_p(j)$ is the signature factor

$$\xi_p(j) = -\frac{e^{-i\pi j} + p}{\sin \pi j}. \tag{7.449}$$

Respectively, the nonsinglet contribution to the quark-antiquark scattering amplitudes with colour singlet and octet quantum numbers ($r = 1, 8$) in the t-channel can be presented in DLA in the form

$$A^r(s; \mu^2) = \int_{\sigma-i\infty}^{\sigma+i\infty} \frac{dj}{2\pi i} \left(\frac{2pp'}{\mu^2}\right)^j \left(\xi_+(j) \, f_+^r(j) + \xi_-(j) \, f_-^r(j)\right). \tag{7.450}$$

Taking into account the corresponding colour projectors

$$P_{i_2,k_1}^{i_1,k_2}(1, qq) = \frac{\delta_{i_2}^{i_1} \delta_{k_1}^{k_2}}{N_c}, \quad P_{i_2,k_1}^{i_1,k_2}(8, qq) = 2 \, \vec{T}_{i_2}^{i_1} \, \vec{T}_{k_1}^{k_2}, \tag{7.451}$$

one obtains for the t-channel partial waves in the Born approximation

$$f_\pm^B(\gamma) = -\frac{\pi}{2} \frac{1}{\gamma}, \quad f_\pm^{(1)B} = -\frac{\pi}{2} \frac{N_c^2 - 1}{2N_c}, \quad f_\pm^{(8)B} = \frac{\pi}{2} \frac{1}{2N_c}. \tag{7.452}$$

Note that for $\mu^2 > Q^2$ the amplitude $F(s, Q^2; \mu^2)$ does not depend on Q^2 and equals

$$F^\pm(s, Q^2; \mu^2)\big|_{\mu^2 \gg Q^2} = F_B^\pm \, A_\pm^{(1)}(s; \mu^2) \frac{2N_c}{N_c^2 - 1}. \tag{7.453}$$

This relation can be considered as an initial condition for the evolution of $F^\pm(s, Q^2; \mu^2)$ in the infrared cut-off $\mu^2 < Q^2$.

The Sudakov parameters α_i and β_i of the quark momenta in the t-channel ladder diagrams

$$k_i = \beta_i \, p + \alpha_i \, q' + k_i^\perp \tag{7.454}$$

are strongly ordered in DLA

$$1 \gg \beta_1 \gg \beta_2 \gg \ldots \gg \beta_n \gg \frac{|k^\perp|^2}{s}, \quad 1 \gg \alpha_n \gg \alpha_{n-1} \gg \ldots \gg \alpha_1 \gg \frac{|k^\perp|^2}{s}$$

(7.455)

and their transverse components

$$|k_i^\perp|^2 \sim s\alpha_i \beta_i$$

(7.456)

are large and can exceed the photon virtuality Q^2 in the deep inelastic scattering. In fact, the double-logarithmic terms are obtained as a result of logarithmic integrations over the parameters α_i and β_i.

Extracting in the ladder the quark and antiquark with the smallest transverse momentum $|k_\perp|^2$ one can write the following relation between $F(s, Q^2; \mu^2)$ and $A(s; \mu^2)$

$$F(s, Q^2; \mu^2) = F_B + \int \frac{d^2 k_\perp}{i(2\pi)^4} \frac{d(-s\alpha) d(s\beta)}{2 |s|} A^{(1)}(-s\alpha; |k_\perp|^2)$$

$$\times \frac{\not k_\perp \ldots \not k_\perp}{(s\alpha\beta - |k_\perp|^2 + i\varepsilon)^2} F(s\beta, Q^2; |k_\perp|^2), \quad |k_\perp|^2 > \mu^2,$$

(7.457)

substituting by $|k_\perp|^2$ in the inner amplitudes the infrared cut-off μ^2. Here the notation $\not k_\perp \ldots \not k_\perp$ means that the corresponding γ-matrices act on different spinors. The spinor structure is simplified after the integration over the azimuthal angles in the two-dimensional transverse subspace with the use of the relation

$$\gamma_\sigma^\perp \not k_\perp \gamma_\mu^\perp \ldots \gamma_\nu^\perp \not k_\perp \gamma_\sigma^\perp \to -\frac{|k_\perp|^2}{2} \gamma_\sigma^\perp \gamma_\rho^\perp \gamma_\mu^\perp \ldots \gamma_\nu^\perp \gamma_\rho^\perp \gamma_\sigma^\perp = -2|k_\perp|^2 \gamma_\mu^\perp \ldots \gamma_\nu^\perp.$$

(7.458)

Further, if one will present $A(-s\alpha; |k_\perp|^2)$ and $F(s\beta, Q^2; |k_\perp|^2)$ as a sum of two contributions with signatures $p = \pm 1$, the signatures of A and F should coincide and be equal to the signature of the result of integration, which can be verified by changing the sign of the invariant s and signs of both Sudakov variables α and β. Therefore, it is enough to consider only the positive values of s and β and multiply the result by the factor 2. The contour of the integration over $-s\alpha$ can be enclosed around the singularities of $A(-s\alpha; |k_\perp|^2)$ situated in the lower semiplane at $-s\alpha > 0$, which gives a possibility to express the integral in terms of its discontinuity $\Delta A = 2i \Im A$. In such way, we obtain for $F(s, Q^2; \mu^2)$ the equation

$$F_\pm(s, Q^2; \mu^2) = F_B^\pm A_\pm^{(1)}(s; Q^2) \frac{2N_c}{N_c^2 - 1} - 2g^2 \int_0^\infty \frac{dy}{y} \int_{Q^2}^{sy/\mu^2} \frac{d(s\beta)}{(2\pi)^4 s}$$

$$\times \int_{\mu^2}^{Q^2} \frac{2\pi |k_\perp|^2 d|k_\perp|^2}{(y + |k_\perp|^2)^2} \int_{\sigma-i\infty}^{\sigma+i\infty} \frac{dj'}{2\pi i} \left(\frac{y}{|k_\perp|^2 \beta} \right)^{j'} f_\pm(j') F_\pm(s\beta, Q^2; |k_\perp|^2),$$

(7.459)

where we introduced the new variable $y = -s\alpha\beta$ instead of $-s\alpha$ and used for $\Delta A_\pm(-s\alpha; |k_\perp|^2)$ the t-channel partial wave representation. The upper limit of integration over $|k_\perp|^2$ was taken to be equal to Q^2, because the contribution from the integration region

$|k_\perp|^2 > Q^2$ was included in the new inhomogeneous term $A_\pm^{(1)}(s; Q^2)$ in an accordance with the above discussion. The amplitudes $F_\pm(s\beta, Q^2; |k_\perp|^2)$ and $A_\pm^{(1)}(s; Q^2)$ have the signatures $p = \pm$. The equation for $A_\pm(s; Q^2)$ will be obtained below. One can perform the integration over y taking into account that in DLA $y \ll |k_\perp|^2$

$$F_\pm(s, Q^2; \mu^2) = \frac{F_B}{2} A_\pm^{(1)}(s; Q^2) \frac{2N_c}{N_c^2 - 1}$$

$$-2g^2 \int_{Q^2}^s \frac{d(s\beta)}{s} \int_{\mu^2}^{Q^2} \frac{d|k_\perp|^2}{(2\pi)^3|k_\perp|^2} \int_{\sigma-i\infty}^{\sigma+i\infty} \frac{dj'}{2\pi i j'} \left(\frac{1}{\beta}\right)^{j'} f_\pm^{(1)}(j') F_\pm \left(s\beta, Q^2; |k_\perp|^2\right).$$

Finally, inserting the Mellin representation for $A_\pm(s; Q^2)$ and $F_\pm(s\beta, Q^2; |k_\perp|^2)$ in the form of an integral over j and γ and calculating subsequently the integrals over $|k_\perp|^2$, $s\beta$, and j' we obtain

$$\gamma f_\pm(j, \gamma) = f_\pm^{(1)}(j) \frac{2N_c}{N_c^2 - 1} - \frac{1}{j} \frac{\alpha_s}{\pi^2} f_\pm^{(1)}(j) f_\pm^{(1)}(j, \gamma), \quad \alpha_s = \frac{g^2}{4\pi}. \tag{7.460}$$

The solution of this equation is

$$f_\pm(j, \gamma) = \frac{f_\pm^{(1)}(j) \frac{2N_c}{N_c^2-1}}{\gamma + \frac{1}{j} \frac{\alpha_s}{\pi^2} f_\pm^{(1)}(j)}. \tag{7.461}$$

On the other hand, one can obtain the similar equation for $f_\pm(j)$ by extracting from the ladder diagrams the quark pair with the smallest transverse momentum

$$f_\pm^{(1)}(j) = f_\pm^{(1)B} - \frac{1}{j^2} \frac{\alpha_s}{\pi^2} \left(f_\pm^{(1)}(j)\right)^2, \quad f_\pm^{(1)B} = -\frac{\pi}{2} \frac{N_c^2 - 1}{2N_c}. \tag{7.462}$$

We shall discuss in the next subsection additional contributions from the soft Sudakov gluons [10]. The neglection of the Sudakov terms is valid for the positive signature, where the solution for $f_+(j, \gamma)$ can be presented in the form

$$f_+(j, \gamma) = -\frac{\gamma(j)/j}{\gamma - \gamma(j)} \frac{\pi^2}{\alpha_s} \frac{2N_c}{N_c^2 - 1}, \tag{7.463}$$

where the position of the pole in γ satisfies the algebraic equation

$$\gamma(j)(\gamma(j) - j) = -\frac{\alpha_s}{2\pi} \frac{N_c^2 - 1}{2N_c}. \tag{7.464}$$

For the negative signature there are corrections to this equation from the soft-gluon contributions.

It is important that even for the singlet-structure functions, the two-gluon exchange also can lead to the corrections $\sim s^0$ related to the double-logarithmic asymptotics [94]. Indeed, let us consider the amplitude of the deep inelastic scattering with a two-gluon intermediate state in the t-channel. In the Feynman gauge, the propagators of these gluons contain the tensor structure

$$L_{\mu_1\nu_1; \mu_2\nu_2} = \delta_{\mu_1\nu_1} \delta_{\mu_2\nu_2}. \tag{7.465}$$

We assume that the tensor L is contracted in the indices μ_1 and μ_2 with the left (L) blob having particles with large Sudakov parameters β_i along the initial momentum p and in the indices ν_1 and ν_2 with the right (R) blob with large parameters α_i along q'. In this case, the leading asymptotic contribution appears from its longitudinal (so-called nonsense) tensor components

$$L_{\mu_1\nu_1;\mu_2\nu_2} = \frac{q'_{\mu_1} p_{\nu_1}}{pq'} \frac{q'_{\mu_2} p_{\nu_2}}{pq'} + \Delta L_{\mu_1\nu_1;\mu_2\nu_2}, \tag{7.466}$$

where the small correction $\Delta L_{\mu_1\nu_1;\mu_2\nu_2}$ with good accuracy can be written as follows

$$\Delta L_{\mu_1\nu_1;\mu_2\nu_2} = \frac{q'_{\mu_1} p_{\nu_1}}{pq'} \delta^{\perp}_{\mu_2\nu_2} + \frac{q'_{\mu_2} p_{\nu_2}}{pq'} \delta^{\perp}_{\mu_1\nu_1}. \tag{7.467}$$

We consider the forward scattering, where the momenta of two gluons are k and $-k$. Due to the colour current conservation $k_{\mu_1} f^L_{\mu_1} = k_{\nu_1} f^R_{\nu_1} = 0$ for the amplitudes describing the left and right blobs, one can substitute

$$\Delta L_{\mu_1\nu_1;\mu_2\nu_2} \to pq' \frac{k^{\perp}_{\mu_1} k^{\perp}_{\nu_1}}{kp \, kq'} \delta^{\perp}_{\mu_2\nu_2} + pq' \frac{k^{\perp}_{\mu_2} k^{\perp}_{\nu_2}}{kp \, kq'} \delta^{\perp}_{\mu_1\nu_1}. \tag{7.468}$$

After averaging over the azimuthal angle in the transverse plane we can present this tensor as a sum of two spin structures factorized in the t-channel

$$\Delta L_{\mu_1\nu_1;\mu_2\nu_2} \to -pq' \frac{|k^{\perp}|^2}{kp \, kq'} \left(\delta^{\perp}_{\mu_1\mu_2} \delta^{\perp}_{\nu_1\nu_2} + \varepsilon^{\perp}_{\mu_1\mu_2} \varepsilon^{\perp}_{\nu_1\nu_2} \right). \tag{7.469}$$

This means that one can consider the forward-scattering amplitudes only with the transverse polarizations. Note that for the two-gluon intermediate state we obtained the factor $|k^{\perp}|^2$ in the numerator similar to the nonsinglet case where the product of the quark and antiquark propagators in the t-channel contain the numerator $k^{\perp} \ldots k^{\perp}$. The presence of this factor leads to the logarithmic divergency of the integral over $|k^{\perp}|^2$ and, as a result, to the appearence of double-logarithmic contributions.

To derive the Bethe–Salpeter equation for the amplitude in DLA for the singlet case, we begin with the DGLAP evolution equation. It is equivalent to the renormalization group equation for the twist-2 operators. The tensor structure $\delta^{\perp}_{\mu_1\mu_2}$ appearing in $L_{\mu_1\nu_1;\mu_2\nu_2}$ corresponds to the operators

$$O^g_{\mu_1\mu_2\ldots\mu_j} = S \, G_{\mu_1\sigma} D_{\mu_2} D_{\mu_3} \ldots D_{\mu_{j-1}} G_{\mu_j\sigma}, \tag{7.470}$$

where S means the symmetrization in indices $\mu_1, \ldots, \mu_j$ combined with the trace subtraction and D_μ are the covariant derivatives. Analogously, $\varepsilon^{\perp}_{\mu_1\mu_2}$ corresponds to the operators

$$\tilde{O}^g_{\mu_1\mu_2\ldots\mu_j} = S \, G_{\mu_1\sigma} D_{\mu_2} D_{\mu_3} \ldots D_{\mu_{j-1}} \tilde{G}_{\mu_j\sigma} \tag{7.471}$$

where $\tilde{G}_{\mu\sigma}$ is the dual tensor of the gluon field. Note that due to the positive-charge parity of the two-photon state, the operators $O^g_{\mu_1\mu_2\ldots\mu_j}$ and $\tilde{O}^g_{\mu_1\mu_2\ldots\mu_j}$ should have, respectively, even and odd values of the Lorentz spin j. It means that the small-x asymptotics of the structure functions $F_{1,2}(x)$ and $g_1(x)$ are related to the t-channel partial waves with the positive and negative signatures, respectively. The anomalous dimension matrix for the gluon operators $O^g_{\mu_1\mu_2...\mu_j}$ and $\tilde{O}^g_{\mu_1\mu_2...\mu_j}$ together with the corresponding quark operators

$$O^g_{\mu_1\mu_2...\mu_j} = S\bar{\psi}\gamma_{\mu_1}\nabla_{\mu_2}\nabla_{\mu_3}\ldots\nabla_{\mu_j}\psi, \tag{7.472}$$

$$\tilde{O}^q_{\mu_1\mu_2...\mu_j} = S\bar{\psi}\gamma_5\gamma_{\mu_1}D_{\mu_2}D_{\mu_3}\ldots D_{\mu_j}\psi \tag{7.473}$$

is well known. We write the evolution equations in the Born approximation for their matrix elements in the form

$$\frac{d}{d\ln Q^2}\begin{pmatrix} n^g_j \\ n^q_j \end{pmatrix} = \frac{\alpha_s(Q^2)}{2\pi j} K \begin{pmatrix} n^g_j \\ n^q_j \end{pmatrix}, \quad K = \begin{pmatrix} -2N_c & -\frac{N_c^2-1}{N_c} \\ n_f & \frac{N_c^2-1}{2N_c} \end{pmatrix}, \tag{7.474}$$

$$\frac{d}{d\ln Q^2}\begin{pmatrix} \tilde{n}^g_j \\ \tilde{n}^q_j \end{pmatrix} = \frac{\alpha_s(Q^2)}{2\pi j} \tilde{K} \begin{pmatrix} \tilde{n}^g_j \\ \tilde{n}^q_j \end{pmatrix}, \quad \tilde{K} = \begin{pmatrix} 4N_c & \frac{N_c^2-1}{N_c} \\ -n_f & \frac{N_c^2-1}{2N_c} \end{pmatrix}, \tag{7.475}$$

taking into account only the pole contributions at $j = 0$, responsible for the appearence of double logarithms. These relations can be considered as differential forms of the Bethe–Salpeter equations for the parton distibutions at small x

$$n(|k|^2, \beta) = n_0(\beta) + \int_\beta^1 \frac{d\beta'}{\beta'} \int_{\mu^2}^{|k|^2} \frac{d|k'|^2}{|k'|^2} \frac{\alpha_s(|k|^2)}{2\pi} K\, n(Q^2, \beta'), \tag{7.476}$$

$$\tilde{n}(|k|^2, \beta) = \tilde{n}_0(\beta) + \int_\beta^1 \frac{d\beta'}{\beta'} \int_{\mu^2}^{|k|^2} \frac{d|k'|^2}{|k'|^2} \frac{\alpha_s(|k|^2)}{2\pi} \tilde{K}\, \tilde{n}(Q^2, \beta'), \tag{7.477}$$

where

$$n_0(\beta) = \delta(1-\beta)\begin{pmatrix} a^g \\ a^q \end{pmatrix}, \quad \tilde{n}_0(\beta) = \delta(1-\beta)\begin{pmatrix} \tilde{a}^g \\ \tilde{a}^q \end{pmatrix} \tag{7.478}$$

and a^r and $\tilde{a}^r$ describe the content of the initial parton. For example, for the longitudinally polarized quark in the initial state we have $a^g = \tilde{a}^g = 0$, $a^q = \tilde{a}^q = 1$.

The amplitudes in the double-logarithmic approximation (for the ladder diagrams) satisfy the same integral equations, but with a different region of integration. Namely, one should order not only the Sudakov variables β_i, but also the variables $\alpha_i = |k_i|^2/(\beta_i s)$. As a result, we obtain the new condition for the region of integration over k'

$$\frac{|k'|^2}{\beta'} < \frac{|k|^2}{\beta} \tag{7.479}$$

instead of the old inequality $|k'|^2 < |k|^2$. In the double-logarithmic region $|k'|^2$ can be larger then $|k|^2$ for $\beta' > \beta$. Note that for the pomeron the region of logarithmically small $|k|^2$ for the gluons inside the BFKL ladder gives a negligible contribution due to the colour-current conservation. It is important only in the asymptotics $g^2 \ln s \to \infty$, which is related to the diffusion in the variable $\ln |k|^2$ at large energies.

Thus, the Bethe–Salpeter equations for the amplitude in DLA are

$$n(k^2, \beta) = n_0(\beta) + \int_\beta^1 \frac{d\beta'}{\beta'} \int_{\mu^2}^{|k|^2 \frac{\beta'}{\beta}} \frac{d|k'|^2}{|k'|^2} \frac{\alpha_s}{2\pi} K \, n(|k'|^2, \beta'), \tag{7.480}$$

$$\tilde{n}(k^2, \beta) = \tilde{n}_0(\beta) + \int_\beta^1 \frac{d\beta'}{\beta'} \int_{\mu^2}^{|k|^2 \frac{\beta'}{\beta}} \frac{d|k'|^2}{|k'|^2} \frac{\alpha_s}{2\pi} \tilde{K} \, \tilde{n}(|k'|^2, \beta'). \tag{7.481}$$

In the leading approximation, the strong coupling fine structure constant α_s can be considered as a quantity independent from $|k|^2$.

Searching the solution of these equations in the form of the Mellin transformation

$$n(Q^2, \beta) = \int_{\sigma-i\infty}^{\sigma+i\infty} \frac{dj}{2\pi i} \left(\frac{1}{\beta}\right)^j \int_{-i\infty}^{i\infty} \frac{d\gamma}{2\pi i} \left(\frac{Q^2}{\mu^2}\right)^\gamma G(j, \gamma) \begin{pmatrix} a^g \\ a^q \end{pmatrix}, \tag{7.482}$$

$$\tilde{n}(Q^2, \beta) = \int_{\sigma-i\infty}^{\sigma+i\infty} \frac{dj}{2\pi i} \left(\frac{1}{\beta}\right)^j \int_{-i\infty}^{i\infty} \frac{d\gamma}{2\pi i} \left(\frac{Q^2}{\mu^2}\right)^\gamma \tilde{G}(j, \gamma) \begin{pmatrix} \tilde{a}^g \\ \tilde{a}^q \end{pmatrix}, \tag{7.483}$$

we obtain for the operators $G(j, \gamma)$ and $\tilde{G}(j, \gamma)$ the following expressions

$$G(j, \gamma) = \frac{j \chi(j)/K}{\gamma - \frac{\alpha_s}{2\pi} \chi(j)}, \qquad \tilde{G}(j, \gamma) = \frac{j \tilde{\chi}(j)/\tilde{K}}{\gamma - \frac{\alpha_s}{2\pi} \tilde{\chi}(j)}, \tag{7.484}$$

where the t-channel partial waves $\chi(j)$ and $\tilde{\chi}(j)$ satisfy the matrix relations

$$j \chi(j) = K + \frac{\alpha_s}{2\pi} (\chi(j))^2, \qquad j \tilde{\chi}(j) = \tilde{K} + \frac{\alpha_s}{2\pi} (\tilde{\chi}(j))^2. \tag{7.485}$$

The last relations are obtained by comparing the coefficients in front of $(k^2/\mu^2)^{\gamma(j)}$ in the left- and right-hand sides of the above equations for $n(k^2, \beta)$ and $\tilde{n}(k^2, \beta)$. The terms independent from k^2 are cancelled in their right-hand sides. The above-derived algebraic relations for the nonsinglet partial waves $f_\pm^{(1)}(j)$ are particular cases of these general equations.

We can calculate the eigenvalues $\lambda_r, \tilde{\lambda}_r$ and eigenfunctions $a_r^i, \tilde{a}_r^i$ of the operators K and $\tilde{K}$

$$\lambda_r \begin{pmatrix} a_r^g \\ a_r^q \end{pmatrix} = K \begin{pmatrix} a_r^g \\ a_r^q \end{pmatrix}, \qquad \tilde{\lambda}_r \begin{pmatrix} \tilde{a}_r^g \\ \tilde{a}_r^q \end{pmatrix} = \tilde{K} \begin{pmatrix} \tilde{a}_r^g \\ \tilde{a}_r^q \end{pmatrix}. \tag{7.486}$$

They are solutions of the corresponding secular equations

$$\lambda_r^2 + \frac{3N_c^2 + 1}{2 N_c} \lambda_r + \frac{N_c^2 - 1}{N_c} (n_f - N_c) = 0, \qquad \frac{a_r^q}{a_r^g} = -\frac{N_c^2 - 1}{N_c(\lambda_r + 2N_c)}, \tag{7.487}$$

$$\tilde{\lambda}_r^2 + \frac{-9N_c^2 + 1}{2N_c} \tilde{\lambda}_r + \frac{N_c^2 - 1}{N_c} (n_f + 2N_c) = 0, \qquad \frac{\tilde{a}_r^q}{\tilde{a}_r^g} = \frac{N_c^2 - 1}{N_c(\lambda_r - 4N_c)}. \tag{7.488}$$

The eigenvalues of the operators $\chi(j)$ and $\tilde{\chi}(j)$ satisfy the algebraic equations [12]

$$j \chi_r(j) = \lambda_r + \frac{\alpha_s}{2\pi} (\chi_r(j))^2, \qquad j \tilde{\chi}_r(j) = \tilde{\lambda}_r + \frac{\alpha_s}{2\pi} (\tilde{\chi}_r(j))^2. \tag{7.489}$$

The eigenvalues are not changed after the similarity transformation of the matrices K and $\widetilde{K}$

$$K \to K' = U^{-1}KU, \quad \widetilde{K} \to \widetilde{K}' = \widetilde{U}^{-1}\widetilde{K}\widetilde{U}. \tag{7.490}$$

For example, using the diagonal matrices U and $\widetilde{U}$ one can transform K and $\widetilde{K}$ to the symmetric forms

$$K' = \begin{pmatrix} -2N_c & i\sqrt{\frac{N_c^2-1}{N_c}}n_f \\ i\sqrt{\frac{N_c^2-1}{N_c}}n_f & \frac{N_c^2-1}{2N_c} \end{pmatrix}, \quad \widetilde{K}' = \begin{pmatrix} 4N_c & i\sqrt{\frac{N_c^2-1}{N_c}}n_f \\ i\sqrt{\frac{N_c^2-1}{N_c}}n_f & \frac{N_c^2-1}{2N_c} \end{pmatrix}. \tag{7.491}$$

One can write the corresponding t-channel partial waves for the parton distributions in DLA as follows

$$G(j,\gamma)\begin{pmatrix} a^g \\ a^q \end{pmatrix} = \sum_{r=1}^{2} \frac{j\,\chi_r(j)\,c_r/\lambda_r}{\gamma - \frac{\alpha_s}{2\pi}\chi_r(j)}\begin{pmatrix} a_r^g \\ a_r^q \end{pmatrix}, \quad \begin{pmatrix} a^g \\ a^q \end{pmatrix} = \sum_{r=1}^{2} c_r \begin{pmatrix} a_r^g \\ a_r^q \end{pmatrix}, \tag{7.492}$$

$$\widetilde{G}(j,\gamma)\begin{pmatrix} \widetilde{a}^g \\ \widetilde{a}^q \end{pmatrix} = \sum_{r=1}^{2} \frac{j\,\widetilde{\chi}_r(j)\,\widetilde{c}_r/\widetilde{\lambda}_r}{\gamma - \frac{\alpha_s}{2\pi}\widetilde{\chi}_r(j)}\begin{pmatrix} \widetilde{a}_r^g \\ \widetilde{a}_r^q \end{pmatrix}, \quad \begin{pmatrix} \widetilde{a}^g \\ \widetilde{a}^q \end{pmatrix} = \sum_{r=1}^{2} \widetilde{c}_r \begin{pmatrix} \widetilde{a}_r^g \\ \widetilde{a}_r^q \end{pmatrix}. \tag{7.493}$$

It turns out that there is an additional contribution in the equation for $\widetilde{\chi}(j)$ from the soft Sudakov gluon with the smallest transverse momentum. This contribution will be constructed below.

7.14.2 Sudakov term in evolution equations

According to the Gribov arguments, the main contribution from the diagrams with the virtual gluon having the smallest transverse momentum $|k^\perp|$ appears from its insertion in the external lines for a scattering amplitude on the mass shell. The inner amplitude has the new infrared cut-off $|k^\perp|$ instead of initial one μ. In LLA for the Regge kinematics these Sudakov-type terms can be presented as follows [12]

$$\Delta A_{i_2 r_1}^{i_1 r_2}(s,\mu^2) = 2\frac{g^2}{(2\pi)^4}\int \frac{\pi\,d\,|k_\perp|^2\,|s|\,d\alpha\,d\beta\,(-2s)}{2i\left(s\alpha\beta - |k_\perp|^2 + i\varepsilon\right)\left(-s(1-\beta)\alpha - |k_\perp|^2 + i\varepsilon\right)}$$

$$\times \left(\frac{\overrightarrow{T}_{i_1'}^{i_1}\,\overrightarrow{T}_{r_1'}^{r_1}\,A_{i_2 r_1'}^{i_1' r_2}(s,|k_\perp|^2)}{s(1+\alpha)\beta - |k_\perp|^2 + i\varepsilon} + \frac{\overrightarrow{T}_{i_1'}^{i_1}\,\overrightarrow{T}_{r_2'}^{r_2}\,A_{i_2 r_1}^{i_1' r_2'}(s,|k_\perp|^2)}{-s(1-\alpha)\beta - |k_\perp|^2 + i\varepsilon}\right). \tag{7.494}$$

The factor 2 in front of the integral corresponds to the contribution of two symmetric Feynman diagrams in which the gluon connects the lines with initial and final particle momenta $\pm p_1$ and $\pm p_2$. The diagrams in which the soft gluon connects the lines with opposite momenta $\pm p_1$ or $\pm p_2$ are not essential at high energies. The operators $\overrightarrow{T}_{i_1'}^{i_1}$ are the generators of the gauge group in the fundamental and adjoint representations for the quark and gluon lines, respectively. We use the Sudakov variables

$$k = \alpha p_2 + \beta p_1 + k_\perp, \quad d^4k = \frac{d^2k_\perp \, |s| \, d\alpha \, d\beta}{2}, \quad p_2^2 = p_1^2 = 0, \tag{7.495}$$

where p_1 and p_2 are the initial particle momenta. The factor $(-2s)$ in the integrand appears as a result of the simplification of the spin structure at high energies. It is convenient to present the amplitudes ΔA and A as a superposition of the states with the various colour quantum numbers in the t-channel. For this purpose, we introduce the projectors on the singlet (1) and octet (8_f), (8_d) states for gluons

$$P^{a_1,b_2}_{a_2,b_1}(1,gg) = \frac{\delta_{a_1,a_2} \, \delta_{b_1,b_2}}{N_c^2 - 1}, \quad P^{a_1,b_2}_{a_2,b_1}(8_f,gg) = \frac{f_{a_1,a_2,c} \, f_{b_1,b_2,c}}{N_c}, \tag{7.496}$$

$$P^{a_1,b_2}_{a_2,b_1}(8_d,gg) = \frac{N_c \, d_{a_1,a_2,c} \, d_{b_1,b_2,c}}{N_c^2 - 4}, \tag{7.497}$$

quark-antiquark pairs

$$P^{i_1,k_2}_{i_2,k_1}(1,qq) = \frac{\delta^{i_1}_{i_2} \, \delta^{k_2}_{k_1}}{N_c}, \quad P^{i_1,k_2}_{i_2,k_1}(8,qq) = 2 \, \vec{T}^{i_1}_{i_2} \, \vec{T}^{k_2}_{k_1}, \tag{7.498}$$

and for matrix elements describing transitions between quarks and gluons

$$P^{a_1,k_2}_{a_2,k_1}(1,gq) = \frac{\delta_{a_1,a_2} \, \delta^{k_2}_{k_1}}{\sqrt{N_c(N_c^2 - 1)}}, \quad P^{a_1,k_2}_{a_2,k_1}(8_f,gq) = \frac{-if_{a_1,a_2,c} \, (T^c)^{k_2}_{k_1}}{\sqrt{N_c/2}}, \tag{7.499}$$

$$P^{a_1,k_2}_{a_2,k_1}(8_d,gq) = \frac{d_{a_1,a_2,c} \, (T^c)^{k_2}_{k_1}}{\sqrt{(N_c^2 - 4)/(2N_c)}}. \tag{7.500}$$

The scattering amplitudes ΔA and A for various processes with quarks and gluons can be expanded over these projectors

$$A^{i_1 k_2}_{i_2 k_1}(s,\mu^2) = \sum_{r=1,8} P^{i_1,k_2}_{i_2,k_1}(1,qq) \, A^{(r)}_{qq}(s,\mu^2), \tag{7.501}$$

$$A^{a_1,b_2}_{a_2,b_1}(s,\mu^2) = \sum_{r=1,8_f,8_d,\ldots} P^{a_1,b_2}_{a_2,b_1}(r,gg) \, A^{(r)}_{gg}(s,\mu^2), \tag{7.502}$$

$$A^{a_1,k_2}_{a_2,k_1}(s,\mu^2) = \sum_{r=1,8_f,8_d} P^{a_1,k_2}_{a_2,k_1}(r,gq) \, A^{(r)}_{gq}(s,\mu^2). \tag{7.503}$$

In particular for the colour matrices appearing in the Born amplitudes, we have the following decomposition

$$f_{b_1,a_1,c} \, f_{b_2,a_2,c} = N_c \, P^{a_1,b_2}_{a_2,b_1}(1,gg) + \frac{N_c}{2} \, P^{a_1,b_2}_{a_2,b_1}(8_f,gg) + \frac{N_c}{2} \, P^{a_1,b_2}_{a_2,b_1}(8_d,gg) + \cdots, \tag{7.504}$$

$$\vec{T}_{i_2}^{k_2}\vec{T}_{k_1}^{i_1} = \frac{N_c^2 - 1}{2 N_c} P_{i_2 k_1}^{i_1 k_2}(1, qq) - \frac{1}{2N_c} P_{i_2 k_1}^{i_1 k_2}(8, qq), \tag{7.505}$$

$$\sqrt{n_f}\,(T^{a_1} T^{a_2})_{k_1}^{k_2} = \frac{1}{2}\sqrt{\frac{N_c^2 - 1}{N_c}} n_f P_{a_2, k_1}^{a_1, k_2}(1, gq) + \frac{\sqrt{2 N_c n_f}}{4} P_{a_2, k_1}^{a_1, k_2}(8_f, gq)$$

$$+ \frac{\sqrt{2(N_c - \frac{4}{N_c}) n_f}}{4} P_{a_2, k_1}^{a_1, k_2}(8_d, gq). \tag{7.506}$$

The matrix elements of the matrices K' and $\widetilde{K}'$ are proportional to the corresponding coefficients in front of the projectors $P(1)$ to the singlet states for the decomposition of the Born colour structures. It is natural to generalize these matrices to other colour states $l = 8_d, 8_f$ in accordance with the Born-term decomposition

$$K'_{8_d} = \begin{pmatrix} -N_c & i\sqrt{\frac{N_c^2 - 4}{2N_c/n_f}} \\ i\sqrt{\frac{N_c^2 - 4}{2N_c/n_f}} & -\frac{1}{2N_c} \end{pmatrix}, \quad \widetilde{K}'_{8_d} = \begin{pmatrix} 2N_c & i\sqrt{\frac{N_c^2 - 4}{2N_c/n_f}} \\ i\sqrt{\frac{N_c^2 - 4}{2N_c/n_f}} & -\frac{1}{2N_c} \end{pmatrix},$$

$$K'_{8_f} = \begin{pmatrix} -N_c & i\sqrt{\frac{N_c n_f}{2}} \\ i\sqrt{\frac{N_c n_f}{2}} & -\frac{1}{2N_c} \end{pmatrix}, \quad \widetilde{K}'_{8_f} = \begin{pmatrix} 2N_c & i\sqrt{\frac{N_c n_f}{2}} \\ i\sqrt{\frac{N_c n_f}{2}} & -\frac{1}{2N_c} \end{pmatrix}. \tag{7.507}$$

In this case, for the ladder diagrams in DLA the nonlinear equations for the t-channel partial waves $\chi_l(j), \widetilde{\chi}_l(j)$ $(l = 1, 8_f, 8_d)$ are universal [12]

$$j\,\chi_l(j) = K'_l + \frac{\alpha_s}{2\pi}(\chi_l(j))^2, \quad j\,\widetilde{\chi}_l(j) = \widetilde{K}'_l + \frac{\alpha_s}{2\pi}(\widetilde{\chi}_l(j))^2. \tag{7.508}$$

Note that signatures of the amplitudes χ_{8_f} and $\widetilde{\chi}_{8_d}$ are negative and positive, respectively.

The soft-gluon contribution expressed in the terms of the colour octet and colour singlet amplitudes is simplified after the integration over the Sudakov variables α and β

$$\Delta A^{(r)}(s, \mu^2) = -2 \frac{g^2}{(2\pi)^3} \int_{\mu^2}^{s} \frac{\pi\,d\,|k_\perp|^2}{|k_\perp|^2}$$

$$\times \left(a_s^{t\to r} \ln \frac{-s}{|k_\perp|^2} - a_u^{t\to r} \ln \frac{s}{|k_\perp|^2} \right) A^{(t)}(s, |k_\perp|^2). \tag{7.509}$$

The coefficients a_s and a_u depend on the transition type (qq), (gg), or (qg) and the colour quantum numbers of the amplitudes $\Delta A^{(r)}$ and $A^{(t)}$ in the t-channel. For ΔA in the singlet state A can be only in the octet state. We obtain the following coefficients

$$a_s^{8\to 1} = a_u^{8\to 1} = \frac{N_c^2 - 1}{2N_c} \tag{7.510}$$

for the qq transitions,

$$a_s^{8_f\to 1} = a_u^{8_f\to 1} = N_c, \quad a_s^{8_d\to 1} = a_u^{8_d\to 1} = 0 \tag{7.511}$$

for the gg transitions, and

$$a_s^{8_f \to 1} = a_u^{8_f \to 1} = \sqrt{\frac{N_c^2 - 1}{2}}, \quad a_s^{8_d \to 1} = a_u^{8_d \to 1} = 0 \qquad (7.512)$$

for the transitions $q \to g$ and $g \to q$.

In the case when $\Delta_g A$ is in the octet state, A can be in the singlet state, where

$$a_s^{1 \to 8} = a_u^{1 \to 8} = \frac{1}{2N_c} \qquad (7.513)$$

for the qq transitions,

$$a_s^{1 \to 8_f} = a_u^{1 \to 8_f} = \frac{N_c}{N_c^2 - 1}, \quad a_s^{1 \to 8_d} = a_u^{1 \to 8_d} = 0 \qquad (7.514)$$

for the gg transitions, and

$$a_s^{1 \to 8_f} = a_u^{1 \to 8_f} = \frac{1}{\sqrt{2(N_c^2 - 1)}}, \quad a_s^{1 \to 8_d} = a_u^{1 \to 8_d} = 0 \qquad (7.515)$$

for the transitions $q \to g$ and $g \to q$.

When $\Delta_g A$ is in the octet state 8_f or 8_d, A can also be in other colour states. We consider below only transitions from 8_f and 8_d states. The coefficients a_s and a_u equal

$$a_s^{8 \to 8} = \frac{N_c^2 - 2}{2N_c}, \quad a_u^{8 \to 8} = -\frac{1}{N_c}, \qquad (7.516)$$

for the qq transitions,

$$a_s^{8_f \to 8_f} = -a_u^{8_f \to 8_f} = \frac{N_c}{4}, \quad a_s^{8_f \to 8_d} = a_u^{8_f \to 8_d} = \frac{N_c}{4},$$

$$a_s^{8_d \to 8_f} = a_u^{8_d \to 8_f} = \frac{N_c}{4}, \quad a_s^{8_d \to 8_d} = a_u^{8_d \to 8_d} = \frac{N_c}{4} \qquad (7.517)$$

for the gg transitions, and

$$a_s^{8_f \to 8_f} = -a_u^{8_f \to 8_f} = \frac{N_c}{4}, \quad a_s^{8_f \to 8_d} = a_u^{8_f \to 8_d} = \frac{\sqrt{N_c^2 - 4}}{4}$$

$$a_s^{8_d \to 8_f} = a_u^{8_d \to 8_f} = \frac{\sqrt{N_c^2 - 4}}{4}, \quad a_s^{8_d \to 8_d} = a_u^{8_d \to 8_d} = \frac{N_c^2 - 4}{4N_c}. \qquad (7.518)$$

for the qg transitions.

The Sudakov contribution can be written in the differential form

$$\frac{\partial}{\partial \ln \frac{1}{\mu^2}} \Delta A^r(s, \mu^2) = -\frac{g^2}{4\pi^2} R^{t \to r}(s) A^t(s, \mu^2), \qquad (7.519)$$

where

$$R^{t \to r}(s) = a_s^{t \to r} \ln \frac{-s}{\mu^2} - a_u^{t \to r} \ln \frac{s}{\mu^2}$$

$$= (a_s^{t\to r} - a_u^{t\to r})\frac{\ln(-s/\mu^2) + \ln(s/\mu^2)}{2} + (a_s^{t\to r} + a_u^{t\to r})\frac{-i\pi\,\varepsilon(s)}{2}. \qquad (7.520)$$

Here $\varepsilon(s) = s/|s|$ is the sign function.

Let us consider the scattering amplitudes with the definite signature $p = \pm 1$

$$A(s) = A_+(s) + A_-(s), \quad A_p(s) = \int_{\sigma-i\infty}^{\sigma+i\infty} \frac{dj}{2\pi i} \left(\frac{s}{\mu^2}\right)^j \xi_p(j)\, f_p(j), \quad \sigma > 0, \qquad (7.521)$$

where $\xi_p(j)$ is the signature factor.

$$\xi_p(j) = -\frac{e^{-i\pi j} + p}{\sin \pi j}. \qquad (7.522)$$

The inverse transformation is

$$f_p(j) = \int_{\mu^2}^\infty \frac{ds}{s} \left(\frac{s}{\mu^2}\right)^{-j} \mathrm{Im}_s\, A_p(s), \qquad (7.523)$$

where $\mathrm{Im}_s\, A_p$ is the s-channel imaginary part of the amplitude symmetric ($p = 1$) or antisymmetric ($p = -1$) under the transformation $s \to -s$.

Let us calculate the product of $R^{i\to r}(s)$ with the scattering amplitudes

$$R^{t\to r}(s)\, A_p^i(s) = \int_{\sigma-i\infty}^{\sigma+i\infty} \frac{dj}{2\pi i} \left(\frac{s}{\mu^2}\right)^j \times \Big((a_s^{t\to r} - a_u^{t\to r})\xi_p(j)\,\varphi_p^t(j)$$

$$+ (a_s^{t\to r} + a_u^{t\to r})\xi_{-p}(j)\,\varphi_{-p}^t(j)\Big), \qquad (7.524)$$

where

$$\varphi_p^t(j) = \int_{\mu^2}^\infty \frac{ds}{s}\left(\frac{s}{\mu^2}\right)^{-j} \int_{\sigma'-i\infty}^{\sigma'+i\infty} \frac{dj'}{2\pi i} \left(\ln\frac{s}{\mu^2} + \frac{\pi}{2}\frac{\cos\pi j' + p}{\sin\pi j'}\right)\left(\frac{s}{\mu^2}\right)^{j'} f_p^t(j')$$

$$= \int_{\sigma'-i\infty}^{\sigma'+i\infty} \frac{dj'}{2\pi i(j-j')}\left(\frac{1}{j-j'} + \frac{\pi}{2}\frac{\cos\pi j' + p}{\sin\pi j'}\right) f_p^t(j') \qquad (7.525)$$

and

$$\varphi_{-p}^t(j) = \int_{\mu^2}^\infty \frac{ds}{s}\left(\frac{s}{\mu^2}\right)^{-j} \int_{\sigma'-i\infty}^{\sigma'+i\infty} \frac{dj'}{2\pi i}\left(\frac{\pi}{2}\frac{\cos\pi j' + p}{\sin\pi j'}\right)\left(\frac{s}{\mu^2}\right)^{j'} f_p^t(j')$$

$$= \int_{\sigma'-i\infty}^{\sigma'+i\infty} \frac{dj'}{2\pi i(j-j')}\left(\frac{\pi}{2}\frac{\cos\pi j' + p}{\sin\pi j'}\right) f_p^t(j'). \qquad (7.526)$$

In the above expressions, the contour of integration over j' lies to the left from the contour of integration over j ($\sigma > \sigma'$). Therefore, we can enclose the integration contour in the variable j' around the pole in the point $j = j'$ and calculate the integral by residues

$$\varphi_p^t(j) = \left(-\frac{d}{dj} + \frac{\pi}{2}\frac{\cos\pi j + p}{\sin\pi j}\right) f_p^t(j), \quad \varphi_{-p}^t(j) = \left(\frac{\pi}{2}\frac{\cos\pi j + p}{\sin\pi j}\right) f_p^t(j). \qquad (7.527)$$

Here we neglected contributions of the poles contained in the factor $(\cos \pi j' + p)/\sin \pi j'$ because they do not lead to the singularities of $\varphi_{\pm}^i(j)$ in the left half of the j-plane related to the asymptotic behaviour of amplitudes at large s. As a result, we can write the Sudakov contribution to the evolution equation in μ^2 as follows

$$
j f_p^r(j)_{Sud} = -\frac{g^2}{4\pi^2}(a_s^{t \to r} - a_u^{t \to r})\left(-\frac{d}{dj} + \frac{\pi}{2}\frac{\cos \pi j + p}{\sin \pi j}\right) f_p^t(j)
$$
$$
-\frac{g^2}{4\pi^2}(a_s^{t \to r} + a_u^{t \to r})\left(\frac{\pi}{2}\frac{\cos \pi j - p}{\sin \pi j}\right) f_{-p}^t(j). \tag{7.528}
$$

One can simplify the brackets in the above equation near $j = 0$

$$
-\frac{d}{dj} + \frac{\pi}{2}\frac{\cos \pi j + p}{\sin \pi j} = -\frac{d}{dj} + \frac{1+p}{2j}, \quad \frac{\pi}{2}\frac{\cos \pi j - p}{\sin \pi j} = \frac{1-p}{2j}. \tag{7.529}
$$

7.14.3 Equations for the partial waves in DLA

As we have shown above, the amplitude for the virtual photon scattering off the quark with nonsinglet quantum numbers in the t-channel is expressed in DLA in terms of the quark-antiquark scattering amplitude on mass shell. The $q\bar{q}$-amplitude satisfies the infrared evolution equation which contains generally the soft quark and Sudakov contributions. These terms were found in the previous sections. The partial waves $f_{\pm}^r(j)$ for quark pairs with the colour quantum numbers r are related to the matrix elements of the more general operators χ^r and $\tilde{\chi}^r$ with the relations of the type

$$
f_+^{(1)}(j) = -\frac{\pi}{2} j \chi_{qq}^{(1)}(j), \quad f_-^{(1)}(j) = -\frac{\pi}{2} j \tilde{\chi}_{qq}^{(1)}(j), \tag{7.530}
$$

In DLA, the singlet operator $\chi_+^{(1)}(j)$ with the positive signature satisfies the equation [12]

$$
j \chi^{(1)}(j) = K' + \frac{\alpha_s}{2\pi}\left(\chi^{(1)}(j)\right)^2. \tag{7.531}
$$

The anomalous dimensions of the twist-2 operators with even j are expressed in terms of two eigenvalues $\chi_{1,2}^{(1)}(j)$ of its solution

$$
\gamma_{1,2} = \frac{\alpha_s}{2\pi}\chi_{1,2}^{(1)}(j), \quad \chi^{(1)}(j) = \frac{\pi}{\alpha_s}\left(j - \sqrt{j^2 - 2\frac{\alpha_s}{\pi}K'}\right). \tag{7.532}
$$

As for the partial wave $\tilde{\chi}^{(1)}(j)$ with the negative signature, its equation contains also the contribution from the soft gluons [12]

$$
j \tilde{\chi}^{(1)}(j) = \tilde{K}' + \frac{\alpha_s}{2\pi}\left(\tilde{\chi}^{(1)}(j)\right)^2
$$
$$
- 2\frac{\alpha_s}{\pi}\frac{1}{j}\left(\begin{array}{cc} N_c \tilde{\chi}_{gg}^{(8_f)}(j) & \sqrt{\frac{N_c^2-1}{2}}\tilde{\chi}_{gq}^{(8_f)}(j) \\ \sqrt{\frac{N_c^2-1}{2}}\tilde{\chi}_{qg}^{(8_f)}(j) & \frac{N_c^2-1}{2N_c}\tilde{\chi}_{qq}^{(8_f)}(j) \end{array}\right), \tag{7.533}
$$

where the operator $\tilde{\chi}^r(j)$ with the octet quantum numbers $r = (8_f)$ and the positive signature satisfies the more complicated Ricatti-type equation

$$j\tilde{\chi}^{(8_f)}(j) = K'_{8_f} + \frac{\alpha_s}{2\pi}\left(\tilde{\chi}^{(8_f)}(j)\right)^2 + \frac{\alpha_s}{2\pi}N_c\,\frac{d}{dj}\,\tilde{\chi}^{(8_f)}(j). \qquad (7.534)$$

In the Sudakov term with the double-logarithmic accuracy we neglected the contributions from the negative-signature amplitudes $\tilde{\chi}^r$ ($r = 1, 8_d, 10, \overline{10}, 27$).

One can search the solution in the form

$$\tilde{\chi}^{(8_f)}(j) = N_c\,\frac{\partial}{\partial j}\,\ln\psi(j), \qquad (7.535)$$

where $\psi(j)$ satisfies the linear equation

$$\left(\frac{\alpha_s}{2\pi}N_c\,\frac{\partial^2}{(\partial j)^2} - j\frac{\partial}{\partial j} + \frac{1}{N_c}K'_{8_f}\right)\psi(j) = 0. \qquad (7.536)$$

By extracting from $\psi(j)$ the factor $\exp(\pi j^2/(2N_c\alpha_s))$ one can reduce it to the Schrödinger equation for the harmonic oscillator, which allows us to find its solution in terms of the parabolic-cylinder function $D_p(z)$ satisfying the equation

$$\left(\frac{\partial^2}{(\partial z)^2} + v + \frac{1}{2} - \frac{z^2}{4}\right)D_v(z) = 0, \qquad (7.537)$$

where

$$z = j\sqrt{\frac{2\pi}{N_c\alpha_s}}, \qquad v = \frac{1}{N_c}K'_{8_f}. \qquad (7.538)$$

Thus, we obtain for $\tilde{\chi}^{(8_f)}(j)$ [12]

$$\tilde{\chi}^{(8_f)}(j) = N_c\,\frac{\partial\ln\left(e^{\frac{z^2}{4}}D_v(z)\right)}{\partial j} = N_c\,\frac{\partial\ln\int_0^\infty\frac{dt}{t^{1+v}}e^{-\frac{t^2}{2}-zt}}{\partial j}. \qquad (7.539)$$

This solution is a matrix. One can diagonalize it

$$\tilde{\chi}^{(8_f)}(j) = L^{-1}N_c\,\frac{\partial}{\partial j}\begin{pmatrix} \ln\left(e^{\frac{z^2}{4}}D_{v_1}(z)\right) & 0 \\ 0 & \ln\left(e^{\frac{z^2}{4}}D_{v_2}(z)\right) \end{pmatrix}L, \qquad (7.540)$$

where the operator L is constructed from the eigenfunctions of the matrix K'_{8_f}/N_c and its eigenvalues v_r

$$L = \begin{pmatrix} (L_1)_1 & (L_2)_1 \\ (L_1)_2 & (L_2)_2 \end{pmatrix}, \qquad \frac{1}{N_c}K'_{8_f}L_r = v_r L_r. \qquad (7.541)$$

After that the matrix

$$\widetilde{S}(j) = K'_{8_f} - 2\frac{\alpha_s}{\pi}\frac{1}{j} \begin{pmatrix} N_c\,\widetilde{\chi}_{gg}^{(8_f)}(j) & \sqrt{\frac{N_c^2-1}{2}}\,\widetilde{\chi}_{gq}^{(8_f)}(j) \\ \sqrt{\frac{N_c^2-1}{2}}\,\widetilde{\chi}_{qg}^{(8_f)}(j) & \frac{N_c^2-1}{2N_c}\,\widetilde{\chi}_{qq}^{(8)}(j) \end{pmatrix} \tag{7.542}$$

is constructed in an explicit form. The anomalous dimensions of the operators with the negative signature are expressed in terms of the eigenvalues of the matrix $\widetilde{\chi}^{(1)}(j)$

$$\gamma_{1,2} = \frac{\alpha_s}{2\pi}\,\widetilde{\chi}_{1,2}^{(1)}(j), \tag{7.543}$$

satisfying the algebraic equation

$$j\widetilde{\chi}_{1,2}^{(1)}(j) = \widetilde{s}_{1,2}(j) + \frac{\alpha_s}{2\pi}\left(\widetilde{\chi}_{1,2}^{(1)}(j)\right)^2, \tag{7.544}$$

where $\widetilde{s}_{1,2}(j)$ are the eigenvalues of the matrix $\widetilde{S}(j)$.

For completeness, we present also the equation for the amplitude with the octet quantum numbers and the negative signature [12]

$$j\,\chi^{(8_f)}(j) = \phi^{(8_f)}(j) + \frac{\alpha_s}{2\pi}\left(\chi^{(8_f)}(j)\right)^2 + \frac{\alpha_s}{2\pi}N_c\frac{d}{dj}\,\widetilde{\chi}^{(8_f)}(j), \tag{7.545}$$

where the inhomogeneous term is given below

$$\phi^{(8_f)}(j) = K'_{(8_f)}$$

$$-\frac{\alpha_s}{\pi}\frac{1}{j}\begin{pmatrix} \frac{2N_c}{N_c^2-1}\chi_{gg}^{(1)}(j) & \sqrt{\frac{2}{N_c^2-1}}\chi_{gq}^{(1)}(j) \\ \sqrt{\frac{2}{N_c^2-1}}\chi_{qg}^{(1)}(j) & \frac{1}{N_c}\chi_{qq}^{(1)} \end{pmatrix}$$

$$-\frac{\alpha_s}{\pi}\frac{1}{j}\begin{pmatrix} \frac{N_c}{2}\,\widetilde{\chi}_{gg}^{(8_d)}(j) & \frac{\sqrt{N_c^2-4}}{2}\,\widetilde{\chi}_{gq}^{(8_d)}(j) \\ \frac{\sqrt{N_c^2-4}}{2}\,\widetilde{\chi}_{qg}^{(8_d)}(j) & \frac{N_c^2-4}{2N_c}\,\widetilde{\chi}_{qq}^{(8)} \end{pmatrix}. \tag{7.546}$$

In principle, writing exactly the signature factors, one can derive more accurate equations taking into account the corrections $\sim j$.

References

[1] Fritzsch, H., Gell-Mann, M. and Leutwyler, H., *Phys. Lett.* **B47** (1973) 365.

[2] Bjorken, J. D. in: Proc. of Intern. School of Physics "Enrico Fermi", Course 41, ed. by J. Steinberger, Academic Press Inc., New York, 1968; Bjorken, J. D. and Paschos, E. A., *Phys. Rev.* **185** (1969) 1975.

[3] Feynman, P. R., *Phys. Rev. Lett.* **23** (1969) 1415.

[4] Ioffe B. L., Khoze, V. A. and Lipatov, L. N., Hard processes, North Holland, Amsterdam, 1984.

[5] Bjorken, J. D., *Phys. Rev.* **179** (1969) 1547.

[6] Ioffe, B. L., *Phys. Lett.* **B30** (1969) 123.

[7] Wilson, K. G., *Phys. Rev.* **179** (1969) 1499.

[8] Muta, T., Foundations of Quantum Chromodynamics, World Scientific, 1987.

[9] Yndurain, E. J., The Theory of Quark and Gluon Interactions, Second Edition, Springer-Verlag, 1992.

[10] Sudakov, V. V., *Sov. Phys.-JETP* **3** (1956) 115.

[11] Gorshkov, V. G., Gribov, V. N., Lipatov, L. N. and Frolov, G. V., *Sov. J. Nucl. Phys.* **6** (1967) 95, 262.

[12] Kirschner, R. and Lipatov, L. N., *Sov. Phys. JETP* **83** (1982) 488, *Phys. Rev.* **D26** (1983) 1202; *Nucl. Phys.* **B123** (1983) 122.

[13] Lipatov, L. N., *Sov. J. Nucl. Phys.* **23** (1976) 642; Fadin, V. S., Kuraev, E. A. and Lipatov, L. N., *Phys. Lett.* **B60** (1975) 50; *Sov. Phys. JETP* **44** (1976) 443; **45** (1977) 199; Balitsky, Ya. Ya. and Lipatov, L. N., *Sov. J. Nucl. Phys.* **28** (1978) 822.

[14] Gribov, V. N., Lipatov, L. N. and Frolov, G. V., *Phys. Lett.* **31B** (1970) 34; Cheng, H. and Wu, T. T., *Phys. Rev.* **D1** (1970) 2775.

[15] Drell, S. D., Levy, D. J. and Yan, T. M., *Phys. Rev.* **187** (1969) 2159; **D1** (1970) 1035, 1617, 2402.

[16] Drell, S. D. and Yan, T. M., *Phys. Rev. Lett.* **24** (1970) 181.

[17] Callan, C. and Gross, D. J., *Phys. Rev. Lett.* **22** (1969) 156.

[18] Gribov, V. N., Ioffe, B. L. and Pomeranchuk, I. Ya., *Sov. J. Nucl. Phys.* **2** (1966) 549.

[19] Polyakov, A. M., Lecture at Intern. School of High Energy Physics, Erevan, Nov. 23–Dec. 4, 1971, Chernogolovka preprint, 1972.

[20] Gross, D. J., Wilczek, F., *Phys. Rev. Lett.* **30** (1973) 1343; Politzer, H. D., *Phys. Rev. Lett.* **30** (1973) 1346.

[21] Gross, D. J. and Wilczek, F. *Phys. Rev.* **D8** (1973) 3633.

[22] Politzer, H. D., *Phys. Rep.* **14C** (1974) 129; Georgi, H., Politzer, H. D., *Phys. Rev.* **D9** (1974) 416.

[23] Gribov, V. N. and Lipatov, L. N., *Sov. J. Nucl. Phys.* **18** (1972) 438, 675; Lipatov, L. N., *Sov. J. Nucl. Phys.* **20** (1975) 93; Altarelli, G. and Parisi, G., *Nucl. Phys.* **B126** (1977) 298; Dokshitzer, Yu. L., *Sov. Phys. JETP* **46** (1977) 641.

[24] Bukhvostov, A. P., Lipatov, L. N., Popov, N. P., *Sov. J. Nucl. Phys.* **20** (1975) 287.

[25] Fishbane, P. M. and Sullivan, J. D., *Phys. Rev.* **D6** (1972) 3568.

[26] Collins, J. C, Soper, D. E. and Sterman, G. in: Perturbative Quantum Chromodynamics, ed. by A. H. Mueller, World Scientific, 1989, p.1.

[27] Floratos, E. G., Ross, D. A., Sachrajda, C. T., *Nucl. Phys.* **B129** (1977) 66, Nucl. Phys. **B139** (1978) 545; Gonzales-Arroyo, A., Lopez C., Yndurain, F. J., *Nucl. Phys.* **B153** (1979) 161; Curci, G., Furmanski, W., Petronzio, R., *Nucl. Phys.* **B175** (1980) 27; Floratos, E. G., Lacaze, R., Kounnas, C., *Phys. Lett.* **B98** (1981) 89.

[28] Floratos, E. G., Ross, D. A., Sachrajda, C. T., *Nucl. Phys.* **B152** (1979) 493; Gonzales-Arroyo, A., Lopez C., *Nucl. Phys.* **B166** (1980) 429; Furmanski, W., Petronzio, R., *Phys. Lett.* **B97** (1980) 437; Floratos, E. G., Lacaze, R., Kounnas, C., *Nucl. Phys.* **D192** (1981) 417, *Phys. Lett.* **B98** (1981) 285; Hamberg, R., van Neerven, W. L., *Nucl. Phys.* **B379** (1992) 143; Ellis, R. K., Vogelsang, W., hep-ph/9602356.

[29] Glück, M., Reya, E. and Vogt, A., *Z. Phys.* **C53** (1992) 651.

[30] Dokshitzer, Yu. L., Khoze, V. A., Troyan, S. I., *Nucl. Phys.* **D53** (1996) 89; Dokshitzer, Yu. L., Marchesini, G.,Salam, G. P., hep-th/0511302; Dokshitzer, Yu. L., Marchesini, G., hep-th/0612248

[31] Drell, S. D., Levy, D. J., Yan, T. M., *Phys. Rev.* **D1** (1970) 1035, 1617.

[32] Zijlstra, E. B. and Neerven, W. L., *Phys. Lett.* **B272** (1991) 127; ibid **B273** (1991) 476; ibid **B297** (1992) 377; *Nucl. Phys.* **B383** (1992) 525; Moch, S., Vermaseren, J. A. M. and Vogt, A., *Phys. Lett.* **B606** (2005) 123; Vermaseren, J. A. M., Vogt, A. and Moch, S., *Nucl. Phys.* **B724** (2005) 3.

[33] Martin, A. D., Stirling, W. J., Thorne, E. S. and Watt, G., *Phys. Lett.* **B652** (2007) 292.

[34] Lai, H. L., *et al*, *JHEP* 0704 (2007) 089.

[35] Jimenez-Delgado, P., Reya, E., arXiv: 0810.4274.

[36] Alekhin, S. I., *Phys. Rev.* **D68** (2003) 014002; *JETP Lett.* **82** (2005) 628.

[37] Tung, W.-K., in: Handbook of QCD, Boris Ioffe Festschrift, v. 2, World Scientific, 2001, p.888.

[38] Martin, A. D., Roberts, R. G., Stirling, W. J. and Throne, R. S., *Eur. Phys. J.* **C28** (2003) 455; ibid **C35** (2004) 325.

[39] Buras, A. J., Fermilab-Pub.-79/17 - THY (1979).

[40] Hagiwara, K. *at al*, Review of Partical Physics, *Phys. Rev.* **D66** (2002) 1.

[41] Vogt, A., Moch, S. and Vermaseren, J. A. M., *Nucl. Phys.* **B688** (2004) 101; ibid **B691** (2004) 129.

[42] Vogt, A., Moch, S. and Vermaseren, J. A. M., *Phys. Lett.* **B606** (2005) 123; *Nucl. Phys.* **B724** (2005) 3.

[43] Martin, A. D., Stirling, W. J. and Thorne, R. S., *Phys. Lett.* **B635** (2006) 305.

[44] Shuryak, E. V. and Vainstein, A. I., *Nucl. Phys.* **B199** (1992) 451.

[45] Alasia, D. *et al*, *Z. Phys.* **C28** (1985) 321.

[46] Broadhurst, D. J., Kataev, A. L. and Maxwell, C. J., Proc. of Quarks-2004, ed. by D. G. Levkov, V. A. Matveev and V. A. Rubakov, INR, 2005, p. 218.

[47] Arneodo, M. *et al*, *Phys. Rev.* **D50** (1994) 1.

[48] Towell, *et al.*, FNAL/Nusea Collaboration, *Phys. Rev.* **D64** (2001) 052002.

[49] Gorishny, S. G. and Larin, S. A., *Phys. Lett.* **B172** (1986) 109; Larin, S. A. and Vermaseren, J. A. M., *Phys. Lett.* **B259** (1991) 345.

[50] Chetyrkin, K. G., Gorishny, S. G., Larin, S. A. and Tkachov, F. V., *Phys. Lett.* **B137** (1984) 230. Larin, S. A., Tkachov, F. V. and Vermaseren, J. A. M., *Phys. Rev. Lett.* **66** (1991) 862.

[51] Kataev, A. L., *Phys. Rev.* **D50** (1994) 5469.

[52] Balitsky, I. I., Braun, V. M. and Kolesnichenko, A. V., *Phys. Lett.* **B242** (1990) 245; Errata **B318** (1993) 648.

[53] Ioffe, B. L., *Phys. At. Nucl.* **58** (1995) 1408.

[54] Brooks, P. M. and Maxwell, C. J., hep-ph/0608339 (2007).

[55] Koba, Z., Nielsen, H. B. and Olesen, P., *Nucl. Phys.* **B40** (1972) 317.

[56] Kodaira, J., Matsuda, S., Sasaki, K. and Uematsu, T., *Nucl. Phys.* **B159** (1979) 99.

[57] Shuryak, E. V. and Vainstein, A. I., *Nucl. Phys.* **B201** (1982) 141.

[58] Bukhvostov, A. P., Kuraev, E. A. and Lipatov, L. N., *Sov. J. Nucl. Phys.* **38** (1983) 263; *Sov. J. Nucl. Phys.* **39** (1984) 121; *Sov. J. Phys. JETP* **60** (1984) 22.

[59] Bukhvostov, A. P., Kuraev, E. A. and Lipatov, L. N., *"Parton model and operator expansion in the quantum field theory"*, Proceedings of the 11 ITEP School on Elementary particles, (1984) (in Russian).

[60] Lampe, B., Reya, E., *Phys. Rep.* **332** (2001) 1.

[61] Bass, S. D., The spin structure of the proton, World Scientific, 2008.

[62] Bass, S. D., Ioffe, B. L., Nikolaev, N. N. and Thomas, A. W., *J. Moscow Phys. Soc.* **1** (1991) 317.

[63] Carlitz, R. D., Collins, J. C. and Mueller, A. H., *Phys. Lett.* **B214** (1988) 229.

[64] Anselmino, M., Efremov, A. and Leader, E., *Phys. Rep.* **261** (1995) 1.

[65] Zijstra, E. B. and van Neerven, W. L., *Nucl. Phys.* **B417** (1994) 61.

[66] Mertig, R. and van Neerven, W. L., *Z. Physik* **C70** (1996) 637.

[67] Vogelsang, W., *Phys. Rev.* **D54** (1996) 2023.

[68] Stoesslein, U., *Acta Phys. Polon.* **B33** (2002) 2813.

[69] Antony, P. L. *et al*, *Phys. Lett.* **B493** (2000) 19.

[70] Adeva, B. *at al*, *Phys. Rev.* **D58** (1998) 112002.

[71] Burkhardt, H. and Cottingham, W. N., *Ann. Phys. (N.Y.)* **56** (1970) 453.

[72] Antony, P. L. *at al*, *Phys. Lett.* **B553** (2003) 18.

[73] Amarian, M. *at al*, *Phys. Rev. Lett.* **92** (2004) 022301.

[74] Larin, S. A., van Ritbergen, T. and Vermaseren, J. A. M., *Phys. Lett.* **B404** (1997) 153.

[75] Hsueh, S. Y. *et al*, *Phys. Rev.* **D38** (1988) 2056.

[76] Anselmino, M., Ioffe, B. L. and Leader, E., *Yad. Fiz.* **49** (1989) 214.

[77] Burkert, V. and Ioffe, B. L., *Phys. Lett.* **B296** (1992) 223; JETP; **78** (1994) 619.

[78] Ioffe, B. L., *Yad. Fiz.* **60** (1997) 1866.

[79] Ralston, J. and Soper, D. E., *Nucl. Phys.* **B152** (1979) 109.

[80] Jaffe, R. L. and Ji, X., *Phys. Rev. Lett.* **67** (1991) 552; Nucl. Phys. **B375** (1992) 527.

[81] Artu, X. and Mekhi, M., *Z. Phys.* **C 45** (1990) 669.

[82] Ioffe, B. L. and Khodjamirian, A., *Phys. Rev.* **D51** (1995) 3373.

[83] Soffer, J., *Phys. Rev. Lett.* **74** (1995) 1292.

[84] Goldstein, G., Jaffe, R. L. and Ji, X., *Phys. Rev.* **D52** (1995) 5006.

[85] Altarelli, G., Forte, S. and Ridolfi, G., *Nucl. Phys.* **B534** (1998) 277.

[86] Bukhvostov, A. P., Frolov, G. V., Kuraev, E. A. and Lipatov, L. N., *Nucl. Phys.* **B258** (1985) 601.

[87] Belitsky, A. V. and Radyushkin, A. V., hep-ph/0504030.

[88] Efremov, A. V. and Radyushkin, A. V., *Phys. Lett.* **B94** (1980) 245; *Theor. Math. Phys.* **42** (1980) 97.

[89] Chernyak, V. L., Zhitnitsky, A. R. and Serbo, V. G., *JETP Lett.* **26** (1977) 594.

[90] Brodsky, S. J. and Lepage, G. P., *Phys. Lett.* **B87** (1979) 359; *Phys. Rev.* **D22** (1980) 2157.

[91] Farrar, G. R. and Jackson, D. R., *Phys. Rev. Lett.* **43** (1979) 246.

[92] Makeenko, Yu. M., *Sov. J. Nucl. Phys.* **33** (1981) 440; Ohrndorf, Th., *Nucl. Phys.* **B198** (1982) 26.

[93] Cvitanovic, P., *Group theory*, Nordita notes (1984).

[94] Ermolaev, B. I., Manaenkov, S. I., Ryskin, M. G., *Z. Phys.* **C69** (1996) 259; Bartels, J., Ermolaev, B. I., S. I., Ryskin, M. G., *Z. Phys.* **C70** (1996) 273, **C72** (1996) 627; Ermolaev, B. I., Greco, M., Troyan, S. I., *Nucl. Phys.* **B571** (2000) 137, **B594** (2001) 71, *Phys. Part. Nucl.* **35** (2004) S105; Bartels, J., Lublinsky, M., *JHEP* **0309** (2003) 076.

8

QCD jets

8.1 Total cross section of e^+e^--annihilation into hadrons

The total cross section $\sigma_t \equiv \sigma_{e^+e^- \to hadrons}$ of e^+e^--annihilation into hadrons is the simplest process for theoretical analysis. The asymptotic freedom of QCD allows one to perform its theoretical calculation reliably. In one-photon approximation, the matrix element of the process $e^+e^- \to X$, where X is any hadronic state, is written as

$$\mathcal{M} = -e^2 \bar{v}(k_+) \gamma_\mu u(k_-) \frac{1}{q^2} \langle X | J_\mu^{el}(0) | 0 \rangle, \tag{8.1}$$

where k_- and k_+ are the electron and positron momenta, $q = k_- + k_+$ is the virtual photon 4–momentum, and J_μ^{el} is the electromagnetic current. Neglecting the electron mass, we have for unpolarized e^+e^- beams:

$$\sigma_t = \frac{e^4}{2s^3} \left(k_{-\mu} k_{+\nu} + k_{+\mu} k_{-\nu} - \frac{q^2}{2} \delta_{\mu\nu} \right) R_{\mu\nu}, \tag{8.2}$$

where $s = q^2$ and

$$R_{\mu\nu} = (2\pi)^4 \sum_X \langle 0 | J_\nu^{el}(0) | X \rangle \langle X | J_\mu^{el}(0) | 0 \rangle \, \delta(q - p_X)$$

$$= \int d^4x \, e^{iqx} \langle 0 | \left[J_\nu^{el}(x) J_\mu^{el}(0) \right] | 0 \rangle. \tag{8.3}$$

Note that only the first term in the commutator gives a contribution for $q^0 > 0$. As in Section 7.1, use of the commutator is helpful to clarify the space-time picture of the process. In the c.m.s., where $q^0 = \sqrt{s}$, the condition of absence of oscillations $|qx| \leq 1$ implies that the main contribution to the cross section comes from the region $|x^0| \leq 1/\sqrt{s}$. Since the commutator is different from 0 only at $x^2 \geq 0$, we obtain similar restrictions on all components of x. Therefore, in contrast to the amplitude (7.2), the amplitude (8.3) is dominated by the domain near the central point $x = 0$. Using the expansion (6.1.1) for the products of the currents in (8.3), we find that operators of dimension d make contributions of the order of $s^{d_J-4-d/2}$, i.e. $s^{-1-d/2}$ since $d_J = 3$. The main contribution of the order of s^{-1} comes from the unit operator and therefore can be calculated in perturbation theory.

Due to current conservation $R_{\mu\nu}$ can be written as

$$R_{\mu\nu} = -\frac{1}{6\pi}\left(q^2\delta_{\mu\nu} - q_\mu q_\nu\right)R, \tag{8.4}$$

where

$$R = -\frac{2\pi}{q^2}R_{\mu\mu} = -\frac{2\pi}{q^2}\int d^4x\, e^{iqx}\langle 0|\left[J_\mu^{el}(x)J_\mu^{el}(0)\right]|0\rangle, \tag{8.5}$$

so that

$$\sigma_t = \frac{4\pi\alpha^2}{3s}R. \tag{8.6}$$

Now, since

$$\frac{4\pi\alpha^2}{3s} = \sigma_{e^+e^-\to\mu^+\mu^-}, \tag{8.7}$$

where $\sigma_{e^+e^-\to\mu^+\mu^-}$ is the total cross section of the process $e^+e^- \to \mu^+\mu^-$ in the Born approximation, we have

$$R = \frac{\sigma_{e^+e^-\to hadrons}}{\sigma_{e^+e^-\to\mu^+\mu^-}}. \tag{8.8}$$

Note that

$$\sigma_{e^+e^-\to\mu^+\mu^-} = \frac{86\cdot 10^{-33}}{s(\text{GeV}^2)}cm^2. \tag{8.9}$$

The variable R is related to the hadronic contribution to the photon vacuum polarization operator $\mathcal{P}$. If $\mathcal{P}$ is defined by

$$\mathcal{P}_{\mu\nu} = i\int d^4x\, e^{iqx}\langle 0|T\left(J_\mu^{el}(x)J_\nu^{el}(0)\right)|0\rangle = \left(q_\mu q_\nu - q^2\delta_{\mu\nu}\right)\mathcal{P}(q^2), \tag{8.10}$$

then

$$R = 12\pi\,\text{Im}\mathcal{P}(s)\Theta(s - s_{th}), \tag{8.11}$$

where s_{th} is the hadron-production threshold.

In the operator expansion for $J_\mu^{el}(x)J_\nu^{el}(0)$, the contribution nondecreasing with s gives only the unit operator (in perturbation theory only this operator makes a contribution). Retaining only this contribution and taking into account that the anomalous dimension of a conserved current is zero, we obtain at $m_q = 0$

$$\left(\mu\frac{\partial}{\partial\mu} + g\beta(g)\frac{\partial}{\partial g}\right)R\left(\frac{s}{\mu^2}, g\right) = 0, \tag{8.12}$$

where $g \equiv g(\mu)$, μ is the renormalization point, and we have made use of R being dimensionless. This means that R is a function of the invariant charge only:

$$R\left(\frac{s}{\mu^2}, g\right) = R\left(1, g(\sqrt{s})\right). \tag{8.13}$$

This function is known to the first three orders in α_s:

$$R = N_c \sum_i Q_i^2 \left[1 + A_1 \frac{\alpha_s}{4\pi} + A_2 \left(\frac{\alpha_s}{4\pi}\right)^2 + A_3 \left(\frac{\alpha_s}{4\pi}\right)^3 \right], \tag{8.14}$$

where the sum is over quark flavours, Q_f is the electric charge of flavour f. The coefficient A_1 is scheme independent. It was found ([1], [2]) long ago:

$$A_1 = 3C_F, \quad C_F = \frac{N_c^2 - 1}{2N_c} = \frac{4}{3}. \tag{8.15}$$

Indeed, up to the colour coefficient it coincides with the QED result [3]. For massive quarks, the first-order correction has also been calculated [4],[5].

The coefficients A_2 and A_3 are scheme dependent. In the $\overline{\text{MS}}$ scheme, the first of them is ([6]–[8]):

$$A_2^{\overline{MS}} = -\frac{3}{2}C_F^2 + C_F C_A \left(\frac{123}{2} - 44\zeta(3)\right) + C_F n_f \left(-11 + 8\zeta(3)\right), \tag{8.16}$$

where $\zeta(n)$ is the Rieman zeta-function, $\zeta(3) \simeq 1.202$. The latter coefficient, A_3, is ([9],[10]):

$$\begin{aligned}
A_3^{\overline{MS}} = &-\frac{69}{2}C_F^3 + C_F^2 C_A \left(-127 - 572\zeta(3) + 880\zeta(5)\right) \\
&+ C_F C_A^2 \left(\frac{90445}{54} - \frac{10948}{9}\zeta(3) - \frac{440}{3}\zeta(5)\right) \\
&+ C_F^2 n_f \left(-\frac{29}{2} + 152\zeta(3) - 160\zeta(5)\right) \\
&+ C_F C_A n_f \left(-\frac{15520}{27} + \frac{3584}{9}\zeta(3) + \frac{80}{3}\zeta(5)\right) \\
&+ C_F n_f^2 \left(\frac{1208}{27} - \frac{304}{9}\zeta(3)\right) - \pi^2 C_f \left(\frac{11}{3}C_A - \frac{2}{3}n_f\right)^2 \\
&+ \frac{(\sum_i Q_i)^2}{\sum_i Q_i^2} \frac{(N_c^2 - 1)(N_c^2 - 4)}{16N_c^2} \left(\frac{176}{3} - 128\zeta(3)\right),
\end{aligned} \tag{8.17}$$

$\zeta(5) \simeq 1.037$. Numerically at $N_c = 3$

$$\begin{aligned}
R = 3 \sum_i Q_i^2 &\left[1 + \frac{\alpha_s}{\pi} + \left(\frac{\alpha_s}{\pi}\right)^2 (1.986 - 0.115 n_f) \right. \\
&\left. + \left(\frac{\alpha_s}{\pi}\right)^3 \left(-6.637 - 1.200 n_f - 0.005 n_f^2\right) \right] - 1.240 \left(\sum_i Q_i\right)^2 \left(\frac{\alpha_s}{\pi}\right)^3. \tag{8.18}
\end{aligned}$$

For five active flavours it is

$$R = \frac{11}{3} \left[1 + \frac{\alpha_s}{\pi} + 1.409 \left(\frac{\alpha_s}{\pi}\right)^2 - 12.805 \left(\frac{\alpha_s}{\pi}\right)^3 \right]. \tag{8.19}$$

One could think that the relatively large value of the last coefficient is related to the asymptotic nature of the perturbation series. However, this seems implausible since the value of this coefficient comes mainly from the negative π^2 terms arising due to the analytical continuation to the physical region of energies [9].

Experimental values of R are used to determine α_s. The disadvantage of this method of determination of α_s is that R is only weakly sensitive to a variation of α_s, while the advantage is the model independence.

8.2 Jet production

The variable R gives only a quite general characteristic of the process. It is very desirable to be able to calculate more detailed properties and to have at least a qualitative picture of hadron production.

A more or less common picture here is the following: A quark q and antiquark $\bar{q}$ are created at small distances in accordance with QCD perturbation theory and then fly apart with relativistic velocities. Perturbation theory remains valid for some time during their propagation. In this time, gluon bremsstrahlung and conversion of radiated gluons into further $q\bar{q}$ pairs take place. These processes are called parton branching. While relative transverse momenta of partons produced in the branching are sufficiently large, these processes are described by perturbation theory, since the transverse momenta determine the scale of the coupling constant. However, perturbation theory stops being applicable at certain transverse momenta. Then the colour fields become deformed in such a way that they turn out to be concentrated in tubes connecting colour charges. Such localization of the fields takes place due to the nonperturbative condensate of gluon fields, analogously to the Meissner effect in a superconductor. In the simplest case of quark and antiquark flying apart, a "flux tube" or "gluon string" is stretched between them. All colour field energy E turns out to be concentrated in this string, with a constant linear energy density k, which is estimated to be $k \sim 0.2$ GeV^2, and the total energy is $E = kr$, r being the distance between the quark and antiquark. Here the strength of the quark and antiquark attraction becomes independent of r. At the subsequent flying apart of the quark and antiquark more $q\bar{q}$ pairs are created from the vacuum (the string breaks into smaller strings with $q\bar{q}$ pairs at their ends). Subsequent repeat of the process leads to blanching and hadronization. The flying apart quark and antiquark turn into hadron jets. In the case when at the first (perturbative) stage additional gluons and quark-antiquark pairs are produced, a complicated structure of chromomagnetic field develops instead of the flux tube and several jets are created.

It is possible also to speak about the last stage using the usual language of perturbation theory, without appealing to condensates and to strings. When the distance between the quark and antiquark reaches ~ 1 fm, the bremsstrahlung of gluons with $k \sim 1$ fm^{-1}, with a subsequent conversion into $q\bar{q}$ pairs occurs with a probability ~ 1 since at such k the effective coupling $\alpha_s(k)$ is large. The creation of soft gluons and $q\bar{q}$ pairs stops only when the flying apart clusters (jets) become colourless.

Evidently, the four-momentum of the jet cannot coincide with the four-momentum of the quark producing the jet. But at high energies the main difference is the result of perturbative emission of gluons and their conversion into $q\bar{q}$ pairs at the stage of parton branching, so that it is described by perturbation theory. Just parton branching determines the mean value $< k_\perp >$ of hadron transverse momenta in jets which can be estimated as

$$< k_\perp >= \frac{1}{\sigma} \int k_\perp \frac{d\sigma}{dk_\perp} dk_\perp \sim \int_0^{\sqrt{s}} k_\perp \alpha_s(k_\perp) \frac{dk_\perp}{k_\perp} \sim \frac{\sqrt{s}}{\ln\left(\frac{s}{\Lambda_{QCD}^2}\right)}. \quad (8.20)$$

As for nonperturbative effects, it is considered that they lead to transverse momenta of hadrons in jets of ~ 300 MeV.

Measurement of the probabilities of jet production makes it possible to understand the parton stage of the process. But in order to measure such probabilities, one needs to define explicitly the notion of jets and to give an algorithm of extraction of jets from data. The JADE algorithm [11] used by many experimental groups is based on the limit of the jet invariant mass: $M_{jet} \leq M_{cut}$. The procedure of extraction of jets consists of sequentially finding pairs a, b of clusters of particles (at the initial step each cluster consists of one particle) with the lowest invariant mass M_{ab} and replacing this pair by cluster c with momentum $p_c = p_a + p_b$ if $M_{ab} < M_{cut}$. Such replacements are repeated until the lowest invariant mass becomes greater than M_{cut}. The final clusters are called jets. The Durham algorithm [12] uses the values $2 \min(E_m^2, E_n^2)(1 - \cos\theta_{mn})$ instead of the invariant mass squared.

8.3 Two-jet events

It follows from the qualitative picture discussed above that at sufficiently high energies most of the hadronic final states in e^+e^--annihilation have a two-jet structure. Such structure was discovered at the SLAC electron-positron colliding beam facility SPEAR [13], [14]. The data taken at different incident beam energies from 3.0 to 7.4 GeV were used to obtain sphericity distributions. The sphericity S,

$$S = 3 \frac{\min \sum_i |p_i \times n|^2}{2 \sum_i p_i^2}, \quad (8.21)$$

served as a measure of the jet-like character of the event. In (8.21), the sum is over all hadrons produced in the event, p_i is the hadron momentum, and the minimum is found by varying the direction of the unit vector n. S is bounded between 0 and 1; events with small S are jet-like. The data show increasing with energy peak at small values of S in the sphericity distributions and decreasing mean sphericity, which is evidence for the two-jet structure. Moreover, the angular distribution of the jet axis indicates that the jets are produced by spin 1/2 partons.

At high energies, cross sections for jet production can be calculated in perturbative QCD, although it operates with massless quarks and gluons instead of massive hadrons. Of course, the cross sections must be infrared, stable and free of collinear divergences in each order of perturbation theory. This can be achieved by summing over degenerate states [15]–[17], so that the cross sections must be inclusive.

In [18], the two-jet events are defined as those in which all but a fraction $\Delta \ll 1$ of the total energy $\sqrt{s}$ is emitted within a pair of oppositely directed cones of half-angle $\delta \ll 1$, lying within two fixed cones of solid angle Ω ($\pi \delta^2 \ll \Omega \ll 1$) at an angle θ to the beam line. In Born approximation, the jet angular distribution is given by

$$\left(\frac{d\sigma}{d\Omega}\right)_B = \frac{\alpha^2}{4s} N_c \sum_f Q_f^2 \left(1 + \cos^2\theta\right), \tag{8.22}$$

where θ is the angle of the axis of the cones to the beam line. In the first order in α_s

$$\left(\frac{d\sigma}{d\Omega}\right) = \left(\frac{d\sigma}{d\Omega}\right)_B + \left(\frac{d\sigma}{d\Omega}\right)_v + \left(\frac{d\sigma}{d\Omega}\right)_r, \tag{8.23}$$

where the subscript v (r) denotes virtual (real) corrections. The virtual part is determined by the single electromagnetic quark formfactor $f_1(q^2)$ surviving in the massless case

$$\langle q(p_+)\bar{q}(p_-)|J_\mu^{el}(0)|0\rangle = \bar{u}(p_+)\gamma_\mu f_1(q^2)v(p_-). \tag{8.24}$$

Using the dimension $D = 4 - 2\epsilon$ for the regularization of both ultraviolet and infrared singularities, one has in one-loop approximation

$$f_1(q^2) = 1 - \frac{\alpha_s}{2\pi} C_F \left(\frac{4\pi\mu^2}{-q^2 - i0}\right)^\epsilon \Gamma(1+\epsilon) \frac{\Gamma^2(1-\epsilon)}{\Gamma(2-2\epsilon)} \left(\frac{1}{\epsilon^2} - \frac{1-2\epsilon}{2\epsilon}\right), \tag{8.25}$$

so that

$$\left(\frac{d\sigma}{d\Omega}\right)_v = \left(\frac{d\sigma}{d\Omega}\right)_B \frac{\alpha_s C_F}{\pi} \left(\frac{4\pi\mu^2}{s}\right)^\epsilon \Gamma(1+\epsilon) \left(-\frac{1}{\epsilon^2} - \frac{3}{2\epsilon} - 4 + \frac{2\pi^2}{3} + \mathcal{O}(\epsilon)\right). \tag{8.26}$$

Here and below, the terms singular at $\epsilon = 0$ are concerned with gluons which are either soft or collinear with the quark or antiquark. The $1/\epsilon^2$ term comes from soft-collinear emission. Note that, for consistency, one has to take $\left(\frac{d\sigma}{d\Omega}\right)_B$ also at the dimension D, i.e.

$$\left(\frac{d\sigma}{d\Omega}\right)_B = \frac{\alpha^2}{4s} N_c \sum_f Q_f^2 (1 - 2\epsilon + \cos^2\theta) \frac{(4\pi)^\epsilon}{\Gamma(1-\epsilon)} \left(\frac{s}{4\mu^2} \sin^2\theta\right)^{-\epsilon}. \tag{8.27}$$

Here we have used that $\delta_\mu^\mu = 4 - 2\epsilon$ and in the system with $q = 0$

$$(2\pi)\delta((q-p)^2) \frac{d^{D-1}p}{2\epsilon(2\pi)^{D-1}} = \frac{d\Omega}{32\pi^2} \frac{(4\pi)^\epsilon}{\Gamma(1-\epsilon)} \left(\frac{s}{4} \sin^2\theta\right)^{-\epsilon}. \tag{8.28}$$

Thus the total cross section in Born approximation is

$$\sigma_t^B = \frac{4\pi\alpha^2}{3s} N_c \sum_f Q_f^2 \left(\frac{s}{4\pi\mu^2}\right)^{-\epsilon} \frac{3(1-\epsilon)}{(3-2\epsilon)} \frac{\Gamma(2-\epsilon)}{\Gamma(2-2\epsilon)}. \tag{8.29}$$

The contribution of real gluon emission can be divided into two parts:

$$\left(\frac{d\sigma}{d\Omega}\right)_r = \left(\frac{d\sigma}{d\Omega}\right)_s + \left(\frac{d\sigma}{d\Omega}\right)_c, \tag{8.30}$$

where the first part takes into account emission of gluons with energy $\omega \leq \Delta\sqrt{s}$ (soft emission) and the second is related to emission of gluons with $\omega > \Delta\sqrt{s}$ in cones with half-angles δ (collinear emission). Using the well-known expression for the soft emission probability

$$dW_s(k) = -4\pi\alpha_s C_F \left(\frac{p_-\mu}{p_-k} - \frac{p_+\mu}{p_+k}\right)^2 \frac{\mu^{2\epsilon}d^{D-1}k}{2\omega(2\pi)^{D-1}}, \tag{8.31}$$

where p_+ and p_- are the quark and antiquark momenta, and noting that

$$\frac{\mu^{2\epsilon}d^{D-1}k}{2\omega(2\pi)^{D-1}} = \frac{\omega^2}{8\pi^2}\left(\frac{4\pi\mu^2}{s}\right)^\epsilon \frac{2^{2\epsilon}}{\Gamma(1-\epsilon)}\frac{dx}{x^{1+2\epsilon}}\frac{d\cos\theta_k}{\sin^{2\epsilon}\theta_k}, \tag{8.32}$$

where $x = 2\omega/\sqrt{s}$ and θ_k is the emission angle, we obtain

$$\left(\frac{d\sigma}{d\Omega}\right)_s = \left(\frac{d\sigma}{d\Omega}\right)_B \frac{\alpha_s}{\pi}C_F\left(\frac{4\pi\mu^2}{s}\right)^\epsilon \frac{2^{1+2\epsilon}}{\Gamma(1-\epsilon)}\int_0^{2\Delta}\frac{dx}{x^{1+2\epsilon}}\int_0^\pi \frac{d\theta_k}{\sin^{1+2\epsilon}\theta_k}$$

$$\simeq \left(\frac{d\sigma}{d\Omega}\right)_B \frac{\alpha_s}{\pi}C_F\left(\frac{4\pi\mu^2}{s}\right)^\epsilon \Gamma(1+\epsilon)\left(\frac{(2\Delta)^{-2\epsilon}}{\epsilon^2} - \frac{\pi^2}{3}\right). \tag{8.33}$$

The collinear part can be easily calculated using the quasireal electron method [19]. Note, however, that the emission probabilities introduced in [19] must be taken with account of the colour factor and at $D = 4 - 2\epsilon$. For emission along the quark momenta, we have

$$\frac{2\omega(2\pi)^{D-1}dW_c(k)}{\mu^{2\epsilon}d^{D-1}k} = \frac{4\pi\alpha_s C_f}{1-x}\frac{1}{4(p_+k)^2}\frac{1}{2}\sum_{pol}|\bar{u}(p_+)\,\not{\epsilon}^*(k)u(p)|^2$$

$$\simeq 4\pi\alpha_s C_f\frac{1}{p_+k}\frac{1+(1-x)^2-\epsilon x^2}{x(1-x)}, \tag{8.34}$$

where $p = p_+ + k = -p_-$, the sum goes over polarizations, and only physical gluon polarizations are used. The factor $(1-x)^{-1}$ takes into account that in the quark phase-space volume $\epsilon_+ \simeq (1-x)\epsilon$. The integration must be performed with respect to $\theta_k \equiv \theta_{(k,\,p)}$. Since

$$x\theta_{(k,\,p)} = (1-x)\theta_{(p_+,\,p)}, \quad \theta_{(k,\,p_+)} = \theta_{(k,\,p)} + \theta_{(p_+,\,p)} = \frac{\theta_{(k,\,p)}}{(1-x)}, \tag{8.35}$$

we obtain

$$\left(\frac{d\sigma}{d\Omega}\right)_c = \left(\frac{d\sigma}{d\Omega}\right)_B \frac{\alpha_s}{\pi}C_F\left(\frac{4\pi\mu^2}{s}\right)^\epsilon \frac{2^{1+2\epsilon}}{\Gamma(1-\epsilon)}$$

$$\times \int_{2\Delta}^1 \frac{dx\left(1+(1-x)^2-\epsilon x^2\right)}{x^{1+2\epsilon}}\int_0^{2(1-x)\delta}\frac{d\theta_k}{\theta_k^{1+2\epsilon}}$$

$$\simeq \left(\frac{d\sigma}{d\Omega}\right)_B \frac{\alpha_s}{\pi}C_F\left(\frac{4\pi\mu^2}{s}\right)^\epsilon \Gamma(1+\epsilon)$$

$$\times \left(\frac{\delta^{-2\epsilon}}{\epsilon}\left(2\ln(2\Delta) + \frac{3}{2}\right) - 2\ln^2(2\Delta) - \frac{2\pi^2}{3} + \frac{13}{2}\right), \tag{8.36}$$

so that the real emission gives

$$
\left(\frac{d\sigma}{d\Omega}\right)_r = \left(\frac{d\sigma}{d\Omega}\right)_B \frac{\alpha_s}{\pi} C_F \left(\frac{4\pi\mu^2}{s}\right)^\epsilon \Gamma(1+\epsilon)
$$
$$
\times \left(\frac{1}{\epsilon^2} + \frac{3}{2\epsilon} - 4\ln\delta\ln 2\Delta - 3\ln\delta - \pi^2 + \frac{13}{2}\right). \tag{8.37}
$$

The final result is

$$
\left(\frac{d\sigma}{d\Omega}\right) = \left(\frac{d\sigma}{d\Omega}\right)_B \left(1 - \frac{\alpha_s}{\pi} C_F \left(4\ln\delta\ln 2\Delta + 3\ln\delta + \frac{\pi^2}{3} - \frac{5}{2}\right)\right). \tag{8.38}
$$

Here the term with $\ln\delta\ln\Delta$ is connected with the suppression of the soft-collinear emission. Its coefficient can be easily obtained using the soft-emission probability (8.31):

$$
4\frac{\alpha_s}{\pi} C_F = 2\frac{dW_s(k)}{d\ln x\, d\ln\theta_k}\bigg|_{\theta_k \ll 1}. \tag{8.39}
$$

The infrared and collinear singularities are cancelled in accordance with the Kinoshita–Lee–Nauenberg theorem [15]–[17]. At high energies and not too small Δ and δ the two-jet cross section is insensitive to large distances and can be calculated order by order in perturbation theory. Evidently, the angular distribution of jets coincides with that of the primary partons. Experimental investigations of the angular distribution therefore elucidate the nature of the partons.

The Sterman–Weinberg definition of two-jet events is not the only possible jet definition. Another definition [20] is based on an invariant mass constraint. Two-jet events are defined as those in which all produced particles can be combined into two clusters with invariant masses less than $y_0 s$.

The calculation of the two-jet cross section can be simplified by integration over all angular correlations of the produced jets with the initial particles. Taking account of the α_s-correction we have

$$
\sigma_{2-jet} = \sigma_{2-jet}(q\bar{q}) + \sigma_{2-jet}(q\bar{q}g), \tag{8.40}
$$

where $\sigma_{2-jet}(q\bar{q})$ coincides with the total cross section of quark-antiquark production and $\sigma_{2-jet}(q\bar{q}g)$ is the cross section of quark-antiquark-gluon production integrated over the two-jet region. Using (8.25) we have for the former term

$$
\sigma_{2-jet}(q\bar{q}) = \sigma_t^B \left(1 + \frac{\alpha_s C_F}{\pi}\left(\frac{4\pi\mu^2}{s}\right)^\epsilon \Gamma(1+\epsilon)\left(-\frac{1}{\epsilon^2} - \frac{3}{2\epsilon} - 4 + \frac{2\pi^2}{3} + \mathcal{O}(\epsilon)\right)\right). \tag{8.41}
$$

For the cross section of $q\bar{q}g$ production, we have

$$
\sigma_{2-jet}(q\bar{q}g) = \frac{e^4}{2s^3}\left(k_{-\mu}k_{+\nu} + k_{+\mu}k_{-\nu} - \frac{q^2}{2}\delta_{\mu\nu}\right) R_{\mu\nu}^{q\bar{q}g}, \tag{8.42}
$$

where $R_{\mu\nu}^{q\bar{q}g}$ is obtained from (8.3) if the sum over X contains only $q\bar{q}g$ states with momenta in the two-jet region:

$$R_{\mu\nu}^{q\bar{q}g} = (2\pi)^D \sum_{q\bar{q}g} \langle 0|J_\nu^{el}(0)|q\bar{q}g\rangle \langle q\bar{q}g|J_\mu^{el}(0)|0\rangle \, \delta(q - p_+ - p_- - k). \qquad (8.43)$$

Here p_+, p_-, and k are the quark, antiquark, and gluon momenta, $\sum$ means the sum over their polarizations, colours, and flavours, as well as integration over the phase space

$$d\rho_{q\bar{q}g} = \frac{d^{D-1}p_+}{(2\pi)^{D-1}2\epsilon_+} \frac{d^{D-1}p_-}{(2\pi)^{D-1}2\epsilon_-} \frac{d^{D-1}k}{(2\pi)^{D-1}2\omega} \qquad (8.44)$$

with the constraint

$$x_\pm = 2p_\pm q/s \geq 1 - y_0, \quad x = 2kq/s = 2 - x_+ - x_- \geq 1 - y_0,$$
$$q = k_+ + k_-, \quad q^2 = s, \qquad (8.45)$$

and

$$\langle 0|J_\mu^{el}(0)|q\bar{q}g\rangle = Q_f \, g\mu^\epsilon \bar{u}(\mathbf{p}_+)t^a \left[\frac{\displaystyle{\not}\epsilon^*(k)(\displaystyle{\not}p_+ + \displaystyle{\not}k)\gamma_\mu}{2(p_+k)} - \frac{\gamma_\mu(\displaystyle{\not}p_- + \displaystyle{\not}k)\,\displaystyle{\not}\epsilon^*(k)}{2(p_-k)} \right] v(\mathbf{p}_-),$$
$$(8.46)$$

where e is the gluon-polarization vector and a is the colour index. Since the constraint (8.45) is Lorentz invariant and depends only on q, we have from current conservation

$$R_{\mu\nu}^{q\bar{q}g} = \frac{1}{D-1}\left(\delta_{\mu\nu} - \frac{q_\mu q_\nu}{q^2}\right) R_{\rho\rho}. \qquad (8.47)$$

Using the expression

$$(2\pi)^D \int \delta(q - p_+ - p_- - k)d\rho_{q\bar{q}g} = \left(\frac{4\pi}{s}\right)^{2\epsilon} \int \frac{s dx_+ dx_- \left((1-x)(1-x_+)(1-x_-)\right)^{-\epsilon}}{2(4\pi)^3\Gamma(2-2\epsilon)}$$
$$(8.48)$$

for the phase-space element and

$$\sum \langle 0|J_\mu^{el}(0)|q\bar{q}g\rangle \langle q\bar{q}g|J_\mu^{el}(0)|0\rangle = Q_f^2 g^2 \mu^{2\epsilon} N_c C_F \frac{-8}{(1-x_+)(1-x_-)}$$
$$\times \left[x_+^2 + x_-^2 - 2\epsilon \left((1-x_+)^2 \right.\right.$$
$$\left.\left. +(1-x_-)^2 + x_+ x_-\right) + \epsilon^2 x^2 \right] \quad (8.49)$$

for the sum over polarizations and colours, we have

$$\sigma_{2-jet}(q\bar{q}g) = \sigma_t^B \frac{\alpha_s C_F}{\pi} \left(\frac{4\pi\mu^2}{s}\right)^\epsilon \frac{3(1-\epsilon)}{2(3-2\epsilon)\Gamma(2-2\epsilon)} \int \frac{dx_+ dx_- (1-x)^{-\epsilon}}{((1-x_+)(1-x_-))^{1+\epsilon}}$$

$$\times \left[x_+^2 + x_-^2 - 2\epsilon\left((1-x_+)^2 + (1-x_-)^2 + x_+ x_-\right) + \epsilon^2 x^2\right], \quad (8.50)$$

where the integration region is defined by (8.45). In the limit $\epsilon \to 0$ (8.50) turns into the result [24],[25]

$$d\sigma_{e^+ e^- \to q\bar{q}g} = \sigma_t^B \frac{\alpha_s}{2\pi} C_F \frac{dx_+ dx_- (x_+^2 + x_-^2)}{(1-x_+)(1-x_-)}. \quad (8.51)$$

Now, performing the integration over x_+ and x_- with the constraint (8.45), one obtains in the limit $\epsilon \to 0$ for $y_0 \ll 1$

$$\sigma_{2-jet}(q\bar{q}g) = \sigma_t^B \frac{\alpha_s C_F}{\pi} \left(\frac{4\pi\mu^2}{s}\right)^\epsilon \Gamma(1+\epsilon) \left(\frac{1}{\epsilon^2} + \frac{3}{2\epsilon} - \ln^2 y_0 - \frac{3}{2}\ln y_0 + \frac{7}{2} - \frac{\pi^2}{2}\right). \quad (8.52)$$

Putting together (8.41) and (8.52), we come to the result [20]

$$\sigma_{2-jet} = \sigma_t^B \left[1 - \frac{\alpha_s C_F}{\pi} \left(\ln^2 y_0 + \frac{3}{2}\ln y_0 + \frac{1}{2} - \frac{\pi^2}{6}\right)\right]. \quad (8.53)$$

Note that this result can be reached with much less effort if the total cross section σ_t is known with one-loop accuracy. Indeed, in α_s-order the two-jet cross section is given by the difference of the total cross section and the cross section for $q\bar{q}g$ production integrated over the region complementary to the two-jet region defined by (8.45). The latter cross section contains neither soft nor collinear singularities and can be calculated directly with $D = 4$ using the result (8.51).

The two-jet cross section to α_s-order does not require renormalization and, strictly speaking, cannot be used to determine the coupling constant or the QCD scale parameter Λ_{QCD}. Knowledge of the α_s^2-corrections is very useful from this point of view. It is required also to understand the convergence of the perturbation series.

The α_s^2-corrections to the two-jet cross section were calculated in [21],[22] for both the Sterman–Weinberg [18] and the invariant mass [20] definitions of two-jet events. In both cases, the angular correlations with $e^+ e^-$ beams were integrated out. All calculations were performed in the Feynman gauge using dimensional regularization for both ultraviolet and infrared divergences and $\overline{\text{MS}}$ renormalization scheme. The α_s^2-corrections are given by the sum of the three parts: the two-loop contribution to $\sigma_{2-jet}(q\bar{q})$, the one-loop contribution to $\sigma_{2-jet}(q\bar{q}g)$, and the four-parton contributions $\sigma_{2-jet}(q\bar{q}gg)$ and $\sigma_{2-jet}(q\bar{q}q\bar{q})$. The calculation of $\sigma_{2-jet}(q\bar{q})$ was performed also in [23]. The one-loop corrections to the differential cross section $d\sigma(q\bar{q}g)$ were calculated previously [26]–[35] in the limit $\epsilon \to 0$; but to find $\sigma_{2-jet}(q\bar{q}g)$ with α_s^2-accuracy the order ϵ terms in $d\sigma(q\bar{q}g)$ are necessary, and therefore $d\sigma(q\bar{q}g)$ was recalculated in [21],[22] taking account of these terms. The cross sections for the processes with four-parton final states were found in [36],[37],[27],[28].

In the case of the invariant mass definition of two-jet events, the result for the α_s^2-correction in the $\overline{\text{MS}}$ renormalization scheme with $\mu^2 = s$ is of the following form:

$$\sigma_{2-jet}^{\alpha_s^2} = \sigma_t^B \left(\frac{\alpha_s}{\pi}\right)^2 C_F \left\{ C_F \left[2\ln^4 y_0 + 6\ln^3 y_0 + \left(\frac{13}{2} - 6\zeta(2)\right)\ln^2 y_0 \right.\right.$$
$$\left. + \left(\frac{9}{4} - 3\zeta(2) - 12\zeta(3)\right)\ln y_0 + \frac{1}{8} - \frac{51}{4}\zeta(2) + 11\zeta(3) + 4\zeta(4)\right]$$
$$+ N_c \left[-\frac{11}{3}\ln^3 y_0 + \left(2\zeta(2) - \frac{169}{26}\right)\ln^2 y_0 \right.$$
$$\left. + \left(6\zeta(3) - \frac{57}{4}\right)\ln y_0 + \frac{31}{9} + \frac{32}{3}\zeta(2) - 13\zeta(3) + \frac{45}{2}\zeta(4)\right]$$
$$\left. + \frac{n_f}{2}\left[-\frac{4}{3}\ln^3 y_0 + \frac{11}{9}\ln^2 y_0 + 5\ln y_0 + \frac{19}{9} - \frac{38}{9}\zeta(2)\right]\right\}. \qquad (8.54)$$

The α_s^2-correction is positive and not large. For $\alpha_s = 0.12$ and $n_f = 5$ at $y_0 = 0.05$ it amounts to 4%, whereas the α_s-correction is about 19%.

8.4 Three-jet events

Existence of three-jet events due to hard single-gluon bremsstrahlung off quark or anti-quark at small distances before hadronization was predicted [1] even before the discovery of the two-jet structure of hadronic final states in e^+e^--annihilation. Soon after its discovery, it was pointed out [24] that it is possible to calculate cross sections for multijet production within perturbative QCD. Experimentally the three-jet events were observed by several groups [38],[39],[40] at PETRA. All groups found a good qualitative agreement of observed jet properties with the lowest-order QCD perturbation theory. The rate and shape of the three-jet events were compared with the distribution (8.51). Peaks at $x_{\pm} = 1$ in this distribution are distinctive for spin 1. Comparison of the data with models has given the evidence of spin 1 for the intermediary of the strong interactions. This observation is now called gluon discovery.

But higher-order corrections must be included for a quantitative comparison and for a determination of α_s. In the next order in α_s, one needs to know one-loop corrections to $q\bar{q}g$ production (calculated in [26]–[35]) as well as the cross sections for processes with four-parton final states (found in [36],[37],[27],[28]). The analytic calculation of the three-jet cross section to order α_s^2 with all angular correlations integrated out was performed in [26],[30]–[32],[34]. As in the two-jet case, various theoretical definitions of three-jet events are possible. In [26],[31] the Sterman–Weinberg type definition was used. The three-jet cross section was defined as the cross section of events which have all but a fraction $\vartheta/2$ of the total energy distributed within three separate cones of full opening angle δ. The corrections are small for not too small ϑ, δ and negative, as one would have expected. The results were used to fit the data from the detector PLUTO at the storage ring PETRA: at $Q = \sqrt{s} = 30, \vartheta = 0.2, \delta = 45°$. In the $\overline{\text{MS}}$ scheme this gives $\alpha_s = 0.17$.

The three-jet cross section for invariant mass cut-off was calculated in [30],[32],[34]. In [34] the three-jet events were defined as those which consist of three clusters, each having an invariant mass smaller than y_{0s}. The "dressed variables" x_1, x_2, and x_3 were introduced as the ratios of twice the energy of the parent quark, antiquark and gluon cluster, respectively, to the total energy. This definition leads to a simpler expression for the cross section than that of the Sterman–Weinberg type:

$$\frac{d\sigma_{3-jet}}{dx_1 dx_2} = \sigma_t^B \frac{\alpha_s(s)}{2\pi} C_F \left\{ \frac{(x_1^2 + x_2^2)}{(1 - x_1)(1 - x_2)} \left[1 + \frac{\alpha_s(s)}{2\pi} (J_1 + J_2 + J_3) \right] \right. $$
$$\left. + \frac{\alpha_s(s)}{2\pi} f(x_1, x_2) \right\}, \tag{8.55}$$

where

$$J_1 = C_F \left[-2 \ln^2 \frac{y_0}{1 - x_3} - 3 \ln y_0 - 1 + 2\zeta(2) + \frac{2y_0}{1 - x_3} \ln \frac{y_0^2}{1 - x_3} \right], \tag{8.56}$$

$$J_2 = N_C \left[\ln^2 \frac{y_0}{1 - x_3} - \ln^2 \frac{y_0}{1 - x_1} - \ln^2 \frac{y_0}{1 - x_2} - \frac{11}{6} \ln y_0 + \frac{67}{18} \right.$$
$$\left. + \zeta(2) - \frac{y_0}{1 - x_3} \ln \frac{y_0^2}{1 - x_3} + \frac{y_0}{1 - x_1} \ln \frac{y_0^2}{1 - x_1} + \frac{y_0}{1 - x_2} \ln \frac{y_0^2}{1 - x_2} \right], \tag{8.57}$$

$$J_3 = \frac{n_f}{2} \left[\frac{2}{3} \ln y_0 - \frac{10}{9} \right], \tag{8.58}$$

$$f(x_1, x_2) = C_F \left[(1 - x_3) \left(\frac{2}{x_3} - \frac{1 - x_3}{x_1(1 - x_2)} \right) - \frac{1 - x_2}{1 - x_1} \right] + N_c \frac{1 - x_1}{x_3(1 - x_2)}$$
$$+ \ln(1 - x_2) \left[N_c \frac{1 - x_2}{x_2} + C_F \left(2 - \frac{(1 - 3x_1)}{x_2} - \frac{1 - x_1}{x_2^2} \right) \right]$$
$$+ (2C_F - N_c) \frac{1 - x_3^2}{x_3^2} \ln(1 - x_3) - N_c \frac{x_1^2}{(1 - x_1)(1 - x_2)} r(x_1, x_2)$$
$$- (2C_F - N_c) \frac{(1 - x_3)^2 + x_1^2}{(1 - x_1)(1 - x_2)} r(x_1, x_3) + (x_1 \leftrightarrow x_2), \tag{8.59}$$

with

$$r(x, y) = \ln(1 - x) \ln(1 - y) + Li_2(x) + Li_2(y) - \frac{\pi^2}{6},$$

$$Li_2(x) = -\int_0^1 \frac{dt}{t} \ln(1 - xt). \tag{8.60}$$

The corrections are negative and small for not too small y_0.

As mentioned before, the first observation of three-jet events is considered to be the discovery of the gluon. Measurement of probabilities of four-jet events provides data on the gluon self-interaction. The limits on C_A/C_F were obtained from an analysis of the ratio of the probability of four-jet events, when two of the jets are of gluon origin, to the squared probability of three-jet events. The limits on T_R/C_F (T_R is defined by the relation $\text{Tr}\, t^a t^b = 1/2 \delta^{ab} T_R$) were obtained from the ratio of the probability of events with four-quark jets to the squared probability of three-jet events.

8.5 Event shape

Two factors affecting the event shape are obvious: the shape of events at the parton level and the hadronization process. Experimental study of hadron-event shapes provide data on both the parton stage of a process and on hadronization. Evidently, we cannot define completely the shape of hadronic events, since we cannot describe the hadronization process. However, it is possible to calculate some inclusive characteristics of the shape. Doing so, the calculations are carried out on the parton level. It is supposed that the hadronization process does not affect them.

A typical value used to describe the hadron event shape is the thrust T [41] defined by

$$T = max \frac{\sum |\boldsymbol{pn}|}{\sum |\boldsymbol{p}|}, \tag{8.61}$$

where the sum is taken over all hadrons produced in a given event, and the maximum is found by varying the direction of the unit vector $\boldsymbol{n}$.[1] Evidently, $T = 1$ for ideal two-jet events; for three-jet events $2/3 < T < 1$, and for spherically symmetric events $T = 1/2$.

The thrust T belongs to the observables called "infrared and collinear stable," which means that they are insensitive to the emission of gluons of small frequencies and/or at small angles. Singular contributions from the infrared and collinear regions, existing separately in virtual radiative corrections and in radiative corrections owing to real gluon emission, cancel in these variables. Therefore, the cross section

$$\sigma(\tau) = \int_{1-\tau}^{1} \frac{d\sigma}{dT} dT, \tag{8.62}$$

is finite order by order in perturbation theory and can be calculated at the parton level. In one-loop approximation, only quark-antiquark and quark-antiquark-gluon production contribute to (8.62). Actually, it is more convenient to calculate the quantity

$$r(\tau) = 1 - \frac{\sigma(\tau)}{\sigma_t} = \frac{1}{\sigma_t} \int_0^{1-\tau} \frac{d\sigma}{dT} dT. \tag{8.63}$$

The latter equality follows from $\sigma(1) = \sigma_t$. In the first order in α_s, one has to know only the quark-antiquark-gluon production cross section (8.51). It is easy to understand that for the process $e^+ e^- \to q \bar{q} g$ we have

$$T = max\{x_+, \ x_-, \ x\}, \tag{8.64}$$

---

[1] Actually, in [41], the variable $d = max \sum(\boldsymbol{pn})\theta(\boldsymbol{pn})/\sum |\boldsymbol{p}|$ was used, which is smaller than T by a factor of two.

such that

$$
\begin{aligned}
r(\tau) &= \frac{\alpha_s}{2\pi} C_F \int dx_- dx_+ \frac{x_+^2 + x_-^2}{(1 - x_+)(1 - x_-)} \\
&\quad \times \theta(1 - \tau - x_-)\theta(1 - \tau - x_+)\theta(x_- + x_+ - 1 - \tau) \\
&= \frac{\alpha_s}{2\pi} C_F \int_{2\tau}^{(1-\tau)} \frac{dx_+}{1 - x_+} \left((1 + x_+^2) \ln\left(\frac{x_+ - \tau}{\tau} \right) + 4\tau - 2x_+ - \tau x_+ + \frac{x_+^2}{2} \right).
\end{aligned}
$$

(8.65)

In the next order in α_s, one needs to know one-loop corrections to the three-parton production process as well as the cross sections for the processes with four-parton final states. The order α_s-correction to the cross section $d\sigma_{e^+e^- \to q\bar{q}g}$ was found in [27],[28],[31]; the cross sections $d\sigma_{e^+e^- \to q\bar{q}gg}$ and $d\sigma_{e^+e^- \to q\bar{q}q\bar{q}}$ were calculated in [37],[36],[27],[28]. The thrust distribution to order α_s^2 has been found in [42] (see also [27],[28],[30],[29]). The α_s^2-corrections are large and positive, in contrast to small negative corrections to two- and three-jet cross sections. As was emphasized in [31],[34], the large corrections indicate sensitivity of the thrust to emission of soft and collinear gluons. (This applies also to sphericity, acoplanarity, and most other "bare" shape variables, i.e. variables calculated using parton momenta.) It was advocated to use "dressed" shape variables calculated using the jet cross sections and jet momenta instead of the parton ones. In [34], it was demonstrated explicitly that the large α_s^2-corrections to bare thrust-like distributions arise primarily from rather soft and collinear partons. This was done adding three- and four-jet cross sections and taking the zero limit of the jet-resolution parameter. In this limit the large corrections were recovered.

At $\tau \to 0$, the expansion parameter in the thrust distribution is $\alpha_s(Q) \ln^2 \tau$. The double-logarithmic terms have the same nature as the Sudakov formfactor, i.e. they are related to the suppression of soft-collinear radiation. The double-logarithmic terms are easily resummed. It is possible to resum not only these but all terms of the type $\alpha_s^n \ln^m (1 - T)$ with $m \geq n$ [43]. The resummed result gives better agreement with the data at large values of T.

8.6 Inclusive spectra

The variable R is the most general (totally inclusive) property of the process. Next in degree of specification is the inclusive cross section $d\sigma_h$ of the hadron h-production process $e^+e^- \to h + X$. The corresponding matrix element is obtained from (8.1) by extraction of the hadron h:

$$
\mathcal{M} = -e^2 \bar{v}(p_+) \gamma_\mu u(p_-) \frac{1}{q^2} \langle hX | J_\mu^{el}(0) | 0 \rangle.
$$

(8.66)

Denoting the hadron momentum by p, we obtain

$$
\epsilon_p \frac{d^3 \sigma_h}{d^3 p} = \frac{4\alpha^2}{\pi s^3} \left(k_{-\mu} k_{+\nu} + k_{+\mu} k_{-\nu} - \frac{q^2}{2} \delta_{\mu\nu} \right) \overline{W}_{\mu\nu},
$$

(8.67)

where the hadronic tensor $\overline{W}_{\mu\nu}$ is defined in Chapter 7, Eq.(7.26). Note that in contrast to the tensor $W_{\mu\nu}$ for deep inelastic scattering of electrons (Chapter 7, Eq.(9)), $\overline{W}_{\mu\nu}$ is not expressed in terms of the matrix element of the current product, since the sum goes not over all states but only over states containing the hadron h with fixed momentum p. If $\overline{W}_{\mu\nu}$ and $W_{\mu\nu}$ were related by crossing (i.e. if they were values of one analytic function in different kinematical regions), then we would have the relations

$$\overline{W}_{\mu\nu}(p, q) = W_{\mu\nu}(-p, q), \quad \overline{F}_i(\nu, q^2) = F_i(-\nu, q^2). \tag{8.68}$$

Here, if we consider the s–channel physical region, then the F_i in (8.68) are outside their region of definition, so that they need to be defined. If crossing could be applied, then this would be simply analytic continuation. But actually crossing connects amplitudes of processes and not their structure functions, the definition of which includes complex conjugation, so that in the relation (8.68) the right-hand member cannot be obtained simply by analytic continuation. This is clear considering that the functions acquire imaginary parts under analytic continuation.

A possible variant of the definition is the analytic continuation with truncation of the imaginary parts. This is correct in the leading logarithmic approximation. More refined methods are required beyond the framework of this approximation.

The cross section differential in the fraction $x = 2(pq)/q^2$ of energy carried by the hadron h and in the solid angle Ω of its flight is given by

$$\frac{d\sigma_h}{x\,dx\,d\Omega} = \frac{\alpha^2}{s}\beta\left(\overline{F}_1 + \frac{x\beta^2\sin^2\theta}{4}\overline{F}_2\right), \tag{8.69}$$

where $\beta = \sqrt{1 - 4M^2/sx^2}$ is the velocity of h in the c.m.s. of the initial beams.

In the parton picture, the cross section is represented in the form of

$$\frac{d\sigma_h}{dx\,d\Omega} = \sum_a \int_x^1 \frac{dz}{z}\,\bar{f}_a^h\left(\frac{x}{z}\right)\frac{d\sigma_a}{dz\,d\Omega}, \tag{8.70}$$

where $d\sigma_a$ is the cross section for the production of type a partons, $\bar{f}_a^h$ is the fragmentation function of parton a to produce hadron h. For partons with spin $1/2$ and electric charge Ze in Born approximation we have

$$\frac{d\sigma_{1/2}}{dx\,d\Omega} = \frac{z^2\alpha^2}{4s}(2 - \beta^2\sin^2\theta)\delta(1 - x), \tag{8.71}$$

while for the partons with spin 0 the result is

$$\frac{d\sigma_0}{dx\,d\Omega} = \frac{z^2\alpha^2}{8s}\beta^3\sin^2\theta\,\delta(1 - x). \tag{8.72}$$

Experiment gives evidence of spin $1/2$ partons.

The representation (8.70) is valid in QCD, but the fragmentation functions depend on s. Evolution of the fragmentation functions in $\ln s$ has the same form as the evolution of parton distributions in hadrons [44]–[48]:

$$\frac{d\bar{f}_a^h(x, q^2)}{d\ln q^2} = \frac{\alpha_s(q^2)}{2\pi} \int_x^1 \frac{dz}{z} \bar{P}_a^b\left(\frac{x}{z}\right) \bar{f}_b^h(z, q^2),\tag{8.73}$$

where $\bar{P}_a^b(x)$ are the splitting functions for the time-like evolution. In leading order they coincide with the splitting functions $P_a^b(x)$ for the space-like evolution (Gribov–Lipatov reciprocity):

$$\bar{P}_b^a(z) = P_b^a(z),\tag{8.74}$$

such that "the number of partons in a hadron is equal to number of hadrons in a parton." The reciprocity is broken beyond the leading order [49]. However, there is hope [50] for getting simpler splitting functions satisfying the reciprocity relations (8.74) as a result of reformulation of the evolution equations (8.73) by generalizing the structure on the right-hand side. It is discussed in more detail in Chapter 7.

At sufficiently small x, when $\ln(1/x) \sim \ln(q^2/\Lambda_{QCD}^2)$, the expansion parameter in the parton distribution becomes double logarithmic. Then the evolution equations (8.73) must be modified. The reason is that in their derivation it was assumed that the angular integrations give $\ln s$. This is evidently not correct in the case of small x. Naively, one could think that the equation can be improved by changing the expansion parameter to $\ln(q^2 x^2/\Lambda_{QCD}^2)$ instead of $\ln(q^2/\Lambda_{QCD}^2)$. But this is not the case. At small x a new phenomenon called colour coherence becomes important.

8.7 Colour coherence

The simplest process where the colour coherence phenomenon becomes apparent is the two-gluon emission in the one-photon annihilation $e^+e^- \to \gamma^* \to q\bar{q}$. Let us consider this process in the center of mass system. To elucidate typical properties of the process, we restrict the consideration to the region giving the main (double-logarithmic) contribution to the cross section. In this region, which is called soft-collinear region, all gluon energies as well as their emission angles are strongly ordered. The matrix element of any hard process where the particle of momentum p is emitting a gluon with momentum k in the soft-collinear region contains the factor $1/|k_\perp|$, where the subscript $\perp$ denotes perpendicularity to p; $|k_\perp| = \omega\theta_{kp}$. The double logarithm appears as a result of integration over the energy ω and the emission angles θ_{kp} of the radiated gluon. All this is quite analogous to the QED case. However, the analogy does not work further. In QED, the soft-emission probability is given by the product of independent factors for the accompanying bremsstrahlung. In QCD, the gluon self-interaction leads to violation of the Poisson distribution for soft emission [51].

Denote the quark, antiquark, and gluon momenta by p_-, p_+, k_i, $i = 1, 2$, respectively, and the gluon-polarization vectors and colour indices by e_i and a_i. It is convenient to use some physical gauge both for the produced gluons (i.e. for their polarization vectors) and for the intermediate ones (propagators). A choice of an appropriate gauge is very helpful for the analysis. Note that one can use different gauges in different kinematical regions. A great advantage of physical gauges is the possibility for estimating matrix elements of separate diagrams. In the Feynman gauge, these matrix elements contain artificial singularities which cancel in the calculation of cross sections due to gauge invariance. These artificial singularities are related to wrong gluon helicities present in the Feynman gauge. In the massless limit, due to helicity and angular momentum conservation, gluon-emission vertices for physical polarizations vanish at collinear momenta. This is not so for unphysical polarizations. As a result, collinear singularities in matrix elements for unphysical polarizations are too strong: $1/\theta^2$ instead of $1/\theta$ for physical polarizations. The $1/\theta^2$ singularities related to the presence of unphysical polarizations in gluon propagators cancel due to gauge invariance of the total matrix element represented by the set of Feynman diagrams. This cancellation makes the consideration of matrix elements for separate diagrams meaningless even for physical polarizations of external gluons. The $1/\theta^2$ singularities related to unphysical polarizations of external gluons remain even after this. Absence of $1/\theta^4$ singularities (or suppression of terms $1/(\theta^2 + m^2/\epsilon_p^2)^2$ by factors of m^2/ϵ_p^2 for massive quarks) in cross sections after Feynman summation over polarizations is determined by cancellation of contributions of time-like and longitudinal polarizations. The main remaining $1/\theta^2$ contributions in the cross sections come from interference of different diagrams, which evidently makes a physical interpretation impossible.

Use of a physical gauge and of physical polarization vectors makes it possible to consider separately gluon emission by quark and antiquark neglecting interference. In the case when both gluons are emitted along the quark momentum, the process is depicted by the diagrams in Fig. 8.1. The matrix elements corresponding to the diagrams Fig. 8.1a and Fig. 8.1b can be easily written for any relation between ω_2 and ω_1:

$$M_a = M_\alpha^{(0)} g^2 \frac{e_2 p_-}{k_2 p_-} \frac{e_1 p_-}{(k_1 + k_2) p_-} \langle \beta | t^{a_2} t^{a_1} | \alpha \rangle,$$

$$M_b = M_\alpha^{(0)} g^2 \frac{e_1 p_-}{k_1 p_-} \frac{e_2 p_-}{(k_1 + k_2) p_-} \langle \beta | t^{a_1} t^{a_2} | \alpha \rangle, \qquad (8.75)$$

where $M_\alpha^{(0)}$ is the matrix element of the hard production of a quark in the colour state $|\alpha\rangle$, t^a are the group generators in the fundamental representation, and $|\beta\rangle$ is the final quark colour state. Of course, the sum over α is implied.

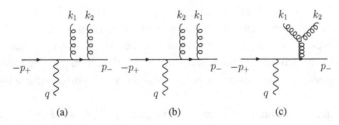

Fig. 8.1. Emission of two gluons by a quark of momentum p_- produced in e^+e^--annihilation.

To estimate the matrix element for the diagram Fig. 8.1c let us take for definiteness $\omega_2 \ll \omega_1$. Besides this, we have to specify the physical gauge for the intermediate gluon propagator. It is convenient to use the light-cone gauge with the gauge-fixing vector p_+, so that

$$d_{\mu\nu}(k) = -\delta_{\mu\nu} + \frac{k_\mu p_{+\nu} + p_{+\mu} k_\nu}{p_+ k}. \tag{8.76}$$

Let us use the decomposition

$$d_{\mu\nu}(k) = \sum_{\lambda=1}^{3} e^*_{\lambda\mu}(k) e_{\lambda\nu}(k), \tag{8.77}$$

where for $i = 1, 2$ the polarization vectors e_i are transverse to the (p_+, k)–plane and are orthonormal:

$$(p_+ e_i(k)) = (k e_i(k)) = 0, \quad (e^*_1(k) e_2(k)) = 0, \quad (e^*_1(k) e_1(k)) = (e^*_2(k) e_2(k)) = -1, \tag{8.78}$$

and $e_3(k) = p_+ \sqrt{k^2}/(k p_+)$.

Consider the convolutions of the three-gluon vertex

$$\gamma_{\mu\nu\rho}(k, -k_1, -k_2) = -\delta_{\mu\nu}(k + k_1)_\rho + \delta_{\mu\rho}(k + k_2)_\nu - \delta_{\nu\rho}(k_2 - k_1)_\mu, \tag{8.79}$$

where $k = k_1 + k_2$, with the corresponding polarization vectors, assuming that the gluon momenta are close to p_-, i.e. $(k_i p_-) \ll (k_i p_+)$ and that their energies are strongly ordered: $(k p_+) \simeq (k_1 p_+) \gg (k_2 p_+)$. Consider the most general case, when all momenta k_i are off mass-shell. Using the notation

$$\gamma_{\lambda\lambda_1\lambda_2}(k; k_1, k_2) = e_{\lambda\mu}(k) \gamma_{\mu\nu\rho}(k, -k_1, -k_2) e^*_{\lambda_1\nu}(k_1) e^*_{\lambda_2\rho}(k_2), \tag{8.80}$$

and choosing the transverse polarization vectors such that they coincide in the collinear limit, we find that for the transverse polarizations within the accuracy up to corrections suppressed either by ω_2/ω_1 or by θ_{12} we have

$$\gamma_{i i_1 i_2} \simeq \delta_{i i_1} 2(e^*_{i_2}(k_2) k_1) \sim \omega_1 \theta_{12}, \tag{8.81}$$

where θ_{12} is the angle between k_1 and k_2. It is easy to see that $\gamma_{\lambda\lambda_1\lambda_2}$ is nonzero only if no more than one polarization is not transverse; then

$$\gamma_{3 i_1 i_2} \simeq -\delta_{i_1 i_2} \sqrt{k^2}, \quad \gamma_{i 3 i_2} \simeq \delta_{i i_2} \sqrt{k_1^2}, \quad \gamma_{i i_1 3} \simeq -\delta_{i i_1} 2\sqrt{k_2^2}. \tag{8.82}$$

In the case under consideration, only the first of these terms is different from zero. Since $k^2 \simeq \omega_1 \omega_2 \theta_{12}^2$, it is easy to see that this contribution is negligible compared with (8.81), so that the matrix element corresponding to the diagram Fig. 8.1c can be written as

$$M_c = M_\alpha^{(0)} g^2 \frac{e^*_2 k_1}{k_2 k_1} \frac{e^*_1 p_-}{(k_1 + k_2) p_-} \langle \beta | [t^{a_1}, t^{a_2}] | \alpha \rangle. \tag{8.83}$$

As $2k_i p_- \simeq \epsilon \omega_i \theta_i^2$, where θ_i is the angle between $\mathbf{k}_i$ and $\mathbf{p}_-$, the region where M_c gives the double-logarithmic contribution is restricted by the inequalities

$$\frac{\omega_1}{\omega_2}\theta_1^2 \gg \theta_2^2 \gtrsim \theta_1^2. \tag{8.84}$$

Indeed, for $\theta_2^2 \ll \theta_1^2$ the denominators in (8.83) do not depend on the emission angle of the second gluon, so that integration with respect to it does not give logarithms. Similarly, when $\omega_2\theta_2^2 \gtrsim \omega_1\theta_1^2$, then integration with respect to the emission angle of the first gluon becomes nonlogarithmic.

The total contribution of the diagrams of Fig. 8.1 in the case $\omega_2 \ll \omega_1$ is given by the sum

$$M = M_a + M_b + M_c, \tag{8.85}$$

where M_a and M_b are given by (8.75). The regions where they give double-logarithmic contributions satisfy the inequalities

$$\theta_1^2 \gg \frac{\omega_2}{\omega_1}\theta_2^2 \tag{8.86}$$

and

$$\theta_2^2 \gg \frac{\omega_1}{\omega_2}\theta_1^2, \tag{8.87}$$

respectively.

Despite the simple form of M_a, M_b, and M_c, it is not suitable to use (8.85) because of the interference of M_a, M_b, and M_c. The contributions of separate Feynman diagrams interfere since the regions, where their contributions are significant, are overlapping. From (8.86) and (8.84) it is easy to see that the double-logarithmic regions for diagrams Fig. 8.1a and Fig. 8.1c indeed overlap. Recall that we consider the case $\omega_1 \gg \omega_2$; evidently, in the opposite case, it would be necessary to account for the interference of the contributions of Fig. 8.1b and Fig. 8.1c.

It is possible, however, to simplify the matrix element (8.85) dividing the momentum space into nonoverlapping regions according to angular ordering of the emitted gluons [52]. For the case of two-gluon emission in the cone along p_- there are three such regions, regions I, II, and III:

$$\text{I. } \theta_1 \gg \theta_2,$$

$$\text{II. } \theta_{12} \ll \theta_2 \simeq \theta_1,$$

$$\text{III. } \theta_2 \gg \theta_1. \tag{8.88}$$

In each of these regions, M acquires the simple form of M_0 times corresponding factors of the accompanying bremsstrahlung.

Indeed, it is easy to see that in region I only M_a is essential; taking into account that in this region $k_1 p_- \gg k_2 p_-$, we find

$$M_I = M_\alpha^{(0)} g^2 \frac{e_2^* p_-}{k_2 p_-} \frac{e_1^* p_-}{k_1 p_-} \langle \beta | t^{a_2} t^{a_1} | \alpha \rangle. \tag{8.89}$$

Evidently, in region II the whole amplitude is given by M_c; since in this region also $k_1 p_- \gg k_2 p_-$, we have

$$M_{II} = M_\alpha^{(0)} g^2 \frac{e_2^* k_1}{k_2 k_1} \frac{e_1^* p_-}{k_1 p_-} T_{a_1 c}^{a_2} \langle \beta | t^c | \alpha \rangle. \tag{8.90}$$

Region III requires a more complicated consideration. Here we must consider two cases:

$$\text{(i)} \quad \theta_2^2 \gg \frac{\omega_1}{\omega_2} \theta_1^2 \gg \theta_1^2, \quad \text{(ii)} \quad \frac{\omega_1}{\omega_2} \theta_1^2 \gg \theta_2^2 \gg \theta_1^2. \tag{8.91}$$

Note that the region $\omega_2 \theta_2^2 \sim \omega_1 \theta_1^2$ does not require consideration because it cannot give double logarithms. In case i, only the diagram Fig. 8.1b contributes; in this case, $k_2 p_- \gg k_1 p_-$, and we have:

$$M_{III} = M_\alpha^{(0)} g^2 \frac{e_1^* p_-}{k_1 p_-} \frac{e_2^* p_-}{k_2 p_-} \langle \beta | t^{a_1} t^{a_2} | \alpha \rangle. \tag{8.92}$$

In case ii, two other diagrams of Fig. 8.1 (Fig. 8.1a and Fig. 8.1c) contribute. However, since here $\theta_{12} \simeq \theta_2$, we have

$$\frac{e_2^* k_1}{k_2 k_1} \simeq \frac{e_2^* p_-}{k_2 p_-}, \tag{8.93}$$

so that taking account of the inequality $k_1 p_- \gg k_2 p_-$ and of the commutation relations for t^a, the sum $M_a + M_c$ is given by (8.92) also in this case. Therefore, (8.92) is valid in the whole region III.

Thus, Eqs. (8.89), (8.90), and (8.92) determine the two-gluon emission amplitude in double-logarithmic approximation (DLA) for the case when gluons are emitted along the quark direction. Note that, strictly speaking, in this derivation the gauge-fixing vector p_+ was assumed for the gluon-polarization vectors e_i. It is clear, however, that the results are valid in a wide class of physical gauges. The change of the gauge-fixing vector $p_+ \to n$ means the change of the polarization vectors $e_i \to e_i - k_i(e_i n)/(k_i n)$. If the momentum of the emitting particle is p, then this change does not affect the matrix elements until

$$\left| \frac{e_i p}{k_i p} \right| \gg \left| \frac{e_i n}{k_i n} \right|. \tag{8.94}$$

This means, in particular, that we can take the gauge-fixing vector as $n = q$, where q is the virtual photon momentum. The convenience of this choice is the symmetry between quark and antiquark, which permits us to use it in any kinematical region. This choice is assumed in the following. Then the matrix elements for gluon emission by antiquarks are obtained from the corresponding matrix elements for emission by quarks using the replacement $p_- \to p_+$, $t_{\alpha\beta}^a \to -t_{\beta\alpha}^a$.

Let us emphasize that in case ii (8.91) the result (8.92) appears as the sum of the contributions of the diagrams Fig. 8.1b and Fig. 8.1c, which implies the importance of their interference. As was noted in [52], there is no gauge where the two-gluon production matrix element in region III is given by the contribution of only one Feynman diagram. This means, in particular, that the diagrams of Fig. 8.2a and Fig. 8.2b must be taken into account when calculating the total cross section and the inclusive spectrum in the DLA.

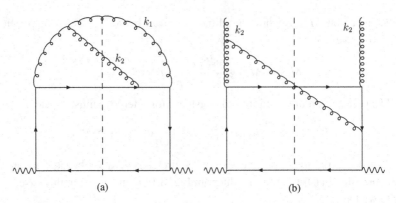

Fig. 8.2. The interference contributions in the total cross section (a) and in the inclusive spectrum (b).

At first sight this looks strange, since it is known that in physical gauges only planar diagrams contribute to the inclusive spectra in leading logarithmic approximation (see, for instance, [53]). Indeed, this is correct when the inclusive spectra are calculated at $x = \omega/\epsilon \sim 1$. At such x the case ii (8.91) does not exist at all. But studying phenomena sensitive to the small x region (such as multiplicity distributions) it is impossible to use familiar ideas of the leading logarithmic approximation and consider only planar Feynman diagrams. The importance of nonplanar Feynman diagrams for the evaluation of jet multiplicity was realized in [54] in a three-loop calculation in an axial gauge.

The fact that the matrix element M_{III} in case ii is given by the sum of the contributions of the Feynman diagrams Fig. 8.1a and Fig. 8.1c is related to an important physical property of soft gluon emission – colour coherence. The discussion above shows explicitly that in this case the second gluon is emitted coherently by the quark and the first gluon. This is clearly seen from the colour structure of the matrix element (8.92): the generator t^{a_2} corresponding to emission of the second gluon, is enclosed between the colour states $\langle\beta|t^{a_1}$ and $|\alpha\rangle$. In general, the colour structure of the matrix elements (8.89) and (8.92) confirms that the gluon flying out at a large angle is emitted by a jet of particles flying out at smaller angles. This statement is almost trivial for region I, as well for region III in case i. Indeed, since a typical time τ for the emission of a gluon with energy ω and emission angle θ is $\tau \sim 1/(\omega\theta^2)$, therefore in region I the first gluon is emitted much earlier than the second one, so that the colour factor $\langle\beta|t^{a_2}t^{a_1}|\alpha\rangle$ is quite natural. In region III for case i, we have the opposite situation: the second gluon is emitted much earlier than the first one, and here the natural colour factor is $\langle\beta|t^{a_1}t^{a_2}|\alpha\rangle$. But in case ii, the sequence of emission is the same as in region I, i.e. it is opposite to case i, whereas the colour factor has the same form as in case i. It takes this form only due to the fact that the second gluon is emitted coherently by the quark and the first gluon, or by a jet consisting of them.

The crucial difference between soft-gluon emission in QCD and soft-photon emission in QED is determined by the fact that gluons are coloured. This appears strikingly in the existence of region II, which is absent in QED. This region shows that the first gluon creates its own jet. As for the regions I and III, they are the same as in QED; the difference is only that the QCD matrix elements have the colour factors.

8.8 Soft-gluon approximations

The matrix elements (8.75) are written in the eikonal approximation. In this approach, the matrix elements for gluon bremsstrahlung off the quark are obtained neglecting gluon momenta in the numerators of quark propagators and keeping in the denominators only terms linear in gluon momenta. After that the matrix elements are simplified using the anti-commutation relations for the γ matrices and the Dirac equation. As a result, the Feynman rules in the eikonal approximation have the following form:

$$\beta, j \quad \underline{\quad p + \sum k_i \quad} \quad \alpha, i \quad = i\delta_{ij} \frac{\delta^{\alpha\beta}}{2p \sum k_i + i0}, \tag{8.95}$$

$$\beta, j \underline{\quad\quad}_{p+k} \overset{c, \rho}{\underset{p}{\frown}} \alpha, i = -ig(t^c)_{ij}\delta^{\alpha\beta}2p^\rho. \tag{8.96}$$

Since soft emission is classical, and is therefore has a spin-independent effect, the Feynman rules (8.8), (8.8) differ only by a factor of $\delta^{\alpha\beta}$ from the corresponding rules for the case of a scalar quark. One can easily write also the Feynman rules for soft-gluon emission by a hard gluon:

$$\nu, b \quad \overset{p + \sum k_i}{\overline{\underset{}{\frown}}} \quad \mu, a \quad = i\delta^{ab} \frac{-\delta_{\mu\nu}}{2p \sum k_i + i0}, \tag{8.97}$$

$$\nu, b \overset{}{\underset{p+k}{\frown}} \overset{p, c}{\underset{p}{\frown}} \mu, a = -ig(T^c)_{ab}(-\delta_{\mu\nu})2p^\rho. \tag{8.98}$$

Four-gluon vertices evidently can be omitted in the soft-collinear region since they give neither a collinear nor a soft singularity.

The eikonal rules (8.8)–(8.8) give the matrix elements (8.75) and (8.83). Actually, they can be used in the DLA for the production of an arbitrary number of gluons. Although this looks quite natural, a strict proof of this statement requires some work. This was done in [59] using physical gauges both for the final and intermediate gluons. The simplicity of the rules (8.8)–(8.8) appears as a great advantage; but this advantage is accompanied by a substantial shortcoming: the matrix elements obtained according to these rules are not gauge invariant. Actually, the rules suppose physical gluon-polarization vectors.

Note that the rules (8.8)–(8.8) differ from the so-called soft insertion rules, where only diagrams obtained by successive insertions of lines of a softer gluon in external lines of diagrams with harder particles are kept, and the dependence on momentum of this gluon is retained only in denominators of propagators adjacent to the emission vertices.

In contrast to the eikonal rules, the soft-insertion rules produce matrix elements possessing a QED-like gauge invariance with respect to soft gluons: the matrix elements turn to zero at the substitutions $e_i \rightarrow k_i$ for any gluon independently of the polarizations of other gluons.

Using the soft-insertion rules is evidently correct in the region where the most important parameters are ratios of gluon energies. However, in DLA the ratios of emission angles θ_i are equally important. Here the kinematical regions where the values $\omega_i \theta_i^2$ are ordered opposite to ω_i become important. At first sight, this makes use of the soft-insertion rules doubtful.

But it happens that they give correct results for the emission of arbitrary numbers of gluons in all double-logarithmic regions. Let us demonstrate this in the case of two-gluon emission. Using the soft-insertion rules, we obtain instead of (8.75) and (8.83)

$$M_a^{(s.i.)} = M_\alpha^{(0)} g^2 \frac{e_2^* p_-}{k_2 p_-} \frac{e_1^* p_-}{k_1 p_-} \langle \beta | t^{a_2} t^{a_1} | \alpha \rangle,$$

$$M_b^{(s.i.)} = 0,$$

$$M_c^{(s.i.)} = M_\alpha^{(0)} g^2 \frac{e_2^* k_1}{k_2 k_1} \frac{e_1^* p_-}{k_1 p_-} \langle \beta | [t^{a_1}, t^{a_2}] | \alpha \rangle. \tag{8.99}$$

For $M = M_a + M_b + M_c$, the difference between (8.75), (8.83), and (8.99) is

$$\Delta M = M_\alpha^{(0)} g^2 \left(\frac{e_2^* p_-}{k_1 p_-} - \frac{e_2^* k_1}{k_2 k_1} \frac{k_2 p_-}{k_1 p_-} \right) \frac{e_1^* p_-}{(k_1 + k_2) p_-} \langle \beta | [t^{a_1}, t^{a_2}] | \alpha \rangle. \tag{8.100}$$

Both terms in (8.100) must be kept only in the region $k_2 p_- \geq k_1 p_-$; however, in this region they cancel each other.

This means that the soft-insertion rules can be used in a much larger region than one could naively expect. This phenomenon has the same nature as the expansion of the region of applicability of the usual formulas for soft photon accompanying bremsstrahlung in high-energy hadron scattering, discovered by Gribov [55]; he has shown that at large c.m.s. energy $\sqrt{s} = \sqrt{(p_A + p_B)^2}$ of colliding hadrons with momenta p_A and p_B this region is restricted by the inequalities

$$\frac{2 p_A k}{s} \ll 1, \quad \frac{2 p_B k}{s} \ll 1,$$

$$k_\perp^2 \approx \frac{2(p_A k) \cdot 2(p_B k)}{s} \ll \mu^2, \tag{8.101}$$

where $k_\perp$ is transverse to the (p_A, p_B)-plane photon momentum, and μ is a typical hadron mass. Before the paper [55], it was generally accepted (see, for example, Ref. [56]) that for the applicability of the accompanying bremsstrahlung formulas it is necessary to have

$$2 p_A k \ll \mu^2, \quad 2 p_B k \ll \mu^2. \tag{8.102}$$

Therefore, Gribov proved that the applicability region of these formulas is considerably extended at large energies. Indeed, the conditions (8.102) are much more stringent than the conditions (8.101), if $s \gg \mu^2$.

The formulas for soft photons accompanying bremsstrahlung are obtained when only those Feynman diagrams are kept where the photon is attached to external charged particles. Furthermore, in calculating the contributions of these diagrams one has to keep the nonradiative part of the amplitude on the mass shell, i.e. one must neglect virtualities of radiating particles. This means nothing else than using the soft-insertion rules for photon-emission amplitudes. Gribov pointed out that these rules give amplitudes which are correct everywhere when they are large. This statement is quite nontrivial and is determined by the gauge invariance of the emission amplitudes [55].

At first sight, the Gribov theorem cannot be extended to QCD, despite the fact that QCD is a gauge theory like QED. An evident obstacle is the masslessness of particles having colour charge. In other words, the typical mass μ in Eq. (8.101) is equal to zero in QCD. The main point in the proof of the Gribov theorem is the smallness of the transverse momentum $k_\perp$ of the emitted quantum of the gauge field (photon or gluon) in comparison with the essential transverse momenta of the other particles. In massive theories, the latter momenta are of the order of (or greater than) μ. Conversely, in QCD essential transverse momenta of virtual particles can be arbitrarily small (this appears as infrared and collinear divergences).

Nevertheless, the soft-insertion rules can be used in QCD. At the Born level, considered so far, the problem of arbitrarily small momenta of virtual particles does not exist at all. It appears when virtual corrections must be considered. In this case, one has to introduce the infrared cut-off μ for transverse momenta of virtual gluons. In the DLA, because of the strong ordering of transverse momenta, one can find the dependence of QCD amplitudes on this cut-off, using the infrared evolution equations ([57],[58]) in different regions of ordering of μ and of soft-gluon transverse momenta of soft real gluons.

The amplitudes obtained by soft-insertion rules are correct since the dependence on the virtuality of matrix elements that correspond to diagrams with photon emission off external lines cancels the matrix elements that correspond to diagrams with photon emission off internal lines (inner emission). It is instructive to follow this cancellation in the case of emission of two gluons. Here the cancellation takes place between the dependence on $k_2 p_-$ of the denominator $(k_1 + k_2) p_-$ in M_a and M_c and the matrix element of the inner emission M_b. The former effect is

$$M_a + M_c - M_a^{(s.i.)} - M_c^{(s.i.)} = - M_\alpha^{(0)} g^2 \left(\frac{e_2^* p_-}{k_2 p_-} \langle \beta | t^{a_2} t^{a_1} | \alpha \rangle + \frac{e_2^* k_1}{k_2 k_1} \langle \beta | [t^{a_1}, t^{a_2}] | \alpha \rangle \right)$$
$$\times \frac{e_1^* p_-}{(k_1 + k_2) p_-} \frac{k_2 p_-}{k_1 p_-}, \tag{8.103}$$

and the latter one is given by M_b (8.75). Both of these become important at $k_2 p_- \gtrsim k_1 p_-$, but just in this region they cancel each other as is seen from their sum given by (8.100).

The proof of the Gribov theorem is based on dispersion relations and is reduced to the proof of the smallness of all singularities of bremsstrahlung amplitudes in the variables related to the emitted photon apart from those related to poles at $k = 0$. The smallness appears as a result of cancellation of different contributions due to gauge invariance. The proof of validity of the soft-insertion rules in QCD with the infrared cut-off μ can be constructed similarly. In the example under consideration, this means cancellations in the amplitude of the process $q(p'_-) \to q(p_-) + g(k_1) + g(k_2)$, analytically continued to the point $p'^2_- = (p_- + k_1 + k_2)^2 = 0$ (for the massless case). In the case of soft gluons with $\omega_2 \ll \omega_1$, this amplitude, presented by the diagrams of Fig. 8.1 with omitted photon line and $-p_+$ replaced by p'_-, contains the factor

$$e^*_1 p_- \left(\frac{e^*_2 k_1}{k_2 k_1} - \frac{e^*_2 p_-}{k_2 p_-} \right). \tag{8.104}$$

For $\omega_2 \ll \omega_1$, the condition $k_2 p_- + k_1 p_- = 0$ can be fulfilled only in the region where k_1 is much closer to p_- than k_2. In this region, the two terms in the above equation cancel each other. This cancellation is the cancellation of residues in the pole $k_2 p_- + k_1 p_- = 0$ in the sum of (8.75) and (8.83).

Thus, amplitudes for the emission of any number of gluons with strongly ordered energies can be obtained by successive use of the soft insertion technique. The amplitudes obtained in this way are gauge invariant. However, it will be convenient afterwards to give up the gauge invariance and use physical polarization vectors for the produced gluons, whence the colour coherence effect becomes apparent. This means that in some kinematical regions the amplitudes are given by contributions of several diagrams. In these regions, it is convenient to find the sum of the diagrams. The colour coherence presents a perfect guide to this.

To describe n-gluon production it is convenient to use the jet terminology. In the DLA, both the gluon energies and emission angles are strongly ordered. Let for definiteness

$$\omega_n \ll \omega_{n-1} \ll \ldots \omega_1 \ll \epsilon_\pm \simeq \epsilon = \frac{\sqrt{q^2}}{2}. \tag{8.105}$$

In each jet there is a particle carrying almost the entire energy of the jet. Let us call that particle the leading particle, and its momentum the jet momentum, and say that the leading particle creates the jet. Jets are called quark or gluon jets according to the kind of leading particles. Let us call the emission angles of the leading particles jet emission angles, and the largest possible angle between momenta of particles constituting the jet the jet-opening angle. Jets can differ by their energies and opening angles, so that there is a sequence of jets, starting from the quark jet, which includes all produced gluons until "elementary jets" consisting of only one gluon or quark. Besides energies and opening angles, jets are characterized by colour states, triplet and octet for quark and gluon jets, respectively. Let us emphasize that the jet colour states coincide with colour states of leading particles only in the case of elementary jets.

Distribution of produced particles into jets can be done recursively. Here it is possible to start both from the largest jet or from elementary ones. Let us start from the largest quark jet. Then the distribution looks like jet branching. The largest quark jet includes all outgoing gluons at small angles with respect to the quark momentum. Let its opening angle be $\theta_0 < 1$ (here and in the following we use simple inequalities instead of strong ones which should be used in the DLA) and the angles between momenta k_l and k_m, $l, m = 0, 1, 2, \ldots n$, $k_0 \equiv p_-$, are θ_{lm}. Provided $\theta_1 = max\ \theta_{j0}$, $j = 1, \ldots n$, all gluons can be divided into two jets, quark and gluon ones, both with opening angles θ_1. The first of these includes gluons flying with respect to the quark at smaller angles than θ_1 (in the DLA it means much smaller). The second one consists of gluons flying with respect to the quark at angles $\simeq \theta_1$. The hardest of these gluons is the leading gluon which creates the jet. We will say that this jet (as well as its leading gluon) is emitted by the quark jet with opening angle θ_1. This is the first step of branching. At the second step, each of the jets obtained after the first branching is divided similarly into two even-smaller jets. The only difference is that the gluon jet is divided into two jets which are both gluon ones (and not quark and gluon ones). They have equal opening angles (small compared with θ_1), but different energies. By definition, the jet with smaller energy is emitted by the jet with greater energy. The jet branching proceeds until each jet contains only one particle (quark or gluon).

The matrix elements of the processes with soft gluons are given by the products of the hard parts and the jet factors describing soft-gluon emission by the particles produced in the hard processes. The jet factors are independent. Unfortunately, they cannot be written explicitly. However, the rules for writing them are easily formulated recursively. It is sufficient to present the rules for quark and gluon jet branching. They differ only by colour structures. Each branching looks like the emission of a gluon jet. Let the emitted jet have colour index c and momentum k, and let the polarization vector of the leading gluon in the jet be e. If it is emitted by the quark jet with momentum p, then the emission is described by the factor

$$g \frac{(e^* p)}{(kp)} \langle \beta | t^c | \alpha \rangle, \tag{8.106}$$

where $|\alpha\rangle$ and $|\beta\rangle$ are the colour states of the quark jet before and after emission. The case when the emitting jet (with the same momentum p) is a gluon jet differs from (8.106) only by the replacement

$$\langle \beta | t^c | \alpha \rangle \rightarrow T^c_{ba}, \tag{8.107}$$

where a and b are the colour indices of the emitting jet before and after emission. As already noted, in the antiquark case $t^c_{\beta\alpha} \rightarrow -t^c_{\alpha\beta}$. Summation over all intermediate colour states is implied. These rules permit one to write martix elements for the production of arbitrary numbers of gluons.

In contrast to the QED case, the jet factors cannot be written explicitly, but they can be represented as matrix elements of jet-production operators. This representation is similar to that well known in QED, but unlike the case of QED the jet-production operators are not given in an explicit form and are defined by operator relations. Recall that in QED the operator for a jet with opening angle θ created by a particle of charge e and momentum p is given by

$$J^\dagger(p, \theta) = \exp\left(e \int \frac{d^3 k \Theta^k_{p,\theta}}{(2\pi)^3 2\omega} a^\dagger_\lambda(k) \frac{e^*_\lambda p}{kp} \right), \tag{8.108}$$

where

$$\Theta_{p,\theta}^k = \theta(|\boldsymbol{p}| - |\boldsymbol{k}|)\theta(\theta - \theta_{kp}), \tag{8.109}$$

$a_\lambda^\dagger(k)$ are the photon-creation operators with normalization

$$\left[a_{\lambda_2}(k_2)a_{\lambda_1}^\dagger(k_1)\right] = (2\pi)^3 2\omega_1 \delta_{\lambda_1\lambda_2}\delta(\boldsymbol{k}_1 - \boldsymbol{k}_2), \tag{8.110}$$

k and λ are the photon momentum and polarization, e_λ are the polarization vectors, and summation over λ is assumed. The factors corresponding to emission of n photons with momenta k_i and polarizations vectors e_i are given by the matrix elements of this operator between vacuum $|0\rangle$ and the n-photon states $\langle \mathcal{N}_\gamma|$,

$$\langle \mathcal{N}_\gamma| = \langle k_1, \lambda_1; \ldots; k_n, \lambda_n| = \langle 0| \prod_{i=1}^n a_{\lambda_i}(k_i). \tag{8.111}$$

In QCD, the jet-production operators are matrices in colour space. The fact that gluons are coloured leads to two important differences: first, the gluon-creation operators in the exponent must be accompanied by operators for the production of gluon jets, and second, because of the colour coherence the exponents must be angular ordered. Thus, for the quark jet, omitting as usual the quark colour indices, we have

$$J_q^\dagger(p, \theta) = \hat{P}_{\theta_{kp}} \exp\left(g \int \frac{d^3 k \Theta_{p,\theta}^k}{(2\pi)^3 2\omega} a_\lambda^{\dagger a}(k) \left(J_g^\dagger(k, \theta_{kp})\right)_{ab} t^b \frac{e_\lambda^* p}{kp}\right), \tag{8.112}$$

where the angular-ordering operator $\hat{P}_\theta$ orders exponents from left to right by way of increasing θ, $J_g^\dagger(k, \theta_{kp})$ is the production operator of the gluon jet with opening angle θ_{kp} and momentum k. For this operator, we have

$$\left(J_g^\dagger(p, \theta)\right)_{ab} = \left[\hat{P}_{\theta_{kp}} \exp\left(g \int \frac{d^3 k \Theta_{p,\theta}^k}{(2\pi)^3 2\omega} a_\lambda^{\dagger c}(k) \left(J_g^\dagger(k, \theta_{kp})\right)_{cd} T^d \frac{e_\lambda^{*\mu} p_\mu}{kp}\right)\right]_{ab}. \tag{8.113}$$

The factors corresponding to emission of n soft gluons accompanying production of partons P, $P = q, g$, with momentum p in any hard processes is given by the matrix elements of the operators $J_P^\dagger(p, \theta)$ between vacuum $|0\rangle$ and the n-gluon states $\langle \mathcal{N}_g|$,

$$\langle \mathcal{N}_g| = \langle k_1, \lambda_1, c_1; \ldots; k_n, \lambda_n, c_n| = \langle 0| \prod_{i=1}^n a_{\lambda_i}^{c_i}(k_i). \tag{8.114}$$

These factors can be represented in terms of functional derivatives of the jet generating functionals. The relations $\langle k, \lambda, ac| = \langle 0|a_\lambda^c(\boldsymbol{k})$, $[a_\lambda^c(\boldsymbol{k}), a_{\lambda'}^{\dagger c'}(\boldsymbol{k}')] = (2\pi)^3 2\omega_1 \delta(\boldsymbol{k} - \boldsymbol{k}')\delta_{\lambda\lambda'}\delta_{cc'}$ and $a_\lambda^c(\boldsymbol{k})|0\rangle = 0$, $\langle 0|a_\lambda^{\dagger c}(\boldsymbol{k}) = 0$, allow one to write

$$\langle \mathcal{N}_g|J_P^\dagger(p, \theta)|0\rangle = \frac{\delta}{\delta u_{\lambda_1}^{a_1}(k_1)} \cdots \frac{\delta}{\delta u_{\lambda_n}^{a_n}(k_n)} J_P(p, \theta|u),$$

where $J_P(p, \theta|u)$ is the generating functional for jets created by partons P,

$$J_q(p,\theta|u) = \hat{P}_{\theta kp} \exp\left(g \int d^3k \Theta^k_{p,\theta} u^a_\lambda(k) \left(J_g(k,\theta kp|u)\right)_{ab} t^b \frac{e^*_\lambda p}{kp}\right), \qquad (8.115)$$

and $J_g(k,\theta kp|u)$ is the generating functional for gluon jets defined by the equation

$$\left(J_g(p,\theta|u)\right)_{ab} = \left[\hat{P}_{\theta kp} \exp\left(g \int_{(p,\theta)} d^3k \Theta^k_{p,\theta} u^c_\lambda(k) \left(J_g(k,\theta kp|u)\right)_{cd} T^d \frac{e^{*\mu}_\lambda p_\mu}{kp}\right)\right]_{ab}.$$
$$(8.116)$$

The above discussion does not concern virtual corrections caused by soft gluons. Inclusion of these corrections in the DLA does not present a problem and can be done quite analogously to QED. In fact, the virtual corrections are determined by the requirement of their cancellation with real corrections in inclusive cross sections and locality of this cancellation in momentum space. Taking account of the virtual corrections, the generating functionals become

$$\mathcal{J}_q(p,\theta|u) = e^{-\frac{1}{2}w_q(p,\theta)} J_q(p,\theta|u), \quad \mathcal{J}_g(p,\theta|u) = e^{-\frac{1}{2}w_g(p,\theta)} J_g(p,\theta|u), \qquad (8.117)$$

where $w_q(p,\theta)$ and $w_g(p,\theta)$ are the Born probabilities of soft-gluon emission by quark and gluon respectively inside a cone of half-angle θ :

$$w_q(p,\theta) = \int \frac{2C_F\alpha_s}{\pi} \frac{d^3k \Theta^k_{p,\theta}}{4\pi\omega^2} \frac{\epsilon_p}{kp}, \quad w_g(p,\theta) = \int \frac{2C_A\alpha_s}{\pi} \frac{d^3k \Theta^k_{p,\theta}}{4\pi\omega^2} \frac{\epsilon_p}{kp}, \qquad (8.118)$$

where $C_F = (N_c^2 - 1)/(2N_c)$ and $C_A = N_c$ are the values of the Casimir operators in the fundamental and in the adjoint representations.

It follows from (8.117) that

$$\langle \mathcal{N}_g | \mathcal{J}^\dagger_P(p,\theta)|0\rangle = \langle \mathcal{N}_g | J^\dagger_P(p,\theta)|0\rangle F(P,\mathcal{N}_g), \qquad (8.119)$$

where $F(P,\mathcal{N}_g)$ is the double-logarithmic formfactor accounting for the virtual corrections,

$$F(P,\mathcal{N}_g) = \exp\left[-\frac{1}{2}w_P(p,\theta) - \frac{1}{2}\sum_{i-1}^n w_g(k_i,\theta_i)\right], \qquad (8.120)$$

and θ_i are emission angles of gluons with momenta k_i. This form of the virtual correction was conjectured in [59] and proved afterwards in [60].

8.9 Soft-gluon distributions

The parton spectra are inclusive characteristics. In principle, calculation of such quantities does not require knowledge of amplitudes with fixed numbers of participating particles; moreover, cancellation of various contributions, virtual and real, to these quantities can be used without explicit evaluation of such contributions. The exclusive approach, which requires knowledge of all amplitudes, is much more detailed. In DLA, the parton (gluon) distributions can be described on a completely exclusive level.

In the following summation over colours and polarizations of soft gluons is assumed. Their distributions can be represented in a simple form using the generating functionals for jet production [61]–[63]. Let us define the generating functional $\Phi_{\{P_a\}}(\{p_a\}, \{\theta_a\}|u)$ for jets with momenta p_a and opening angles θ_a created by partons P_a produced in the hard process with cross section $d\sigma^{(0)}$ by the relations

$$d\sigma^{(n)} = d\sigma^{(0)} \prod_{i=1}^{n} \frac{\delta}{\delta u(k_i)} \Phi_{\{P_a\}}(\{p_a\}, \{\theta_a\}|u)\Big|_{u=0}, \qquad (8.121)$$

where $d\sigma^{(n)}$ is the cross section of the process with n soft gluons in the accompanying bremsstrahlung and k_i are the gluon momenta. The gluon distributions in different jets are independent, so that

$$\Phi_{\{P_a\}}(\{p_a\}, \{\theta_a\}|u) = \prod_a \Phi_{P_a}(p_a, \theta_a|u), \qquad (8.122)$$

where $\Phi_{P_a}(p_a, \theta_a|u)$ are the generating functionals for separate jets. Having Eqs. (8.115), (8.116), and (8.117), it is easy to write the corresponding equations for $\Phi_P(p, \theta|u)$:

$$\Phi_q(p, \theta|u) = e^{-w_q(p,\theta)} \exp\left(\int \frac{2C_F \alpha_s}{\pi} \frac{d^3k \Theta_{p,\theta}^k}{4\pi\omega^2} \frac{\epsilon_p}{kp} u(k) \Phi_g(k, \theta_{kp}|u)\right), \qquad (8.123)$$

$$\Phi_g(p, \theta|u) = e^{-w_g(p,\theta)} \exp\left(\int \frac{2C_A \alpha_s}{\pi} \frac{d^3k \Theta_{p,\theta}^k}{4\pi\omega^2} \frac{\epsilon_p}{kp} u(k) \Phi_g(k, \theta_{kp}|u)\right), \qquad (8.124)$$

where $\Theta_{p,\theta}^k$ is given by (8.109), $w_q(p, \theta)$ and $w_g(p, \theta)$ by (8.118). As one can see from (8.123), (8.124),

$$\Phi_q(p, \theta|u) = \left(\Phi_g(p, \theta|u)\right)^{C_F/C_A}. \qquad (8.125)$$

From (8.123), (8.124), it also follows that the functionals satisfy the boundary conditions

$$\Phi_P(p, \theta|u)\Big|_{u=0} = e^{-w_P(p,\theta)}, \quad \Phi_P(p, \theta|u)\Big|_{u=1} = 1. \qquad (8.126)$$

Sometimes it is convenient to use the representation

$$\Phi_g(p, \theta|u) = \exp[-w_g(p, \theta) + W(p, \theta|u)], \qquad (8.127)$$

where $W(p, \theta|u)$ is defined to be the perturbative solution of the following equation:

$$W(p, \theta|u) = \int \frac{2C_A \alpha_s}{\pi} \frac{d^3k \Theta_{p,\theta}^k}{4\pi\omega^2} \frac{\epsilon_p}{kp} u(k) \exp[-w(k, \theta_{kp}) + W(k, \theta_{kp}|u)], \qquad (8.128)$$

and has the boundary conditions

$$W(p, \theta|u)\Big|_{u=0} = 1, \quad W(p, \theta|u)\Big|_{u=1} = w_g(p, \theta). \qquad (8.129)$$

The generating functionals allow one to write easily the inclusive as well as exclusive cross sections. Thus, the totally inclusive cross section (cross section with production of arbitrary numbers of gluons) can be written as (compare with (8.121))

$$
d\sigma_{incl} = d\sigma^{(0)} \exp\left(\int d^3k \frac{\delta}{\delta u(k)}\right) \Phi_{\{P_a\}}(\{p_a\}, \{\theta_a\}|u)\Big|_{u=0}
$$

$$
= d\sigma^{(0)} \Phi_{\{P_a\}}(\{p_a\}, \{\theta_a\}|u)\Big|_{u=1} = d\sigma^{(0)}. \tag{8.130}
$$

The last equality follows from (8.126). For the inclusive production of n gluons, we have

$$
d\sigma_{incl}^{(n)} = d\sigma^{(0)} \prod_{i=1}^{n} \frac{\delta}{\delta u(k_i)} \exp\left(\int d^3k \frac{\delta}{\delta u(k)}\right) \Phi_{\{P_a\}}(\{p_a\}, \{\theta_a\}|u)\Big|_{u=0}
$$

$$
= d\sigma^{(0)} \prod_{i=1}^{n} \frac{\delta}{\delta u(k_i)} \Phi_{\{P_a\}}(\{p_a\}, \{\theta_a\}|u)\Big|_{u=1}. \tag{8.131}
$$

These equations provide a simple qualitative picture of soft-gluon emission and have a clear probabilistic interpretation. Hard-production cross sections are not affected by soft-collinear gluons if emission of any number of such gluons is permitted. In this case under quite crude angular resolution, each hard parton P accompanied by soft-collinear gluons looks like a jet created by this parton. A finer resolution shows that this jet contains nonoverlapping gluon jets with ordered emission angles. The angular size of each of these jets is small compared with its emission angle. The jets are emitted by the parton independently with probabilities equal to the Born probability of production of the leading gluons in the jets. The probability of fixed configurations of jets is suppressed by the Sudakov formfactor $\exp[-w_P(p, \theta)]$. In turn, each gluon jet with leading gluon having momentum k consists of jets emitted independently by the leading gluon; the probability of fixed configurations of these jets is suppressed by the Sudakov formfactor $\exp[-w_g(k, \theta_{kp})]$, and so on.

The functional method is easily generalized to account for the running of the coupling constant. A plausible choice of the argument of α_s is $k_\perp^2$, where $k_\perp$ is the gluon momentum transverse to the momentum of the radiating particle.

The mean gluon multiplicity in the parton P jet of angular size θ and momentum p is

$$
n_P(p, \theta) = \int d^3k \frac{\delta}{\delta u(k)} \Phi_P(p, \theta|u)\Big|_{u=1}; \tag{8.132}
$$

with

$$
n_q(p, \theta) = \frac{C_F}{C_A} n_g(p, \theta), \tag{8.133}
$$

$$
n_g(p, \theta) = \int d^3k \frac{\delta}{\delta u(k)} W(p, \theta|u)\Big|_{u=1} \equiv n(p, \theta). \tag{8.134}
$$

It follows from (8.128) that $n(p, \theta)$ obeys the equation

$$
n(p, \theta) = \int \frac{2C_A \alpha_s(k_\perp^2)}{\pi} \frac{d^3k \Theta_{p,\theta}^k}{4\pi\omega^2} \frac{\epsilon_p}{kp} [1 + n(k, \theta_{kp})]. \tag{8.135}
$$

Evidently, this equation contains infrared and collinear divergences, which must be regularized. A natural way of regularization is the infrared cut-off $k_\perp^2 > Q_0^2$. Then the integration region is

$$\epsilon^2\theta^2 \geq \omega^2\theta^2 \geq k_\perp^2 \geq Q_0^2. \qquad (8.136)$$

The integral in (8.135) can be rewritten as

$$\int \frac{d^3k\Theta_{p,\theta}^k}{2\pi\omega^2}\frac{\epsilon_p}{kp} = \int_{Q_0^2}^{\epsilon^2\theta^2}\frac{dk_\perp^2}{k_\perp^2}\int_{k_\perp}^{\epsilon\theta}\frac{d(\omega\theta)}{\omega\theta}, \qquad (8.137)$$

such that $n(p,\theta)$ actually depends only on the product $\epsilon\theta$. Thus, equation (8.135) is reduced to

$$n(p,\theta) = \int_{Q_0^2}^{\epsilon^2\theta^2}\frac{dk_\perp^2}{k_\perp^2}\frac{C_A\alpha_s(k_\perp^2)}{2\pi}\ln\left(\frac{\epsilon^2\theta^2}{k_\perp^2}\right)[1 + n(k,\theta_{kp})]. \qquad (8.138)$$

Using the variables $Y = \ln\epsilon\theta/Q_0$ and denoting $n(p,\theta) = n(Y)$, we can rewrite (8.138) as

$$n(Y) = \int_0^Y dy\,\frac{2C_A\alpha_s(k_\perp^2)}{\pi}(Y - y)[1 + n(y)], \qquad (8.139)$$

where $k_\perp^2 = Q_0^2 e^{2y}$. This equation can be reduced to the differential one

$$\frac{d^2}{dY^2}n(Y) = \frac{2C_A\alpha_s(\epsilon^2\theta^2)}{\pi}[1 + n(Y)], \qquad (8.140)$$

supplemented by the initial conditions $n(0) = 0$, $n'(0) = 0$. With the one-loop running coupling in the form of

$$\alpha_s(\epsilon^2\theta^2) = \frac{2\pi}{b(Y + \lambda)}, \quad \lambda = \ln\left(\frac{Q_0}{\Lambda_{QCD}}\right), \quad b = \frac{11}{3}N_c - \frac{2}{3}n_f, \qquad (8.141)$$

the solution of (8.140) is written as

$$n(Y) = a\sqrt{Y + \lambda}\left(I_1(a\sqrt{Y+\lambda})K_0(a\sqrt{\lambda}) + K_1(a\sqrt{Y+\lambda})I_0(a\sqrt{\lambda})\right) - 1, \qquad (8.142)$$

where K_i and I_i are the modified Bessel functions and $a = \sqrt{16C_A/b}$. In the perturbative region $\lambda \gg 1$, so that

$$n(Y) \simeq \left(\frac{Y+\lambda}{\lambda}\right)^{1/4}\cosh\left[a\left(\sqrt{Y+\lambda} - \sqrt{\lambda}\right)\right] - 1. \qquad (8.143)$$

In the region $Y \ll \lambda$, where the running of α_s is negligible, we have

$$n(Y) \simeq \cosh\left(\sqrt{\frac{2C_A}{\pi}\alpha_s}\,Y\right) - 1. \qquad (8.144)$$

Asymptotically, for $Y \gg \lambda$,

$$n(Y) \sim \left(\frac{Y}{\lambda}\right)^{1/4}\exp\sqrt{\frac{16C_A}{b}Y}. \qquad (8.145)$$

The exponent $\sqrt{(16C_A/b)Y}$ in (8.145) differs by a factor of $1/\sqrt{2}$ from the results obtained in [64]–[68] by summation of contributions of planar diagrams in axial gauges, i.e. without account of the coherence effect.

According to (8.132), (8.133) the gluon multiplicity in e^+e^--annihilation is given by $n(Y)$ with the factor $2C_F/C_A$. Note, however, that it was assumed in (8.133) that the quark mass M does not exceed Q_0 significantly. In the case $M_Q \gg Q_0$, the large quark mass suppresses emission of gluon jets because of the angular distribution $d\theta^2/(\theta^2 + M_Q^2/\epsilon_Q^2)$, reducing therefore the mean multiplicity, which becomes equal to [62]

$$n_{Q\bar{Q}}(2\epsilon) \simeq n_{q\bar{q}}(2\epsilon) - n_{q\bar{q}}(2M_Q). \tag{8.146}$$

Actually, one should expect that it is even smaller, since the effective energy for the evolution of the gluon cascade is [69]

$$\epsilon_{eff} \simeq \epsilon(1- <x_Q>), \quad <x_Q> \simeq \left(\frac{\alpha_s(\epsilon^2)}{\alpha_s(M_Q^2)}\right)^{\frac{32}{81}}. \tag{8.147}$$

Note that the exponent in (8.145) does not depend on λ, i.e. is infrared safe and so is calculable in perturbation theory. Therefore, one can affirm that it determines the asymptotic behaviour of the mean hadronic multiplicity $< N_h(s) >$ in e^+e^--annihilation

$$\ln < N_h(s) > \simeq \sqrt{\frac{8C_A}{b} \ln s}, \tag{8.148}$$

for any hadron.

8.10 Hump-backed shape of parton spectra

Quite analogously to the mean multiplicity, the inclusive one-gluon distribution D with respect to the gluon energy ω in a jet of energy ϵ and opening angle θ depends only on the variables y and Y defined by

$$y = \ln\left(\frac{\omega\theta}{Q_0}\right), \quad Y = \ln\left(\frac{\epsilon\theta}{Q_0}\right). \tag{8.149}$$

Normalizing the distribution so that

$$n(Y) = \int_0^Y dy D(y, Y), \tag{8.150}$$

we have, from (8.131) and (8.127),

$$D(y, Y) = \int d^3k\delta \left(1 - \frac{k_0}{\omega}\right) \frac{\delta}{\delta u(k)} W(p, \theta|u)\Big|_{u=1}. \tag{8.151}$$

With the help of (8.122) and (8.125), the inclusive gluon spectra for any hard process can be expressed in terms of D. In particular, for e^+e^--annihilation

$$\frac{1}{\sigma} x \frac{d\sigma}{dx} = 2\frac{C_F}{C_A} D(y, Y), \tag{8.152}$$

where y, Y are given by (8.149) with $\theta \sim 1$ and $\epsilon = \epsilon_{\pm}$.

Using Eqs. (8.128) and (8.129), we come from (8.151) to the equation for D. Including in $D(y, Y)$ the term $\delta(y - Y)$ to account for the leading gluon, we obtain

$$D(y, Y) = \delta(y - Y) + \frac{2C_A}{\pi} \int_0^y dy_1 \int_{y_1}^{Y-y+y_1} dY_1 \alpha_s(k_\perp^2) D(y_1, Y_1), \quad (8.153)$$

where $k_\perp^2 = Q_0^2 e^{2Y_1}$. Taking $\alpha_s(k_\perp^2)$ in the one-loop approximation, and defining

$$G(y, z) = \frac{4C_A}{b(z + y + \lambda)} D(y, z + y), \quad (8.154)$$

we have

$$\frac{b}{4C_A}(\lambda + y + z)G(y, z) = \delta(z) + \int_0^y dy_1 \int_0^z dz_1 G(y_1, z_1), \quad (8.155)$$

where $z = \ln(\epsilon/\omega), \quad y = \ln(\omega\theta/Q_0)$. For the Mellin image of $G(y, z)$

$$L(\alpha, \beta) = \int_0^\infty dz \int_0^\infty dy e^{-\alpha z - \beta y} G(y, z) \quad (8.156)$$

we obtain

$$\left(\lambda - \frac{\partial}{\partial \alpha} - \frac{\partial}{\partial \beta}\right) L(\alpha, \beta) = \frac{a^2}{4\alpha\beta} L(\alpha, \beta) + \frac{a^2}{4\beta}, \quad (8.157)$$

where $a^2 = 16C_A/b$. The perturbative solution of this equation can be written as

$$L(\alpha, \beta) = \int_0^\infty ds \exp\left[-s\left(\lambda - \frac{\partial}{\partial \alpha} - \frac{\partial}{\partial \beta} - \frac{a^2}{4\alpha\beta}\right)\right] \frac{a^2}{4\beta}. \quad (8.158)$$

Denoting

$$f(s) = \exp\left[s\left(\frac{\partial}{\partial \alpha} + \frac{\partial}{\partial \beta} + \frac{a^2}{4\alpha\beta}\right)\right] \exp\left[-s\left(\frac{\partial}{\partial \alpha} + \frac{\partial}{\partial \beta}\right)\right], \quad (8.159)$$

we have

$$\frac{df(s)}{ds} = f(s) \exp\left[s\left(\frac{\partial}{\partial \alpha} + \frac{\partial}{\partial \beta}\right)\right] \frac{a^2}{4\alpha\beta} \exp\left[-s\left(\frac{\partial}{\partial \alpha} + \frac{\partial}{\partial \beta}\right)\right] = f(s) \frac{a^2}{4(s + \alpha)(s + \beta)}, \quad (8.160)$$

such that

$$f(s) = \exp\left[\frac{a^2}{4(\alpha - \beta)} \ln\left(\frac{\alpha(\beta + s)}{\beta(\alpha + s)}\right)\right], \quad (8.161)$$

and hence

$$L(\alpha, \beta) = \frac{a^2}{4} \int_0^\infty \frac{ds}{\beta + s} \exp\left[-s\lambda + \frac{a^2}{4(\alpha - \beta)} \ln\left(\frac{\alpha(\beta + s)}{\beta(\alpha + s)}\right)\right]. \quad (8.162)$$

Thus we come to the following representation for $D(y, Y)$:

$$D(y, Y) = (Y + \lambda) \int_{\gamma - i\infty}^{\gamma + i\infty} \frac{d\alpha}{2\pi i} \int_{\delta - i\infty}^{\delta + i\infty} \frac{d\beta}{2\pi i} e^{\alpha(Y - y) + \beta y}$$

$$\times \int_0^\infty \frac{ds}{\beta + s} e^{-s\lambda} \left(\frac{\alpha(\beta + s)}{\beta(\alpha + s)} \right)^{\frac{a^2}{4(\alpha - \beta)}}. \tag{8.163}$$

Evaluating the integral by the steepest-descent method and neglecting the pre-exponential factor gives

$$D(y, Y) \simeq e^{\phi(\alpha_0, \beta_0, y, Y)}, \tag{8.164}$$

where

$$\phi(\alpha, \beta, y, Y) = \alpha(Y - y) + \beta y + \frac{a^2}{4(\alpha - \beta)} \ln \left(\frac{\alpha(\beta + s_0)}{\beta(\alpha + s_0)} \right) - \lambda s_0,$$

$$s_0 = \frac{1}{2} \left(\sqrt{\frac{a^2}{\lambda^2} + (\alpha - \beta)^2} - (\alpha + \beta) \right), \tag{8.165}$$

and α_0, β_0 are defined by the equations

$$\frac{\partial}{\partial \alpha} \phi(\alpha, \beta, y, Y) = 0, \quad \frac{\partial}{\partial \beta} \phi(\alpha, \beta, y, Y) = 0. \tag{8.166}$$

It is suitable to introduce new variables μ, ν by the relations

$$\frac{2\sqrt{Y + \lambda}}{a} \alpha_0 = \left(\frac{2\sqrt{Y + \lambda}}{a} \beta_0 \right)^{-1} = e^\mu, \quad \frac{2\sqrt{\lambda}(\alpha_0 + s_0)}{a} = \left(\frac{2\sqrt{\lambda}(\beta_0 + s_0)}{a} \right)^{-1} = e^\nu. \tag{8.167}$$

In these variables we have

$$\ln D(y, Y) \simeq a \left(\sqrt{Y + \lambda} - \sqrt{\lambda} \right) \frac{\mu - \nu}{\sinh \mu - \sinh \nu}, \tag{8.168}$$

where μ, ν satisfy the equations

$$\frac{2y - Y}{Y} = \frac{(\sinh(2\mu) - 2\mu) - (\sinh(2\nu) - 2\nu)}{2(\sinh^2 \mu - \sinh^2 \nu)},$$

$$\frac{\sinh \nu}{\sqrt{\lambda}} = \frac{\sinh \mu}{\sqrt{Y + \lambda}}. \tag{8.169}$$

It is easy to see that D has a maximum at $y = Y/2$ ($\sqrt{\lambda}\, \mu = \sqrt{Y + \lambda}\, \nu = 0$). Expanding the exponent in (8.164) around this point and restoring the pre-exponential factor from the normalization condition (8.150), we obtain

$$D(y, Y) = n(Y) \left(\frac{c}{\pi} \right)^{1/2} \exp \left(-c \left(y - \frac{Y}{2} \right)^2 \right), \tag{8.170}$$

where

$$c = \sqrt{\frac{16N_c}{b}} \frac{3}{2\left((Y+\lambda)^{\frac{3}{2}} - \lambda^{\frac{3}{2}}\right)}, \qquad b = \frac{11}{3}N_c - \frac{2}{3}n_f. \tag{8.171}$$

Note that a fixed-coupling constant α_s corresponds to the limit $Y \ll \lambda$. In this case, it is easy to obtain an exact expression for $D(y, Y)$:

$$D(y, Y) = \left(\frac{2C_A \alpha_s y}{\pi(Y-y)}\right)^{1/2} I_1\left(\sqrt{\frac{8C_A \alpha_s}{\pi}} y(Y-y)\right), \tag{8.172}$$

where I_1 is a modified Bessel function. The most striking feature of $D(y, Y)$ is the maximum at $y \simeq Y/2$ and the collapse in the region of small energies (instead of the usually assumed plateau), see Fig. 8.3. This is a direct consequence of the coherence. Note that the position of the maximum is approximately the same for fixed and running α_s; the width of the distribution is $\propto Y^{1/2}$ in the former case and $\propto Y^{3/4}$ in the latter (at large $Y \gg \lambda$).

Using the fact that the distribution D in the gluon jet with energy ϵ and opening angle θ depends only on the variables $\epsilon\theta/Q_0$ and $\omega\theta/Q_0$, one can obtain the double-differential gluon distribution (with respect to the gluon energy ω and the angle θ between the flight direction and the jet axis). Thus in the process of e^+e^--annihilation we have

$$\frac{1}{\sigma} \frac{d\sigma}{dy\, d\ln\theta} = \frac{C_F}{C_A} \frac{\partial}{\partial \ln\theta} D(y + \ln\theta, Y + \ln\theta), \tag{8.173}$$

where $y = \ln\omega/Q_0$, $Y = \ln\epsilon/Q_0$. Similarly, for the angular distribution we get

$$\frac{1}{\sigma} \frac{d\sigma}{d\ln\theta} = \frac{C_F}{C_A} \frac{\partial}{\partial \ln\theta} n(Y + \ln\theta). \tag{8.174}$$

The functional method described above makes it possible to obtain also more complicated inclusive distributions. As an example, let us present without derivation the following two-particle distribution:

$$\frac{\omega_1\omega_2}{\sigma} \frac{d\sigma}{d\omega_1 d\omega_2} \simeq \frac{\omega_1}{\sigma} \frac{d\sigma}{d\omega_1} \frac{\omega_2}{\sigma} \frac{d\sigma}{d\omega_2} \left[1 + \frac{C_F}{6C_A} \frac{1}{1 + \frac{4}{3}\sinh^2\left(\frac{\mu_1-\mu_2}{2}\right)}\right], \tag{8.175}$$

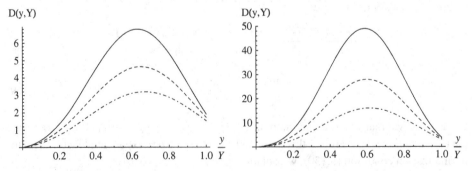

Fig. 8.3. The distributions $D(y, Y)$ (8.172) at $\alpha_s = 0.1$ (left) and $\alpha_s = 0.2$ (right). Dot-dashed, dashed, and solid lines correspond to $Y = 8$, $Y = 9$ and $Y = 10$, respectively.

where $\mu_{1,2}$ are the solutions of Eq.(8.169) at $y = y_{1,2}$ respectively. The factor in square brackets describes damping of two-particle correlations with increasing $|\eta|$, $\eta = \ln \omega_1/\omega_2$. At $\eta = 0$ $(\mu_1 = \mu_2)$ it coincides with the value of the multiplicity correlator

$$\frac{\langle n(n-1)\rangle_{q\bar{q}}}{\langle n\rangle_{q\bar{q}}^2} = \left(1 + \frac{C_A}{6C_F}\right). \tag{8.176}$$

Fixing only the ratio of gluon energies in the region of maximal contribution, it is easy to obtain

$$\frac{1}{\sigma}\frac{d\sigma}{d\eta} \simeq \langle n(n-1)\rangle_{q\bar{q}}\sqrt{\frac{c}{2\pi}}e^{-c\eta^2/2}, \tag{8.177}$$

where c is defined in (8.171). It is clear from the above statements that the coherence phenomenon leads to striking qualitative predictions for inclusive distributions: appearance of a maximum in the rapidity distribution and the collapse in the region of small rapidities, dependence of the distribution on the jet-transverse momentum instead of the total jet energy, characteristic behaviour of the double-inclusive spectra, and so on. These predictions are confirmed experimentally, which supports the soft-blanching hypothesis [53],[70], affirming the similarity of hadron and parton spectra.

8.11 Multiplicity distributions and KNO scaling

As we have seen, the coherence leads to substantial reduction of multiplicity with respect to predictions based on extension to the low x region of LLA selection rules for Feynman diagrams. The question arises: what is the influence of coherence on multiplicity moments $n_k = \langle n(n-1)\ldots(n-k+1)\rangle$. To investigate this question it is possible to take $u(k) = u$ independent of k in the generating functionals, so that the functionals become functions of u. For the gluon jet, the function $\Phi(Y, u)$ is defined by the equation

$$\ln \Phi(Y, u) = \int_0^Y dy(Y - y)\frac{2C_A\alpha_s(k_\perp^2)}{\pi}\left(u\Phi(y, u) - 1\right), \tag{8.178}$$

where $Y = \ln(\epsilon\theta/Q_0)$, $k_\perp^2 = Q_0^2e^{2y}$. If there are several jets with energies and opening angles of the same order, then the corresponding function is

$$\Phi^{(n_q,n_g)}(Y, u) = [\Phi(Y, u)]^{\frac{C_F}{C_A}n_q+n_g}, \tag{8.179}$$

where n_q and n_g are the numbers of quark (antiquark) and gluon jets, respectively.

The probabilities of exclusive production of k gluons are given by the coefficients of the expansion of $\Phi(Y, u)$ in powers of u, whereas the moments n_k are given by the coefficients of the expansion in powers of $(u - 1)$:

$$\Phi(Y, u) = \sum_{k=0}^{\infty} P_k(Y)u^k = \sum_{k=0}^{\infty} n_k(Y)\frac{(u-1)^k}{k!},$$

$$n_0(Y) = 1, \quad n_1(Y) = n(Y) \equiv \bar{n}(Y). \tag{8.180}$$

Using equation (8.178), one can show [63] that although the coherence has a substantial influence on the energy dependence of each n_k, it does not change the normalized ratios $n_k/\bar{n}^k$, which were found in [65],[66], without taking account of coherence. In particular, the Koba–Nielsen–Olesen (KNO) [71] scaling function

$$f(x) = \lim_{Y \to \infty, \, n \to \infty} [\bar{n}(Y) P_n(Y)], \tag{8.181}$$

where the limit is taken at fixed $x = n/\bar{n}(Y)$, is also not affected. Here $f(x)$, as well as the ratios $n_k/\bar{n}^k$ which appear as the moments of f, do not depend on the interaction at all. At $x \ll 1$ (i.e. $n \ll \bar{n}$), we have

$$f(x) \sim \frac{1}{x} e^{-\frac{1}{2} \ln^2 x}, \tag{8.182}$$

which is related to the Sudakov suppression of small multiplicity events. At $x \gg 1$ (i.e. $n \gg \bar{n}$)

$$f(x) \simeq e^{\beta_0 x} 2\beta_0 \left(\beta_0 x + 1 + \frac{1}{3\beta_0 x} + \dots \right), \tag{8.183}$$

where $\beta_0 < 0$,

$$\ln |\beta_0| = \int_1^\infty \frac{dx}{x} \left(\frac{1}{\sqrt{2(x - 1 - \ln x)}} - \frac{1}{x - 1} \right),$$

$$\beta_0 \simeq -2.552. \tag{8.184}$$

The expression (8.183) differs from that obtained in [67] (without account of the colour coherence) because of the approximations used there.

8.12 Moments of fragmentation functions at small $j - 1$

The hadron multiplicity in e^+e^--annihilation is defined by the relation

$$< N_h(s) > = \frac{1}{\sigma_t} \int_0^1 dx \frac{d\sigma_h}{dx}, \tag{8.185}$$

where $\sigma_t \equiv \sigma_{e^+e^- \to hadrons}$ is the total cross section of e^+e^--annihilation into hadrons, $d\sigma_h \equiv d\sigma_{e^+e^- \to h+X}$ is the inclusive cross section for hadron h-production, $x = \epsilon_h/\epsilon_\pm$ is the fraction of the electron (positron) energy carried by the hadron. According to (8.70), the cross section $d\sigma_h/(dx)$ can be written as

$$\frac{d\sigma_h}{dx} = \int d\Omega \frac{d\sigma_h}{dx d\Omega} = \sum_q \int_x^1 \frac{dz}{z} \bar{f}_q^h \left(\frac{x}{z}, s \right) \int d\Omega \frac{d\sigma_{e^+e^- \to q\bar{q}}}{dz d\Omega}$$

$$= \sigma_t \frac{\sum_q e_q^2 \bar{f}_q^h(x, s)}{\sum_q e_q^2}, \tag{8.186}$$

where we have used that

$$\frac{d\sigma_{e^+e^- \to q\bar{q}}}{dz d\Omega} = \frac{\alpha^2 e_q^2 N_c}{4s} (1 + \cos^2 \theta) \delta(1 - z). \tag{8.187}$$

Therefore, we can write

$$< N_h(s) > \frac{\sum_q e_q^2 n_q^h(s)}{\sum_q e_q^2}, \tag{8.188}$$

where

$$n_q^h(Q^2) = \int_0^1 dx \, \bar{f}_q^h(x, Q^2) \tag{8.189}$$

is the hadron multiplicity in the quark jet. This multiplicity can be considered to be the limit of the Mellin image M_q of the fragmentation function $\bar{f}_q^h$,

$$M_q(j, Q^2) = \int_0^1 dx x^{j-1} \bar{f}_q^h(x, Q^2) \tag{8.190}$$

at $j \to 1$. The fragmentation functions obey the equations (8.73). For the Mellin images, the equations is of the following form:

$$\frac{dM_a(j, Q^2)}{d \ln Q^2} = \frac{\alpha_s(Q^2)}{2\pi} A_a^b(j) M_b(j, Q^2), \tag{8.191}$$

where

$$A_a^b(j) = \int_0^1 \frac{dz}{z} z^j \bar{P}_a^b(z). \tag{8.192}$$

At small x, the leading contributions in (8.73) come from the splitting functions $\bar{P}_q^g$ and $\bar{P}_g^g$ with $1/x$ singularities:

$$\bar{P}_q^g(x) \simeq \frac{2C_F}{x}, \quad \bar{P}_q^g(x) \simeq \frac{2C_A}{x}. \tag{8.193}$$

The small x region is related to $j \simeq 1$. Correspondingly, the functions $A_q^g(j)$ and $A_g^g(j)$ are singular at $j = 1$:

$$A_q^g(j) \simeq \frac{2C_F}{j-1}, \quad A_g^g(j) \simeq \frac{2C_A}{j-1}. \tag{8.194}$$

In perturbation theory, the Mellin transforms M_a are defined in the half-plane Re $j > 1$. Thus we have for the terms most singular at $j = 1$:

$$\frac{dM_q(j, Q^2)}{d \ln Q^2} = \frac{\alpha_s(Q^2)}{2\pi} A_q^g(j) M_g(j, Q^2), \quad \frac{dM_g(j, Q^2)}{d \ln Q^2} = \frac{\alpha_s(Q^2)}{2\pi} A_g^g(j) M_g(j, Q^2), \tag{8.195}$$

such that

$$M_g(j, Q^2) = \exp\left(\int_{Q_0^2}^{Q^2} \frac{dQ^2}{Q^2} \frac{\alpha_s(Q^2)}{2\pi} A_g^g(j) \right) M_g(j, Q_0^2). \tag{8.196}$$

This means that in the one-loop approximation the leading anomalous dimension is divergent at $j \to 1$, such as

$$\gamma_j^{(1)}(\alpha_s) \simeq \frac{C_A \alpha_s}{\pi} \frac{1}{j-1}. \tag{8.197}$$

The existence of the double logarithms means that in the n-loop approximation

$$\gamma_j^{(n)}(\alpha_s) \sim \left(\frac{C_A \alpha_s}{2\pi}\right)^n \frac{1}{(j-1)^{2n+1}}. \tag{8.198}$$

Thus, in perturbation theory the anomalous dimension is singular at $j = 1$, and the singularity increases with powers of α_s. But from (8.148) it follows that the sum of the singular fixed-order contributions must give the following finite result:

$$\gamma_1(\alpha_s) = \lim_{j \to 1} \sum_{n=1}^{\infty} \gamma_j^{(n)}(\alpha_s) = \sqrt{\frac{C_A \alpha_s}{2\pi}}. \tag{8.199}$$

To analyze the behaviour of $\gamma_j(\alpha_s)$ near $j = 1$ let us rewrite Eq. (8.153) in terms of the variables $x = \omega/\epsilon$ and $Q^2 = \epsilon^2 \theta^2$,

$$Y = \frac{1}{2}\ln(Q^2/Q_0^2), \quad y = \ln x + Y, \tag{8.200}$$

$$D(x, Q^2) = \delta(x-1) + \frac{C_A}{\pi}\int_x^1 \frac{dz}{z}\int_{Q_0^2 z^2/x^2}^{Q^2 z^2} \frac{dk_\perp^2}{k_\perp^2}\alpha_s(k_\perp^2)D(\frac{x}{z}, k_\perp^2), \tag{8.201}$$

Taking account of the difference in the normalization (8.189) and (8.150), we have

$$M(j, Q^2) = \int_0^1 \frac{dx}{x}x^{j-1}D(x, Q^2). \tag{8.202}$$

so that

$$\begin{aligned}
Q^2\frac{dM(j, Q^2)}{dQ^2} &= \int_0^1 \frac{dz}{z}\int_0^z \frac{dx}{x}x^{j-1}\frac{C_A}{\pi}\alpha_s(Q^2z^2)D(\frac{x}{z}, Q^2z^2) \\
&= \int_0^1 \frac{dz}{z}z^{j-1}\frac{C_A}{\pi}\alpha_s(Q^2z^2)M(j, Q^2z^2) \\
&= \int_0^{Q^2} \frac{dk_\perp^2}{k_\perp^2}\left(\frac{k_\perp^2}{Q^2}\right)^{\frac{j-1}{2}}\frac{C_A}{2\pi}\alpha_s(k_\perp^2)M(j, k_\perp^2). \tag{8.203}
\end{aligned}$$

The second derivative gives

$$\left(Q^2\frac{d}{dQ^2}\right)^2 M(j, Q^2) = -\frac{(j-1)}{2}Q^2\frac{d}{dQ^2}M(j, Q^2) + \frac{C_A\alpha_s(Q^2)}{2\pi}M(j, Q^2). \tag{8.204}$$

The leading singularities of γ_j are not affected by the running of α_s. At fixed α_s, looking for solutions of this equation with $M_j(Q^2) \propto (Q^2)^{\gamma_j}$, we obtain

$$\gamma_j^2 + \frac{(j-1)}{2}\gamma_j - \frac{C_A\alpha_s}{2\pi} = 0, \tag{8.205}$$

which gives for the leading anomalous dimension

$$\gamma_j = \frac{1}{4}\left(\sqrt{(j-1)^2 + \frac{8C_A\alpha_s}{\pi}} - (j-1)\right). \tag{8.206}$$

At $j = 1$ we come to (8.199).

8.13 Modified Leading Logarithmic Approximation (MLLA)

The double-logarithmic approximation (DLA) discussed above gives the qualitative picture of particle production in hard interactions with relatively small momenta. But this approximation is too rough for reasonable quantitative predictions. In order to obtain reliable predictions one needs to take into account nonleading logarithms. Evidently, there is no hope to sum all of such logarithms. What one can hope for is to understand and identify those which are essential.

At first sight this task looks hopeless. In order to take into account nonleading logarithms one has to consider a huge number of interference contributions. But then it looks impossible to maintain a probabilistic interpretation.

Therefore, it seems striking that a scheme, taking into account essential single logarithms and based on the parton-shower picture, does exist. Moreover, it looks like a simple generalization of the DLA. This scheme, called the modified leading logarithmic approximation (MLLA), was developed in Refs. [72]–[73].

The results of this approach agree with the renormalization-group approach developed in Refs. [74]–[76]. The undoubted advantage of the latter approach is its rigor and the possibility to study higher-order corrections. However, it has a limited application region and is not as transparent as the approach based on parton branching. The MLLA has the advantage that the most important contributions come from small angles. To find subleading logarithms coming from this region one needs to take into account the running of α_s and to consider hard-parton decays.

The DLA can be improved by the generalization of Eqs. (8.123), (8.124) for the generating functionals. These equations can be rewritten in the form of

$$Z_P(p, \theta|u) = e^{-w_P(p,\theta)} \left(u(p) + \int \frac{2C_P\alpha_s}{\pi} \frac{d^3k\Theta_{p,\theta}^k}{4\pi\omega^2} \frac{\epsilon_p}{kp} Z_g(k, \theta_{kp}|u) e^{w_P(p,\theta_{kp})} Z_P(p, \theta_{kp}|u) \right),$$

(8.207)

where

$$Z_P(p, \theta|u) = u(p)\Phi_P(p, \theta|u).$$

(8.208)

Taking the derivative with respect to θ and using $z = \omega/\epsilon$,

$$\frac{d^3k}{4\pi\omega^2} \frac{\epsilon_p}{kp} u(k) = \frac{dz}{z} \frac{d\theta_{kp}}{\theta_{kp}}, \qquad dw_P(p,\theta) = \frac{2C_P\alpha_s}{\pi} \frac{dz}{z} \frac{d\theta}{\theta},$$

(8.209)

we obtain

$$\theta\frac{d}{d\theta} Z_P(p, \theta|u) = \frac{2C_P\alpha_s}{\pi} \int_0^1 \frac{dz}{z} \left(Z_g(zp, \theta|u) - 1 \right) Z_P(p, \theta|u).$$

(8.210)

The forms (8.207), (8.210) resemble the equations of jet calculus [65],[77]. These equations could be considered as more general than (8.207), since they are written for partons with $x \sim 1$. However, they do not take into account the colour coherence. Therefore, the evolution parameter in these equations is the jet invariant mass, whereas at small x it should be the jet opening angle. At large x the difference in the evolution parameters disappear.

Therefore, the MLLA generalization of (8.210) looks like the jet calculus equations, where the jet-opening angle is taken as the evolution parameter:

$$\theta \frac{d}{d\theta} Z_a(p, \theta|u) = \sum_{b,c} \int_0^1 dz \frac{\alpha_s(k_\perp^2)}{\pi} P_a^{bc}(z) \left(Z_b(zp, \theta|u) Z_c((1-z)p, \theta|u)\right)$$
$$- Z_a(p, \theta|u)). \tag{8.211}$$

Here, as before, $Z_a(p, \theta|u)$ is the generating functional for the jet of opening angle θ created by parton a with momentum p; but it describes now the production of partons with any energy fraction and of any species. Therefore, now u is the set of the functions $\{u_P(k)\}$ for all species.

In (8.211), $k_\perp^2 = z(1-z)\epsilon^2\theta^2$ and the splitting functions $P_a^{bc}(z)$ are defined by the following relations (see Chapter 7)

$$\theta^2 \frac{d^2 w_a^{bc}}{dz d\theta^2} = \frac{\alpha_s}{2\pi} P_a^{bc}(z), \tag{8.212}$$

where w_a^{bc} are the Born probabilities of the decays $a \to bc$ with $\epsilon_b/\epsilon_a \le z$ and $\theta_{ba} \le \theta$. (The splitting functions were presented in Chapter 7.) Equation (8.211) requires initial conditions. They are imposed at $\theta = Q_0/p$:

$$Z_a(p, \theta|u)|_{p\theta=Q_0} = u(p). \tag{8.213}$$

Together with Eq. (8.211), these conditions provide for the generating functionals the same boundary values as in the double-logarithmic approximation:

$$Z_a(p, \theta|u)|_{u=1} = 1. \tag{8.214}$$

For a system of jets produced in some hard interaction, the generating functional is given by the product of the generating functionals for separate jets, analogously to (8.122).

For the inclusive distributions

$$D_a^b(x, Q^2) = \int d^3 k \delta\left(\frac{k_0}{\epsilon} - x\right) \frac{\delta}{\delta u(k)} Z_a(p, \theta|u)\Big|_{u=1}, \tag{8.215}$$

where $Q^2 = \epsilon^2\theta^2$, taking account of (8.214) we obtain from (8.211)

$$Q^2 \frac{d}{dQ^2} D_a^b(x, Q^2) = \sum_C \int_x^1 \frac{dz}{z} \frac{\alpha_s(k_\perp^2)}{2\pi} P_a^c\left(\frac{x}{z}\right) D_c^b\left(z, Q^2 z^2\right). \tag{8.216}$$

Here $P_a^b(z)$ are the regularized splitting functions:

$$P_a^b(z) = \sum_c P_a^{bc}(z) - \delta(1-z)\delta_a^b \frac{1}{2} \sum_{c,d} \int_0^1 dz' P_a^{cd}(z'), \tag{8.217}$$

which were given by Eqs. (7.135) in Chapter 7.

Since the region of small $(1 - z)$ is not essential in the integral (8.216), $k_\perp^2 = Q^2 z^2 (1-z)^2$ in the argument of α_s can be changed for $Q^2 z^2$. After that, for the Mellin transforms

$$M_a^b(j, Q^2) = \int_0^1 dx\, x^{j-1} D_a^b(x, Q^2).\tag{8.218}$$

Eqs. (8.216) give

$$Q^2 \frac{d}{dQ^2} M_a^b(j, Q^2) = \sum_c \int_0^1 dz\, z^{j-1} \frac{\alpha_s(Q^2 z^2)}{2\pi} P_a^c(z) M_c^b\left(j, Q^2 z^2\right).\tag{8.219}$$

In the case of j not close to 1, which corresponds to values of x less than 1 but of the order of 1, we have in the essential integration region $z \sim 1$. Neglecting the z-dependence of M and of α_s, we come from (8.219) to the usual DGLAP equations (8.191). But for $j \to 1$, the region of small z becomes important for the terms $\propto 1/z$ in $P_a^c(z)$. For these terms it is not correct to neglect the z-dependence of M and of α_s. They can be considered in the same way as in (8.203), (8.204). Note that such terms are present only in $P_a^g(z)$ and are equal to $2C_a/z$. In the integrals with remaining terms of $P_a^b(z)$, the z-dependence of M and of α_s can be neglected as well as for j not close to 1. Thus for the gluon distributions we come to the following equations (compare with (8.204) and (8.191)):

$$\frac{d^2 M_a(j, Q^2)}{d(\ln Q^2)^2} = -\frac{(j-1)}{2} \frac{dM_g(j, Q^2)}{d\ln Q^2} + \frac{C_a \alpha_s(Q^2)}{2\pi} M_g(j, Q^2)$$
$$+ \frac{\alpha_s(Q^2)}{2\pi} \sum_c \left(A_a^c(j) - \delta_g^c \frac{2C_a}{j-1}\right)\left(\frac{(j-1)}{2} + \frac{d}{d\ln Q^2}\right) M_c(j, Q^2),\tag{8.220}$$

where $A_a^c(j)$ is given by (8.192). In the derivation of (8.220), we neglect

$$\frac{d\ln \alpha_s(Q^2)}{d\ln Q^2} \sim \alpha_s(Q^2)\tag{8.221}$$

in comparison with

$$\frac{d\ln M_a(j, Q^2)}{d\ln Q^2} \sim \sqrt{\alpha_s(Q^2)}.\tag{8.222}$$

At small $j - 1$, the last term in this equation can be considered as a small correction. It determines the correction to the leading anomalous dimension (8.206). Taking (8.220) with $a = g$ and considering that according to (8.219) in the leading approximation, we have

$$\frac{dM_q(j, Q^2)}{d\ln Q^2} = \frac{C_F}{C_A} \frac{dM_g(j, Q^2)}{d\ln Q^2},\tag{8.223}$$

and that in the limit $j - 1 = 0$,

$$A_g^g(j) - \frac{2C_A}{j-1} = -\frac{11}{6} N_c - \frac{n_f}{3}, \quad A_g^q(j) = \frac{1}{3},\tag{8.224}$$

we obtain instead of (8.205) the following:

$$\gamma_j^2\left(\alpha_s(Q^2)\right) + \frac{(j-1)}{2}\gamma_j\left(\alpha_s(Q^2)\right) - \frac{C_A\alpha_s(Q^2)}{2\pi} + \frac{d\gamma_j\left(\alpha_s(Q^2)\right)}{d\ln Q^2}$$

$$+ \frac{\alpha_s(Q^2)}{2\pi}\left(\frac{j-1}{2} + \gamma_j(\alpha_s(Q^2))\right)\left(\frac{11}{6}N_c + \frac{n_f}{3N_c^2}\right) = 0. \qquad (8.225)$$

Using the relation

$$\frac{d\gamma_j\left(\alpha_s(Q^2)\right)}{d\ln Q^2} = -\beta_0\frac{\alpha_s^2(Q^2)}{4\pi}\frac{d\gamma_j\left(\alpha_s(Q^2)\right)}{d\alpha_s(Q^2)}, \quad \beta_0 = \frac{11}{3}N_c - \frac{2n_f}{3}, \qquad (8.226)$$

and denoting $(11/6)N_c + n_f/(3N_c^2)$ by a, one gets

$$\gamma_j(\alpha_s) = \gamma_j^{DLA} + \frac{\alpha_s}{4\pi}\left[-a\left(1 + \frac{j-1}{\sqrt{(j-1)^2 + 16\gamma_0^2}}\right) + \beta_0\frac{4\gamma_0^2}{(j-1)^2 + 16\gamma_0^2}\right], \qquad (8.227)$$

where γ_j^{DLA} is given by (8.206) and $\gamma_0 = \sqrt{C_A\alpha_s/(2\pi)}$.

The change of the anomalous dimension means, in particular, the change of the hump-backed x-spectrum and the multiplicity. At large Y, $Y \gg \ln^2(Q_0/\Lambda_{QCD})$,

$$n(Y) \propto Y^{-\frac{B}{2}+\frac{1}{4}}\exp\sqrt{\frac{16N_c}{\beta_0}Y}, \qquad (8.228)$$

where $B = 2a/\beta_0$. In terms of the running coupling,

$$\ln n(Y) = \sqrt{\frac{32\pi N_c}{\alpha_s(Q^2)}\frac{1}{\beta_0}} + \left(\frac{B}{2} - \frac{1}{4}\right)\ln\alpha_s(Y) + \mathcal{O}(1). \qquad (8.229)$$

The MLLA effect shifts the peak in the fragmentation function from the DLA position $\ln\frac{1}{x} = \frac{Y}{2}$, $Y = \ln(Q/Q_0)$, to the larger value

$$\ln\frac{1}{x_{max}} = Y\left(\frac{1}{2} + a\sqrt{\frac{\alpha_s(Q^2)}{8N_c\pi}} - a^2\frac{\alpha_s(Q^2)}{8N_c\pi} + \dots\right), \qquad (8.230)$$

thus softening the spectrum. The shape of the peak is also affected by the MLLA corrections: the peak is narrowed and skewed towards higher values of x.

References

[1] Appelquist, T. and Georgi, H., *Phys. Rev.* **D8** (1973) 4000.
[2] Zee, A., *Phys. Rev.* **D8** (1973) 4038.
[3] Jost, R. and Luttinger, J.M., *Helv. Phys. Acta* **23** (1950) 201.
[4] Poggio, E. C., Quinn, H. R. and Weinberg, S., *Phys. Rev.* **D13** (1976) 1958.
[5] Barnett, R. M., Dine, M. and McLerran, L. D., *Phys. Rev.* **D22** (1980) 594.
[6] Chetyrkin, K. G., Kataev, A. L. and Tkachov, F. V., *Phys. Lett.* **B85** (1979) 277.

[7] Dine, M. and Sapirstein, J. R., *Phys. Rev. Lett.* **43** (1979) 668.

[8] Celmaster, W. and Gonsalves, R. J., *Phys. Rev.* **D21** (1980) 3112.

[9] Gorishny, S. G., Kataev, A. L. and Larin, S. A., *Phys. Lett.* **B259** (1991) 144.

[10] Surguladze, L. R. and Samuel, M. A., *Phys. Rev. Lett.* **66** (1991) 560 [Erratum-ibid. **66** (1991) 2416].

[11] Bethke, S. *et al.* [JADE Collaboration], *Phys. Lett.* **B213** (1988) 235.

[12] Bethke, S., Kunszt, Z., Soper, D. E. and Stirling, W. J., *Nucl. Phys.* **B370** (1992) 310 [Erratum-ibid. **B523** (1998) 681].

[13] Schwitters, R., SLAC-PUB-1666, Nov 1975. 20pp. Invited paper presented at the Intern. Symp. on Lepton and Photon Interactions at High Energy, Stanford, California, Aug 21–27, 1975. Published in: *Lepton-Photon Symp.* 1975:5 (QCD161:I71:1975)

[14] Hanson, G. *et al.*, *Phys. Rev. Lett.* **35** (1975) 1609.

[15] Kinoshita, T., *J. Math. Phys.* **3** (1962) 650.

[16] Lee, T. D. and Nauenberg, M., *Phys. Rev.* **B133** (1964) 1549;

[17] Lee, T. D. and Nauenberg, M., *Ann. Phys.* **77** (1973) 570.

[18] Sterman, G. and Weinberg, S., *Phys. Rev. Lett.* **39** (1977) 1436.

[19] Baier, V. N., Fadin, V. S. and Khoze, V. A., *Nucl. Phys.* **B65** (1973) 381.

[20] Kramer, G., *Springer Tracts Mod. Phys.* **102** (1984) 1.

[21] Lampe, B. and Kramer, G., *Prog. Theor. Phys.* **76** (1986) 1340.

[22] Kramer, G. and Lampe, B., *Fortsch. Phys.* **37** (1989) 161.

[23] Gonsalves, R. J., *Phys. Rev.* **D28** (1983) 1542.

[24] Ellis, J. R., Gaillard, M. K. and Ross, G. G., *Nucl. Phys.* **B111** (1976) 253 [Erratum-ibid. **B130** (1977) 516].

[25] Hoyer, P., Osland, P., Sander, H. G., Walsh, T. F. and Zerwas, P. M., *Nucl. Phys.* **B161** (1979) 349.

[26] Fabricius, K., Schmitt, I., Schierholz, G. and Kramer, G., *Phys. Lett.* **B97** (1980) 431.

[27] Ellis, R. K., Ross, D. A. and Terrano, A. E., *Phys. Rev. Lett.* **45** (1980) 1226.

[28] Ellis, R. K., Ross, D. A. and Terrano, A. E., *Nucl. Phys.* **B178** (1981) 421.

[29] Vermaseren, J. A. M., Gaemers, K. J. F. and Oldham, S. J., *Nucl. Phys.* **B187** (1981) 301.

[30] Kunszt, Z., *Phys. Lett.* **B99** (1981) 429.

[31] Fabricius, K., Schmitt, I., Kramer, G. and Schierholz, G., *Z. Phys.* **C11** (1981) 315.

[32] Kunszt, Z., *Phys. Lett.* **B107** (1981) 123.

[33] Ali, A., *Phys. Lett.* **B110** (1982) 67.

[34] Gutbrod, F., Kramer, G. and Schierholz, G., *Z. Phys.* **C21** (1984) 235.

[35] Lampe, B. and Kramer, G., *Commun. Math. Phys.* **97** (1985) 257.

[36] Gaemers, K. J. F. and Vermaseren, J. A. M., *Z. Phys.* **C7** (1980) 81.

[37] Ali, A., Korner, J. G., Kunszt, Z., Pietarinen, E., Kramer, G., Schierholz, G. and Willrodt, J., *Nucl. Phys.* **B167** (1980) 454.

[38] Brandelik, R. *et al.* [TASSO Collaboration], *Phys. Lett.* **B86** (1979) 243.

[39] Barber, D. P. *et al.*, *Phys. Rev. Lett.* **43** (1979) 830.

[40] Bartel, W. *et al.* [JADE Collaboration], *Phys. Lett.* **B91** (1980) 142.

[41] Farhi, E., *Phys. Rev. Lett.* **39** (1977) 1587.

[42] Ellis, R. K. and Ross, D. A., *Phys. Lett.* **B106** (1981) 88.

[43] Catani, S., Turnock, G., Webber, B. R. and Trentadue, L., *Phys. Lett.* **B263** (1991) 491.

[44] Gribov, V. N. and Lipatov, L. N., *Sov. J. Nucl. Phys.* **15** (1972) 438 [*Yad. Fiz.* **15** (1972) 781].

[45] Gribov, V. N. and Lipatov, L. N., *Sov. J. Nucl. Phys.* **15** (1972) 675 [*Yad. Fiz.* **15** (1972) 1218].

[46] Lipatov, L. N., *Sov. J. Nucl. Phys.* **20** (1975) 94 [*Yad. Fiz.* **20** (1974) 181].

[47] Altarelli, G. and Parisi, G., *Nucl. Phys.* **B126** (1977) 298.

[48] Dokshitzer, Y. L., *Sov. Phys. JETP* **46** (1977) 641 [*Zh. Eksp. Teor. Fiz.* **73** (1977) 1216].

[49] Curci, G., Furmanski, W. and Petronzio, R., *Nucl. Phys.* **B175**(1980) 27.

[50] Dokshitzer, Yu. L., Marchesini, G. and Salam, G. P., *Phys. Lett.* **B634** (2006) 504 [arXiv:hep-ph/0511302].

[51] Kuraev, E. A. and Fadin, V. S., *Yad. Fiz.* **27** (1978) 1107.

[52] Ermolaev, B. I. and Fadin, V. S., *JETP Lett.* **33** (1981) 269 [*Pisma Zh. Eksp. Teor. Fiz.* **33** (1981) 285].

[53] Dokshitzer, Y. L., Diakonov, D. and Troyan, S. I., *Phys. Rept.* **58** (1980) 269.

[54] Mueller, A. H., *Phys. Lett.* **B104** (1981) 161.

[55] Gribov, V. N., *Sov. J. Nucl. Phys.* **5** (1967) 280 [*Yad. Fiz.* **5** (1967) 399].

[56] Low, F. E., *Phys. Rev.* **110** (1958) 974.

[57] Kirschner, R. and Lipatov, L. N., *Nucl. Phys.* **B213** (1983) 122.

[58] Kirschner, R. and Lipatov, L. N., *Sov. Phys. JETP* **56** (1982) 266 [*Zh. Eksp. Teor. Fiz.* **83** (1982) 488].

[59] Fadin, V. S., *Yad. Fiz.* **37** (1983) 408.

[60] Ermolaev, B. I., Lipatov, L. N. and Fadin, V. S., *Yad. Fiz.* **45** (1987) 817.

[61] Dokshitzer, Y. L., Fadin, V. S. and Khoze, V. A., *Phys. Lett.* **B115** (1982) 242.

[62] Dokshitzer, Y. L., Fadin, V. S. and Khoze, V. A., *Z. Phys.* **C15** (1982) 325.

[63] Dokshitzer, Y. L., Fadin, V. S. and Khoze, V. A., *Z. Phys.* **C18** (1983) 37.

[64] Furmanski, W., Petronzio, R. and Pokorski, S., *Nucl. Phys.* **B155** (1979) 253.

[65] Konishi, K., Ukawa, A. and Veneziano, G., *Nucl. Phys.* **B157** (1979) 45.

[66] Konishi, K., (Rutherford). RL-79-035, May 1979. 18 pp.

[67] Bassetto, A., Ciafaloni, M. and Marchesini, G., *Nucl. Phys.* **B163** (1980) 477.

[68] Amati, D., Bassetto, A., Ciafaloni, M., Marchesini, G. and Veneziano, G., *Nucl. Phys.* **B173** (1980) 429.

[69] Azimov, Ya. I., Dokshitzer, Y. L. and Khoze, V. A., Materials of XVI LNPI Winter School, (1981) pp. 26–53.

[70] Azimov, Ya. I., Dokshitzer, Y. L. and Khoze, V. A., Materials of XVII LNPI Winter School, (1982) pp. 162–225.

[71] Koba, Z., Nielsen, H. B. and Olesen, P. *Nucl. Phys.* **B40** (1972) 317.

[72] Dokshitzer, Y. L. and Troyan, S. I., Proceedings of the XIX Winter School of the LNPI, volume 1, page 144. Leningrad, 1984.

[73] Dokshitzer, Y. L., Khoze, V. A. and Troyan, S. I., In *Perurbative QCD*, ed. A. H. Mueller, World Scientific, Singapore, 1989, pp. 241–410.

[74] Mueller, A. H., *Nucl. Phys.* **B213** (1983) 85.

[75] Mueller, A. H., *Nucl. Phys.* **B228** (1983) 351.

[76] Mueller, A. H., *Nucl. Phys.* **B241** (1984) 141.

[77] Cvitanovic, P., Hoyer, P. and Konishi, K., *Phys. Lett.* **B85** (1979) 413.

9
BFKL approach

9.1 Introduction

The Balitsky–Fadin–Kuraev–Lipatov (BFKL) equation [1]–[4] became famous owing to the prediction of the rapid growth of the $\gamma^* p$ cross section at increasing energy, subsequently discovered experimentally. Therefore, the BFKL equation is usually associated with the evolution of the unintegrated gluon distribution. The parton distributions serve now as the inherent part in the theoretical description of hard QCD processes. In hadron collisions, cross sections of processes with a hard scale Q^2 are given by the convolution

$$d\sigma_{AB}(s) = \sum_{a,b} \int_0^1 dx_a \int_0^1 dx_b \, f_A^a(x_a, Q^2) f_B^b(x_b, Q^2) \hat{\sigma}_{ab}(x_a x_b s, Q^2), \qquad (9.1)$$

where s is the squared total energy in the centre of mass system, $f_A^a(x, Q^2)$ is the density of the probability to find the parton a in the hadron A carrying a fraction x_a of its momentum, and $\hat{\sigma}_{ab}(x_a x_b s, Q^2)$ is the partonic cross section. Evolution of the parton distributions with $\tau = \ln\left(Q^2/\Lambda_{QCD}^2\right)$ is determined by the DGLAP [5]–[9] equations discussed in Chapter 7. The DGLAP equations permit to sum the terms strengthened in each order of the perturbation series by powers of $\ln Q^2$. These logarithms are called collinear since they are picked up from the region of small angles between parton momenta. There are logarithms of another kind, which are called soft ones, arising at integration over ratios of parton energies. These logarithms are present both in parton distributions and in partonic cross sections. At small values of the ratio $x = Q^2/s$, the soft logarithms appear to be even more important than the collinear ones.

At small x the gluons are the dominant partons. The unintegrated gluon density $\mathcal{F}(x, \mathbf{k})$ is defined in such a way that the gluon distribution $f^g(x, Q^2)$ is given by the integral $\int \mathcal{F}(x, \mathbf{k}) dk/(\pi k^2)\theta(Q^2 - k^2)$. The equation describing evolution of $\mathcal{F}(x, \mathbf{k})$ as a function of $\ln(1/x)$ has the structure

$$\frac{\partial \mathcal{F}}{\partial \ln(1/x)} = \mathcal{K} \otimes \mathcal{F}, \qquad (9.2)$$

where $\mathcal{K}$ is the BFKL kernel and $\otimes$ means convolution with respect to transverse momenta. Originally, the kernel was found in the NLA. The equation with this kernel permits to sum the leading terms $(\alpha_S \ln(1/x))^n$. The summation leads to rising cross sections

$$\sigma \sim s^{\omega_P}, \quad \omega_P = 4N_c \frac{\alpha_s}{\pi} \ln 2. \tag{9.3}$$

Just this result brought fame to the BFKL equation since the sharp rise of the proton structure function with decreasing x was discovered in the experiments on deep inelastic $e - p$ scattering at the collider HERA [10].

But the region of applicability of the BFKL approach is much wider. The approach gives the description of QCD-scattering amplitudes in the region of large s and fixed momentum transfer t, $s \gg |t|$ (Regge region), with various colour states in the t channel. The evolution equation for the unintegrated gluon distribution appears in this approach as a particular result for the imaginary part of the forward-scattering amplitude ($t = 0$ and vacuum quantum numbers in the t channel). It is worthwhile to add that the approach was developed, and is more suitable, for the description of processes with only one hard scale, such as $\gamma^*\gamma^*$-scattering with both photon virtualities of the same order, where the DGLAP evolution (i.e. evolution in Q^2) is not appropriate. The forward BFKL kernel can carry only restrictive information about the BFKL dynamics. Moreover, the nonforward case has an advantage of smaller sensitivity to large-distance contributions, since the diffusion in the infrared region is limited by $\sqrt{|t|}$.

The leading logarithmic approximation (LLA) can provide only qualitative predictions, because it does fix neither the scale of energy nor the scale of significant transverse momenta which determine the value of the coupling α_s. They can be determined in the next-to-leading approximation (NLA), when the terms $\alpha_S(\alpha_S \ln(1/x))^n$ are resummed. Therefore, the normalization of cross sections and the exponent ω_P in (9.3), called pomeron intercept, can be fixed only in the NLA.

Evidently, the power growth (9.3) of cross sections violate the Froissart bound [11]

$$\sigma_{tot} < const (\ln s)^2 \tag{9.4}$$

which follows from unitarity. The violation of the Froissart bound cannot be removed by calculation of radiative corrections at any fixed $NNN\ldots NL$ order and requires other methods. Most popular at present are nonlinear generalizations of the BFKL equation, related to the idea of saturation of parton densities [12]. A general approach to the unitarization problem is the reformulation of QCD in terms of a gauge-invariant effective field theory for the reggeized gluon interactions [13],[14]. It is discussed in Chapter 10.

Another well-known equation for the unintegrated gluon distribution is the Ciafaloni–Catani–Fiorani–Marchesini (CCFM) equation [15]–[17] based on an angular ordering, which follows from colour coherence effects discussed in Chapter 8. This equation interpolates between DGLAP and BFKL and unifies descriptions of the small and large x domains. It reduces to the DGLAP formalism at moderate x and give the same results as the BFKL approach at small x [18]–[20].

9.2 Gluon reggeization

The basis of the BFKL approach is the gluon reggeization. The notion reggeization of elementary particles in perturbation theory was introduced in [21]. In terms of the relativistic partial wave amplitude $A_j(t)$, analytically continued to complex j values, this means that the nonanalytic terms (proportional to Kronecker delta symbols), arising in the Born approximation on account of one-particle exchanges in the t channel, disappear as a consequence of radiative corrections. In other words, reggeization of an elementary particle with spin j_0 and mass m means that at large s and fixed t Born amplitudes with exchange of this particle in the t channel acquire a factor $s^{j(t)-j_0}$, with $j(m^2) = j_0$, as a result of radiative corrections. This phenomenon was discovered originally in QED in backward Compton scattering [21]. It was called reggeization because just such form of amplitudes is given by the Regge poles – moving poles in the complex angular momentum plane (j-plane) introduced by Regge [22]; therefore, these poles are called Regge poles (reggeons), and the functions $j(t)$ are Regge trajectories. The value $j(0)$ is the intercept, and the derivative $j'(0)$ is the slope of the trajectory.

For relativistic particles, the theory of complex angular momenta was developed by Gribov. This theory had outstanding significance in elementary particle physics. In the 1960s it was the main and almost unique tool of the theoretical analysis of strong interactions.

Recall that, as compared with ordinary particles, reggeons possess an additional quantum number, called a signature. The signature means parity with respect to the substitution $\cos \theta_t \leftrightarrow -\cos \theta_t$, where θ_t is the t-channel-scattering angle. At large s, this substitution is equivalent to the substitution $s \leftrightarrow u$, and at the same time $s = -u$. The reason for the appearance of this quantum number is that t-channel partial waves $f_j(t)$ cannot be represented at all integer j as values of analytical functions $f(j, t)$ with the required properties in the complex j-plane. Such representation is possible only separately for even (positive signature) and odd (negative signature) values of angular momentum. Therefore, for any $2 \rightarrow 2$ process, amplitudes with definite signatures are introduced in the Gribov–Regge theory as even and odd with respect to the substitution $\cos \theta_t \leftrightarrow -\cos \theta_t$ parts of the amplitude of this process. In the following, for brevity, we will use the words "process with definite signature" assuming the corresponding amplitude.

The fundamental role in the Gribov theory belongs to the reggeon with vacuum quantum numbers and positive signature, which is called pomeron after I. Ya. Pomeranchuk. The pomeron determines behaviour of total cross sections at high energy. Originally, it was introduced (with intercept equal to one) [23],[24] to provide constant cross sections at asymptotically high energies. Very important and intriguing is another reggeon, differing from the pomeron by C and P-parity and called odderon [25],[26]. The odderon is responsible for the difference of particle-particle and particle-antiparticle cross sections.

QCD is a unique theory where all elementary particles reggeize. In contrast to QED, where the electron does reggeize in perturbation theory [21], but the photon remains elementary [27], in QCD the gluon does reggeize [28]–[30],[1],[2] as well as the quark [31],[32],[33]. The reggeization is very important for the theoretical description of

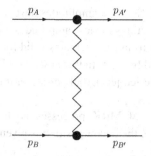

Fig. 9.1. The process $A + B \rightarrow A' + B'$ with colour octet in the t channel and negative signature. The zigzag line represent reggeized gluon exchange.

high-energy processes with fixed momentum transfer. Especially important is the gluon reggeization, because gluon exchanges provide nondecrease of cross sections with energy. In each order of perturbation theory, amplitudes with negative signature do dominate, owing to cancellation of the leading logarithmic terms in amplitudes with positive signature (these amplitudes turn out to be pure imaginary in the LLA due to this cancellation). Therefore, the primary reggeon in QCD turns out to be the reggeized gluon, which has negative signature. The pomeron and the odderon emerge as compound states of two and three reggeized gluons, respectively.

The gluon reggeization determines the form of QCD amplitudes at large energies and limited transverse momenta. For the process $A + B \rightarrow A' + B'$, amplitudes with a colour octet t-channel exchange and negative signature can be depicted by the diagram of Fig. 9.1 and have the Regge form

$$\mathcal{A}_{AB}^{A'B'} = \frac{s}{t} \Gamma_{A'A}^{c} \left[\left(\frac{-s}{-t} \right)^{\omega(t)} + \left(\frac{s}{-t} \right)^{\omega(t)} \right] \Gamma_{B'B}^{c} , \qquad (9.5)$$

which is valid in the NLA as well as in the LLA. In (9.5), $s = (p_A + p_B)^2$; $\omega(t)$ is called gluon trajectory (in fact, the trajectory is $j(t) = 1 + \omega(t)$), c is a color index, and $\Gamma_{P'P}^{c}$ are the particle–particle–reggeon (PPR) vertices (we also will call them scattering vertices) which do not depend on s. Note that the form (9.5) represents correctly the analytical structure of the scattering amplitude, which is quite simple in the elastic case.

The gluon reggeization determines the form not only of elastic, but also inelastic, amplitudes in the multi-Regge kinematics (MRK), which is the most important at high energy. We call MRK the kinematics where all particles have limited transverse momenta and are combined into jets with limited invariant mass of each jet and large (increasing with s) invariant mass of any pair of jets. This kinematics gives dominant contributions to cross sections of QCD processes. In the LLA, each jet is actually one particle. In the NLA, one of the jets can contain a couple of particles. Such kinematics is called also quasi-multi-Regge kinematics (QMRK). We use the notion of jets and extend the notion of MRK, so that it includes QMRK, in order to unify our considerations.

In perturbation theory, the MRK amplitudes are determined by gluon exchanges in channels with fixed momentum transfer. Despite a great number of contributing Feynman diagrams, it turns out that in MRK these amplitudes have a simple factorized form. Quite uncommonly, the radiative corrections to these amplitudes do not destroy this form but give only simple Regge factors. The form (9.5) remains valid for the case when any of the particles A', B', or A, B is replaced by a jet. In this case, the PPR vertices $\Gamma^c_{A'A}$ and $\Gamma^c_{B'B}$ are the effective vertices for particle-jet, jet-particle or jet-jet transitions owing to interactions with the reggeized gluons.

A suitable tool for analysis of MRK processes is the Sudakov decomposition of momenta. For any momentum p, the decomposition is given by

$$p = \beta p_1 + \alpha p_2 + p_\perp, \tag{9.6}$$

where p_1 and p_2 are light-like vectors,

$$(p_1 + p_2)^2 = 2p_1 p_2 = s, \quad p^2 = s\alpha\beta + p_\perp^2 = s\alpha\beta - \boldsymbol{p}^2. \tag{9.7}$$

Here and below to emphasize the Euclidean properties of components of momenta transverse to the p_1, p_2 plane, we will use vector notation. We also will use fixed (not connected with s) light-cone momenta n_1 and n_2 along p_1 and p_2, respectively, with $n_1^2 = n_2^2 = 0$, $(n_1 n_2) = 1$, and denote $(pn_2) \equiv p^+$, $(pn_1) \equiv p^-$.

Let us consider the amplitude $\mathcal{A}_{2 \to n+2}$ of the process $A + B \to A' + J_1 + \ldots + J_n + B'$ of production of n jets (see Fig. 9.2). We assume that the initial momenta p_A and p_B have dominant components p_A^+ and p_B^-. Generally, it is not assumed that the components $p_{A\perp}$, $p_{B\perp}$ transverse to the (p_1, p_2) plane are zero:

$$p_A = p_1 + \frac{p^2 A - p^2 A\perp}{s} p_2 + p_{A\perp}, \quad p_B = p_2 + \frac{p^2 B - p^2 B\perp}{s} p_1 + p_{B\perp},$$

$$(p_A + p_B)^2 \simeq s = 2p_1 p_2. \tag{9.8}$$

Moreover, A and B, as well as A' and B', can represent jets. Denoting momenta of the final jets k_i, $i = 0 \ldots, n+1$,

$$k_i = \beta_i p_1 + \alpha_i p_2 + k_{i\perp}, \quad s\alpha_i \beta_i = k_i^2 - k_{i\perp}^2 = k_i^2 + \boldsymbol{k}_i^2, \tag{9.9}$$

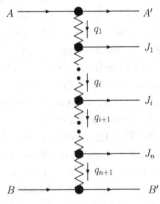

Fig. 9.2. Schematic representation of the process $A + B \to A' + J_1 + \cdots + J_n + B'$ in MRK. The zigzag lines represent reggeized gluon exchange; the black circles denote the reggeon vertices; q_i are reggeon momenta.

we have

$$\frac{1}{s} \sim \alpha_0 \ll \alpha_1 \cdots \ll \alpha_n \ll \alpha_{n+1} \simeq 1, \qquad \frac{1}{s} \sim \beta_{n+1} \ll \beta_n \cdots \ll \beta_1 \ll \beta_0 \simeq 1. \quad (9.10)$$

Eqs. (9.9) and (9.10) ensure the squared invariant masses of neighbouring jets,

$$s_i = (k_{i-1} + k_i)^2 \approx s\beta_{i-1}\alpha_i = \frac{\beta_{i-1}}{\beta_i} \left(k_i^2 + k_i^{\,2} \right), \quad (9.11)$$

to be large compared with the squared transverse momenta:

$$s_i \gg k_i^2 \sim | t_i | =| q_i^2 |, \quad (9.12)$$

where

$$t_i = q_i^2 \approx q_{i\perp}^2 = -\boldsymbol{q}_i^{\,2}, \quad (9.13)$$

and their product is proportional to s:

$$\prod_{i=1}^{n+1} s_i = s \prod_{i=1}^{n} (k_i^2 + k_i^2). \quad (9.14)$$

Amplitudes dominant in each order of perturbation theory can be represented by Fig. 9.2. Multiparticle amplitudes have a complicated analytical structure. They are not simple even in MRK (see, for instance, [34]–[36], [37]). Fortunately, only real parts of these amplitudes are used in the BFKL approach in NLA as well as in LLA. Restricting ourselves to the real parts (although it is not explicitly indicated below), we can write (see [38] and references therein)

$$\mathcal{A}_{AB}^{A'B'+n} = 2s\Gamma_{A'A}^{c_1} \left[\prod_{i=1}^{n} \frac{1}{t_i} \gamma_{c_i c_{i+1}}^{J_i}(q_i, q_{i+1}) \left(\frac{s_i}{s_i^0} \right)^{\omega(t_i)} \right] \frac{1}{t_{n+1}} \left(\frac{s_{n+1}}{s_{n+1}^0} \right)^{\omega(t_{n+1})} \Gamma_{B'B}^{c_{n+1}}, \quad (9.15)$$

where the vertices $\Gamma_{A'A}^a$ and $\Gamma_{B'B}^b$ are the same as in (9.5), s_i^0, $i = 1, \ldots, n+1$ are energy scales, and $\gamma_{c_i c_{i+1}}^{J_i}(q_i, q_{i+1})$ are the reggeon–reggeon–particle (RRP) vertices, i.e. the effective vertices for production of jets J_i with momenta $k_i = q_i - q_{i+1}$ in collisions of reggeons with momenta q_i and $-q_{i+1}$ and colour indices c_i and c_{i+1}. Actually, in the LLA, only one gluon can be produced in the RRP vertex; in the NLA, a jet can contain two gluons or a $q\bar{q}$ pair. Note that we have taken definite energy scales in the Regge factors in Eq. (9.15) as well as in Eq. (9.5). In the LLA the energy scales are not important at all. In the NLA we could take, in principle, an arbitrary scale s_R; in this case the PPR and RRP vertices would become dependent on s_R. Of course, physical results must be scale independent. We will use the following scales:

$$s_1^0 = \sqrt{\boldsymbol{q}_1^2 \boldsymbol{k}_1^2}, \quad s_j^0 = \sqrt{\boldsymbol{k}_{j-1}^2 \boldsymbol{k}_j^2}, \quad j = 2, \ldots n, \quad s_{n+1}^0 = \sqrt{\boldsymbol{k}_n^2 \boldsymbol{q}_{n+1}^2}. \quad (9.16)$$

For brevity in the following, we call the forms (9.5) and (9.15) reggeized forms, and when speaking about gluon reggeization we mean these forms. The gluon-reggeization hypothesis is extremely powerful since an infinite number of amplitudes is expressed in terms of the gluon Regge trajectory and several reggeon vertices. The gluon reggeization was proved in [39] (see also [40]) in the LLA and recently also in the NLA (see [41] and references therein). The proof is based on "bootstrap" relations, required by compatibility of the gluon reggeization with s-channel unitarity. It turns out that fulfillment of all these relations ensures the reggeized form of energy-dependent radiative corrections order by order in perturbation theory. It is quite nontrivial that an infinite number of bootstrap relations for the multiparticle production amplitudes can be fulfilled if the reggeon vertices and trajectory satisfy several bootstrap conditions. All these conditions are derived and are proved to be satisfied.

9.3 Reggeon vertices and trajectory

9.3.1 Reggeon vertices in Born approximation

The idea of gluon reggeization arose as the result of the fixed-order calculations in non-Abelian gauge theories with spontaneously broken gauge invariance [2],[30]. The dispersion method of calculations based on unitarity and analyticity suggested in [30] and developed in [2] appeared to be very effective. It greatly simplifies calculations even in the case of Born amplitudes, which can be easily found using t-channel unitarity. For elastic-scattering amplitudes $\mathcal{A}_{AB}^{A'B'}(s,t)$ the t-channel discontinuities are shown by Fig. 9.3, where the dashed line representing the t-channel discontinuity means the substitution

$$\frac{1}{t+i0} \longrightarrow -2\pi i \delta(t) \tag{9.17}$$

in the gluon propagator. The straight lines can represent either quarks or gluons.

Note that in order to preserve usual analyticity properties of the amplitude we have to use covariant gauges for the intermediate gluons. We will use the Feynman gauge. In the Regge region, it is convenient to exploit the substitution

$$g_{\mu\nu} \longrightarrow \frac{2}{s} p_{2\mu} p_{1\nu} \tag{9.18}$$

for the tensor $g_{\mu\nu}$ in the numerator of the gluon propagator between vertices with indices μ and ν and momenta close to p_1 and p_2, respectively. This is possible because in the decomposition

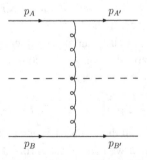

Fig. 9.3. The t-channel discontinuity of the amplitude of the process $A + B \rightarrow A' + B'$.

$$g_{\mu\nu} = \frac{2}{s}\left(p_{2\mu}p_{1\nu} + p_{1\mu}p_{2\nu}\right) + g_{\mu\nu}^{\perp} \tag{9.19}$$

the last two terms give negligible contributions. Therefore, we obtain

$$2i\,\mathrm{Im}_t\,\mathcal{A}_{AB}^{A'B'}(s,t) = -4\pi is\delta(t)\,\Gamma_{A'A}^c\,\Gamma_{B'B}^c, \tag{9.20}$$

where Im_t denotes the t-channel imaginary part and the vertices $\Gamma_{A'A}^c$ and $\Gamma_{B'B}^c$ are the interaction vertices of the t-channel gluon with the gluon colour index c and polarization vectors ip_2/s and ip_1/s, respectively. Renormalizability of the theory ensures against existence of terms of order s not decreasing with t, so that (9.20) determines the amplitude unambiguously:

$$\mathcal{A}_{AB}^{A'B'}(s,t) = \frac{2s}{t}\Gamma_{A'A}^c\Gamma_{B'B}^c = 2p_A^+\Gamma_{A'A}^c\frac{1}{t}\,2p_B^-\Gamma_{B'B}^c. \tag{9.21}$$

From comparison with (9.5), one can see that in fact the vertices $\Gamma_{A'A}^c$ and $\Gamma_{B'B}^c$ are the RRP vertices in the leading order, i.e. the RRP vertices can be easily found assuming the form (9.5). In the helicity basis, all these vertices have identical form:

$$\Gamma_{P'P}^c = gT_{P'P}^c\delta_{\lambda_{P'}\lambda_P}, \tag{9.22}$$

where $T_{P'P}^c$ represents now the matrix elements of the colour-group generators in the corresponding representations and λ are parton helicities. Except for a common coefficient, the vertices (9.22) can be written down without calculation because they are given by forward-matrix elements of the conserved current. Note that Eq. (9.22) implies a definite choice of the relative phase of spin-wave functions of particles P' and P. Evidently, the phase is zero at $t = 0$. In (9.22), the s-channel helicity conservation is exhibited explicitly. Note that for gluons and for massive quarks it is valid only in the leading order.

It is easy to rewrite (9.22) in terms of Dirac spinors for quarks and of physical polarization vectors for gluons. The vertex for the $q(p) \to q(p')$ quark transition with momenta p and p' having dominant components along p_1 can be represented as

$$\Gamma_{Q'Q}^c = g\bar{u}(p')t^c\frac{\slashed{p}_2}{2pp_2}u(p), \quad 2p^+\Gamma_{Q'Q}^c = g\bar{u}(p')t^c\gamma^+u(p), \tag{9.23}$$

where t^c are the colour-group generators in the fundamental representation; for antiquarks we have correspondingly

$$\Gamma_{\bar{Q}'\bar{Q}}^c = -g\bar{v}(p)t^c\frac{\slashed{p}_2}{2pp_2}v(p'), \quad 2p^+\Gamma_{\bar{Q}'\bar{Q}}^c = -g\bar{v}(p)t^c\gamma^+v(p'). \tag{9.24}$$

Evidently, the vertices for the quark and antiquark transitions with momenta having dominant components along p_2 are obtained from (9.23) and (9.24) by the replacement $1 \leftrightarrow 2$, $+ \leftrightarrow -$. Eqs. (9.21), (9.23), (9.24) make obvious that in the leading order the reggeon acts as a gluon with polarization vector $in_{2,1}$ when it interacts with a particle with momentum p that has a large component along $p_{1,2}$, respectively. Then, because of the factor $2s$ in (9.21), the reggeon vertices are obtained from gluon vertices with polarization vectors $ip_{2,1}/(2pp_{2,1})$.

The vertices for gluon transitions acquire a simple form in physical gauges. For gluons G and G' having momenta k and k' with predominant components along p_1 we will take their polarization vectors e and e' in the light-cone gauge: $(ep_2) = 0$, $(e'p_2) = 0$, so that

$$e = e_\perp - \frac{(e_\perp k_\perp)}{kp_2}p_2, \quad e' = e'_\perp - \frac{(e'_\perp k'_\perp)}{k'p_2}p_2, \tag{9.25}$$

and

$$\Gamma^c_{G'G} = -g(e'^*_\perp e_\perp)\, T^c_{G'G}, \tag{9.26}$$

with the colour generators in the adjoint representation. Conversely, for gluons with predominant components along p_2 we will take their polarization vectors in the gauge $(ep_1) = 0$, $(e'p_1) = 0$, i.e.

$$e = e_\perp - \frac{(e_\perp k_\perp)}{kp_1}p_1, \quad e' = e'_\perp - \frac{(e'_\perp k'_\perp)}{k'p_1}p_1. \tag{9.27}$$

The form (9.26) of the gluon–gluon–reggeon (GGR) vertex remains unchanged. Note, however, that in different gauges the transverse parts of the polarization vectors are different. Therefore, to be rigorous one should indicate somehow the gauge that is used. We do not do this to avoid overloading the notation. It is easy to see that the polarization vectors in the gauges (9.25) and (9.27) are connected by a gauge transformation:

$$e \rightarrow e - 2\frac{(e_\perp k_\perp)}{k_\perp^2}k, \quad e' \rightarrow e' - 2\frac{(e'_\perp k'_\perp)}{k_\perp'^2}k'. \tag{9.28}$$

For transverse components, this means

$$e_{\perp\mu} \rightarrow \Omega_{\mu\nu}e_{\perp\nu}, \quad e'_{\perp\mu} \rightarrow \Omega'_{\mu\nu}e'_{\perp\nu}, \tag{9.29}$$

where

$$\Omega_{\mu\nu} = \Omega_{\nu\mu} = \delta^\perp_{\mu\nu} - 2\frac{k_{\perp\mu}k_{\perp\nu}}{k_\perp^2}, \quad \Omega_{\mu\nu}\Omega_{\nu\rho} = \delta_{\mu\rho}. \tag{9.30}$$

The dispersion approach requires knowledge of production amplitudes in the MRK. Again in the Born approximation they can be calculated in the leading order without great effort using the t-channel unitarity. In the leading order, only gluons can be produced. The amplitudes $\mathcal{A}^{A'GB'}_{AB}$ are calculated using the t_1- and t_2-channel discontinuities. Schematically, they are represented in Fig. 9.4, where the reggeized form (9.5) of the $2 \rightarrow 2$ amplitudes must be taken in the lowest order, so that it is given by (9.21), evidently with the replacements $s \rightarrow s_2 = (p'_B + k)^2 = (p_B + q_1)^2$, $t \rightarrow t_2$ in the case of Fig. 9.4a and $s \rightarrow s_1 = (p'_A + k)^2 = (p_A - q_2)^2$, $t \rightarrow t_1$ in the case of Fig. 9.4b.

Here we meet a complication. Until now the gluon–gluon–reggeon vertices were defined only for physical gluon polarizations. But in order to use the Feynman gauge in the amplitudes of Fig. 9.4 it is necessary to have the vertices in gauge invariant form. It is easy to see that the required form can be obtained from (9.26) by the replacement

$$(e'^*_\perp e_\perp) \longrightarrow (e'^*e) - \frac{(p_i e'^*)(p'e)}{(p_i p')} - \frac{(pe'^*)(p_i e)}{(p_i p)} + \frac{(pp')}{(p_i p)(p_i p')}(p_i e'^*)(p_i e), \tag{9.31}$$

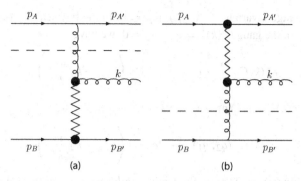

Fig. 9.4. Schematic representation of the discontinuities of the $A + B \to A' + G + B'$ amplitude in the t_1 (a) and t_2 (b) channels.

where $p_i = p_2$ ($p_i = p_1$) for the gluons with dominant components of momenta along p_1 (p_2), such that $(p_i p) \simeq (p_i p') \gg (p p')$. Using the vertices in the covariant form (9.31), one can easily find the contributions $\mathcal{A}_a$ and $\mathcal{A}_b$ with discontinuities corresponding to the diagrams Fig. 9.4a and Fig. 9.4b:

$$\mathcal{A}_a = 2s\Gamma^{c_1}_{A'A}\frac{1}{t_1}gT^c_{c_1 c_2}e^*_\mu(k)\left(-q_{1\mu}-q_{2\mu}+2p_{1\mu}\frac{kp_2}{p_1 p_2}-p_{2\mu}\left(\frac{q_2^2}{kp_2}+2\frac{kp_1}{p_1 p_2}\right)\right)\frac{1}{t_2}\Gamma^{c_2}_{B'B},$$
(9.32)

$$\mathcal{A}_b = 2s\Gamma^{c_1}_{A'A}\frac{1}{t_1}gT^c_{c_1 c_2}e^*_\mu(k)\left(-q_{1\mu}-q_{2\mu}+p_{1\mu}\left(\frac{q_1^2}{kp_2}+2\frac{kp_2}{p_1 p_2}\right)-2p_{2\mu}\frac{kp_1}{p_1 p_2}\right)\frac{1}{t_2}\Gamma^{c_2}_{B'B}.$$
(9.33)

Here $k = q_1 - q_2$, $e(k)$, and c are the gluon momentum, polarization vector, and colour index, respectively. It is easy to see that the amplitude $\mathcal{A}^{A'GB'}_{AB}$,

$$\mathcal{A}^{A'GB'}_{AB} = 2s\Gamma^{c_1}_{A'A}\frac{1}{t_1}\gamma^c_{c_1 c_2}(q_1, q_2)\frac{1}{t_2}\Gamma^{c_2}_{B'B},$$
(9.34)

with

$$\gamma^c_{c_1 c_2}(q_1, q_2) = gT^c_{c_1 c_2}e^*_\mu(k)C_\mu(q_2, q_1),$$
(9.35)

$$C_\mu(q_2, q_1) = -q_{1\mu}-q_{2\mu}+p_{1\mu}\left(\frac{q_1^2}{kp_1}+2\frac{kp_2}{p_1 p_2}\right)-p_{2\mu}\left(\frac{q_2^2}{kp_2}+2\frac{kp_1}{p_1 p_2}\right)$$

$$= -q_{1\perp\mu}-q_{2\perp\mu}-\frac{p_{1\mu}}{2(kp_1)}\left(k_\perp^2-2q_{1\perp}^2\right)+\frac{p_{2\mu}}{2(kp_2)}\left(k_\perp^2-2q_{2\perp}^2\right), \quad (9.36)$$

has the correct discontinuities both in the t_1- and t_2-channels. This means that it is the correct amplitude, because contributions $\sim s$ without singularities in the t_1- and t_2-channels are forbidden by renormalizability. Therefore, the vertex (9.35) is in fact the RRG vertex. In the leading order it is the only RRP vertex.

The vertex is gauge invariant: $C_\mu(q_2, q_1)k_\mu = 0$. In the physical light-cone gauges the vertex simplifies. In the gauge $e(k)k = e(k)p_2 = 0$, we have

$$e_\mu^*(k)C_\mu(q_2, q_1) = -2e_\perp^*(k)\left(q_{1\perp} - k_\perp \frac{q_{1\perp}^2}{k_\perp^2}\right), \tag{9.37}$$

and in the gauge $e(k)k = e(k)p_1 = 0$

$$e_\mu^*(k)C_\mu(q_2, q_1) = -2e_\perp^*(k)\left(q_{2\perp} + k_\perp \frac{q_{2\perp}^2}{k_\perp^2}\right). \tag{9.38}$$

Thus, *assuming* gluon reggeization, the leading order reggeon vertices are found without noticeable effort.

9.3.2 One-loop gluon trajectory

It is quite easy also to find the gluon trajectory in the leading order. To do this it is sufficient to find the lowest-order contribution to the s-channel discontinuity of some elastic amplitude with negative signature and to compare it with (9.5). Of course, the trajectory must be process independent. This requirement serves as a check of self-consistency of the reggeization hypothesis.

In the lowest order, only two-particle intermediate states contribute in the unitarity relation:

$$\mathrm{Im}_s \mathcal{A}_{AB}^{A'B'} = \frac{1}{2}\sum_{\tilde{A},\tilde{B}}\int \mathcal{A}_{AB}^{\tilde{A}\tilde{B}}\left(\mathcal{A}_{A'B'}^{\tilde{A}\tilde{B}}\right)^* d\Phi_{\tilde{A}\tilde{B}}, \tag{9.39}$$

where Im_s stands for the s-channel imaginary part, $\sum_{\tilde{A},\tilde{B}}$ means the sum over discrete quantum numbers of the intermediate particles, and $d\Phi_{\tilde{A},\tilde{B}}$ is the element of their phase space:

$$d\Phi_{\tilde{A}\tilde{B}} = (2\pi)^D\delta^D(p_A + p_B - p_{\tilde{A}} - p_{\tilde{B}})\frac{d^{D-1}p_{\tilde{A}}}{(2\pi)^{D-1}2\epsilon_{\tilde{A}}}\frac{d^{D-1}p_{\tilde{B}}}{(2\pi)^{D-1}2\epsilon_{\tilde{B}}}. \tag{9.40}$$

Here and in the following in this Chapter $D = 4+2\epsilon$ is the space-time dimension taken different from 4 to regularize the infrared, collinear, and ultraviolet divergences. Supposing, without any loss of generality, that $\beta_{\tilde{A}} \gg \beta_{\tilde{B}}$, using (9.21) and the equalities

$$d^D p = \frac{s}{2}d\alpha d\beta d^{D-2}p_\perp, \quad \delta^D(p) = \frac{2}{s}\delta(\alpha)\delta(\beta)\delta^{D-2}(p_\perp),$$

$$\frac{d^{D-1}p}{(2\pi)^{D-1}2\epsilon_p} = \delta(s\alpha\beta + p_\perp^2 - m^2)d^D p = \frac{d\beta}{2\beta}d^{D-2}p_\perp = \frac{d\alpha}{2\alpha}d^{D-2}p_\perp, \tag{9.41}$$

we obtain

$$d\Phi_{\tilde{A}\tilde{B}} = \frac{d^{D-2}q_{1\perp}}{2s(2\pi)^{(D-2)}}, \tag{9.42}$$

where $q_1 = p_A - p_{\tilde{A}}$, and

$$\text{Im}_s \, \mathcal{A}_{AB}^{A'B'} = s \sum_{\tilde{A}} \Gamma_{\tilde{A}A}^c \Gamma_{A'\tilde{A}}^{c'} \sum_{\tilde{B}} \Gamma_{\tilde{B}B}^c \Gamma_{A'\tilde{A}}^{c'} \frac{1}{(2\pi)^{(D-2)}} \int \frac{d^{D-2}q_1}{q_1^2 (q - q_1)^2}. \qquad (9.43)$$

Eq. (9.43) gives the s-channel discontinuity of the amplitude of the process $A + B \to A' + B'$. Usually this amplitude does not have a definite signature. If we want to study reggeization, we need to perform "signaturization," i.e. to construct amplitudes of definite parity with respect to the replacement $\cos\theta_t \leftrightarrow -\cos\theta_t$, where θ_t is the t-channel scattering angle. In general, the signaturization is not a simple task. It requires partial-wave decomposition of amplitudes in the t-channel with subsequent symmetrization (antisymmetrization) in $\cos\theta_t$ and analytical continuation into the s-channel. This procedure is relatively simple only in the case of elastic scattering of spin-zero particles. But, generally speaking, even in this case the amplitudes of definite signature cannot be expressed in terms of physical amplitudes related by crossing, because the substitution $\cos\theta_t \leftrightarrow -\cos\theta_t$ does not necessarily mean transition to the cross channel. Fortunately, the signaturization can be easily done in the Regge kinematics. As well as crossing properties, it is more convenient to formulate the signaturization in terms of truncated amplitudes, i.e. amplitudes with omitted wave functions (polarization and colour vectors and Dirac spinors). In Regge kinematics, the substitution $\cos\theta_t \leftrightarrow -\cos\theta_t$ means $s \leftrightarrow -s$. The crucial point is that $s = 2p_A^+ p_B^-$ is determined by the longitudinal components of momenta, which are conserved in the scattering: $p_{A'}^+ = p_A^+$, $p_{B'}^- = p_B^-$. Therefore, for the truncated amplitudes the substitution $s \leftrightarrow -s$ is equivalent to the change of signs of the longitudinal components of p_A and $p_{A'}$ (or, equivalently, p_B and $p_{B'}$) without changing the transverse components. At the same time, all particles remain on their mass-shell, so that the substitution is equivalent to the transition into the cross channel.

Recall here that the amplitudes of $\bar{Q}' \to \bar{Q}$ transitions are not obtained from amplitudes of $Q \to Q'$ transitions simply by crossing (i.e. they are not given by the same analytical functions in different regions of kinematical variables), but acquire an additional factor of -1 because of Fermi statistics. This must be taken into account in the definition of signaturization. Therefore, the truncated amplitudes $\mathcal{A}_{AB}^{A'B'}$ are signaturized by the operator

$$\hat{S}_\tau = \frac{1}{2} \left(1 + \tau \hat{R}_A \hat{R}_A' \right), \qquad (9.44)$$

where $\tau = \pm 1$ is the signature and $\hat{R}_P$ transfers the particle P from initial (final) state into the antiparticle $\bar{P}$ in the final (initial) state with the same longitudinal and opposite transverse momenta (which is equivalent to changing of the signs of the longitudinal momenta and accounting for Fermi statistics). From this definition, it follows that the s-channel discontinuity of the signaturized amplitude

$$\mathcal{A}_{AB}^{A'B'}(\tau) = \hat{S}_\tau \, \mathcal{A}_{AB}^{A'B'}, \qquad \mathcal{A}_{AB}^{A'B'} = \sum_{\tau = \pm 1} \mathcal{A}_{AB}^{A'B'}(\tau), \qquad (9.45)$$

for $\tau = +1$ ($\tau = -1$) is equal to half the sum (half the difference) of the s- and u-channel discontinuities of $\mathcal{A}_{AB}^{A'B'}$. Moreover, the s- and u-channel discontinuity of the amplitude (9.45) coincide (differs by sign), so one needs to find only one of them.

Using the vertices (9.23), (9.24), and (9.26), we derive that the s-channel discontinuity of the amplitude with negative signature is obtained from (9.43) by antisymmetrization in the colour indices in the first (or second) sum. In fact, it could be anticipated. Indeed, from the t-channel point of view, negative signature means antisymmetry with respect to exchange of momenta of gluons interacting with particles AA' (BB'). In the Regge region, these gluons have identical polarizations; therefore, due to Bose statistics the antisymmetry with respect to exchange of momenta implies antisymmetry with respect to exchange of colour indices. The commutation relations of the colour-group generators give

$$\frac{1}{2}\left(\Gamma^{c}_{\tilde{A}A}\Gamma^{c'}_{A'\tilde{A}} - \Gamma^{c'}_{\tilde{A}A}\Gamma^{c}_{A'\tilde{A}}\right) = -i\frac{g}{2}f_{cc'a}\Gamma^{a}_{A'A}, \quad -i\frac{g}{2}f_{cc'a}\Gamma^{c}_{\tilde{B}B}\Gamma^{c'}_{B'\tilde{B}} = -g^{2}\frac{N_{c}}{4}\Gamma^{a}_{B'B},$$

(9.46)

where $f_{cc'a}$ are the group-structure constants, $f_{cc'a}f_{cc'b} = N_{c}$, N_{c} in the number of colours. Comparing (9.43) with (9.5) and taking account of (9.46) we obtain

$$\omega(t) = \frac{g^{2}N_{c}t}{2(2\pi)^{D-1}}\int\frac{d^{D-2}q_{1}}{q_{1}^{2}(q-q_{1})^{2}}$$

$$= -g^{2}\frac{N_{c}\Gamma(1-\epsilon)}{(4\pi)^{D/2}}\frac{\Gamma^{2}(\epsilon)}{\Gamma(2\epsilon)}(q^{2})^{\epsilon} \simeq -g^{2}\frac{N_{c}\Gamma(1-\epsilon)}{(4\pi)^{2+\epsilon}}\frac{2}{\epsilon}(q^{2})^{\epsilon}.$$

(9.47)

Thus we see that, assuming the gluon reggeization, all the Reggeon vertices and the Regge trajectory can be easily obtained in the leading order. To find the vertices, it is sufficient to calculate elastic and one-gluon production amplitudes in the Born approximation; to find the trajectory, it is enough to calculate in the lowest-order the s-channel discontinuity of some signaturized elastic-scattering amplitude.

Originally, the reggeized form of elastic amplitudes was established in the leading order in two loops [30]. The three-loop calculations [2] confirmed this form and permitted formulation of the reggeization hypothesis for inelastic amplitudes.

9.3.3 Next-to-leading order Reggeon vertices

In the NLA, the gluon trajectory and the reggeon vertices emerging already in the LLA must be taken in the next-to-leading order. Again, assuming the gluon reggeization, they can be extracted from the fixed-order results: the vertices from the elastic and the one-gluon production amplitudes in the one-loop approximation and the trajectory from the two-loop s-channel discontinuities of elastic amplitudes. Of course, neither the calculations, nor the results are as simple as in the leading order. Therefore, we do not present here the calculations and give below simplified versions of the results.

To find the PPR vertices, one has to calculate nonlogarithmic terms in the one-loop elastic amplitudes. Various processes can be used; the results must be process independent. The QQR vertex can be extracted both from quark-quark and quark–gluon scattering amplitudes. It was obtained in [42] at arbitrary space-time dimension $D = 4 + 2\epsilon$ and quark masses; in the case of massless quarks, it can be written in the form of

$$\Gamma^a_{Q'Q} = \Gamma^{a(B)}_{Q'Q} \left(1 + \frac{\omega^{(1)}(t)}{2} \left[\frac{1}{\epsilon} + \psi(1-\epsilon) + \psi(1) - 2\psi(1+\epsilon) + \frac{2+\epsilon}{2(1+2\epsilon)(3+2\epsilon)} \right. \right.$$

$$\left. \left. - \frac{1}{N_c^2} \left(\frac{1}{\epsilon} - \frac{(3-2\epsilon)}{2(1+2\epsilon)} \right) - \frac{n_f}{N_c} \frac{(1+\epsilon)}{(1+2\epsilon)(3+2\epsilon)} \right] \right), \tag{9.48}$$

where the superscripts (B) and (1) mean the Born and the one-loop approximations, $\psi(x)$ is the logarithmic derivative of the Eueler Γ-function and n_f is the number of quark flavours. Evidently, the helicity of massless quarks is conserved, so that the one-loop vertex is proportional to the Born vertex. This is not so in the case of massive quarks [42].

To find the GGR vertex, both quark-gluon and gluon-gluon scattering can be used. This vertex was also found ([37],[43]) for arbitrary D and quark masses; in the massless case, it takes the form of

$$\Gamma^a_{G'G} = \Gamma^{a(B)}_{G'G} \left\{ 1 + \frac{\omega^{(1)}(t)}{2} \left[\frac{2}{\epsilon} + \psi(1) + \psi(1-\epsilon) - 2\psi(1+\epsilon) \right. \right.$$

$$\left. \left. - \frac{9(1+\epsilon)^2+2}{2(1+\epsilon)(1+2\epsilon)(3+2\epsilon)} + \frac{n_f}{N_c} \frac{(1+\epsilon)^3+\epsilon^2}{(1+\epsilon)^2(1+2\epsilon)(3+2\epsilon)} \right] \right\} \tag{9.49}$$

$$+ gT^a_{G'G} e'^*_{\perp\mu} e_{\perp\nu} \left(\delta^\perp_{\mu\nu} - (D-2)\frac{q_{\perp\mu} q_{\perp\nu}}{q_\perp^2} \right)$$

$$\times \frac{\epsilon \, \omega^{(1)}(t)}{2(1+\epsilon)^2(1+2\epsilon)(3+2\epsilon)} \left(1 + \epsilon - \frac{n_f}{N_c} \right).$$

The last term here exhibits violation of helicity conservation.

The PPR vertices presented above are the most simple Reggeon vertices. The gluon-production vertex $\gamma^G_{ab}(q_1, q_2)$ is much more complicated. The simplest piece of $\gamma^G_{ab}(q_1, q_2)$ – the quark part of the vertex – was found immediately at arbitrary D [44], just as the vertices (9.48) and (9.49). The calculation of the gluon part, which is much more involved, has a long history. First, only the terms not vanishing at $\epsilon \to 0$ were found [37]. But in the process of calculation of the next-to-leading order BFKL kernel it was realized that at small transverse momentum k of the produced gluon the RRG vertex must be known at arbitrary D. After this, the vertex at small k was calculated [45] at arbitrary D. Later the results of [37],[45] were obtained by another method in [46]. But then it became clear that for verification of the bootstrap condition for the BFKL kernel the vertex must be known at arbitrary D in a wider kinematical region. Finally, it was calculated at arbitrary D in [47]. Unfortunately, a complete expression for the vertex at arbitrary D is very complicated. We present it here in the form where in the gluon part only terms singular at small k are given at arbitrary D, but the other terms in the limit $\epsilon \to 0$. It can be written as

$$\gamma^G_{c_1c_2}(q_1, q_2) = \gamma^{G(B)}_{c_1c_2}(q_1, q_2) \left[1 + \frac{2g^2 N_c \Gamma(1-\epsilon)}{(4\pi)^{2+\epsilon}} (f_1 + f_2) \right] + gT^a_{c_1c_2} \frac{g^2 N_c \Gamma(1-\epsilon)}{(4\pi)^{2+\epsilon}}$$

$$\times \left[f_3 - (2k^2 - q_1^2 - q_2^2)f_2 \right] \left(\frac{p_A}{(kp_A)} - \frac{p_B}{(kp_B)} \right)_\mu e^*_\mu(k), \tag{9.50}$$

where $\gamma_{ab}^{G(B)}(q_1, q_2)$ is the Born vertex (9.35). This form is explicitly gauge invariant. As compared with the Born vertex (9.35) it contains new vector structure. The functions f_n, $n = 1, 2, 3$, are given by the sums of the quark and gluon parts:

$$f_n = \frac{n_f}{N_c} \frac{r^2(1+\epsilon)}{(-\epsilon)(4+2\epsilon)} f_n^Q + f_n^G, \tag{9.51}$$

where

$$f_1^Q = \frac{(q_1^2 + q_2^2)}{(q_1^2 - q_2^2)}(1+\epsilon)^2\phi_0, \qquad f_2^Q = \frac{k^2}{(q_1^2 - q_2^2)^3}\left[\epsilon\phi_2 - q_1^2 q_2^2(2+\epsilon)\phi_0\right],$$

$$f_3^Q = \frac{1}{(q_1^2 - q_2^2)}\left[q_1^2 q_2^2(2+\epsilon)^2\phi_0 + \epsilon k^2\phi_1\right], \quad \phi_n = (q_1^2)^{n+\epsilon} - (q_2^2)^{n+\epsilon}. \tag{9.52}$$

For the complete functions f_n, taking the limit $\epsilon \to 0$, but admitting $\epsilon \ln(1/k^2) \sim 1$, we have:

$$2f_1 = \left(\frac{11}{6} - \frac{n_f}{3N_c}\right)\frac{(q_1^2 + q_2^2)}{(q_1^2 - q_2^2)}\ln\left(\frac{q_1^2}{q_2^2}\right) - \frac{1}{2}\ln^2\left(\frac{q_1^2}{q_2^2}\right) - (k^2)^\epsilon\left(\frac{1}{\epsilon^2} - 3\zeta(2) + 2\epsilon\zeta(3)\right),$$

$$2f_2 = \left(1 - \frac{n_f}{N_c}\right)\frac{k^2}{3(q_1^2 - q_2^2)^2}\left[q_1^2 + q_2^2 - 2\frac{q_1^2 q_2^2}{(q_1^2 - q_2^2)}\ln\left(\frac{q_1^2}{q_2^2}\right)\right],$$

$$f_3 = \left(\frac{11}{3} - \frac{2n_f}{3N_c}\right)\frac{q_1^2 q_2^2}{(q_1^2 - q_2^2)}\ln\left(\frac{q_1^2}{q_2^2}\right) + \left(1 - \frac{n_f}{N_c}\right)\frac{k^2}{6}, \tag{9.53}$$

where $\zeta(n)$ is the Riemann zeta-function. The overall factor of the new vector structure in (9.50) is infrared finite, as it should be.

9.3.4 Gluon trajectory in two loops

The next-to-leading order correction $\omega^{(2)}(t)$ to the gluon trajectory was calculated in Refs. [48]–[52]. This correction can be determined from the s-channel discontinuity of the two-loop scattering amplitude of any the elastic-scattering processes (quark-quark, gluon-gluon, or quark-gluon) with colour octet and negative signature in the t-channel. The coincidence of the results served as a check of the reggeization.

With the assumption of gluon reggeization, the elastic scattering amplitudes with colour octet and negative signature are given by (9.5). In the helicity basis, the vertices $\Gamma_{P'P}^c$ can be represented as

$$\Gamma_{P'P}^c = g\, T_{P'P}^c\left[\delta_{\lambda_{P'}, \lambda_P}\left(1 + \Gamma_{PP}^{(+)}(t)\right) + \delta_{\lambda_P, -\lambda_{P'}}\Gamma_{PP}^{(-)}(t)\right], \tag{9.54}$$

where $\Gamma_{PP}^{(+)}(t)$ are the radiative corrections to the helicity conserving leading order vertices (9.22), $\Gamma_{PP}^{(-)}(t)$ are the helicity nonconserving next-to-leading order parts. To calculate the two-loop correction to the trajectory $\omega^{(2)}(t)$, we can consider only the part of the scattering amplitude (9.5) conserving helicities of each of the colliding particles. Let us write the two-loop contribution to the s-channel discontinuity of this part in the following form:

$$\text{disc}_s \, \mathcal{A}_{AB}^{A'B'} = -2\pi i g^2 \, T_{A'A}^c T_{B'B}^c \, \frac{s}{t} \, \Delta_{AB}. \tag{9.55}$$

Calculating this contribution from Eq. (9.5) with the help of Eq. (9.54) we find

$$\omega^{(2)}(t) = \Delta_{AB} - \left(\omega^{(1)}(t)\right)^2 \ln\left(\frac{s}{-t}\right) - \left[\Gamma_{AA}^{(+)}(t) + \Gamma_{BB}^{(+)}(t)\right]\omega^{(1)}(t), \tag{9.56}$$

where $\omega^{(1)}(t)$ is the one-loop contribution (9.47). Having the one-loop corrections to the PPR vertices, one can obtain $\omega^{(2)}(t)$ calculating the discontinuity (9.55) for any of the elementary scattering processes in QCD. Of course, the trajectory cannot depend on the colliding particles; therefore, comparing the results of the calculation one can verify the gluon reggeization.

With the correction $\omega^{(2)}(t)$ found in this way, the trajectory can be written at arbitrary D in terms of integrals over transverse momenta:

$$\omega(t) = \frac{g^2 N_c t}{2(2\pi)^{D-1}} \int \frac{d^{(D-2)}q_1}{q_1^2 (q_1 - q)^2}\left(1 + f(q_1, q) - 2f(q_1, q_1)\right), \tag{9.57}$$

where

$$f(q_1, q) = -\frac{g^2 N_c q^2}{2(2\pi)^{D-1}} \int \frac{d^{(D-2)}q_2}{q_2^2 (q_2 - q)^2}$$

$$\times \left[\ln\left(\frac{q^2}{(q_1 - q_2)^2}\right) - 2\psi(1 + 2\epsilon) - \psi(1 - \epsilon) + 2\psi(\epsilon) + \psi(1)\right.$$

$$\left. + \frac{1}{1 + 2\epsilon}\left(\frac{1}{\epsilon} + \frac{1 + \epsilon}{2(3 + 2\epsilon)}\right)\right] + \frac{4g^2 n_f \Gamma(2 - \frac{D}{2})\Gamma^2\left(\frac{D}{2}\right)}{(4\pi)^{\frac{D}{2}}\Gamma(D)}(q^2)^{(D/2 - 1)}. \tag{9.58}$$

The integrals in (9.57) can be expressed in terms of elementary functions [52] only in the limit $\epsilon \to 0$, where they give

$$\omega(t) = \omega^{(1)}(t)\left\{1 + \frac{\omega^{(1)}(t)}{4}\left[\frac{11}{3} + \left(\frac{\pi^2}{3} - \frac{67}{9}\right)\epsilon\right.\right.$$

$$\left.\left. + \left(\frac{404}{27} - 2\zeta(3)\right)\epsilon^2 - \frac{2n_f}{3N_c}\left(1 - \frac{5}{3}\epsilon + \frac{28}{9}\epsilon^2\right)\right]\right\}. \tag{9.59}$$

The remarkable fact which occurred during the calculation is the cancellation of the third-order poles in ϵ arising in separate contributions in (9.57). This cancellation is necessary for the absence of infrared divergences in the corrections to the BFKL equation for the colour singlet channel. As a result of this cancellation, the gluon and quark contributions to $\omega^{(2)}(t)$ have similar infrared behaviour. Moreover, the coefficient of the leading singularity in ϵ is proportional to the coefficient of the one-loop β function. This means that infrared divergences are strongly correlated with the ultraviolet ones.

Equation (9.59) gives the most economic representation of the trajectory. Note, however, that in order to retain terms finite at $\epsilon \to 0$ one has to conserve the terms of order ϵ in $\omega^{(1)}$ taking the square of $\omega^{(1)}$. This implies that one cannot use the last expression in (9.47).

The result (9.59) was confirmed in [53]. Then it was obtained quite independently by taking the high-energy limit of the two-loop amplitudes for parton–parton scattering [54].

Up to now we have systematically used the perturbative expansion in terms of the bare coupling, so that g is the bare-coupling constant related to the renormalized coupling g_μ in the $\overline{\text{MS}}$ scheme by the relation

$$g = g_\mu \mu^{-\epsilon} \left(1 + \beta_0 \frac{\bar{g}_\mu^2}{2\epsilon}\right), \quad \beta_0 = \left(\frac{11}{3} - \frac{2}{3}\frac{n_f}{N_c}\right), \quad \bar{g}_\mu^2 = \frac{g_\mu^2 N_c \Gamma(1-\epsilon)}{(4\pi)^{2+\epsilon}}. \quad (9.60)$$

We will use also the notation

$$\bar{g}^2 = \frac{g^2 N_c \Gamma(1-\epsilon)}{(4\pi)^{2+\epsilon}}, \quad \bar{g}^2 = \bar{g}_\mu^2 \mu^{-2\epsilon}\left(1 + \beta_0 \frac{\bar{g}_\mu^2}{\epsilon}\right). \quad (9.61)$$

As well as the cross sections, the reggeon vertices and trajectories themselves must not be renormalized, therefore the renormalization procedure for them is reduced to the transition from the bare-coupling constant g to the the renormalized one g_μ using (9.60). For the gluon trajectory in two-loop approximation, we obtain

$$\omega(t) = -\bar{g}_\mu^2 \left(\frac{q^2}{\mu^2}\right)^\epsilon \frac{2}{\epsilon} \left\{1 + \frac{\bar{g}_\mu^2}{\epsilon}\left[\left(\frac{11}{3} - \frac{2}{3}\frac{n_f}{N}\right)\left(1 - \frac{\pi^2}{6}\epsilon^2\right) - \left(\frac{q^2}{\mu^2}\right)^\epsilon \right.\right.$$

$$\times \left(\frac{11}{6} + \left(\frac{\pi^2}{6} - \frac{67}{18}\right)\epsilon + \left(\frac{202}{27} - \frac{11\pi^2}{18} - \zeta(3)\right)\epsilon^2\right)$$

$$\left.\left. - \frac{n_f}{3N}\left(1 - \frac{5}{3}\epsilon + \left(\frac{28}{9} - \frac{\pi^2}{3}\right)\epsilon^2\right)\right)\right]\right\}. \quad (9.62)$$

The correlation of the infrared and ultraviolet divergences is unique in the sense that it provides the independence on q^2 of singular contributions to $\omega(t)$. Indeed, expanding Eq. (9.62) we have

$$\omega(t) = -\bar{g}_\mu^2 \left(\frac{2}{\epsilon} + 2\ln\left(\frac{q^2}{\mu^2}\right)\right) - \bar{g}_\mu^4 \left[\left(\frac{11}{3} - \frac{2}{3}\frac{n_f}{N}\right)\left(\frac{1}{\epsilon^2} - \ln^2\left(\frac{q^2}{\mu^2}\right)\right)\right.$$

$$\left. + \left(\frac{67}{9} - \frac{\pi^2}{3} - \frac{10}{9}\frac{n_f}{N}\right)\left(\frac{1}{\epsilon} + 2\ln\left(\frac{q^2}{\mu^2}\right)\right) - \frac{404}{27} + 2\zeta(3) + \frac{56}{27}\frac{n_f}{N}\right]. \quad (9.63)$$

Eq. (9.63) exhibits explicitly all singularities of the trajectory in the two-loop approxima-tion and gives its finite part in the limit $\epsilon \to 0$. In the BFKL equation for the colour singlet channel, the singularities of the trajectory must be cancelled by the infrared singularities of the part related to real particle production.

9.3.5 *Vertices in quasi-multi-Regge kinematics*

The vertices and the trajectory presented above allow us to describe in the NLA the production of particles strongly ordered in rapidity space, i.e. in the same kinematics which contributes in the LLA. To obtain production amplitudes in this kinematics in the NLA it is sufficient to take one of the vertices or the trajectory in (9.15) in the next-to-leading order. But in the NLA another kinematics becomes important: one of the produced jets can contain two particles. Such a jet can be produced either in the fragmentation regions of the initial particles or in the central region, i.e. with rapidities far away from those of the colliding particles. Production amplitudes are given by (9.15), where the two-particle jet is in the first case either A' or B' and in the second case is one of the J_i. Note that because any two-particle jet in the unitarity relations leads to the loss of a large logarithm, the energy scales in (9.15) are unimportant in the NLA; moreover, the trajectory and the vertices are needed there only in the leading order.

Let us begin with vertices for the production of two-particle jets in the fragmentation region. To be definite, we take the fragmentation region of particle A. If particle A is a quark, only a quark–gluon jet can be produced; but if it is a gluon, then the jet can contain either two gluons or a $q\bar{q}$ pair. In all three cases, for generality, we take p_A as

$$p_A \equiv k = \beta p_1 + \frac{k^2 + m_A^2}{\beta s} p_2 + k_\perp, \tag{9.64}$$

the momenta of produced particles as

$$k_1 = \beta_1 p_1 + \frac{k_1^2 + m_1^2}{\beta_1 s} p_2 + k_{1\perp}, \quad k_2 = \beta_2 p_1 + \frac{k_2^2 + m_2^2}{\beta_2 s} p_2 + k_{2\perp}, \tag{9.65}$$

and we use the notation

$$\beta_1 = x_1 \beta, \quad \beta_2 = x_2 \beta, \quad x_1 + x_2 = 1. \tag{9.66}$$

In all three cases, we take polarization vectors of the participating gluons in the light-cone gauge (9.25).

In the case of quark–gluon production, let k_1 and k_2 be the momenta of the final quark and gluon respectively, such that $m_A = m_1 = m$, $m_2 = 0$. Then one has [55]

$$\Gamma^c_{\{QG\}Q} = (t^a t^c)_{i_1 i_2} \left(A_b \big((x_2 k_1 - x_1 k_2)_\perp \big) - A_b \big((k_1 - x_1 k)_\perp \big) \right)$$
$$- \left(t^c t^a \right)_{i_1 i_2} \left(A_b \big((-k_2 + x_2 k)_\perp \big) - A_b \big((k_1 - x_1 k)_\perp \big) \right), \tag{9.67}$$

where i_1 and i_2 are the colour indices of the outgoing and incoming quarks, a is the colour index of the produced gluon G, and the amplitudes A_b have the form:

$$A_b(p_\perp) = -\frac{g^2}{p_\perp^2 - x_2^2 m^2} \bar{u}(k_1) \frac{\not{p}_2}{\beta s} \left(x_1 \not{e}_\perp^* \not{p}_\perp + \not{p}_\perp \not{e}_\perp^* + \not{e}_\perp^* x_2^2 m \right) u(k). \tag{9.68}$$

Here e is the polarization vector of the produced gluon.

The reggeon vertex for $q\bar{q}$ production was found in [56]. It is of the form

$$\Gamma^c_{\{Q\bar{Q}\}G} = (t^a t^c)_{i_1 i_2} \Big(\mathcal{A}_p((k_1 - x_1 k)_\perp) - \mathcal{A}_p((x_2 k_1 - x_1 k_2)_\perp) \Big)$$
$$- (t^c t^a)_{i_1 i_2} \Big(\mathcal{A}_p((-k_2 + x_2 k)_\perp) - \mathcal{A}_p((x_2 k_1 - x_1 k_2)_\perp) \Big), \quad (9.69)$$

where i_1, i_2 are now the quark and antiquark colour indices, and a is the colour index of the incoming gluon G. The amplitudes $\mathcal{A}_p(p_\perp)$ in the light-cone gauge (9.25) are:

$$\mathcal{A}_p(p_\perp) = \frac{g^2}{p_\perp^2 - m^2} \bar{u}(k_1) \frac{\not{p}_2}{\beta s} \Big(x_1 \not{e}_\perp \not{p}_\perp - x_2 \not{p}_\perp \not{e}_\perp - \not{e}_\perp m \Big) v(k_2). \quad (9.70)$$

The amplitudes $\mathcal{A}_p$ and $\mathcal{A}_b$ are related by crossing.

The vertex $\Gamma^c_{\{G_1 G_2\}G}$ for two-gluon production can be represented in the same form as (9.69), with the difference that k_1 and k_2 now are the momenta of the produced gluons, i_1 and i_2 are their colour indices. Taking their polarization vectors e_1 and e_2 in the light-cone gauge (9.25), we have [56]

$$\Gamma^c_{\{G_1 G_2\}G} = (T^a T^c)_{i_1 i_2} \Big(\mathcal{A}((k_1 - x_1 k)_\perp) - \mathcal{A}((x_2 k_1 - x_1 k_2)_\perp) \Big)$$
$$- (T^c T^a)_{i_1 i_2} \Big(\mathcal{A}((-k_2 + x_2 k)_\perp) - \mathcal{A}((x_2 k_1 - x_1 k_2)_\perp) \Big), \quad (9.71)$$

where the amplitudes $\mathcal{A}(p_\perp)$ now have the form

$$\mathcal{A}(p_\perp) = \frac{2g^2}{p_\perp^2} \Big[x_1 x_2 \left(e^*_{1\perp} e^*_{2\perp} \right) (e_\perp p_\perp) - x_1 \left(e^*_{1\perp} e_\perp \right) \left(e^*_{2\perp} p_\perp \right) - x_2 \left(e^*_{2\perp} e_\perp \right) \left(e^*_{1\perp} p_\perp \right) \Big].$$
$$(9.72)$$

In the central region, jets are produced by two reggeons. Denoting reggeon momenta by q_1 and q_2, we can put

$$q_1 = \beta p_1 + q_{1\perp}, \quad q_2 = -\alpha p_2 + q_{2\perp}, \quad \beta = \beta_1 + \beta_2, \quad \alpha = \alpha_1 + \alpha_2, \quad (9.73)$$

where β_i and α_i are the Sudakov parameters for the produced particles, $i = 1, 2$. The particles can be either $q\bar{q}$ or two gluons. For simplicity, we discuss below the case of massless quarks, although the massive case can be considered quite analogously. Then, for momenta k_1 and k_2 of the produced particles, we have:

$$k = k_1 + k_2 = q_1 - q_2, \quad k_i = \beta_i p_1 + \alpha_i p_2 + k_{i\perp},$$
$$s \alpha_i \beta_i = -k_{i\perp}^2 = k_i^2, \quad \beta_i = x_i \beta, \quad x_1 + x_2 = 1. \quad (9.74)$$

The reggeon vertex for quark-antiquark production in reggeon-reggeon collisions was obtained in [57]–[59]. If k_1 and k_2 are the quark and antiquark momentum respectively, then the vertex has the form

$$\gamma^{Q\bar{Q}}_{c_1 c_2}(q_1, q_2) = \frac{1}{2} g^2 \bar{u}(k_1) \Big[t^{c_1} t^{c_2} a(q_1; k_1, k_2) - t^{c_2} t^{c_1} \overline{a(q_1; k_2, k_1)} \Big] v(k_2), \quad (9.75)$$

where $a(q_1; k_1, k_2)$ and $\overline{a(q_1; k_2, k_1)}$ can be written [58] in the following way:

$$a(q_1; k_1, k_2) = \frac{4 \not{p}_1 \not{Q}_1 \not{p}_2}{s \tilde{t}_1} - \frac{1}{k^2} \not{\Gamma}, \quad \overline{a(q_1; k_2, k_1)} = \frac{4 \not{p}_2 \not{Q}_2 \not{p}_1}{s \tilde{t}_2} - \frac{1}{k^2} \not{\Gamma}, \quad (9.76)$$

with

$$\tilde{t}_1 = (q_1 - k_1)^2, \quad \tilde{t}_2 = (q_1 - k_2)^2, \quad Q_1 = q_{1\perp} - k_{1\perp}, \quad Q_2 = q_{1\perp} - k_{2\perp},$$

$$\Gamma = 2\left[(q_1 + q_2)_\perp - \beta p_1 \left(1 - 2\frac{q_1^2}{s\alpha\beta}\right) + \alpha p_2 \left(1 - 2\frac{q_2^2}{s\alpha\beta}\right)\right]. \tag{9.77}$$

Using the notation $D(p, q)$ and $d(p, q)$,

$$D(p, q) = x_1 p_\perp^2 + x_2 q_\perp^2, \quad d(p, q) = (x_1 p_\perp - x_2 q_\perp)^2, \tag{9.78}$$

for the denominators in the vertex, and noting that for arbitrary $p_\perp$ we have

$$\bar{u}(k_1)\not{p}_\perp v(k_2) = \bar{u}(k_1)\frac{\not{p}_2}{s\beta}\left(\frac{\not{k}_{1\perp}\not{p}_\perp}{x_1} + \frac{\not{p}_\perp\not{k}_{2\perp}}{x_2}\right)v(k_2), \tag{9.79}$$

we can represent $a(q_1; k_1, k_2)$ and $\overline{a(q_1; k_2, k_1)}$ as

$$a(q_1; k_1, k_2) = \frac{4}{s\beta}\not{p}_2 b(q_1; k_1, k_2), \quad \overline{a(q_1; k_2, k_1)} = \frac{4}{s\beta}\not{p}_2 \overline{b(q_1; k_2, k_1)}, \tag{9.80}$$

where

$$b(q_1; k_1, k_2) = \frac{\not{k}_{1\perp}(\not{k}_{1\perp} - \not{q}_{1\perp})}{D(k_1 - q_1, k_1)} - \frac{x_1 x_2}{d(k_2, k_1)}\left(\frac{q_{1\perp}^2 \not{k}_{1\perp}\not{k}_{2\perp}}{D(k_2, k_1)}\right.$$
$$\left. - \frac{\not{k}_{1\perp}\not{q}_{1\perp}}{x_1} - \frac{\not{q}_{1\perp}\not{k}_{2\perp}}{x_2} - q_{1\perp}^2 + 2(q_{2\perp}(k_1 + k_2)_\perp)\right) - 1,$$

$$\overline{b(q_1; k_2, k_1)} = \frac{(\not{k}_{2\perp} - \not{q}_{1\perp})\not{k}_{2\perp}}{D(k_2, k_2 - q_1)} - \frac{x_1 x_2}{d(k_2, k_1)}\left(\frac{q_{1\perp}^2 \not{k}_{1\perp}\not{k}_{2\perp}}{D(k_2, k_1)}\right.$$
$$\left. - \frac{\not{k}_{1\perp}\not{q}_{1\perp}}{x_1} - \frac{\not{q}_{1\perp}\not{k}_{2\perp}}{x_2} - q_{1\perp}^2 + 2(q_{1\perp}(k_1 + k_2)_\perp)\right) - 1. \tag{9.81}$$

The vertex for two-gluon production was obtained in [60]. It possesses QED-like gauge invariance, i.e. it turns into zero when the polarization vector of one of the gluons is replaced by its momentum, independently of the polarizations of other particles. For the one-gluon vertex, such property could be considered as the consequence of the factorization of MRK amplitudes, although the matter is not quite simple. One could think that this property is not worthy of notice, because one can always obtain vertices with QED-like gauge invariance starting from noninvariant ones by addition of terms giving zero at physical polarizations. Let us consider, for example, the vertex $e_{1\mu}^* e_{2\nu}^* C_{\mu\nu}$, where $C_{\mu\nu}$ does not possesses QED-like gauge invariance. Let $k_{1\mu} C_{\mu\nu} = k_{2\nu} f_1$, $k_{2\nu} C_{\mu\nu} = k_{1\mu} f_2$, in accordance with usual QCD gauge invariance, which requires vanishing of $k_{1\mu} C_{\mu\nu} e_{2\nu}^*$ and $e_{1\mu}^* C_{\mu\nu} k_{2\nu}$ at physical e_1 and e_2. Physical results do not change under the replacement

$$C_{\mu\nu} \rightarrow \bar{C}_{\mu\nu} = C_{\mu\nu} - \frac{k_{2\mu} k_{2\nu}}{(k_1 k_2)} f_1 - \frac{k_{1\mu} k_{1\nu}}{(k_1 k_2)} f_2, \tag{9.82}$$

where $\bar{C}_{\mu\nu}$ is QED-like gauge invariant. But, generally speaking, $\bar{C}_{\mu\nu}$ contains unphysical singularities because of $(k_1 k_2)$ in the denominator and does not correspond to Feynman diagrams. In general terms, it contains singularities in overlapping channels and contradicts the Steinmann rule [61]. But it is not so nether for the one-gluon vertex (9.35), (9.36), not for the two-gluon vertex obtained in [60]. One can say that for them the QED-like gauge invariance is natural.

In the light-cone gauge (9.25) for both gluons the vertex takes the form [62]:

$$\gamma_{c_1 c_2}^{G_1 G_2}(q_1, q_2) = 4g^2 (e_{1\perp}^*)_\alpha (e_{2\perp}^*)_\beta$$

$$\times \left[\left(T^{i_1} T^{i_2} \right)_{c_1 c_2} b_{\alpha\beta}(q_1; k_1, k_2) + \left(T^{i_2} T^{i_1} \right)_{c_1 c_2} b_{\beta\alpha}(q_1; k_2, k_1) \right], \qquad (9.83)$$

where $e_{1,2}$ are the polarization vectors of the produced gluons, $i_{1,2}$ are their colour indices, and

$$b_{\alpha\beta}(q_1; k_1, k_2) = \frac{1}{2} g_{\alpha\beta}^\perp \left[\frac{x_1 x_2}{d(k_2, k_1)} \left(2q_{1\perp}(x_1 k_2 - x_2 k_1)_\perp + q_{1\perp}^2 \left(x_2 - \frac{x_1 k_{2\perp}^2}{D(k_2, k_1)} \right) \right) \right.$$

$$\left. - x_2 \left(1 - \frac{k_{1\perp}^2}{D(q_1 - k_1, k_1)} \right) \right] - \frac{x_2 k_{1\perp\alpha} q_{1\perp\beta} - x_1 q_{1\perp\alpha}(q_1 - k_1)_{\perp\beta}}{D(q_1 - k_1, k_1)}$$

$$- \frac{x_1 q_{1\perp}^2 k_{1\perp\alpha}(q_1 - k_1)_{\perp\beta}}{k_{1\perp}^2 D(q_1 - k_1, k_1)} - \frac{x_1 q_{1\perp\alpha}(x_1 k_2 - x_2 k_1)_{\perp\beta} + x_2 q_{1\perp\beta}(x_1 k_2 - x_2 k_1)_{\perp\alpha}}{d(k_2, k_1)}$$

$$+ \frac{x_1 q_{1\perp}^2 k_{1\perp\alpha} k_{2\perp\beta}}{k_{1\perp}^2 D(k_2, k_1)} + \frac{x_1 x_2 q_{1\perp}^2}{d(k_2, k_1) D(k_2, k_1)}$$

$$\times \left((x_1 k_2 - x_2 k_1)_{\perp\alpha} k_{2\perp\beta} + k_{1\perp\alpha}(x_1 k_2 - x_2 k_1)_{\perp\beta} \right). \qquad (9.84)$$

Here we use the notation (9.78). Note that one can come to (9.84) starting from the vertex in the gauge $e(k_1)p_1 = 0$, $e(k_2)p_2 = 0$ [63]. Our $b_{\alpha\beta}(q_1; k_1, k_2)$ can be obtained from $c_{\alpha\beta}(k_1, k_2)$ defined in [63] by the gauge transformation

$$b_{\alpha\beta}(q_1; k_1, k_2) = \left(g_{\alpha\gamma}^\perp - 2 \frac{k_{1\perp\alpha} k_{1\perp\gamma}}{k_{1\perp}^2} \right) c_{\gamma\beta}(k_1, k_2). \qquad (9.85)$$

The vertex (9.83) describes production of a two-gluon jet. The squared invariant mass of the jet

$$(k_1 + k_2)^2 = \frac{(x_2 k_1 - x_1 k_2)^2}{x_1 x_2} \qquad (9.86)$$

becomes large at small x_1 or x_2. Let us consider the behaviour of the vertex (9.83) in the limits $x_1 \to 1$, $x_2 \to 0$ and $x_1 \to 0$, $x_2 \to 1$. The former corresponds to the case when the first gluon is much closer in rapidity space to particle A than the second gluon. Therefore, in this limit, the two-gluon production vertex must be factorized as

$$\gamma_{ij}^{G_1 G_2}(q_1, q_2) = \gamma_{il}^{G_1}(q_1, q_1 - k_1) \frac{1}{(q_1 - k_1)_\perp^2} \gamma_{lj}^{G_2}(q_1 - k_1, q_2), \qquad (9.87)$$

where $\gamma_{ij}^G(q_1, q_2)$ is the one-gluon production vertex. Indeed, at $x_1 = 1$, $x_2 = 0$ we have $D(p, q) = d(p, q) = p_\perp^2$, which gives

$$b_{\beta\alpha}(q_1; k_2, k_1)|_{x_1=1} = 0, \tag{9.88}$$

such that

$$\gamma_{ij}^{G_1 G_2}(q_1, q_2) = 4g^2 e^*_{1\perp\alpha} e^*_{2\perp\beta} \left(T^{i_1} T^{i_2}\right)_{ij} b_{\alpha\beta}(q_1; k_1, k_2)|_{x_1=1}, \tag{9.89}$$

where

$$b_{\alpha\beta}(q_1; k_1, k_2)|_{x_1=1} = \frac{1}{(q_1-k_1)_\perp^2} \left[q_{1\perp} - \frac{q_{1\perp}^2}{k_{1\perp}^2} k_{1\perp}\right]_\alpha \left[q_{1\perp} - k_{1\perp} - \frac{(q_{1\perp}-k_{1\perp})^2}{k_{2\perp}^2} k_{2\perp}\right]_\beta. \tag{9.90}$$

Comparing with the one-gluon production vertex (9.35), (9.37), we see that the factorization property (9.87) is fulfilled. In the second limit, i.e. $x_1 = 0$, $x_2 = 1$, we get $D(p, q) = d(p, q) = q_\perp^2$, $b_{\alpha\beta}(q_1; k_1, k_2)|_{x_2=1} = 0$,

$$b_{\beta\alpha}(q_1; k_2, k_1)|_{x_2=1} = \frac{1}{(q_1-k_2)_\perp^2} \left[q_{1\perp} - \frac{q_{1\perp}^2}{k_{2\perp}^2} k_{2\perp}\right]_\beta \left[q_{1\perp} - k_{2\perp} - \frac{(q_{1\perp}-k_{2\perp})^2}{k_{1\perp}^2} k_{1\perp}\right]_\alpha \tag{9.91}$$

and

$$\gamma_{ij}^{G_1 G_2}(q_1, q_2) = \gamma_{il}^{G_2}(q_1, q_1 - k_2) \frac{1}{(q_1-k_2)_\perp^2} \gamma_{lj}^{G_1}(q_1 - k_2, q_2). \tag{9.92}$$

9.4 BFKL equation

9.4.1 BFKL kernel and impact factors

The gluon reggeization determines amplitudes with colour octet states and negative signature in the t-channels. Amplitudes with other quantum numbers are found in the BFKL approach using s-channel unitarity. In the unitarity relations, the contribution of order s, in which we are interested, is given by the MRK. Large logarithms come from integration over longitudinal momenta of the produced jets. For elastic amplitudes the s-channel unitarity relation gives (see Fig. 9.5)

$$\text{Im}_s \mathcal{A}_{AB}^{A'B'} = \frac{1}{2} \sum_{n=0}^{\infty} \sum_{\{f\}} \int \mathcal{A}_{AB}^{\tilde{A}\tilde{B}+n} \left(\mathcal{A}_{A'B'}^{\tilde{A}\tilde{B}+n}\right)^* d\Phi_{\tilde{A}\tilde{B}+n}, \tag{9.93}$$

where $\sum_{\{f\}}$ means sum over discrete quantum numbers of intermediate particles, the amplitudes $\mathcal{A}_{AB}^{\tilde{A}\tilde{B}+n}$ and $\left(\mathcal{A}_{A'B'}^{\tilde{A}\tilde{B}+n}\right)$ are defined by (9.15) and $d\Phi_{\tilde{A}\tilde{B}+n}$ is the phase-space element of the produced jets. In the LLA, where production of each additional particle must give a large logarithm, each jet is in fact a gluon, so that

$$d\Phi_{\tilde{A}\tilde{B}+n} = (2\pi)^D \delta^D \left(p_A + p_B - \sum_{i=0}^{n+1} k_i\right) \prod_{i=0}^{n+1} \frac{d^{D-1}k_i}{(2\pi)^{D-1} 2\epsilon_i}. \tag{9.94}$$

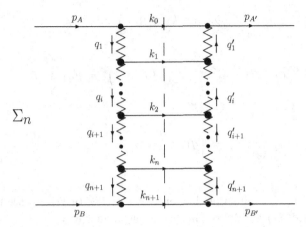

Fig. 9.5. Schematic representation of the s-channel discontinuity of the amplitude of the process $A + B \to A' + B'$.

In the MRK, with account of the strong ordering (9.10) and equalities (9.41), we have

$$d\Phi_{\tilde{A}\tilde{B}+n} = \frac{\pi}{s} \prod_{i=1}^{n} \frac{d\beta_i}{2\beta_i} \prod_{i=1}^{n+1} \frac{d^{D-2}q_{i\perp}}{(2\pi)^{D-1}}, \qquad (9.95)$$

where $q_i = p_A - \sum_{l=0}^{i-1} k_l$.

The integrand in (9.93) contains products of the Regge factors and convolutions of the reggeon vertices, which must be integrated over momenta of particles in intermediate states. In order to represent the result in a compact way it is convenient to use the operator notation and to define operators $\hat{\omega}$ for the gluon trajectory and $\hat{\mathcal{K}}_r$ for the convolutions of the production vertices. We will also use notation which accumulates all reggeon quantum numbers. Let $|\mathcal{G}\rangle$ denote the reggeized gluon state with transverse momentum $q_\perp$ and colour index c, and let us use the normalization

$$\langle \mathcal{G}|\mathcal{G}'\rangle = -q_\perp^2 \delta(q_\perp - q_\perp')\delta_{cc'}. \qquad (9.96)$$

The operator $\hat{\omega}$ is naturally defined as

$$\hat{\omega} = \omega\left(\hat{q}_\perp^2\right), \qquad (9.97)$$

such that

$$\hat{\omega}|\mathcal{G}\rangle = \omega(q_\perp^2)|\mathcal{G}\rangle, \qquad (9.98)$$

where $\omega(t)$ is the gluon trajectory. Then, $\langle \mathcal{G}_1\mathcal{G}_2|$ and $|\mathcal{G}_1\mathcal{G}_2\rangle$ are *bra*- and *ket*-vectors for the t-channel states of two reggeized gluons with transverse momenta $q_{1\perp}$ and $q_{2\perp}$ and colour indices c_1 and c_2, respectively. It is convenient to distinguish the states $|\mathcal{G}_1\mathcal{G}_2\rangle$ and $|\mathcal{G}_2\mathcal{G}_1\rangle$. We will associate the former with the case when the reggeon $\mathcal{G}_1$ is contained in the amplitude with initial particles (in the left part of Fig. 9.5), and the latter with the case when it is contained in the amplitude with final particles (in the right part of Fig. 9.5). The states $|\mathcal{G}_1\mathcal{G}_2\rangle$ form a complete set of states. If their scalar products are defined according to (9.96), i.e.

$$\langle \mathcal{G}_1\mathcal{G}_2|\mathcal{G}_1'\mathcal{G}_2'\rangle = q_{1\perp}^2 q_{2\perp}^2 \delta(q_{1\perp} - q_{1\perp}')\delta(q_{2\perp} - q_{2\perp}')\delta_{c_1 c_1'}\delta_{c_2 c_2'}, \qquad (9.99)$$

then the completeness means

$$\langle \Psi|\Phi\rangle = \int \frac{d^{D-2}q_{1\perp}d^{D-2}q_{2\perp}}{q_{1\perp}^2 q_{2\perp}^2}\langle \Psi|\mathcal{G}_1\mathcal{G}_2\rangle\langle \mathcal{G}_1\mathcal{G}_2|\Phi\rangle, \qquad (9.100)$$

where summation over colours c_1 and c_2 is implied. In the following, we will also use the letters $\mathcal{G}_i$ instead of c_i.

The operator $\hat{\mathcal{K}}_r$ is defined by the relations

$$\langle \mathcal{G}_1\mathcal{G}_2|\hat{\mathcal{K}}_r|\mathcal{G}_1'\mathcal{G}_2'\rangle = \delta(q_{1\perp} + q_{2\perp} - q_{1\perp}' - q_{2\perp}')\mathcal{K}_r(q_{1\perp}, q_{1\perp}'; q_\perp), \qquad (9.101)$$

$$\mathcal{K}_r(q_{1\perp}, q_{1\perp}'; q_\perp) = \frac{1}{2(2\pi)^{D-1}}\sum_G \gamma_{c_1 c_1'}^G(q_1, q_1')\left(\gamma_{c_2 c_2'}^G(-q_2, -q_2')\right)^*, \qquad (9.102)$$

where $q_\perp = q_{1\perp} + q_{2\perp} = q_{1\perp}' + q_{2\perp}'$ is the total t-channel momentum, the vertices $\gamma_{cc'}^G$ are defined in (9.35) and the sum is taken over spin and colour states of the gluon G. To carry out the summation, it is convenient to use one of the light-cone gauges (9.37) and (9.38). The result is

$$\mathcal{K}_r(q_{1\perp}, q_{1\perp}'; q_\perp) = T_{c_1 c_1'}^a (T_{c_2 c_2'}^a)^* \frac{2g^2}{(2\pi)^{D-1}}\left(q_{1\perp} - k_\perp \frac{q_{1\perp}^2}{k_\perp^2}\right)\left(q_{2\perp} + k_\perp \frac{q_{2\perp}^2}{k_\perp^2}\right)$$

$$= -T_{c_1 c_1'}^a (T_{c_2 c_2'}^a)^* \frac{g^2}{(2\pi)^{D-1}}\left(\frac{q_{1\perp}^2 q_{2\perp}'^2 + q_{2\perp}^2 q_{1\perp}'^2}{k_\perp^2} - q_\perp^2\right), \qquad (9.103)$$

where $k_\perp = q_{1\perp} - q_{1\perp}' = q_{2\perp}' - q_{2\perp}$. Note that $\mathcal{K}_r(q_{1\perp}, q_{1\perp}'; q_\perp)$ remains a matrix in colour space. To simplify formulas we do not indicate the matrix indices explicitly.

We also have to describe the transitions $A \rightarrow A'$ and $B \rightarrow B'$, shown by the upper and lower lines of Fig. 9.5. To do this, we introduce the impact factors $\Phi_{A'A}$ and $\Phi_{B'B}$. For amplitudes with signature τ, they are defined by

$$\Phi_{A'A}(q_1, q_2) = \sum_{\tilde{A}} \left(\Gamma_{\tilde{A}A}^{c_1}\Gamma_{A'\tilde{A}}^{c_2} + \tau\Gamma_{\tilde{A}A}^{c_2}\Gamma_{A'\tilde{A}}^{c_1}\right), \qquad (9.104)$$

where $q_1 = p_A - p_{\tilde{A}}$, $q_2 = p_{\tilde{A}} - p_{A'}$, and

$$\Phi_{B'B}(q_1, q_2) = \sum_{\tilde{B}} \left(\Gamma_{\tilde{B}B}^{c_1}\Gamma_{B'\tilde{B}}^{c_2} + \tau\Gamma_{\tilde{B}B}^{c_2}\Gamma_{B'\tilde{B}}^{c_1}\right), \qquad (9.105)$$

where $q_1 = p_{\tilde{B}} - p_B$, $q_2 = p_{B'} - p_{\tilde{B}}$. For simplicity, we omit the signature τ and the colour indices of the impact-factor symbols. Finally, we define the states $|\bar{B}'B\rangle$ and $\langle A'\bar{A}|$ by the relations

$$\langle \mathcal{G}_1\mathcal{G}_2|\bar{B}'B\rangle = \delta(q_{B\perp} - q_{1\perp} - q_{2\perp})\Phi_{B'B}(q_1, q_2),$$

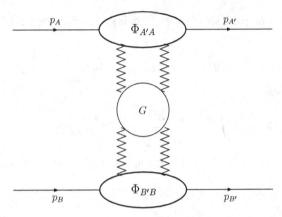

Fig. 9.6. Schematic representation of the process $A + B \rightarrow A' + B'$.

$$\langle A'\bar{A}|\mathcal{G}_1\mathcal{G}_2\rangle = \delta(q_{A\perp} - q_{1\perp} - q_{2\perp})\Phi_{A'A}(q_1, q_2), \qquad (9.106)$$

where $q_B = p_{B'} - p_B$, $q_A = p_A - p_{A'}$.

With the definitions given above, the s-channel discontinuities of signaturized scattering amplitudes for the processes $A + B \rightarrow A' + B'$ can be represented in the form (see Fig. 9.6)

$$-2i(2\pi)^{D-2}\delta(\boldsymbol{q}_A - \boldsymbol{q}_B)\mathrm{disc}_s\mathcal{A}_{AB}^{A'B'} = s\langle A'\bar{A}|\hat{G}(Y)|\bar{B}'B\rangle, \qquad (9.107)$$

where $Y = \ln(s/s_0)$, s_0 is an appropriate energy scale (which cannot be specified in the LLA), and $\hat{G}(Y)$ represents the Green function of two interacting reggeized gluons. It is given by the series

$$\hat{G}(Y) = e^{\hat{\Omega}Y} + \sum_{n=1}^{\infty}\int_0^Y e^{\hat{\Omega}(Y-y_1)}\hat{\mathcal{K}}_r dy_1 \int_0^{y_1} dy_2 e^{\hat{\Omega}(y_1-y_2)}\hat{\mathcal{K}}_r \ldots \int_0^{y_{n-1}} dy_n e^{\hat{\Omega}(y_{n-1}-y_n)}\hat{\mathcal{K}}_r e^{\hat{\Omega}y_n}. \qquad (9.108)$$

Here $\hat{\Omega} = \hat{\omega}_1 + \hat{\omega}_2$, $y_i = Y + \ln\beta_i$, and we have used the relations

$$s_i = s_0\frac{\beta_i - 1}{\beta_i} = s_0 e^{y_{i-1}-y_i}, \quad y_0 = Y, \qquad (9.109)$$

which are fulfilled in the LLA.

From (9.108) it is easy to see that $\hat{G}(Y)$ obeys the equation

$$d\hat{G}(Y)/dY = \hat{\mathcal{K}}\hat{G}(Y), \qquad (9.110)$$

where

$$\hat{\mathcal{K}} = \hat{\Omega} + \hat{\mathcal{K}}_r = \hat{\omega}_1 + \hat{\omega}_2 + \hat{\mathcal{K}}_r, \qquad (9.111)$$

with the initial condition $\hat{G}(0) = 1$. Therefore,

$$\hat{G}(Y) = e^{\hat{\mathcal{K}}Y}. \qquad (9.112)$$

Eq. (9.110) is the operator form of the BFKL equation, and (9.111) is the BFKL kernel. As is seen from (9.111), it consists of two parts; the former contains the gluon trajectory (this part is called virtual), and the latter is expressed in terms of effective vertices for particle production in reggeon–reggeon interaction (this part is called real). An unfolded form of the BFKL equation in momentum representation and of the corresponding initial condition is written as

$$\frac{d}{dY} G\left(Y; q_{1\perp}, q'_{1\perp}; q_{\perp}\right) = \int \frac{d^{D-2}r_{\perp}}{r_{\perp}^{2}(q_{\perp}-r_{\perp})^{2}} \mathcal{K}\left(q_{1\perp}, r_{\perp}; q_{\perp}\right) G\left(Y; r_{\perp}, q'_{1\perp}; q_{\perp}\right),$$

(9.113)

$$G\left(0; q_{1\perp}, q'_{1\perp}; q_{\perp}\right) = q_{1\perp}^{2} q_{2\perp}^{2} \delta^{(D-2)}\left(q_{1\perp}-q'_{1\perp}\right),$$

(9.114)

where $G\left(Y; q_{1\perp}, q'_{1\perp}; q_{\perp}\right)$ is defined by the relation

$$\langle \mathcal{G}_{1}\mathcal{G}_{2}|\hat{G}(Y)|\mathcal{G}'_{1}\mathcal{G}'_{2}\rangle = \delta(q_{1\perp}+q_{2\perp}-q'_{1\perp}-q'_{2\perp})G\left(Y; q_{1\perp}, q'_{1\perp}; q_{\perp}\right), \quad q_{\perp}=q_{1\perp}+q_{2\perp},$$

(9.115)

and the BFKL kernel is

$$\mathcal{K}\left(q_{1\perp}, q'_{1\perp}; q_{\perp}\right) = q_{1\perp}^{2} q_{2\perp}^{2} \delta^{(D-2)}\left(q_{1\perp}-q'_{1\perp}\right)\left(\omega(q_{1\perp}^{2})+\omega(q_{2\perp}^{2})\right)$$
$$+ \mathcal{K}_{r}\left(q_{1\perp}, q'_{1\perp}; q_{\perp}\right).$$

(9.116)

Here the real part $\mathcal{K}_{r}$ of the kernel is defined in (9.101), (9.103).

9.4.2 Representation of scattering amplitudes

Scattering amplitudes $A(s,t)$ with definite signatures τ are expressed in terms of their s–channel discontinuities. Defining the Mellin transforms $f_{\omega}(t)$ of the discontinuities as

$$f_{\omega}(t) = \int_{0}^{\infty} dY e^{-\omega Y} \frac{\mathrm{Im}_{s} A(s,t)}{s},$$

(9.117)

one has

$$A(s,t) = A^{B}(s,t) + \frac{s}{2\pi i} \int_{\delta-i\infty}^{\delta+i\infty} \frac{d\omega}{\sin(\pi\omega)} \left(\tau \left(\frac{s}{s_{0}}\right)^{\omega} - \left(\frac{-s}{s_{0}}\right)^{\omega}\right) f_{\omega}(t),$$

(9.118)

where $A^{B}(s,t)$ is the Born contribution and the integration contour lies on the right of the point $\omega = 0$. Eq. (9.118) is the high-energy limit of the analytic continuation of the t-channel partial-wave expansion, and $f_{\omega}(t)$ is the partial wave with $j = 1 + \omega$. The limit $s \to \infty$ means the limit $j \to 1$, or $\omega \to 0$; the LLA means summation of terms $(\alpha_{s}/\omega)^{n}$. In our case, the Born contribution exists only for the gluon exchange (colour octet and negative signature). The gluon reggeization means that one can omit this contribution shifting the integration contour to the left from the point $\omega = 0$. It will be accepted in the following. The proof of the gluon reggeization in the NLA as well as in the LLA is discussed in Section 9.6.

For the Mellin transform $\hat{G}_\omega$ of $\hat{G}(Y)$,

$$\hat{G}_\omega = \int\limits_0^\infty dY e^{-\omega Y} \hat{G}(Y), \quad \hat{G}(Y) = \frac{1}{2\pi i} \int\limits_{\delta-i\infty}^{\delta+i\infty} d\omega e^{\omega Y} \hat{G}_\omega, \tag{9.119}$$

we have from (9.112)

$$\hat{G}_\omega = \frac{1}{\omega - \hat{\mathcal{K}}}. \tag{9.120}$$

In momentum representation, the Mellin transform $G_\omega\left(q_{1\perp}, q'_{1\perp}; q_\perp\right)$ is determined by the equation:

$$\omega G_\omega\left(q_{1\perp}, q'_{1\perp}; q_\perp\right) = q^2_{1\perp} q^2_{2\perp} \delta^{(D-2)}\left(q_{1\perp} - q'_{1\perp}\right)$$
$$+ \int \frac{d^{D-2}r_\perp}{r^2_\perp (q_\perp - r_\perp)^2} \mathcal{K}\left(q_{1\perp}, r_\perp; q_\perp\right) G_\omega\left(r_\perp, q'_{1\perp}; q_\perp\right). \tag{9.121}$$

From (9.107), (9.120) we obtain for the amplitudes with signature τ at $s > 0$

$$\delta(q_A - q_B)\mathcal{A}^{A'B'}_{AB} = \frac{s}{4(2\pi)^{D-2}} \int\limits_{\delta-i\infty}^{\delta+i\infty} \frac{d\omega}{2\pi i} \frac{e^{\omega Y}(\tau - e^{-i\pi\omega})}{\sin(\pi\omega)} \langle A'\bar{A}| \frac{1}{\omega - \hat{\mathcal{K}}} |\bar{B}'B\rangle, \tag{9.122}$$

where $q_A = p_A - p_{A'}$, $q_B = p_{B'} - p_B$, $Y = \ln(s/s_0)$, s_0 is an energy scale, $\langle A'\bar{A}|$ and $|\bar{B}'B\rangle$ are the t-channel states representing impact factors of the colliding particles, and $\hat{\mathcal{K}}$ is the BFKL kernel.

9.4.3 Colour decomposition

So far we allowed all possible colour states in the t-channel, so that $\hat{\mathcal{K}}$ is an operator in colour space. Due to colour conservation, one can write

$$\hat{\mathcal{K}} = \sum_R \hat{\mathcal{K}}^{(R)} \hat{\mathcal{P}}_R, \tag{9.123}$$

where $\hat{\mathcal{P}}_R$ are the projection operators of the two-reggeon colour states on the irreducible representations R of the colour group, and $\hat{\mathcal{K}}^{(R)}$ are the operators in momentum space (but not operators in colour space any more). Then, we have

$$\hat{\mathcal{P}}_R = \sum_\lambda \hat{\mathcal{P}}^\lambda_R, \tag{9.124}$$

where λ enumerates basis states of the representation R. Defining

$$|\bar{B}'B\rangle^\lambda_R = \hat{\mathcal{P}}^\lambda_R |\bar{B}'B\rangle, \quad \langle A'\bar{A}|^\lambda_R = \langle A'\bar{A}|\hat{\mathcal{P}}^\lambda_R, \tag{9.125}$$

we finally obtain from (9.122)

$$\delta(\boldsymbol{q}_A - \boldsymbol{q}_B)\mathcal{A}_{AB}^{A'B'} = \frac{s}{4(2\pi)^{D-2}} \int\limits_{\delta-i\infty}^{\delta+i\infty} \frac{d\omega}{2\pi i} \frac{e^{\omega Y}(\tau - e^{-i\pi\omega})}{\sin(\pi\omega)} \sum_{R,\lambda} \langle A'\bar{A}|_R^\lambda \frac{1}{\omega - \hat{K}^{(R)}} |\bar{B}'B\rangle_R^\lambda.$$

(9.126)

Since the virtual part of $\hat{K}$ is the unit matrix in colour space, we can write

$$\hat{K}^{(R)} = \hat{\Omega} + \hat{K}_r^{(R)},$$

(9.127)

where $\hat{K}^{(R)}$ is determined by the decomposition of the colour structure of $\hat{K}_r$. It is seen from (9.101)–(9.103) that $\hat{K}_r$ can be written as

$$\hat{K}_r = \frac{\hat{T}_1^a (\hat{T}_2^a)^*}{N_c} \hat{K}_r,$$

(9.128)

where $\hat{T}_i^a$, $i = 1, 2$ are the colour-group generators for the i-th reggeon, and $\hat{K}_r$ is defined as

$$\langle q_{1\perp} q_{2\perp} | \hat{K}_r | q'_{1\perp} q'_{2\perp} \rangle = \delta(q_{1\perp} + q_{2\perp} - q'_{1\perp} - q'_{2\perp}) K_r(q_{1\perp}, q'_{1\perp}; q_\perp),$$

$$K_r(q_{1\perp}, q'_{1\perp}; q_\perp) = -\frac{g^2 N_c}{(2\pi)^{D-1}} \left(\frac{q_{1\perp}^2 (q - q'_1)_\perp^2 + (q - q_1)_\perp^2 q'^2_{1\perp}}{(q_1 - q'_1)_\perp^2} - q_\perp^2 \right). \quad (9.129)$$

Here the remarkable properties of the kernel

$$K_r(0, q'_{1\perp}; q_\perp) = K_r(q_{1\perp}, 0; q_\perp) = K_r(q_\perp, q'_{1\perp}; q_\perp) = K_r(q_{1\perp}, q_\perp; q_\perp) = 0, \quad (9.130)$$

and

$$K_r(q_{1\perp}, q'_{1\perp}; q_\perp) = K_r(q_\perp - q_{1\perp}, q_\perp - q'_{1\perp}; q_\perp) = K_r(q'_{1\perp}, q_{1\perp}; q_\perp) \quad (9.131)$$

are explicitly exhibited. They remain valid also in the next-to-leading order. The properties (9.130) mean that the kernel turns to zero at zero transverse momentum of any the reggeons; they follow from gauge invariance; equalities (9.131) are the consequences of the symmetry of the reggeon–reggeon scattering amplitude.

Using the decomposition

$$T_{c_1 c'_1}^a (T_{c_2 c'_2}^a)^* = N_c \sum_R c_R \langle c_1 c_2 | \hat{P}_R | c'_1 c'_2 \rangle,$$

(9.132)

we obtain

$$\hat{K}^{(R)} = \hat{\Omega} + c_R \hat{K}_r.$$

(9.133)

The most interesting representations are the singlet (pomeron) and antisymmetrical octet (reggeized gluon) representations. For the former we have

$$\langle c_1 c_2 | \hat{P}_1 | c'_1 c'_2 \rangle = \frac{\delta_{c_1 c_2} \delta_{c'_1 c'_2}}{N_c^2 - 1}$$

(9.134)

and for the latter

$$\langle c_1 c_2 | \hat{P}_8 | c'_1 c'_2 \rangle = \frac{f_{ac_1 c_2} f_{ac'_1 c'_2}}{N_c}.$$

(9.135)

Using the identities of Section 1.13 one can easily find

$$c_1 = 1, \quad c_{8a} = \frac{1}{2}. \tag{9.136}$$

The projection operators $\hat{\mathcal{P}}_R$ for other representations are given in Section 1.13. Using these operators and the identities presented there, one can find

$$c_{8_s} = c_{8_a} = \frac{1}{2}, \quad c_{10} = c_{\overline{10}} = 0, \quad c_{27} = -c_{N_c>3} = -\frac{1}{4N_c}. \tag{9.137}$$

Here it is worth noting the equality $c_{8_s} = c_{8_a}$. Due to this equality, the total kernels (9.133) coincide for antisymmetric and symmetric octet representations. It was already mentioned that amplitudes with definite colour symmetry have definite symmetry in the t-channel scattering angle, i.e. they have definite signatures (see section 9.3.2). Therefore, equality of the kernels for antisymmetric and symmetric octet representations means equality of the octet kernels for odd and even signatures, i.e. the same energy behaviour of the s-channel imaginary parts and therefore the same partial waves. This phenomenon is called degeneracy in signature. In the limit of large N_c ($N_c \to \infty$, $g^2 \to 0$, $g^2 N_c$ is fixed), this phenomenon is not restricted by the LLA framework, but appears to be exact. As is well known, in the large N_c limit only planar Feynman diagrams contribute to scattering amplitudes. These diagrams have no simultaneous s- and u-channel discontinuities. The signature degeneracy for the diagrams with only s- (or u-) channel discontinuities is evident.

9.5 BFKL pomeron

9.5.1 BFKL kernel in the pomeron channel

With the definition

$$\langle q_{1\perp} q_{2\perp} | f \rangle = f(q_{1\perp}, q_{2\perp}), \tag{9.138}$$

we have in the singlet (pomeron) channel (in this section we consider only this channel and omit the superscript (1) of the singlet representation)

$$\langle q_{1\perp} q_{2\perp} | \hat{K} | f \rangle = \left(\omega(q_{1\perp}^2) + \omega(q_{2\perp}^2) \right) f(q_{1\perp}, q_{2\perp}) + \int \frac{d^{D-2}k_\perp}{(q_1 - k)_\perp^2 (q_2 + k)_\perp^2}$$
$$\times K_r(q_{1\perp}, (q_1 - k)_\perp; q_\perp) f((q_1 - k)_\perp, (q_2 + k)_\perp), \tag{9.139}$$

where $q_\perp = q_{1\perp} + q_{2\perp}$. Using the integral representation (9.47) for the gluon trajectory and (9.129) for K_r, we obtain:

$$\langle q_{1\perp} q_{2\perp} | \hat{K} | f \rangle = \frac{g^2 N_c}{(2\pi)^{3+2\epsilon}} \int d^{D-2}k_\perp \left[\left(q_\perp^2 - \frac{q_{1\perp}^2 q_{2\perp}'^2 + q_{2\perp}^2 q_{1\perp}'^2}{k_\perp^2} \right) \frac{f(q_{1\perp}', q_{2\perp}')}{q_{1\perp}'^2, q_{2\perp}'^2} \right.$$
$$\left. + \left(\frac{q_{1\perp}^2}{2k_\perp^2 q_{1\perp}'^2} + \frac{q_{2\perp}^2}{2k_\perp^2 q_{2\perp}'^2} \right) f(q_{1\perp}, q_{2\perp}) \right], \tag{9.140}$$

where $q_{1\perp}' = (q_1 - k)_\perp$, $q_{2\perp}' = (q_2 + k)$.

The important fact, which follows from (9.140), is the cancellation of the infrared singularities. Recall that the real kernel turns to zero at zero transverse momentum of any the reggeons. Therefore, it gives the singular at $\epsilon \to 0$ contribution only at $k = 0$. It is easy to see that this contribution is completely cancelled by the singular contribution of the virtual part. Therefore, we can take in (9.140) the limit $\epsilon \to 0$. Moreover, we can simply put $\epsilon = 0$, regularizing the singularities at $k_\perp^2 = 0$, $(q_1 - k)_\perp^2 = 0$ and $(q_2 + k)_\perp^2 = 0$ by limiting the integration region $|k_\perp^2| > \lambda^2$, $|(q_1 - k)_\perp^2| > \lambda^2$, $|(q_2 + k)_\perp^2| > \lambda^2$ with $\lambda \to 0$, or in an equivalent way (for example, changing $k_\perp^2 \to k_\perp^2 - \lambda^2$, $(q_1 - k)_\perp^2 \to (q_1 - k)_\perp^2 - \lambda^2$, $(q_2 + k)_\perp^2 \to (q_2 + k)_\perp^2 - \lambda^2$). In the following such kind of regularization is implied.

In the particular case of forward scattering, $q_\perp = q_{1\perp} + q_{2\perp} = 0$, we have

$$K(q_1, q_1') \equiv \frac{K_r(q_{1\perp}, q_{1\perp}'; 0)}{q_{1\perp}^2 q_{1\perp}'^2} + 2\,\omega(q_{1\perp}^2)\delta(q_{1\perp} - q_{1\perp}')$$

$$= \frac{g^2 N_c}{(2\pi)^3} \left(\frac{2}{k^2} - \int \frac{q_1^2 dl}{l^2 (q_1 - l)^2} \delta(k) \right), \qquad (9.141)$$

where $k = q_1 - q_1'$. Until we consider azimuthal correlations, we can average over the angle ϕ between q_1' and q_1. Writing

$$\int \frac{dl}{l^2 (q_1 - l)^2} = \int \frac{dl}{l^2 + (q_1 - l)^2} \left(\frac{1}{(q_1 - l)^2} + \frac{1}{l^2} \right) = 2 \int \frac{dl}{(l^2 + (q_1 - l)^2)(q_1 - l)^2}$$

$$= 2 \int \frac{dl}{l^2} \left(\frac{1}{(q_1 - l)^2} - \frac{1}{l^2 + (q_1 - l)^2} \right) \qquad (9.142)$$

and using

$$\int_0^\pi \frac{d\phi}{a + b \cos \phi} = \frac{\pi}{\sqrt{a^2 - b^2}}, \qquad (9.143)$$

we arrive at the form introduced in [3] and widely used in the literature:

$$\int dl\, K(q, l) f(l^2) = \frac{N_c \alpha_s}{\pi} \int dl^2 \left[\frac{f(l^2)}{|q^2 - l^2|} - f(q^2)\frac{q^2}{l^2} \left(\frac{1}{|q^2 - l^2|} - \frac{1}{\sqrt{q^4 + 4l^4}} \right) \right]. \qquad (9.144)$$

Evidently, the representation used here for the virtual part (coefficient of $f(q^2)$) is not unique. One can easily see that it is possible to put this coefficient equal to $g(l^2/q^2)|l^2 - q^2|^{-1}$ with any function g satisfying the requirements $g(1) = 1$ and

$$\int_0^\infty \frac{dx}{|x - 1|} g(x)\, \theta(|x - 1| - \delta) = 2 \ln \frac{1}{\delta} \qquad (9.145)$$

at $\delta \to 0$. A possible choice [64] is

$$\int dl \, K(q,l) f(l^2) = \frac{N_c \alpha_s}{\pi} \int \frac{dl^2}{|q^2 - l^2|} \left[f(l^2) - 2 \frac{\min(q^2, l^2)}{q^2 + l^2} f(q^2) \right]. \tag{9.146}$$

Of course, the representations (9.144) and (9.146) are equivalent. Both of them make explicit the scale invariance of the kernel, due to which its eigenfunctions are powers of l^2. Usually they are written as $l^{2(\gamma-1)}$, and the corresponding eigenvalues as

$$\omega(\gamma) = \frac{N_c \alpha_s}{\pi} \chi(\gamma). \tag{9.147}$$

From (9.146) it is easy to obtain

$$\chi(\gamma) = \int_0^1 \frac{dx}{1-x} \left(x^{\gamma-1} + x^{-\gamma} - 2 \right) = 2\psi(1) - \psi(\gamma) - \psi(1-\gamma), \tag{9.148}$$

where $\psi(z) = \Gamma'(z)/\Gamma(z)$.

The values $\gamma = 1/2 + i\nu$, $-\infty < \nu < \infty$ are especially important for investigations of the high-energy behaviour of cross sections because the functions $(l^2)^{-1/2+i\nu}$ form a complete set. At that, the eigenvalues of the kernel

$$\omega\left(\frac{1}{2} + i\nu\right) = \frac{N_c \alpha_s}{\pi} \sum_{k=0}^{\infty} \frac{k + \frac{1}{2} - 2\nu^2}{(k+1)((k+\frac{1}{2})^2 + \nu^2)} \tag{9.149}$$

form a continuous spectrum in the region $-\infty < \omega(\nu) < \omega_P^B$ with the maximal value

$$\omega_P^B = 4N_c \frac{\alpha_s}{\pi} \ln 2 \simeq 2.77 N_c(\alpha_s/\pi) \tag{9.150}$$

which is reached at $\nu = 0$ (see Fig. 9.7).

Generally, amplitudes with zero momentum transfer can contain azimuthal correlations, so that one can be interested in eigenfunctions and eigenvalues of the nonaveraged kernel (9.141). They were found in [4]. The kernel $K(q,l)$ is invariant under the scale

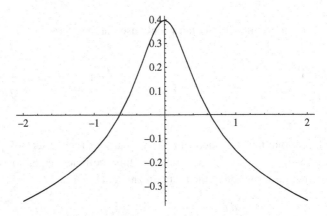

Fig. 9.7. Plot of $\omega(1/2 + i\nu)$ dependence on ν for $\alpha_s = 0.15$.

transformations and rotation in the transverse momentum plane. Therefore, the eigenfunctions can be taken in the form $e^{i n\phi} l^{2(\gamma-1)}$, where ϕ is the azimuthal angle of the
momentum l. Denoting the corresponding eigenvalues by

$$\omega(\gamma, n) = \frac{\alpha_c N_c}{\pi} \chi(\gamma, n) \tag{9.151}$$

and using

$$\int_0^{2\pi} \frac{d\phi e^{in\phi}}{(q - l)^2} = \frac{2\pi}{|q^2 - l^2|} \left(\frac{\min(q^2, l^2)}{\max(q^2, l^2)} \right)^{\frac{|n|}{2}}, \tag{9.152}$$

we obtain

$$\chi(\gamma, n) = \chi(\gamma, |n|) = \int_0^1 \frac{dx}{1 - x} \left(x^{\frac{|n|}{2}+\gamma-1} + x^{\frac{|n|}{2}-\gamma} - 2 \right)$$

$$= 2\psi(1) - \psi\left(\gamma + \frac{n}{2}\right) - \psi\left(1 - \gamma + \frac{n}{2}\right). \tag{9.153}$$

At $\gamma = 1/2 + i\nu$, we have

$$\omega(\gamma, n) = \frac{N_c \alpha_s}{\pi} \sum_{k=0}^{\infty} \frac{(1 - |n|)(k + \frac{|n|+1}{2}) - 2\nu^2}{(k + 1)((k + \frac{|n|+1}{2})^2 + \nu^2)}, \tag{9.154}$$

the maximal value is reached at $\nu = 0$. As it should be, these eigenvalues are less than ω_P^B
by the amount of $\Delta(n)$,

$$\Delta(n) = \frac{N_c \alpha_s}{\pi} \sum_{k=0}^{\infty} \frac{4|n|}{(1 + 2k)(1 + 2k + |n|)}. \tag{9.155}$$

9.5.2 Rising cross section and diffusion

According to (9.126), the total cross section

$$\sigma_{AB}(s) = \frac{\mathrm{Im} \mathcal{A}_{AB}^{AB}}{s} \tag{9.156}$$

is represented in the form of

$$\sigma_{AB}(s) = \frac{1}{4(2\pi)^2} \int_{\delta-i\infty}^{\delta+i\infty} \frac{d\omega}{2\pi i} e^{\omega Y} \int \frac{dq}{q^2} \int \frac{dl}{l^2} \Phi_A(q) G_\omega(q, l) \Phi_B(l), \tag{9.157}$$

where $\Phi_A(q) = \Phi_{A'A}(q, -q)$, $\Phi_B(q) = \Phi_{B'B}(q, -q)$ are the colour-singlet impact
factors and G_ω is the Mellin transform of the Green function for forward scattering. G_ω
has the spectral representation

$$G_\omega(q, l) = \frac{1}{2\pi^2} \sum_{n=-\infty}^{+\infty} \int_{-\infty}^{+\infty} \frac{d\nu}{\omega - \omega(\gamma, n)} e^{in\phi} q^{2(-\frac{1}{2}+i\nu)} l^{2(-\frac{1}{2}-i\nu)}, \quad \gamma = \frac{1}{2} + i\nu. \tag{9.158}$$

As is seen from this representation, G_ω as a function of ω has a cut from $-\infty$ to ω_P^B. This means that ω_P^B is the leading singularity of the colour-singlet partial waves with positive signature at $t = 0$, which is called the pomeron intercept. The singularity is a branch point. Moreover, it is a fixed branch point (i.e. not depending on t). Actually, in the LLA it can be predicted from dimensional reasons. Indeed, the positions of singularities of partial waves in the complex momentum plane can depend on t, but not on the colliding particles. Therefore, in the LLA there are no dimensionless parameters to describe singularities moving with t. On the other hand, fixed singularities can be only branch points and not poles.

At large s, the main contribution to the forward Green function

$$G(Y; q, l) = \int_{\delta - i\infty}^{\delta + i\infty} \frac{d\omega}{2\pi i} e^{\omega Y} G_\omega(q, l) = \frac{1}{2\pi^2} \sum_{n=-\infty}^{+\infty} \int_{-\infty}^{+\infty} dv e^{\omega(\gamma, n)Y} e^{in\phi} q^{2(-\frac{1}{2}+iv)} l^{2(-\frac{1}{2}-iv)}$$

(9.159)

comes from $n = 0$ and the region of small v. In this region,

$$\omega(\gamma) \simeq \omega_P^B - Dv^2 + O(v^4),$$ (9.160)

where ω_P^B is given by (9.150) and

$$D = 14\zeta(3)N_c(\alpha_s/\pi) \simeq 16.8N_c(\alpha_s/\pi).$$ (9.161)

Therefore,

$$G(Y; q, l) \simeq \frac{1}{2\pi\sqrt{\pi Y D q^2 l^2}} \exp\left[\omega_P^B Y - \frac{\left(\ln q^2 - \ln l^2\right)^2}{4DY}\right].$$ (9.162)

This equation demonstrates explicitly two main problems of the approximation. First, since the impact factors are energy independent, one can see from (9.162) and (9.157) that

$$\sigma_{AB}(s) \sim \frac{s^{\omega_P^B}}{\sqrt{\ln s}}.$$ (9.163)

This is just the famous result (9.3), where $\omega_P = \omega_P^B$ (9.150), which signifies violation of the Froissart bound [11]. The second problem is the broadening with Y of the width of the $|\ln q^2|$ distribution. This broadening means the appearance of large contributions from small transverse momenta at sufficiently high energies. In other words, it is diffusion into the large distance region where perturbative calculations may be unreliable. Actually, $G(Y; q, l)$ in (9.162) is up to the coefficient $\exp[\omega_P Y]/(\pi|q||l|)$, the Green function for the diffusion equation, where $\ln q^2$, Y and D are respectively the coordinate, time and diffusion coefficient. The diffusion property was recognized already in [3]. Numerically, its importance was analyzed in [65].

The singularity in the ω-plane is changed if one considers the running of the coupling constant. Strictly speaking, account of the running oversteps the limits of the LLA. Nevertheless, it is not unreasonable to take it into account just in the leading order BFKL kernel in order to understand qualitative effects of the running (see [12],[66]–[73]). One important effect ([66]) is the conversion of the cut $[0, \omega_P^B]$ into an infinite series of moving poles with limiting point at $\omega = 0$. The pomeron trajectories found in ([66]) are represented in the form

$$\omega_k(t) = \frac{c}{k + \eta(q^2) + 1/4}, \quad k = 0, \pm 1, \pm 2 \ldots, \tag{9.164}$$

where c is the calculable constant. The function $\eta(q^2)$ is determined by large distances, but is limited by the inequalities $-1 \le 4\eta \le 3$ at small $|t|$. In the region of large $|t|$ ($\alpha_s(|t|) \ll 1$), the family of poles is approximated by the moving cut with the branch point

$$\omega(t) = \frac{4N_c}{\pi} \ln 2 \, \alpha_s(|t|). \tag{9.165}$$

9.5.3 Resummation of terms $\alpha_s^n/(j-1)^n$ in anomalous dimensions

The remarkable result of the BFKL approach is the resummation of the most singular at $j = 1$ terms in the anomalous dimension of the twist-2 operators [74]. Let us consider deep inelastic scattering, and let particle A be a photon with virtuality Q^2; consequently particle B represents a hadron target. Then the Q^2-dependence of the deep inelastic moments

$$M_\omega(Q^2) = \int\limits_{Q^2}^{\infty} \frac{ds}{s} \left(\frac{s}{Q^2}\right)^{-\omega} \sigma_{AB}(s) \tag{9.166}$$

is determined by the anomalous dimensions of the twist-2 operators. At small ω, we have from (9.157)

$$M_\omega(Q^2) = \frac{1}{4(2\pi)^2} \int \frac{dq}{q^2} \int \frac{dl}{l^2} \Phi_A(q) G_\omega(q, l) \Phi_B(l). \tag{9.167}$$

Essential regions of integrations (in 9.167) are $q^2 \sim Q^2$ and $l^2 \sim M_t^2$, where M_t is the target mass. Therefore, the large Q^2 behaviour is determined by $G_\omega(q, l)$ in the region $Q^2 \sim q^2 \gg l^2$.

After averaging over ϕ, we obtain from (9.158)

$$G_\omega(q^2, l^2) = \frac{1}{2\pi^2 q^2} \int\limits_{1/2-i\infty}^{1/2+i\infty} \frac{d\gamma}{\omega - \frac{\alpha_s N_c}{\pi} \chi(\gamma)} \exp\left[\gamma(\ln q^2 - \ln l^2)\right]. \tag{9.168}$$

One can shift the integration contour in the γ plane to the left, picking up contributions of singularities remaining on the right. At large q^2, the main contribution

$$G_\omega(q^2, l^2) \sim \left(q^2\right)^{\gamma_\omega} \tag{9.169}$$

comes from the rightmost singularity, which is the pole at γ_ω defined by the equation

$$\omega = \frac{N_c \alpha_s}{\pi} \chi(\gamma_\omega). \tag{9.170}$$

Eq. (9.169) gives just the renormalization group behaviour for fixed α_s, and γ_ω appears as the resummed anomalous dimension. Expanding γ_ω in powers of $\bar{\alpha}/\omega$, where $\bar{\alpha} = \alpha_s N_c/\pi$, we obtain from (9.170)

$$\gamma_\omega = \frac{\bar{\alpha}_s}{\omega} + 2.4 \left(\frac{\bar{\alpha}_s}{\omega}\right)^4 + 7 \left(\frac{\bar{\alpha}_s}{\omega}\right)^6 + \dots \qquad (9.171)$$

9.5.4 Colour-dipole model and colour glass condensate (CGC)

The power growth of cross sections (9.163) conflicts with unitarity and should be modified. Starting from pioneer paper [12], this modification of the growth is known as gluon saturation. In deep inelastic scattering (DIS), the saturation is reached when the density becomes so large that gluons "overlap" with each other, i.e. their interaction cross section $\sim \alpha_s/Q^2$ multiplied by the density is comparable with the geometrical cross section of the target. At fixed $x = Q^2/s$, it determines the saturation scale $Q_s(x)$ [12]:

$$\frac{\alpha_s(Q_s^2)xG(x, Q_s^2)}{Q_s^2} \sim \pi R_t^2, \quad xG(x, Q^2) = \int \mathcal{F}(x, k^2)dk^2\theta(Q^2-k^2)/k^2. \qquad (9.172)$$

Here $G(x, Q^2) = f^g(x, Q^2)$ is the gluon distribution function and $\mathcal{F}(x, k^2)$ is the unintegrated gluon density. A more refined definition of the saturation scale takes into account dependence of their interaction cross section on gluon transverse momentum and is formulated in terms of $\mathcal{F}(x, k^2)$:

$$\frac{\mathcal{F}(x, Q_s^2)}{Q_s^2 \pi R_t^2} \sim \frac{1}{\alpha_s(Q_s^2)}. \qquad (9.173)$$

Here the left-hand side is the gluon occupation number. Therefore, this definition shows that the saturation is reached at large occupation numbers, admitting (semi)classical description.

It follows from (9.163) that $Q_s^2(x) \sim Q_0^2 x^{-\omega_P^B}$. The (x, Q^2) plane is divided by the curve $Q_s^2(x)$ into two parts. In the low-density domain $Q^2 > Q_s^2(x)$, the gluon-density evolution is described by the DGLAP and BFKL equations; in the region $Q^2 < Q_s^2(x)$, the high density leads to large nonlinear effects which restrict the growth of the density and produce some equilibrium (saturated) state of partons. The first equation taking into account the nonlinear effects was obtained in the double-logarithmic approximation [12]. It is called Gribov–Levin–Ryskin (GLR) equation and looks as follows [12],[75]:

$$\frac{\partial^2(xG(x, Q^2))}{\partial \ln(1/x)\partial \ln Q^2} = \frac{\alpha_s N_c}{\pi}xG(x, Q^2) - \frac{4\alpha_s^2 N_c}{3C_F R^2}\frac{1}{Q^2}[xG(x, Q^2)]^2, \qquad (9.174)$$

where R is the hadron radius. The last term in this equation takes into account not only ladder-type diagrams, representing the BFKL pomeron, but also so-called fan diagrams, describing the pomeron splitting into two pomerons.

Further development of the nonlinear generalization of the BFKL equation is related to the colour-dipole model. This model gives a simple physical picture of γ^* interactions with nucleus at small x allowing to derive the BFKL equation in impact-parameter space

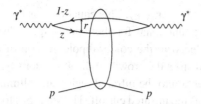

Fig. 9.8. Schematic representation of γ^* interaction with nucleus.

[76]–[79]. In the colour-dipole picture γ^* interaction with a nucleus is described as follows (see Fig. 9.8): in the reference frame in which the virtual photon carries sufficiently large energy it dissociates into a $q\bar{q}$ pair (colour dipole) much earlier than the pair scatters off the gluon field in the nucleus. Therefore, one has for the total cross section

$$\sigma_{\gamma^*}(Y, Q^2) = \int d^2r \int_0^1 dz |\Psi_{\gamma^*}(r, z, Q^2)|^2 \sigma_{dp}(r, Y), \tag{9.175}$$

where $\Psi_{\gamma^*}(r, z, Q^2)$ is the $q\bar{q}$ component of the γ^* wave function, z is the quark momentum fraction, $r = r_1 - r_2$, r_1 and r_2 are the quark-pair transverse coordinates, $Y = \log(s/Q^2)$ and $\sigma_{dp}(r, Y)$ is the dipole cross section. The important property of the interaction at high energy is the conservation of transverse coordinates of quark-antiquark pairs. The cross section $\sigma_{dp}(r, Y)$ is expressed in terms of the dipole elastic scattering S-matrix:

$$\sigma_{dp}(r, Y) = \int d^2b \left(1 - S(r_1, r_2; Y)\right), \tag{9.176}$$

where $b = (r_1 + r_2)/2$ is the impact parameter, $S(r_1, r_2; Y)$ is the S-matrix element, which can be presented as the average of the product of two Wilson lines:

$$S(r_1, r_2; Y) = \langle \operatorname{tr}\left(U^n(r_1)U^{n\dagger}(r_2)\right)\rangle / N_c, \tag{9.177}$$

where

$$U^n(x_\perp) = P \exp\left[ig \int_{-\infty}^{+\infty} du\, n_\mu A_\mu^c(un + x_\perp)t^c\right]. \tag{9.178}$$

Here P stands for ordering along the straight line, n is the particle velocity.

In Ref. [80], the equation for evolution of the product $U^n U^{n\dagger}$ with respect to the slope of the straight line was derived. However, it is not a closed equation, because it contains products of four Wilson lines. In general, there is the infinite hierarchy of coupled equations with increasing number of $U^n U^{n\dagger}$. In the approximation of factorization, when the four-line average is replaced by the product of two-line averages (which is justified when a target is a large nucleus) in the limit of large N_c the evolution takes a simple form of the Balitsky–Kovchegov (BK) equation

$$\frac{\partial}{\partial Y} N(r_1, r_2; Y) = \frac{\alpha_s N_c}{2\pi^2} \int dr_0 \frac{(r_1 - r_2)^2}{(r_1 - r_0)^2 (r_2 - r_0)^2}$$

$$\times (N(r_1, r_0; Y) + N(r_0, r_2; Y) - N(r_1, r_2; Y) - N(r_1, r_0; Y)N(r_0, r_2; Y)), \tag{9.179}$$

where $N(r_1, r_2; Y) = 1 - S(r_1, r_2; Y)$. This equation was derived independently in [81] with use of the generating functional for the colour-dipole distribution [78]. For small N, equation (9.179) is reduced to the colour-dipole version of the BFKL equation. The nonlinear effects not only restrict the growth of the gluon density, but also diminish the diffusion of transverse momenta into the infrared region and eliminate the infrared problem; $Q_s(x)$ appears effectively as the infrared cut-off. The solutions to the BK equation have the remarkable property [82] called geometrical scaling [83]: $N(r_1, r_2; Y) = f(r^2 Q_s^2(x; b))$. In fact, this property holds not only in the saturation region $Q^2 < Q_s^2(x)$, but in the much wider kinematical region [82]–[85] up to $Q^2 \sim Q_s^4(x)/\Lambda_{QCD}^2$, where $1 - S(r_1, r_2, Y) \simeq (r^2 Q_s^2(x, b))^\gamma$, with $\gamma \simeq 0.64$.

An evident imperfection of the BK equation is its target-projectile asymmetry. Like the GLR equation, it allows only summing of the fan diagrams. The equation contains the pomeron splitting, but does not contain the pomeron fusion, i.e. pomeron loops which are necessary from the point of view of the Gribov reggeon calculus.

Another approach to the saturation problem, which was much developed in the past decade (see for instance [86]–[105]), is called colour glass condensate (CGC) [92]–[94]. It is worthwhile to note that the saturation phenomenon is not necessary related to the growth of the gluon density at small x. For a nucleus with $A \gg 1$, it can be reached even at $x \sim 1$. In the model of independent sources in Born approximation

$$\mathcal{F}(x, Q^2) = AN_c \frac{\alpha_s C_F}{\pi}, \tag{9.180}$$

so that using $R_A \sim A^{1/3} M_N^{-1}$ one obtains from (9.173)

$$Q_s^2 \sim A^{1/3} \left(\frac{\alpha_s(Q_s^2) N_c}{\pi} \right)^2 M_N^2. \tag{9.181}$$

At sufficiently large A, we can have $\alpha_s(Q_s^2) \ll 1$, so that the use of perturbation theory is justified [86]–[88]. At small x summation of the term $(\alpha_s \ln(1/x))^n$ gives in the right-hand side of (9.181) the factor $(1/x)^{\omega_P^B}$.

The CGC approach is based on a model for the small-x hadronic wavefunction of a fast hadron. The small-x short-lived gluons are radiated semiclassically by a "frozen" configuration of faster partons (glass) with a random distribution $W_Y[\rho]$ of the colour charge $\rho_a(r)$. For the gluonic modes with $|k_\perp| < Q_s(x)$ the occupation numbers are $\sim 1/\alpha_s$ (condensate), corresponding to strong classical fields $A_a^\mu[\rho] \sim 1/g$. Evolution of $W_Y[\rho]$ with Y is governed by the functional evolution equation which was derived [89]–[94], using the Wilson renormalization group. It is known as Jalilian–Marian–Iancu–McLerran–Weigert–Leonidov–Kovner (JIMWLK) equation. The JIMWLK equation includes the BK equation and is equivalent to the Balitsky hierarchy. Nevertheless, the JIMWLK equation is incomplete, as was recognized in Refs. [97]–[101], because it misses the effects of gluon-number fluctuations. After this, new equations have been proposed [101]–[105], which can be interpreted [103]–[105] as an effective theory which involves the pomeron loops.

For virtual photon interaction cross section (9.157) can be rewritten in the form (9.175) of the colour-dipole model. Evidently σ_{AB} (9.157) makes a sense only if the impact factors vanish at zero momentum transfer; otherwise it tends to infinity. In fact, it is infinite for scattering of colour particles, as it should be, and finite for colourless ones, because for these we have

$$\Phi_P(q^2) \sim q^2 \tag{9.182}$$

at small q^2 due to the gauge invariance. Physically, this means that long-wave gluons do not interact with a system of zero total colour. A remarkable example of impact factors for scattering of colourless particles is Φ_{γ^*} – the virtual photon impact factor. Having an important physical implication, this impact factor can be unambiguously calculated in perturbation theory.

If particle A is a photon with virtuality Q^2, then the upper line in Fig. 9.5 represents a quark-antiquark pair. Let the quark and antiquark momenta will be l_1 and l_2, respectively,

$$p_A = p_1 - \frac{Q^2}{s}p_2, \quad l_1 + l_2 = k = p_1 + \frac{k^2 + \mathbf{k}^2}{s}p_2 + k_\perp, \quad l_i = x_i p_1 + \frac{l_i^2 + m^2}{s x_i}p_2 + l_{i\perp}. \tag{9.183}$$

In this case, we have to replace in the phase-space element (9.94)

$$\frac{d^{D-1}k}{(2\pi)^{D-1}2\epsilon} \rightarrow \delta^D(k - l_1 - l_2)d^D k \prod_{i=1}^{2} \frac{d^{D-1}l_i}{(2\pi)^{D-1}2\epsilon_i} = \frac{d^{D-1}k}{(2\pi)^{D-1}2\epsilon_k} \frac{dx d^{D-2}l_\perp}{2x(1-x)(2\pi)^{D-1}}, \tag{9.184}$$

where $x \equiv x_1$, $l_\perp \equiv l_{1\perp}$. Accordingly, the last factor appears in the impact-factor definition (9.104). For the case of forward scattering, we obtain

$$\Phi_{\gamma^*}(q) = \frac{2}{\sqrt{N_C^2 - 1}} \sum_{\{a\}} \int |\Gamma^c_{\gamma^*R \to q\bar{q}}|^2 \frac{dx d^{D-2}l_\perp}{2x(1-x)(2\pi)^{D-1}}, \tag{9.185}$$

where $\Gamma^c_{\gamma^*R \to q\bar{q}}$ is the amplitude of $q\bar{q}$ production in the γ^*R collision, and the sum $\{a\}$ is over all discrete quantum numbers of the produced pair. As was already mentioned, in the leading order the reggeized gluon acts as an ordinary gluon, so that $\Gamma^c_{\gamma^*R \to q\bar{q}}$ is given by the amplitude $\gamma^*g \to q\bar{q}$ with the gluon-colour index c and the polarization vector p_2/s. Therefore, we have (see Fig. 9.9)

$$\Gamma^c_{\gamma^*R \to q\bar{q}} = \langle t^c \rangle \, (\Gamma_a + \Gamma_b), \tag{9.186}$$

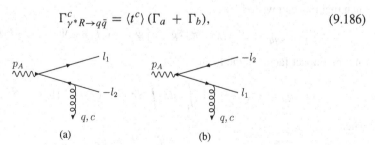

Fig. 9.9. The diagrams corresponding to the vertex $\gamma^*R \to q\bar{q}$.

where $\langle t^c \rangle$ is the colour matrix element, and

$$\Gamma_a = \frac{e_q g}{s} \bar{u}(x_1, l_{1\perp}) \frac{k_{\gamma^*}(\slashed{l}_2 + \slashed{q} - m) \slashed{p}_2}{m^2 - (p_A - l_1)^2} v(x_2, l_{2\perp}), \qquad (9.187)$$

$$\Gamma_b = -\frac{e_q g}{s} \bar{u}(x_1, l_{1\perp}) \frac{\slashed{p}_2(\slashed{l}_1 + \slashed{q} + m) k_{\gamma^*}}{m^2 - (p_A - l_1)^2} v(x_2, l_{2\perp}). \qquad (9.188)$$

Here e_q is the electric charge of the quark, e_{γ^*} is the polarisation vector of the photon, $l_{1\perp} + l_{2\perp} + q_\perp = 0$, $u(x_1, l_{1\perp}) = u(l_1)$, $v(x_2, l_{2\perp}) = v(l_2)$ are usual Dirac spinors. Using the equalities

$$m^2 - (p_A - l_i)^2 = \frac{l_i^2 + m^2 + x_1 x_2 Q^2}{x_i},$$

$$(\slashed{l}_2 + \slashed{q} - m) \slashed{p}_2 = (x_2 \slashed{p}_1 - \slashed{l}_{1\perp} - m) \slashed{p}_2 = \sum_\lambda v^\lambda(x_2, -l_{1\perp}) \bar{v}^\lambda(x_2, -l_{1\perp}) \slashed{p}_2,$$

$$\slashed{p}_2(\slashed{l}_1 + \slashed{q} + m) = \slashed{p}_2(x_1 \slashed{p}_1 - \slashed{l}_{2\perp} + m) = \slashed{p}_2 \sum_\lambda u^\lambda(x_1, -l_{2\perp}) \bar{u}^\lambda(x_2, -l_{1\perp}), \qquad (9.189)$$

and

$$\bar{v}^\lambda(x_2, -l_{1\perp}) \slashed{p}_2 v^{\lambda'}(x_2, l_{2\perp}) \simeq x_2 s \delta_{\lambda\lambda'}, \quad \bar{u}^\lambda(x_1, l_{1\perp}) \slashed{p}_2 u^{\lambda'}(x_1, -l_{2\perp}) \simeq x_1 s \delta_{\lambda\lambda'}, \qquad (9.190)$$

we obtain

$$\Gamma^c_{\gamma^* R \to q\bar{q}} = e_q g x_1 x_2 \left[\frac{\bar{u}(x_1, l_1) \, k_{\gamma^*} t^c v(x_2, -l_1)}{l_1^2 + x_1 x_2 Q^2} - \frac{\bar{u}(x_1, -l_2) \, k_{\gamma^*} t^c v(x_2, l_2)}{l_2^2 + x_1 x_2 Q^2} \right]. \qquad (9.191)$$

This representation of the vertex appears most naturally in the old-fashioned or light-cone perturbation theory. It makes obvious that the vertex turns to zero at $q_\perp = l_{1\perp} + l_{2\perp} = 0$. Moreover, this representation permits us to write the BFKL cross section for virtual photon interaction σ_{γ^*} in the same form as the colour-dipole model. In momentum representation, the $q\bar{q}$ component of the γ^* wave function is

$$\Psi_{\gamma^*}(x, k) = \frac{e_q}{(2\pi)^{3/2}} \sqrt{\frac{x(1-x)}{2}} \frac{\bar{u}(x, k) \, k_{\gamma^*} v(1-x, -k)}{k^2 + x(1-x)Q^2}, \qquad (9.192)$$

such that one can write Γ^c as

$$\Gamma^c_{\gamma^* R \to q\bar{q}} = \sqrt{2x_1 x_3} (2\pi)^{\frac{3}{2}} g \langle t^c \rangle \left[\Psi_{\gamma^*}(x_1, l_1) - \Psi_{\gamma^*}(x_1, -l_2) \right] \qquad (9.193)$$

and the impact factor in the form

$$\Phi_{\gamma^*}(q) = g^2 \sqrt{N_c^2 - 1} \int_0^1 dx \int dr \, |\Psi_{\gamma^*}(x, r)|^2 \, |1 - e^{-iqr}|^2, \qquad (9.194)$$

where

$$\Psi_{\gamma^*}(x, r) = \int \frac{dk}{(2\pi)^{\frac{3}{2}}} e^{ikr} \Psi_{\gamma^*}(x, k). \qquad (9.195)$$

Using Eq. (9.194), one can rewrite the BFKL representation of the cross section (see (9.157)) in the colour-dipole form, identifying the dipole cross section with the convolution of the Green function with the target impact factor:

$$\sigma_{dp}(r, Y) = \int \frac{d^2 q}{2\pi q^2} |1 - e^{-iqr}|^2 \mathcal{F}(q), \qquad (9.196)$$

where

$$\mathcal{F}(q) = \frac{g^2 \sqrt{N_c^2 - 1}}{4} \int_{\delta - i\infty}^{\delta + i\infty} \frac{d\omega}{2\pi i} \int \frac{d^2 l}{2\pi l^2} \left(\frac{s}{Q^2} \right)^\omega G_\omega(q, l) \Phi_B(l). \qquad (9.197)$$

Eq. (9.194) can be easily generalized to the nonforward case, when instead of (9.185) we have

$$\Phi_{\gamma^{*'} \gamma^*}(q_1, q_2) = \frac{1}{\sqrt{N_C^2 - 1}} \sum_{\{a\}} \int \frac{dx d^{D-2} l_\perp}{2x(1-x)(2\pi)^{D-1}} \Gamma_{\gamma^* R \to q\bar{q}}^c$$

$$\times \left(\Gamma_{\gamma^{*'} R' \to q\bar{q}}^c \right)^* + (q_1 \leftrightarrow q_2), \qquad (9.198)$$

where $\Gamma_{\gamma^* R \to q\bar{q}}^c$ is given by (9.191) at $l_1 + l_2 + q_1 = 0$ and $\Gamma_{\gamma^{*'} R' \to q\bar{q}}^c$ is obtained by the substitution $q_1 \to -q_2$, $Q^2 \to Q'^2$; Q'^2 is the $\gamma^{*'}$ virtuality. Then it is easy to see that instead of (9.194) the impact factor takes the form

$$\Phi_{\gamma^{*'} \gamma^*}(q_1, q_2) = \frac{g^2}{2} \sqrt{N_c^2 - 1} \int_0^1 dx \int dr \left(\Psi_{\gamma^*}(x, r) \Psi_{\gamma^{*'}}^*(x, r) + \Psi_{\gamma^{*'}}(x, r) \Psi_{\gamma^*}^*(x, r) \right)$$

$$\times \left(1 - e^{-iq_1 r} \right) \left(1 - e^{-iq_2 r} \right), \qquad (9.199)$$

and the scattering amplitude $\gamma^* B \to \gamma^{*'} B'$ (see (9.122)) with the momentum transfer q is written as

$$\mathcal{A}_{\gamma^* B}^{\gamma^{*'} B'} = s \int d^2 r \int_0^1 dx \Psi_{\gamma^*}(x, r) \Psi_{\gamma^{*'}}^*(x, r)$$

$$\times \int \frac{d^2 q_1}{2\pi q_1^2 (q - q_1)^2} \left(1 - e^{-iq_1 r} \right) \left(1 - e^{-i(q - q_1)r} \right) \mathcal{F}(q_1, q - q_1), \qquad (9.200)$$

where

$$\mathcal{F}(q_1, q - q_1) = \frac{g^2 \sqrt{N_c^2 - 1}}{4} \int_{\delta - i\infty}^{\delta + i\infty} \frac{d\omega}{2\pi i} \frac{(1 - e^{-i\pi\omega})}{\sin \pi \omega}$$

$$\int \frac{d^2 q_1'}{2\pi q_1'^2 (q - q_1')^2} \left(\frac{s}{Q^2} \right)^\omega G_\omega(q_1, q_1'; q) \Phi_{B'B}(q_1', q - q_1'). \qquad (9.201)$$

As was already mentioned, the nonforward case has an advantage of smaller sensitivity to large-distance contributions, since the diffusion in the infrared region is limited by

$\sqrt{|t|}$. The momentum transfer acts as an effective infrared cut-off which ensures that the dominant contribution arises from short-distance physics [106].

9.5.5 BFKL kernel in Möbius representation

It was found that for scattering of colourless particles the leading order BFKL equation can be solved in a general form not only for the forward case, considered above, but for general momentum transfer [66]. This striking fact is related to the remarkable property of the BFKL equation discovered in [66]: the two-dimensional conformal invariance in the impact parameter space.

To discuss this question it is convenient to change the definition of the kernel and the state normalization. We will use the following notation: q_i' and q_i, $i = 1, 2$, represent the transverse momenta of reggeons in initial and final t-channel states, while r_i' and r_i are the corresponding conjugate coordinates. The state normalization is

$$\langle q | q' \rangle = \delta(q - q'), \quad \langle r | r' \rangle = \delta(r - r'), \tag{9.202}$$

so that

$$\langle r | q \rangle = \frac{e^{iq\,r}}{(2\pi)^{1+\epsilon}}. \tag{9.203}$$

We will also use the notation $q = q_1 + q_2$, $q' = q_1' + q_2'$; $k = q_1 - q_1' = q_2' - q_2$ and, for brevity, we will usually write $p_{ij'} = p_i - p_j'$. The s-channel discontinuities (9.107) are written in the form

$$- 2i(2\pi)^{D-2}\delta(q_A - q_B)\text{disc}_s \mathcal{A}_{AB}^{A'B'} = s\,(A'\bar{A}|e^{Y\hat{\mathcal{K}}}|\bar{B}'B), \tag{9.204}$$

where (cf. (9.106))

$$\langle q_1, q_2 | \bar{B}'B \rangle = \delta(q_B - q_1 - q_2)\frac{\Phi_{B'B}(q_1, q_2)}{q_1^2 q_2^2},$$

$$(A'\bar{A}|q_1, q_2) = \delta(q_A - q_1 - q_2)\Phi_{A'A}(q_1, q_2), \tag{9.205}$$

and

$$\langle q_1, q_2 | \hat{\mathcal{K}} | q_1', q_2' \rangle = \delta(q - q')\frac{1}{q_1^2 q_2^2}K(q_1, q_1'; q), \tag{9.206}$$

where $K(q_1, q_1'; q)$ is defined by (9.116). For colourless particles the impact factors $\Phi_{A'A}(q_1, q_2)$ turn into zero at zero transverse momentum of one of the reggeons, i.e. $\Phi_{A'A}(0, q) = \Phi_{A'A}(q, 0) = 0$. This means that $(A'\bar{A}|\Psi) = 0$ for any $|\Psi\rangle$ if in the coordinate space $\langle r_1, r_2|\Psi\rangle$ does not depend either on r_1 or on r_2. As it can be seen from (9.130), the BFKL kernel conserves this property, which permits the change in (9.204) $|\bar{B}'B\rangle$ for $|\bar{B}'B)_d$, where the subscript d means the dipole property $\langle r, r|\bar{B}'B)_d = 0$. Then, one can omit in the matrix elements $\langle r_1, r_2|\hat{\mathcal{K}}_d|r_1', r_2'\rangle$ the terms proportional to $\delta(r_{1'2'})$, and turn $\langle r, r|\hat{\mathcal{K}}_d|r_1', r_2'\rangle$ into zero adding terms not depending on either r_1 or r_2. This form of the kernel is called dipole or Möbius form.

It was indicated after Eq. (9.140) that we can put there $\epsilon = 0$ due to the cancellation of the infrared singularities. Using (9.140), we obtain

$$\frac{1}{q_1^2 q_2^2} K(q_1, q_1'; q) = \frac{g^2 N_c}{(2\pi)^3} \left[\frac{2}{k^2} - \frac{2(kq_1)}{k^2 q_1^2} + \frac{2(kq_2)}{k^2 q_2^2} - \frac{2(q_1 q_2)}{q_1^2 q_2^2} \right.$$
$$\left. - \delta(k) \int dl \left(\frac{2}{l^2} + \frac{(l(q_1 - l))}{l^2 (q_1 - l)^2} - \frac{(l(q_2 + l))}{l^2 (q_2 + l)^2} \right) \right], \quad (9.207)$$

where the cancellation is obvious. The kernel in the coordinate representation is given by the integral

$$\langle r_1, r_2 | \hat{\underline{K}} | r_1', r_2' \rangle = \int \frac{dq_1 \, dq_2 \, dk}{(2\pi)^4} \frac{1}{q_1^2 q_2^2} K(q_1, q_1'; q) \, e^{i(q_1 r_{11'} + q_2 r_{22'} + k r_{1'2'})}.$$
$$(9.208)$$

The last term in the real part of the kernel (9.207) gives here $\delta(r_{1'2'})$ and can be omitted in the Möbius form. The first terms in the real and virtual parts give the contribution $g^2 N_c \delta(r_{11'}) \delta(r_{22'}) V(r_{12})/(2\pi)^3$, where

$$V(r_{12}) = 2 \int dk \frac{1}{k^2} \left(e^{i k r_{12}} - 1 \right) \quad (9.209)$$

contains ultraviolet divergence. This divergence is artificial: it appears as a result of the separation of both real and virtual parts of the kernel (9.207) into several pieces. To make evident cancellation of ultraviolet singularities of separate terms in (9.208) we represent $V(r_{12})$ in the integral form:

$$V(r_{12}) = -\int \frac{dk_1 \, dk_2 \, dr_0}{(2\pi)^2} \frac{(k_1 k_2)}{k_1^2 k_2^2} \left(2 e^{i(k_1 r_{10} + k_2 r_{20})} - e^{i(k_1 + k_2)r_{10}} - e^{(k_1 + k_2)r_{20}} \right)$$
$$= -\int dr_0 \frac{r_{12}^2}{r_{10}^2 r_{20}^2}. \quad (9.210)$$

The second terms in the real and virtual parts give

$$-\frac{g^2 N_c}{(2\pi)^3} \delta(r_{22'}) \int \frac{dq_1 \, dk}{(2\pi)^2} \frac{(k q_1)}{k^2 q_1^2} e^{i q_1 r_{11'}} \left(2 e^{-i k r_{21'}} + e^{i k r_{11'}} \right)$$

$$= \frac{g^2 N_c}{(2\pi)^3} \delta(r_{22'}) \left(\frac{r_{12}^2}{r_{11'}^2 r_{21'}^2} - \frac{1}{r_{21'}^2} \right). \quad (9.211)$$

In the Möbius form the last term must be omitted. The contribution of the third terms is obtained by the substitution $1 \to 2$. Finally, we have

$$\langle r_1, r_2 | \hat{K}_M | r_1', r_2' \rangle = \frac{\alpha_s N_c}{2\pi^2} \int dr_0 \frac{r_{12}^2}{r_{10}^2 r_{20}^2} \left(\delta(r_{11'}) \delta(r_{02'}) \right.$$
$$\left. + \delta(r_{22'}) \delta(r_{01'}) - \delta(r_{11'}) \delta(r_{22'}) \right), \quad (9.212)$$

where the subscript M denotes the Möbius form. Remarkably, this form exactly coincides with the kernel of the dipole approach (see (9.179)). Moreover, in the forward scattering case, there is a remarkable functional identity between the Möbious form and the BFKL kernel in the momentum representation. Defining for this case $\langle r|\hat{\mathcal{K}}_M|r'\rangle = \int d\vec{r}_0 \left(r, \vec{0}|\hat{\mathcal{K}}_M|r' + \vec{r}_0, \vec{r}_0 \rangle \right)$ we have the equality

$$\frac{\vec{r}'^2}{\vec{r}^2}\langle \vec{r}|\hat{\mathcal{K}}_M|\vec{r}^{prime}\rangle = \frac{\vec{q}^2}{\vec{q}'^2}K(\vec{q},\vec{q}')\Big|_{\vec{q}\to\vec{r},\vec{q}'\to\vec{r}'},$$

where $K(\vec{q},\vec{q}')$ is defined in (9.141). It is worth noting that the ultraviolet singularities of separate terms in (9.212) cancel in their sum with account of the dipole property of the target impact factors.

The transformations of the Möbius group in the two-dimensional space $r = (x, y)$ can be written as

$$z \to \frac{az+b}{cz+d}, \tag{9.213}$$

where $z = x + iy$, a, b, c, d are complex numbers, with $ad - bc \neq 0$. Under these transformations, one has

$$z_1 - z_2 \to \frac{z_1 - z_2}{(cz_1 + d)(cz_2 + d)}(ad - bc),$$

$$dzdz^* \to dzdz^* \frac{|ad - bc|^2}{\left|(cz + d)^2\right|^2}, \tag{9.214}$$

so that the conformal invariance of (9.212) is evident.

The eigenfunctions of the kernel $\hat{\mathcal{K}}_M$ can be obtained from the eigenfunctions of the forward scattering kernel by transformations of the Möbius group, in particular, by inversion and shift. In [66], they were chosen in the form

$$E_{n,\nu}(r_{10}, r_{20}) = (-1)^n \left(\frac{z_{12}}{z_{10}z_{20}}\right)^{\frac{1}{2}+i\nu-\frac{n}{2}} \left(\frac{z_{12}^*}{z_{10}^*z_{20}^*}\right)^{\frac{1}{2}+i\nu+\frac{n}{2}}, \tag{9.215}$$

where the vector r_0 is introduced for indexing of the wave functions, $r_{ij} = r_i - r_j$, n is integer and ν is real. The eigenvalues coincide with the eigenvalues of the forward kernel $\omega(\gamma, n)$, $\gamma = 1/2 + i\nu$ (9.154). One can see it by noticing that the eigenvalues do not depend on r_0 and that integration over r_0 gives

$$\int dr_0 E_{n,\nu}(r_{10}, r_{20}) = C_{n,\nu}(z_{12})^{\frac{1}{2}-i\nu+\frac{n}{2}}(z_{12}^*)^{\frac{1}{2}-i\nu-\frac{n}{2}} = C_{n,\nu}\, e^{i\,n\phi_{12}}r_{12}^{2(1-\gamma)}, \tag{9.216}$$

where ϕ_{12} is the azimuthal angle of the vector r_{12}. Then the problem is reduced to the calculations of the integrals (9.152), (9.153) of the forward case. The form of the integral (9.216) can be easily obtained using the change of the integration variables $z_{10} = z_{12}z$; the coefficient $C_{n,\nu}$ is irrelevant for the calculation of the eigenvalues. In fact, it was found in [66]:

$$C_{n,\nu} = \frac{\pi 2^{4i\nu}}{|n| - 2i\nu} \frac{\Gamma(\kappa)}{\Gamma(\kappa^*)} \frac{\Gamma(\kappa^* - 1/2)}{\Gamma(\kappa - 1/2)}, \qquad \kappa = \frac{1}{2} - i\nu + \frac{|n|}{2}. \tag{9.217}$$

One can easily show that

$$z_{12}^2 \frac{\partial}{\partial z_1} \frac{\partial}{\partial z_2} E_{n,\nu} = \lambda_{n,\nu} E_{n,\nu}, \quad z_{12}^{*2} \frac{\partial}{\partial z_1^*} \frac{\partial}{\partial z_2^*} E_{n,\nu} = \lambda_{n,-\nu} E_{n,\nu},$$

$$\lambda_{n,\nu} = \frac{1}{4} - \left(\frac{n}{2} - i\nu\right)^2. \tag{9.218}$$

The completeness condition derived in [66] has the form

$$4 \sum_{n=-\infty}^{+\infty} \int_{-\infty}^{+\infty} d\nu \int d^2 r_0 \frac{n^2 + 4\nu^2}{r_{12}^2 r_{1'2'}^2} E_{n,\nu}(r_{10}, r_{20}) E_{n,\nu}^*(r_{1'0}, r_{2'0}) = (2\pi)^4 \delta(r_{11'}) \delta(r_{22'}). \tag{9.219}$$

Using (9.218) and (9.219) one obtains

$$\left\langle r_1, r_2 \left| \frac{1}{\omega - \hat{\mathcal{K}}_M} \frac{1}{\hat{q}_1^2 \hat{q}_2^2} \right| r_1', r_2' \right\rangle$$

$$= 4 \sum_{n=-\infty}^{+\infty} \int_{-\infty}^{+\infty} d\nu \int d^2 r_0 \frac{(n^2 + 4\nu^2) E_{n,\nu}(r_{10}, r_{20}) E_{n,\nu}^*(r_{1'0}, r_{2'0})}{[(n+1)^2 + 4\nu^2][(n-1)^2 + 4\nu^2](\omega - \omega(\gamma, n))}. \tag{9.220}$$

For $n = \pm 1$ the integral over ν here must be taken in the sense of its principal value.

9.6 Bootstrap of the gluon reggeization

9.6.1 Multi-Regge form of QCD amplitudes

The BFKL pomeron discussed in the preceding section contributes to amplitudes with positive signature, $\tau = +1$. It gives the main contribution to total cross sections. But in each order of perturbation theory its contribution to scattering amplitudes is subleading, because of cancellation of leading terms in the positive signature. This cancellation is evident from (9.126). In this section, we will consider MRK amplitudes with $\tau = -1$ in all t_i-channels. These amplitudes are leading in each order. They are especially interesting because their investigation permits us to check the compatibility of the gluon reggeization with s–channel unitarity. Moreover, it lets us to prove the reggeization hypothesis, i.e. the reggeized form (9.15) of QCD amplitudes in the MRK.

First, we have to define the signaturization procedure. As already mentioned, although in general signaturization is not a simple problem, it is greatly simplified in Regge kinematics. For $2 \to 2$ amplitudes the signaturization procedure was defined in Section 9.2. It can be easily generalized for MRK amplitudes, due to the fact that in the MRK all energy invariants $s_{i,j} = (k_i + k_j)^2$ are large and determined only by the longitudinal components of momenta ($s_{i,j} = 2k_i^+ k_j^-$, $i < j$). Due to largeness of $s_{i,j}$ signaturization in the q_l channel means symmetrization (antisymmetrization) with respect to the substitution $s_{i,j} \leftrightarrow -s_{i,j}, i < l \le j$. Since $s_{i,j}$ are determined by longitudinal components, this substitution is equivalent to the replacement $k_i^\pm \leftrightarrow -k_i^\pm, i < l$, $p_A^\pm \leftrightarrow -p_A^\pm$ (or, equivalently, $k_j^\pm \leftrightarrow -k_j^\pm, j \ge l$, $p_B^\pm \leftrightarrow -p_B^\pm$) in the truncated amplitudes without change of transverse components. Note that such substitution does not violate the total momentum conservation due to the strong ordering of the longitudinal components. At the same time, all particles remain on their mass-shell, so that the substitution is equivalent to the transition to the cross-channel. Therefore, signaturization of the truncated amplitudes $\mathcal{A}_{2 \to n+2}$ of the process $A + B \to A' + J_1 + \ldots + J_n + B'$ in the t_l-channel is performed by the operator

$$\hat{S}_{\tau_l} = \frac{1}{2}\left(1 + \tau_l \hat{R}_A \hat{R}'_A \prod_{i=1}^{l-1} \hat{R}_{J_i}\right), \tag{9.221}$$

with the operator $\hat{R}$ introduced in (9.44). Note that in the NLA it can act on jets; in this case, it transfers each particle of the jet from the initial (final) state into the antiparticle in the final (initial) state with the same longitudinal and opposite transverse momenta. The amplitude $\mathcal{A}_{2 \to 2+n}$ is given by the sum of the signaturized amplitudes over signatures in all t_i channels:

$$\mathcal{A}_{2 \to 2+n} = \sum_{\tau_i = \pm 1} \mathcal{A}_{2 \to 2+n}(\tau_1, \tau_2, \ldots \tau_{n+1}), \tag{9.222}$$

where

$$\mathcal{A}_{2 \to 2+n}(\tau_1, \tau_2, \ldots \tau_{n+1}) = \prod_{i=1}^{n+1} \hat{S}_{\tau_i} \mathcal{A}_{2 \to 2+n}. \tag{9.223}$$

At $\tau_i = +1$ ($\tau_i = -1$) the amplitude (9.223) has s_i-channel discontinuities equal to (differing by sign from) the u_i-channel ones and equal to the half-sum (half-difference) of the s_i- and u_i-channel discontinuities of $\mathcal{A}_{2 \to n+2}$. Recall that the signaturization is performed for truncated amplitudes.

Repeat that we use light-cone momenta n_1 and n_2, with $n_1^2 = n_2^2 = 0$, $(n_1 n_2) = 1$, denote $(pn_2) = p^+$, $(pn_1) = p^-$ and assume that the initial momenta p_A and p_B have predominant components p_A^+ and p_B^-. For generality it is not assumed that the components $p_{A\perp}$, $p_{B\perp}$ that transverse to the (n_1, n_2) plane are zero. Moreover, A and B, as well as A' and B', can represent jets. We define rapidities of the final jets J_i, $i = 1, 2, \ldots n$ with momenta k_i as $y_i = \frac{1}{2}\ln\left(k_i^+/k_i^-\right)$. It is supposed that they decrease with i: $y_0 > y_1 > \cdots > y_n > y_{n+1}$. As for y_0 and y_{n+1}, it is convenient to define them as $y_0 = y_A \equiv \ln\left(\sqrt{2}p_A^+/|q_{1\perp}|\right)$ and $y_{n+1} = y_B \equiv \ln\left(|q_{(n+1)\perp}|/\sqrt{2}p_B^-\right)$. Let us rewrite (9.15) in the form

$$\mathcal{A}_{2 \to n+2}^R = 4(p_A p_B)\Gamma_{A'A}^{R_1}\left(\prod_{i=1}^{n} \frac{e^{\omega_i(y_{i-1} - y_i)}}{q_{i\perp}^2} \gamma_{R_i R_{i+1}}^{J_i}\right)\frac{e^{\omega_{n+1}(y_n - y_{n+1})}}{q_{(n+1)\perp}^2}\Gamma_{B'B}^{R_{n+1}}. \tag{9.224}$$

Here $\omega_i \equiv \omega(q_{i\perp}^2)$, $\Gamma_{B'B}^R$ and $\Gamma_{A'A}^R$ are the scattering vertices, $\gamma_{R_i R_{i+1}}^{J_i}$ are the production vertices. For particles and reggeons we use notation which accumulates all their parameters. All reggeon vertices, as well as the gluon trajectory, are known with the next-to-leading order accuracy and are presented in Section 9.2.

We will call (9.224) the multi-Regge form. Our aim is to prove both in the LLA and in the NLA that this form is correct for real parts of the $\mathcal{A}_{2 \to n+2}$ amplitudes in the MRK. Let us emphasize that only the real parts have such a simple form, and only these parts are given by the reggeized gluon contributions. As for imaginary parts, they come into amplitudes both from the parts with positive and negative signatures. They can be calculated using the unitarity relations and the real parts of the $\mathcal{A}_{2\to n+2}$. As one can see from preceding section, they are complicated even for the LLA elastic amplitudes. Fortunately, both in the LLA and in the NLA only real parts of the MRK amplitudes contribute to the unitarity relations. Note that the real parts of amplitudes with positive signature even in one of t_i-channels must be taken into account only in the next-to-next-to leading approximation. Therefore, with our accuracy only amplitudes with negative signature in all t_i-channels must be considered. In the following the negative signature is assumed.

The proof is based on the bootstrap relations. These relations are derived from the requirement of the compatibility of the multi-Regge form (9.224) with s-channel unitarity. We show that the fulfilment of the bootstrap relations guarantees this form. Then we demonstrate that an infinite set of the bootstrap relations are fulfilled if several conditions imposed on the reggeon vertices and the trajectory (bootstrap conditions) hold true. All these conditions are proved to be satisfied.

Our aim is to prove the form (9.224) for real parts of the $\mathcal{A}_{2\to n+2}$ amplitudes, whereas s-channel unitarity determines the imaginary parts. Evidently, to derive the bootstrap relations we need connection between the real and imaginary parts. It is well known that because of analyticity these parts are connected by dispersion relations. These relations have an integral form and are very complicated in the inelastic case. Fortunately, in the high-energy limit the real and imaginary parts are connected also by differential relations (differential analyticity), which will be used below.

9.6.2 Bootstrap for elastic amplitudes

Let us begin with the $2 \to 2$ amplitude. To derive the differential analyticity relation here it is sufficient to note that with the next-to-leading order accuracy

$$\frac{1}{-2\pi i}\mathrm{disc}_s\ (\ln^n(-s) + \ln^n s) = \frac{1}{2}\frac{\partial}{\partial \ln s}\mathrm{Re}\left[\ln^n(-s) + \ln^n s\right], \tag{9.225}$$

where disc_s denotes the s-channel discontinuity. Eq. (9.225) gives the differential relation:

$$\frac{1}{-2\pi i s}\mathrm{disc}_s\ \mathcal{A}_{2\to 2s} = \frac{1}{2}\frac{\partial}{\partial \ln s}\mathrm{Re}\ \mathcal{A}_{2\to 2}, \tag{9.226}$$

Using in the right-hand side the Regge form (9.5) we come to the bootstrap relation:

$$\frac{1}{-2\pi i}\mathrm{disc}_s\mathcal{A}_{2\to 2} = \frac{1}{2}\omega(t)\mathcal{A}^R_{2\to 2}. \tag{9.227}$$

Note that it is not exact relation; we can not demand it in approximations higher than the NLA.

The important point is that the left-hand side of (9.227) can be calculated using the amplitudes (9.224) in the unitarity condition. Since the amplitudes are expressed through the gluon–Regge trajectory and the vertices of reggeon interactions, the relation (9.227) imposes strong restrictions on the trajectory and the vertices. Thus, using for the discontinuity in (9.227) Eqs. (9.107) and (9.112) and for $\mathcal{A}^R_{2\to 2}$ Eq. (9.224), one obtains

$$\frac{1}{2(2\pi)^{D-1}} \langle A'\bar{A}|e^{\hat{\mathcal{K}}Y}|\bar{B}'B\rangle = -\delta(q_{A\perp}-q_{B\perp})\Gamma^c_{A'A}\frac{\omega(q_\perp^2)}{q_\perp^2}e^{\omega(q_\perp^2)Y}\,\Gamma^c_{B'B}, \qquad (9.228)$$

where $q_\perp = q_{A\perp} = q_{B\perp}$, $q_{A\perp} = p_{A\perp} - p_{A'\perp}$, $q_{B\perp} = p_{B'\perp} - p_{B\perp}$. In the leading order this relation requires fulfilment of the following bootstrap conditions. The impact factors of the colliding particles must contain as coefficients the corresponding reggeon vertices:

$$|\bar{B}'B\rangle = g\Gamma^R_{B'B}|R_\omega(q_{B\perp})\rangle, \qquad \langle A'\bar{A}| = g\Gamma^R_{A'A}\langle R_\omega(q_{A\perp})|, \qquad (9.229)$$

where $\langle R_\omega(q_\perp)|$ and $|R_\omega(q_\perp)\rangle$ are the *bra-* and *ket-*vectors of the universal (process independent) colour-octet state (the sum over colour indices R is implied) with the total t-channel momentum $q_\perp$. This state must be an eigenstate of the kernel $\hat{\mathcal{K}}$ with the eigenvalue $\omega(q_\perp^2)$,

$$\hat{\mathcal{K}}|R_\omega(q_\perp)\rangle = \omega(q_\perp^2)|R_\omega(q_\perp)\rangle, \qquad \langle R_\omega(q_\perp)|\hat{\mathcal{K}} = \langle R_\omega(q_\perp)|\omega(q_\perp^2). \qquad (9.230)$$

And finally, it must have the normalization

$$\frac{g^2 q_\perp^2}{2(2\pi)^{D-1}}\langle R_\omega(q'_\perp)|R_\omega(q_\perp)\rangle = -\delta(q_\perp - q'_\perp)\delta^{RR'}\omega(q_\perp^2). \qquad (9.231)$$

Eqs. (9.230) and (9.229) are called respectively the first and the second bootstrap conditions. In the leading order it is not difficult to see that all these conditions are satisfied. Indeed, fulfilment of (9.229) is evident from (9.104), (9.105), and (9.46). Moreover, it follows from these equations, that

$$\langle \mathcal{G}_1\mathcal{G}_2|R_\omega(q_\perp)\rangle = T^R_{\mathcal{G}_1\mathcal{G}_2}\delta(q_\perp - q_{1\perp} - q_{2\perp}), \qquad \langle R_\omega(q_\perp)|\mathcal{G}_1\mathcal{G}_2\rangle = T^R_{\mathcal{G}_1\mathcal{G}_2}\delta(q_\perp - q_{1\perp} - q_{2\perp}). \qquad (9.232)$$

With account of the scalar product definition (9.100) and the explicit form of the trajectory (9.47) one can see that (9.232) ensures the fulfilment of the normalization condition (9.231). Then, using (9.232) and the kernel definition (9.111), (9.101), (9.103), the first bootstrap condition (9.230) is written as

$$\frac{g^2 N_c}{2(2\pi)^{D-1}}\int \frac{d^{D-2}k_\perp}{q_{1\perp}'^2 q_{2\perp}'^2}\left(\frac{q_{1\perp}^2 q_{2\perp}'^2 + q_{2\perp}^2 q_{1\perp}'^2}{k_\perp^2} - q_\perp^2\right) = \omega(q_{1\perp}^2) + \omega(q_{2\perp}^2) - \omega(q_\perp^2),$$

$$\qquad (9.233)$$

where $q_{1\perp}' = q_{1\perp} - k_\perp$, $q_{2\perp}' = q_{2\perp} + k_\perp$, $q_\perp = q_{1\perp} + q_{2\perp}$. This equality is evidently satisfied by taking into account (9.47).

9.6.3 Bootstrap relations

Fulfilment of the bootstrap conditions (9.229)–(9.231) guarantees that the Regge form (9.5) is reproduced by calculations using (9.15) and s-channel unitarity. Clearly, it gives only a limited though important consistency check of the reggeization hypothesis. Actually, this check carried out in [2] was the first successful test of the BFKL equation. However, for consistency of the whole approach, we must be sure that the amplitudes (9.15) are reproduced not only in the case of $2 \rightarrow 2$ processes but for $2 \rightarrow 2+n$ processes with arbitrary n.

This can be checked using the unitarity conditions in all $s_{i,j}$-channels, $s_{i,j} = (k_i + k_j)^2$. These conditions provide us with the discontinuities $\mathrm{disc}_{s_{i,j}}$ of the amplitudes $\mathcal{A}_{2\rightarrow n+2}$ (recall that here we consider signaturized amplitudes with negative signature in all t_i-channels). Note that, generally speaking, in the NLA these discontinuities are not pure imaginary, since a discontinuity in one of the channels can have, in turn, a discontinuity in another channel. But it is clear that these double discontinuities are sub-subleading, so that we will neglect them in the following.

Turning to amplitudes with $n > 0$ we meet a complication: analytical properties of the production amplitudes are very intricate even in the MRK [34]–[36]. But, fortunately, if in the MRK we confine ourselves to the NLA, these properties are greatly simplified and allow us to express partial derivatives $\partial/\partial y_j$ of the amplitudes, considered as functions of rapidities y_j ($j = 0, \ldots, n+1$) and transverse momenta, in terms of the discontinuities of the signaturized amplitudes [107]:

$$\frac{1}{-\pi i} \left(\sum_{l=j+1}^{n+1} \mathrm{disc}_{s_{j,l}} - \sum_{l=0}^{j-1} \mathrm{disc}_{s_{l,j}} \right) \mathcal{A}_{2\rightarrow n+2} \Big/ (p_A p_B) = \frac{\partial}{\partial y_j} \left(\mathcal{A}_{2\rightarrow n+2} \Big/ (p_A p_B) \right).$$

(9.234)

Note that taking the sum of the equations (9.234) over j from 0 to $n+1$ it is easy to see that $\mathcal{A}_{2\rightarrow n+2}$ depends only on differences of the rapidities y_i, as it must be. The division by $(p_A p_B)$ is performed in Eq. (9.234) in order to differentiate the rapidity dependence of radiative corrections only.

Equalities (9.234) can be easily proved using the Steinmann theorem [61], or, more definitely, the statement [34] that the amplitudes can be represented as sums of contributions corresponding to various sets of the $n+1$ nonoverlapping channels s_{i_k, j_k}, $i_k < j_k$, $k = 1, \ldots, n+1$; then each of the contributions can be written as a signaturized series in logarithms of the energy variables s_{i_k, j_k} with coefficients which are real functions of the transverse momenta. Recall that two channels s_{i_1, j_1} and s_{i_2, j_2} are called overlapping if either $i_1 < i_2 \le j_1 < j_2$, or $i_2 < i_1 \le j_2 < j_1$. Since scattering amplitudes enter in the relations (9.234) linearly and uniformly, it is sufficient to prove these relations separately for the contribution of one of the sets. Two observations now are important. First, the coefficients which depend only on transverse momenta enter equally in both sides of (9.234), therefore we can omit them calculating both the discontinuities and the derivatives with respect to y_j. Second, the energy variables s_{i_k, j_k} entering in each set are independent, i.e. there are no relations between the differences $y_{i_k} - y_{j_k}$ for nonoverlapping channels s_{i_k, j_k}; this means, in particular, that we need to consider only leading and next-to leading orders in these variables.

Therefore, it is sufficient to prove the equalities (9.234) with next-to-leading order accuracy for the symmetrized products

$$SP = \hat{S} \prod_{i < j=1}^{n+1} \left(\frac{s_{i,j}}{|k_{i\perp}| \, |k_{j\perp}|} \right)^{\alpha_{ij}}$$

(9.235)

instead of $\mathcal{A}_{2\to n+2}/(p_A p_B)$. Here the exponents $\alpha_{ij} \sim \alpha_S$ are different from zero only for some set of nonoverlapping channels and are arbitrary in all other respects; $\hat{S}$ means symmetrization with respect to simultaneous change of signs of all $s_{i,j}$ with $i < k \le j$, performed independently for each $k = 1, \ldots, n+1$. Indeed, due to the above-mentioned arbitrariness of α_{ij} the fulfilment of the equalities (9.234) for SP guarantees it for any logarithmic series.

Since we consider only real parts of discontinuities in the invariants $s_{i,j}$, calculating the discontinuity of SP in one of the $s_{i,j}$ at real $\alpha_{ij} \sim \alpha_S$, we can neglect the signs of the other invariants not only in the LLA, but also in the NLA, so that we have

$$\frac{1}{-\pi i}\left(\sum_{l=j+1}^{n+1}\mathrm{disc}_{s_{j,l}} - \sum_{l=0}^{j-1}\mathrm{disc}_{s_{l,j}}\right)SP = \left(\sum_{l=j+1}^{n+1}\alpha_{jl} - \sum_{l=0}^{j-1}\alpha_{lj}\right)SP. \tag{9.236}$$

On the other hand, taking into account that

$$\left(\frac{s_{i,j}}{|k_{i\perp}|\,|k_{j\perp}|}\right)^{\alpha_{ij}} = e^{\alpha_{ij}(y_i - y_j)}, \tag{9.237}$$

we have for the real part

$$SP = e^{\sum_{i<j=1}^{n+1}\alpha_{ij}(y_i - y_j)}\left(1 + \mathcal{O}(\alpha_S^2)\right), \tag{9.238}$$

such that, within the NLA accuracy

$$\frac{\partial}{\partial y_j}SP = \left(\sum_{l=j+1}^{n+1}\alpha_{jl} - \sum_{l=0}^{j-1}\alpha_{lj}\right)SP. \tag{9.239}$$

It is clear from Eqs. (9.236) and (9.239) that the equalities (9.234) are fulfilled.

The important point is that the relations (9.234) give a possibility to find in the next-to-leading order all MRK amplitudes to all orders of the coupling constant if they are known (for all n) in the one-loop approximation. Indeed, these relations express all partial derivatives of the real parts for some number of loops in terms of the discontinuities, which can be calculated using the s-channel unitarity in terms of amplitudes with a smaller number of loops; moreover, in the NLA only the MRK is important and only real parts of the amplitudes contribute in the unitarity relations. To find $\mathcal{A}_{2\to n+2}$ in addition to the derivatives determined by Eq. (9.234) suitable initial conditions are required, but since they can be taken at fixed y_i, they are needed in the NLA only with one-loop accuracy. Therefore, the relations (9.234) together with the one-loop approximation for the MRK amplitudes unambiguously determine all $\mathcal{A}_{2\to n+2}$.

Thus, in order to prove the multi-Regge form (9.224) in the NLA it is sufficient to know that it is valid in the one-loop approximation and satisfies the equalities (9.234), where the discontinuities are calculated using this form in the unitarity relations.

Substituting Eq. (9.224) in the right-hand side of Eq. (9.234), we obtain the bootstrap relations

$$\frac{1}{-\pi i}\left(\sum_{l=j+1}^{n+1}\mathrm{disc}_{s_{j,l}} - \sum_{l=0}^{j-1}\mathrm{disc}_{s_{l,j}}\right)\mathcal{A}_{2\to n+2} = \left(\omega(t_{j+1}) - \omega(t_j)\right)\mathcal{A}_{2\to n+2}^R. \tag{9.240}$$

The discontinuities in these relations must be calculated using the s-channel unitarity and the multi-Regge form of the amplitudes (9.15). Evidently, there is an infinite number of bootstrap relations, because there is an infinite number of amplitudes $\mathcal{A}_{2 \to n+2}$. At first sight, it seems a miracle to satisfy all of them, since all these amplitudes are expressed through several reggeon vertices and the gluon Regge trajectory. Moreover, it is quite nontrivial to satisfy even some definite bootstrap relation for a definite amplitude, because it connects two infinite series in powers of y_i, and therefore it leads to an infinite number of equalities between coefficients of these series.

In fact, two miracles must occur in order to satisfy all the bootstrap relations: first, for each particular amplitude $\mathcal{A}_{2 \to n+2}$ it must be possible to reduce the bootstrap relation to a limited number of restrictions (bootstrap conditions) on the gluon trajectory and the reggeon vertices, and second, starting from some $n = n_0$ these bootstrap conditions must be the same as those obtained for amplitudes with $n < n_0$. Finally, all bootstrap conditions must be satisfied by the known expressions for the trajectory and the vertices.

It is necessary to add here that the amplitude in the right-hand side of Eq. (9.240) contains only colour octets in each of the q_i channels. The discontinuities in the left-hand side, taken separately, along with the colour octet hold other representations of the colour group, which cancel in the sum.

9.6.4 Calculation of s-channel discontinuites

Thus, we have to calculate the discontinuities in any of $s_{i,j}$ channel. They must be found using multi-Regge form (9.224). It worth noting that the calculation given below can be applied in the case of a positive signature as well, with the only difference in symmetrization of the impact factors. Each of the $s_{i,j}$-channel discontinuities, being expressed with the help of the s-channel unitarity through the product of amplitudes (9.224), contains two reggeons in the channels q_l at $i < l \leq j$. As an example, the $s_{j,n+1}$-channel discontinuity is represented schematically in Fig. 9.6. Large blobs are meant to account for the signaturization.

In order to represent the discontinuities in a compact way, we will use the operator notation introduced in Section 9.4. Our goal is to write them as matrix elements of operator expressions, consisting of the operators $\hat{\mathcal{J}}_i$ for jet-J_i production and of the operator $\hat{\mathcal{K}}$ for the reggeon–reggeon interaction kernel, between *bra-* and *ket-*vectors, describing either particle-particle or reggeon-particle transitions (actually "particle" here can denote a jet, as was already mentioned) due to interaction with reggeized gluons ($R_j \to J_j$ and $B \to B'$ transitions in Fig. 9.10). We will call these states particle-particle or reggeon-particle impact factors.

To calculate the discontinuity we need to convolute reggeon vertices with account of the signaturization and to integrate over momenta of particles in intermediate states. Since the convolutions of the reggeon vertices depend on the transverse components of momenta only, the signaturization is reduced to antisymmetrization with respect to the attached reggeon lines. In order to escape double counting in the next-to-leading order, we introduce an auxiliary parameter $\Delta \gg 1$ which constrains the difference in rapidities of particles

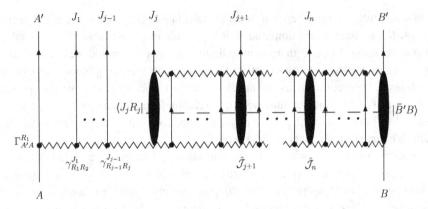

Fig. 9.10. Schematic representation of the $s_{j,n+1}$−channel discontinuity. The zigzag lines depict reggeized gluon exchanges. The right and left blobs represent the $B \to B'$ and $R_j \to J_j$ transitions respectively. The intermediate blobs depict jet productions.

belonging to one jet. Note that the largeness of Δ is numerical, but not parametrical (related to s), so that terms of order $\alpha_S \Delta$ are considered as subleading. Of course, the final answer must not depend on Δ.

We denote the momenta of the intermediate jets l_α, their rapidities z_α, with $z_\alpha = \frac{1}{2} \ln(l_\alpha^+ / l_\alpha^-)$. Rapidities of the intermediate jets related neither to jet-jet or reggeon-jet transitions, nor to the jet production (in the example depicted in Fig. 9.10 not contained in the blobs) are confined in the intervals $[y_{k+1} + \Delta, y_k - \Delta]$, where for the $s_{i,j}$-channel discontinuity k takes values from i to $j - 1$. In each interval, we need to perform integration over rapidities and summation over number of jets from 0 to ∞. Denoting by $r_{1\perp}$ and $r_{2\perp}$ the momenta of reggeons between the jets l_α and $l_{\alpha+1}$ we write all corresponding Regge factors in the same form $\exp[(\omega(r_{1\perp}^2) + \omega(r_{2\perp}^2))(z_\alpha - z_{\alpha+1})]$. Instead, the Regge factors for the reggeons interacting either with the colliding particles or reggeons or with the produced particles (in Fig. 9.10 attached to the blobs) are not uniform. In order to unify them, we include uniformity-violating multipliers in the definitions of jet-production operators and impact factors for particle-particle and reggeon-particle transitions. After that, the two-reggeon exchange in the q_j-channel is represented by the operator (compare with (9.108))

$$
\hat{G}(Y_j)^\Delta = \sum_{n=0}^{\infty} \int_{y_j+n\Delta}^{y_{j-1}-\Delta} e^{\hat{\Omega}(y_{j-1}-z_1)} \hat{\mathcal{K}}_r^\Delta dz_1 \int_{y_j+(n-1)\Delta}^{z_1-\Delta} dz_2 e^{\hat{\Omega}(z_1-z_2)} \hat{\mathcal{K}}_r^\Delta \cdots
$$

$$
\int_{y_i+\Delta}^{z_{n-1}-\Delta} dz_n e^{\hat{\Omega}(z_{n-1}-z_n)} \hat{\mathcal{K}}_r^\Delta e^{\hat{\Omega}(z_n-y_j)}, \tag{9.241}
$$

where the term for $n = 0$ is equal to $e^{\hat{\Omega}Y_j}$, with $Y_j = y_{j-1} - y_j$, $\hat{\Omega} = \omega(\hat{r}_1^2) + \omega(\hat{r}_2^2)$. The operator $\hat{\mathcal{K}}_r^\Delta$ takes into account production of the intermediate jets J with intervals of particle rapidities Δ_J in them less than Δ:

$$\langle \mathcal{G}_1\mathcal{G}_2|\hat{\mathcal{K}}_r^\Delta|\mathcal{G}_1'\mathcal{G}_2'\rangle = \delta(r_{1\perp}+r_{2\perp}-r_{1\perp}'-r_{2\perp}')\sum_J \int \gamma_{\mathcal{G}_1\mathcal{G}_1'}^J \gamma_J^{\mathcal{G}_2\mathcal{G}_2'}\frac{d\phi_J}{2(2\pi)^{D-1}}\theta(\Delta-\Delta_J).$$

(9.242)

Here the sum is taken over all discrete quantum numbers; $\gamma_J^{\mathcal{G}_2\mathcal{G}_2'}$ is the effective vertex for absorption of the jet J in the reggeon transition $\mathcal{G}_2' \to \mathcal{G}_2$. It is related to $\gamma_{\mathcal{G}_2\mathcal{G}_2'}^{\bar{J}}$ by the crossing described above (i.e. by the change of signs of longitudinal momenta and the corresponding change of wave functions). Then, we have

$$d\phi_J = \frac{dk_J^2}{2\pi}(2\pi)^D\delta^D(k_J-\sum_i p_i)\prod_i \frac{d^{D-1}p_i}{(2\pi)^{D-1}2\epsilon_i}$$

(9.243)

for a jet J with total momentum k_J consisting of particles with momenta p_i. The integration limits in Eq. (9.241) correspond to the limits on the intervals of particle rapidities in Eq. (9.242).

As already mentioned, terms of order $\alpha_S\Delta$ are subleading, therefore we need to retain in Eq. (9.241) only terms linear in Δ with coefficients of order α_S. With this accuracy we can write $\hat{G}(Y_j)^\Delta = (1-\hat{\mathcal{K}}_r^B\Delta)\hat{G}(Y_j)(1-\hat{\mathcal{K}}_r^B\Delta)$; the superscript B here and below denotes leading order, so that $\mathcal{K}_r^B$ is given by $\mathcal{O}(\alpha_S)$ terms in Eq. (9.242), and $\hat{G}(Y_j)$ is obtained from Eq. (9.241) by the omission of Δ in the integration limit and the replacement $\hat{\mathcal{K}}_r^\Delta \to \hat{\mathcal{K}}_r$, where

$$\hat{\mathcal{K}}_r = \hat{\mathcal{K}}_r^\Delta - \hat{\mathcal{K}}_r^B\hat{\mathcal{K}}_r^B\Delta.$$

(9.244)

We include the factors $(1-\hat{\mathcal{K}}_r^B\Delta)$ in the definitions of jet-production operators and impact factors for particle–particle and reggeon–particle transitions. Then the two-reggeon exchange in the q_j channel is represented by the operator $\hat{G}(Y_j)$. It is easy to see that it obeys the same equation (9.110) as in the LLA, with the same form of the kernel (9.111) and the same initial condition $\hat{G}(0) = 1$, so that $\hat{G}(Y_j) = \exp(\hat{\mathcal{K}}Y_j)$.

With account of the terms discussed before Eq. (9.241) and after Eq. (9.244) the impact factor for the $B \to B'$ transition is defined as

$$|\bar{B}'B\rangle = |\bar{B}'B\rangle^\Delta - \left(\omega(\hat{r}_{1\perp}^2)\ln\left|\frac{\hat{r}_{1\perp}}{q_{B\perp}}\right| + \omega(\hat{r}_{2\perp}^2)\ln\left|\frac{\hat{r}_{2\perp}}{q_{B\perp}}\right| + \hat{\mathcal{K}}_r^B\Delta\right)|\bar{B}'B\rangle^B,$$ (9.245)

where $\omega(t)$ is the one-loop trajectory (9.47),

$$\langle \mathcal{G}_1\mathcal{G}_2|\bar{B}'B\rangle^\Delta = \delta(q_{B\perp}-r_{1\perp}-r_{2\perp})\sum_{\bar{B}}\int\left(\Gamma_{\bar{B}B}^{\mathcal{G}_1}\Gamma_{B'\bar{B}}^{\mathcal{G}_2}-\Gamma_{\bar{B}B}^{\mathcal{G}_2}\Gamma_{B'\bar{B}}^{\mathcal{G}_1}\right)d\phi_{\bar{B}}\prod_l\theta\left(\Delta-(z_l-y_B)\right).$$

(9.246)

Here $q_B = p_{B'} - p_B$ and z_l are the rapidities of particles in intermediate jets. The terms with ω in Eq. (9.245) take into account the difference of the Regge factors related to the reggeons interacting with the particles B and B' and the "uniform" factors used in the series (9.241) for $\hat{G}(Y_{n+1})^\Delta$. The term with $\hat{\mathcal{K}}_r^B$ in Eq. (9.245) comes from the relation between $\hat{G}(Y_{n+1})^\Delta$ and $\hat{G}(Y_{n+1})$. Note that in the case when B or B' is a two-particle jet, only the first term must be kept in Eq. (9.245); moreover, only the Born approximation for this term must be taken in Eq. (9.246).

It is clear that for the impact factor of the $A \to A'$ transition we have

$$\langle A'\bar{A}| = \langle A'\bar{A}|^\Delta - \langle A'\bar{A}|^B \left(\omega(\hat{r}_{1\perp}^2) \ln \left|\frac{\hat{r}_{1\perp}^2}{q_{A\perp}}\right| + \omega(\hat{r}_{2\perp}^2) \ln \left|\frac{\hat{r}_{2\perp}^2}{q_{A\perp}}\right| + \hat{\mathcal{K}}_r^B \, \Delta \right), \quad (9.247)$$

$$\langle A'\bar{A}|\mathcal{G}_1\mathcal{G}_2\rangle^\Delta = \delta(q_{A\perp}-r_{1\perp}-r_{2\perp}) \sum_{\tilde{A}} \int \left(\Gamma_{\tilde{A}A}^{\mathcal{G}_1}\Gamma_{A'\tilde{A}}^{\mathcal{G}_2} - \Gamma_{\tilde{A}A}^{\mathcal{G}_2}\Gamma_{A'\tilde{A}}^{\mathcal{G}_1} \right) d\phi_{\tilde{A}} \prod_l \theta\left(\Delta-(y_A-z_l)\right), \quad (9.248)$$

where $q_A = p_A - p_{A'}$.

The antisymmetrization with respect to the permutation $\mathcal{G}_1 \leftrightarrow \mathcal{G}_2$ in Eqs. (9.246) and (9.248) takes into account the signaturization. The important fact is that due to the signaturization only the antisymmetric colour octet survives of all possible colour states of the two reggeons $\mathcal{G}_1$ and $\mathcal{G}_2$. For quark and gluon impact factors this follows from results of Refs. [55],[56]. For the case when some state is a two-particle state this can be seen from results presented in Ref. [108].

Accordingly, the reggeon-particle impact factors are defined as

$$|\bar{J}_i R_{i+1}\rangle = |\bar{J}_i R_{i+1}\rangle_s - |\bar{J}_i R_{i+1}\rangle_u, \quad |\bar{J}_i R_{i+1}\rangle_s = |\bar{J}_i R_{i+1}\rangle_s^\Delta$$

$$-\left(\left(\omega\left(q_{(i+1)\perp}^2\right) - \omega\left(\hat{r}_{1\perp}^2\right) \right) \ln \left|\frac{k_{i\perp}}{q_{(i+1)\perp} - \hat{r}_{1\perp}}\right| - \omega(\hat{r}_{2\perp}^2) \ln \left|\frac{k_{i\perp}}{\hat{r}_{2\perp}}\right| + \hat{\mathcal{K}}_r^B \Delta \right) |\bar{J}_i R_{i+1}\rangle_s^B, \quad (9.249)$$

$$\langle \mathcal{G}_1\mathcal{G}_2|\bar{J}_i R_{i+1}\rangle_u = \langle \mathcal{G}_1\mathcal{G}_2|\bar{J}_i R_{i+1}\rangle_s, \quad \langle \mathcal{G}_1\mathcal{G}_2|\bar{J}_i R_{i+1}\rangle_s^\Delta = \delta\left(q_{(i+1)\perp} + k_{i\perp} - r_{1\perp} - r_{2\perp}\right)$$

$$\times \sum_J \int \gamma_{\mathcal{G}_1 R_{i+1}}^{\bar{J}} \Gamma_{J_i J}^{\mathcal{G}_2} d\phi_J \prod_l \theta\left(\Delta - (z_l - y_i)\right), \quad (9.250)$$

and

$$\langle J_i R_i| = \langle J_i R_i|_s - \langle J_i R_i|_u, \quad \langle J_i R_i|_s = \langle J_i R_i|_s^\Delta$$

$$-\langle J_i R_i|_s^B \left(\left(\omega\left(q_{i\perp}^2\right) - \omega\left(\hat{r}_{1\perp}^2\right) \right) \ln \left|\frac{k_{i\perp}}{q_{i\perp} - \hat{r}_{1\perp}}\right| - \omega\left(\hat{r}_{2\perp}^2\right) \ln \left|\frac{k_{i\perp}}{\hat{r}_{2\perp}}\right| + \hat{\mathcal{K}}_r^B \Delta \right), \quad (9.251)$$

$$\langle J_i R_i|\mathcal{G}_1\mathcal{G}_2\rangle_u = \langle J_i R_i|\mathcal{G}_1\mathcal{G}_2\rangle_s, \quad \langle J_i R_i|\mathcal{G}_1\mathcal{G}_2\rangle_s^\Delta = \delta(r_{1\perp} + r_{2\perp} - q_{i\perp} + k_{i\perp})$$

$$\times \sum_J \int \gamma_{R_i \mathcal{G}_1}^{J} \Gamma_{J_i J}^{\mathcal{G}_2} d\phi_J \prod_l \theta\left(\Delta - (y_i - z_l)\right). \quad (9.252)$$

Finally, the operators $\hat{\mathcal{J}}_i$ for production of jets J_i are defined as

$$\hat{\mathcal{J}}_i = \hat{\mathcal{J}}_i^A - \left(\hat{\mathcal{K}}_r^B \hat{\mathcal{J}}_i^B + \hat{\mathcal{J}}_i^B \hat{\mathcal{K}}_r^B\right)\Delta, \quad \langle \mathcal{G}_1 \mathcal{G}_2 | \hat{\mathcal{J}}_i^A | \mathcal{G}_1' \mathcal{G}_2' \rangle = \delta(r_{1\perp} + r_{2\perp} - k_{i\perp} - r_{1\perp}' - r_{2\perp}')$$

$$\times \left[\gamma_{\mathcal{G}_1 \mathcal{G}_1'}^{J_i} \delta(r_{2\perp} - r_{2\perp}')r_{2\perp}^2 \delta_{\mathcal{G}_2 \mathcal{G}_2'} + \gamma_{\mathcal{G}_2 \mathcal{G}_2'}^{J_i} \delta(r_{1\perp} - r_{1\perp}')r_{1\perp}^2 \delta_{\mathcal{G}_1 \mathcal{G}_1'}\right.$$

$$\left. + \sum_G \int_{y_i - \Delta}^{y_i + \Delta} \frac{dz_G}{2(2\pi)^{D-1}} \left(\gamma_{\mathcal{G}_1 \mathcal{G}_1'}^{\{J_i G\}} \gamma_{\mathcal{G}_G}^{\mathcal{G}_2 \mathcal{G}_2'} + \gamma_{\mathcal{G}_1 \mathcal{G}_1'}^{G} \gamma_{J_i G}^{\mathcal{G}_2 \mathcal{G}_2'}\right)\right]. \tag{9.253}$$

Here the last term appears only in the case when $J_i \equiv G_i$ is a single gluon, the sum in this term goes over quantum numbers of the intermediate gluon G and the vertices must be taken in the Born approximation. At that $\gamma_{\mathcal{G}_1 \mathcal{G}_1'}^{\{J_i G\}}$ is the vertex for production of the jet consisting of the gluons G_i and G, $\gamma_{\mathcal{G}_i G}^{\mathcal{G}_2 \mathcal{G}_2'}$ is the vertex for absorption of gluon G and production of gluon G_i in the $\mathcal{G}_2 \to \mathcal{G}_2'$ transition; it can be obtained from $\gamma_{\mathcal{G}_2 \mathcal{G}_2'}^{\{G_i G\}}$ by crossing with respect to the gluon G.

With the definitions given above, we obtain

$$-4i(2\pi)^{D-2}\delta\left(q_{i\perp} - q_{(j+1)\perp} - \sum_{l=i}^{l=j} k_{l\perp}\right) \mathrm{disc}_{s_{i,j}} \mathcal{A}_{2\to n+2} = 2s\Gamma_{A'A}^{R_1} \frac{e^{\omega_1(y_0 - y_1)}}{q_{1\perp}^2}$$

$$\times \left(\prod_{l=2}^{i} \gamma_{R_{l-1} R_l}^{J_{l-1}} \frac{e^{\omega_l(y_{l-1} - y_l)}}{q_{l\perp}^2}\right) \langle J_i R_i | \left(\prod_{l=i+1}^{j-1} e^{\hat{\mathcal{K}}(y_{l-1} - y_l)} \hat{\mathcal{J}}_l\right) e^{\hat{\mathcal{K}}(y_{j-1} - y_j)} |\bar{J}_j R_{j+1}\rangle$$

$$\times \left(\prod_{l=j+1}^{n} \frac{e^{\omega_l(y_{l-1} - y_l)}}{q_{l\perp}^2} \gamma_{R_l R_{l+1}}^{J_l}\right) \frac{e^{\omega_{n+1}(y_n - y_{n+1})}}{q_{(n+1)\perp}^2} \Gamma_{B'B}^{R_{n+1}}, \tag{9.254}$$

where $\omega_i = \omega(q_{i\perp}^2)$. If $i = 0$ we must omit all factors to the left of $\langle J_0 R_0|$ and replace $\langle J_0 R_0|$ by $\langle A' \bar{A}|$, $k_0 - q_0$ by $p_{A'} - p_A$; in the case $j = n+1$ we must omit all factors to the right of $|\bar{J}_{n+1} R_{n+2}\rangle$ and perform the substitutions $|\bar{J}_{n+1} R_{n+2}\rangle \to |\bar{B}'B\rangle$, $k_{n+1} + q_{n+2} \to p_{B'} - p_B$.

9.6.5 Proof of the gluon reggeization

Using the representation (9.254) for the discontinuities, we prove below that an infinite number of the bootstrap relations (9.240) are satisfied if the following bootstrap conditions are fulfilled: the impact factors for colliding particles satisfy equations (9.229), where $\langle R_\omega(q_\perp)|$ and $|R_\omega(q_\perp)\rangle$ are the *bra*- and *ket*-vectors of the process independent eigenstate of the kernel $\hat{\mathcal{K}}$ with the eigenvalue $\omega(q_\perp^2)$ (see (9.230)), the normalization condition (9.231) is fulfilled and the reggeon-gluon impact factors and the gluon-production vertices satisfy the equations

$$\hat{\mathcal{J}}_i |R_\omega(q_{(i+1)\perp})\rangle \, g \, q_{(i+1)\perp}^2 + |\bar{J}_i R_{i+1}\rangle = |R_\omega(q_{i\perp})\rangle \, g \, \gamma_{R_i R_{i+1}}^{J_i},$$

$$g \, q_{i\perp}^2 \langle R_\omega(q_{i\perp})| \hat{\mathcal{J}}_i + \langle J_i R_i| = g \, \gamma_{R_i R_{i+1}}^{J_i} \langle R_\omega(q_{(i+1)\perp})|, \tag{9.255}$$

where $q_{(i+1)\perp} = q_{i\perp} - k_{i\perp}$. Actually, the second of Eqs. (9.229), (9.230) and (9.255) are not independent since *bra*- and *ket*-vectors are related to each other by the change of $+$ and $-$ momentum components.

To prove that the bootstrap conditions (9.229)–(9.231), (9.255) provide the fulfilment of the entire infinite set of bootstrap relations (9.240), consider first the terms with $l = n$ and $l = n + 1$ in Eq. (9.240). Using the representation (9.254) for the discontinuities and applying the bootstrap conditions (9.229) and (9.230) to the $s_{k,n+1}$-channel discontinuity, we find that the sum of the discontinuities in the channels $s_{k,n}$ and $s_{k,n+1}$ contains

$$g\,\hat{\mathcal{J}}_n\,|R_\omega(q_{(n+1)\perp})\rangle + |\bar{\mathcal{J}}_n R_{n+1}\rangle \frac{1}{q^2_{(n+1)\perp}} = |R_\omega(q_{n\perp})\rangle\,g\,\gamma^{J_n}_{R_n R_{n+1}} \frac{1}{q^2_{(n+1)\perp}}. \tag{9.256}$$

The equality here follows from the bootstrap condition (9.255). Now the procedure can be repeated: we can apply to this sum the bootstrap condition (9.230), and to the sum of the obtained result with the $s_{k,n-1}$-channel discontinuity again Eq. (9.255). Thus the entire sum over l from $j + 1$ to $n + 1$ in the representation (9.240) is reduced to one term. A quite analogous procedure (with the use of the bootstrap conditions for *bra*-vectors) can be applied to the sum over l from 0 to $j - 1$. As a result, we find that the left part of the representation (9.240) with the coefficient $-2(2\pi)^{D-1}\delta(q_{j\perp} - q_{(j+1)\perp} - k_{j\perp})$, where $q_{(j+1)\perp} = p_{B'\perp} - p_{B\perp} + \sum_{l=j+1}^{l=n} k_{l\perp}$ and $q_{j\perp} = p_{A\perp} - p_{A'\perp} - \sum_{l=1}^{l=j-1} k_{l\perp}$, can be obtained from the right-hand side of the multi-Regge form (9.224) by the replacement

$$\gamma^{J_j}_{R_j R_{j+1}} \longrightarrow \langle J_j R_j | R_\omega(q_{(j+1)\perp}) \rangle g q^2_{(j+1)\perp} - g q^2_{j\perp} \langle R_\omega(q_{j\perp}) | \bar{J}_j R_{j+1} \rangle. \tag{9.257}$$

Taking the difference of the first equality in the condition (9.255) for $i = j$ multiplied by $g q^2_{j\perp} \langle R_\omega(q_{j\perp}) |$ and the second equality multiplied by $|R_\omega(q_{(j+1)\perp})\rangle g q^2_{(j+1)\perp}$ and using the normalization (9.231), we obtain

$$\langle J_j R_j | R_\omega(q_{j+1}) \rangle\,g\,q^2_{(j+1)\perp} - g\,q^2_{j\perp}\,\langle R_\omega(q_j) | \bar{J}_j R_{j+1} \rangle$$
$$= -2(2\pi)^{D-1}\delta(q_{j\perp} - q_{(j+1)\perp} - k_{j\perp})\left(\omega(q^2_{(j+1)\perp}) - \omega(q^2_{j\perp})\right)\gamma^{J_j}_{R_j R_{j+1}}. \tag{9.258}$$

That concludes the proof.

Thus, the fulfilment of the bootstrap conditions (9.229)–(9.231), (9.255) guarantees the implementation of the entire infinite set of the bootstrap relations (9.240).

In the leading order there is only one production vertex, the reggeon–reggeon–gluon (RRG) vertex. The definition of the reggeon–gluon impact factor $\langle GR|$ and the gluon production operator $\hat{\mathcal{G}}$ are reduced to the following:

$$\langle GR|\mathcal{G}_1\mathcal{G}_2\rangle = \delta(q_\perp - r_{1\perp} - r_{2\perp} - k_\perp)\sum_{G'}\left(\gamma^{G'}_{R\mathcal{G}_1}\Gamma^{\mathcal{G}_2}_{GG'} - \gamma^{G'}_{R\mathcal{G}_2}\Gamma^{\mathcal{G}_1}_{GG'}\right), \tag{9.259}$$

where q, r_1, r_2 and k are the momenta of the reggeons $R, \mathcal{G}_1, \mathcal{G}_2$ and the gluon G;

$$\langle \mathcal{G}'_1\mathcal{G}'_2|\hat{\mathcal{G}}|\mathcal{G}_1\mathcal{G}_2\rangle = \delta(r_{1\perp} + r_{2\perp} + k_\perp - r'_{1\perp} - r'_{2\perp})$$
$$\times \left[\gamma^G_{\mathcal{G}'_1\mathcal{G}_1}\delta(r_{2\perp} - r'_{2\perp})r^2_{2\perp}\delta_{\mathcal{G}_2\mathcal{G}'_2} + \gamma^G_{\mathcal{G}'_2\mathcal{G}_2}\delta(r_{1\perp} - r'_{1\perp})r^2_{1\perp}\delta_{\mathcal{G}_1\mathcal{G}'_1}\right]. \tag{9.260}$$

Using the reggeon vertices in the form (9.26), (9.35), (9.37), we have

$$\langle GR|\mathcal{G}_1\mathcal{G}_2\rangle = -\delta(q_\perp - r_{1\perp} - r_{2\perp} - k_\perp)2g^2 e_\perp^* \times \left[T_{RG_1}^{G'} T_{GG'}^{\mathcal{G}_2} \left(q_\perp - q_\perp^2 \frac{q_\perp - r_{1\perp}}{(q_\perp - r_{1\perp})^2} \right) \right.$$
$$\left. - T_{RG_2}^{G'} T_{GG'}^{\mathcal{G}_1} \left(q_\perp - q_\perp^2 \frac{q_\perp - r_{2\perp}}{(q_\perp - r_{2\perp})^2} \right) \right], \tag{9.261}$$

$$\langle \mathcal{G}_1' \mathcal{G}_2' | \hat{\mathcal{G}} | \mathcal{G}_1 \mathcal{G}_2\rangle = -\delta(r_{1\perp} + r_{2\perp} + k_\perp - r_{1\perp}' - r_{2\perp}')2g e_\perp^*$$
$$\times \left[T_{\mathcal{G}_1' \mathcal{G}_1}^G \left(r_{1\perp}' - k_\perp \frac{r_{1\perp}'^2}{k_\perp^2} \right) \delta(r_{2\perp} - r_{2\perp}') r_{2\perp}^2 \delta_{\mathcal{G}_2 \mathcal{G}_2'} + T_{\mathcal{G}_2' \mathcal{G}_2}^G \right.$$
$$\left. \times \left(r_{2\perp}' - k_\perp \frac{r_{2\perp}'^2}{k_\perp^2} \right) \delta(r_{1\perp} - r_{1\perp}') r_{1\perp}^2 \delta_{\mathcal{G}_1 \mathcal{G}_1'} \right]. \tag{9.262}$$

Taking account of (9.232) gives

$$\langle R_\omega(q_\perp) | \hat{\mathcal{G}} | \mathcal{G}_1 \mathcal{G}_2\rangle = \delta(q - r_{1\perp} - r_{2\perp} - k_\perp)2g e_\perp^* \tag{9.263}$$
$$\times \left[T_{\mathcal{G}_1 \mathcal{G}_2}^R T_{\mathcal{G}_1' \mathcal{G}_1}^G \left(\frac{k_\perp}{k_\perp^2} - \frac{r_{1\perp} + k_\perp}{(r_{1\perp} + k_\perp)^2} \right) + T_{\mathcal{G}_1 \mathcal{G}_2'}^R T_{\mathcal{G}_2' \mathcal{G}_2}^G \left(\frac{k_\perp}{k_\perp^2} - \frac{r_{2\perp} + k_\perp}{(r_{2\perp} + k_\perp)^2} \right) \right],$$

such that

$$g q_\perp^2 \langle R_\omega(q_\perp) | \hat{\mathcal{G}} | \mathcal{G}_1 \mathcal{G}_2\rangle + \langle GR|\mathcal{G}_1\mathcal{G}_2\rangle = - \delta(q_\perp - r_{1\perp} - r_{2\perp} - k_\perp)2g^2 e_\perp^* T_{\mathcal{G}_1 \mathcal{G}_2}^{R_1}$$
$$\times T_{RR_1}^G \left(q_\perp - q_\perp^2 \frac{k_\perp}{k_\perp^2} \right). \tag{9.264}$$

Comparing the right-hand side of this equation with the right side of the bootstrap condition (9.255) and taking account of (9.232), (9.35) and (9.37) one sees that the bootstrap condition is fulfilled.

In the next-to-leading order the bootstrap conditions (9.229) and (9.230) have been known for a long time [38],[109]–[111] and have been proved to be satisfied in Refs. [55, 56],[111]–[115]. The bootstrap relations for elastic amplitudes require only a weak form of the conditions (9.229) and (9.230), namely only the projection of these conditions on $|R_\omega\rangle$. It was recognized [107] that in addition to the conditions (9.229) and (9.230), the bootstrap relations for one-gluon production amplitudes require also a weak form of the condition (9.255). Thus, the bootstrap relations for one-gluon production amplitudes play a twofold role: they strengthen the conditions imposed by the elastic bootstrap and give a new one. One could expect that history will repeat itself upon addition of each next gluon in the final state. If this were so, we would have to consider the bootstrap relations for production of an arbitrary number of gluons and would obtain an infinite number of bootstrap conditions. Fortunately, the history is repeated only partly: it was shown [116] that already the bootstrap relations for two-gluon production require the strong form of the last condition (i.e. Eq. (9.255)) and do not require new conditions.

The bootstrap conditions with two-particle jets are required in the NLA only with the reggeon vertices taken in the Born approximation. They were checked and proved to be satisfied in Refs. [62] and [108]. After that only the condition (9.255) remained not evident. Its fulfilment was proved recently [117]. Thus, now it is shown that all bootstrap conditions are fulfilled.

9.7 Next-to-leading order BFKL

9.7.1 One-gluon contribution

The correctness of the multi-Regge form (9.224) means that the representation (9.126) for elastic amplitudes and the decomposition (9.127) for the BFKL kernel remain valid in the NLA. The difference is that the gluon trajectory is taken in the next-to-leading order (9.63), and the real part includes the contributions, coming from one-gluon, two-gluon, and quark-antiquark pair production:

$$\hat{\mathcal{K}}_r = \hat{\mathcal{K}}_G + \hat{\mathcal{K}}_{Q\bar{Q}} + \hat{\mathcal{K}}_{GG}. \tag{9.265}$$

The former part is present already in the LLA (see (9.102)). Here we must take account of radiative corrections. It is easy to do substituting the one-loop production vertex (9.50) in (9.102). Then the quark contribution to $\hat{\mathcal{K}}_G$ can be obtained for arbitrary D. Writing $\hat{\mathcal{K}}_G = \hat{\mathcal{K}}_G^Q + \hat{\mathcal{K}}_G^G$, we obtain for the representation R of the colour group

$$K_G^{(R)Q}\left(\boldsymbol{q}_1, \boldsymbol{q}_1'; \boldsymbol{q}\right) = c_R \frac{g^4 n_f N_c}{(2\pi)^{D-1}} \frac{\Gamma(-\epsilon)}{(4\pi)^{2+\epsilon}} \frac{[\Gamma(1+\epsilon)]^2}{\Gamma(4+2\epsilon)}$$

$$\times \left\{ \left[k^2 \left(2k^2 - \boldsymbol{q}_1^2 - 2\boldsymbol{q}_2^2 - \boldsymbol{q}_1'^2 - 2\boldsymbol{q}_2'^2 + 2\boldsymbol{q}^2 \right) + (\boldsymbol{q}_1^2 - \boldsymbol{q}_1'^2)(\boldsymbol{q}_2^2 - \boldsymbol{q}_2'^2) \right] \right.$$

$$\times \frac{[2(1+\epsilon)\boldsymbol{q}_1^2\boldsymbol{q}_1'^2\phi_0 - \epsilon(\boldsymbol{q}_1^2 + \boldsymbol{q}_1'^2)\phi_1]}{(\boldsymbol{q}_1^2 - \boldsymbol{q}_1'^2)^3} + \frac{(k^2 - \boldsymbol{q}_2^2 - \boldsymbol{q}_2'^2)}{(\boldsymbol{q}_1^2 - \boldsymbol{q}_1'^2)} \epsilon\phi_1 + 2(1+\epsilon)^2$$

$$\times \left(\frac{\boldsymbol{q}_1^2\boldsymbol{q}_2'^2 - \boldsymbol{q}_2^2\boldsymbol{q}_1'^2}{k^2} + \frac{2\boldsymbol{q}_1^2\boldsymbol{q}_1'^2 - \boldsymbol{q}^2(\boldsymbol{q}_1^2 + \boldsymbol{q}_1'^2)}{(\boldsymbol{q}_1^2 - \boldsymbol{q}_1'^2)} \right) \phi_0 + (1 \longleftrightarrow 2) \right\}. \tag{9.266}$$

where $\phi_n = (\boldsymbol{q}_1^2)^{n+\epsilon} - (\boldsymbol{q}_1'^2)^{n+\epsilon}$. The substitution $1 \longleftrightarrow 2$ here and below means $\boldsymbol{q}_1 \leftrightarrow \boldsymbol{q}_2$, $\boldsymbol{q}_1' \leftrightarrow \boldsymbol{q}_2'$ (and then $k \leftrightarrow -k$ because $k = \boldsymbol{q}_1 - \boldsymbol{q}_1' = \boldsymbol{q}_2' - \boldsymbol{q}_2$). Since the one-loop vertex (9.50) has the same colour structure as the Born vertex, the coefficients c_R here are given by (9.137).

The gluon contribution is very complicated at arbitrary D. Of course, for physical applications the limit $\epsilon \to 0$ must be considered. But here one needs to recognize that it would be wrong to take this limit in $\mathcal{K}_G$. The reason is the singularity of the real part of the kernel at $k^2 = 0$. One has to retain all terms giving nonvanishing contributions in the limit $\epsilon \to 0$ after integration over $d^{2+\epsilon}k$. Within this accuracy, one has

$$K_G^{(R)}(\boldsymbol{q}_1, \boldsymbol{q}'; \boldsymbol{q}) = \frac{g^2 N_c c_R}{(2\pi)^{D-1}} \left\{ \left(\frac{q_1^2 q_2'^2 + q_2^2 q_1'^2}{k^2} - q^2 \right) \right.$$

$$\times \left[\frac{1}{2} + \frac{\bar{g}^2}{2} \left(-(k^2)^\epsilon \left(\frac{2}{\epsilon^2} - \pi^2 + 4\epsilon\,\zeta(3) \right) - \ln^2 \left(\frac{q_1^2}{q_1'^2} \right) \right) \right]$$

$$+ \frac{\bar{g}^2}{6} \left[\left(11 - 2\frac{n_f}{N_c} \right) \left(\frac{2q_1^2 q_1'^2}{q_1^2 - q_1'^2} + \frac{q_1^2 q_2'^2 - q_1'^2 q_2^2}{k^2} - \frac{q_1^2 + q_1'^2}{q_1^2 - q_1'^2} q^2 \right) \ln \left(\frac{q_1^2}{q_1'^2} \right) \right.$$

$$+ \left(1 - \frac{n_f}{N_c} \right) \left(\left(\frac{(q_2^2 - q_2'^2)}{(q_1^2 - q_1'^2)} - \frac{k^2}{(q_1^2 - q_1'^2)^2} \left(q_1^2 + q_1'^2 + 4q_2 q_2' - 2q^2 \right) \right) \right.$$

$$\left. \left. \times \left(\frac{2q_1^2 q_1'^2}{q_1^2 - q_1'^2} \ln \left(\frac{q_1^2}{q_1'^2} \right) - q_1^2 - q_1'^2 \right) - 2q_2 q_2' \right) \right] + (1 \longleftrightarrow 2) \right\}. \tag{9.267}$$

For arbitrary D this part of the kernel can be found in [47].

9.7.2 Two-particle production

The new contributions which appear in the next-to-leading order are $\hat{\mathcal{K}}_{Q\bar{Q}}$ and $\hat{\mathcal{K}}_{GG}$. They have the form (9.244), where $\Delta \gg 1$ is the auxiliary parameter and Δ_J are the intervals of particle rapidities in the jets. Passing from the limitation Δ on rapidity intervals to the limitation s_Λ on invariant masses of the jets, one obtains

$$\mathcal{K}_{GG}(\boldsymbol{q}_1, \boldsymbol{q}_1'; \boldsymbol{q}) + \mathcal{K}_{Q\bar{Q}}(\boldsymbol{q}_1, \boldsymbol{q}_1'; \boldsymbol{q}) = \mathcal{K}_{GG}^\Lambda(\boldsymbol{q}_1, \boldsymbol{q}_1'; \boldsymbol{q}) + \mathcal{K}_{Q\bar{Q}}^\Lambda(\boldsymbol{q}_1, \boldsymbol{q}_1'; \boldsymbol{q})$$

$$- \frac{1}{2} \int \frac{d^{D-2}r}{r^2 (q-r)^2} \mathcal{K}_r^B(\boldsymbol{q}_1, \boldsymbol{r}; \boldsymbol{q}) \, \mathcal{K}_r^B(\boldsymbol{r}, \boldsymbol{q}_1'; \boldsymbol{q}) \ln \left(\frac{s_\Lambda^2}{(q_1 - r)^2 (q_1' - r)^2} \right), \tag{9.268}$$

where for the representation R of the colour group the nonsubtracted kernel $\mathcal{K}_J^\Lambda$ is

$$K_J^{(R)\Lambda}(\boldsymbol{q}_1, \boldsymbol{q}_1'; \boldsymbol{q}) = \frac{\langle b, b' | \hat{\mathcal{P}}_R | a a' \rangle}{n_R} \sum_J \int \gamma_{ab}^J \left(q_1, q_1' \right) \left(\gamma_{a'b'}^J \left(-q_2, -q_2' \right) \right)^* \frac{d\phi_J}{2(2\pi)^{D-1}}. \tag{9.269}$$

Here J is GG or $Q\bar{Q}$, $\hat{\mathcal{P}}_R$ projects two-gluon colour states on the representation R; a, a' and b, b' are the reggeon colour indices; n_R is the number of independent states in R; $\gamma_{ab}^J (q_1, q_2)$ is the effective vertex for production of the state J in the collision of reggeons with momenta $q_1 = \beta p_1 + q_{1\perp}$, $q_1' = -\alpha p_2 + q_{1\perp}'$; $d\phi_J$ is the corresponding phase space element; the sum is over all discrete states in J. The operators $\hat{\mathcal{P}}_R$ are defined in Section 1.13 for any $\mathcal{R}$; in the most important cases of singlet and antisymmetric octet they are given in (9.134) and (9.135), respectively. For a jet J consisting of particles with momenta l_i, $i = 1, 2$, with total momentum $k = q_1 - q_1'$, we have

$$d\phi_J = \frac{dk^2}{2\pi} \theta(s_\Lambda - k^2)(2\pi)^D \delta^D(k - \sum_i l_i) \prod_i \frac{d^{D-1} l_i}{(2\pi)^{D-1} 2\epsilon_i}$$

$$= \frac{dx_1 dx_2}{2x_1 x_2} \delta(1 - x_1 - x_2) \frac{d^{D-2} l_1 d^{D-2} l_2}{(2\pi)^{(D-1)}} \delta^{D-2}(k_\perp - l_{1\perp} - l_{2\perp}) \theta(s_\Lambda - k^2), \quad (9.270)$$

where x_i are fractions of the jet longitudinal momentum k^+. Here the "+" component is taken only for definiteness; evidently, of course, Eq. (9.270) remains valid with the replacement $x_i \rightarrow y_i$, where y_i are the fractions of k^-. In the case of two-gluon jet $d\phi_J$ contains an additional factor of $1/2!$ to accounting for the identity. The intermediate parameter s_Λ in Eq. (9.268) must be taken tending to infinity. The second term on the right-hand side of Eq. (9.268) serves to subtract the large k^2 contribution, in order to avoid double counting of this region. Such contribution is important only in the case of two-gluon production, so that this term must be included in $\mathcal{K}_{GG}$.

The vertices $\gamma_{ab}^{Q\bar{Q}}$ and γ_{ab}^{GG} are given in (9.75), (9.80), (9.81) and (9.83), (9.84) respectively. Each of them contains two colour structures; accordingly, $K_{Q\bar{Q}}^{(R)}$ and $K_{GG}^{(R)}$ are written as sums of two terms with coefficients depending on R. We can write

$$K_J^{(R)\Lambda}(\boldsymbol{q}_1, \boldsymbol{q}_1'; \boldsymbol{q}) = 8g^4 N_c^2 \int \left(a_R F_a^J(l_1, l_2) + b_R F_b^J(l_1, l_2) \right) \frac{d\phi_J}{(2\pi)^{D-1}}, \quad (9.271)$$

where a_R and b_R are the group coefficients. If they are defined as

$$a_R = \frac{\langle b, b' | \hat{\mathcal{P}}_R | aa' \rangle}{2N_c^2 n_R} \text{Tr} \left(T^{a'} T^a T^b T^{b'} + T^a T^{a'} T^{b'} T^b \right),$$

$$b_R = \frac{\langle b, b' | \hat{\mathcal{P}}_R | aa' \rangle}{2N_c^2 n_R} \text{Tr} \left(T^a T^b T^{a'} T^{b'} + T^b T^a T^{b'} T^{a'} \right), \quad (9.272)$$

where $T_{ij}^a = -if_{aij}$ for gluons, and $T_{\alpha\beta}^a = t_{\alpha\beta}^a$ for quarks, then

$$F_a^{Q\bar{Q}}(l_1, l_2) = n_f \frac{x_1 x_2}{8} \text{Tr} \left[-b(q_1; l_1, l_2) \overline{b(-q_2; l_1, l_2)} - b(q_1; l_2, l_1) \overline{b(-q_2; l_2, l_1)} \right],$$

$$F_b^{Q\bar{Q}}(l_1, l_2) = n_f \frac{x_1 x_2}{8} \text{Tr} \left[b(q_1; l_1, l_2) \overline{b(-q_2; l_2, l_1)} + b(q_1; l_2, l_1) \overline{b(-q_2; l_1, l_2)} \right],$$

$$F_a^{GG}(l_1, l_2) = b^{\alpha\beta}(q_1; l_1, l_2) b_{\alpha\beta}(-q_2; l_1, l_2) + b^{\alpha\beta}(q_1; l_2, l_1) b_{\alpha\beta}(-q_2; l_2, l_1),$$

$$F_b^{GG}(l_1, l_2) = b^{\alpha\beta}(q_1; l_1, l_2) b_{\beta\alpha}(-q_2; l_2, l_1) + b^{\alpha\beta}(q_1; l_2, l_1) b_{\beta\alpha}(-q_2; l_1, l_2), \quad (9.273)$$

where n_f are the number of quark flavours; the matrices b and $\bar{b}$ for quarks and tensors $b^{\alpha\beta}$ for gluons are defined in (9.81) and (9.84), respectively. The explicit forms of the functions $F_{a,b}^{Q\bar{Q}}$ and $F_{a,b}^{GG}$ are given in [113] and [118],[119], respectively.

Using the commutation relations for colour generators and the trace normalization

$$\text{Tr} \left(T^a T^b \right) = T \delta^{ab}, \quad (9.274)$$

$T = T_F = 1/2$ for quarks and $T = T_A = N_c$ for gluons, and remembering that the leading order colour coefficient c_R is given by

$$c_R = \frac{\langle b, b' | \hat{\mathcal{P}}_R | aa' \rangle}{N_c n_R} f_{ab}^c f_{a'b'}^c, \quad (9.275)$$

we obtain from (9.272)

$$b_R = a_R - \frac{T}{2N_c} c_R.$$

(9.276)

The coefficients a_R are easily found. The colour structure of these coefficients correspond to the $Q\bar{Q}$ and GG t-channel states. In the first case, only singlet and octet representations are admitted. A simple colour algebra calculation gives

$$a_1^Q = \frac{N_c^2 - 1}{4N_c^3}, \quad a_{8_a}^Q = \frac{1}{8N_c}, \quad a_{8_s}^Q = \frac{N_c^2 - 4}{8N_c^3}, \quad b_1^Q = \frac{-1}{4N_c^3}, \quad b_{8_a}^Q = 0, \quad b_{8_s}^Q = \frac{-4}{8N_c^3}.$$

(9.277)

Note that the coefficients b_R^Q are suppressed in the limit of large N_c, because they correspond to nonplanar diagrams. Further, the coefficients for symmetric and antisymmetric colour octets coincide only in this limit; this means that the signature degeneracy is broken by terms nonleading in N_c. And, finally, an important fact: the equality $b_{8_a}^Q = 0$. This equality plays a crucial role in the proof of the gluon reggeization.

As for the coefficients a_R in the gluon case, one can easily see from comparison of the first equality in (9.272) with (9.275) that $a_R^G = c_R^2$. This relation is important for the cancellation of the s_Λ-dependence in the kernel (9.268). Taking account of (9.276) we have finally:

$$a_R^G = c_R^2, \quad b_R^G = a_R^G - \frac{1}{2}c_R = c_R \left(c_R - \frac{1}{2} \right).$$

(9.278)

With the results (9.136), (9.137) this gives, in particular,

$$a_1^G = 1, \quad a_{8_a}^G = a_{8_s}^G = \frac{1}{4}, \quad b_1^G = \frac{1}{2}, \quad b_{8_a}^G = b_{8_s}^G = 0.$$

(9.279)

Therefore, for both symmetric and antisymmetric colour-octet representations the coefficients b_R^G are zero. This is especially important for the antisymmetric case, since the vanishing of b_{8_a} is crucial for the gluon reggeization. The vanishing of $b_{8_s}^G$ means that in pure gluodynamics the signature degeneracy exists in the NLA as well as in the LLA.

9.7.3 Colour-octet kernel

The vanishing of b_{8_a} simplifies drastically the calculations of the kernel $K_J^{(8_a)}$. The $Q\bar{Q}$ contribution to this kernel was obtained in [113] at arbitrary D:

$$K_{Q\bar{Q}}^{(8_a)}(q_1, q_1'; q) = \frac{g^2 n_f}{(2\pi)^{D-1}} \frac{\bar{g}^2}{2\epsilon} \frac{[\Gamma(1+\epsilon)]^2}{\Gamma(4+2\epsilon)}$$

$$\times \left\{ 2(1+\epsilon)^2 \left(\frac{(k^2)^\epsilon}{k^2} (q_1^2 q_2'^2 + q_2^2 q_1'^2) + (q^2)^{1+\epsilon} \right) \right.$$

$$\left. + \left[k^2 \left(2k^2 - q_1^2 - 2q_2^2 - q_1'^2 - 2q_2'^2 + 2q^2 \right) + (q_1^2 - q_1'^2)(q_2^2 - q_2'^2) \right] \right.$$

$$\times \frac{[2(1+\epsilon)q_1^2 q_1'^2 \phi_0 - \epsilon(q_1^2 + q_1'^2)\phi_1]}{(q_1^2 - q_1'^2)^3} + \frac{\epsilon(k^2 - q_2^2 - q_2'^2) - 4(1+\epsilon)^2 q^2}{(q_1^2 - q_1'^2)}\phi_1$$

$$+ \ 4(1+\epsilon)^2 \frac{q_1^2 q_1'^2}{(q_1^2 - q_1'^2)^2}\phi_0 + \ (1 \longleftrightarrow 2)\Bigg\}, \qquad (9.280)$$

where the functions ϕ_n are defined after (9.266).

The quark part of the real kernel contains the sum of $K_{Q\bar{Q}}$ and the virtual quark contribution to K_G, which is given by (9.266). The sum is greatly simplified in comparison with (9.280) and (9.266). Denoting it by $K_Q^{(8a)}$ we have

$$K_Q^{(8a)}(q_1, q_1'; q) = \frac{g^2 n_f}{(2\pi)^{D-1}} \frac{\bar{g}^2}{\epsilon} \frac{[\Gamma(2+\epsilon)]^2}{\Gamma(4+2\epsilon)}$$

$$\times \ \Bigg\{ 2\frac{(k^2)^\epsilon}{k^2}(q_1^2 q_2'^2 + q_2^2 q_1'^2) + q^2\left(2q^{2\epsilon} - q_1^{2\epsilon} - q_2^{2\epsilon} - q_1'^{2\epsilon} - q_2'^{2\epsilon}\right)$$

$$- \ \frac{(q_1^2 q_2'^2 - q_2^2 q_1'^2)}{k^2}\left(q_1^{2\epsilon} - q_2^{2\epsilon} - q_1'^{2\epsilon} + q_2'^{2\epsilon}\right)\Bigg\}. \qquad (9.281)$$

This simplification indicates that the splitting of the next-to-leading order corrections into real and virtual ones is not the best way of doing the calculations. Now the method of finding the total contribution without complicated calculations of separate pieces is known [120].

The kernel $K_{GG}^{(8)}$ was calculated in [121],[122]. At arbitrary D it can be found in [122]. Here we present the kernel in the limit $\epsilon \to 0$, retaining, as always, the terms giving nonvanishing contributions after integration over $d^{2+\epsilon}k$:

$$K_{GG}^{(8)}(q_1, q_1'; q) = \frac{g^2 N_c}{(2\pi)^{D-1}} \frac{\bar{g}^2}{2}\Bigg\{(k^2)^{\epsilon-1}\left(q_1^2 q_2'^2 + q_2^2 q_1'^2 - q^2 k^2\right)\left(\frac{1}{\epsilon^2} - \frac{11}{6\epsilon} + \frac{67}{18}\right.$$

$$-4\zeta(2) + \epsilon\left(-\frac{202}{27} + 9\zeta(3) + \frac{11}{6}\zeta(2)\right)\Bigg) + q^2\left[\frac{11}{6}\left(\ln\left(\frac{q_1^2 q_1'^2}{k^2 q^2}\right) + \frac{(q_1^2 + q_1'^2)}{(q_1^2 - q_1'^2)}\right.\right.$$

$$\times \ \ln\left(\frac{q_1^2}{q_1'^2}\right)\Bigg) + \ \frac{1}{2}\ln\left(\frac{q_1^2}{q^2}\right)\ln\left(\frac{q_1'^2}{q^2}\right) - \frac{1}{8}\ln^2\left(\frac{q_1^2}{q_2^2}\right) - \frac{1}{8}\ln^2\left(\frac{q_1'^2}{q_2'^2}\right)\Bigg] - \frac{q_1 q_1'}{3}$$

$$- \ \frac{11}{6}(q_1^2 + q_1'^2) - \frac{1}{6}\left(11 - \frac{k^2}{(q_1^2 - q_1'^2)^2}\left(q_1^2 + q_1'^2 + 4(q_2 q_2') - 2q^2\right)\right)$$

$$+ \ \frac{q_2^2 - q_2'^2}{(q_1^2 - q_1'^2)}\Bigg)\left(\frac{2q_1^2 q_1'^2}{(q_1^2 - q_1'^2)}\ln\left(\frac{q_1^2}{q_1'^2}\right) - q_1^2 - q_1'^2\right) - \frac{q_1^2 q_2'^2 - q_2^2 q_1'^2}{4k^2}\ln\left(\frac{q_1^2}{q_1'^2}\right)$$

$$\times \ \ln\left(\frac{q_1^2 q_1'^2}{k^4}\right) + \frac{1}{2}I(q_1^2, q_1'^2; k^2)[q^2(k^2 - q_1^2 - q_1'^2)$$

$$
+ 2q_1^2 q_1'^2 - q_1^2 q_2'^2 - q_2^2 q_1'^2 + \frac{q_1^2 q_2'^2 - q_2^2 q_1'^2}{k^2} (q_1^2 - q_1'^2)] \Big\}
$$

$$
+ \frac{g^2 N_c}{(2\pi)^{D-1}} \frac{\bar{g}^2}{2} \left\{ 1 \longleftrightarrow 2 \right\}, \tag{9.282}
$$

where

$$
I(a, b, c) = \int_0^1 \frac{dx}{a(1-x) + bx - cx(1-x)} \ln \left(\frac{a(1-x) + bx}{cx(1-x)} \right). \tag{9.283}
$$

The integral $I(a, b, c)$ is invariant with respect to any permutation of its arguments, as it can be seen from the representation [115]

$$
I(a, b, c) = \int_0^1 \int_0^1 \int_0^1 \frac{dx_1 dx_2 dx_3 \delta(1 - x_1 - x_2 - x_3)}{(ax_1 + bx_2 + cx_3)(x_1 x_2 + x_1 x_3 + x_2 x_3)}. \tag{9.284}
$$

Combining (9.282) with the gluon part of (9.267) and taking account of the equality $c_8 = 1/2$, we obtain the complete real part of the kernel in gluodynamics:

$$
K_G^{(8)}(q_1, q'; q) = \frac{g^2 N_c}{2(2\pi)^{D-1}} \left\{ \left(\frac{q_1^2 q_2'^2 + q_2^2 q_1'^2}{k^2} - q^2 \right) \left[\frac{1}{2} + \bar{g}^2 \left(k^2 \right)^\epsilon \left(-\frac{11}{6\epsilon} + \frac{67}{18} \right) \right. \right.
$$

$$
\left. - \zeta(2) + \epsilon \left(-\frac{202}{27} + 7\zeta(3) + \frac{11}{6} \zeta(2) \right) \right) \right] + \bar{g}^2 \left[q^2 \left(\frac{11}{6} \ln \left(\frac{q_1^2 q_1'^2}{k^2 q^2} \right) \right) \right.
$$

$$
+ \frac{1}{4} \ln \left(\frac{q_1^2}{q^2} \right) \ln \left(\frac{q_2^2}{q^2} \right) + \frac{1}{4} \ln \left(\frac{q_1'^2}{q^2} \right) \ln \left(\frac{q_2'^2}{q^2} \right) + \frac{1}{4} \ln^2 \left(\frac{q_1^2}{q_2^2} \right) - \frac{q_1^2 q_2'^2 + q_2^2 q_1'^2}{2k^2}
$$

$$
\times \ln^2 \left(\frac{q_1^2}{q_1'^2} \right) + \frac{q_1^2 q_2'^2 - q_2^2 q_1'^2}{k^2} \ln \left(\frac{q_1^2}{q_1'^2} \right) \left(\frac{11}{6} - \frac{1}{4} \ln \left(\frac{q_1^2 q_1'^2}{k^4} \right) \right) + \frac{1}{2} I(q_1^2, q_1'^2; k^2)
$$

$$
\times \left[q^2(k^2 - q_1^2 - q_1'^2) + 2q_1^2 q_1'^2 - q_1^2 q_2'^2 - q_2^2 q_1'^2 \right.
$$

$$
\left. \left. + \frac{q_1^2 q_2'^2 - q_2^2 q_1'^2}{k^2} (q_1^2 - q_1'^2) \right] \right\} + \frac{g^2 N_c}{2(2\pi)^{D-1}} \left\{ 1 \longleftrightarrow 2 \right\}. \tag{9.285}
$$

Note that the $1/\epsilon^2$ terms in (9.282) and (9.267) cancel in the sum. As we will see, this cancellation is crucial for the infrared safety of the colour-singlet BFKL kernel.

9.7.4 Colour-singlet kernel

As we have seen, the kernel is relatively simple for amplitudes with colour octet and negative signature. This simplicity is evidently related to the gluon reggeization. Technically, the reason for the simplicity is the absence of nonplanar diagrams, which are represented by the F_b term (9.271). But most important for physical applications is the singlet representation, and the F_b term is necessary in this case. It turns out, however, that it is much better to consider not F_b alone, but the combination $F = F_a + F_b$. This combination appears naturally in the singlet kernel. Indeed, looking at the colour coefficients (9.277), (9.279) and at the nonsubtracted kernels (9.271) one can obtain the relation

$$K_J^{(0)\Lambda} = 2K_J^{(8a)\Lambda} + K_J^{(s)\Lambda}, \tag{9.286}$$

where $K_J^{(s)\Lambda}$ is given by (9.271)

$$K_J^{(s)\Lambda}(q_1, q_1'; q) = 8g^4 N_c^2 d_J \int F^J(l_1, l_2) \frac{d\phi_J}{(2\pi)^{D-1}}, \tag{9.287}$$

and $F^J(l_1, l_2) = F_a^J(l_1, l_2) + F_b^J(l_1, l_2)$, $d_{Q\bar{Q}} = -1/(4N_c^3)$ and $d_{GG} = 1/2$. Furthermore, $\hat{K}_G$ in (9.265) and the leading order kernels of the subtraction term in (9.268) have the colour coefficients $c_1 = 1$, $c_8 = 1/2$. Remembering that the $Q\bar{Q}$ contribution does not require subtraction, one can therefore write:

$$K_r^{(0)} = 2K_r^{(8a)} + K^{(s)}, \quad K^{(s)} = K_{Q\bar{Q}}^{(s)} + K_{GG}^{(s)}, \quad K_{Q\bar{Q}}^{(s)} = K_{Q\bar{Q}}^{(s)\Lambda}, \tag{9.288}$$

and

$$K_{GG}^{(s)}(q_1, q_1'; q) = K_{GG}^{(s)\Lambda}(q_1, q_1'; q) - \frac{1}{4}\int \frac{d^{D-2}r}{r^2(q-r)^2} K_r^{B(0)}(q_1, r; q)$$

$$\times K_r^{B(0)}(r, q_1'; q) \ln\left(\frac{s_\Lambda^2}{(q_1-r)^2(q_1'-r)^2}\right). \tag{9.289}$$

It turns out, that F^J has a simpler form in comparison with F_a^J and F_b^J. Even more important is the infrared safety of $K^{(s)}$, whereas F_a^J and F_b^J are infrared singular. Finally, $K^{(s)}$ has no ultraviolet singularities, in contrast to $K^{(8)}$.

In the case of massless quarks, we have [113]:

$$F^{Q\bar{Q}}(l_1, l_2) = \frac{n_f}{4} x_1 x_2 \left\{ x_1 x_2 \left(\frac{2(q_1 l_1) - q_1^2}{\sigma_{11}} + \frac{2(q_1 l_2) - q_1^2}{\sigma_{21}} \right) \right.$$

$$\times \left(\frac{2(q_2 l_1) + q_2^2}{\sigma_{12}} + \frac{2(q_2 l_2) + q_2^2}{\sigma_{22}} \right) + \frac{x_1 q^2 (2(q_1 l_1) - q_1^2)}{2\sigma_{11}} \left(\frac{1}{\sigma_{22}} - \frac{1}{\sigma_{12}} \right)$$

$$+ \frac{x_1 q^2 (2(q_2 l_1) + q_2^2)}{2\sigma_{12}} \left(\frac{1}{\sigma_{11}} - \frac{1}{\sigma_{21}} \right) + \frac{1}{\sigma_{11}\sigma_{22}} \left(-2(q_1 l_1)(q_2 q_2') \right)$$

$$\left. - 2(q_2 l_1)(q_1 q_1') + (q_2^2 - q_1^2)(l_1 k) + q_1^2 q_2'^2 - \frac{k^2 q^2}{2} \right) \right\}, \tag{9.290}$$

where

$$\sigma_{ij} = (l_i + (-1)^j x_i q_j)^2 + x_1 x_2 q_j^2. \tag{9.291}$$

It is easy to see that $F^{Q\bar{Q}}$ decreases as l_i^{-4} at $l_i^2 \to \infty$, such that the integral (9.287) is well convergent in the ultraviolet region. This is not surprising because (9.287) describes real particle production. Less evident is the fact exhibited by (9.290) that $F^{Q\bar{Q}}$ does not contain infrared singularities. However, this fact is also predictable, because up to a numerical coefficient $K_{Q\bar{Q}}^{(s)\Lambda}$ coincides with the imaginary part of virtual photon-scattering amplitudes.

Therefore $K^{(s)}_{Q\bar{Q}}$ is called the abelian part of the quark contribution to the colour-singlet kernel, in contrast to the non-Abelian part $2K^{(8_a)}_{Q\bar{Q}}$ (see (9.280)).

In spite of the fact that $F^{Q\bar{Q}}$ has nice ultraviolet and infrared behaviour, and the integration in (9.271) can be performed at $D = 4$, the result is very complicated. The complexity is evidently related to nonplanar diagrams as is well known. In fact, $K^{(s)}_{Q\bar{Q}}$ was calculated many years ago [123],[124] within the framework of QED and can be obtained from the results found there. We have

$$K^{(s)}_{Q\bar{Q}}(q_1, q_1'; q) = \frac{\alpha_s^2}{2\pi^3}\left(-\frac{n_f}{N_c^3}\right)K_1\left(q_1 - \frac{q}{2}, -q_1' - \frac{q}{2}\right), \qquad (9.292)$$

with the function K_1 given by Eq. (A39) of Ref. [124], where on the right-hand side the substitutions

$$q \rightarrow q_1 - \frac{q}{2}, \quad q' \rightarrow q_1' - \frac{q}{2}, \quad r \rightarrow \frac{q}{2}, \quad Q \rightarrow q_1 - q(1-y), \quad Q' \rightarrow q_1' - q(1-x)$$
$$(9.293)$$

must be done. Note that the Abelian part of the quark contibution is suppressed at large N_c. It is worthwhile to note that right side of Eq. (9.292) contains a nonzero fermion mass and, at first sight, has a logarithmic singularity when the mass tends to zero; but the singularity is spurious because of cancellations among various terms.

The piece $K^{(s)}_{GG}$ of the gluon contribution to the singlet kernel is called the symmetric part. It also has neither ultraviolet nor infrared singularities. For the former ones the reason is evident: $K^{(s)}_{GG}$, as well as $K^{(s)}_{Q\bar{Q}}$, describes real particle production. But absence of the infrared singularities is achieved in a much more intricate way. Indeed, F^{GG} is infrared singular; in $K^{(s)}_{GG}$ the singularities are absent due to the cancellation between $K^{(s)\Lambda}_{GG}$ and the subtraction term in (9.289). It is possible, however, to perform the cancellation before integration ([119]). The matter is that due to the factorization property (9.87) and (9.92) of the two-gluon production vertex one has

$$\left(\frac{2g^2 N_c c_R}{(2\pi)^{D-1}}\right)^2 F_a^{GG}(l_1, l_2)\bigg|_{x_1=1} = \frac{K_r^{(R)B}\left(q_1, q_1 - l_1; q\right)K_r^{(R)B}\left(q_1 - l_1, q_1'; q\right)}{(q_1 - l_1)^2(q_2 + l_1)^2}.$$
$$(9.294)$$

and

$$\left(\frac{2g^2 N_c c_R}{(2\pi)^{D-1}}\right)^2 F_a^{GG}(l_1, l_2)\bigg|_{x_2=1} = \frac{K_r^{(R)B}\left(q_1, q_1 - l_2; q\right)K_r^{(R)B}\left(q_1 - l_2, q_1'; q\right)}{(q_1 - l_2)^2(q_2 + l_2)^2}.$$
$$(9.295)$$

From the expression (9.84) one can see that the tensors $b^{\alpha\beta}(q; l_1, l_2)$ drop as $1/l_i^2$ for $l_i^2 \rightarrow \infty$ at fixed x_i. Therefore, the integral over l_i in Eq. (9.271) is well convergent in the ultraviolet region, so that the restrictions imposed by the theta-function can be written as

$$x_i \geq \frac{l_i^2}{s_\Lambda}, \qquad (9.296)$$

and the subtraction term in (9.268) acquires the form

$$\frac{g^4 N_c^2 c_R^2}{(2\pi)^{D-1}} \int \frac{d^{2+2\epsilon} l_1}{(2\pi)^{D-1}} \left(F_a^{GG}(l_1, l_2)|_{x_1=1} + F_a^{GG}(l_1, l_2)|_{x_2=1} \right) \ln \left(\frac{s_\Lambda^2}{l_1^2 l_2^2} \right)$$

$$= 8g^4 N_c^2 a_R \int \frac{d\phi_{GG}}{(2\pi)^{D-1}} \left(x_1 F_a^{GG}(l_1, l_2)|_{x_1=1} + x_2 F_a^{GG}(l_1, l_2)|_{x_2=1} \right)$$

$$+ \frac{g^4 N_c^2 a_R}{(2\pi)^{D-1}} \int \frac{d^{2+2\epsilon} l_1}{(2\pi)^{D-1}} \left(F_a^{GG}(l_1, l_2)|_{x_1=1} - F_a^{GG}(l_1, l_2)|_{x_2=1} \right) \ln \left(\frac{l_2^2}{l_1^2} \right). \quad (9.297)$$

Here the equality $a_R = c_R^2$, the expression (9.270) for the phase-space element with account of the identity factor $1/2!$ and the restrictions given by the inequality (9.296) on x_i were taken into account. Note that the second integral on the right-hand side of Eq. (9.297) is completely antisymmetric with respect to the substitution $q_1 \leftrightarrow -q_1'$, $q \leftrightarrow -q$. Therefore, the subtraction term can be obtained by symmetrization of the first integral. Consequently, using the definition

$$\left(\frac{f(x)}{x(1-x)} \right)_+ \equiv \frac{1}{x}[f(x) - f(0)] + \frac{1}{(1-x)}[f(x) - f(1)], \quad (9.298)$$

we can write the two-gluon contribution to the kernel (9.268) in the limit $s_\Lambda \to \infty$ in the form of

$$K_{GG}^{(R)}(q_1, q_1'; q) = \frac{2g^4 N_c^2}{(2\pi)^{D-1}} \hat{S} \int_0^1 dx \int \frac{d^{2+2\epsilon} l_1}{(2\pi)^{D-1}} \left(\frac{a_R F_a^{GG}(l_1, l_2) + b_R F_b^{GG}(l_1, l_2)}{x(1-x)} \right)_+, \quad (9.299)$$

where $x \equiv x_1$ and the operator $\hat{S}$ symmetrizes with respect to the substitution $q_1 \leftrightarrow -q_1'$, $q \leftrightarrow -q$. We have used here that $F_b^{GG}(l_1, l_2)|_{x_1=0} = F_b^{GG}(l_1, l_2)|_{x_1=1} = 0$, according to the definition (9.273) and the properties $b_{\alpha\beta}(q_1; l_1, l_2)|_{x_1=0} = b_{\beta\alpha}(q_1; l_2, l_1)|_{x_1=1} = 0$. Consequently, for the symmetric part, one has

$$K_{GG}^{(s)}(q_1, q_1'; q) = \frac{g^4 N_c^2}{(2\pi)^{D-1}} \hat{S} \int_0^1 dx \int \frac{d^{2+2\epsilon} l_1}{(2\pi)^{D-1}} \left(\frac{F^{GG}(l_1, l_2)}{x(1-x)} \right)_+, \quad (9.300)$$

where the function $F^{GG}(l_1, l_2) = F_a^{GG}(l_1, l_2) + F_b^{GG}(l_1, l_2)$ is given by the convolution

$$F^{GG}(l_1, l_2) = \left(b_{\alpha\beta}(q_1; l_1, l_2) + b_{\beta\alpha}(q_1; l_2, l_1) \right) \left(b_{\alpha\beta}(q_1'; l_1, l_2) + b_{\beta\alpha}(q_1'; l_2, l_1) \right). \quad (9.301)$$

The calculation of $K_{GG}^{(s)}(q_1, q_1'; q)$ is more convenient because the sum

$$b_{\alpha\beta}(q_1; l_1, l_2) + b_{\beta\alpha}(q_1; l_2, l_1) = \frac{q_{1\perp}^2 l_{1\perp\alpha} l_{2\perp\beta}}{l_{1\perp}^2 l_{2\perp}^2}$$

$$+ \frac{1}{2} g_{\alpha\beta}^\perp x_1 x_2 \frac{q_{1\perp}^2 - 2q_{1\perp} l_{1\perp}}{\sigma_{11}} + \frac{x_2 l_{1\perp\alpha} q_{1\perp\beta} - x_1 q_{1\perp\alpha}(q_1 - l_1)_{\perp\beta}}{\sigma_{11}} + x_1 \frac{q_{1\perp}^2 l_{1\perp\alpha}(q_1 - l_1)_{\perp\beta}}{k_{1\perp}^2 \sigma_{11}}$$

$$+ \frac{1}{2} g_{\alpha\beta}^{\perp} x_1 x_2 \frac{q_{1\perp}^2 - 2q_{1\perp} l_{2\perp}}{\sigma_{12}} + \frac{x_1 q_{1\perp\alpha} l_{2\perp\beta} - x_2 (q_1 - l_2)_{\perp\alpha} q_{1\perp\beta}}{\sigma_{12}} + x_2 \frac{q_{1\perp}^2 (q_1 - l_2)_{\perp\alpha} l_{2\perp\beta}}{l_{2\perp}^2 \sigma_{12}},$$

$$(9.302)$$

looks simpler than $b_{\alpha\beta}(q_1; l_1, l_2)$ and $b_{\beta\alpha}(q_1; l_2, l_1)$ taken separately. Nevertheless, it evidently contains infrared singularities. Therefore, in contrast to the quark case, the function F^{GG} contains the piece F_{sing}^{GG} which is infrared singular. But one can check that this piece does not depend on x. On the other hand, the boundary values $F^{GG}|_{x=0}$ and $F^{GG}|_{x=1}$ do not contain new (i.e. different from F_{sing}^{GG}) nonintegrable infrared singularities in the limit $\epsilon \to 0$. Therefore, such singularities are absent in $\left(F^{GG}/[x(1-x)]\right)_+$, such that the integration in (9.300) can be performed at $D = 4$, as in the quark case. Nevertheless, because of the contribution of nonplanar diagrams the result is very complicated. As could be expected, the result is more complicated than in the quark case, because besides contributions of cross-box diagrams it contains contributions of cross-hexagon and cross-octagon diagrams. The result is given in the Appendix.

9.7.5 Möbius representation of the next-to-leading order kernel

The singlet kernel is strongly simplified in the Möbius representation, i.e. in the space of impact parameters r_1, r_2 and functions vanishing at $r_1 = r_2$ [125]–[127]. In the next-to-leading order, a general form of the kernel in this representation is (compare with (9.213)):

$$\langle r_1 r_2 | \hat{K}_M | r_1' r_2' \rangle = \delta(r_{11'}) \delta(r_{22'}) \int dr_0 g_0(r_1, r_2; r_0)$$

$$+ \delta(r_{11'}) g_1(r_1, r_2; r_2') + \delta(r_{22'}) g_1(r_2, r_1; r_1') + \frac{1}{\pi} g_2(r_1, r_2; r_1', r_2') \qquad (9.303)$$

with the functions g_i turning into zero when their first two arguments coincide. The first three terms here contain ultraviolet singularities which cancel in their sum, as well as in the leading order, with account of the dipole property of the target impact factors. The coefficient of $\delta(r_{11'}) \delta(r_{22'})$ is written in the integral form in order to make the cancellation evident. The function g_2 is absent in the leading order because the leading order kernel in the momentum space does not contain terms depending on all three independent momenta simultaneously.

Note that the next-to-leading order kernel is ambiguious. In the BFKL approach, scattering amplitudes are invariant (see, for example, (9.123)) under the operator transformation of the kernel

$$\hat{K} \to \hat{O}^{-1} \hat{K} \hat{O} \qquad (9.304)$$

accompanied by corresponding transformations of the impact factors. Even if the kernel is fixed in the leading order by the requirement of the conformal invariance of its Möbius form, transformations with $\hat{O} \to 1 - \hat{O}$, where $\hat{O} \sim g^2$, are still possible. Then

$$\hat{K} \to \hat{K} - [\hat{K}^B \hat{O}], \qquad (9.305)$$

where $\hat{\mathcal{K}}^B$ is the leading order kernel. These transformations rearrange next-to-leading order corrections to the kernel and impact factors and can be used for simplification of the kernel. The operator adopted in [125]–[127] is:

$$\hat{O} = -\frac{\alpha_s}{8\pi}\beta_0 \ln\left(\hat{q}_1^{\,2}\hat{q}_2^{\,2}\right), \qquad \beta_0 = \left(\frac{11}{3}N_c - \frac{2n_f}{3}\right). \tag{9.306}$$

At first sight, there is one more ambiguity of the next-to-leading order kernel, related to the choice of the energy scale. But it was shown [128] that changes of the energy scale can be compensated by corresponding redefinitions of the impact factors. It means that the ambiguity in the energy scale gives nothing new as compared with (9.305).

With the transformation (9.305), (9.306) one has

$$g_1(r_1, r_2; r_2') = \frac{\alpha_s(4e^{-2\gamma}/r_{12}^2)N_c}{2\pi^2} \frac{r_{12}^2}{r_{22'}^2 r_{12'}^2}\left[1 + \frac{\alpha_s N_c}{2\pi}\left(\frac{67}{18} - \zeta(2) - \frac{5n_f}{9N_c}\right.\right.$$

$$\left.\left. + \frac{\beta_0}{2N_c}\frac{r_{12'}^2 - r_{22'}^2}{\vec{r}_{12}^2}\ln\left(\frac{r_{22'}^2}{\vec{r}_{12'}^2}\right) - \frac{1}{2}\ln\left(\frac{r_{12}^2}{r_{22'}^2}\right)\ln\left(\frac{r_{12}^2}{r_{12'}^2}\right) + \frac{r_{12'}^2}{2r_{12}^2}\ln\left(\frac{r_{12'}^2}{r_{22'}^2}\right)\ln\left(\frac{r_{12}^2}{r_{12'}^2}\right)\right)\right], \tag{9.307}$$

where $\gamma \simeq 0.577216$ is Euler's constant,

$$\alpha_s(q^2) = \alpha_s(\mu^2)\left(1 - \beta_0\frac{\alpha_s(\mu^2)}{4\pi}\ln\left(\frac{q^2}{\mu^2}\right)\right). \tag{9.308}$$

The function g_0 can be taken in different forms giving the same integral contribution. One can take

$$g_0(r_1, r_2; r_0) = -g(r_1, r_2; r_0) + \frac{\alpha_s^2 N_c^2}{8\pi r^3}\frac{r_{12}^2}{r_{10}^2 r_{10}^2}\ln\left(\frac{r_{10}^2}{r_{12}^2}\right)\ln\left(\frac{r_{20}^2}{r_{12}^2}\right). \tag{9.309}$$

The function g_2 is not so simple (although it is incomparably simpler than the kernel in the momentum representation):

$$g_2(r_1, r_2; r_1', r_2') = \frac{\alpha_s^2 N_c^2}{4\pi^3}\left[\frac{1}{2r_{1'2'}^4}\left(\frac{r_{12'}^2 r_{21'}^2}{d}\ln\left(\frac{r_{12'}^2 r_{21'}^2}{r_{11'}^2 r_{22'}^2}\right) - 1\right)\left(1 + \frac{n_f}{N_c^3}\right)\right.$$

$$- \left(\frac{(4 + n_f/N_c^3)\,r_{12}^2 r_{1'2'}^2}{4r_{1'2'}^4} - \frac{1}{4r_{11'}^2 r_{22'}^2}\left(\frac{r_{12}^4}{d} - \frac{r_{12}^2}{r_{1'2'}^2}\right)\right)\ln\left(\frac{r_{12'}^2 r_{21'}^2}{r_{11'}^2 r_{22'}^2}\right)$$

$$+ \frac{\ln\left(\frac{r_{12}^2}{r_{1'2'}^2}\right)}{4r_{11'}^2 r_{22'}^2} + \frac{\ln\left(\frac{r_{12}^2 r_{1'2'}^2}{r_{11'}^2 r_{22'}^2}\right)}{2r_{12'}^2 r_{21'}^2}\left(\frac{r_{12}^2}{2r_{1'2'}^2} + \frac{1}{2} - \frac{r_{22'}^2}{r_{1'2'}^2}\right) + \frac{r_{12}^2\ln\left(\frac{r_{12'}^2 r_{1'2'}^2}{r_{12'}^2 r_{21'}^2}\right)}{4r_{11'}^2 r_{22'}^2 r_{1'2'}^2}$$

$$\begin{aligned}
&+ \frac{\ln\left(\frac{r_{22'}^2}{r_{12}^2}\right)}{2r_{11'}^2 r_{12'}^2} + \frac{\ln\left(\frac{r_{12}^2 r_{1'2'}^2}{r_{12'}^2 r_{22'}^2}\right)}{2r_{11'}^2 r_{1'2'}^2} + \frac{\ln\left(\frac{r_{12}^2 r_{11'}^2}{r_{22'}^2 r_{1'2'}^2}\right)}{2r_{12'}^2 r_{1'2'}^2} \\[2em]
&+ \frac{r_{12}^2 \ln\left(\frac{r_{11'}^2}{r_{1'2'}^2}\right)}{2r_{11'}^2 r_{12'}^2 r_{22'}^2} + \frac{r_{21'}^2 \ln\left(\frac{r_{21'}^2 r_{1'2'}^2}{r_{12}^2 r_{11'}^2}\right)}{2r_{11'}^2 r_{22'}^2 r_{1'2'}^2} + (1 \leftrightarrow 2) \Bigg],
\end{aligned} \tag{9.310}$$

where $d = r_{12'}^2 r_{21'}^2 - r_{11'}^2 r_{22'}^2$. The quark part of $\hat{\mathcal{K}}_M$ agrees with the results obtained in Refs. [129],[130] by the direct calculation of the quark contribution to the dipole kernel in the coordinate representation.

Evidently, the conformal invariance is violated in the next-to-leading order by renormalization. It is seen, however, that the renormalization is not the only source of the violation. Note that the abelian part of the quark contribution to the dipole kernel in the coordinate representation. But the gluon part of the dipole kernel, which was found then in [131], disagrees with Eq. (9.310). However, it was shown [132] that with account of the correction made in [133] the transformations (9.305) permit to match these parts also. Moreover, these transformations allow to present the kernel in the form where the conformal invariance is violated only by renormalization. It is especially interesting for the QED Pomeron and Yang-Mills theories with N = 4 extended supersymmetry.

9.7.6 Impact factors

In the BFKL approach, scattering amplitudes are given by the convolution of the impact factors of interacting particles with the Green's function of two interacting reggeized gluons. All energy dependence is defined by the universal (i.e. process independent) Green's function, which is determined by the BFKL kernel. The impact factors describing the scattering of particles by the reggeized gluons contain all the dependence on the nature of the particles and are energy independent. For a consistent description of scattering amplitudes in the BFKL approach one needs to know the impact factors with the same accuracy as the kernel.

On the parton level, the next-to-leading order impact factors have been calculated for quarks and gluons [55],[56],[134] and for forward-jet production [135],[136]. As a rule, the next-to-leading order impact factors are rather complicated and we don't present them here. The exception is the impact factors for partons (quarks and gluons) in scattering amplitudes with a colour-octet t-channel exchange and negative signature. Because of the gluon reggeization, they are expressed in terms of the reggeon vertices and the process-independent eigenstate $|R_\omega(q_\perp)\rangle$ of the colour-octet kernel (see (9.230)). The reggeon vertices for quarks and gluons are given by Eqs. (9.48) and (9.48) correspondingly. In the leading order the matrix elements

$\langle \mathcal{G}_1 \mathcal{G}_2 | R_\omega(q_\perp) \rangle$ are defined in (9.233). In the next-to-leading order they acquire the factor [111]:

$$R_\omega(q_1, q_2) = 1 + \frac{\omega(t)}{2} \left[\tilde{K}_1 + \left(\left(\frac{q_1^2}{q^2} \right)^\epsilon + \left(\frac{q_2^2}{q^2} \right)^\epsilon - 1 \right) \right.$$

$$\times \left\{ \frac{1}{2\epsilon} + \psi(1+2\epsilon) - \psi(1+\epsilon) + \frac{11+7\epsilon}{2(1+2\epsilon)(3+2\epsilon)} - \frac{n_f}{N_c} \frac{(1+\epsilon)}{(1+2\epsilon)(3+2\epsilon)} \right\}$$

$$\left. - \frac{1}{2\epsilon} + \psi(1) + \psi(1+\epsilon) - \psi(1-\epsilon) - \psi(1+2\epsilon) \right], \qquad (9.311)$$

where $q = q_1 + q_2$, $\omega(t)$ is the one-loop gluon trajectory,

$$\tilde{K}_1 = \frac{(4\pi)^{2+\epsilon} \Gamma(1+2\epsilon)\epsilon \left(q^2 \right)^{-\epsilon}}{4\Gamma(1-\epsilon)\Gamma^2(1+\epsilon)} \int \frac{d^{D-2}k}{(2\pi)^{D-1}} \ln\left(\frac{q^2}{k^2} \right) \frac{q^2}{(k-q_1)^2(k+q_2)^2} \, . \qquad (9.312)$$

The result of integration in (9.312), in form of expansion in ϵ, can be found in Ref. [56].

Among the impact factors of colourless objects, the most important one from the phenomenological point of view is the virtual-photon impact factor, because it is determined from the first principles in perturbative QCD and opens the way to predictions of the $\gamma^* \gamma^*$ total cross section. Its calculation turned out to be a very complicated problem, which is not yet completely solved after long-standing efforts [137]–[142]. Recently, important progress was reached in solution of a related problem: the next-to-leading order impact factor for the transition of a virtual photon to a light-vector meson was found as a closed analytical expression in the case of $t = 0$ and longitudinal polarizations [143],[144],[143]. The knowledge of the $\gamma^* \rightarrow V$ impact factor allows determination completely within perturbative QCD and with NLA accuracy the amplitude $\gamma^* \gamma^* \rightarrow VV$ [145]–[147].

9.7.7 Next-to-leading order BFKL pomeron at zero momentum transfer

The kernel simplifies considerably in the case of forward scattering. Actually, just this case was considered at first. The quark-antiquark pair production contribution was calculated in Refs. [57],[58],[59],[148]–[151], for massive as well as for massless quarks. Denoting, as well in the leading order, $K(q_1, q_1') = K(q_1, q_1'; 0)q_1^{-2}q_1'^{-2}$, we have for massless quark flavours

$$K_{Q\bar{Q}}(q_1, q_2) = \frac{4\bar{g}_\mu^4 \mu^{-2\epsilon} n_f}{\pi^{1+\epsilon}\Gamma(1-\epsilon)N_c^3} \left\{ N_c^2 \left[\frac{1}{k^2} \left(\frac{k^2}{\mu^2} \right)^\epsilon \frac{2}{3} \left(\frac{1}{\epsilon} - \frac{5}{3} + \epsilon \left(\frac{28}{9} - \frac{\pi^2}{6} \right) \right) \right. \right.$$

$$\left. + \frac{1}{(q_1^2 - q_2^2)} \left[1 - \frac{k^2(q_1^2 + q_2^2 + 4q_1 q_2)}{3(q_1^2 - q_2^2)^2} \right] \ln\frac{q_1^2}{q_2^2} + \frac{k^2}{(q_1^2 - q_2^2)^2} \right.$$

$$\times \left(2 - \frac{k^2(q_1^2 + q_2^2)}{3q_1^2 q_2^2}\right) + \frac{(2k^2 - q_1^2 - q_2^2)}{3q_1^2 q_2^2}\right] + \left[-1 - \frac{(q_1^2 - q_2^2)^2}{4q_1^2 q_2^2}\right.$$

$$+ \left(\frac{(q_1 q_2)^2}{q_1^2 q_2^2} - 2\right)\frac{(2q_1^2 q_2^2 - 3q_1^4 - 3q_2^4)}{16q_1^2 q_2^2}\right]\int_0^\infty \frac{dx}{(q_1^2 + x^2 q_2^2)}\ln\left|\frac{1+x}{1-x}\right|$$

$$+ \frac{(3(q_1 q_2)^2 - 2q_1^2 q_2^2)}{16q_1^4 q_2^4}\left[(q_1^2 - q_2^2)\ln\left(\frac{q_1^2}{q_2^2}\right) + 2(q_1^2 + q_2^2)\right]\right\}, \tag{9.313}$$

where $k = q_1 - q_2$. The right-hand side of Eq. (9.313) exhibits several terms with unphysical singularities at $q_1^2 = q_2^2$. It is not difficult to see that the only real singularity is at $k^2 = 0$. All others are spurious and singular terms cancel each other. As for the singularity at $k^2 = 0$, the region of such small k^2 where $\ln(1/k^2) \sim 1/\epsilon$ is essential in the subsequent integration over k. In order to retain all terms which give nonzero contributions in the physical case $\epsilon \to 0$ we must not expand $(k^2/\mu^2)^\epsilon$ in powers of ϵ in such terms and have to keep the terms of order ϵ in the coefficient.

Investigation of the two-gluon production contribution was started in [60]; the next step was done in [57]. The final result obtained in [63] is:

$$K_{GG}(q_1, q_2) = \frac{4\bar{g}_\mu^4 \mu^{-2\epsilon}}{\pi^{1+\epsilon}\Gamma(1-\epsilon)q_1^2 q_2^2}\left\{\frac{2q_1^2 q_2^2}{k^2}\left(\frac{k^2}{\mu^2}\right)^\epsilon\left[\frac{1}{\epsilon^2} - \frac{11}{6\epsilon} - \frac{2\pi^2}{3} + \frac{67}{18}\right.\right.$$

$$+ \epsilon\left(\frac{11\pi^2}{36} - \frac{202}{27} + 9\zeta(3)\right)\right] - \frac{(q_1^2 + q_2^2)(2q_1^2 q_2^2 - 3(q_1 q_2)^2)}{8q_1^2 q_2^2}$$

$$- \left(\frac{11}{3}\frac{q_1^2 q_2^2}{(q_1^2 - q_2^2)} + \frac{(2q_1^2 q_2^2 - 3(q_1 q_2)^2)}{16q_1^2 q_2^2}(q_1^2 - q_2^2)\right)\ln\left(\frac{q_1^2}{q_2^2}\right) - \frac{2}{3}\frac{q_1^2 q_2^2}{(q_1^2 - q_2^2)^3}$$

$$\times \left[\left(1 - \frac{2(q_1 q_2)^2}{q_1^2 q_2^2}\right)\left(q_1^4 - q_2^4 - 2q_1^2 q_2^2 \ln\left(\frac{q_1^2}{q_2^2}\right)\right) + (q_1 q_2)\right.$$

$$\times \left(2q_1^2 - - 2q_2^2 - (q_1^2 + q_2^2)\ln\left(\frac{q_1^2}{q_2^2}\right)\right)\right] + \frac{2q_1^2 q_2^2(k(q_1 + q_2))}{k^2(q_1 + q_2)^2}$$

$$\times \left[\frac{1}{2}\ln\left(\frac{q_1^2}{q_2^2}\right)\ln\left(\frac{q_1^2 q_2^2 k^4}{(q_1^2 + q_2^2)^4}\right) - Li_2\left(-\frac{q_1^2}{q_2^2}\right) + Li_2\left(-\frac{q_2^2}{q_1^2}\right)\right]$$

$$- \left[4q_1^2 q_2^2 + \frac{(q_1^2 - q_2^2)^2}{4} + (2q_1^2 q_2^2 - 3q_1^4 - 3q_2^4)\frac{(2q_1^2 q_2^2 - (q_1 q_2)^2)}{16q_1^2 q_2^2}\right]$$

$$\times \int_0^\infty \frac{dx}{(q_1^2 + x^2 q_2^2)}\ln\left|\frac{1+x}{1-x}\right| - q_1^2 q_2^2\left(1 - \frac{(k(q_1 + q_2))^2}{k^2(q_1 + q_2)^2}\right)$$

$$\times \left(\int_0^1 - \int_1^\infty\right)dz\frac{\ln((zq_1)^2/q_2^2)}{(q_2 - zq_1)^2}\right\}, \tag{9.314}$$

where $Li_2(z) = -\int_0^z \frac{dt}{t} \ln(1-t)$ is the dilogarithm and $\zeta(n) = \sum_{k=1}^{\infty} k^{-n}$ is the Riemann ζ-function.

From (9.265), using $K_G^{(R)}$ (9.267) for the forward case (singlet representation and $q = 0$) we obtain the total real part of the forward kernel $K_r(q_1, q_2)$ [152]:

$$
K_r(q_1, q_2) = \frac{4\bar{g}_\mu^2 \, \mu^{-2\epsilon}}{\pi^{1+\epsilon} \, \Gamma(1-\epsilon)} \frac{1}{k^2} \left\{ 1 + \bar{g}_\mu^2 \left[\frac{\beta_0}{N_{c\epsilon}} \left(1 - \left(\frac{k^2}{\mu^2}\right)^\epsilon \left(1 - \epsilon^2 \frac{\pi^2}{6}\right) \right) \right. \right.
$$

$$
+ \left(\frac{k^2}{\mu^2}\right)^\epsilon \left(\frac{67}{9} - \frac{\pi^2}{3} - \frac{10}{9}\frac{n_f}{N_c} + \epsilon\left(-\frac{404}{27} + 14\zeta(3) + \frac{56}{27}\frac{n_f}{N_c}\right) \right) - \left(\ln \frac{q_1^2}{q_2^2}\right)^2 \right]
$$

$$
+ \frac{4\bar{g}_\mu^4 \, \mu^{-2\epsilon}}{\pi^{1+\epsilon} \, \Gamma(1-\epsilon)} [f_1(q_1, q_2) + f_2(q_1, q_2)], \tag{9.315}
$$

where $\beta_0 = (11/3)N_c - 2n_f/3$ is the first coefficient of the β-function,

$$
f_1(q_1, q_2) = -\frac{2(q_1^2 - q_2^2)}{k^2(q_1 + q_2)^2} \left(\frac{1}{2}\ln\left(\frac{q_2^2}{q_1^2}\right) \ln\left(\frac{q_1^2 q_2^2 k^4}{(q_1^2 + q_2^2)^4}\right) + Li_2\left(-\frac{q_1^2}{q_2^2}\right) \right.
$$

$$
\left. - Li_2\left(-\frac{q_2^2}{q_1^2}\right) \right) - \left(1 - \frac{(q_1^2 - q_2^2)^2}{k^2(q_1 + q_2)^2}\right) \left(\int_0^1 - \int_1^\infty\right) \frac{dz \, \ln \frac{(zq_1)^2}{(q_2)^2}}{(q_2 - zq_1)^2}, \tag{9.316}
$$

$$
f_2(q_1, q_2) = -\left(1 + \frac{n_f}{N_c^3}\right) \frac{2q_1^2 q_2^2 - 3(q_1 q_2)^2}{16 q_1^2 q_2^2} \left(\frac{2}{q_2^2} + \frac{2}{q_1^2}\right)
$$

$$
+ \left(\frac{1}{q_2^2} - \frac{1}{q_1^2}\right) \ln \frac{q_1^2}{q_2^2} - \left(3 + \left(1 + \frac{n_f}{N_c^3}\right)\left(1 - \frac{(q_1^2 + q_2^2)^2}{8 q_1^2 q_2^2}\right)\right)
$$

$$
- \frac{2q_1^2 q_2^2 - 3q_1^4 - 3q_2^4}{16 q_1^4 q_2^4}(q_1 q_2)^2\right) \int_0^\infty \frac{dx \, \ln\left|\frac{1+x}{1-x}\right|}{q_1^2 + x^2 q_2^2}. \tag{9.317}
$$

As usual, the terms $\sim \epsilon$ are taken into account in the coefficient at k^{-2} in order to save all contributions nonvanishing in the limit $\epsilon \to 0$ after the integrations. The remarkable fact exhibited by Eq. (9.315) is the cancellation of the $1/\epsilon^2$ terms which are present in K_G and K_{GG}. After cancellation of the terms $\sim 1/\epsilon^2$ the leading singularity of the kernel is $1/\epsilon$. Because of the singular behaviour of the kernel at $k^2 = 0$ it turns again into $\sim 1/\epsilon^2$ after subsequent integrations of the kernel and cancels with the singularity in the virtual part of the kernel.

There is the representation of the gluon trajectory [127] which allows the cancellation explicitly. Taking into account the charge renormalization (9.60) in Eqs. (9.57), (9.58) and retaining in the integrand only terms giving nonvanishing contribution to the trajectory in the limit $\epsilon \to 0$, one has

$$\omega(-q^2) = -\frac{\bar{g}_\mu^2 \, q^2}{\pi^{1+\epsilon}\Gamma(1-\epsilon)} \int \frac{d^{2+2\epsilon}k \, \mu^{-2\epsilon}}{k^2(k-q)^2} \left(1 + \bar{g}_\mu^2 f_\omega(k, k-q)\right), \qquad (9.318)$$

where

$$f_\omega(k_1, k_2) = \frac{\beta_0}{\epsilon N_c} - f(k_1, k_2) + f(k_1, 0) + f(0, k_2) = \frac{\beta_0}{\epsilon N_c}$$

$$+ \left[\frac{\beta_0}{N_c} - \frac{67}{9} + 2\zeta(2) + \frac{10\,n_f}{9\,N_c} + \epsilon\left(\frac{404}{27} - \frac{\beta_0}{\epsilon N_c}\zeta(2) - 6\zeta(3) - \frac{56\,n_f}{27\,N_c}\right)\right]$$

$$\times \left[\left(\frac{k_{12}^2}{\mu^2}\right)^\epsilon - \left(\frac{k_1^2}{\mu^2}\right)^\epsilon - \left(\frac{k_2^2}{\mu^2}\right)^\epsilon\right] - \ln\left(\frac{k_{12}^2}{k_1^2}\right)\ln\left(\frac{k_{12}^2}{k_2^2}\right), \qquad (9.319)$$

and $k_{12} = k_1 - k_2$. The representation (9.318)–(9.319) is extremely convenient, since it enables to get easily the expression (9.63) for the trajectory in the limit $\epsilon \to 0$. But its main advantage is that it gives the possibility to perform explicitly the cancellation of the infrared singularities and writing of the kernel at the physical space-time dimension $D = 4$. Let us introduce the cut-off $\lambda \to 0$, making it tending to zero after taking the limit $\epsilon \to 0$, and divide the integration region in the integral representation of the trajectory (9.318) into three domains. In two of them either $k^2 \le \lambda^2$, or $(k - q_i)^2 \le \lambda^2$, and in the third one both $k^2 > \lambda^2$ and $(k - q_i)^2 > \lambda^2$. Then in the third domain we can take the limit $\epsilon = 0$ in (9.319) and put $f_\omega(k_1, k_2) = f_\omega^{(0)}(k_1, k_2)$, where

$$f_\omega^{(0)}(k_1, k_2) = \frac{67}{9} - 2\zeta(2) - \frac{10\,n_f}{9\,N_c} - \frac{\beta_0}{N_c}\ln\left(\frac{k_1^2 k_2^2}{\mu^2 k_{12}^2}\right) - \ln\left(\frac{k_{12}^2}{k_1^2}\right)\ln\left(\frac{k_{12}^2}{k_2^2}\right). \qquad (9.320)$$

In the first domain we have

$$f_\omega(k, k-q) = \frac{\beta_0}{\epsilon N_c} - \left(\frac{k^2}{\mu^2}\right)^\epsilon \left[\frac{\beta_0}{\epsilon N_c} - \frac{67}{9} + 2\zeta(2) + \frac{10\,n_f}{9\,N_c}\right.$$

$$\left. + \epsilon\left(\frac{404}{27} - \frac{\beta_0}{N_c}\zeta(2) - 6\zeta(3) - \frac{56\,n_f}{27\,N_c}\right)\right] \qquad (9.321)$$

and in the second domain we have the same expression with the substitution $k^2 \to (k - q)^2$. Comparing (9.315) with (9.320), we see that the contribution of (9.315) from the region $k^2 < \lambda^2$ cancels almost completely the contributions of the regions $k^2 \le \lambda^2$ and $(k - q_i)^2 \le \lambda^2$ in the doubled trajectory $\omega(-q^2)$. The only piece which remains uncanceled at $\epsilon \to 0$ in the trajectory is

$$\frac{\bar{g}_\mu^4}{\pi^{1+\epsilon}\Gamma(1-\epsilon)} \int \frac{d^{2+2\epsilon}k \, \mu^{-2\epsilon}}{k^2} 16\epsilon\zeta(3)\left(\frac{k^2}{\mu^2}\right)^\epsilon \theta(\lambda^2 - k^2) = \frac{\alpha_s^2(\mu)N_c^2}{2\pi^2}\zeta(3). \qquad (9.322)$$

On account of this cancellation and using the equality

$$\int \frac{d^2k}{4\pi} \frac{q^2}{k^2(k-q)^2} \ln\left(\frac{k^2}{q^2}\right)\ln\left(\frac{(k-q)^2}{q^2}\right) = \zeta(3), \qquad (9.323)$$

we can put

$$K(q,l) = \frac{\alpha_s(\mu^2)N_c}{2\pi^2} \left[\frac{2}{(q-l)^2} - \delta(q-l) \int \frac{dk\, q^2}{(q-k)^2 k^2} \right]$$

$$\left[1 + \frac{\alpha_s(\mu^2)N_c}{4\pi} \left(\frac{67}{9} - 2\zeta(2) - \frac{10}{9}\frac{n_f}{N_c} \right) \right] + \frac{\alpha_s^2(\mu^2)N_c^2}{4\pi^3}$$

$$\left[\frac{1}{(q-l)^2} \left(\frac{\beta_0}{N_c} \ln\left(\frac{\mu^2}{(q-l)^2} \right) - \ln^2\left(\frac{q^2}{l^2} \right) \right) + f_1(q,l) + f_2(q,l) \right.$$

$$\left. + \delta(q-l)\left(\frac{\beta_0}{2N_c} \int \frac{dk\, q^2}{(q-k)^2 k^2} \ln\left(\frac{(q-k)^2 k^2}{\mu^2 q^2} \right) + 6\pi\zeta(3) \right) \right]. \qquad (9.324)$$

This representation facilitates calculation of the eigenvalues of the kernel. Strictly speaking, because of the charge renormalization the eigenfunctions of the leading order kernel not are any more eigenfunctions of the next-to-leading order kernel. But with required accuracy we can write

$$\alpha_s(\mu^2) = \alpha_s(q^2)\left(1 + \frac{\beta_0}{4\pi}\alpha_s(q^2)\ln\left(\frac{q^2}{\mu^2} \right) \right), \qquad (9.325)$$

and obtain [152],[153]

$$\int d^2l\, K(q,l) \left(\frac{l^2}{q^2} \right)^{\gamma-1} = \omega(q^2,\gamma) = \frac{\alpha_s(q^2)N_c}{\pi}\chi(\gamma),$$

$$\chi(\gamma) = \chi_B(\gamma) + \frac{\alpha_s N_c}{\pi}\chi^{(1)}(\gamma). \qquad (9.326)$$

Here $\chi_B(\gamma)$ is given by (9.148) and the correction $\chi^{(1)}(\gamma)$ is:

$$\chi^{(1)}(\gamma) = -\frac{1}{4}\left[\frac{\beta_0}{2N_c}\left(\chi_B^2(\gamma) + \chi_B'(\gamma) - 6\zeta(3) \right) \right.$$

$$+ \frac{\pi^2\cos(\pi\gamma)}{\sin^2(\pi\gamma)(1-2\gamma)}\left(3 + \left(1 + \frac{n_f}{N_c^3} \right)\frac{2+3\gamma(1-\gamma)}{(3-2\gamma)(1+2\gamma)} \right)$$

$$\left. - \left(\frac{67}{9} - \frac{\pi^2}{3} - \frac{10}{9}\frac{n_f}{N_c} \right)\chi_B(\gamma) + \chi_B''(\gamma)) - \frac{\pi^3}{\sin(\pi\gamma)} + 4\phi(\gamma) \right], \qquad (9.327)$$

where

$$\phi(\gamma) = -\int_0^1 \frac{dx}{1+x}\left(x^{\gamma-1} + x^{-\gamma} \right)\int_x^1 \frac{dt}{t}\ln(1-t)$$

$$= \sum_{n=0}^\infty (-1)^n \left[\frac{\psi(n+1+\gamma) - \psi(1)}{(n+\gamma)^2} + \frac{\psi(n+2-\gamma) - \psi(1)}{(n+1-\gamma)^2} \right]. \qquad (9.328)$$

Note that almost all the terms in $\chi^{(1)}(\gamma)$ except the contribution $-\beta_0/(8N_c)\chi_B'(\gamma)$ are symmetric with respect to the transformation $\gamma \leftrightarrow 1-\gamma$. It is possible to remove this contribution [152] redefining the function $l^{2(\gamma-1)}$ by including in it the logarithmic factor $(\alpha_s(l^2)/\alpha_s(\mu^2))^{-1/2}$.

For the relative correction $r(\gamma)$ defined by $\chi^{(1)}(\gamma) = -r(\gamma)\chi_B(\gamma)$ in the symmetrical point $\gamma = 1/2$, corresponding to the largest eigenvalue of the leading order kernel, we have

$$
r\left(\frac{1}{2}\right) = \left(\frac{11}{6} - \frac{n_f}{3N_c}\right)\ln 2 - \frac{67}{36} + \frac{\pi^2}{12} + \frac{5}{18}\frac{n_f}{N_c} + \frac{1}{\ln 2}\left[\int_0^1 \arctan(\sqrt{t})\ln\left(\frac{1}{1-t}\right)\frac{dt}{t}\right.
$$

$$
\left. + \frac{11}{8}\zeta(3) + \frac{\pi^3}{32}\left(\frac{27}{16} + \frac{11}{16}\frac{n_f}{N_c^3}\right)\right] \simeq 6.46 + 0.05\frac{n_f}{N_c} + 0.96\frac{n_f}{N_c^3}. \tag{9.329}
$$

This shows that the correction is very large.

The great value of the correction became the subject of intensive discussion (see, for example, [154]–[162]) just after appearance of the result. Some authors paid attention to various problems related to the next-to-leading order BFKL (negative values for the gluon-splitting function $P_{gg}(x, \alpha_s)$ [154], which was considered an incompatibility of the BFKL approach with the Q^2 evolution of structure functions and absence of any phenomenological relevance in the $\ln(1/x)$ summation [155]; appearance of a cut in the complex j plane along the whole real axis, which makes questionable the use of the complex angular momentum variable [156]). There was an attempt [157] to overcome the problem of the large and negative correction to the pomeron intercept using the fact that the corrected eigenvalue function (9.326) completely changes its ν-dependence (see Fig. 9.11). In the leading order, the point $\nu = 0$ corresponds to the maximal eigenvalue of the kernel (pomeron intercept). In the next-to-leading order, this is true only at very small α_s ($\alpha_s \leq 0.05$). For values of α_s above 0.5 instead of having a single maximum at $\nu = 0$ the function $\omega(q^2, \gamma)$, $\gamma = 1/2 + i\nu$, has a local minimum here. Near $\nu = 0$, we have

$$
\omega(q^2, \gamma) \simeq \omega_0 + a\nu^2 - b\nu^4, \tag{9.330}
$$

where for $\alpha_s = 0.15$ $\omega_0 = 0.021$, $a = 4.19$, $b = 47.4$. The maximum of the eigenfunction is at $\nu^2 = a/2b$ and

$$
\sigma \sim s^{(\omega_0 + a^2/4b)} \simeq s^{0.12}. \tag{9.331}
$$

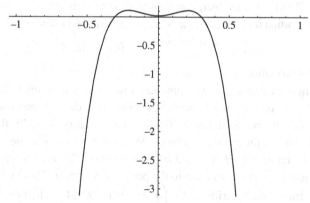

Fig. 9.11. Dependence $\omega(q^2, 1/2 + i\nu)$ on ν for $\alpha_s = 0.15$.

However, the solution obtained in [157] contains oscillations which can lead to negative cross sections. The oscillation were confirmed in [158] with the conclusion that the next-to-leading order BFKL has a serious pathology.

In [159], the iteration formalism for the complete next-to-leading order kernel was developed and the resummed Green's function was obtained. It was discovered that the running coupling part of the kernel leads to a non-Regge term in the energy dependence of high-energy hard scattering:

$$G(q_1, q_2) \simeq \frac{1}{2\pi \sqrt{\pi D q_1^2 q_2^2 Y}} \exp\left[\omega_P Y - \frac{(\ln q_1^2 - \ln q_2^2)^2}{4DY} + \frac{D}{3}\left(\alpha_s(\mu^2)\omega_P^B\right)^2 Y^3\right],$$

(9.332)

where D is the diffusion coefficient (9.161), so that the Regge asymptotics is valid only in a limited region of energy $Y \leq (\alpha_s)^{-5/3}$.

Other authors tried to make the next-to-leading order BFKL sensible. It is necessary for this to reduce the relative value of the next-to-leading order correction, i.e. to rearrange the perturbation expansion in some way. One of the first attempts [160] to do it was based on a suggestion of Lipatov to limit from below the relative rapidities of produced particles in the LLA. The idea was to include in the LLA only the interval of rapidity where correlations in the hadron production processes become unimportant. Evidently it reduces the leading order intercept and consequently the next-to-leading order correction. Unfortunately, this approach did not gain further development.

There is a possibility to reduce the correction by the choice of an appropriate renormalization scheme and scale setting [161]. As is well known (see Chapter 1), any physical value calculated in a fixed order of perturbation theory depends on a renormalization scheme and scale taken for a coupling. In the eigenvalue function (9.326) $\chi^{(1)}(\gamma)$ is scheme and scale dependent. In the non-Abelian physical renormalization schemes (such as the Υ-scheme) with the BLM scale setting

$$\chi_1^{BLM}(\gamma) = \chi^{(1)}(\gamma) - (\beta_0\text{-dependent terms})$$

(9.333)

the corresponding corrections are not large; this gives an opportunity to apply the next-to-leading order BFKL to high-energy phenomenology (such as $\gamma^*\gamma^*$ scattering) [161]. Moreover, the conformal invariance of the NLA is approximately conserved in this scheme:

$$\omega_P^{next-to-leading\,order} \simeq 0.13 \div 0.18$$

(9.334)

in a wide range of virtualities.

To date, the most extended are two approaches to small x resummation. One of them [162]–[169] is based on the BFKL framework and uses the renormalization group to improve the BFKL eigenvalue function. The approach was initiated by the observation [162] that a large part of the next-to-leading order correction to ω_P comes from collinear region. Indeed, the terms with $\psi''(\gamma)$ and $\psi''(1 - \gamma)$ in (9.327) are responsible for about half the next-to-leading order correction to the pomeron intercept. These terms are associated with double transverse logarithms $(\ln^2\left(q^2/l^2\right)$ in (9.324)). Their resummation with the help of the renormalization group considerably improves properties of the kernel.

Another approach [170]–[176] is based on the DGLAP framework with use of the BFKL results for improvement of the anomalous dimensions in the region of j close to 1. Now there is a complete agreement between these two approaches.

Until recently, the small x resummation was not necessary for DIS phenomenology, because next-to-leading order DGLAP description of data was quite satisfactory. The reason is the absence of the second and third order terms in the expansion (9.171). However, with the advent of next-to-next-to-leading order [177] the resummation became necessary.

As well as in the leading order (see subsection 9.5.3), the eigenvalue function $\omega(q^2\gamma)$ (9.326) can be applied to the calculation of anomalous dimensions of the twist-2 operators in the vicinity of the point $\omega = j - 1 = 0$. However, here we meet a complication. In the LLA, the energy scale s_0 entering in the Green's function (9.112) (recall that $Y = \ln(s/s_0)$) and in the discontinuity (9.107) is optional. It is not so in the NLA. As it is seen from the general representation (9.254) at $n = 0$, $i = 0$, $j = n+1$, in the NLA $s_0 = Q_A Q_B$, where Q_A and Q_B are typical momenta for the impact factors Φ_A and Φ_B. Therefore for the deep inelastic moments (9.166) instead of (9.167) we obtain

$$M_\omega(Q^2) = \frac{1}{4(2\pi)^2} \int \frac{dq}{q^2} \int \frac{dl}{l^2} \Phi_A(q) e^{\omega \ln(Q/Q_B)} G_\omega(q,l) \Phi_B(l), \qquad (9.335)$$

where G_ω after averaging over the azimuthal angle takes the form (9.168), with $\alpha_s = \alpha_s(q^2)$ and $\chi(\gamma)$ given by (9.326). As the result, for the anomalous dimension $\gamma_\omega(\alpha_s)$ one obtains

$$\gamma_\omega = \frac{\omega}{2} + \chi^{-1}\left(\frac{\pi\omega}{N_c \alpha_s}\right), \qquad (9.336)$$

where χ^{-1} is the inverse function. In other words, the anomalous dimensions $\gamma_\omega(\alpha_s)$ of the twist-2 operators near the point $\omega = 0$ are determined from the solution of the equation

$$\omega = \frac{\alpha_s N_c}{\pi} \chi\left(\gamma - \frac{\omega}{2}\right) \simeq \frac{\alpha_s N_c}{\pi} \chi(\gamma) - \frac{\alpha_s^2 N_c^2}{2\pi^2} \chi(\gamma)\chi'(\gamma)$$

$$\simeq \frac{\alpha_s N_c}{\pi} \frac{1}{\gamma} - \frac{\alpha_s^2 N_c^2}{4\pi^2}\left(\frac{11 + 2n_f/N_c^3}{3\gamma^2} + \frac{n_f(10 + 13/N_c^2)}{9\gamma N_c}\right.$$

$$\left. + \frac{395}{27} - 2\zeta(3) - \frac{11}{3}\frac{\pi^2}{6} + \frac{n_f}{N_c^3}\left(\frac{71}{27} - \frac{\pi^2}{9}\right) + \mathcal{O}(\gamma)\right) \qquad (9.337)$$

for $\gamma \to 0$. This equation was derived in [152] and used to reproduce the known results and predict the higher-loop correction for $\omega \to 0$:

$$\gamma \simeq \frac{\alpha_s N_c}{\pi}\left(\frac{1}{\omega} - \frac{11}{12} - \frac{n_f}{6N_c^3}\right) - \left(\frac{\alpha_s}{\pi}\right)^2 \frac{n_f N_c}{6\omega}\left(\frac{5}{3} + \frac{13}{6N_c^2}\right)$$

$$- \frac{1}{4\omega^2}\left(\frac{\alpha_s N_c}{\pi}\right)^3\left(\frac{395}{27} - 2\zeta(3) - \frac{11}{3}\frac{\pi^2}{6} + \frac{n_f}{N_c^3}\left(\frac{71}{27} - \frac{\pi^2}{9}\right)\right). \qquad (9.338)$$

The result for the three-loop correction was confirmed later in [177],[178].

The representation (9.324) is useful for finding all eigenvalues of the kernel. Defining

$$\int d^2l \, K(q,l) \left(\frac{l^2}{q^2}\right)^{\gamma-1} e^{in(\phi_l-\phi_q)} = \frac{\alpha_s(q^2) N_c}{\pi} \left(\chi_B(\gamma,n) + \frac{\alpha_s N_c}{\pi} \chi^{(1)}(\gamma,n)\right), \tag{9.339}$$

where $\chi_B(\gamma,n)$ is given by (9.153), and using the integrals

$$\int \frac{dl}{\pi(q-l)^2} \left[\left(\frac{l^2}{q^2}\right)^{\gamma-1} e^{in(\phi_l-\phi_q)} - \frac{q^2}{l^2}\right] \ln\left(\frac{q^2}{(q-l)^2}\right) = -\frac{1}{2}\left(\chi_B^2(\gamma,n) + \chi_B'(\gamma,n)\right), \tag{9.340}$$

$$\int \frac{dl}{\pi(q-l)^2} \left(\frac{l^2}{q^2}\right)^{\gamma-1} e^{in(\phi_l-\phi_q)} \ln^2\left(\frac{q^2}{l^2}\right) = \chi_B''(\gamma,n)), \tag{9.341}$$

$$\int \frac{dl}{\pi} \left(\frac{l^2}{q^2}\right)^{\gamma-1} e^{in(\phi_l-\phi_q)} f_2(q,l) = F(n,\gamma), \tag{9.342}$$

where

$$F(n,\gamma) = \frac{\pi^2 \cos(\pi\gamma)}{\sin^2(\pi\gamma)(1-2\gamma)} \left[\frac{\gamma(1-\gamma)(\delta_{n,2}+\delta_{n,-2})}{2(3-2\gamma)(1+2\gamma)}\right.$$
$$\times \left(1+\frac{n_f}{N_c^3}\right) - \left(\frac{3\gamma(1-\gamma)+2}{(3-2\gamma)(1+2\gamma)}\left(1+\frac{n_f}{N_c^3}\right)+3\right)\delta_{n,0}\bigg], \tag{9.343}$$

and

$$\int \frac{dl}{\pi} \left(\frac{l^2}{q^2}\right)^{\gamma-1} e^{in(\phi_l-\phi_q)} f_1(q,l) = -2\left(\Phi(n,\gamma) + \Phi(n,1-\gamma)\right), \tag{9.344}$$

where

$$\Phi(n,\gamma) = \int_0^1 \frac{dt}{1+t} t^{\gamma-1+n/2} \left\{\frac{\pi^2}{12} - \frac{1}{2}\psi'\left(\frac{n+1}{2}\right) - Li_2(t) - Li_2(-t)\right.$$
$$- \left(\psi(n+1) - \psi(1) + \ln(1+t) + \sum_{k=1}^{\infty} \frac{(-t)^k}{k+n}\right) \ln t - \sum_{k=1}^{\infty} \frac{t^k}{(k+n)^2}\left[1-(-1)^k\right]\bigg\}, \tag{9.345}$$

we come to the correction obtained in [179]:

$$4\chi^{(1)}(\gamma,n) = -\frac{\beta_0}{2N_c}\left(\chi_B^2(\gamma,n) + \chi_B'(\gamma,n)\right) + 6\zeta(3) - \chi_B''(\gamma,n)$$

$$+ \left(\frac{67}{9} - \frac{\pi^2}{3} - \frac{10}{9}\frac{n_f}{N_c}\right)\chi_B(\gamma,n) - 2\Phi(n,\gamma) - 2\Phi(n,1-\gamma) + F(n,\gamma). \tag{9.346}$$

The integrals (9.340)–(9.345) can be calculated using the expansions

$$\frac{1-t^2}{1-2tz+t^2}\bigg|_{|t|<1} = 1 + 2\sum_{n=1}^{\infty} t^n T_n(z), \quad \ln\left(1-2tz+t^2\right)\bigg|_{|t|<1} = -2\sum_{n=1}^{\infty} \frac{t^n}{n} T_n(z), \tag{9.347}$$

where $T_n(z)$ are the Chebyshev polynomials with the properties

$$2T_n(z)T_m(z) = T_{n+m}(z) + T_{|n-m|}(z), \quad T_0(z) = 1, \quad T_{-n}(z) = T_n(z),$$

$$\int_{-\pi}^{\pi} \frac{d\phi}{\pi} e^{in\phi} T_m(\cos\phi) = 2\int_0^{\pi} \frac{d\phi}{\pi} e^{in\phi} T_m(\cos\phi) = \delta_{nm}(1 + \delta_{n0}). \tag{9.348}$$

9.7.8 Appendix

The result of integration in (9.300) is written [119] in the form:

$$K_{GG}^{(s)}(\boldsymbol{q}_1, \boldsymbol{q}_1'; \boldsymbol{q}) = \frac{\alpha_s^2 N_c^2}{4(2\pi)^3} \left\{ \left[\left(q_2^2 \left(\frac{4q_1^2 q_1'^2 - (k^2 - q_1^2 - q_1'^2)^2}{k^2} I(k^2, q_1^2, q_1'^2) \right. \right. \right. \right.$$

$$- \frac{k^2 + q_1'^2 - q_1^2}{k^2} \ln\left(\frac{k^2}{q_1'^2}\right) \ln\left(\frac{q_1^2}{q_1'^2}\right) \right) + q^2 \ln\left(\frac{q_1^2}{q^2}\right) \ln\left(\frac{q_2^2}{q^2}\right)$$

$$- (q_1^2 + q_2^2 - q^2) \left(\frac{100}{9} - 2\zeta(2) \right) + \frac{11}{3} \left(q_1^2 \ln\left(\frac{q_1^2}{k^2}\right) + q_2^2 \ln\left(\frac{q_2^2}{k^2}\right) - q^2 \ln\left(\frac{q^2}{k^2}\right) \right)$$

$$\left. - 4J(\boldsymbol{q}_1, \boldsymbol{q}_1'; \boldsymbol{q}) \right) + \left(q_i \leftrightarrow -q_i' \right) \right] + \left[1 \leftrightarrow 2 \right] \right\}, \tag{9.349}$$

where $1 \leftrightarrow 2$ means $\boldsymbol{q}_1 \leftrightarrow \boldsymbol{q}_2$, $\boldsymbol{q}_1' \leftrightarrow \boldsymbol{q}_2'$ (and consequently $\boldsymbol{k} \leftrightarrow -\boldsymbol{k}$; remind that $\boldsymbol{k} = \boldsymbol{q}_1 - \boldsymbol{q}_1' = -\boldsymbol{q}_2 + \boldsymbol{q}_2'$); the integral $I(a, b, c)$ is defined in (9.283), (9.284), and

$$J(\boldsymbol{q}_1, \boldsymbol{q}_1'; \boldsymbol{q}) = \int_0^1 dx \int_0^1 dz \left\{ q_1 q_2 \left((x_1 x_2 - 2) \ln\left(\frac{Q^2}{k^2}\right) + \frac{2}{x_1} \ln\left(\frac{Q^2}{Q_0^2}\right) \right) \right.$$

$$- \frac{1}{2Q^2} x_1 x_2 (q_1^2 - 2q_1 p_1)(q_2^2 + 2q_2 p_2) + \frac{2}{x_1} \left[\left(x_2 q_1 q_2 (p_1(q_2 + p_2)) - q_2^2 q_1 p_2 \right) \frac{1}{Q^2} \right.$$

$$+ \left. \left(q_2^2 (z q_1 k - (1-z) q_1 q_2) - z(1-z) q_2'^2 q_1 q_2 \right) \frac{1}{Q_0^2} \right] - \frac{1}{Q^2} \left(q_2^2 q_1 (p_1 + 2q_2) \right.$$

$$- 4x_1 q_1^2 (q_2 p_2) + q_2 q_1 (q_2 q_1 - q_2 p_1 + q_1 p_2) - 2(q_2 p_1)(q_1 p_2) + 2(q_2 p_2)(q_1 p_1) \right)$$

$$+ q_2^2 \left[\frac{1}{\mu_2^2 Q^2} \left(2\frac{x_2}{x_1} (q_1 p_2) q_2 k + x_2 (q_2 p_2)(q_1'^2 - k^2) - 2(q_1' p_2) q_1 q \right) \right.$$

$$- \frac{2}{\mu_0^2 Q_0^2} \frac{1}{x_1} (q_1 p_0) q_2 k + \frac{q_1(q_2 - k)}{x_1} \left(\frac{x_2}{p_2^2} \ln\left(\frac{Q^2}{\mu_2^2}\right) - \frac{1}{p_0^2} \ln\left(\frac{Q_0^2}{\mu_0^2}\right) \right)$$

$$+ \frac{1}{p_2^2} \left(\frac{1}{p_2^2} \ln\left(\frac{Q^2}{\mu_2^2}\right) + \frac{1}{Q^2} \right) \left(2\frac{x_2}{x_1} (q_1 p_2)(k - q_2) p_2 + 2((x_2 q_2 - q_1') p_2) q_1 p_2 \right)$$

$$-\frac{1}{p_0^2}\left(\frac{1}{p_0^2}\ln\left(\frac{Q_0^2}{\mu_0^2}\right)+\frac{1}{Q_0^2}\right)\left(2\frac{1}{x_1}(q_1p_0)(k-q_2)p_0\right)-\frac{(x_2q_2-q_1')q_1}{p_2^2}\ln\left(\frac{Q^2}{\mu_2^2}\right)$$

$$-\frac{q_1^2}{d}\left((q_1'k)(q_2'k)\left(\frac{Q^2}{d}\mathcal{L}-\frac{1}{k^2}\right)+(q_1'p_2)(q_2'k)\left(\frac{1}{\mu_2^2}-\frac{\mu_1^2}{d}\mathcal{L}\right)+(q_1'k)(q_2'p_1)\right.$$

$$\left.\times\left(\frac{1}{\mu_1^2}-\frac{\mu_2^2}{d}\mathcal{L}\right)+(q_1'p_2)(q_2'p_1)\left(\frac{k^2}{d}\mathcal{L}-\frac{1}{Q^2}\right)+\frac{(q_1'q_2')}{2}\mathcal{L}\right)\right\}.\qquad(9.350)$$

Here we use the following notations:

$$p_1=zxq_1+(1-z)(xk+(1-x)q_2'),\quad p_2=z((1-x)k-xq_1')-(1-z)(1-x)q_2;$$

$$p_1+p_2=k,\ Q^2=x(1-x)(q_1^2z+q_2^2(1-z))+z(1-z)(q_1'^2x+q_2'^2(1-x)-q^2x(1-x)),$$

$$\mu_i^2=Q^2+p_i^2,\ p_0=zk-(1-z)q_2;\quad Q_0^2=z(1-z)q_2'^2,\quad \mu_0^2=zk^2+(1-z)q_2^2,$$

$$d=\mu_1^2\mu_2^2-k^2Q^2=z(1-z)x(1-x)\left((k^2-q_1^2-q_2'^2)(k^2-q_1'^2-q_2^2)+k^2q^2\right)$$

$$+q_1^2q_1'^2xz(x+z-1)+q_2^2q_2'^2(1-x)(1-z)(1-x-z),\ \mathcal{L}=\ln\left(\frac{\mu_1^2\mu_2^2}{k^2Q^2}\right).\quad(9.351)$$

In pure gluodynamics, the real part of the colour-singlet kernel $K_r^{(0)}$ can be written in the limit $D=4+2\epsilon\to4$ as the sum of two parts:

$$K_r=K_r^{sing}+K_r^{(reg)}.\qquad(9.352)$$

Here the first contains all singularities:

$$K_r^{sing}(q_1,q_2;\vec{q})=\frac{2\bar{g}_\mu^2\mu^{-2\epsilon}}{\pi^{1+\epsilon}\Gamma(1-\epsilon)}\left(\frac{q_1^2q_2'^2+q_1'^2q_2^2}{k^2}-q^2\right)\left\{1+\bar{g}_\mu^2\left[\frac{11}{3\epsilon}\right.\right.$$

$$\left.\left.+\left(\frac{k^2}{\mu^2}\right)^\epsilon\left\{-\frac{11}{3\epsilon}+\frac{67}{9}-2\zeta(2)+\epsilon\left(-\frac{404}{27}+14\zeta(3)+\frac{11}{3}\zeta(2)\right)\right\}\right]\right\},\qquad(9.353)$$

and the second, putting $\epsilon=0$ and $\bar{g}_\mu^2=\alpha_s(\mu^2)N_c/(4\pi)$, is given by

$$K_r^{reg}(q_1,q_1';q)=\frac{\alpha_s^2(\mu^2)N_c^2}{16\pi^3}\left[2(q_1^2+q_1'^2-q^2)\left(\zeta(2)-\frac{50}{9}\right)+\frac{11}{3}\left(q_1^2\ln\left(\frac{q_1^2}{k^2}\right)\right)\right.$$

$$+q_1'^2\ln\left(\frac{q_1'^2}{k^2}\right)+q^2\ln\left(\frac{q_1^2q_1'^2}{q^4}\right)+\frac{q_1^2q_2'^2-q_1'^2q_2^2}{k^2}\ln\left(\frac{q_1^2}{q_1'^2}\right)\right)+q^2\left(\ln\left(\frac{q_1^2}{q^2}\right)\ln\left(\frac{q_2^2}{q^2}\right)\right.$$

$$+\ln\left(\frac{q_1'^2}{q^2}\right)\ln\left(\frac{q_2'^2}{q^2}\right)+\frac{1}{2}\ln^2\left(\frac{q_1^2}{q_1'^2}\right)\right)+\ln\left(\frac{q_1^2}{q_1'^2}\right)\left(\frac{q_2^2}{2}\ln\left(\frac{q_1'^2}{k^2}\right)-\frac{q_2'^2}{2}\ln\left(\frac{q_1^2}{k^2}\right)\right.$$

$$-\frac{q_1^2 q_2'^2 + q_1'^2 q_2^2}{2k^2} \ln\left(\frac{q_1^2}{q_1'^2}\right) + \frac{q_2^2(q_1^2 - 3q_1'^2)}{2k^2} \ln\left(\frac{k^2}{q_1'^2}\right) + \frac{q_2'^2(3q_1^2 - q_1'^2)}{2k^2} \ln\left(\frac{k^2}{q_1^2}\right)\right)$$

$$+ \left(q^2(k^2 - q_1^2 - q_1'^2) + 2q_1^2 q_1'^2 - \frac{(q_1^2 - q_1'^2)(q_1^2 + q_1'^2)(q_2^2 - q_2'^2)}{2k^2} + q_1^2 q_2^2 + q_1'^2 q_2'^2\right.$$

$$\left. - \frac{k^2}{2}(q_2^2 + q_2'^2)\right) I(k^2, q_1'^2, q_1^2) - 2J(q_1, q_1'; q) - 2J(-q_1', -q_1; -q)\right] + \left\{1 \leftrightarrow 2\right\}.$$

$$(9.354)$$

References

[1] Fadin, V. S., Kuraev, E. A. and Lipatov, L. N., *Phys. Lett.* **B60** (1975) 50.

[2] Kuraev, E. A., Lipatov, L. N. and Fadin, V. S., *Sov. Phys. JETP* **44** (1976) 443 [*Zh. Eksp. Teor. Fiz.* **71** (1976) 840].

[3] Kuraev, E. A., Lipatov, L. N. and Fadin, V. S., *Sov. Phys. JETP* **45** (1977) 199 [*Zh. Eksp. Teor. Fiz.* **72** (1977) 377].

[4] Balitsky, I. I. and Lipatov, L. N., *Sov. J. Nucl. Phys.* **28** (1978) 822 [*Yad. Fiz.* **28** (1978) 1597].

[5] Gribov, V. N. and Lipatov, L. N. , *Sov. J. Nucl. Phys.* **15** (1972) 438 [*Yad. Fiz.* **15** (1972) 781].

[6] Gribov, V. N. and Lipatov, L. N. , *Sov. J. Nucl. Phys.* **15** (1972) 675 [*Yad. Fiz.* **15** (1972) 1218].

[7] Lipatov, L. N. , *Sov. J. Nucl. Phys.* **20** (1975) 94 [*Yad. Fiz.* **20** (1974) 181].

[8] Altarelli, G. and Parisi, G. , *Nucl. Phys.* **B126** (1977) 298.

[9] Dokshitzer, Y. L. , *Sov. Phys. JETP* **46** (1977) 641 [*Zh. Eksp. Teor. Fiz.* **73** (1977) 1216].

[10] I. Abt, I. *et al.* [H1 Collaboration], *Nucl. Phys.* **B407** (1993) 515.

[11] Froissart, M., *Phys. Rev.* **123** (1961) 1053.

[12] Gribov, L. V., Levin, E. M. and Ryskin, M. G., *Phys. Rept.* **100** (1983) 1.

[13] Lipatov, L. N., *Nucl. Phys.* **B452** (1995) 369.

[14] Lipatov, L. N., *Phys. Rept.* **286** (1997) 131.

[15] Ciafaloni, M., *Nucl. Phys.* **B296** (1988) 49.

[16] Catani, S., Fiorani, F. and Marchesini, G., *Phys. Lett.* **B234** (1990) 339.

[17] Catani, S., Fiorani, F. and Marchesini, G., *Nucl. Phys.* **B336** (1990) 18.

[18] Marchesini, G., *Nucl. Phys.* **B445** (1995) 49.

[19] Salam, G. P., *JHEP* **9903** (1999) 009.

[20] Salam, G. P., *Nucl. Phys. Proc. Suppl.* **79** (1999) 426.

[21] Gell-Mann, M., Goldberger, M. L., Low, F. E., Marx, E. and Zachariasen, F., *Phys. Rev.* **B133** (1964) 145.

[22] Regge, T., *Nuovo Cim.* **14** (1959) 951.

[23] Gribov, V. N., *Sov. Phys. JETP* **14** (1962) 1395 [*Zh. Eksp. Teor. Fiz.* **41** (1961) 1962].

[24] Chew, G. F. and Frautschi, S. C., *Phys. Rev. Lett.* **7** (1961) 394.

[25] Gribov, V. N., Mur, V. D., Kobzarev, I. Y., Okun, L. B. and Popov, V. S., *Sov. J. Nucl. Phys.* **13** (1971) 381 [*Yad. Fiz.* **13** (1971) 670].

[26] Lukaszuk, L. and Nicolescu, B., *Lett. Nuovo Cim.* **8** (1973) 405.

[27] Mandelstam, S., *Phys. Rev.* **B137** (1964) 949.

[28] Grisaru, M. T. and Schnitzer, H. J., *Phys. Lett.* **30** (1973) 811.

[29] Grisaru, M. T., Schnitzer, H. J. and Tsao, H. S., *Phys. Rev.* **D8** (1973) 4498 .

[30] Lipatov, L. N., *Sov. J. Nucl. Phys.* **23** (1976) 338 [*Yad. Fiz.* **23** (1976) 642].

[31] Fadin, V. S. and Sherman, V. E., *Zh. Eksp. Teor. Fiz. Pis'ma* **23** (1976) 599.

[32] Fadin, V. S. and Sherman, V. E., *Zh. Eksp. Teor. Fiz.* **72** (1977) 1640.

[33] Bogdan, A. V., Del Duca, V., Fadin, V. S. and Glover, E. W. N., *JHEP* **0203** (2002) 032.

[34] Bartels, J., *Phys. Rev.* **D11** (1975) 2977.

[35] Bartels, J., *Phys. Rev.* **D11** (1975) 2989.

[36] Bartels, J., *Nucl. Phys.* **B175** (1980) 365.

[37] Fadin, V. S. and Lipatov, L. N., *Nucl. Phys.* **B406** (1993) 259.

[38] Fadin, V. S. and Fiore, R. *Phys. Lett.* **B440** (1998) 359.

[39] Balitskii, Ya.Ya., Lipatov, L.N. and Fadin, V.S., in *Materials of IV Winter School of LNPI* (Leningrad, 1979) p. 109.

[40] Lipatov, L. N., *Adv. Ser. Direct. High Energy Phys.* **5** (1989) 411.

[41] Fadin, V. S., Fiore, R., Kozlov, M. G. and Reznichenko, A. V., *Phys. Lett.* **B639** (2006) 74.

[42] Fadin, V. S., Fiore, R., and Quartarolo, A., *Phys. Rev.* **D50** (1994) 2265.

[43] Fadin, V. S., and Fiore, R., *Phys. Lett.* **B294** (1992) 286.

[44] Fadin, V. S., Fiore, R., and Quartarolo, A., *Phys. Rev.* **D50** (1994) 5893.

[45] Fadin, V. S., Fiore, R., and Kotsky, M. I., *Phys. Lett.* **B389** (1996) 737.

[46] Del Duca, V. and Schmidt, C. R., *Phys. Rev.* **D59** (1999) 074004.

[47] Fadin, V. S., Fiore, R. and Papa, A., *Phys. Rev.* **D63** (2001) 034001.

[48] Fadin, V. S., *JETP Lett.* **61** (1995) 346 [*Pisma Zh. Eksp. Teor. Fiz.* **61** (1995) 342].

[49] Fadin, V. S., Fiore, R., and Quartarolo, A., *Phys. Rev.* **D53** (1996) 2729.

[50] Kotsky, M. I. and Fadin, V. S., *Phys. Atom. Nucl.* **59** (1996) 1035 [Yad. Fiz. **59** (1996) 1080].

[51] Fadin, V. S., Kotsky, M. I., and Fiore, R., *Phys. Lett.* **B359** (1995) 181.

[52] Fadin, V. S., Fiore, R. and Kotsky, M. I., *Phys. Lett.* **B387** (1996) 593.

[53] Blumlein, J., Ravindran, V. and van Neerven, W. L., *Phys. Rev.* **D58** (1998) 091502.

[54] Del Duca, V. and Glover, E. W. N., *JHEP* **0110** (2001) 035 [arXiv:hep-ph/0109028].

[55] Fadin, V. S., Fiore, R., Kotsky, M. I. and Papa, A., *Phys. Rev.* **D61** (2000) 094006.

[56] Fadin, V. S., Fiore, R., Kotsky, M. I. and Papa, A., *Phys. Rev.* **D61** (2000) 094005.

[57] Fadin, V. S. and Lipatov, L. N., *Nucl. Phys.* **B477** 767 (1996).

[58] Fadin, V. S., Kotsky, M. I., Fiore, R. and Flachi, A., *Phys. Lett.* **422B** (1998) 287.

[59] Fadin, V. S., Kotsky, M. I., Fiore, R. and Flachi, A., *Phys. Atom. Nucl.* **62** (1999) 999 [Yad. Fiz. **62** (1999) 1066]

[60] Fadin, V. S. and Lipatov, L. N., *Zh. Eksp. Teor. Fiz. Pis'ma* **49** (1989) 311 [*Sov. Phys. JETP Lett.* **49** (1989) 352]; *Yad. Fiz.* **50** (1989) 1141 [*Sov. J. Nucl. Phys.* **50** (1989) 712].

[61] Steinmann, O., *Helv. Phys. Acta* **33** (1960) 33 .

[62] Fadin, V. S., Kozlov, M. G. and Reznichenko, A. V., *Phys. Atom. Nucl.* **67** (2004) 359 [*Yad. Fiz.* **67** (2004) 377].

[63] Fadin, V. S., Kotsky, M. I. and Lipatov, L. N., *Phys. Lett.* **B415**(1997) 97 ; *Yad. Fiz.* **61** (1998) 716.

[64] Fadin, V. S. arXiv:hep-ph/9807528.

[65] Bartels, J. and Lotter, H., *Phys. Lett.* **B30B** (1993) 400.

[66] Lipatov, L. N., *Sov. Phys. JETP* **63** (1986) 904 [*Zh. Eksp. Teor. Fiz.* **90** (1986) 1536].

[67] Kwiecinski, J., *Z. Phys.* **C29** (1985) 561.

[68] Collins, J. C. and Kwiecinski, J., *Nucl. Phys.* **B316** (1989) 307.

[69] Hancock, R. E. and Ross, D. A., *Nucl. Phys.* **B383** (1992) 575.

[70] Haakman, L. P. A., Kancheli, O. V. and Koch, J. H., *Nucl. Phys.* **B518** (1998) 275.

[71] Camici, G. and Ciafaloni, M., *Phys. Lett.* **B395** (1997) 118.

[72] Thorne, R. S., *Phys. Rev.* **D60** (1999) 054031.

[73] Ciafaloni, M., Taiuti, M. and Mueller, A. H., *Nucl. Phys.* **B616** (2001) 349.

[74] Jaroszewicz, T., *Phys. Lett.* **B116** (1982) 291.

[75] Mueller, A. H. and Qiu, J. w., *Nucl. Phys.* **B 268** (1986) 427.

[76] Nikolaev, N. N. and Zakharov, B. G., *Z. Phys.* **C64** (1994) 631.

[77] Nikolaev, N. N., Zakharov, B. G. and Zoller, V. R., *JETP Lett.* **59** (1994) 6.

[78] Mueller, A. H., *Nucl. Phys.* **B415** (1994) 373.

[79] Mueller, A. H. and Patel, B., *Nucl. Phys.* **B425** (1994) 471.

[80] Balitsky, I., *Nucl. Phys.* **B463** (1996) 99.

[81] Kovchegov, Yu., *Phys. Rev.* **D60** (1999) 034008.

[82] Levin, E. and Tuchin, K., *Nucl. Phys.* **A691** (2001) 779.

[83] Stasto, A. M., Golec-Biernat, K. J. and Kwiecinski, J., *Phys. Rev. Lett.* **86** (2001) 596.

[84] Iancu, E., Itakura, K. and McLerran, L., *Nucl. Phys.* **A708** (2002) 327.

[85] Mueller, A. H. and Triantafyllopoulos, D. N., *Nucl. Phys.* **B640** (2002) 331.

[86] McLerran, L. D. and Venugopalan, R., *Phys. Rev.* **D49** (1994) 2233.

[87] McLerran, L. D. and Venugopalan, R., *Phys. Rev.* **D49** (1994) 3352.

[88] McLerran, L. D. and Venugopalan, R., *Phys. Rev.* **D50** (1994) 2225.

[89] Jalilian-Marian, J., Kovner, A., Leonidov, A. and Weigert, H., Nucl. Phys. B **504** (1997) 415 *Nucl. Phys.* **B504** (1997) 415.

[90] Jalilian-Marian, J., Kovner, A., Leonidov, A. and Weigert, H., *Phys. Rev.* **D59** (1999) 014015.

[91] Jalilian-Marian, J., Kovner, A., Leonidov, A. and Weigert, H., *Phys. Rev.* **D59** (1999) 014014.

[92] Iancu, E., Leonidov, A. and McLerran, L., *Nucl. Phys.* **A692** (2001) 583.

[93] Iancu, E., Leonidov, A. and McLerran, L., *Phys. Lett.* **B510** (2001) 133.

[94] Ferreiro, E., Iancu, E. Leonidov, A. and McLerran, L., *Nucl. Phys.* **A703** (2002) 489.

[95] Kovner, A., Milhano, J. G. and Weigert, H., *Phys. Rev.* **D62** (2000) 114005.

[96] Weigert, H., *Nucl. Phys. Rev.* **A703** (2000) 823.

[97] Iancu E. and Mueller, A. H., *Nucl. Phys.* **A730** (2004) 494.

[98] Mueller, A. H. and Shoshi, A. I., *Nucl. Phys.* **B692** (2004) 175.

[99] Mueller, A. H., Shoshi, A. I. and Wong, S. M. H., *Nucl. Phys.* **B715** (2005) 440.

[100] Iancu E., Mueller, A. H. and Munier, S., *Phys. Lett.* **B606** (2005) 342.

[101] Iancu, E. and Triantafyllopoulos, D. N., *Nucl. Phys.* **A756** (2005) 419.

[102] Iancu, E. and Triantafyllopoulos, D. N., *Phys. Lett.* **B610** (2005) 253.

[103] Levin, E. and Lublinsky, M., *Nucl. Phys.* **A763** (2005) 172.

[104] Blaizot, J. P., Iancu, E., Itakura, K. and Triantafyllopoulos, D. N., *Phys. Lett.* **B615** (2005) 221.

[105] Iancu, E., Soyez, D. N. and Triantafyllopoulos, D. N., *Nucl. Phys* **A768** (2006) 194.

[106] Forshaw, J. R. and Sutton, P. J., *Eur. Phys. J.* **C1** (1998) 285.

[107] Fadin, V. S. *Prepared for NATO Advanced Research Workshop on Diffraction 2002, Alushta, Ukraine, 31 Aug - 6 Sep 2002* Published in *Alushta 2002, Diffraction 2002*, Ed. by R. Fiore *et al.*, NATO Science Series, Vol. 101, pp.235-245.

[108] Fadin, V. S., *Phys. Atom. Nucl.* **66** (2003) 2017.

[109] Braun, M. and Vacca, G. P., *Phys. Lett.* **B454** (1999) 319.

[110] Braun, M. A., arXiv:hep-ph/9901447.

[111] Fadin, V. S., Fiore, R., Kotsky, M. I. and Papa, A., *Phys. Lett.* **B495** (2000) 329.

[112] Braun, M. and Vacca, G. P., *Phys. Lett.* **B477** (2000) 156.

[113] Fadin, V. S., Fiore, R. and Papa, A., *Phys. Rev.* **D60** (1999) 074025.

[114] Fadin, V. S., Fiore, R. and Kotsky, M. I., *Phys. Lett.* **B494** (2000) 100.

[115] Fadin, V. S. and Papa, A., *Nucl. Phys.* **B640** (2002) 309.

[116] Bartels, J., Fadin, V. S. and Fiore, R., *Nucl. Phys.* **B672** (2003) 329.

[117] Fadin, V. S., Kozlov, M. G. and Reznichenko, A. V., to be pubidhed.

[118] Fadin, V. S. and Fiore, R., *Phys. Lett.* **B610** (2005) 61 [Erratum-ibid. **B621** (2005) 61].

[119] Fadin, V. S. and Fiore, R., *Phys. Rev.* **D72** (2005) 014018.

[120] Gerasimov, R. E. and Fadin, V. S., *preprint* BUDKER-INP-2008-36, 2008, 27 pp.

[121] Fadin, V. S. and Gorbachev, D. A., *JETP Lett.* **71** (2000) 222 [*Pisma Zh. Eksp. Teor. Fiz.* **71** (2000) 322].

[122] Fadin, V. S. and Gorbachev, D. A., *Phys. Atom. Nucl.* **63** (2000) 2157 [*Yad. Fiz.* **63** (2000) 2253].

[123] Gribov, V. N., Lipatov, L. N. and Frolov, G. V., *Sov. J. Nucl. Phys.* **12** (1971) 543 [*Yad. Fiz.* **12** (1970) 994].

[124] Cheng, H. and Wu, T. T., *Phys. Rev.* **D1** (1970) 2775.

[125] Fadin, V. S., Fiore, R. and Papa, A., *Nucl. Phys.* **B769** (2007) 108.

[126] Fadin, V. S., Fiore, R. and Papa, A., Phys. Lett. **B647** (2007) 179.

[127] Fadin, V. S., Fiore, R., Grabovsky, A. V. and Papa, A., *Nucl. Phys.* **B784** (2007) 49.

[128] Fadin, V. S., arXiv:hep-ph/9807527.

[129] Kovchegov, Y. V., Weigert, H., *Nucl. Phys.* **A784** (2007) 188.

[130] Balitsky, I., *Phys. Rev.* **D75** (2007) 014001.

[131] Balitsky, I. and Chirilli, G. A., *Phys. Rev.* **D77** (2008) 014019.

[132] Fadin, V. S., Fiore, R. and Grabovsky, A. V., *Nucl. Phys.* **B820** (2009) 334; to be published.

[133] Balitsky, I. and Chirilli, G. A., *Nucl. Phys.* **B822** (2009) 45.

[134] Ciafaloni, M. and Rodrigo G. , *JHEP* **0005** (2000) 042.

[135] Bartels, J., Colferai, D. and Vacca, G. P., *Eur. Phys. J.* **C24** (2002) 83.

[136] Bartels, J., Colferai, D. and Vacca, G. P., *Eur. Phys. J.* **C29** (2003) 235.

[137] Bartels, J., Gieseke, S. and Qiao, C. F., *Phys. Rev.* **D63** (2001) 056014 [Erratum-*ibid.* **D65** (2002) 079902].

[138] Bartels, J., Gieseke, S. and Kyrieleis, A., *Phys. Rev.* **D65** (2002) 014006.

[139] Bartels, J., Colferai, D., Gieseke, S. and Kyrieleis, A., *Phys. Rev.* **D66** (2002) 094017.

[140] Bartels, J., and Kyrieleis, A., *Phys. Rev.* **F70** (2004) 114003.

[141] Fadin, V. S., Ivanov, D. and Kotsky, M., *Phys. Atom. Nucl.* **65** (2002) 1513 [*Yad. Fiz.* **65** (2002) 1551].

[142] Fadin, V. S., Ivanov, D. and Kotsky, M., *Nucl. Phys.* **B658** (2003) 156.

[143] Ivanov, D., Kotsky, M. and Papa, A., *Eur. Phys. J.* **C38**, 195 (2004).

[144] Ivanov, D., Kotsky, M. and Papa, A., *Nucl. Phys. Proc. Suppl.* **146** (2005) 117.

[145] Ivanov, D. and Papa, A., *Nucl. Phys.* **B732** (2006) 183.

[146] Ivanov, D. and Papa, A., *Eur. Phys. J.* **C49** (2007) 947.

[147] Ivanov, D. and Papa, A., *Acta Phys. Polon.* **B39** (2008) 2391.

[148] Catani, S., Ciafaloni, M. and Hautmann, F., *Phys. Lett.* **B242** (1990) 97.

[149] Catani, S., Ciafaloni, M. and Hautmann, F., *Nucl. Phys.* **B366** (1991) 135.

[150] Camici, G. and Ciafaloni, M., *Phys. Lett.* **B386** (1996) 341.

[151] Camici, G. and Ciafaloni, M., *Nucl. Phys.* B **496** (1997) 305 [Erratum-ibid. **B607** (2001) 431].

[152] Fadin, V. S. and Lipatov, L. N., *Phys. Lett.* **B429** (1998) 127.

[153] Ciafaloni, M. and Camici, G., *Phys. Lett.* **B430** (1998) 349.

[154] Blumlein, J., Ravindran, V., van Neerven, W. L. and Vogt, A., arXiv:hep-ph/9806368.

[155] Ball, R. D. and Forte, S., arXiv:hep-ph/9805315.

[156] Armesto, N., Bartels J. and Braun, M. A., *Phys. Lett.* **B442** (1998) 459.

[157] Ross, D. A., *Phys. Lett.* **B431** (1998) 161.

[158] Levin, E., arXiv:hep-ph/9806228.

[159] Kovchegov, Y. V. and Mueller, A. H., *Phys. Lett.* **B439** (1998) 428.

[160] Schmidt, C. R., *Phys. Rev.* **D60** (1999) 074003.

[161] Brodsky, S. J., Fadin, V. S., Kim, V. T., Lipatov, L. N. and Pivovarov, G. B., *JETP Lett.* **70** (1999) 155.

[162] Salam, G. P., *JHEP* **9807** (1998) 019.

[163] Ciafaloni, M., Colferai, D. and Salam, G. P., *Phys. Rev.* **D60** (1999) 114036.

[164] Ciafaloni, M., Colferai, D. and Salam, G. P., *JHEP* **0007** (2000) 054.

[165] Ciafaloni, M., Colferai, D., Salam, G. P. and Stasto, A. M., *Phys. Rev.* **D66** (2002) 054014.

[166] Ciafaloni, M., Colferai, D., Salam, G. P. and Stasto, A. M., *Phys. Lett.* **B576** (2003) 143.

[167] Ciafaloni, M., Colferai, D., Salam, G. P. and Stasto, A. M., *Phys. Rev.* **D68** (2003) 114003.

[168] Ciafaloni, M., Colferai, D., Salam, G. P. and Stasto, A. M., *Phys. Lett.* **B587** (2004) 87.

[169] Ciafaloni, M., Colferai, D., Salam, G. P. and Stasto, A. M., *JHEP* **0708** (2007) 046.

[170] Ball, R. D. and Forte, S., *Phys. Lett.* **B465** (1999) 271.

[171] Altarelli, G., Ball, R. D. and Forte, S., *Nucl. Phys.* **B575** (2000) 313.

[172] Altarelli, G., Ball, R. D. and Forte, S., *Nucl. Phys.* **B599** (2001) 383.

[173] Altarelli, G., Ball, R. D. and Forte, S., *Nucl. Phys.* **B621** (2002) 359.

[174] Altarelli, G., Ball, R. D. and Forte, S., *Nucl. Phys.* **B674** (2003) 459.

[175] Altarelli, G., Ball, R. D. and Forte, S., *Nucl. Phys.* **B742** (2006) 1.

[176] Ball, R. D. and Forte, S., *Nucl. Phys.* **B742** (2006) 158.

[177] Vogt, A., Moch, S. and Vermaseren, J. A. M., *Nucl. Phys.* **B691** (2004) 129.

[178] Kotikov, A. V., Lipatov, L. N., Onishchenko, A. I. and Velizhanin, V. N., *Phys. Lett.* **B595** (2004) 521.

[179] Kotikov, A. V. and Lipatov, L. N., *Nucl. Phys.* **B582** (2000) 19.

[180] Gerasimov, R. E. and Fadin, V. S., *preprint* BUDKER-INP-2008-36, 2008, 27 pp.

[181] Balitsky, I. and Chirilli, G. A., *Phys. Rev.* **D77** (2008) 014019.

[182] Fadin, V. S., Fiore, R. and Grabovsky, A. V., *Nucl. Phys.* **B820** (2009) 334; to be published.

[183] Balitsky, I. and Chirilli, G. A., *Nucl. Phys.* **B822** (2009) 45.

10

Further developments in high-energy QCD

10.1 Effective-action approach

10.1.1 Amplitudes with the multi-Regge unitarity

In the Born approximation of QCD, the amplitude for a two-coloured particle scattering is factorized in the Regge kinematics $s \gg -t$ (see Fig. 10.1):

$$M_{AB}^{A'B'}(s,t)|_{Born} = \Gamma_{A'A}^c \frac{2s}{t} \Gamma_{B'B}^c, \qquad \Gamma_{A'A}^c = g\, T_{A'A}^c\, \delta_{\lambda_{A'}\lambda_A}, \qquad (10.1)$$

where T^c are the generators of the colour-group $SU(N_c)$ in the corresponding representation and λ_r are the helicities of the colliding and final-state particles (see Chapter 9). Helicity conservation is related to the fact that at small $t = -q^2$ the t-channel gluon interacts with the conserved colour charge. As a result, the matrix elements of this operator between different states vanish at $t = 0$ and do not depend on the helicities.

In LLA, the scattering amplitude has the Regge form [1] (see Fig. 10.2 and Chapter 9):

$$M_{AB}^{A'B'}(s,t) = M_{AB}^{A'B'}(s,t)|_{Born}\, s^{\omega(t)}, \qquad \alpha_s \ln s \sim 1, \qquad (10.2)$$

where the gluon Regge trajectory is

$$\omega(-|q|^2) = -\int \frac{d^2k}{4\pi^2} \frac{\alpha_s N_c\, |q|^2}{|k|^2|q-k|^2} \approx -\frac{\alpha_s N_c}{2\pi} \ln \frac{|q^2|}{\lambda^2}. \qquad (10.3)$$

Here the fictitious gluon mass λ is introduced to regularize the infrared divergence. This trajectory was calculated also in a two-loop approximation in QCD [2] and in supersymmetric gauge theories [3].

The gluon production at high energies can be investigated in the multi-Regge kinematics (see Fig. 10.3)

$$s \gg s_1, \quad s_2, \ldots, \quad s_{n+1} \gg -t_1, \quad -t_2, \ldots, \quad -t_{n+1}, \qquad (10.4)$$

where s_r is the square of the sum of neighbouring particle momenta k_{r-1}, k_r, and $-t_r$ is the square of the momentum transfer q_r, $r = 1, 2, \ldots, n+1$. Furthermore, the gluon-production amplitude in multi-Regge kinematics can be written in factorized form [1]:

$$M_{2 \to 1+n} = 2s\, \Gamma_{A'A}^{c_1} \frac{s_1^{\omega_1}}{|q_1|^2}\, g T_{c_2 c_1}^{d_1} C_{\mu_1}(q_2, q_1)\, e_{\mu_1} \ldots C_{\mu_{n-1}}(q_n, q_{n-1})\, e_{\mu_{n-1}} \frac{s_n^{\omega_n}}{|q_n|^2}\, \Gamma_{B'B}^{c_n}, \qquad (10.5)$$

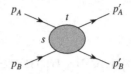

Fig. 10.1. Elastic amplitude.

Fig. 10.2. Regge pole exchange.

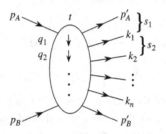

Fig. 10.3. Multi-particle production.

where the gluon–gluon–reggeon (GGR) vertices Γ contain the Kronecker symbols $\delta_{\lambda\lambda'}$ corresponding to helicity conservation for each of the colliding particles. The reggeon–reggeon–gluon (RRG) vertex is proportional to the vector

$$C(q_2, q_1) = -(q_2 + q_1)_\perp + P_A \left(\frac{kP_B}{P_A P_B} + \frac{q_1^2}{kP_A} \right) - P_B \left(\frac{kP_A}{P_A P_B} + \frac{q_2^2}{kP_B} \right). \qquad (10.6)$$

This vector is gauge invariant: $C(q_2, q_1)k_1 = 0$. Therefore, we can use arbitrary gauges for the polarization vector $e(k_1)$. There are two convenient light-cone gauges – left-hand (L) and right-hand (R) ones. The corresponding polarization vectors satisfy the conditions $P_A e_L = P_B e_R = 0$ and are expressed in terms of their transverse components

$$e_L(k) = e_L^\perp - \frac{e_L k_\perp}{P_A k} P_A, \quad e_R(k) = e_R^\perp - \frac{e_R k_\perp}{P_B k} P_B, \quad ke_L = ke_R = 0. \qquad (10.7)$$

The gauge transformation relating these two polarization vectors is given by

$$e_R = e_L - 2\frac{k_\perp e_L}{k_\perp^2} k, \quad k_\perp e_R = -k_\perp e_L. \qquad (10.8)$$

In particular, this gives the following relation between their transverse components

$$e_R^i = \left(\delta^{ir} - 2\frac{k^i k^r}{k^2} \right) e_L^r. \qquad (10.9)$$

We can introduce complex coordinates for two-dimensional vectors a

$$a = a_1 + ia_2, \quad a^* = a_1 - ia_2, \quad a_\mu^\perp b_\mu^\perp = -\frac{1}{2}(ab^* + a^*b). \tag{10.10}$$

In these complex notations, one obtains for the RRG vertex

$$C_\mu e_L^\mu(k_1) = C_L^*(q_2, q_1) e^L + C_L(q_2, q_1) e^{L*}, \quad C_L(q_2, q_1) = \frac{q_1^* q_2}{k_1^*} \tag{10.11}$$

and

$$C_\mu e_R^\mu(k_1) = C_R^*(q_2, q_1) e^R + C_R(q_2, q_1) e^{R*}, \quad C_R(q_2, q_1) = -\frac{q_1 q_2^*}{k_1^*}. \tag{10.12}$$

The above expressions are in agreement with the gauge transformations

$$e_R(k) = -\frac{k}{k^*} e_L^*(k), \quad C_L(q_2, q_1) = -\frac{k_1}{k_1^*} C_R^*(q_2, q_1). \tag{10.13}$$

It is obvious that in multi-Regge kinematics the production amplitude is significantly simplified. Moreover, with the use of the above relations and the s- and t-channel unitarity conditions one can easily construct the equations of Balitsky–Fadin–Kuraev–Lipatov (BFKL) [1] and Bartels–Kwiecinskii–Praszalowicz (BKP) [4] in a generalized LLA. However, in each step of this iterative procedure the imaginary part of the production amplitude is neglected in comparison with its real part, because one of the energy logarithms is replaced by $i\pi$ in the corresponding contribution. As a result, the Froissart bound $\sigma_t < c \ln^2 s$ for the total cross section is violated in the region $g^2 \ln s \gg 1$.

A natural idea to restore the s-channel unitarity in multi-Regge kinematics is to take account of the terms $g^2 \ln s$ and $ig^2\pi$ on equal footing in the above iteration procedure. This so-called $i\pi$ approximation was constructed explicitly for amplitudes with quasi-elastic unitarity [5].

Another possibility to unitarize the scattering amplitudes is to use an effective field theory in which the Feynman vertices coincide with the above reggeon–gluon effective vertices Γ and C [6] (see also Ref. [8]). For this purpose one can introduce new fields $A^\pm$:

$$A^\pm = A_\mp = A^0 \pm A^3, \quad P_A^\mu A_\mu = \frac{\sqrt{s}}{2} A^-, \quad P_B^\mu A_\mu = \frac{\sqrt{s}}{2} A^+, \tag{10.14}$$

which describe the production and annihilation of the reggeized gluons in the t-channel. In multi-Regge kinematics the gluon momentum components α_r, β_r of the Sudakov decomposition

$$q_r = \beta_r P_A + \alpha_r P_B + q_r^\perp$$

are strongly ordered:

$$\beta_{r-1} \gg \beta_r, \quad \alpha_r \gg \alpha_{r-1}. \tag{10.15}$$

This means that we can neglect the Sudakov component β of the reggeized gluon in propagators and vertices of the particles which emit this t-channel gluon and the Sudakov component α in corresponding quantities for the particles which absorb it. This kinematical constraint can be expressed by the relations

$$\partial_- A_+ = \partial_+ A_- = 0, \quad A_\pm = A_0 \pm A_3, \quad \partial_\pm = \partial_1 \mp \partial_2. \tag{10.16}$$

Because in propagators of the reggeized gluons one can neglect longitudinal momenta, their free Green function has the form

$$\langle A_+^a(\rho, z) A_-^b(0, 0) \rangle = -i \frac{\delta^{ab}}{2\pi} \ln \frac{1}{|\rho|^2} \delta^2(z), \quad \delta^2(z) = \delta(x_3) \delta(x_0), \tag{10.17}$$

where a, b are colour indices and ρ is the transverse relative coordinate. We shall assume also the ordering in the rapidity $y_+ - y_- = \Delta y > 0$ for the fields A^+ and A^-.

To describe the physical gluons with two helicities living in the s-channel intermediate states it is convenient to introduce a complex scalar field ϕ in such a way that the corresponding transverse gluon fields $V = V_1 + iV_2$ and $V^* = V_1 - iV_2$ are expressed in terms of ϕ as follows:

$$V_R = i\partial^*\phi, \quad V_L = i\partial^*\phi^*, \quad \partial = \frac{\partial}{\partial\rho}, \quad \partial^* = \frac{\partial}{\partial\rho^*}. \tag{10.18}$$

These definitions are in agreement with the gauge relation $\partial V_R = -\partial^* V_L^*$ discussed above. The extraction of the transverse derivatives from the fields V_R and V_L is related to the fact that the above RRG vertex is nonlocal but becomes local in terms of the fields ϕ, ϕ^*.

The effective action that describes the interaction of the physical and reggeized gluons in multi-Regge kinematics is the sum of four contributions [6]:

$$S_{eff}^{mR} = S_{free} + S_{int}^s + S_{int}^p + S_{ext}. \tag{10.19}$$

Here S_{free} corresponds to kinetic terms for the introduced fields:

$$S_{free} = \int d^4x \left(\frac{1}{2}(\partial_+\partial\phi^a)(\partial_-\partial^{a*}\phi^*) + 2(\partial A_+^a)(\partial^* A_-^a) \right). \tag{10.20}$$

From this expression one can derive the propagators of the reggeized and physical gluons. Note that the kinetic term for the field ϕ contains four derivatives, which leads to the propagator being a product of two usual propagators for the particle that lives independently in the longitudinal and transverse subspaces. The propagator in the longitudinal subspace is obtained after neglecting transverse momenta of the produced gluons in the multi-Regge kinematics. The second propagator in the transverse subspace appears due to the factors ∂, ∂^* that arise in the transition from V_R and V_L to the fields ϕ^* and ϕ.

The term S_{int}^s

$$S_{int}^s = \int d^4x \left(j_+^a(x) A_-^a(x) + j_-^a(x) A_+^a(x) \right) \tag{10.21}$$

with

$$j_+^a = g (\partial^*\phi^*) T^a i \partial_+ \partial\phi, \quad j_-^a = g (\partial\phi) T^a i \partial_- \partial^*\phi \tag{10.22}$$

describes the physical gluon scattering caused by the emission or absorption of a reggeized gluon. Due to the kinematical constraint on the fields $A_\pm$ the derivatives $\partial_\pm$ can be applied also to the fields ϕ^* by integration by parts. Note that helicity conservation of the fields V is transformed here into charge conservation for the complex field ϕ.

The term S^p_{int}

$$S^p_{int} = \int d^4x \left(j^a(x)\, \phi^{*a}(x) + j^{*a}(x)\, \phi^a(x) \right) \tag{10.23}$$

with the currents

$$j^a = g\,(\partial^* A_-) T^a \partial A_+, \quad j^{*a} = g\,(\partial^* A_+) T^a \partial A_- \tag{10.24}$$

describes the gluon production in the collision of two reggeized gluons.

Finally, the term S_{ext} describes the production and annihilation of the reggeized gluons by external particles:

$$S_{ext} = \int d^4x \left(j^{ext}_+(P_B, x) A_-(x) + j^{ext}_-(P_A, x) A_+(x) + \dots \right), \tag{10.25}$$

where $j^{ext}_\pm$ are the colour currents generated by quarks and gluons existing inside the colliding hadrons.

The S-matrix constructed with the use of the effective action S^{mr}_{eff} is a generalization of LLA amplitudes in QCD because it has multi-Regge unitarity in all subchannels. Other approaches to unitarization of the amplitudes do not have this property. Note that it is difficult to formulate this effective theory in a gauge-invariant way. However, such formulation is possible in a more general approach developed in the next subsection.

10.1.2 Gauge-invariant effective action for reggeized gluons

Initially, calculations of scattering amplitudes in Regge kinematics were performed by an iterative method based on analyticity, unitarity, and renormalizability of the theory [1] (see Chapter 9). The s-channel unitarity was incorporated partly in the form of bootstrap equations for the amplitudes generated by the gluon exchanges. But later it turned out that for this purpose one can also use an effective field theory for reggeized gluons [7],[9].

The high-energy effective action is written below. It is valid for interactions of particles having their rapidities y in a certain interval η around the rapidity y_0,

$$y = \frac{1}{2} \ln \frac{\epsilon_k + |k|}{\epsilon_k - |k|}, \quad |y - y_0| < \eta, \quad \eta \ll \ln s. \tag{10.26}$$

The corresponding gluon and quark fields are

$$v_\mu(x) = -i T^a v^a_\mu(x), \quad \psi(x), \quad \bar{\psi}(x), \quad [T^a, T^b] = i f_{abc} T^c. \tag{10.27}$$

In the case of supersymmetric models, one should take into account also the gluinos and scalars with known Yang–Mills and Yukawa interactions. Let us introduce the fields describing the production and annihilation of reggeized gluons

$$A_\pm(x) = -i T^a A^a_\pm(x). \tag{10.28}$$

Under global colour-group rotations, the introduced fields are transformed in the standard way:

$$\delta v_\mu(x) = [v_\mu(x), \chi], \quad \delta\psi(x) = -\chi\,\psi(x), \quad \delta A(x) = [A(x), \chi], \tag{10.29}$$

but under local gauge transformations with $\chi(x) \to 0$ for $x \to \infty$ we have [7]

$$\delta v_\mu(x) = \frac{1}{g}[D_\mu, \chi(x)], \quad \delta\psi(x) = -\chi(x)\,\psi(x), \quad \delta A_\pm(x) = 0. \tag{10.30}$$

This means that the reggeon fields are gauge invariant.

In quasi-multi-Regge kinematics, the particles are produced in groups (clusters) with fixed invariant masses. These clusters have significantly different rapidities in accordance with their multi-Regge ordering. Similar to the previous subsection, one obtains the following kinematical constraints on the reggeon fields:

$$\partial_\mp A_\pm(x) = 0, \quad \partial_\pm = n_\pm^\mu \partial_\mu, \tag{10.31}$$

where $n_\pm^\mu = \delta_0^\mu \pm \delta_3^\mu$. In QCD, the gauge-invariant effective action local in the rapidity y is of the following form [7]:

$$S_{eff} = \int d^4x\,(L_0 + L_{ind}), \tag{10.32}$$

where L_0 is the usual Yang–Mills Lagrangian

$$L_0 = i\bar\psi\,\slashed{\nabla}\psi + \frac{1}{2}\operatorname{Tr} G_{\mu\nu}^2, \quad D_\mu = \partial_\mu + gv_\mu, \quad G_{\mu\nu} = \frac{1}{g}[D_\mu, D_\nu] \tag{10.33}$$

and the induced contribution is given by

$$L_{ind} = \operatorname{Tr}(L_{ind}^k + L_{ind}^{GR}), \quad L_{ind}^k = 2\,\partial_\mu A_+ \partial_\mu A_-. \tag{10.34}$$

Here the gluon–reggeon interaction can be represented in terms of the Wilson P exponents and their perturbative expansions

$$
\begin{aligned}
L_{ind}^{GR} &= -\frac{1}{g}\partial_+\,P\exp\left(-\frac{g}{2}\int_{-\infty}^{x^+} v_+(x'^+)dx'^+\right)\partial_\sigma^2 A_- \\
&\quad -\frac{1}{g}\partial_-\,P\exp\left(-\frac{g}{2}\int_{-\infty}^{x^-} v_-(x'^-)dx'^-\right)\partial_\sigma^2 A_+ \\
&= \left(v_+ - gv_+\frac{1}{\partial_+}v_+ + g^2 v_+\frac{1}{\partial_+}v_+\frac{1}{\partial_+}v_+ - \dots\right)\partial_\sigma^2 A_- \\
&\quad + \left(v_- - gv_-\frac{1}{\partial_-}v_- + g^2 v_-\frac{1}{\partial_-}v_-\frac{1}{\partial_-}v_- - \dots\right)\partial_\sigma^2 A_+.
\end{aligned}
\tag{10.35}
$$

Note that the contribution of L_{ind}^{GR} to S_{eff} is gauge invariant and real, which can be verified by taking its Hermitian conjugate with subsequent integration by parts. It is important also that the nonlocal factors $\partial_\pm^{-1}$ appear together with the Laplacians ∂_σ^2 applied to the reggeon fields $A_\mp$ which are functions of coordinates of gluons situated on the light-cone integration paths in the P exponents. The Laplacians cancel the nearest reggeon propagators in momentum space and the P exponents can be interpreted as a coherent contribution of the diagrams in which the gluons belonging to the given interval $|y - y_0|$ of rapidities are emitted by particles with other rapidities. In principle, these physical properties of L_{ind}^{GR} are sufficient for its restoration.

The classical equations of motion for the above effective action can be written as follows [7]:

$$\left[D_\mu, G_{\mu\sigma}^\perp\right] = 0, \quad [D_\mu, G_{\mu\pm}] = j_\pm^{ind}, \tag{10.36}$$

where the induced gauge-invariant currents are

$$j_\pm^{ind}(x) = S(v_\mp)\partial_\sigma^2 A_\pm(x)S^+(v_\mp) \tag{10.37}$$

and

$$S(v_\mp) = P\exp\left(-\frac{g}{2}\int_{-\infty}^{x^\mp} v_\mp(x')dx'^\mp\right) = 1 - \frac{g}{\partial_\mp}v_\mp + \left(\frac{g}{\partial_\mp}v_\mp\right)^2 - \dots. \tag{10.38}$$

Note that the two equations (10.36) for the light-cone components of the Yang–Mills current j_σ can be written in another form:

$$[D_\mp, j_\pm] = \left[D_\mp, [D_\mu, G_{\mu\pm}]\right] = 0. \tag{10.39}$$

As a result, the fields $A_\pm$ can be considered as functions which parametrize their classical solutions [7]. Indeed, the equations of motion have nontrivial solutions

$$v_\pm = A_\pm + V_\pm, \tag{10.40}$$

where the correction $V_\pm$ is of the order of g and can be calculated within the framework of perturbation theory. This means that $A_\pm$ can be considered as a classical value of the field $v_\pm$ for weak couplings. Inserting the above decomposition of $v_\pm$ in the effective action we can expand it in the fields $A_\pm$ and $V_\pm$ [7]:

$$S_{eff} = \text{Tr}\int d^4x\left(j_+A_- + j_-A_+ + L_2(A_+, A_-, V) + O(g^2)\right), \tag{10.41}$$

where

$$\frac{1}{g}j_\pm = -\partial_\sigma^2\left(V_\pm\frac{1}{\partial_\pm}V_\pm\right) + [V_\nu, \partial_\pm V_\nu - 2\partial_\nu V_\pm] + [V_\pm, \partial_\nu V_\nu] \tag{10.42}$$

and

$$\frac{1}{g}L_2(A_+, A_-, V) = -\left(\partial_\sigma^2 A_-\right)\left[V_+, \frac{1}{\partial_+}A_+\right] - \left(\partial_\sigma^2 A_+\right)\left[V_-, \frac{1}{\partial_-}A_-\right]$$
$$+ (\partial_+V_- - \partial_-V_+)[A_-, A_+] + V_\nu\left([A_-, \partial_\nu A_+] + [A_+, \partial_\nu A_-]\right). \tag{10.43}$$

One can verify that under the abelian gauge transformation

$$V_\sigma \longrightarrow V_\sigma + \partial_\sigma\chi \tag{10.44}$$

the contribution $L_2(A_+, A_-, V)$ is invariant and therefore it is possible to use an arbitrary gauge for V_μ. In the left-hand light-cone gauge $V_+ = 0$ the fields describing the gluon on its mass-shell and having a physical polarization satisfy the equations

$$\left(\partial_+\partial_- - \partial_i^2\right) V_\mu = \frac{1}{2}\partial_+ V_- - \partial V = 0. \tag{10.45}$$

With the use of these constraints the production contribution to the action can be transformed to the form of

$$S_{eff}^p = 2g \, \text{Tr} \int d^4x \left((\partial^{*-1}V_L)[\partial^* A_-, \partial A_+] + (\partial^{-1}V_L^*)[\partial A_-, \partial^* A_+]\right), \tag{10.46}$$

where we used the complex notation $\partial = (\partial_1 - i\partial_2)/2$ and ∂^* for derivatives in the transverse subspace. Correspondingly, in the right-hand light-cone gauge $V_- = 0$ we obtain

$$S_{eff}^p = -2g \, \text{Tr} \int d^4x \left((\partial^{-1}V_R^*)[\partial A_-, \partial^* A_+] + (\partial^{-1}V_R)[\partial^* A_-, \partial A_+]\right). \tag{10.47}$$

By going to the new fields ϕ and ϕ^* defined by Eq. (10.18) one can derive the production term (10.23) in the effective action for multi-Regge processes.

Now let us consider the contribution linear in the reggeon fields $A_\pm$. This term is not invariant under the above gauge transformation of the field V_μ. But such invariance appears provided that the corresponding gluons are on mass-shell and have physical polarization ($\partial_\sigma^2 V_\mu = \partial_\mu V_\mu = 0$). In this case, it is convenient to choose the left-hand gauge $V_+ = 0$ for the gluon field interacting with the reggeon field A_- and the right-hand gauge $V_- = 0$ for the field interacting with A_+. The corresponding contribution to the action is simplified as follows:

$$S_{eff}^s = g \, \text{Tr} \int d^4x \left(- A_-[V_L, \partial_+ V_L^*] - A_+[V_R, \partial_- V_R^*]\right). \tag{10.48}$$

After the transition to the fields ϕ and ϕ^*, the above expression gives the corresponding term (10.21) in the action for the amplitudes with multi-Regge unitarity discussed in the previous subsection.

The kinetic contribution to the effective action for the field V appears from the free term in the Yang–Mills action for the field v. The corresponding expressions for the fields $A_\pm$ include also the bilinear contribution arising from the free action for the fields v_σ after the shift $v_\sigma = A_\sigma + V_\sigma$ (cf. (10.20)):

$$S_{free} = -\text{Tr} \int d^4x \left(\partial_+ V^* \partial_- V + 4(\partial A_+)(\partial^* A_-)\right). \tag{10.49}$$

Thus, the effective action for the amplitudes with multi-Regge unitarity can be derived from the more general gauge-invariant effective action for the quasi-multi-Regge processes.

From this effective field theory it is possible also to obtain the Bartels–Kwiecinskii–Praszalowicz (BKP) equation for the colourless composite states of n reggeized gluons [4]. In this equation only the pairwise interaction of gluons is taken into account. The effective action for the interaction of k reggeized gluons is the sum of the effective actions for pairs of gluons in arbitrary colour states. As in the case of the BFKL equation, here the infrared divergences are cancelled between the virtual contributions corresponding to the gluon Regge trajectories and the terms appearing from the real gluon emission. We shall discuss this equation in the next sections.

One can formulate the Feynman rules for the above effective theory directly in momentum space [9]. It is important to take into account the gluon-momentum conservation in induced vertices

$$k_0^{\pm} + k_1^{\pm} + \ldots + k_r^{\pm} = 0. \tag{10.50}$$

Some simple examples of the induced reggeon–gluon vertices are

$$\Delta_{a_0 c}^{\nu_0 +} = q_{\perp}^2 \, \delta_{a_0 c} \, (n^+)^{\nu_0}, \quad \Delta_{a_0 a_1 c}^{\nu_0 \nu_1 +} = q_{\perp}^2 \, T_{a_1 a_0}^c \, (n^+)^{\nu_1} \frac{1}{k_1^+} (n^+)^{\nu_0}, \tag{10.51}$$

$$\Delta_{a_0 a_1 a_2 c}^{\nu_0 \nu_1 \nu_2 +} = q_{\perp}^2 \, (n^+)^{\nu_0} (n^+)^{\nu_1} (n^+)^{\nu_2} \left(\frac{T_{a_2 a_0}^a \, T_{a_1 a}^c}{k_1^+ k_2^+} + \frac{T_{a_2 a_1}^a \, T_{a_0 a}^c}{k_0^+ k_2^+} \right). \tag{10.52}$$

They can be used to construct the effective vertices of various gluon–reggeon interactions [7]. In the general case, the induced vertices are factorized in the form

$$\Delta_{a_0 a_1 \ldots a_r c}^{\nu_0 \nu_1 \ldots \nu_r +} = (-1)^r q_{\perp}^2 \prod_{s=0}^{r} (n^+)^{\nu_s} \, 2 \, \mathrm{Tr} \left(T^c G_{a_0 a_1 \ldots a_r} \right), \tag{10.53}$$

where T^c are the colour-group generators in the fundamental representation. In more details, $G_{a_0 a_1 \ldots a_r}$ can be written as [9]

$$G_{a_0 a_1 \ldots a_r} = \sum_{\{i_0, i_1, \ldots, i_r\}} \frac{T^{a_{i_0}} T^{a_{i_1}} T^{a_{i_2}} \ldots T^{a_{i_r}}}{k_{i_0}^+ (k_{i_0}^+ + k_{i_1}^+) \ldots (k_{i_0}^+ + k_{i_1}^+ + \ldots + k_{i_{r-1}}^+)}. \tag{10.54}$$

The induced vertices satisfy the recurrence relations (Ward identities) [7]

$$- k_r^+ \, \Delta_{a_0 a_1 \ldots a_r c}^{\nu_0 \nu_1 \ldots \nu_r +} (k_0^+, \ldots, k_r^+)$$

$$= n^{+\nu_r} \sum_{i=0}^{r-1} i f_{a a_r a_i} \, \Delta_{a_0 \ldots a_{i-1} a a_{i+1} \ldots a_{r-1} c}^{\nu_0 \ldots \nu_{r-1} +} \left(k_0^+, \ldots, k_{i-1}^+, k_i^+ + k_r^+, k_{i+1}^+, \ldots \right). \tag{10.55}$$

With the use of the effective theory developed above one can calculate tree amplitudes for the production of a cluster of three gluons or of a gluon and a pair of fermions or scalar particles (in the case of an extended supersymmetric model) in the collision of two reggeized gluons [9]. It is possible also to derive the gluon-production vertex in the collision of three reggeons [10]) and the signature structure of production amplitudes [11]. The square of the amplitude of three-particle production integrated over the momenta of these particles is one of the main ingredients for the BFKL kernel in the next-to-next-to-leading approximation. Using this action, it is also possible to calculate loop corrections to various reggeon-particle vertices. For $N = 4$ SUSY one can apply for this purpose also the results obtained by Bern–Dixon–Smirnov [12].

10.2 BFKL dynamics and integrability

10.2.1 Möbius invariance

Because the production amplitudes in QCD are factorized in the multi-Regge kinematics, one can write a Bethe–Salpeter-type equation for the total cross section σ_t in LLA. Using also the optical theorem the result can be presented as the BFKL equation for the pomeron wave function [1] (see Chapter 9)

$$E\,\Psi(\rho_1,\rho_2) = H_{12}\,\Psi(\rho_1,\rho_2), \quad \omega = -\frac{\alpha_s N_c}{2\pi}\,E, \quad \Delta = \max\omega, \tag{10.56}$$

where the intercept Δ enters in the expression for the total cross-section $\sigma_t \sim s^{\Delta}$ and the BFKL Hamiltonian in the coordinate representation is given by [13] (see Chapter 9)

$$H_{12} = \ln|p_1 p_2|^2 + \frac{1}{p_1 p_2^*}\left(\ln|\rho_{12}|^2\right) p_1 p_2^*$$

$$+ \frac{1}{p_1^* p_2}(\ln|\rho_{12}|^2) p_1^* p_2 - 4\psi(1), \quad \rho_{12} = \rho_1 - \rho_2. \tag{10.57}$$

Here the kinetic term $\ln|p_1 p_2|^2$ is the contribution of two-gluon Regge trajectories and the potential term corresponds to the Fourier transform of the product of two effective RRG vertices C. Note that the infrared divergence in H_{12} is cancelled. We used here complex notations for the transverse coordinates and their canonically conjugate momenta:

$$\rho_r = x_r + iy_r, \quad \rho_r^* = x_r - iy_r, \quad p_r = i\partial_r, \quad p_r^* = i\partial_r^*. \tag{10.58}$$

The Hamiltonian H_{12} is invariant under the Möbius transformation [14]

$$\rho_k \longrightarrow \frac{a\rho_k + b}{c\rho_k + d}. \tag{10.59}$$

The corresponding generators for n reggeized gluons are

$$M = \sum_r M_{r=1}^n, \quad M_r^3 = \rho_r \partial_r, \quad M_r^+ = \partial_r, \quad M_r^- = -\rho_r^2 \partial_r. \tag{10.60}$$

As a consequence of the Möbius invariance, solutions f of the above Schrödinger equation for the pomeron are also eigenfunctions of two Casimir operators

$$M^2 = \left(\sum_r M_r\right)^2 = -\sum_{r<r'} \rho_{rr'}^2 \partial_r \partial_{r'}, \quad M^{*2} = \left(M^2\right)^*, \tag{10.61}$$

$$M^2 f_{m,\tilde{m}} = m(m-1) f_{m,\tilde{m}}, \quad M^{*2} f_{m,\tilde{m}} = \tilde{m}(\tilde{m}-1) f_{m,\tilde{m}}. \tag{10.62}$$

Here m and $\tilde{m}$ are the conformal weights defined by the relations [14]

$$m = \gamma + n/2, \quad \tilde{m} = \gamma - n/2, \quad \gamma = 1/2 + i\nu \tag{10.63}$$

for the principal series of unitary representations of the Möbius group. The quantity γ is the anomalous dimension of the twist-2 operators and the integer n is their conformal spin. The eigenfunctions of the Casimir operators are well known [14]:

$$\Psi_{m,\tilde{m}}(\boldsymbol{\rho}_1, \boldsymbol{\rho}_2) = \left(\frac{\rho_{12}}{\rho_{10}\rho_{20}}\right)^m \left(\frac{\rho_{12}^*}{\rho_{10}^*\rho_{20}^*}\right)^{\tilde{m}}, \tag{10.64}$$

where ρ_0 in the differences ρ_{10} and ρ_{20} can be considered the pomeron coordinate. The eigenvalues E of the Hamiltonian H_{12} have the holomorphic separability property [15]

$$E_{m,\tilde{m}} = \epsilon_m + \epsilon_{\tilde{m}}, \quad \epsilon_m = \psi(m) + \psi(1-m) - 2\psi(1), \tag{10.65}$$

where the Euler function $\psi(m) = (\ln \Gamma(m))'$ can be written in the integral form

$$\psi(m) - \psi(1) = \int_0^1 dx \, \frac{1 - x^{m-1}}{1 - x}. \tag{10.66}$$

The ground-state energy is negative

$$E_0 = E_{\frac{1}{2},\frac{1}{2}} = -8 \ln 2 \tag{10.67}$$

and therefore the intercept $\Delta = g^2 N_c \ln 2 / \pi^2$ of the BFKL pomeron in LLA is positive ($\Delta \approx 0.5$ for $\alpha_s \approx 0.2$). This means that the Froissart theorem in this approximation is not fulfilled. In the next-to-leading approximation [16] the value of the intercept, obtained with the use of the BLM procedure, is significantly smaller: $\Delta \approx 0.2$ [17]. For the unitarization of scattering amplitudes we should take into account also the diagrams with many reggeized gluons which will be considered below.

10.2.2 Large-N_c limit and holomorphic factorization

The Bartels–Kwiecinski–Praszalowicz (BKP) equation for colourless composite states of several reggeized gluons has the following form [4]:

$$E \, \Psi(\boldsymbol{\rho}_1, \ldots, \boldsymbol{\rho}_n) = H \, \Psi(\boldsymbol{\rho}_1, \ldots, \boldsymbol{\rho}_n), \quad H = \sum_{k<l} \frac{T_k T_l}{-N_c} H_{kl}. \tag{10.68}$$

Here H_{kl} is the BFKL Hamiltonian. The intercept Δ of the corresponding colourless t-channel state giving the contribution $\sim s^\Delta$ to the total cross section σ_t is proportional to the ground-state energy of this equation:

$$\Delta_{m,\tilde{m}} = -\frac{g^2 N_c}{8\pi^2} E_{m,\tilde{m}}, \tag{10.69}$$

where m and $\tilde{m}$ are the conformal weights.

To simplify the structure of the equation for composite states of n reggeized gluons, we consider the multicolour limit $N_c \to \infty$ [15]. According to 't Hooft only planar diagrams are essential in multicolour QCD. It is convenient to describe the colour structure of the gluon r by an Hermitian matrix $A_a T_r^a$ of rank N_c with its Green function represented by a pair of quark and antiquark lines. At $N_c \to \infty$ only cylinder-type diagrams for the colourless t-channel exchange survive. It is enough to study an irreducible case in which the wave function has the following colour structure:

$$\Psi_{m,\tilde{m}}(\boldsymbol{\rho}_1, \ldots, \boldsymbol{\rho}_n; \boldsymbol{\rho}_0) = \sum_{\{i_1, \ldots, i_n\}} f_{m,\tilde{m}}(\boldsymbol{\rho}_{i_1}, \ldots, \boldsymbol{\rho}_{i_n}; \boldsymbol{\rho}_0) \, \mathrm{Tr}\left(T^{a_{i_1}} \ldots T^{a_{i_n}}\right), \tag{10.70}$$

where the summation is performed over all noncyclic permutations $\{i_1 \ldots i_n\}$ of gluons $1, 2, \ldots, n$. Note that we consider solutions with fixed values of the conformal weights $m, \tilde{m}$. At large N_c each term in the sum satisfies the Schrödinger equation and therefore for the function $f_{m,\tilde{m}}(\rho_1, \rho_2, \ldots, \rho_n; \rho_0)$ symmetric under the cyclic permutations

$$f_{m,\tilde{m}}(\rho_1, \rho_2, \ldots \rho_n; \rho_0) = f_{m,\tilde{m}}(\rho_n, \rho_1, \ldots \rho_{n-1}; \rho_0) \quad (10.71)$$

one can derive the following simplified BKP equation

$$E_{m,\tilde{m}} \, f_{m,\tilde{m}} = H \, f_{m,\tilde{m}}, \quad H = \frac{1}{2} \sum_{r=1}^{n} H_{r,r+1}. \quad (10.72)$$

It is obvious that for $N_c \to \infty$ only neighbouring gluons interact with each other and the factor $1/2$ is related to the fact that in this case, the pair of neighbouring gluons is in an adjoint representation of the gauge group. We implied also that $H_{n,n+1} = H_{1,n}$.

Remarkable is that the Hamiltonian H in multicolour QCD has the property of holomorphic separability [15]:

$$H = \frac{1}{2}(h + h^*), \quad [h, h^*] = 0, \quad (10.73)$$

where the holomorphic and antiholomorphic Hamiltonians

$$h = \sum_{k=1}^{n} h_{k,k+1}, \quad h^* = \sum_{k=1}^{n} h^*_{k,k+1} \quad (10.74)$$

are expressed in terms of the corresponding BFKL contributions (see (10.57)) [13]

$$H_{k,k+1} = h_{k,k+1} + h^*_{k,k+1},$$
$$h_{k,k+1} = \ln(p_k \, p_{k+1}) + p_k^{-1} \ln(\rho_{k,k+1}) \, p_k + p_{k+1}^{-1} \ln(\rho_{k,k+1}) \, p_{k+1} + 2\gamma. \quad (10.75)$$

Owing to the holomorphic separability of H, the wave function $f_{m,\tilde{m}}$ has the property of holomorphic factorization [15]:

$$f_{m,\tilde{m}}(\rho_1, \ldots, \rho_n; \rho_0) = \sum_{r,l} c_{r,l} \, f_m^r(\rho_1, \ldots, \rho_n; \rho_0) \, f_{\tilde{m}}^l(\rho_1^*, \ldots, \rho_n^*; \rho_0^*), \quad (10.76)$$

where r and l enumerate degenerate solutions of the Schrödinger equation in the holomorphic and antiholomorphic subspaces:

$$\epsilon_m \, f_m = h \, f_m, \quad \epsilon_{\tilde{m}} \, f_{\tilde{m}} = h^* \, f_{\tilde{m}}, \quad E_{m,\tilde{m}} = \frac{1}{2}(\epsilon_m + \epsilon_{\tilde{m}}). \quad (10.77)$$

Similarly to the case of two-dimensional conformal field theories, the coefficients $c_{r,l}$ are fixed by the single-valuedness condition for the wave function $f_{m,\tilde{m}}(\rho_1, \rho_2, \ldots, \rho_n; \rho_0)$ in the two-dimensional ρ-space [18]. Note that in these conformal models the holomorphic factorization of the Green functions is a consequence of the invariance of the operator algebra under the infinitely dimensional Virasoro group [18].

10.2.3 Integrability of the BKP equation

One can easily verify that the pair Hamiltonian can be written in another form:

$$h_{k,\,k+1} = \rho_{k,k+1} \ln(p_k\, p_{k+1})\, \rho_{k,k+1}^{-1} + 2\ln(\rho_{k,k+1}) + 2\,\gamma. \tag{10.78}$$

As a result, there are two different normalization conditions for the wave function [13]:

$$\|f\|_1^2 = \int \prod_{r=1}^{n} d^2\rho_r \left| \prod_{r=1}^{n} \rho_{r,r+1}^{-1}\, f \right|^2, \quad \|f\|_2^2 = \int \prod_{r=1}^{n} d^2\rho_r \left| \prod_{r=1}^{n} p_r\, f \right|^2 \tag{10.79}$$

compatible with the hermicity properties of H. This is related to the fact that the transposed Hamiltonian h^t is related to h by two different similarity transformations [13]

$$h^t = \prod_{r=1}^{n} p_r\, h \prod_{r=1}^{n} p_r^{-1} = \prod_{r=1}^{n} \rho_{r,r+1}^{-1}\, h \prod_{r=1}^{n} \rho_{r,r+1}. \tag{10.80}$$

Therefore h commutes with the differential operator A defined by [13]

$$A = \rho_{12}\rho_{23} \dots \rho_{n1}\, p_1 p_2 \dots p_n, \tag{10.81}$$

i.e. we have

$$[h, A] = 0. \tag{10.82}$$

Furthermore, [19], there is a family $\{q_r\}$ of mutually commuting integrals of motion:

$$[q_r, q_s] = 0, \quad [q_r, h] = 0. \tag{10.83}$$

They are given by

$$q_r = \sum_{i_1 < i_2 < \dots < i_r} \rho_{i_1 i_2}\, \rho_{i_2 i_3} \dots \rho_{i_r i_1}\, p_{i_1}\, p_{i_2} \dots p_{i_r}. \tag{10.84}$$

In particular, q_n is equal to A and q_2 is proportional to the Casimir operator M^2 of the Möbius group.

The generating function for these integrals of motion coincides with the so-called transfer matrix $T(u)$ of the integrable XXX model [19]

$$T(u) = \text{Tr}\left(L_1(u)L_2(u)\dots L_n(u)\right) = \sum_{r=0}^{n} u^{n-r}\, q_r, \tag{10.85}$$

where the L-operators are constructed in terms of the Möbius group generators

$$L_k(u) = \begin{pmatrix} u + \rho_k\, p_k & p_k \\ -\rho_k^2\, p_k & u - \rho_k\, p_k \end{pmatrix}. \tag{10.86}$$

Here the variable u is the spectral parameter.

The transfer matrix is the trace of the so-called monodromy matrix $t(u)$:

$$T(u) = \text{Tr}\left(t(u)\right), \quad t(u) = L_1(u)L_2(u)\dots L_n(u). \tag{10.87}$$

It can be verified from similar relations for the operators $L(u)$ and $L(v)$ that $t(u)$ satisfies the Yang–Baxter (YB) equation [19]

$$t^{s_1}_{r_1}(u)\, t^{s_2}_{r_2}(v)\, l^{r'_1 r'_2}_{r_1 r_2}(v-u) = l^{s_1 s_2}_{s'_1 s'_2}(v-u)\, t^{s'_2}_{r_2}(v)\, t^{s'_1}_{r_1}(u), \tag{10.88}$$

where $l(w)$ is the L-operator of the well-known Heisenberg spin model

$$l^{s_1 s_2}_{s'_1 s'_2}(w) = w\, \delta^{s_1}_{s'_1}\, \delta^{s_2}_{s'_2} + i\, \delta^{s_1}_{s'_2}\, \delta^{s_2}_{s'_1}. \tag{10.89}$$

In this model, the local Hamiltonian is $H = \lambda \sum_k \sigma_k \sigma_{k+1}$, where σ are the Pauli matrices. The commutativity of $T(u)$ and $T(v)$,

$$[T(u), T(v)] = 0 \tag{10.90}$$

is a consequence of the Yang–Baxter equation, which can be easily verified with the use of the property that the trace of an operator is invariant under its similarity transformation.

The BKP Hamiltonian coincides with the local Hamiltonian of the integrable Heisenberg spin model in which spins are generators of the Möbius group [20] (see also [21]). The general method for solving such models was suggested by Sklyanin [22].

10.2.4 Hidden Lorentz symmetry

If one would parametrize $t(u)$ in the form of

$$t(u) = \begin{pmatrix} j_0(u) + j_3(u) & j_-(u) \\ j_+(u) & j_0(u) - j_3(u) \end{pmatrix}, \tag{10.91}$$

then the Yang–Baxter equation would be reduced to the following Lorentz covariant relations for the currents $j_\mu(u)$ [23]:

$$\big[j_\mu(u), j_\nu(v) \big] = \big[j_\mu(v), j_\nu(u) \big] = -\frac{i\, \epsilon_{\mu\nu\rho\sigma}}{2(u-v)} \left(j^\rho(u) j^\sigma(v) - j^\rho(v) j^\sigma(u) \right). \tag{10.92}$$

Here $\epsilon_{\mu\nu\rho\sigma}$ is the antisymmetric tensor ($\epsilon_{0123} = 1$) in the four-dimensional Minkowski space and the metric tensor $\eta^{\mu\nu}$ has the signature $(1, -1, -1, -1)$. This form is compatible with the invariance of the Yang–Baxter equations under Lorentz transformations.

The generators of the spatial rotations coincide with that of the Möbius transformations M. The commutation relations for the Lorentz algebra are given by

$$[M^s, M^t] = i\epsilon_{stu} M^u, \quad [M^s, N^t] = i\epsilon_{stu} N^u, \quad [N^s, N^t] = i\epsilon_{stu} M^u, \tag{10.93}$$

where N are the Lorentz boost generators.

The commutativity of the transfer matrix $T(u)$ with the Hamiltonian h,

$$[T(u), h] = 0 \tag{10.94}$$

is a consequence of the relation

$$\big[L_k(u) L_{k+1}(u), h_{k,k+1} \big] = -i \left(L_k(u) - L_{k+1}(u) \right) \tag{10.95}$$

for the pair Hamiltonian $h_{k,k+1}$, which is easily verified by direct calculations.

In turn this relation follows from the Möbius invariance of $h_{k,k+1}$ and from the identity [23]

$$\left[h_{k,k+1}, \left[\left(M_{k,k+1} \right)^2, N_{k,k+1} \right] \right] = 4 N_{k,k+1}, \qquad (10.96)$$

where

$$M_{k,k+1} = M_k + M_{k+1}, \quad N_{k,k+1} = M_k - M_{k+1} \qquad (10.97)$$

are the Lorentz-group generators for the two-gluon state. To check this identity, one should take into account that the pair Hamiltonian $h_{k,k+1}$ depends only on the Casimir operator $\left(M_{k,k+1} \right)^2$ and is therefore diagonal:

$$h_{k,k+1} \left| m_{k,k+1} \right\rangle = \left(\psi(m_{k,k+1}) + \psi(1 - m_{k,k+1}) - 2\psi(1) \right) \left| m_{k,k+1} \right\rangle \qquad (10.98)$$

in the conformal weight representation:

$$\left(M_{k,k+1} \right)^2 \left| m_{k,k+1} \right\rangle = m_{k,k+1}(m_{k,k+1} - 1) \left| m_{k,k+1} \right\rangle. \qquad (10.99)$$

Further, using the commutation relations of $M_{k,k+1}$ with $N_{k,k+1}$ and taking into account that $(M_k)^2 = 0$, one can verify that the operator $N_{k,k+1}$ has nonvanishing matrix elements only between the states $\left| m_{k,k+1} \right\rangle$ and $\left| m_{k,k+1} \pm 1 \right\rangle$. As a result, the above identity (10.96) for $N_{k,k+1}$ turns out to be a consequence of the well-known recurrence relations for the ψ-functions:

$$\psi(m) = \psi(m-1) + 1/(m-1), \quad \psi(1-m) = \psi(2-m) + 1/(m-1). \qquad (10.100)$$

10.2.5 Algebraic Bethe ansatz

The pair Hamiltonian $h_{k,k+1}$ can be expressed in terms of a small-u asymptotics for the fundamental $\hat{L}$-operator of the integrable Heisenberg model with spins being the generators of the Möbius group [20],[21]

$$\hat{L}_{k,k+1}(u) = P_{k,k+1}(1 + i\,u\,h_{k,k+1} + \ldots). \qquad (10.101)$$

The operator $\hat{L}$ contrary to $L_k(u)$ has the same representations in the basic and auxiliary subspaces and acts as an integral operator on the functions $f(\rho_k, \rho_{k+1})$. The permutation operator $P_{k,k+1}$ is defined by the relation

$$P_{k,k+1}\, f(\rho_k, \rho_{k+1}) = f(\rho_{k+1}, \rho_k). \qquad (10.102)$$

The fundamental operator $\hat{L}_{k,k+1}$ satisfies the linear Yang–Baxter equation

$$L_k(u)\, L_{k+1}(v)\, \hat{L}_{k,k+1}(u - v) = \hat{L}_{k,k+1}(u - v)\, L_{k+1}(v)\, L_k(u). \qquad (10.103)$$

This equation can be solved in terms of Γ functions in a way similar to what was done above for $h_{k,k+1}$, and the proportionality constant is fixed from the triangle Yang–Baxter equation

$$\hat{L}_{13}(u)\, \hat{L}_{23}(v)\, \hat{L}_{12}(u - v) = \hat{L}_{12}(u - v)\, \hat{L}_{23}(v)\, \hat{L}_{13}(u). \qquad (10.104)$$

To find a representation of the operators obeying the Yang–Baxter commutation relations, the algebraic Bethe ansatz can be used [21]. To begin with, in the above parametrization (10.91) of the monodromy matrix $t(u)$ in terms of the currents $j_\mu(u)$, one should construct the pseudovacuum state $|0\rangle$ satisfying the equation

$$j_+(u)\,|0\rangle = 0. \tag{10.105}$$

However, these equations have a nontrivial solution only if the above L-operators are regularized [23] by introducing a small conformal weight $\delta \to 0$ for reggeized gluons:

$$L_k^\delta(u) = \begin{pmatrix} u + \rho_k\,p_k - i\,\delta & p_k \\ -\rho_k^2\,p_k + 2i\,\rho_k\delta & u - \rho_k\,p_k + i\delta \end{pmatrix}. \tag{10.106}$$

Another possibility is to use the dual space corresponding to $\delta = -1$ [20]. For the above regularization, the pseudovacuum state is

$$|\delta\rangle = \prod_{k=1}^n \rho_k^{2\delta}. \tag{10.107}$$

It is also an eigenstate of the transfer matrix:

$$T(u)\,|\delta\rangle = 2\,j_0(u)\,|\delta\rangle = \left((u - i\,\delta)^n + (u + i\,\delta)^n\right)|\delta\rangle. \tag{10.108}$$

Furthermore, excited states are obtained by applying the product of the currents $j_-(v)$ to $|\delta\rangle$

$$|v_1 v_2 \ldots v_k\rangle = j_-(v_1)\,j_-(v_2)\ldots j_-(v_k)\,|\delta\rangle. \tag{10.109}$$

They are eigenfunctions of the transfer matrix $T(u)$ with the eigenvalues:

$$\widetilde{T}(u) = (u + i\delta)^n \prod_{r=1}^k \frac{u - v_r - i}{u - v_r} + (u - i\delta)^n \prod_{r=1}^k \frac{u - v_r + i}{u - v_r}, \tag{10.110}$$

providing that the spectral parameters $v_1, v_2, \ldots, v_k$ are chosen to be solutions of the set of the Bethe equations [21]

$$\left(\frac{v_s - i\delta}{v_s + i\delta}\right)^n = \prod_{r \neq s} \frac{v_s - v_r - i}{v_s - v_r + i} \tag{10.111}$$

for $s = 1, 2 \ldots k$.

One can define the Baxter function

$$Q^{(k)}(u) = \prod_{r=1}^k (u - v_r), \tag{10.112}$$

where v_r are the Bethe roots. Due to above relations, it satisfies the Baxter equation [20],[21]

$$\widetilde{T}(u)\,Q(u) = (u - i\delta)^n\,Q(u + i) + (u + i\delta)^n\,Q(u - i). \tag{10.113}$$

Here $\widetilde{T}(u)$ is an eigenvalue of the transfer matrix $T(u)$. One can slightly simplify the Baxter equation by choosing $\delta = -1$ [20].

The eigenfunctions of h and q_k can be expressed also in terms of the Baxter function $Q^{(k)}(u)$ using the Sklyanin ansatz [22]

$$|v_1 v_2 \ldots v_k\rangle = Q^{(k)}(\widehat{u}_1)\, Q^{(k)}(\widehat{u}_2) \ldots Q^{(k)}(\widehat{u}_{n-1})\,|\delta\rangle, \qquad (10.114)$$

where the integral operators $\widehat{u}_r$ are zeros of the current $j_-(u)$ entering in the monodromy matrix (10.91)

$$j_-(u) = c \prod_{r=1}^{n-1} (u - \widehat{u}_r). \qquad (10.115)$$

Providing that the Baxter function is a polynomial ($k < \infty$), the Baxter–Sklyanin approach is equivalent to the method based on the solution of the Bethe equations. But we need the wave functions belonging to the principal series of unitary representations which cannot be constructed within the framework of the traditional Bethe ansatz. In this case, one should use the Baxter–Sklyanin procedure.

Thus, the problem of finding the wave functions and intercepts of composite states of reggeized gluons is reduced to the search of nonpolynomial solutions of the Baxter equation [20],[21]. We shall consider the Baxter–Sklyanin approach later.

10.2.6 Duality symmetry

The integrals of motion q_r and the Hamiltonian h are invariant under the cyclic permutation of gluon indices $i \to i+1$ ($i = 1, 2 \ldots n$), corresponding to the Bose symmetry of the reggeon wave function at $N_c \to \infty$. It is remarkable that these operators are invariant also under the more general canonical transformation [23]:

$$\rho_{i-1,i} \to p_i \to \rho_{i,i+1}, \qquad (10.116)$$

combined with reversing the order of the operator multiplication.

This duality symmetry is realized as an unitary transformation only for a vanishing total momentum:

$$p = \sum_{r=1}^{n} p_r = 0. \qquad (10.117)$$

The wave function $\psi_{m,\widetilde{m}}$ of the composite state with $p = 0$ can be written in terms of the eigenfunction $f_{m\widetilde{m}}$ of the integrals of motion q_k and q_k^* for $k = 1, 2 \ldots n$ as follows:

$$\psi_{m,\widetilde{m}}(\rho_{12}, \rho_{23}, \ldots, \rho_{n1}) = \int \frac{d^2\rho_0}{2\pi}\, f_{m,\widetilde{m}}(\rho_1, \rho_2, \ldots, \rho_n; \rho_0). \qquad (10.118)$$

Taking into account the hermicity of the total Hamiltonian [19]:

$$H^+ = \prod_{k=1}^{n} |\rho_{k,k+1}|^{-2}\, H \prod_{k=1}^{n} |\rho_{k,k+1}|^{2} = \prod_{k=1}^{n} |p_k|^{2}\, H \prod_{k=1}^{n} |p_k|^{-2}, \qquad (10.119)$$

the solution $\psi_{\tilde{m},m}^{+}$ of the complex conjugate Schrödinger equation for $p = 0$ can be expressed in terms of $\psi_{\tilde{m},m}$ as follows:

$$\psi_{\tilde{m},m}^{+}(\rho_{12}, \rho_{23}, \ldots) = \prod_{k=1}^{n} \left|\rho_{k,k+1}\right|^{-2} \left(\psi_{\tilde{m},m}(\rho_{12}, \rho_{23}, \ldots)\right)^{*}. \qquad (10.120)$$

Because $\psi_{m,\tilde{m}}$ is also an eigenfunction of the integrals of motion $A = q_n$ and A^* with their eigenvalues λ_m and $\lambda_m^* = \lambda_{\tilde{m}}$ [13],

$$A\,\psi_{m,\tilde{m}} = \lambda_m\,\psi_{m,\tilde{m}}, \quad A^*\,\psi_{m,\tilde{m}} = \lambda_{\tilde{m}}\,\psi_{m,\tilde{m}}, \quad A = \rho_{12}\ldots\rho_{n1}\,p_1\ldots p_n, \qquad (10.121)$$

one can verify that the duality symmetry takes the form of the following integral equation for $\psi_{m,\tilde{m}}$ [23]:

$$\frac{\psi_{m,\tilde{m}}(\rho_{12}, \ldots, \rho_{n1})}{|\lambda_m|\,2^n} = \int \prod_{k=1}^{n-1} \frac{d^2\rho_{k-1,k}'}{2\pi} \prod_{k=1}^{n} \frac{e^{i\rho_{k,k+1}\rho_k'}}{\left|\rho_{k,k+1}'\right|^2}\,\psi_{\tilde{m},m}^{*}(\rho_{12}', \ldots, \rho_{n1}'). \qquad (10.122)$$

In the next section we consider the application of the integrability to the three-gluon composite state in QCD.

10.3 The odderon in QCD

10.3.1 Three-reggeon composite states

In the particular case of the odderon, being a composite state of three reggeized gluons with charge parity $C = -1$ and signature $P_j = -1$ [24], the colour factor coincides with the well-known completely symmetric tensor d_{abc}. Therefore, taking into account that for this state each pair of gluons belongs to the adjoint representation, the equation for the function $f_{m,\tilde{m}}$ that multiplies d_{abc} is simplified as follows [4]:

$$E_{m,\tilde{m}}\,f_{m,\tilde{m}} = \frac{1}{2}\,(H_{12} + H_{13} + H_{23})\,f_{m,\tilde{m}}, \quad \Delta_{m,\tilde{m}} = -\frac{g^2 N_c}{8\pi^2}\,E_{m,\tilde{m}}. \qquad (10.123)$$

The eigenvalue of this equation is related to the high-energy behaviour of the difference of the total proton-proton and proton-antiproton cross sections σ_{pp} and $\sigma_{p\bar{p}}$,

$$\sigma_{pp} - \sigma_{p\bar{p}} \sim s^{\Delta_{m,\tilde{m}}}. \qquad (10.124)$$

According to the Pomeranchuck theorem, the odderon intercept should be negative. Below we shall calculate $\Delta_{m,\tilde{m}}$ in the perturbative QCD.

Due to the Bose symmetry the wave function is completely symmetric

$$f_{m,\tilde{m}}(\rho_1, \rho_2, \rho_3; \rho_0) = f_{m,\tilde{m}}(\rho_2, \rho_1, \rho_3; \rho_0) = f_{m,\tilde{m}}(\rho_1, \rho_3, \rho_2; \rho_0). \qquad (10.125)$$

Note that the other solution proportional to the structure constants f_{abc} is completely antisymmetric and describes a state with pomeron quantum numbers $C = P_j = 1$.

In the case of the odderon, the conformal invariance fixes the solution of the Schrödinger equation [14]

$$f_{m,\tilde{m}}(\boldsymbol{\rho}_1, \boldsymbol{\rho}_2, \boldsymbol{\rho}_3; \boldsymbol{\rho}_0) = \left(\frac{\rho_{12} \, \rho_{23} \, \rho_{31}}{\rho_{10}^2 \, \rho_{20}^2 \, \rho_{30}^2} \right)^{m/3} \left(\frac{\rho_{12}^* \, \rho_{23}^* \, \rho_{31}^*}{\rho_{10}^{*2} \, \rho_{20}^{*2} \, \rho_{30}^{*2}} \right)^{\tilde{m}/3} f_{m,\tilde{m}}(x) \qquad (10.126)$$

up to an arbitrary function $f_{m,\tilde{m}}(x)$ of one complex variable where x is the anharmonic ratio of four coordinates

$$x = \frac{\rho_{12} \, \rho_{30}}{\rho_{10} \, \rho_{32}}. \qquad (10.127)$$

Owing to the Bose symmetry, the function $f_{m,\tilde{m}}(x)$ has simple transformation properties under the substitutions $x \to 1 - x$, $x \to 1/x$ [23],[25]. This function is known explicitly for one of the odderon solutions [26].

The wave function $\psi_{m,\tilde{m}}(\boldsymbol{\rho}_{ij})$ at $q = 0$ can be written as

$$\psi_{m,\tilde{m}}(\boldsymbol{\rho}_{ij}) = \left(\frac{\rho_{23}}{\rho_{12}\rho_{31}} \right)^{m-1} \left(\frac{\rho_{23}^*}{\rho_{12}^*\rho_{31}^*} \right)^{\tilde{m}-1} \chi_{m,\tilde{m}}(z), \quad z = \frac{\rho_{12}}{\rho_{32}}, \qquad (10.128)$$

where

$$\chi_{m,\tilde{m}}(z) = \int \frac{d^2x \, f_{m,\tilde{m}}(x)}{2\pi \, |x - z|^4} \left(\frac{(x - z)^3}{x(1 - x)} \right)^{2m/3} \left(\frac{(x^* - z^*)^3}{x^*(1 - x^*)} \right)^{2\tilde{m}/3}. \qquad (10.129)$$

In fact, this function is proportional to $f_{1-m,1-\tilde{m}}(z)$:

$$\chi_{m,\tilde{m}}(z) \sim (x(1 - x))^{2(m-1)/3} \left(x^*(1 - x^*) \right)^{2(\tilde{m}-1)/3} f_{1-m,1-\tilde{m}}(z), \qquad (10.130)$$

which is a certain realization of the linear dependence between two representations $(m, \tilde{m})$ and $(1 - m, 1 - \tilde{m})$. The corresponding reality property for the Möbius group representations can be represented by the integral relation

$$\chi_{m,\tilde{m}}(z) = \int \frac{d^2x}{2\pi} (x - z)^{2m-2} (x^* - z^*)^{2\tilde{m}-2} \chi_{1-m,1-\tilde{m}}(x) \qquad (10.131)$$

for an appropriate choice of phases for the functions $\chi_{m,\tilde{m}}$ and $\chi_{1-m,1-\tilde{m}}$.

10.3.2 Duality equation for the odderon

The duality equation (10.122) for $\chi_{m,\tilde{m}}(z)$ can be written in a pseudodifferential form [23]:

$$|z(1 - z)|^2 (i\partial)^{2-m} (i\partial^*)^{2-\tilde{m}} \varphi_{1-m,1-\tilde{m}}(z) = \left| \lambda_{m,\tilde{m}} \right| \left(\varphi_{1-m,1-\tilde{m}}(z) \right)^*, \qquad (10.132)$$

where

$$\varphi_{1-m,1-\tilde{m}}(z) = (z(1 - z))^{1-m} \left(z^*(1 - z^*) \right)^{1-\tilde{m}} \chi_{m,\tilde{m}}(z) \qquad (10.133)$$

and $\lambda_{m,\tilde{m}}$ is the eigenvalue of the integral of motion A_3.

The following definition of the norm for the odderon wave function

$$\left\| \varphi_{m,\tilde{m}} \right\|_1^2 = \int \frac{d^2x}{|x(1 - x)|^2} \left| \varphi_{m,\tilde{m}}(x) \right|^2 \qquad (10.134)$$

is compatible with the duality symmetry. Another definition, namely

$$\|\varphi_{m,\tilde{m}}\|_2^2 = \int d^2 x \varphi_{m,\tilde{m}}^*(x) \, (-\partial^* \partial) \, \varphi_{m,\tilde{m}}(x), \tag{10.135}$$

is equivalent to the former, providing that $\varphi_{m,\tilde{m}}$ satisfies the duality equations.

For the holomorphic factors $\varphi^{(m)}(x)$, the duality equations are simplified [23]:

$$a_m \varphi^{(m)}(x) = \lambda^m \varphi^{(1-m)}(x), \quad a_{1-m} \varphi^{(1-m)}(x) = \lambda^{1-m} \varphi^{(m)}(x), \tag{10.136}$$

where

$$a_m = x(1-x) \, p^{1+m}, \quad p = i \frac{\partial}{\partial x}. \tag{10.137}$$

As a result, the eigenvalue equation for the integral of motion A can be written in the form of

$$a_{1-m} \, a_m \varphi^{(m)}(x) = \lambda \varphi^{(m)}(x). \tag{10.138}$$

If we consider p as a coordinate and $x - 1/2$ as a momentum, then this equation for the most important case $m = 1/2$ can be reduced to the Schrödinger equation with potential $V(p) = \sqrt{\lambda} \, p^{-3/2}$ [23].

For each eigenvalue λ of the integral of motion A_3 (10.81) there are three independent solutions $\varphi_i^{(m)}(x, \lambda)$ of the third-order ordinary differential equation corresponding to the diagonalization of the operator A [13]:

$$A \varphi = -i x(1-x) \left(x(1-x)\partial^2 + (2-m) \left((1-2x)\partial - 1 + m \right) \right) \partial \varphi = \lambda \, \varphi. \tag{10.139}$$

In the region $x \to 0$, they can be chosen as follows [25]:

$$\varphi_r^{(m)}(x, \lambda) = \sum_{k=1}^{\infty} d_k^{(m)}(\lambda) \, x^k, \quad d_1^{(m)}(\lambda) = 1. \tag{10.140}$$

$$\varphi_s^{(m)}(x, \lambda) = \sum_{k=0}^{\infty} a_k^{(m)}(\lambda) \, x^k + \varphi_r^{(m)}(x, \lambda) \ln x, \quad a_1^{(m)} = 0, \tag{10.141}$$

$$\varphi_f^{(m)}(x, \lambda) = \sum_{k=0}^{\infty} c_{k+m}^{(m)}(\lambda) \, x^{k+m}, \quad c_m^{(m)}(\lambda) = 1. \tag{10.142}$$

Due to the above differential equation (10.139), the coefficients a_k, c_k, and d_k satisfy certain recurrence relations. From the single-valuedness condition near $x = 0$, we obtain the following representation for the wave function in x, x^*-space:

$$\varphi_{m,\tilde{m}}(x, x^*) = \varphi_f^{(m)}(x, \lambda) \, \varphi_f^{(\tilde{m})}(x^*, \lambda^*) + c_2 \, \varphi_r^{(m)}(x, \lambda) \, \varphi_r^{(\tilde{m})}(x^*, \lambda^*)$$
$$+ c_1 \left(\varphi_s^{(m)}(x, \lambda) \, \varphi_r^{(\tilde{m})}(x^*, \lambda^*) + \varphi_r^{(m)}(x, \lambda) \, \varphi_s^{(\tilde{m})}(x^*, \lambda^*) \right) + (\lambda \to -\lambda). \tag{10.143}$$

The complex coefficients c_1, c_2 and the eigenvalues λ are completely fixed by the single-valuedness condition for $f_{m,\tilde{m}}(\rho_1, \rho_2, \rho_3; \rho_0)$ at $\rho_3 = \rho_i$ $(i = 1, 2)$ and the Bose symmetry [25].

With the use of the duality equations (10.136) we obtain [23]

$$|c_1| = |\lambda|. \tag{10.144}$$

Another relation

$$\mathrm{Im}\,\frac{c_2}{c_1} = \mathrm{Im}\,(m^{-1} + \tilde{m}^{-1}) \tag{10.145}$$

can be derived if one takes into account that the complex conjugate representations $\varphi_{m,\tilde{m}}$ and $\varphi_{1-m,1-\tilde{m}}$ of the Möbius group are related by the linear transformation (10.131) discussed above. It is possible to verify from the numerical results of Ref. [25] that both relations for c_1 and c_2 are satisfied [23].

If we introduce for general n the time-dependent pair Hamiltonian $h_{k,k+1}(t)$ by the definition

$$h_{k,k+1}(t) = \exp(i\,T(u)\,t)\,h_{k,k+1}\,\exp(-i\,T(u)\,t), \tag{10.146}$$

then the total holomorphic Hamiltonian h is not changed after the substitution

$$h_{k,k+1} \rightarrow h_{k,k+1}(t) \tag{10.147}$$

due to the commutativity of h and $T(u)$. On the other hand, as a result of the rapid oscillations at $t \rightarrow \infty$ each pair Hamiltonian $h_{k,k+1}(t)$ is diagonalized in the representation where the transfer matrix $T(u)$ is diagonal:

$$h_{k,k+1}(\infty) = f_{k,k+1}(\widehat{q}_2, \widehat{q}_3, \dots \widehat{q}_n). \tag{10.148}$$

This gives a possibility to express h in terms of integrals of motion. Below we shall use another approach for this purpose.

10.3.3 Odderon Hamiltonian

Generally, one can present the holomorphic Hamiltonian for n reggeized gluons in a form explicitly invariant under the Möbius transformations [23]:

$$h = \sum_{k=1}^{n}\left(\log\left(\frac{\rho_{k+2,0}\,\rho_{k,k+1}^2}{\rho_{k+1,0}\,\rho_{k+1,k+2}}\,\partial_k\right) + \log\left(\frac{\rho_{k-2,0}\,\rho_{k,k-1}^2}{\rho_{k-1,0}\,\rho_{k-1,k-2}}\,\partial_k\right) - 2\,\psi(1)\right) \tag{10.149}$$

by introducing the coordinate ρ_0 of the composite state.

In the case of the odderon $h_{k,k+1}(\infty)$ is a function of the total conformal momentum M^2 and of the integral of motion $q_3 = A$, which can be written as follows:

$$A = \frac{i^3}{2}\left[M_{12}^2,\ M_{13}^2\right] = \frac{i^3}{2}\left[M_{23}^2,\ M_{12}^2\right] = \frac{i^3}{2}\left[M_{13}^2,\ M_{23}^2\right]. \tag{10.150}$$

Let us simplify the Schrödinger equation for the odderon using the conformal ansatz for its wave function

$$f_m(\rho_1, \rho_2, \rho_3; \rho_0) = \left(\frac{\rho_{23}}{\rho_{20}\rho_{30}}\right)^m \varphi_m(x), \quad x = \frac{\rho_{12}\rho_{30}}{\rho_{10}\rho_{32}}. \tag{10.151}$$

By applying h to f_m one can derive the following Hamiltonian acting on the function $\varphi_m(x)$ in the space of the anharmonic ratio x [13]

$$h = 6\gamma + \log\left(x^2\partial\right) + \log\left((1-x)^2\partial\right) + \log\left(x^2\left(\partial + \frac{m}{1-x}\right)\right)$$
$$+ \log\left(\partial + \frac{m}{1-x}\right) + \log\left((1-x)^2\left(\partial - \frac{m}{x}\right)\right) + \log\left(\partial - \frac{m}{x}\right). \quad (10.152)$$

It is convenient to introduce the logarithmic derivative $P \equiv x\partial$ as a new momentum. With the use of relations of the type of

$$\log(\partial) = -\log(x) + \psi(-x\partial), \quad \log(x^2\partial) = \log(\partial) + 2\log(x) - \frac{1}{P},$$

and expanding h in a series in x, one can transform the odderon Hamiltonian to the normal order [23]:

$$\frac{h}{2} = -\log(x) + \psi(1-P) + \psi(-P) + \psi(m-P) - 3\psi(1) + \sum_{k=1}^{\infty} x^k f_k(P), \quad (10.153)$$

where

$$f_k(P) = -\frac{2}{k} + \frac{1}{2}\left(\frac{1}{P+k-m} + \frac{1}{P+k}\right) + \sum_{t=0}^{k} \frac{c_t(k)}{P+t} \quad (10.154)$$

and

$$c_t(k) = \frac{(-1)^{k-t}\,\Gamma(m+t)\,((t-k)\,(m+t)+m\,k/2)}{k\,\Gamma(m-k+t+1)\,\Gamma(t+1)\,\Gamma(k-t+1)}. \quad (10.155)$$

10.3.4 Expansion in the inverse integral of motion

The holomorphic Hamiltonian h is a function of $B = iA$ because h and B commute with each other. In particular, for large B this function should be of the form

$$\frac{h}{2} = \log(B) + 3\gamma + \sum_{r=1}^{\infty} \frac{c_r}{B^{2r}}. \quad (10.156)$$

The first two terms of this asymptotic expansion were calculated in [13]. The series is constructed in inverse powers of B^2, because h should be invariant under all modular transformations, including the inversion $x \to 1/x$ under which B changes its sign. The same functional relation should be valid for the eigenvalues $\varepsilon/2$ and $\mu = i\lambda$ of these operators.

For large μ it is convenient to consider the corresponding eigenvalue equations in the P representation, where x is the shift operator

$$x = \exp\left(-\frac{d}{dP}\right), \quad (10.157)$$

after extracting from eigenfunctions (10.151) of B and h the common factor

$$\varphi_m(P) = \Gamma(-P)\,\Gamma(1-P)\,\Gamma(m-P)\,\exp(i\pi P)\,\Phi_m(P). \quad (10.158)$$

The function $\Phi_m(P)$ can be expanded in a series in $1/\mu$

$$\Phi_m(P) = \sum_{n=0}^{\infty} \mu^{-n} \Phi_m^n(P), \quad \Phi_m^0(P) = 1, \tag{10.159}$$

where the coefficients $\Phi_m^n(P)$ turn out to be polynomials of order $4n$ satisfying the following recurrence relation:

$$\Phi_m^n(P) = \sum_{k=1}^{P} (k-1)(k-1-m)\Big((k-m)\Phi_m^{n-1}(k-1) + (k-2)\Phi_{1-m}^{n-1}(k-1-m)\Big)$$

$$- \frac{1}{2} \sum_{k=1}^{m} (k-1)(k-1-m)\Big((k-m)\Phi_m^{n-1}(k-1) + (k-2)\Phi_{1-m}^{n-1}(k-1-m)\Big), \tag{10.160}$$

which is valid due to the duality equations (10.136). These equations are written below after the substitution $x\,\mu \to x$ in (10.136) for a definite choice of the phase of $\Phi_m(P)$:

$$\Phi_{1-m}(P+1-m) - \frac{1}{\mu} P(P-1)(P-m)\,\Phi_{1-m}(P-m) = \Phi_m(P),$$

$$\Phi_m(P+m) - \frac{1}{\mu} P(P-1)(P+m-1)\,\Phi_m(P+m-1) = \Phi_{1-m}(P). \tag{10.161}$$

The recurrence relation (10.160) in difference form can be obtained from Eqs. (10.161) by changing the argument $P \to P - m$ in the second line and adding it to the first line, hence

$$\Phi_m(P) - \Phi_m(P-1) = \frac{1}{\mu}(P-1)(P-1-m)$$

$$\times \Big((P-m)\Phi_m(P-1) + (P-2)\,\Phi_{1-m}(P-m-1)\Big). \tag{10.162}$$

Note that the summation constants $\Phi_m^n(0)$ in (10.160) have the antisymmetry property

$$\Phi_m^n(0) = -\Phi_{1-m}^n(0), \tag{10.163}$$

which guarantees the fulfilment of the relation

$$\Phi_m^n(m) = \Phi_{1-m}^n(0), \tag{10.164}$$

as a consequence of the duality relation. On the other hand, taking into account the last relation and the fact that $r_m = \Phi_{1-m}^n(0) - \Phi_m^n(0)$ is an antisymmetric function, we can choose $\Phi_m^n(0) = -r_m/2$ because adding a symmetric contribution will redefine only the initial condition $\Phi_m^0(P) = 1$ for the recurrence relation (10.160). The most general solution of the duality equation is the function $\Phi_m(P)$ multiplied by an arbitrary constant symmetric to the substitution $m \to 1 - m$.

Note that with the use of the recurrence relation we obtain

$$\Phi_m(1) = \Phi_m(0).$$

10.3.5 Expressions for the odderon energy

The odderon energy can be expressed with the use of (10.153) in terms of $\Phi_m(P)$ as follows [23]

$$\frac{\varepsilon}{2} = \log(\mu) + 3\gamma + \frac{\partial}{\partial P} \log \Phi_m(P)$$

$$+ \sum_{k=1}^{\infty} \mu^{-k} f_k(P-k) \frac{\Phi_m(P-k)}{\Phi_m(P)} \prod_{r=1}^{k} (P-r)(P-r+1)(P-r-m+1) \quad (10.165)$$

and this expression does not depend on P due to the commutativity of h and B.

Since for $P \to 1$ we have

$$f_1(P-1) \to \frac{c_0(1)}{P-1} = \frac{m}{2} \frac{1}{P-1}, \quad f_k(P-k) \neq \infty$$

and $\Phi_m(1) = \Phi_m(0)$, one can obtain a simpler expression for ε:

$$\frac{\varepsilon}{2} = \log(\mu) + 3\gamma + \frac{\Phi'_m(1)}{\Phi_m(1)} + \frac{m(1-m)}{2\mu}. \quad (10.166)$$

It is possible to express ε in terms of the values of the function $\Phi_m(P)$ in other integer points s

$$\frac{\varepsilon}{2} = \log(\mu) + 3\gamma + \frac{\Phi'_m(s)}{\Phi_m(s)}$$

$$+ \frac{(1-m) c_0(k)}{\mu^s} \frac{\Phi_m(0)}{\Phi_m(s)} \prod_{r=1}^{s-1} (s-r)(s-r+1)(s-r-m+1)$$

$$+ \sum_{k=1}^{s-1} \mu^{-k} f_k(s-k) \frac{\Phi_m(s-k)}{\Phi_m(s)} \prod_{r=1}^{k} (s-r)(s-r+1)(s-r-m+1).$$

This representation is equivalent to the previous one due to the recurrence relations for $\Phi_{m,1-m}(P)$ following from the duality equation.

One can fix $\Phi(P)$ at some point in accordance with the duality relation without loss of generality:

$$\Phi_m(1+m) = \Phi_m(m) = \Phi_m(1) = \Phi_m(0) = 1. \quad (10.167)$$

For other integer arguments of Φ_m we have the recurrence relations following from the eigenfunction equation for the integral of motion

$$\Phi_m(2) = \left(1 + \frac{(1-m)(2-m)}{\mu} \right), \quad \Phi_m(s+1)$$

$$= \left(1 + \frac{s(s-m)}{\mu} (2s-m) \right) \Phi_m(s) - \frac{s(s-1)^2 (s-m)^2 (s-m-1)}{\mu^2} \Phi_m(s-1),$$

The solution of these equations is a polynomial in μ^{-1} and m

$$\Phi_m(s) = \sum_{k=0}^{2(s-1)-1} \sum_{l=0}^{2k} c_{kl} \mu^{-k} m^l.$$

Defining

$$\Phi'_m(1) = e(m), \tag{10.168}$$

we can calculate also the derivatives in integer points:

$$\Phi'_m(2) = \left(1 + \frac{(1-m)(2-m)}{\mu}\right) e(m) + \frac{m^2 - 6m + 6}{\mu} \tag{10.169}$$

and

$$\Phi'_m(s+1) = \left(1 + \frac{s(s-m)}{\mu}(2s-m)\right)\Phi'_m(s) - \frac{s(s-1)^2(s-m)^2(s-m-1)}{\mu^2}$$

$$\times \Phi'_m(s-1) + \frac{1}{\mu}\left(s(s-m)(2s-m)\right)' \Phi_m(s)$$

$$- \frac{1}{\mu^2}\left(s(s-1)^2(s-m)^2(s-m-1)\right)' \Phi_m(s-1). \tag{10.170}$$

The parameter $e(m)$ is fixed by the condition that the duality equation for $\Phi_m(s)$ is valid at $P \to \infty$. This requirement can be formulated in a simpler way if one returns to the initial definition of the eigenfunction (10.158) and presents $\phi_m(P)$ as a sum over poles of the first and second order with residues satisfying the recurrence relations obtained from the above relations for $\Phi_m(s)$ and $\Phi'_m(s)$.

It is plausible that the holomorphic energies for the different meromorphic solutions $\Phi_m(P)$ are generally different. Since the odderon wave function constructed as a bilinear combination of these solutions in the holomorphic and antiholomorphic subspaces should have a definite total energy, the quantization of μ should arise as a result of the coincidence of the holomorphic energies for different solutions similar to the case of the Baxter–Sklyanin approach [21],[22],[27].

By solving the recurrence relation (10.160) for $\Phi^n_m(P)$ and putting the result in the expression (10.166) for the energy, we obtain the following asymptotic expansion for $\varepsilon/2$ [23]:

$$\lim_{\mu \to \infty} \frac{\varepsilon}{2} = \log(\mu) + 3\gamma + \left(\frac{3}{448} + \frac{13}{120}(m - 1/2)^2 - \frac{1}{12}(m - 1/2)^4\right)\frac{1}{\mu^2}$$

$$+ \left(-\frac{4185}{2050048} - \frac{2151}{49280}(m - 1/2)^2 + \dots\right)\frac{1}{\mu^4}$$

$$+ \left(\frac{965925}{37044224} + \dots\right)\frac{1}{\mu^6} + \dots. \tag{10.171}$$

This expansion can be used with a certain accuracy even for the smallest eigenvalue $\mu = 0.20526$ that corresponds to the energy $\varepsilon = 0.49434$ [25]. For the first excited state with the same conformal weight $m = 1/2$, where $\varepsilon = 5.16930$ and $\mu = 2.34392$ [9], the energy can be calculated from the above asymptotic series with good precision. The behaviour of the holomorphic energy in the other limit $\mu = 0$ can be obtained from the Baxter equation (see the next section and Ref. [27])

$$\varepsilon|_{\mu=0} = \frac{\pi}{\sin(\pi m)} + \psi(m) + \psi(1-m) - 2\psi(1). \tag{10.172}$$

In particular, for $m = 1/2$ we have a positive value for the energy:

$$\epsilon|_{m=1/2} = \pi - 4\ln 2 = 0.369804 \tag{10.173}$$

and for $m = 0$ and $m = 1$ the result is

$$\epsilon|_{m=0} = \epsilon|_{m=1} = 0. \tag{10.174}$$

The corrections to ϵ of the order of μ^2 have also been calculated [27]. Note, that odderons considered in this subsection have negative intercepts Δ being in an agreement with the Pomeranchuck theorem, but below we shall construct another odderon solution with the intercept equal to zero.

10.3.6 New odderon solution

From the above expression (10.152) for h one can derive a representation of the odderon Hamiltonian in the two-dimensional space x [29]:

$$2H = h + h^* = 12\gamma + \ln\left(|x|^4|\partial|^2\right) + \ln\left(|1 - x|^4|\partial|^2\right)$$
$$+ |(x - 1)^m (x^* - 1)^{\tilde{m}} \left(\ln(|\partial|^2) + \ln(|x|^4|\partial|^2)\right) (x - 1)^{-m} (x^* - 1)^{-\tilde{m}}$$
$$+ (-x)^m (-x^*)^{\tilde{m}} \left(\ln(|1 - x|^4|\partial|^2) + \ln(|\partial|^2)\right) (-x)^{-m} (-x^*)^{-\tilde{m}}. \tag{10.175}$$

The logarithms in this expression can be represented as integral operators with the use of the relation

$$\int \frac{d^2p}{2\pi} \exp(i\,p\,y) \left(2\gamma + \ln\frac{(p)^2}{4}\right) = -2\left(\frac{\theta(|y| - \varepsilon)}{|y|^2} - 2\pi \ln\frac{1}{\varepsilon}\delta^2(y)\right). \tag{10.176}$$

This representation can be used to find the eigenvalue of the Hamiltonian for the following eigenfunction of the integrals of motion B and B^* with vanishing eigenvalues $\mu = \mu^* = 0$:

$$\varphi_{m,\tilde{m}}^{(0)}(x) = 1 + (-x)^m(-x^*)^{\tilde{m}} + (x - 1)^m(x^* - 1)^{\tilde{m}}. \tag{10.177}$$

The corresponding wave function with nonamputated propagators $f_{m,\tilde{m}}(\rho_1, \rho_2, \rho_3; \rho_0)$ is, in fact, a linear combination of the corresponding pomeron wave functions $f_{m,\tilde{m}}(\rho_1, \rho_2; \rho_0)$, $f_{m,\tilde{m}}(\rho_2, \rho_3; \rho_0)$ and $f_{m,\tilde{m}}(\rho_1, \rho_3; \rho_0)$. It is invariant under the cyclic permutation of coordinates $\rho_1 \to \rho_2 \to \rho_3 \to \rho_1$. But $f_{m,\tilde{m}}(\rho_1, \rho_2, \rho_3; \rho_0)$ is symmetric under the permutation $\rho_1 \leftrightarrow \rho_2$ only for even value of the conformal spin $n = \tilde{m} - m$, where the norm $\|\varphi_{m,\tilde{m}}\|_1$ (10.134) is divergent due to the singularities of φ at $x = 0, 1, \infty$. This is the reason why the solution $\phi^{(0)}$ exists only for the case

$$\tilde{m} - m = 2k + 1, \quad k = 0, \pm 1, \pm 2, \ldots, \tag{10.178}$$

where the wave function f is antisymmetric under the permutations of the two coordinates ρ_k. Owing to the Bose symmetry of the wave function, this state corresponds to f-coupling and has positive charge-parity C similar to the pomeron. It could be responsible for the small-x behaviour of the structure function $g_2(x)$ [29]. Using the above representation (10.175) for H, we obtain

$$H \, \varphi_{m,\widetilde{m}}^{(0)}(x) = E_{m,\widetilde{m}}^P \, \varphi_{m,\widetilde{m}}^{(0)}(x), \tag{10.179}$$

where $E_{m,\widetilde{m}}^P$ is the corresponding eigenvalue of the pomeron Hamiltonian

$$E_{m,\widetilde{m}}^P = \epsilon_m^P + \epsilon_{\widetilde{m}}^P \tag{10.180}$$

and

$$\epsilon_m^P = \psi(1-m) + \psi(m) - 2\psi(1). \tag{10.181}$$

The minimal value of $E_{m,\widetilde{m}}^P$ is obtained at $\widetilde{m} - m = \pm 1$ and corresponds to the odderon intercept $\omega = 0$.

For the case of odd $n = \widetilde{m} - m$, the norm $\|\varphi_{m,\widetilde{m}}\|_1$ (10.134) of $\varphi_{m,\widetilde{m}}(x)$ is finite:

$$\int \frac{d^2x \, |\varphi_{m,\widetilde{m}}(x)|^2}{3\pi \, |x(1-x)|^2} = \mathrm{Re} \, (\psi(m) + \psi(1-m) + \psi(\widetilde{m}) + \psi(1-\widetilde{m}) - 4\psi(1)), \tag{10.182}$$

but the other norm $\|\varphi_{m,\widetilde{m}}\|_2$ (10.135) is divergent because the solution is the sum of the pomeron solutions which do not depend on one of the coordinates ρ_i. Therefore, this solution is nonphysical. However, it is possible that the divergence disappears for a more general solution with a nonvanishing value of λ.

Using the duality transformation (10.122) in the form of [26]

$$Q_{m,\widetilde{m}} \, \phi_{m,\widetilde{m}}^{odd} = a_m(x) \, a_{\widetilde{m}}(x^*) \, \phi_{m,\widetilde{m}}^{odd} = \phi_{m,\widetilde{m}}^{(0)}(x, x^*), \tag{10.183}$$

where $a_m(x)$ is defined in Eq. (10.137), one can obtain from the function $\varphi_{m,\widetilde{m}}^{(0)}(x)$ (10.177) a new odderon solution $\phi_{m,\widetilde{m}}^{odd}$ symmetric in the coordinates ρ_k. The function $\phi_{m,\widetilde{m}}^{odd}$ corresponds to the eigenfunction with nonamputated propagators $f_{m,\widetilde{m}}^{odd}$. The amputation of the propagators leads to the result

$$F_{m,\widetilde{m}}^{odd}(\rho_1, \rho_2, \rho_3; \rho_0) = \left| \frac{1}{\rho_{12}\rho_{23}\rho_{31}} \right|^2 |A|^2 \, f_{m,\widetilde{m}}^{odd}(\rho_1, \rho_2, \rho_3; \rho_0)$$

$$= \left| \frac{1}{\rho_{12}\rho_{23}\rho_{31}} \right|^2 \left(\frac{\rho_{23}}{\rho_{20}\rho_{30}} \right)^m \left(\frac{\rho_{23}^*}{\rho_{20}^*\rho_{30}^*} \right)^{\widetilde{m}} \Phi_{m,\widetilde{m}}^{odd}(x, x^*), \tag{10.184}$$

where

$$\Phi_{m,\widetilde{m}}^{odd}(x, x^*) = |Q_{m,\widetilde{m}}|^2 \, \phi_{m,\widetilde{m}}^{odd}. \tag{10.185}$$

One can verify the relation

$$(i\partial)^{2-m} (i\partial^*)^{2-\widetilde{m}}(x, x^*) \sim \delta^2(x) - \delta^2(1-x) + \frac{x^m x^{*\widetilde{m}}}{|x|^6} \delta^2\left(\frac{1}{x}\right). \tag{10.186}$$

Using it we obtain [26]

$$F_{m,\tilde{m}}^{odd}(\boldsymbol{\rho}_1, \boldsymbol{\rho}_2, \boldsymbol{\rho}_3; \boldsymbol{\rho}_0) = \frac{E_{m,\tilde{m}}(\boldsymbol{\rho}_{20}, \boldsymbol{\rho}_{30})}{|\rho_{23}|^4} \delta^2(\rho_{12}) + \frac{E_{m,\tilde{m}}(\boldsymbol{\rho}_{10}, \boldsymbol{\rho}_{20})}{|\rho_{12}|^4} \delta^2(\rho_{31})$$

$$+ \frac{E_{m,\tilde{m}}(\boldsymbol{\rho}_{30}, \boldsymbol{\rho}_{10})}{|\rho_{31}|^4} \delta^2(\rho_{23}) \tag{10.187}$$

where

$$E_{m,\tilde{m}}(\boldsymbol{\rho}_{20}, \boldsymbol{\rho}_{30}) = \left(\frac{\rho_{23}}{\rho_{20}\rho_{30}}\right)^m \left(\frac{\rho_{23}^*}{\rho_{20}^*\rho_{30}^*}\right)^{\tilde{m}} \tag{10.188}$$

is the BFKL wave function. As this is an eigenfunction of the corresponding Casimir operator, the solution $F_{m,\tilde{m}}^{odd}$ can be written as follows:

$$F_{m,\tilde{m}}^{odd}(\boldsymbol{\rho}_1, \boldsymbol{\rho}_2, \boldsymbol{\rho}_3; \boldsymbol{\rho}_0) = \sum_{i,k\neq l} \delta^2(\rho_{li}) |\partial_i|^2 |\partial_k|^2 E_{m,\tilde{m}}(\boldsymbol{\rho}_{i0}, \boldsymbol{\rho}_{k0}). \tag{10.189}$$

In momentum space the solution with the amputated gluon propagators is the sum of the corresponding solutions for the pomeron [26]:

$$F_{m,\tilde{m}}^{odd}(\boldsymbol{k}_1, \boldsymbol{k}_2, \boldsymbol{k}_3) \sim \sum_{i,k\neq l} F_{m,\tilde{m}}(\boldsymbol{k}_i + \boldsymbol{k}_l, \boldsymbol{k}_k), \tag{10.190}$$

where

$$F_{m,\tilde{m}}(\boldsymbol{k}_1, \boldsymbol{k}_2) = (\boldsymbol{k}_1)^2 (\boldsymbol{k}_2)^2 \int \frac{d^2\rho_1 d^2\rho_2}{(2\pi)^2} \exp\left(i\sum_{r=1}^{2}(k_r, \rho_r)\right) E_{m,\tilde{m}}(\boldsymbol{\rho}_1, \boldsymbol{\rho}_2). \tag{10.191}$$

The spectrum of the ω-plane singularities for this solution is given by

$$\omega(\nu, n) = \frac{\alpha_s N_c}{\pi} \left(2\psi(1) - 2\mathrm{Re}\psi\left(\frac{1}{2} + i\nu + \frac{|n|}{2}\right)\right), \tag{10.192}$$

which coincides with the BFKL pomeron spectrum but for odd values of the conformal spin n. The rightmost singularity is situated at

$$\omega(0, 1) = 0, \tag{10.193}$$

which corresponds to an approximate constant behaviour of the difference $\sigma_{pp} - \sigma_{p\bar{p}}$ of the total cross sections for particle–particle and particle–antiparticle interactions. Asymptotically this difference is much smaller then the sum of the corresponding cross sections $\sigma_t \sim s^{\Delta_{BFKL}}$ in an agreement with the Pomeranchuck theorem formulated in the form

$$\frac{\sigma_{pp} - \sigma_{p\bar{p}}}{\sigma_{pp} + \sigma_{p\bar{p}}} \to 0.$$

The solution we have constructed is normalized according to the norm $\|\varphi_{m,\tilde{m}}\|_2$ (10.135) compatible with the hermicity properties of the BFKL Hamiltonian. The normalization constant can be obtained from (10.182) with the use of the duality transformation. Note that the intercept j_0 for the new solution exceeds the intercepts for the solutions vanishing at $\rho_{ij} \to 0$ constructed in Ref. [25]. Solutions of the BKP equation for pomeron and odderon with $n = 1$ have the same energies, which is related to the interpretation of the duality symmetry at large N_c as a symmetry between two reggeons having gluon quantum numbers and opposite signatures [26].

10.4 Baxter–Sklyanin representation

10.4.1 Sklyanin ansatz

According to the previous discussion, the problem of finding solutions of the Schrödinger equation for reggeized gluon interactions is reduced to the search of a representation of the monodromy matrix satisfying the Yang–Baxter bilinear relations (10.92) [19],[21]. For this purpose, it is convenient to work in the conjugated space [20], where the monodromy matrix is parametrized as follows:

$$\widetilde{t}(u) = \widetilde{L}_n(u) \ldots \widetilde{L}_1(u) = \begin{pmatrix} A(u) & B(u) \\ C(u) & D(u) \end{pmatrix} \tag{10.194}$$

and $\widetilde{L}_k(u)$ is given by

$$\widetilde{L}_k(u) = \begin{pmatrix} u + p_k \rho_{k0} & -p_k \rho_{k0}^2 \\ p_k \rho_{k0}^2 & u - p_k \rho_{k0} \end{pmatrix}. \tag{10.195}$$

The pseudovacuum state annihilated by the operators $C(u)$ and $C^*(u)$ has the form of [20]

$$\Psi^{(0)}(\rho_1, \rho_2, \ldots, \rho_n; \rho_0) = \prod_{k=1}^{n} \frac{1}{|\rho_{k0}|^4}. \tag{10.196}$$

To construct the colourless n-reggeon states with physical values of conformal weights $m, \widetilde{m}$ within the framework of the Bethe ansatz (10.109), one can use the Baxter–Sklyanin approach [21],[22]. To begin with, we should introduce the Baxter function satisfying the equation (see [20],[27],[28]) (cf. (10.113))

$$\Lambda^{(n)}(\lambda; \, \mu) \, Q \, (\lambda; \, m, \mu) = (\lambda + i)^n \, Q \, (\lambda + i; \, m, \mu) + (\lambda - i)^n \, Q \, (\lambda - i; \, m, \mu), \tag{10.197}$$

where $\Lambda^{(n)}(\lambda)$ is the eigenvalue of the monodromy matrix

$$\Lambda^{(n)}(\lambda; \, \mu) = \sum_{k=0}^{n} (-i)^k \, \mu_k \, \lambda^{n-k}, \quad \mu_0 = 2, \quad \mu_1 = 0, \quad \mu_2 = m(m - 1). \tag{10.198}$$

Here we assume [27], that the eigenvalues $\mu_k = i^k \, q_k$ of the integrals of motion are real.

The eigenfunctions of the holomorphic Schrödinger equation can be expressed in terms of the Baxter function $Q(\lambda)$ using the Sklyanin ansatz [22] (see (10.114))

$$f(\rho_1, \rho_2, \ldots, \rho_n; \rho_0) = Q \, (\widehat{\lambda}_1; \, m, \mu) \, Q \, (\widehat{\lambda}_2; \, m, \mu) \ldots Q \, (\widehat{\lambda}_{n-1}; \, m, \mu) \, \Psi^{(0)}, \tag{10.199}$$

where $\widehat{\lambda}_r$ are the operator zeros of the matrix element $B(u)$ in $\widetilde{t}(u)$ (10.194)

$$B(u) = -P \prod_{r=1}^{n-1} \left(u - \widehat{\lambda}_r\right), \quad P = \sum_{k=1}^{n} p_k. \tag{10.200}$$

10.4.2 Holomorhic factorization and quantization

In Ref. [27], a unitary transformation was suggested for the transition from the usual coordinate representation to the Baxter–Sklyanin representation. In the latter representation, the operators $\widehat{\lambda}_r$ are diagonal (see also Ref. [28]). As a consequence of the single-valuedness condition for the corresponding integral kernel the arguments of the Baxter functions $Q(\lambda)$ and $Q(\lambda^*)$ in the holomorphic and antiholomorphic subspaces are quantized (see [27],[28]):

$$\lambda = \sigma + i\frac{N}{2}, \quad \lambda^* = \sigma - i\frac{N}{2}, \tag{10.201}$$

where σ and N are real and integer numbers, respectively.

In Ref. [27], a general method of solving the Baxter equation for the n-reggeon composite state was proposed and the wave functions and intercepts of the composite states of three and four reggeons were calculated. It turns out [27] that there is a set of independent Baxter functions $Q^{(t)}$ ($t = 0, 1, \ldots, n - 1$) that have multiple poles simultaneously in the upper ($+$) and lower ($-$) half-λ planes in the points $\lambda = ik$ ($k = 0, \pm 1, \pm 2, \ldots$). The orders of these poles are $n_+ = r$ and $n_- = n - 1 - r$, respectively. Using all n independent functions $Q^{(t)}$ one can construct the normalizable total Baxter function $Q_{m, \widetilde{m}, \boldsymbol{\mu}}(\lambda)$ without the pole at $\sigma = 0$ [27]

$$Q_{m, \widetilde{m}, \boldsymbol{\mu}}(\lambda) = \sum_{t,l} C_{t,l}\, Q^{(t)}\left(\lambda; m, \boldsymbol{\mu}\right) Q^{(l)}\left(\lambda^*; \widetilde{m}, \boldsymbol{\mu}^s\right) \tag{10.202}$$

by adjusting for this purpose the coefficients $C_{t,l}$.

The total energy $E_{m, \widetilde{m}}$ can be expressed in terms of this Baxter function (see Ref. [27])

$$E = i \lim_{\lambda, \lambda^* \to i} \frac{\partial}{\partial \lambda} \frac{\partial}{\partial \lambda^*} \ln\left[(\lambda - i)^{n-1}(\lambda^* - i)^{n-1}|\lambda|^{2n}\, Q_{m, \widetilde{m}, \boldsymbol{\mu}}(\lambda)\right]. \tag{10.203}$$

Let us rewrite the Baxter equation for the n-reggeon composite state in a real form introducing the new variable $x \equiv -i\lambda$,

$$\Omega(x, \boldsymbol{\mu})\, Q(x, \boldsymbol{\mu}) = (x + 1)^n\, Q(x + 1, \boldsymbol{\mu}) + (x - 1)^n\, Q(x - 1, \boldsymbol{\mu}), \tag{10.204}$$

where

$$\Omega(x, \boldsymbol{\mu}) = \sum_{k=0}^{n} (-1)^k\, \mu_k\, x^{n-k} \tag{10.205}$$

and

$$\mu_0 = 2, \quad \mu_1 = 0, \quad \mu_2 = m(m - 1),$$

assuming that the eigenvalues of the integrals of motion μ_k ($k > 2$) are real numbers.

10.4.3 Meromorphic solutions of the Baxter equation

To solve the Baxter equation we introduce a set of the auxiliary functions f_r for $r = 1, 2, \ldots, n - 1$ [27]

$$f_r(x, \mu) = \sum_{l=0}^{\infty} \left[\frac{\tilde{a}_l(\mu)}{(x - l)^r} + \frac{\tilde{b}_l(\mu)}{(x - l)^{r-1}} + \ldots + \frac{\tilde{g}_l(\mu)}{x - l} \right], \tag{10.206}$$

where the coefficients $\tilde{a}_l, \ldots, \tilde{g}_l$ satisfy recurrence relations obtained by inserting f_r instead of $Q(x)$ in the Baxter equation with the following initial conditions:

$$\tilde{a}_0 = 1, \quad \tilde{b}_0 = \ldots = \tilde{g}_0 = 0. \tag{10.207}$$

Note that all functions $f_r(x, \mu)$ are expressed in terms of a subset of pole residues $\tilde{a}_l, \ldots, \tilde{z}_l$ of $f_{n-1}(x, \mu)$, and therefore they can be obtained from that function.

There are n "minimal" independent solutions $Q^{(t)}(x, \mu)$ ($t = 0, 1, 2, \ldots, n - 1$) of the Baxter equation having t-order poles at positive integer x and $(n - 1 - t)$-order poles at negative integer x [27]:

$$Q^{(t)}(x, \mu) = \sum_{r=1}^{t} C_r^{(t)}(\mu) f_r(x, \mu) + \beta^{(t)}(\mu) \sum_{r=1}^{n-1-t} C_r^{(n-1-t)}(\mu^s) f_r(-x, \mu^s), \tag{10.208}$$

where the meromorphic functions $f_r(x, \mu)$ were defined above and $\mu_r^s = (-1)^r \mu_r$. Such form of the solution is related to the invariance of the Baxter equation under the substitution $x \to -x, \mu \to \mu^s$.

The coefficients $C_r^{(t)}(\mu)$, $C_r^{(n-1-t)}(\mu^s)$ and $\beta^t(\mu)$ are obtained by imposing the validity of the Baxter equation at $x \to \infty$,

$$\lim_{x \to \infty} x^{n-2} Q^{(t)}(x, \mu) = 0. \tag{10.209}$$

This leads to a system of $n - 2$ linear equations for the coefficients $C_r^{(t)}$. We normalize $Q^{(t)}(x, \mu)$ by choosing

$$C_t^{(t)}(\mu) = C_{n-1-t}^{(n-1-t)}(\mu) = 1. \tag{10.210}$$

It is important to note that three subsequent solutions $Q^{(r)}$ for $r = 1, 2, \ldots, n - 2$ obey a linear relation [27] similar to the case of orthogonal polynomials:

$$\left[\delta^{(r)}(\mu) + \pi \cot(\pi x) \right] Q^{(r)}(x, \mu) = Q^{(r+1)}(x, \mu) + \alpha^{(r)}(\mu) Q^{(r-1)}(x, \mu). \tag{10.211}$$

Indeed, the left-hand and right-hand sides satisfy the Baxter equation everywhere including $x \to \infty$ and have the same singularities. Therefore, due to the uniqueness of the "minimal" solutions, the quantity $\pi \cot(\pi x) Q^{(r)}(x, \mu)$ can be expressed as a linear combination of $Q^{(r-1)}(x, \mu)$, $Q^{(r)}(x, \mu)$ and $Q^{(r+1)}(x, \mu)$. Furthermore, the coefficient in front of $Q^{(r+1)}(x, \mu)$ is chosen to be unity taking into account our normalization of $Q^{(r)}(x, \mu)$.

The Baxter function in the two-dimensional space $\vec{x}$ is a bilinear combination of holomorphic and antiholomorphic functions $Q^{(r)}$. Therefore, the holomorphic energy expressed in terms of the residues $a_0 = 1, b_0, a_1, b_1$ of the poles closest to zero (cf. (10.203)),

$$\epsilon = \frac{b_1}{a_1} + n = b_0 - \frac{\mu_{n-1}}{\mu_n} \qquad (10.212)$$

should be the same for all solutions $\epsilon^{(0)} = \epsilon^{(1)} = \ldots = \epsilon^{(n)}$. This leads to a quantization of the integrals of motion μ_k and of the energy E [27].

The total energy (10.203) of the composite state of n reggeons is the sum of the holomorphic and antiholomorphic energies

$$E_{m,\tilde{m}} = \epsilon_m(\mu) + \epsilon_{\tilde{m}}(\mu^{s*}). \qquad (10.213)$$

It can be obtained from the Schrödinger equation for the wave function $\phi_{m,\tilde{m}}$ in the Baxter–Sklyanin representation in the limit $\lambda, \lambda^* \to i$ [27]. We can obtain the analogous expression

$$E_{m,\tilde{m}} = \epsilon_m(\mu^s) + \epsilon_{\tilde{m}}(\mu^*) \qquad (10.214)$$

by taking instead another limit $\lambda, \lambda^* \to -i$. These two expressions for energies were derived from the Schrödinger equation with a Hermitian Hamiltonian and should coincide for the quantized values of μ [27]

$$\epsilon_m(\mu) + \epsilon_{\tilde{m}}(\mu^{s*}) = \epsilon_m(\mu^s) + \epsilon_{\tilde{m}}(\mu^*). \qquad (10.215)$$

This gives an additional constraint on the spectrum of the integrals of motion. One of the possible solutions of this constraint is that μ is real or purely imaginary. Note however that, providing that the wave function $Q(x)$ does not contain all possible bilinear combinations of the Baxter functions $Q^{(r)}$ and $Q^{(s)*}$, the quantization conditions should be not so restrictive.

10.4.4 Anomalous dimensions and intercepts of reggeons

The Q^2-dependence of the inclusive probabilities $n_i(x, \ln Q^2)$ to have a parton i with the momentum fraction x inside a hadron with large momentum $|p| \to \infty$ can be found from the evolution equations [30]. The eigenvalues of its integral kernels describing probabilities of inclusive parton transitions $i \to k$ coincide with the matrix elements $\gamma_j^{ki}(\alpha)$ of the anomalous dimension matrix for the twist-2 operators O^j with Lorentz spins $j = 2, 3, \ldots$.

For example, in the case of the pure Yang–Mills theory with the gauge group $SU(N_c)$ we have only one multiplicatively renormalized operator. Similarly in the $N = 4$ supersymmetric gauge theory [31] there is one supermultiplet of twist-2 operators [32]. Its anomalous dimension is singular at the nonphysical point $\omega = j - 1 \to 0$. In this limit, one can calculate the anomalous dimension in all orders of perturbation theory [33],

$$\gamma_{\omega \to 0} = \frac{\alpha N_c}{\pi \omega} - \Psi''(1) \left(\frac{\alpha N_c}{\pi \omega} \right)^4 + \ldots \qquad (10.216)$$

from the eigenvalue of the integral kernel of the BFKL equation in LLA [1] at $n = 0$:

$$\omega_{BFKL} = \frac{\alpha N_c}{\pi} \left[2\Psi(1) - \Psi(\gamma) - \Psi(1 - \gamma) \right]. \qquad (10.217)$$

Using next-to-leading corrections [16] it is possible also to predict the residues of the less singular terms $\sim \alpha^n / \omega^{n-1}$.

Using the BFKL equation, one can find the anomalous dimensions of higher-twist operators by solving the eigenvalue equation near the other singular points $\gamma = -k$ ($k = 1, 2, \ldots$). But for the unitarization program it is more important to calculate the anomalous dimensions of the quasipartonic operators (see Ref. [34]) constructed from several gluonic or quark fields. The simplest operator of such type is the product of the twist-2 gluon operators. In the limit $N_c \to \infty$ this operator is multiplicatively renormalized [35].

Let us return now to the high-energy asymptotics of irreducible Feynman diagrams in which each of n reggeized gluons at $N_c \to \infty$ interacts only with two neighbours. In the Born approximation, the corresponding Green function is a product of free-gluon propagators $\prod_{r=1}^{n} \ln |\rho_r - \rho_r'|^2$. For small coupling constants α_s, the full dimension of the operator related to the composite state of n reggeized gluons is approximately equal to the position of the pole $(m + \tilde{m})/2 \approx n/2$ in the eigenvalue ω of the Schrödinger equation $\omega(m, \tilde{m}; \mu_3, \ldots \mu_n)$ (see [27]):

$$\frac{m + \tilde{m}}{2} = \frac{n}{2} - \gamma^{(n)}, \quad \gamma^{(n)} = c^{(n)} \frac{\alpha_s N_c}{\omega} + O\left(\left[\frac{\alpha_s N_c}{\omega} \right]^2 \right). \qquad (10.218)$$

Here $\gamma^{(n)}$ is the anomalous dimension of the corresponding operator. This expression can be obtained also from the equation for matrix elements of the quasipartonic operators [34] written with double-logarithmic accuracy [35]. For the odderon, we find from the Baxter equation that $c^{(3)} = 0$. Note, however, that in the case of the solution found in Ref. [26] γ has a singularity at $\omega = 0$. In a similar way, a pole singularity was found also for $n = 4$ near $\frac{m + \tilde{m}}{2} = 2$ [27]. Moreover, the anomalous dimensions γ_3 and γ_4 were calculated for arbitrary α/ω (see [27]), which is important for the study of multireggeon contributions to deep inelastic processes at small Bjorken x.

Using the above quantization condition (10.212) for the odderon integral of motion μ, one can calculate the first roots numerically for $m = \tilde{m} = 1/2$ (see [27])

$$\mu_1 = 0.205257506\ldots, \quad \mu_2 = 2.3439211\ldots, \quad \mu_3 = 8.32635\ldots \qquad (10.219)$$

with the corresponding energies (compare the discussion after (10.171))

$$E_1 = 0.49434\ldots, \quad E_2 = 5.16930\ldots, \quad E_3 = 7.70234\ldots \qquad (10.220)$$

in agreement with Ref. [25]. The eigenvalues with odderon quantum numbers have been computed as functions of m for $0 < m < 1$ (see [27]), which corresponds to the analytic continuation to the region of real γ. The energy decreases from $E = E_1$ at $m = 1/2$ in a monotonic way. Only $m = 0$, 1 and $\frac{1}{2}$ are physical values. For other m the function $E(m)$ describes the behaviour of the anomalous dimension of the corresponding higher-twist operators. The energy vanishes at $m = 0, 1$ ($n = \pm 1$), which follows from its explicit expression (10.172) for $\mu = 0$ [27]. Note that $E(m, \mu \equiv 0)$ corresponds to the bilinear solution, in which the function $Q^{(1)}$ having the poles in both semi-planes of the λ-plane is absent and therefore here our general method of quantization does not work.

We obtain numerically at $m \to 0$

$$E(m) = 2.152\,m - 2.754\,m^2 + \ldots, \qquad \mu(m) = 0.375\,\sqrt{m} - 0.0228\,m + \ldots . \qquad (10.221)$$

The state with $m = 1$ and $\tilde{m} = 0$ (or vice versa) is therefore the ground state of the odderon corresponding to $|n| = 1$. It has a vanishing energy for $v \to 0$ and is situated below the eigenstates with $m = \tilde{m} = 1/2$. Note that generally this solution is different from that found in Ref. [26] because for it μ is nonzero. The first eigenstate with $n = 2$ was also investigated [27]. The energy proceeds to decrease with increasing m. The physical eigenstate with $|n| = 2$ is absent on this trajectory because μ is pure imaginary in this interval and vanishes only at $m = 1$ and $m = 2$.

Let us consider now the Baxter equation for the quarteton (four-reggeon state) [27]. A new integral of motion $\mu_4 = q_4$ appears here. The eigenvalues μ and q_4 are assumed to be real, which is compatible with the single-valuedness condition of the wave function in ρ space. Following the general method presented above one can search solutions of the Baxter equation for the quarteton in the form of a series of poles. As a result of our quantization procedure (10.212) we obtain for $m = \tilde{m} = 1/2$ [27]

$$\mu = 0, \quad q_4 = 0.1535892, \quad E = -1.34832.$$

$$\mu = 0.73833, \quad q_4 = -0.3703, \quad E = 2.34105.$$

One finds for the first eigenvalue with $m = 0$, $\tilde{m} = 1$ corresponding to $|n| = 1$

$$\mu = 0, \quad q_4 = 0.12167, \quad E = -2.0799.$$

The state of the quarteton with $|n| = 1$ has $m = 0$, $\tilde{m} = 1$. Its energy is lower than the energy of the above state with $m = \tilde{m} = \frac{1}{2}$.

The eigenvalue with $\mu = 0$ as a function of m in the interval $0 < m < \frac{1}{2}$ was also calculated (see [27]). Contrary to the odderon case, the energy eigenvalue does not vanish for $m = 0$. It decreases with m for $0 < m < \frac{1}{2}$ and takes the value $E = -2.0799$ at $m = 0$.

The state with $m = 3/2$ (corresponding to $n = 2$, $v = 0$) can be considered as the ground state for the quarteton because for it the eigenvalue of q_4 is real. It has a large negative energy $E = -5.863$, lower than the energy $E = -5.545$ of the BFKL pomeron constructed from two reggeized gluons [27]. But to prove that this state is a physical ground state, one should construct a bilinear combination of the corresponding Baxter functions to verify the normalizability of the corresponding solution (cf. [28]).

10.4.5 Pomeron in the thermostat

One of the footprints of the quark–gluon plasma is a decrease of the number of produced ψ mesons due to breaking the confining quark-antiquark potential at large temperature T.

Therefore, it is interesting to investigate the properties of composite states of reggeized gluons, in particular the intercept of the pomeron at a nonzero t channel temperature [36].

The correlators of colourless currents at a finite temperature satisfy an additional symmetry. Namely, they are periodic functions under a shift of the Euclidean time $x_4 \rightarrow x_4 + 1/T$. This periodicity leads to the quantization of the corresponding Euclidean energies $E_l = 2\pi l T$ [37] in the t-channel, where we introduced a nonzero temperature. After analytic continuation of these correlators to the s-channel with the Regge kinematics $s \gg T^2 \sim -t > 0$ one should impose on them the periodicity under the transformation $y \rightarrow y + 1/T$ of an impact parameter coordinate. Correspondingly, its canonically conjugate momentum should be quantized: $k_y^{(l)} = 2\pi l T$. In this cylinder-type topology it is convenient to introduce the rescaled variables ρ and p:

$$\rho = x + iy \rightarrow \frac{1}{2\pi T}\, \rho, \quad p^{(l)} = \frac{p_x^{(l)} - ip_y^{(l)}}{2} \rightarrow \pi T\, p^{(l)} \tag{10.222}$$

with the temperature constraints

$$0 < \mathrm{Im}\,\rho < 2\pi, \quad \mathrm{Im}\,p^{(l)} = \frac{l}{2}, \quad [p, \rho] = i. \tag{10.223}$$

In the case of nonzero temperatures the BFKL equation is modified, but the holomorphic separability remains [36]:

$$H_{12}\Psi = \Psi, \quad H_{12} = h_{12} + h_{12}^*. \tag{10.224}$$

Now the holomorphic Hamiltonian is

$$h_{12} = \sum_{r=1}^{2} \left[\Omega(q_r) + \frac{1}{p_r} G(\rho_{12})\, p_r \right]. \tag{10.225}$$

Here the kinetic energy corresponds to two reggeon trajectories

$$\Omega(q) = \frac{\pi T}{2\lambda} + \frac{1}{2} \left[\psi(1 + iq) + \psi(1 - iq) - 2\psi(1) \right] \tag{10.226}$$

and the potential energy is expressed in terms of the Green function for the cylinder topology

$$G(\rho_{12}) = -\frac{\pi T}{2\lambda} + \ln\left(2\sinh\frac{\rho_{12}}{2} \right). \tag{10.227}$$

It turns out that the BFKL equation at a nonzero temperature can be solved exactly [36]. The reason is that one can find the conformal transformation

$$\rho_r = \ln \rho_r', \tag{10.228}$$

after which the Hamiltonian and the integral of motion take the form, corresponding to zero temperature

$$h_{12} = \ln\left(p_1'\, p_2' \right) + \frac{1}{p_1'} \log\left(\rho_{12}' \right) p_1' + \frac{1}{p_2'} \log\left(\rho_{12}' \right) p_2' - 2\psi(1), \tag{10.229}$$

$$A = -(\rho'_{12})^2 \frac{\partial}{\partial \rho'_1} \frac{\partial}{\partial \rho'_2}. \tag{10.230}$$

To verify it, one can use the following operator identity

$$\frac{1}{2}\left[\psi\left(1 + z\frac{\partial}{\partial z}\right) + \psi\left(-z\frac{\partial}{\partial z}\right)\right] = \ln z + \ln\frac{\partial}{\partial z}. \tag{10.231}$$

Moreover, for the case of n reggeized gluons the Hamiltonian at a finite temperature coincides with the local Hamiltonian of the integrable Heisenberg spin model considered above, but with the spins realizing a different representation of the Möbius group generators [36]

$$M_k = \partial_k, \quad M_+ = e^{-\rho_k}\partial_k, \quad M_- = -e^{\rho_k}\partial_k. \tag{10.232}$$

It is interesting that in the pomeron case the eigenvalue equation for the integral of motion M^2 at $t = 0$ coincides with the Baxter equation (10.197) for the Heisenberg spin model [36].

10.4.6 BFKL pomeron in supersymmetric models and graviton

One can calculate the integral kernel for the BFKL equation also in two loops [16]. Its eigenvalue can be written as follows

$$\omega = 4\,\hat{a}\,\chi(n, \gamma) + 4\,\hat{a}^2\,\Delta(n, \gamma), \quad \hat{a} = g^2 N_c/(16\pi^2), \tag{10.233}$$

where

$$\chi(n, \gamma) = 2\psi(1) - \psi(\gamma + |n|/2) - \psi(1 - \gamma + |n|/2) \tag{10.234}$$

and $\psi(x) = \Gamma'(x)/\Gamma(x)$. The one-loop correction $\Delta(n, \gamma)$ in QCD contains the nonanalytic terms – the Kronecker symbols $\delta_{|n|,0}$ and $\delta_{|n|,2}$ [3] (see Chapter 9).

It is interesting to find a theoretical model in which such nonanalytic terms are absent. Supersymmetric gauge models are considered now as possible generalizations of QCD and the electroweak theory. In these models, a symmetry transformation between bosons and fermions is introduced in such way that these particles can be considered as different components of the same supermultiplet. In the simplest case, such a multiplet includes the gluon and gluino being the Majorana fermion in the adjoint representation of the gauge group. It turns out that in this model the coefficient in front of $\delta_{|n|,2}$ is indeed zero [3]. To cancel the coefficient in front of $\delta_{|n|,0}$, one can consider the N-extended supersymmetric theories in which the supermultiplet includes also scalar particles belonging to the adjoint representation of the gauge group. In the maximally extended $N = 4$ supersymmetric theory (SUSY) we have four gluinos and six scalar particles. Apart from the gauge interaction they participate also in the Yukawa interactions with the gauge-coupling constant g. The Lagrangian of the $N = 4$ model is completely fixed by the supersymmetry. One of the remarkable properties of this model is the absence of the coupling constant renormalization. According to the Maldacena hypothesis, the $N = 4$-supersymmetric gauge model is dual to the superstring theory living in the ten-dimensional anti-de-Sitter space [31].

It turns out that in $N = 4$ SUSY, all Kronecker symbol contributions to ω are cancelled and it is an analytic function of the conformal spin n. The final result for $\Delta(n, \gamma)$ in this model has the hermitially separable form [3],[38]

$$\Delta(n, \gamma) = \phi(M) + \phi(M^*) - \frac{\rho(M) + \rho(M^*)}{2\hat{a}/\omega}, \qquad M = \gamma + \frac{|n|}{2}, \qquad (10.235)$$

$$\rho(M) = \beta'(M) + \frac{1}{2}\zeta(2), \qquad \beta'(z) = \frac{1}{4}\left[\Psi'\left(\frac{z+1}{2}\right) - \Psi'\left(\frac{z}{2}\right)\right]. \qquad (10.236)$$

It is important that all functions entering in these expressions have the property of maximal transcendentality [38]. The maximal transcendentality of an expression means, by definition, that the special functions and numbers with lower complexities do not contribute to it. In particular, $\phi(M)$ can be written as follows:

$$\phi(M) = 3\zeta(3) + \psi''(M) - 2\Phi(M) + 2\beta'(M)\big(\psi(1) - \psi(M)\big), \qquad (10.237)$$

$$\Phi(M) = \sum_{k=0}^{\infty} \frac{(-1)^k}{k+M}\left(\psi'(k+1) - \frac{\psi(k+1) - \psi(1)}{k+M}\right). \qquad (10.238)$$

By definition, $\psi(M)$ has the transcendentality equal to 1, the transcendentalities of $\psi^{(n)}$ and $\zeta(n+1)$ are $n+1$ and the additional poles in the sum over k increase the transcendentality of the function $\Phi(M)$ up to 3. The maximal transcendentality hypothesis is valid also for the anomalous dimensions of twist-2 operators in $N = 4$ SUSY [39],[40], contrary to the case of QCD [41]. These quantities will be discussed in the next subsection.

Generally, the BFKL equation in the diffusion approximation can be written in the simple form [1]

$$j = 2 - \Delta - D\nu^2, \qquad (10.239)$$

where ν is related to the anomalous dimension of the twist-2 operators as follows [16]:

$$\gamma = 1 + \frac{j-2}{2} + i\nu. \qquad (10.240)$$

The parameters Δ and D are functions of the coupling constant $\hat{a}$ and are known up to two loops [3]. For large coupling constants, one can expect [42] that the leading pomeron singularity in $N = 4$ SUSY is moved to the point $j = 2$ and asymptotically the pomeron coincides with the graviton. This assumption is related to the AdS/CFT correspondence, formulated within the framework of the Maldacena hypothesis [31],[43],[44]. It is natural to impose on (10.239) the physical constraint that for the conserved energy-momentum tensor $\vartheta_{\mu\nu}(x)$ having $j = 2$ the anomalous dimension γ is zero. As a result, we find that the parameters Δ and D coincide [40]. In this case, one can calculate γ from the above BFKL equation:

$$\gamma = (j-2)\left(\frac{1}{2} - \frac{1/\Delta}{1 + \sqrt{1 + (j-2)/\Delta}}\right). \qquad (10.241)$$

Using the dictionary developed within the framework of the AdS/CFT correspondence between anomalous dimensions of local operators and energies of superstring states [43], one can present the eigenvalue equation (10.239) in the form of the graviton Regge trajectory [40]

$$j = 2 + \frac{\alpha'}{2} t, \quad t = E^2/R^2, \quad \alpha' = \frac{R^2}{2} \Delta. \tag{10.242}$$

On the other hand, Gubser–Klebanov–Polyakov predicted the following asymptotics of the anomalous dimension at large $\hat{a}$ and j [45]:

$$\gamma_{|\hat{a}, j \to \infty} = -\sqrt{j-2} \, \Delta_{|j \to \infty}^{-1/2} = \sqrt{2\pi j} \, \hat{a}^{1/4}, \quad \hat{a} = \frac{\alpha_s N_c}{4\pi}. \tag{10.243}$$

By comparing this prediction with (10.241), one can obtain the explicit expression for the pomeron intercept at large coupling constants [40],[46]

$$j = 2 - \Delta, \quad \Delta = \frac{1}{2\pi} \hat{a}^{-1/2}. \tag{10.244}$$

10.5 Maximal transcendentality and anomalous dimensions

10.5.1 Anomalous dimensions of twist-2 operators

The anomalous dimension of twist-2 operators in $N = 4$ SUSY in one-loop approximation was calculated comparatively recently [32]. The final expression is proportional to $\psi(j-1) - \psi(j)$. In Ref. [32] it was argued using this result that in this model the evolution equations for the so-called quasipartonic operators [34] are integrable in LLA. Later the integrability for $N = 4$ SUSY was generalized to other operators [47] and to higher loops [48].

The anomalous dimension of twist-2 operators was calculated in two loops in Ref. [39] confirming the result obtained with the use of the maximal transcendentality hypothesis [38].

The universal anomalous dimension for the twist-2 operators was found with the use of the maximal transcendentality hypothesis in $N=4$ SUSY up to three loops [38],[39],[40]

$$\gamma(j) = \hat{\alpha} \gamma_1(j) + \hat{\alpha}^2 \gamma_2(j) + \hat{\alpha}^3 \gamma_3(j) + \ldots, \quad \hat{\alpha} = \frac{\alpha_s N_c}{4\pi}, \tag{10.245}$$

where

$$\gamma_1(j+2) = -4S_1(j), \tag{10.246}$$

$$\frac{\gamma_2(j+2)}{8} = 2S_1 \left(S_2 + S_{-2} \right) - 2S_{-2,1} + S_3 + S_{-3} \tag{10.247}$$

$$\frac{\gamma_3(j+2)}{32} = -12 \left(S_{-3,1,1} + S_{-2,1,2} + S_{-2,2,1} \right)$$

$$+ 6 \left(S_{-4,1} + S_{-3,2} + S_{-2,3} \right) - 3 S_{-5} - 2 S_3 S_{-2} - S_5$$

$$- 2 S_1^2 \left(3 S_{-3} + S_3 - 2 S_{-2,1} \right) - S_2 \left(S_{-3} + S_3 - 2 S_{-2,1} \right)$$

$$+ 24 S_{-2,1,1,1} - S_1 \left(8 S_{-4} + S_{-2}^2 + 4 S_2 S_{-2} + 2 S_2^2 \right)$$

$$- S_1 \left(3 S_4 - 12 S_{-3,1} - 10 S_{-2,2} + 16 S_{-2,1,1} \right). \tag{10.248}$$

The harmonic sums are defined below in a recursive way

$$S_a(j) = \sum_{m=1}^{j} \frac{1}{m^a}, \quad S_{a,b,c,\ldots}(j) = \sum_{m=1}^{j} \frac{1}{m^a} S_{b,c,\ldots}(m),$$

$$S_{-a}(j) = \sum_{m=1}^{j} \frac{(-1)^m}{m^a}, \quad S_{-a,b,\ldots}(j) = \sum_{m=1}^{j} \frac{(-1)^m}{m^a} S_{b,\ldots}(m),$$

$$\overline{S}_{-a,b,c\cdots}(j) = (-1)^j S_{-a,b,\ldots}(j) + S_{-a,b,\cdots}(\infty)\Big(1 - (-1)^j\Big). \tag{10.249}$$

During the past several years there was great progress in the investigation of the N=4 Super Symmetric Yang–Mills (SYM) theory in a framework of the AdS/CFT correspondence [31],[43],[44]. This model at a strong-coupling regime $\alpha_s N_c \to \infty$ is equivalent to classical supergravity on the anti-de Sitter space $AdS_5 \times S^5$. In particular, a very interesting prediction [49] was obtained for the large-j behaviour of the anomalous dimension of twist-2 operators

$$\gamma(j) = a(z) \ln j, \quad z = \frac{\alpha_s N_c}{\pi} \tag{10.250}$$

in the strong coupling regime:

$$\lim_{z \to \infty} a = -\left(\frac{\alpha_s N_c}{\pi}\right)^{1/2} + \ldots . \tag{10.251}$$

Note that in our normalization $\gamma(j)$ contains the extra factor $-1/2$ in comparison with that in Ref. [49].

On the other hand, with the use of the information about the asymptotic behaviour of the two-loop anomalous dimension γ_2 one can formulate a resummation procedure based on the solution of the following algebraic equation for a [39]:

$$\frac{\alpha_s N_c}{\pi} = -\tilde{a} + \frac{\pi^2}{12} \tilde{a}^2. \tag{10.252}$$

Using this equation, the following large-α_s behavior of $\tilde{a}$ can be obtained:

$$\lim_{\alpha_s \to} \tilde{a} \approx -1.1632 \left(\frac{\alpha_s N_c}{\pi}\right)^{1/2} + \ldots \tag{10.253}$$

in rather good agreement with the above prediction based on the AdS/CFT correspondence. Moreover, the small-$\tilde{a}$ expansion of the solution of this equation

$$\tilde{a} = -\frac{\alpha_s N_c}{\pi} + \frac{\pi^2}{12}\left(\frac{\alpha_s N_c}{\pi}\right)^2 - \frac{1}{72}\pi^4\left(\frac{\alpha_s N_c}{\pi}\right)^3 + \ldots \tag{10.254}$$

also coincides with good accuracy with exact calculations up to three loops [40]:

$$a = -\frac{\alpha_s N_c}{\pi} + \frac{\pi^2}{12}\left(\frac{\alpha_s N_c}{\pi}\right)^2 - \frac{11}{720}\pi^4\left(\frac{\alpha_s N_c}{\pi}\right)^3 + \ldots . \tag{10.255}$$

The anomalous dimension is zero at $j = 2$ due to energy-momentum conservation. One can consider its slope $b = \gamma'(2)$ in this point. To resum the perturbation theory for this quantity we use the same procedure as above. Namely, it is possible to write the following algebraic equation [39]:

$$\frac{\pi^2}{6}\frac{\alpha_s N_c}{\pi} = -\tilde{b} + \frac{1}{2}\tilde{b}^2. \tag{10.256}$$

Its perturbative solution

$$\tilde{b} = -\frac{\pi^2}{6}\frac{\alpha_s N_c}{\pi} + \frac{\pi^4}{72}\left(\frac{\alpha_s N_c}{\pi}\right)^2 - \frac{1}{432}\pi^6\left(\frac{\alpha_s N_c}{\pi}\right)^3 + \dots \tag{10.257}$$

is in rather good agreement with the exact result up to three loops [40]:

$$b = -\frac{\pi^2}{6}\frac{\alpha_s N_c}{\pi} + \frac{\pi^4}{72}\left(\frac{\alpha_s N_c}{\pi}\right)^2 - \frac{1}{540}\pi^6\left(\frac{\alpha_s N_c}{\pi}\right)^3 + \dots . \tag{10.258}$$

Therefore, one can attempt to estimate the strong coupling behaviour of b from the above resummation

$$\lim_{\alpha_s\to\infty}\tilde{b} = \frac{\pi}{\sqrt{3}}\sqrt{\frac{\alpha_s N_c}{\pi}}. \tag{10.259}$$

This should be compared with the exact result obtained from AdS/CFT correspondence (see (10.241) and (10.244)) [40]

$$\lim_{\alpha_s\to\infty}b = \frac{\pi}{2}\sqrt{\frac{\alpha_s N_c}{\pi}}. \tag{10.260}$$

It is important also that the behaviour of the anomalous dimension near the singularity at $\omega = j - 1 \to 0$ is in agreement with the prediction of the BFKL equation in the next-to-leading logarithmic approximation [38] with the use of relation (10.240)

$$\lim_{\omega\to0}\gamma(j) = \frac{4}{\omega}\frac{\alpha_s N_c}{\pi} + 0\left(\frac{\alpha_s N_c}{\pi}\right)^2 + \frac{32\zeta(3)}{\omega^2}\left(\frac{\alpha_s N_c}{\pi}\right)^3 + \dots . \tag{10.261}$$

10.5.2 Beisert–Eden–Staudacher equation

Using integrability and maximal transcendentality the integral equation for the anomalous dimension at large j was constructed to all orders of perturbation theory [50],[51]. Its asymptotic behaviour in this region is given by

$$\lim_{j\to\infty}\gamma(j) = -\frac{1}{2}\gamma_K\ln j, \quad \gamma_K = 8g\, f(0), \quad g = \sqrt{\frac{\alpha_s N_c}{4\pi}}. \tag{10.262}$$

Here γ_K is the so-called cusp anomalous dimension introduced by A. M. Polyakov. It is expressed in terms of the solution of the Eden–Staudacher (ES) equation

$$\epsilon\, f(x) = \frac{t}{e^t - 1}\left(\frac{J_1(x)}{x} - \int_0^\infty dx'\, K(x, x')\, f(x')\right), \quad t = \epsilon x, \tag{10.263}$$

$$K(x, y) = \frac{J_1(x)\, J_0(y) - J_1(y)\, J_0(x)}{x - y}, \quad \epsilon = \frac{1}{2g}, \tag{10.264}$$

where $J_n(x)$ are the Bessel functions.

Using the Mellin transformation

$$f(x) = \int_{-i\infty}^{i\infty} \frac{d\,j}{2\pi\,i}\, e^{x\,j}\, \phi(j), \quad \lim_{j\to\infty} \phi(j) = \frac{\gamma_K}{8g\,j}, \tag{10.265}$$

one can represent $\phi(j)$ as the following sum [52]:

$$\phi(j) = \sum_{n=1}^{\infty} \phi_{n,\epsilon}(j)\, (\delta_{n,1} - a_{n,\epsilon}), \quad \gamma_K = 8g^2(1 - a_{1,\epsilon}), \tag{10.266}$$

where

$$\phi_{n,\epsilon}(j) = \sum_{s=1}^{\infty} \frac{\left(\sqrt{(j+s\,\epsilon)^2+1}+j+s\,\epsilon\right)^{-n}}{\sqrt{(j+s\,\epsilon)^2+1}}. \tag{10.267}$$

The coefficients $a_{n,\epsilon}$ satisfy the set of linear algebraic equations [52]

$$a_{n,\epsilon} = \sum_{n'=1}^{\infty} K_{n,n'}(\epsilon)\, (\delta_{n',1} - a_{n',\epsilon}), \tag{10.268}$$

where the kernel is given by the expression

$$K_{n,n'}(\epsilon) = \sum_{R=0}^{\infty} (-1)^R \frac{\zeta(2R+n+n')}{(2\epsilon)^{2R+n+n'}}\, S_{n,n'}^R, \tag{10.269}$$

which allows one to find the perturbative expansion of γ_K. The coefficients $S_{n,n'}^R$

$$S_{n,n'}^R = 2n\, \frac{(2R+n+n'-1)!\,(2R+n+n')!}{R!\,(R+n)!\,(R+n')!\,(R+n+n')!} \tag{10.270}$$

are integer numbers. As a result, the anomalous dimension has the property of maximal transcendentality in all loops

$$\gamma_K(\epsilon) = 8 \sum_{k=1}^{\infty} \left(-\frac{1}{4\epsilon^2}\right)^k \sum_{[s_t]} c_{[s_t]} \prod_r \zeta(s_r), \quad \sum_t s_t = 2k-2 \tag{10.271}$$

with the integer coefficients $c_{[s_t]}$ expressed as sums of products of $S_{n,n'}^R$ [52].

It turns out that the solution of the ES equation does not have a consistent asymptotic behaviour at large coupling constants [52] in accordance with the fact that the correct equation should include effects of the so-called dressing phase. The necessity of these corrections was understood in direct four-loop calculations [53]. Beisert–Eden–Staudacher calculated the dressing phase and constructed a new equation for γ_K [51]. Its perturbative solution for γ_K is different from the solution of the ES equation only by the change of the sign in the contributions in which the zeta-functions with odd integer arguments appear in products twice modulo 4. One can derive from this equation the following asymptotics of the cusp-anomalous dimension at large coupling constants [52],[54]

$$\lim_{\alpha_s N_c \to \infty} \gamma_K = 2 \left(\frac{\alpha_s N_c}{\pi} \right)^{1/2} \tag{10.272}$$

in agreement with the AdS/CFT prediction [43].

10.5.3 Anomalous dimension in four loops

To calculate the anomalous dimension of the twist-2 operators for $N = 4$ SUSY in four loops one can use the integrability approach based on the asymptotic Bethe ansatz [48]. The corresponding equations for the Bethe roots u_k are given below (cf. (10.111))

$$\left(\frac{x_k^+}{x_k^-} \right)^2 = \prod_{r=1}^{j-2} \frac{x_k^- - x_r^+}{x_k^+ - x_r^-} \frac{1 - g^2/x_k^+ x_r^-}{1 - g^2/x_k^- x_r^+} \exp\left(2i\,\theta(u_k, u_r)\right). \tag{10.273}$$

Here we used the notation

$$x_k^{\pm} = \frac{u_k^{\pm}}{2} + \sqrt{\frac{(u_k^{\pm})^2}{4} - g^2}, \quad u^{\pm} = u \pm \frac{i}{2} \tag{10.274}$$

and the dressing phase expansion [51]

$$\theta(u_k, u_j) = 4\,\zeta(3)\,g^6 \big(q_2(u_k)\,q_3(u_j) - q_3(u_k)\,q_2(u_j)\big) + \dots, \tag{10.275}$$

where $q_2(u)$ and $q_3(u)$ are eigenvalues of integrals of motion. The calculated Bethe roots $u_k^{\pm}$ allow one to find the anomalous dimensions:

$$\gamma(g, M) = 2g^2 \sum_{k=1}^{M} \left(\frac{i}{x_k^+} - \frac{i}{x_k^-} \right). \tag{10.276}$$

In particular, for four loops one can obtain [55]

$$\frac{\gamma_4}{256} = 4S_{-7} + 6S_7 + 2\left(S_{-3,1,3} + S_{-3,2,2} + S_{-3,3,1} + S_{-2,4,1}\right)$$
$$+ \dots\dots\dots\dots\dots\dots\dots\dots\dots\dots\dots\dots\dots\dots$$
$$- 80\,S_{1,1,-4,1} - \zeta(3)\,S_1\,(S_3 - S_{-3} + 2S_{-2,1}), \tag{10.277}$$

where the argument of the harmonic sums is $M = j - 2$ and dots mean the omitted terms (their number exceeds 100). All these terms satisfy the maximal transcendentality property. The last term appears from the dressing phase.

It turns out that after the analytic continuation of this expression to the complex j-plane from the first two most singular terms we obtain the pole $\sim 1/\omega^7$ for $\omega = j - 1 \to 0$, which does not agree with the singularity at this point predicted from the BFKL equation (10.233), (10.236)

$$\lim_{j \to 1} \gamma_4(j) = -\frac{32}{\omega^4} \left(32\zeta_3 + \frac{\pi^4}{9}\,\omega \right) + \dots. \tag{10.278}$$

This means that the asymptotic Bethe ansatz should be modified starting from four loops. Namely, one should take into account so-called wrapping effects [55]. The contributions corresponding to the wrapping diagrams were calculated in the paper of Janik with collaborators [56]. It turns out that the total expression for $\gamma_4(j)$ is now in a full agreement with the BFKL prediction (10.278). Thus, the anomalous dimension for twist-2 operators in $N = 4$ SUSY is known exactly up to four loops. Moreover, the method developed in Ref. [57] seems to allow one to calculate γ in all loops.

Interesting results in $N = 4$ SUSY were obtained also for scattering amplitudes [12]. These amplitudes were used in Ref. [58] to construct higher-loop corrections to the BFKL kernel in this model. However, it was shown [58] that the Bern–Dixon–Smirnov (BDS) ansatz [12] does not satisfy correct factorization properties in the multi-Regge kinematics. The two-gluon production amplitude in the multi-Regge kinematics was calculated exactly in LLA [12]. It was shown that in the planar limit the production amlitudes contain generally contributions of the Mandelstam cuts. Moreover, the Schrödinger equation for the composite states of n-reggeized gluons in the adjoint representation is completely integrable and corresponds to the open spin chain [59].

10.6 Discussion of obtained results

We have shown above that the gluon Regge trajectory and various reggeon couplings can be obtained from the effective-action approach. In LLA, the BFKL equation is invariant under the Möbius transformation which allows one to find its exact solution. Moreover, in LLA the BKP equations for reggeized gluon composite states in multicolour QCD have the property of the holomorphic separability and are integrable. Further, the corresponding Hamiltonian is equivalent to the local Hamiltonian of an integrable Heisenberg spin model. In particular, it has the duality symmetry. For the three-reggeon case, the duality equation allows one to construct solutions of the Schrödinger equation and calculate the odderon intercepts in terms of expansions in the inverse eigenvalue λ of the integral of motion. In the next-to-leading approximation for $N = 4$ SUSY, the eigenvalue of the pomeron kernel has remarkable properties including its analyticity in the conformal spin n and the maximal transcendentality. In this model, the pomeron coincides with the reggeized graviton, which gives a possibility to calculate its intercept at large coupling constants. Maximal transcendentality together with integrability allow one to find the anomalous dimensions of twist-2 operators in this model up to four loops. Production amplitudes in the planar limit for this supersymmetric model also have remarkable properties related to the integrability.

References

[1] Lipatov, L. N., *Sov. J. Nucl. Phys.* **23** (1976) 338;
Fadin, V. S., Kuraev, E. A. and Lipatov, L.N., *Phys. Lett.* **B60** (1975) 50;
Kuraev, E. A., Lipatov, L. N. and Fadin, V. S., *Sov. Phys. JETP* **44** (1976) 443; **45** (1977) 199; Balitsky I. I. and Lipatov L. N., *Sov. J. Nucl. Phys.* **28** (1978) 822.

[2] Fadin, V. S., Fiore, R. and Kotsky, M. I., *Phys. Lett.* **B387** (1996) 593.

[3] Kotikov, A. V. and Lipatov, L. N., *Nucl. Phys.* **B582** (2000) 19.

[4] Bartels, J., *Nucl. Phys.* **B175** (1980) 365; Kwiecinskii, J. and Praszalowicz, M., *Phys. Lett.* **B94** (1980) 413.

[5] Lipatov, L. N., *Nucl. Phys.* **B307** (1988) 705.

[6] Lipatov, L. N., *Nucl. Phys.* **B365** (1991) 614.

[7] Lipatov, L. N., *Nucl. Phys.* **B452** (1995) 369; *Phys. Rept.* **286** (1997) 131.

[8] Kirschner, R., Lipatov, L. N., Szymanowski, L., *Nucl. Phys.* **B425** (1994) 579; *Phys. Rev.* **D51** (1995) 838.

[9] Antonov, E. N., Lipatov, L. N., Kuraev, E. A. and Cherednikov, I. O., *Nucl. Phys.* **B721** (2005) 111.

[10] Braun, M. A., Vyazovsky, M. I., *Eur. Phys. J.* **C51** (2007) 103.

[11] Hentschinski, M., Bartels, J., Lipatov, L. N., arXiv:0809.4146v1 [hep-th].

[12] Bern, Z., Dixon, L. J. and Smirnov, V. A., *Phys. Rev.* **D72** (2005) 085001; *Phys. Rev.* **D72** (2005) 085001.

[13] Lipatov, L. N., *Phys. Lett.* **B309** (1993) 394.

[14] Lipatov, L. N., *Sov. Phys. JETP* **63** (1986) 904.

[15] Lipatov, L. N., *Phys. Lett.* B **251** (1990) 284; *Pomeron in QCD*, in "Perturbative QCD", ed. by A. N. Mueller, World Scientific, 1989.

[16] Fadin, V. S. and Lipatov, L. N., *Phys. Lett.* **B429** (1998) 127; Camici, G. and Ciafaloni, M., *Phys. Lett.* **B430** (1998) 349.; Kotikov, A. V. and Lipatov, L. N., *Nucl. Phys.* **B582** (2000).

[17] Brodsky, S. J., Fadin, V. S., Kim, V. T., Lipatov, L. N. and Pivovarov, G. V., *JETP Letters* **70** (1999) 155; **76** (2002) 306.

[18] Belavin, A. A., Zamolodchikov, A. B. and Polyakov, A. M., *Nucl. Phys.* **B241** (1984) 333.

[19] Lipatov, L. N., *High energy asymptotics of multi-colour QCD and exactly solvable lattice models*, Padova preprint DFPD/93/TH/70, hep-th/9311037, unpublished.

[20] Lipatov, L. N., *JETP Lett.* **59** (1994) 596; Faddeev, L. D. and Korchemsky, G. P., *Phys. Lett.* **B342** (1995) 311.

[21] Baxter, R. J., *Exactly Solved Models in Statistical Mechanics*, (Academic Press, New York, 1982); Tarasov, V. O., Takhtajan, L. A. and Faddeev, L. D., *Theor. Math. Phys.* **57** (1983) 163.

[22] Sklyanin, E. K., *Lect. Notes in Phys.* **226**, Springer-Verlag, Berlin, (1985).

[23] Lipatov, L. N., *Nucl. Phys.* **B548** (1999) 328.

[24] Lukaszuk, L. and Nicolescu, B., *Lett. Nuov. Cim.* **8** (1973) 405; Gauron, P., Lipatov, L. N. and Nicolescu, B., *Phys. Lett.* **B304** (1993) 334.

[25] Janik, R. and Wosiek, J., *Phys. Rev. Lett.* **79** (1997) 2935; **82** (1999) 1092.

[26] Bartels, J., Lipatov, L. N. and Vacca, G. P., *Phys. Lett.* **B477**(2000) 178.

[27] de Vega, H. J. and Lipatov, L. N., *Phys. Rev.* **D64** (2001) 114019; **D66**, (2002) 074013-1.

[28] Derkachev, A., Korchemsky, G. and Manashov, A., *Nucl. Phys.* **B617** (2001) 375; Derkachev, A., Korchemsky G., Kotanski, J. and Manashov, A., *Nucl. Phys.* **B645** (2002) 237.

[29] Lipatov, L. N., *Nucl. Phys.* (Proc. Suppl.) **B79** (1999) 207; *Phys. Reports* **320** (1999) 249.

[30] Gribov, V. N. and Lipatov, L. N., *Sov. J. Nucl. Phys.* **18** (1972) 438, 675; Lipatov, L. N., *Sov. J. Nucl. Phys.* **20** (1975) 93; Altarelli, G. and Parisi, G., *Nucl. Phys.* **B126** (1977) 298; Dokshitzer, Yu. L., *Sov. Phys. JETP* **46** (1977) 641.

[31] Maldacena, J., *Adv. Theor. Math. Phys.* **2** (1998) 231.

[32] Lipatov, L. N., talk at "Perspectives in Hadronics Physics," Proceedings of the ICTP conference (World Scientific, Singapore, 1997).

[33] Jaroszevicz, T., *Acta Phys. Polon.* **B11** (1980) 965.

[34] Bukhvostov, A. P., Frolov, G. V., Kuraev, E. A. and Lipatov, L. N., *Nucl. Phys.* **258** (1985) 601.

[35] Gribov, L. V., Levin, E. M. and Ryskin, M. G., Physics Reports, **C100** (1983) 1; Levin, E. M., Ryskin, M. G. and Shuvaev, A. G., *Nucl. Phys.* **B387** (1992) 589; Bartels, J., *Z. Phys.* **60** (1993) 471; Laenen, E., Levin, E. M. and Shuvaev, A. G., *Nucl. Phys.* **B419** (1994) 39; Shuvaev, A. G., hep-ph/9504341, unpublished.

[36] de Vega, H. J. and Lipatov, L. N., *Phys. Lett.* **B578** (2004) 335.

[37] Eletsky, V. L. and Ioffe B. L., *Yad. Fiz.* **48** (1988) 602.

[38] Kotikov, A. V. and Lipatov L. N., *Nucl. Phys.* **B661** (2003) 19.

[39] Kotikov, A. V., Lipatov, L. N. and Velizhanin, V. N., *Phys. Lett.* **B557** (2003) 114.

[40] Kotikov, A. V., Lipatov, L. N., Onishchenko, A. I. and Velizhanin, V. N., *Phys. Lett.* **B595** (2004) 521; [Erratum-ibid. **B632** (2006) 754].

[41] Moch, S., Vermaseren, J. A. M. and Vogt, A., *Nucl. Phys.* **B688** (2004) 101.

[42] Polchinski, J. and Strassler, M. J., hep-th/0209211.

[43] Gubser, S. S., Klebanov, I. R. and Polyakov, A. M., *Phys. Lett.* **B428** (1998) 105.

[44] Witten, E., *Adv. Theor. Math. Phys.* **2** (1998) 253.

[45] Gubser, S. S., Klebanov, I. R. and Polyakov, A. M., *Nucl. Phys.* **B636** (2002) 99.

[46] Brower, R. C., Polchinski, J., Strassler, M. J. and Tan, C. I., *JHEP* **0712** (2007) 005.

[47] Minahan, J. A. and Zarembo, K., *JHEP* **0303** (2003) 013.

[48] Beisert, N. and Staudacher, M, *Nucl. Phys.* B **670** (2003) 439.

[49] Gubser, S. S., Klebanov I. R. and Polyakov, A. M., *Nucl. Phys.* **B636** (2002) 99.

[50] Eden, B. and Staudacher, M., *J. Stat. Mech.* **0611** (2006) P014.

[51] Beisert., N, Eden, B. and Staudacher, M., *J. Stat. Mech.* **0701** (2007) P021.

[52] Kotikov, A. V. and Lipatov, L. N., *Nucl. Phys.* **B769** (2007) 217.

[53] Bern, Z., Czakson, M., Dixon, L. J., Kosower, D. A. and Smirnov, V. A., *Phys. Rev.* bf D75 (2007) 085010.

[54] Benna, M. K., Benvenuti, S., Klebanov, I. R. and Scardicchio, A., *Phys. Rev. Lett.* **98** (2007) 131603; Alday, L. F., Arutyunov, G., Benna, M. K., Eden, B. and Klebanov, I. R., *JHEP* **0704** (2007) 082; Basso, B., Korchemsky, G. P. and Kotanski, J., arXiv:0708.3933 [hep-th]; Kostov, I., Serban, D. and Volin, D., arXiv:0801.2542 [hep-th].

[55] Kotikov, A. V., Lipatov, L. N., Rej, A., Staudacher, M. and Velizhanin, V. N., *J. Stat. Mech.* **0710** (2007) P10003.

[56] Bajnok, Z., Janik, R. A. and Lukowski, T. arXiv:0811.4448v4 [hep-th].

[57] Gromov, N., Kazakov, V., Viera, P., arXiv:0901.3753 [hep-th].

[58] Bartels, J., Lipatov, L. N. and Sabio Vera, A., arXiv:hep-th/0802.2065, arXiv:hep-th/0807.0894

[59] Lipatov, L. N., arXiv:hep-th/0810.3815

Notations

Feynman notation is used in the book: $a_\mu = \{a_0, \boldsymbol{a}\}$, $\delta_{\mu\nu} = (1, -1, -1, -1)$, $\varepsilon_{0123} = 1$ $a_\mu b_\mu = a_0 b_0 - \boldsymbol{ab}$, $\gamma_\mu = \{\beta, \beta\boldsymbol{\alpha}\}$, $\gamma_\mu\gamma_\nu + \gamma_\mu\gamma_\nu = 2\delta_{\mu\nu}$, $\gamma_5 = \begin{pmatrix} 0 & -1 \\ -1 & 0 \end{pmatrix}$, $\hat{a} = a_\mu\gamma_\mu$; the covariant derivative: in fundamental representation—$\nabla_\mu = \partial_\mu + ig\frac{\lambda^n}{2}A_\mu$; in adjoint representation: $D_\mu^{nm} = \partial_\mu\delta^{nm} - gf^{nlm}A_\mu^l$, λ^n—Gell-Mann matrices, f^{nlm}—$SU(3)$ multiplication factors.

Index

active flavours 34, 404
Adler–Bell–Jackiw anomaly 84, 86–87
Adler condition 66
Adler function 208, 220
Adler theorem 66
ALEPH 207, 212, 217, 221
analyticity 92, 284, 454, 493, 537, 575
angular ordering 420, 428, 449
anomalous
    commutator 125
    dimension 28, 30, 32, 42–45, 59, 86, 100, 191,
        198–202, 216, 234, 241, 243, 250–251, 269–271,
        302, 319–320, 326, 328, 330, 340–341, 344–348,
        358, 379, 381, 383, 388, 395, 397, 403, 439–440,
        443–444, 481–482, 522–523, 542, 564, 565,
        569–575
    magnetic moment 146, 269, 370
anomalies
    axial 54, 74, 82–93, 99–100, 123, 124, 132, 199,
        203, 238, 277
    scale 82, 95, 97, 99
α_S (value) 207–212
Appelquist–Carazzone theorem 55
asymptotic
    estimate 146, 152, 154–156, 164–167, 173–174,
        180–183, 185–192, 554, 557, 570–575
    freedom 28, 33, 35, 37, 38, 53, 155, 157, 159, 196,
        309, 340, 346, 402, 405
Atiyah–Singer theorem 130–132
azimuthal correlations 477–478

Banks–Casher relation 142
baryon 52, 54, 59–65, 200, 204, 239, 241, 243–252,
    262, 272–273, 275–276, 286, 330, 343
baryonic charge current 54
Bartels-Kwiecinskii–Praszalowicz (BKP) 535, 540,
    543–546, 560, 575
Baxter equation 548, 549, 557, 562, 563, 565, 566,
    568
Baxter–Sklyanin representation 561–565
Beisert–Eden–Staudacher equation 572–573
β-function 28–30, 32–33, 35, 40, 98, 100, 191–192,
    518

Bethe ansatz 547–549, 561, 574, 575
Bethe–Salpeter equation 387–389
BFKL equation 448–449, 463–464, 469, 471, 473,
    475, 482, 484, 488, 494, 541–542, 565, 567–569,
    572, 574, 575
BFKL kernel 448, 449, 461, 469, 473, 474, 476, 480,
    488, 504, 509, 515, 541, 575
Bianchi identity 119, 256
Bjorken 261, 274, 286, 288, 309, 313, 316, 338, 352,
    353, 360, 368, 370, 379, 565
    scaling 286, 288, 313, 316, 360, 379
    sum rules 261, 274, 353, 368, 370, 379
    variable 286, 288, 313, 338, 565
BLM scale setting 42, 215, 522
bootstrap 454, 461, 491, 493–497, 499, 501–504, 537
Born approximation 43, 324–325, 338, 346, 348,
    361–362, 383–384, 388, 403, 407, 416, 450, 454,
    456, 460, 484, 499, 501, 504, 533, 565
Borel
    parameter 205, 232, 234, 251, 284–285, 287, 291,
        292, 295, 297
    plane 146
    resummation 147, 155, 156, 192
    summation (summable) 145, 146, 150, 154, 155,
        181
    transformation 145, 150, 153, 191, 205–206,
        216–217, 220–223, 231, 234, 236, 238–239, 242,
        244, 247, 250, 252, 260–263, 268–269, 280–282,
        287, 288–289, 291, 295–296

Cabibbo angle 352
Cabibbo–Kobayashi–Maskawa matrix 315
Callan–Symanzik equations 147, 149, 197–198
Callan–Gross relation 313, 314, 350
Characteristic mass of srtong interactions 53–54
charmed quark 223, 225, 226
charmonium 104, 201, 223–224, 228–229, 281
chiral
    currents 54
    effective theory (CET) 54, 58, 65–73, 91, 123, 200
    logarithms 71
    symmetry 53–55, 59–62, 65, 75–76, 87, 107, 123,
        132, 199–202, 218

coefficient functions 195–196, 199, 346, 360, 361
collinear divergence 406, 425, 432
collinear singularities 13, 409, 411, 418
colour
  algebra 45, 47, 507
  coherence 417, 422, 426, 428, 438, 441
  dipole picture 483
  factor 47, 48, 324–325, 328–331, 347, 408, 422, 423, 550
  glass condensate 482, 484
  generator 1, 2, 45, 54, 63, 324, 360, 391, 418, 422, 455, 456, 460, 475, 506, 533, 541
  octet 392, 451, 462, 469, 473, 494, 497, 500, 507, 509, 515
  singlet 37, 384, 392, 463, 464, 479, 480, 509, 510, 526
  space 1, 93, 176, 184, 237, 428, 471, 474–475
complex angular momentum 450, 521
confinement 20, 53, 94, 107, 279, 286
Coulomb corrections 229
Coulomb interaction 12, 100, 228
crossing 84–85, 93, 102, 338, 340, 383, 416, 459, 466, 499, 501
cross-sections 13, 18, 25–26, 100–104, 309–316, 332, 334, 359, 361, 363, 492–403, 406–407, 409, 411–418, 421–422, 429–431, 438, 448–451, 464, 478, 479, 482–484, 486–487, 491, 516, 521, 535, 542–543, 550, 560
counterterms 20, 71, 156, 165, 168, 171, 173, 244
confinement 20, 53, 94, 107, 279, 286
conformal invariance 162, 376–377, 488, 490, 513, 515, 522, 551
Crewther theorem 125, 127
CTEQ Collaboration 299, 349
currents
  baryon 240, 246, 249, 251
  vector 54, 67–68, 71, 83, 85–88, 141, 199, 204, 212, 221, 223, 229–230, 234, 237–238, 256, 259, 265
  axial 16, 53–56, 60–62, 64, 66–67, 74–75, 77, 83–94, 123–126, 141, 199, 203, 207, 211–212, 235–238, 274–278, 280, 281, 289–290, 328
Cutkosky's rule 26, 281

Das–Mathur–Okubo sum rule 76
Dashen theorem 55, 58
Dashen relation 57
decoupling 34
density
  instanton 113, 134–137, 197, 201, 214, 238, 246
  vacuum energy 98–99
DGLAP equations 291, 297, 320–321, 339–340, 346, 350, 383, 387, 443, 448–449, 482, 522, 523
diffusion 368, 449, 479, 480, 484, 487, 522, 569
dilatation 96–99, 165, 167, 169, 174, 338
dilute gas 134, 138, 139, 265
dimensional regularization 14, 16, 18–21, 70–71, 86, 93, 94, 137, 243, 316, 368, 411

dipole 134, 482–490, 513, 515
dispersion representation 225, 259–261
double-dispersion relation 260, 270, 280, 281, 288, 291
double-dispersion representation 259, 260, 263, 281
distributions
  quark — 204, 278, 285–299, 310–316, 322, 331–332, 337–345, 348–350, 359, 364, 367–372, 390, 417
  gluon — 103, 285–286, 316, 322, 331, 337–345, 349–351, 367–374, 390, 417, 429–433, 436, 443, 448, 449, 482
divergency 325, 387
DLA 311, 384–392, 395, 421, 423–429, 441, 444
double logarithms 383, 388, 421, 440
double-logarithmic asymptotic 311, 386
Drell–Levy–Yan relation 338, 348
duality 92, 121, 262, 338, 549–561, 575
Dyson 145, 146, 151, 153, 158, 163, 186, 190

e^+e^- -annihilation 25, 75, 195, 221, 315–316, 335, 348, 402–403, 406, 412, 418, 433, 436, 438
E155 Collaboration 368
eikonal 423
elastic scattering 459–460, 462, 483
effective Lagrangians 20, 34, 54, 67–72, 128, 180
effective action 152, 180, 187, 189, 196, 533, 535–541, 575
effective vertex 147, 499, 505
Euclidean
  action 109, 116
  space-time 9, 110, 111, 115–117, 120, 130, 133, 139, 142, 146, 155, 174, 567
  QCD 115
event shape 414

factorization 161, 201–202, 206, 211, 215, 216, 219, 220, 227, 230, 234, 241, 242, 244, 249, 255, 258, 265–267, 273, 280, 300–301, 318, 343, 345, 368, 467, 469, 483, 511, 543–544, 562, 575
Faddeev–Popov determinant 4, 9, 13
fastest apparent convergence 39
Feynman diagrams 10, 47–48, 56, 93, 145–146, 152, 160–161, 168, 171, 190, 241, 242, 266, 323, 390, 418, 420–422, 425, 437, 476, 565
Feynman parameters 313, 319, 356
Feynman rules 10–11, 282, 423, 541
fermion loop 86, 187
finite energy sum rule (FESR) 205, 219, 264
flavor (flavour)
  singlet 123, 199, 203, 340, 379, 381
  nonsinglet 54, 340, 346
formfactor 61, 76–78, 88, 146, 204, 273, 279–285, 374, 407, 415, 429, 431
fragmentation functions 286, 310, 316, 334–348, 416–417, 438–439, 444
Froissart bound 449, 480, 535

functional integral 3–6, 36, 86, 116, 146, 151,
    159–160, 165–167, 184

gauge
    axial 9, 12, 13, 15, 363, 366, 422, 433
    Coulomb 3, 7, 9, 12, 35, 36
    fixed-point 253, 260
    invariance 1–4, 14, 19, 23–24, 33, 44, 73, 78,
        82–83, 86, 101, 147–148, 186, 253, 312,
        315–316, 333, 339, 360, 418, 423–426, 449, 454,
        456, 458, 462, 467–468, 475, 485, 534, 537–540
    light-cone 13, 317, 324, 329, 334, 363–364, 372,
        419, 456, 458, 465, 466, 468, 471, 534, 540
    planar 9, 12, 13, 20
    transformations 1, 4, 8, 118, 120, 125, 182, 456,
        468, 534–535, 538–540
Gell-Mann matrices 1, 71, 87, 93, 200
Gell-Mann–Low function 28–35, 40, 98, 100, 147,
    156, 159–160, 163–165, 171, 184, 185, 191–192,
    198, 518
Gell-Mann–Oakes–Renner (GMOR) relation 60, 69,
    126, 219, 236
Gell-Mann–Okubo relation 246, 251
generating functional 6, 14, 356–358, 428–431, 437,
    441–442, 484
Glashow–Illiopoulos–Maiani mechanism 82
ghost
    field 3, 5, 12, 25, 135
    propagator 11–12, 19, 29
ghost-gluon- vertex 11–12, 19, 29
gluon
    density 448, 482, 484
    distribution 103, 285–286, 316, 322, 331, 337–345,
        349–351, 367–374, 390, 417, 429–433, 436, 443,
        448, 449, 482
    unintegrated 448–449, 482
Goldberger–Treiman relation 62, 274–275
Goldstone
    theorem 53, 61, 63
    boson 53–55, 63, 65–66, 91, 123–124, 127, 203,
        238, 277, 294
graviton 568–570, 575
Grassmann algebra 5
Grassmann variables 5–6
Gribov
    ambiguity 7, 9
    theorem 425
Gribov–Lipatov relation 338, 348, 417
hard processes 10, 53, 286, 309–311, 317, 335, 338,
    345, 417, 427–428, 430, 433
hard scale 448–449
    Higgs field 153, 180, 183
    holomorphic factorization 161, 543–544
    hyperon 272–274, 369
imaginary time 108–110, 115
impact factors 469, 471, 474, 479–480, 484–488, 490,
    497–502, 513–516, 523

impact parameters 482–483, 488,
    513, 567
inclusive spectra 415, 422, 437
infrared
    divergences 14, 20, 199, 269, 358, 376, 411, 463,
        533, 541, 542
    evolution equation 383, 395, 425
    singularities 71, 243, 407, 464, 477, 489, 510–511,
        513, 519
    stable 191, 406
instanton
    action 134–135, 137
    field 121, 130–132, 134, 138–141, 183, 212–213,
        237–238, 246
    gas model 139, 197, 201–202, 212–213, 222,
        237–238, 246, 271
    liquid model 213
    radius 121, 213, 214, 238
intercept
    pomeron 449–450, 480, 521–522, 542–543, 560,
        567, 570
    odderon 550, 558–560, 575
Ioffe
    current 241, 294
    formula 243
    time 309
Ising model 190, 192

jets
    three-jet events 412–414
    three-jet cross section 412–413, 415
    two-jet events 406–407, 409, 411–412, 414
    Sterman–Weinberg definition 409, 411–413

kernels 310–311, 321–331, 338, 346, 348, 358,
    367–368, 375–377, 381, 448–449, 461, 469,
    473–478, 480, 488–490, 494, 497, 499, 501,
    504–505, 507–523, 526, 541, 562, 564–565, 568,
    573, 575
(splitting kernel) 310–311, 322–331, 346, 348, 358,
    367–368, 375–376
kinematical pole 75, 207–208, 216, 234
Kinoshita–Lee–Nauenberg theorem 409
KNO scaling 355, 437
Kobayashi–Maskawa matrix 208, 315

Landau pole 151
Laplace transformation 205
large N_C limit 122, 219, 476, 483, 507, 543, 565
leading logarithm approximation (LLA) 146, 151,
    192, 197–199, 310, 318, 320, 335, 337–345,
    347–349, 351, 375, 390, 416, 422, 437, 441–444,
    449, 451, 453–454, 460, 465, 469, 472–473, 476,
    480, 492, 499, 504, 507, 522–523, 533, 535, 537,
    542–543, 565, 570, 572, 575
leading order (LO) 70–71, 290–292, 297, 299, 310,
    331–332, 349–351, 367, 372, 417, 455–460, 462,

465, 480, 485, 488, 493–497, 502, 506, 510,
513–515, 520–523
Lehmann–Källen representation 64
light cone 13, 93, 161, 309–310, 317, 324, 327, 329,
332–335, 360–364, 370, 372, 376, 419, 452, 456,
458, 465–466, 468, 471, 486, 492, 534, 536,
539–540
light cone components 539
Liouville interaction 161
longitudinal gluons 100, 103–104
Low theorem 370
low-x 292, 299, 437

magnetic moments 146, 204, 260, 264, 269–273, 370
Majorana fermion 330, 568
mass splitting 58–59, 251
Mellin transformation 355, 358, 389, 434, 439, 443,
473–474, 479, 573
mesons
axial 233–234, 283
pseudoscalar 54–55, 58, 61–66, 71–73, 87, 91, 203,
235–239
meromorphic 557, 563
minimal sensitivity 40
Minkowski space-time 107, 110, 115–116, 127–129,
134, 182
Möbius
form 488–491
group 490, 542, 545–547, 551, 553, 564, 568
invariance 542, 547
representation 488–489, 513–515, 551
moments of
fragmentation function 438–439
parton distribution 339–343
quark distribution 292, 293
structure functions 296, 310
monodromy 545, 548–549, 561
$\overline{\text{MS}}$ scheme, MS scheme. 20–22, 25, 27–28, 33, 42,
44, 59, 135, 137, 191–192, 208, 210, 215, 222,
226, 229, 279, 346, 368, 404, 411–412, 464
MS-like schemes 20–21, 27–29, 32–34
multiplicity 422, 431, 433, 437–439, 444
multi-Regge
kinematics 451–453, 456, 465, 467, 469–470,
491–492, 495–496, 533, 535–536, 538, 542, 575
unitarity 533, 537, 540

next-to-leading
corrections 311, 565
approximation (NLA) 338, 348, 449, 541,
543, 575
NLA 448–449, 451, 453–455, 460, 465, 473,
492–493, 495–496, 499, 504, 507, 516, 523
next-to-leading order 70–72, 103, 209, 292, 299, 311,
349–352, 368–369, 372, 460–462, 465, 475,
492–493, 495–497, 503–517, 519–523, 525, 565
next-to-next-to leading order 349–352, 493,
523, 541

non-Abelian gauge theories 107, 122, 146, 454
nonperturbative effects 61, 147, 150, 193, 196, 406

odderon 450–451, 550–551, 553–559, 561, 565–566,
575
operator product expansion 75–76, 93, 103, 107,
147–148, 195–198, 204–210, 213–224, 227–230,
233–241, 243, 245, 248–250, 252–256, 259,
262–264, 275, 280, 282, 286–287, 291, 294, 297,
302, 310, 313, 332–334, 360–362, 364, 372, 379
optical theorem 542
oscillator potential 110

partial waves 384, 387, 389–390, 392, 395, 450, 476,
480
parton
correlation functions (PCF) 375–383
correlators 310, 353
distribution 286, 310, 315–317, 319, 321–323, 331,
337–343, 345, 348–350, 359, 367–369, 371–372,
390, 417, 448
momenta 316, 415, 448
model 95, 217, 220, 222, 225, 231–233, 278,
290–291, 299, 309–317, 332–333, 342, 350–351,
372
spectra 429, 433, 437
Pauli–Jordan (causal) function 64
Pauli–Villars regularization 14, 86, 135, 137, 185
PCAC 54–55, 66, 74, 76
π^0 width 90
planar diagrams 422, 433, 507, 509, 511, 513, 543
polarization operator 22–23, 25, 35, 74–75, 123–124,
141, 147, 196–197, 203–204, 207, 210–211, 213,
215, 219–220, 223–225, 228, 230, 233–242,
246–250, 252–253, 256–257, 259, 264, 266, 274,
278, 403
Pomeranchuk theorem 550, 558, 560
pomeron
hard (or BFKL, or perturbative QCD) 215,
476–485, 491, 516, 543, 560, 566, 568
intercept 449–450, 480, 521–522, 542–543, 567,
570

QCD at low energies 53, 55, 256
quantization of the QCD Lagrangian 2–5
quantum electrodynamics 3, 10, 14, 18–20, 35, 37, 41,
82, 88, 97, 100–103, 107, 145–147, 150–154,
157, 186, 190, 192, 229, 310–311, 317, 328, 336,
338, 404, 417, 422–425, 427, 429, 450, 511, 515
quantum fluctuations 152, 161, 165, 167–168, 182
quark
condensate 59, 61, 63, 65, 78, 123–124, 127,
137–138, 142, 199–202, 211, 213, 215, 219, 239,
241, 243–246, 255, 257, 265–266, 280, 282, 284,
297, 301
condensate magnetic susceptibility 202, 265, 266,
301
density 334

quark (cont.)
  flavour 1, 6, 23–24, 26, 42, 87, 91, 107, 137–138,
    202, 300, 372, 404, 461, 506, 516
  light-mass 53, 55, 59, 138, 199
  heavy-mass 104, 137
quark-gluon condensate 202, 241–242
quasipartonic operators 310, 367, 372–373, 375, 377,
    379, 565, 570
quasi-multi-Regge 451, 465, 538, 540

rapidity 437, 465, 468, 495, 505, 522,
    536, 538
Regge behaviour 287, 292, 349
Regge region 449, 454, 460
Regge poles 450, 534
Regge trajectory 450, 454, 460, 493, 497, 533,
    541–542, 570, 575
reggeization 450–451, 453–454, 458–460, 462–463,
    469, 473, 491, 493–495, 497, 499, 501, 507, 509,
    515
  gluon — 450–451, 453–454, 458, 460, 462–463,
    469, 473, 491, 493, 495, 497, 499, 501, 507, 509,
    515
reggeon 450–467, 470, 473–475, 477, 484, 488,
    492–494, 497–505, 515, 534–536, 538, 540–541,
    549–550, 561–562, 564–567, 575
  vertices 452, 454–461, 464–470, 492–494, 497,
    503–504, 515
renormalization scheme (see MS, $\overline{\text{MS}}$ scheme)
renormalization group 27, 29, 31, 33, 59, 98, 135,
    147–151, 159, 191–192, 198, 200, 209, 215, 219,
    231, 241, 251, 292, 310, 317, 340, 387, 441, 482,
    484, 522
  renormalization scale 19–21, 27–28
renormalons 146–151, 193
representation
  fundamental 1, 42, 45, 47, 316, 360, 418, 455, 541
  adjoint 2, 45, 58, 176, 253, 330, 373, 391, 429, 456,
    544, 550, 568, 575
  constants 45
  group 551

saddle point 146, 148–150, 152–155, 159–161,
    165–168, 174, 176–177, 180–181, 184, 186, 345
scaling violation 379
Schwinger term 124
s-channel discontinuity 458–460, 462–463, 470, 493
signature 347, 384–386, 388, 392, 394–397, 450–451,
    458–460, 462, 469, 471, 473–474, 476, 480,
    491–493, 495, 497, 507, 509, 515, 541, 546, 550,
    561
Sklyanin ansatz 549, 561
soft insertion 423–426
soft logarithms 448
splitting
  kernels 310–311, 322–331, 346, 348, 358,
    367–368, 375–376
  functions 346, 348–349, 417, 439, 442, 521

sphericity 406, 415
spontaneous violation of chiral symmetry 53, 59, 65,
    107, 200
standard model 82, 147, 205
structure functions 92–93, 95, 103–104, 204, 216,
    219, 240, 250–251, 280–281, 286–288, 290, 296,
    309–313, 316, 345–346, 348, 350–352, 359, 361,
    364, 370, 372, 379, 383, 386–387, 416, 449, 521,
    559
Sudakov (parametrization, variables) 317, 320,
    322–324, 336–337, 363, 366, 375–376, 384–385,
    387–388, 391–392, 466, 535
sum rules
  Adler 352
  Bjorken 261, 274, 368, 370
  Burkhardt–Cottingham 365, 368
  Gerasimov–Drell–Hearn 370
  Gross–Llewellyn–Smith 352
  Gottfried 352–353
  Weinberg 75, 76, 204
supersymmetry 330, 568
symmetry
  of strong interactions 54, 54, 57
  $SU(2)_L \times SU(2)_R \times U(1)$ 54
  $SU(2)_L \times SU(2)_R$ 54, 63, 66–69, 76
  $SU(3)_L \times SU(3)_R$ 54–55, 63, 66,
    71, 199

t-channel 314, 376, 383–392, 395, 450–451, 454–456,
    459–460, 462, 469–471, 473–474, 476, 488, 494,
    507, 515, 533, 535, 543, 567
θ-term 123, 125
't Hooft conjecture (hypothesis) 92
thrust 414–415
topological
  current 115, 122, 134, 139
  susceptibility 203, 276–278
transcendentality 569–575
transversity 370–372
tunneling 107–108, 111, 115, 127–129
twist 42–43, 261, 288, 290, 310, 313, 334, 339–340,
    346, 350, 352–353, 358, 361–362, 364, 366–373,
    376, 379, 381, 387, 395, 481, 523, 542, 564–565,
    569–571, 574–575

ultraviolet divergences 14, 18, 96, 168, 199, 260, 316,
    458, 464, 489
ultraviolet cut-off 14, 37, 86, 135, 147–148, 188, 192,
    204, 264, 269, 316, 340, 354
unitarity 3, 14, 35, 449, 454, 456, 458, 465, 469, 482,
    491–494, 496–497, 533, 535, 537, 540
unitarization 449, 537, 543, 565

vacuum
  energy density 98–99, 201, 405
  structure 107, 133, 137
  topological susceptibility 203, 276–278
  wave function 107, 133

vertex
  particle-particle-reggeon 451
  reggeon-reggeon-particle 453

Wandzura–Wilczek relation 366
Ward identities 17, 19, 65, 72, 122, 278, 541
Weinberg angle 175
Wess–Zumino term 73
Wess–Zumino Lagrangian 73

Wilson approach 190
winding numbers 115, 119, 120, 122, 127, 133

Yang–Baxter 546–548, 561
Yang–Mills theory 175–176, 181–185, 330, 358, 564

zero charge 35, 155–157, 192
zero modes 113, 130, 132–133, 135–137, 140, 142,
  165, 167, 170, 174, 182, 185, 212